CEREBRAL CORTEX

Volume 13
Models of Cortical Circuits

CEREBRAL CORTEX

Series Editors: Edward G. Jones and Alan Peters

Volume 1: Cellular Components of the Cerebral Cortex

Volume 2: Functional Properties of Cortical Cells

Volume 3: Visual Cortex

Volume 4: Association and Auditory Cortices

Volume 5: Sensory-Motor Areas and Aspects of Cortical Connectivity

Volume 6: Further Aspects of Cortical Function, Including Hippocampus

Volume 7: Development and Maturation of Cerebral Cortex

Volume 8A: Comparative Structure and Evolution of Cerebral Cortex, Part I

Volume 8B: Comparative Structure and Evolution of Cerebral Cortex, Part II

Volume 9: Normal and Altered States of Function

Volume 10: Primary Visual Cortex in Primates

Volume 11: The Barrel Cortex of Rodents

Volume 12: Extrastriate Cortex in Primates

Volume 13: Models of Cortical Circuits

Volume 14: Neurodegenerative and Age-Related Changes in Structure and Function of Cerebral Cortex

Series Editors: Edward G. Jones and Alan Peters

CEREBRAL CORTEX

Volume 13
Models of Cortical Circuits

Edited by

PHILIP S. ULINSKI

University of Chicago
Chicago, Illinois

EDWARD G. JONES

University of California, Davis
Davis, California

and

ALAN PETERS

Boston University School of Medicine
Boston, Massachusetts

Springer Science+Business Media, LLC

The Library of Congress cataloged the first volume of this title as follows:

Main entry under title:

Cerebral cortex.

Includes bibliographical references and index.
Contents: v. 1. Cellular components of the cerebral cortex.
1. Cerebral cortex—Collected works. I. Peters, Alan, 1929– II. Jones, Edward G., 1939– . [DNLM: 1. Cerebral cortex—Anatomy and histology. 2. Cerebral cortex—Physiology. WL 307 C4136]

QP383.C45 1984 612'.825 84-1982

ISBN 978-1-4613-7223-3 ISBN 978-1-4615-4903-1 (eBook)
DOI 10.1007/978-1-4615-4903-1

Originally published by Kluwer Academic / Plenum Publishers in 1999
Softcover reprint of the hardcover 1st edition 1999

10 9 8 7 6 5 4 3 2 1

A C.I.P. record for this book is available from the Library of Congress.

Contributors

Lyle J. Borg-Graham Equipe Cognisciences, Institut Alfred Fessard-CNRS, 91198 Gif-sur-Yvette, France

Paul Bush Computational Neurobiology Laboratory, Salk Institute, La Jolla, California 92037; *present address:* Department of Physiology, University of California at San Francisco, San Francisco, California 94143-0444

Matteo Carandini Howard Hughes Medical Institute and Center for Neural Science, New York University, New York, New York 10003; *present address:* Institute of Neuroinformatics, ETH/UNIZ, Zurich 8057, Switzerland

Rodney Douglas Institute of Neuroinformatics, ETH/UNIZ, Zurich 8006, Switzerland; and Computation and Neural Systems Program, California Institute of Technology, Pasadena, California 91125

Michael E. Hasselmo Department of Psychology, Boston University, Boston, Massachusetts 02215

David J. Heeger Department of Psychology, Stanford University, Stanford, California 94305

Christof Koch Computation and Neural Systems Program, California Institute of Technology, Pasadena, California 91125

Christiane Linster Department of Psychology and Program in Neuroscience, Harvard University, Cambridge, Massachusetts 02138

William W. Lytton	Department of Neurology, University of Wisconsin, and William S. Middleton VA Hospital, Madison, Wisconsin 53706
Misha Mahowald	Institute of Neuroinformatics, ETH/UNIZ, Zurich 8006, Switzerland
Kevan Martin	Institute of Neuroinformatics, ETH/UNIZ, Zurich 8006, Switzerland
J. Anthony Movshon	Howard Hughes Medical Institute and Center for Neural Science, New York University, New York, New York 10003
Paul A. Rhodes	Mathematical Research Branch, NIDDK, National Institutes of Health, Bethesda, Maryland 20814
Elizabeth Thomas	Department of Neurology, University of Wisconsin, and William S. Middleton VA Hospital, Madison, Wisconsin 53706
Philip S. Ulinski	Department of Organismal Biology and Anatomy, University of Chicago, Chicago, Illinois 60637
Hugh R. Wilson	Visual Sciences Center, University of Chicago, Chicago, Illinois 60637
Florentin Wörgötter	Institut für Physiologie, Ruhr Universität Bochum, D-44780 Bochum, Germany

Preface

This is the first volume in the *Cerebral Cortex* series devoted to mathematical models of the cortex. It was motivated by the realization that computational models of individual neurons and ensembles of neurons are increasingly used in research on cortical organization and function. This is, in part, because of the now ubiquitous presence of powerful and affordable computers. Suitable machines were formerly rare in research laboratories and required substantial programming expertise to be used in constructing and using neuronal models. However, computers are now routinely used in all areas of neurobiology and a number of software packages allow scientists with minimal computer science and mathematical backgrounds to construct serious neuronal models. A second factor leading to the proliferation of modeling studies is the development of technologies that allow the kinds of data collection needed to develop realistic models of cortical neurons. Characterization of the kinetics of voltage- and ligand-gated channels and receptors had been limited to relatively large neurons. However, the rapid development of slice preparations, patch-clamp methods, and imaging methods based on voltage-sensitive dyes and intracellular calcium indicators has resulted in a significant data base on the biophysical features of cortical neurons.

The scope of modeling approaches to cortical neurons and functions is wide and it seemed necessary to limit the purview of the volume. The focus is on attempts to understand the properties of individual cortical neurons and neuronal circuitry through models that incorporate significant features of cellular morphology and physiology. No attempt was made to include modeling approaches to understanding cortical development and plasticity. Thus, work dealing with the development of ocular dominance columns and the orientation selectivity of neurons in visual cortex is not considered. Similarly, models dealing with the cellular mechanisms underlying long-term plasticity and with approaches to learning and memory based on modification of Hebbian synapses are not considered. Relatively abstract attempts to understand higher level and cognitive processes based on neural nets represent a second, major area of work that is not treated. Models of cognitive processes based on dynamical systems methods in which no attempt is made to include the biophysical features of individual neurons are also not considered.

The ten major chapters fall into three groups. The first group deals with compartmental models of individual cortical neurons. Lyle Borg-Graham provides an introduction to the methods involved in constructing compartmental models and then reviews the existing models of hippocampal pyramidal cells. Because of the effectiveness of hippocampal slice preparations, these neurons have well-characterized biophysical properties. This chapter illustrates how compartmental models can be used to synthesize experimental data and provide an integrative view of the properties of individual neurons. Paul Rhodes continues the theme by focusing on the role of voltage-gated channels located on the dendrites of cortical neurons. This is an area in which technological advances in the visualization of neurons in slice preparations based on infrared microscopy have greatly expanded the information available on dendritic function in just a few years. The chapter both reviews the experimental data on active dendritic conductances and emphasizes their potential functional roles.

The second group of chapters deal with the generation of receptive field properties of neurons within visual cortex. They address issues stemming from the original attempt to understand how the receptive field properties of neurons in cat and monkey primary visual cortex are generated by interactions between geniculate afferents and cortical neurons. The chapter by Florentin Wörgötter evaluates models that have been used to analyze the generation of receptive field properties. Rodney Douglas and his colleagues address a specific set of issues dealing with the role of intracortical excitation mediated by pyramidal cell collaterals. An important feature of this chapter is its relation to attempt to construct fabricated circuits that duplicate the functions of cortical circuits. The chapter by Philip Ulinski focuses on the generation of motion-selective properties in cortical neurons. It seeks to identify cellular mechanisms used by neurons that respond preferentially to visual stimuli moving with particular speeds or directions. Matteo Carandini and his colleagues discuss the feature of cortical neurons, known as gain control, that allows neurons to respond effectively to visual stimuli by pooling information across populations of cortical neurons. The chapter by Hugh Wilson deals with the receptive field properties of extrastriate areas and introduces new work analyzing face-selective neurons.

The final set of chapters consider models of ensembles of thalamic and cortical neurons. The chapter by William Lytton and Elizabeth Thomas uses the theory of dynamical systems to analyze the temporal relationships between thalamic and cortical neurons. An important feature of the interaction between thalamus and cortex is the presence of oscillations that depend in part upon the voltage-gated conductances present on individual neurons and in part on the structure of the overall network. Paul Bush continues this emphasis on oscillations by discussing a model that deals with the generation of synchronized oscillations in visual cortex. Oscillations of this kind have attracted substantial attention in recent years because of their potential role in cognitive processes. The last chapter, by Michael Hasselmo and Christiane Linster, reviews their work on modeling piriform cortex, emphasizing the role of cholinergic mechanisms in modulating the activity of cortical neurons.

An attempt has been made throughout to make the volume accessible to readers with minimal mathematical backgrounds. The volume thus begins with a short history of models of cortical neurons and circuitry that introduces the principal modeling styles. The chapters by Wörgötter and Ulinski contain more extensive introductions to some of the modeling methods that have been used to study visual cortex, and the mathematically challenged reader will find that the chapter by Lytton and Thomas contains a readable introduction to the use of dynamical systems theory in neurobiology.

Philip S. Ulinski
Edward G. Jones

Chicago and Davis

Contents

Chapter 3

Functional Implications of Active Currents in the Dendrites of Pyramidal Neurons

Paul A. Rhodes

Chapter 4

Comparing Different Modeling Approaches of Visual Cortical Cell Characteristics

Florentin Wörgötter

Chapter 5

The Role of Recurrent Excitation in Neocortical Circuits

Rodney Douglas, Christof Koch, Misha Mahowald, and Kevan Martin

Chapter 6

Neural Mechanisms Underlying the Analysis of Moving Visual Stimuli

Philip S. Ulinski

Chapter 7

Linearity and Gain Control in V1 Simple Cells

Matteo Carandini, David J. Heeger, and J. Anthony Movshon

Chapter 8

Non-Fourier Cortical Processes in Texture, Form, and Motion Perception

Hugh R. Wilson

Chapter 9

Modeling Thalamocortical Oscillations

William W. Lytton and Elizabeth Thomas

Chapter 10

Realistic Network Models of Synchronized Oscillations in Visual Cortex

Paul Bush

Chapter 11

Modeling the Piriform Cortex

Michael E. Hasselmo and Christiane Linster

CEREBRAL CORTEX

Volume 13
Models of Cortical Circuits

1

Modeling Cortical Circuitry

A History and Prospectus

PHILIP S. ULINSKI

1. Introduction

Two features of cortical organization have motivated attempts to analyze the structure and function of the cerebral cortex with approaches from computational neuroscience. First, cortical function depends on the properties of the voltage- and ligand-gated receptors and channels distributed on individual neurons as well as the anatomical connections between the thalamus and between cortical areas. It is therefore important to relate information on the membrane biophysics of cortical neurons to their morphology and to the pattern of afferent synapses on their somata and dendrites. Mathematical models of individual neurons provide an effective context in which to examine the relationship of membrane-level factors to the behavior of the whole neuron. This is an area of work that is often called single-neuron computation.

Second, a defining feature of cortical organization is that it contains many interacting elements. A working hypothesis in much of neurobiology is that the pattern of these interactions underlies the generation of behaviors, including the relatively complex behaviors associated with human cognition. Modeling approaches are an effective way to deal with both of these features of cortical organiza-

PHILIP S. ULINSKI • Department of Organismal Biology and Anatomy, University of Chicago, Chicago, Illinois 60637.

tion. Some models include detailed descriptions of the biophysics of individual neurons and allow investigators to relate information between organizational levels. They provide a means, for example, of understanding the significance of particular neurotransmitter systems to the function of cortical circuits. Other models contain only abstract characterizations of individual neurons. They cannot make statements about the functional significance of different types of ligand- or voltage-gated channels, but they do provide insights into the dynamics of large assemblages of cortical neurons. An important class of such models is based on neural nets and these models are called connectionist models. In contrast to single-neuron models, both of these kinds of models are called large-scale models because they involve large numbers of neurons or neuronlike elements.

Since computational or modeling approaches are well suited to studying these two features of cortical organization, it is not surprising that attempts to understand cortical function through quantitative analyses extend back for over a half century. This chapter briefly reviews the history of two approaches that have been used in modeling cortex and its constituent elements. One is most useful for large-scale models with relatively abstract elements. The second is valuable for both single-neuron and computational models that incorporate details of cellular biophysics. After considering these approaches, the chapter speculates on the future of computational approaches to cortex. It does not provide details of various modeling methods, which are subjects of individual chapters in the volume. The chapter also does not attempt to provide a complete discussion of the general field of computational neuroscience. Introductions to various areas in computational neuroscience can be found in recent books (Anderson *et al.*, 1990; Anderson and Rosenfeld, 1988; Arbib, 1995; Churchland and Sejnowski, 1992; Cotterill, 1989; Gluck and Rumelhart, 1990; Haykin, 1994; Koch and Davis, 1994; Koch and Segev, 1989; McKenna *et al.*, 1992; Rieke *et al.*, 1997; Rose and Dobson, 1985; Schwartz, 1990; Ulinski, 1993) and in articles published in *Neural Computation* and the *Journal of Computational Neuroscience.*

2. Lorente de Nó through Dynamical Systems Models

2.1. Lorente de Nó

Although the word "model" often connotes mathematical formulations, a more general usage includes conceptual models that articulate basic relationships within a structure. The drawings of cortical neurons prepared by Santiago Ramón y Cajal (1995) are models in this sense. They accurately depict the morphology of individual neurons, but many go further in suggesting the spatial and synaptic relationships between neurons. For example, Cajal's famous drawing of neurons in rat hippocampus (Fig. 1) includes arrows that suggest the direction of information flow within the hippocampus and thus makes some statement about interactions between hippocampal neurons. This tradition was continued by Cajal's student, Raphael Lorente de Nó, whose paper on the hippocampal formation (Lorente de Nó, 1934) goes beyond descriptions of individual neurons to discuss issues pertaining to the fundamental organization of the nervous system. His paper (Lorente

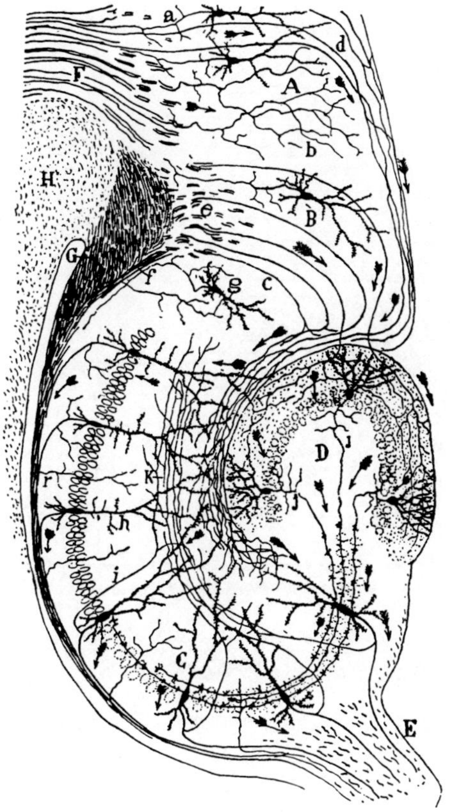

Figure 1. Informal model of hippocampal circuit. This and the next five figures show a progression from relatively anatomical to relatively abstract conceptions of cortical circuitry. This figure is a schematic drawing by Santiago Ramón y Cajal (1995) of neurons in mammalian hippocampus. Notice that inferences about the direction of information flow have been made and are indicated by arrows. From Ramón y Cajal (1995).

de Nó, 1938) on sequences of excitatory neurons (so-called reverbatory circuits) prompted discussions about the neural substrates of memory, and much of the work on cortical anatomy that followed World War II was based on his diagram of neurons in the barrel cortex of mouse somatosensory cortex (Fig. 2A), first published in the 1938 edition of Fulton's textbook of physiology (Fulton, 1938). The diagram shows the spatial relationships of the principal types of neurons found in neocortex, but also introduced the important distinction between specific and nonspecific afferents. The specific afferents were correctly associated with spatially localized inputs from thalamic relay nuclei, while the origin of nonspecific afferents was less certain. Like Cajal, Lorente de Nó made inferences about information flow in the system. His chapter on cortical organization, for example, includes a simplified diagram phrased in terms of chains of neurons in which he indicates information flow by arrows (Fig. 2B).

2.2. Cell Assemblies and Neural Nets

Lorente de Nó's work was one influence on an approach to modeling cortical circuitry that began with the concept of networks of interacting neurons developed by Walter Pitts and Warren McCulloch (McCulloch and Pitts, 1943) and continues into contemporary work on artificial neural nets and dynamical systems. McCulloch was impressed by Rudolph Carnap's development of mathematical logic (Carnap,

Figure 2. Circuit diagrams of mammalian neocortex. These are two drawings made by Lorente de Nó of mammalian neocortex. (A) A schematic diagram that illustrates the principal types of neurons seen in Nissl preparations (at the very left of the drawing) and in Golgi preparations. Cortical neurons are shown in the center of the illustration and afferents are shown to the right. The thalamic afferents labeled a and b are specific afferents that terminate in layer IV. Afferents labeled c–f were called nonspecific afferents.

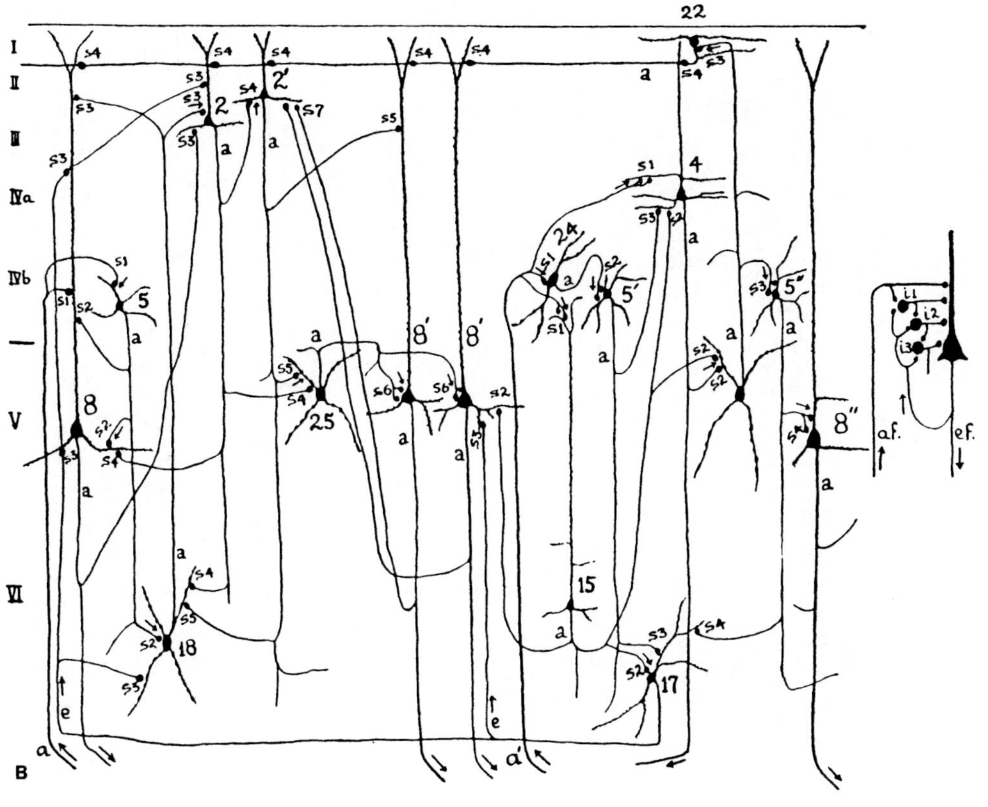

Figure 2. (B) A less realistic sketch shows "chains" of neurons with directions of information flow being indicated by arrows. The small sketch shown in the inset to the far right illustrates neurons reduced to geometric objects. Again, information flow is indicated by arrows. From Lorente de Nó (1938).

1938) and set forth with Pitts to formulate a "logical calculus" for neurons. Individual neurons were viewed as logical devices that were either active or inactive (Fig. 3). Such neurons could be assembled into "nervous nets" that could calculate many logical functions, with the implication that interactions between neurons can form a substrate for complex calculations by the nervous system. Lorente de Nó's ideas are not explicitly mentioned in the McCulloch and Pitts 1943 paper, but a less frequently cited paper (Pitts and McCulloch, 1947) develops a model of cortical circuitry based more directly upon cortical anatomy and includes the distinction between specific and nonspecific afferents. The model attempts to explain how cortical neurons can recognize (Fig. 4) the general features of external objects (such as a square or a triangle) independent of their position in visual space. A key idea was that this cannot be achieved by a simple, topographic representation of the external world. Information about the object was, in some sense, distributed throughout the network of neurons.

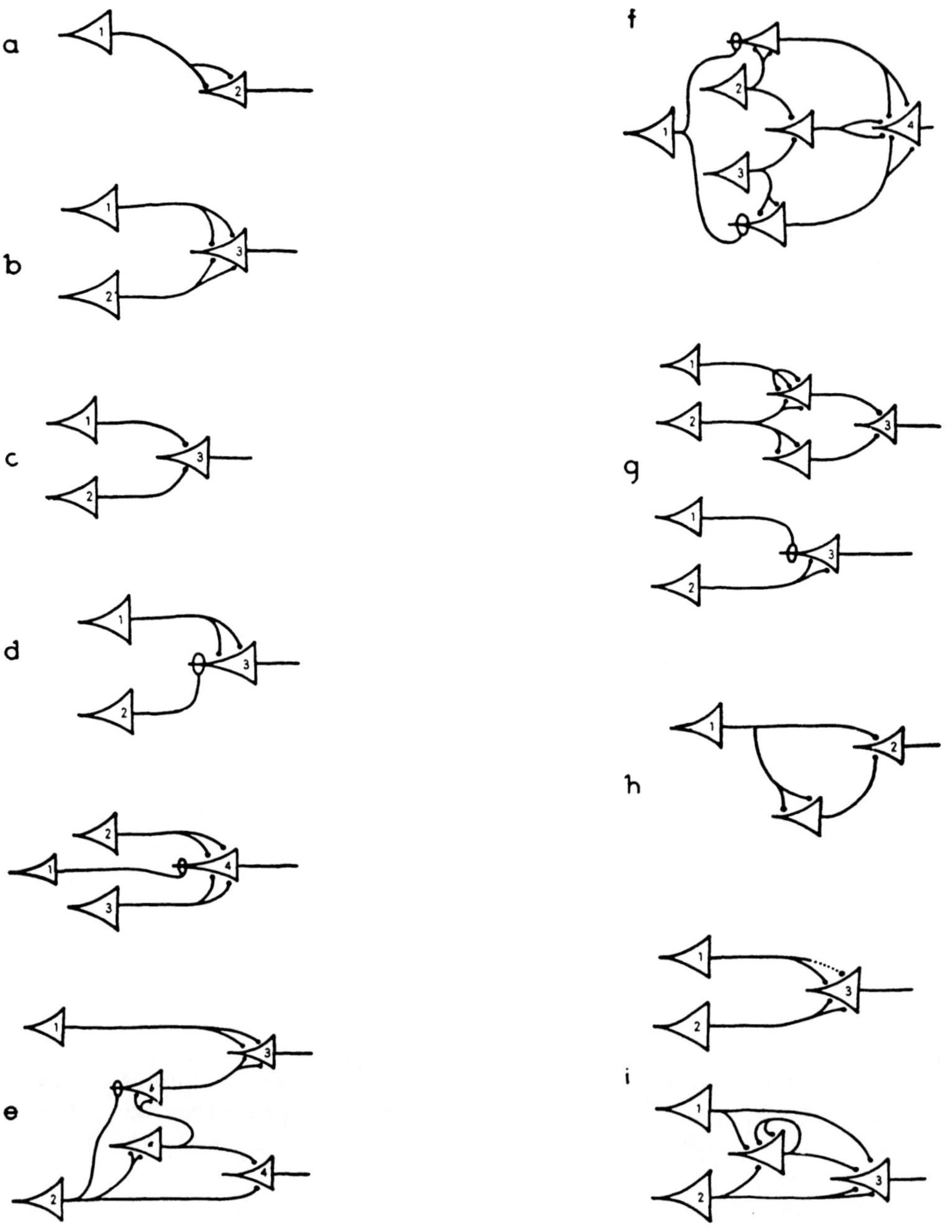

Figure 3. McCulloch–Pitts neurons. Here neuronlike objects have been arranged into nervous nets. The different nets would carry out different logical functions. From McCulloch and Pitts (1943), with permission of the publisher.

Lorente de Nó's work is also explicitly figured by Donald Hebb in *The Organization of Behavior* (Hebb, 1949), which established influential notions about how the neural structure of the nervous system is related to behavior. Hebb shows drawings from Lorente de Nó's paper on reverberating circuits and comments on the obser-

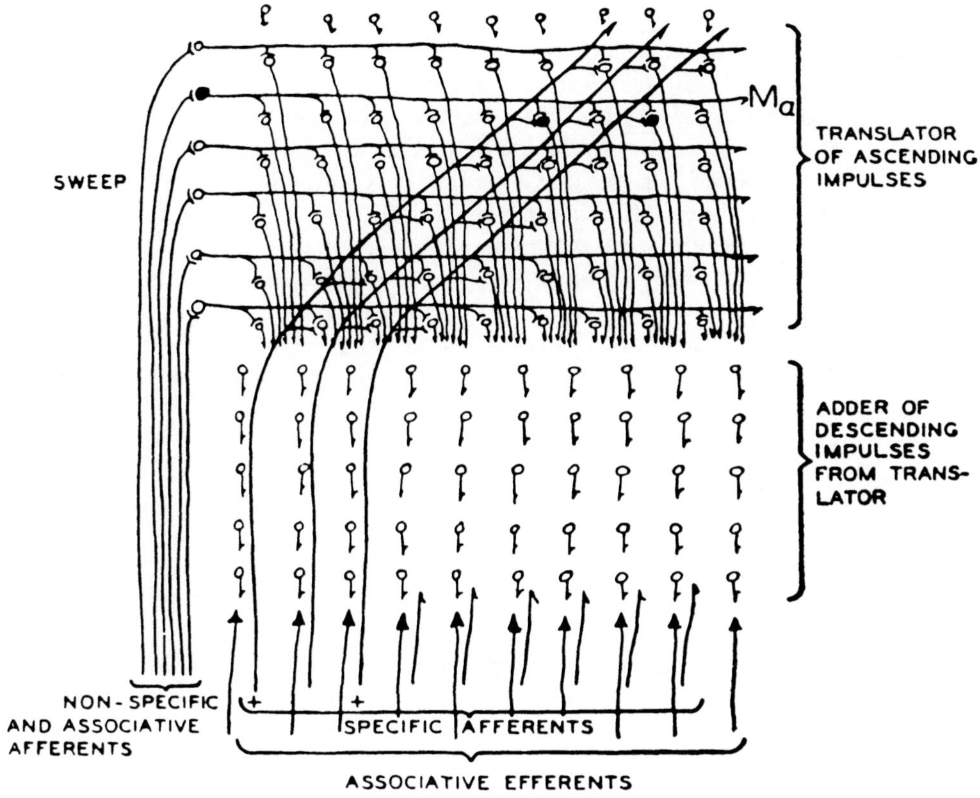

Figure 4. Model of auditory cortex. This model of auditory cortex by Pitts and McCulloch was based on Lorente de Nó's 1938 drawing. It is the second figure in the paper. The first is a Nissl section of the auditory cortex. Notice that Pitts and McCulloch make explicit use of specific and nonspecific afferents. Neurons have been reduced to circles. From Pitts and McCulloch (1947), with permission of the publisher.

vations that axons in the central nervous system branch and end in multiple synaptic knobs (or axon terminals) upon the somata and dendrites of neurons. He notes that the knobs vary significantly in size, and conjectures that large knobs form more effective synapses than do small knobs. The efficacy of synapses could change as the knobs grow or shrink. He further hypothesized that thalamic afferents and cortical neurons, such as those located in the primary and secondary visual cortices, are interconnected into what he called cell assemblies. He thought that the pattern of interconnections and variations in the effectiveness of synapses might be a key to understanding how cortical neurons could respond selectively to specific stimulus patterns. This led to the now famous idea of Hebbian synapses which can be modified by experience and contribute to the selectivity of the cell assemblies. The static nervous nets of McCulloch and Pitts thus became plastic or modifiable assemblies that could learn and adapt.

The idea that groups of cortical neurons function as logical devices was further developed by Frank Rosenblatt (1958), who introduced neural nets called perceptrons. They consisted of an input device, such as the retina, one or two association areas, and an output device. The strength of synapses between elements in the net could be changed so that the synaptic weights within the perceptron would converge over a number of iterations to a stable configuration that allowed the net to peform a variety of tasks. Rosenblatt was an active advocate of his ideas and generated considerable interest in the use of artificial neural networks, but enthusiasm gradually dampened as it became clear that perceptrons had relatively limited computational capabilities (Minsky and Papert, 1969) and could not, for example, solve the problem of telling apart a square and triangle.

A renaissance of interest in neural nets began in the 1980s when the problem of how to adjust the synaptic weights in networks containing an input layer, an output layer, and an intermediate or hidden layer (Fig. 5) was solved (e.g., Rumelhart *et al.* 1986). It had been appreciated that such networks would be more powerful than perceptrons, but there was no clear algorithm for adjusting the synaptic weights at each layer of the network. Rumelhart *et al.* (1986) established a simple method of adjusting the synaptic weights so that the performance of the neural net could change with time. The network could be instructed to match its output to any desired outcome by systematically varying the weights of synapses made by hidden-layer neurons upon the output neurons and, subsequently, the weights of synapses made by input-layer neurons upon hidden-layer neurons. Neural networks can thus be modified, or learn, by propagating changes in synaptic weights "backward" through the net, a process called backpropagation. Introduction of backpropagation algorithms in three-layered structures greatly extended the utility of neural nets and such approaches have a large number of practical applications.

The relationship between artificial neural nets and real neurons is not entirely clear (Gardner, 1993) and neural nets are often not optimal ways to model cortical circuits. They have, however, found some applications in modeling the receptive field properties of cortical neurons (e.g., Zipser and Andersen, 1988; Lehky and Sejnowski, 1988), but they have had an arguably more important influence as a metaphor for distributed systems. The point made by McCulloch and Pitts was that a nervous system that does no more than repeatedly re-represent the external world in a series of topographic maps has very limited computational power. More sophisticated problems such as representing the position of an organism in its environment or representing human speech and logic may employ a mode of information storage involving large populations of neurons. How such representations are achieved by the nervous system is not known, but neural nets are providing a useful, if only metaphorical, way of thinking about this problem that is at the center of much contemporary work in cognitive neuroscience.

2.3. Dynamic Systems Models

Neural net models store information about the external world in a static way by altering synaptic weights or efficiencies. An alternate set of ideas about how information can be represented in populations of neurons comprise a modern version of Lorente de Nó's concept of reverberating circuits. Lorente de Nó's idea was that information (a memory trace, perhaps) could be stored in loops of excitatory neurons. Once one neuron was activated, all of the neurons in the loop would continue to fire in sequence. This idea was eventually abandoned as it became clear that memory traces are too stable to be explained by perpetual reverberations. A 1950 paper by Karl Lashley (Lashley, 1950) played a major role in arguing against the possibility that memory is localized in specific circuits, presenting data that showed that memory tasks were unimpaired by small lesions of the brain. However, the notion that information can be stored in temporal patterns of activity is still under active consideration.

Attempts to develop quantitative models for such phenomena followed from the mathematical description of action potentials by Hodgkin and Huxley (1952). They showed that the mechanisms underlying the production of individual action potentials or sequences of action potentials could be represented in a family of differential equations. However, their work posed a number of practical challenges.

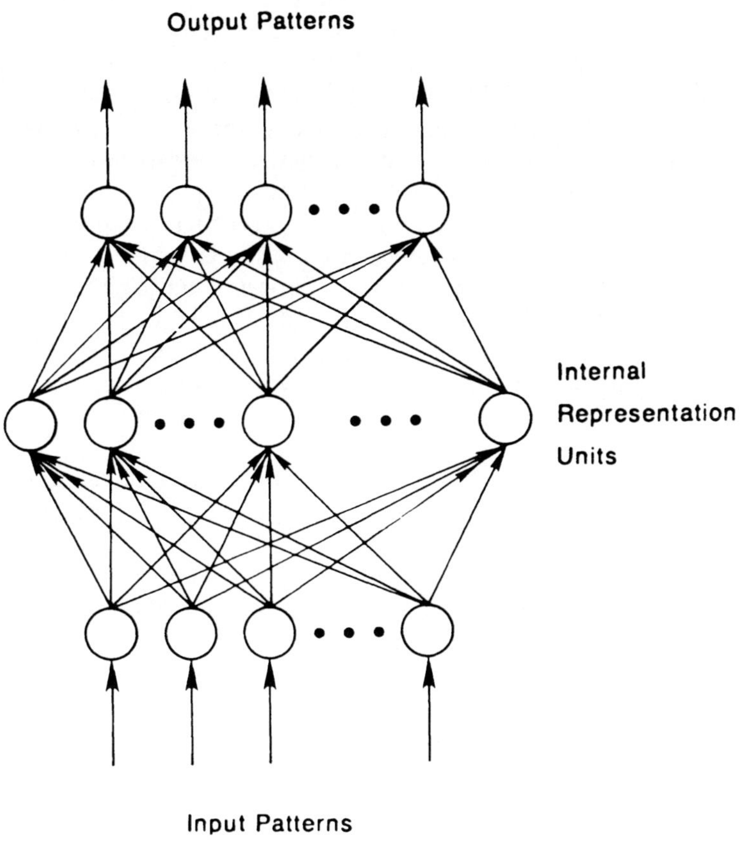

Figure 5. Three-layered neural net. This is a modern neural net based on modifications of the perceptron concept. The first layer receives coded inputs. They could, for example, carry information from different points in visual space. Internal representation units are often called hidden units and receive nontopographic projections from the input layer. They in turn make nontopographic projections to the output units. Arrowheads represent synapses whose weights are adjusted by backpropagation algorithms. From Rumelhart *et al.* (1986), by permission of the publisher.

One was that it took the academic community roughly a decade to absorb and digest what they had done, so there were no experimental data on voltage-gated conductances in other neurons. Another was that the equations developed by Hodgkin and Huxley were mathematically difficult, nonlinear differential equations and the calculations they did took weeks to carry out. Any notion of using the approach to look at the firing patterns of groups of interacting neurons was out of the question.

Several workers developed ways of characterizing the firing patterns of neurons that involved more tractable equations and fewer parameters. Examples of this approach are the relatively simple equations developed by Fitzhugh (1960) and Nagumo *et al.* (1962). Their equations describe "integrate-and-fire" neurons which sum inputs and generate an action potential when the somatic membrane potential crosses a threshold. They are easier to work with and enabled mathematicians to treat general properties of networks of interacting neurons without having to incorporate the biophysical details of synapses and the spike-generating mechanism into the equation. One example of this approach is an analysis of the relative roles of excitation and inhibition in cortical networks carried out by Wilson and Cowan (1972). Another example is the large body of work by Stephen Grossberg and his colleagues (e.g., Grossberg, 1976) in which large groups of neurons are described in terms of differential equations. Formulations of this kind are amenable to analysis by the branch of applied mathematics called nonlinear systems or dynamical systems theory (e.g., Strogatz, 1994). Such systems frequently have stable states in which groups of neurons can fire in complex, but repeatable patterns so that some event in the external world is represented by a particular pattern of activity in a population of neurons. Attempts to develop experimental data for these kinds of representations are found in the work of Walter Freeman and colleagues, who analyzed firing patterns in the olfactory cortex of rats (Freeman, 1975). Freeman argued that complex events in the external world, such as natural patterns of odorants, are represented in the olfactory cortex by stable, dynamic states involving large populations of neurons.

The advantage of using dynamical systems theory to think about cortical function is that it provides a way of conceptualizing the changing patterns of activity involving very large numbers of neurons. There is no doubt that most, if not all, cortical functions involve assemblages of significant numbers of cortical neurons. The difficulty was that there were no effective experimental methods for collecting the kinds of data needed to develop dynamical systems models. This situation is changing with the advent of multiple-electrode arrays that allow investigators to monitor the activity of tens or hundreds of neurons while an animal is carrying out a behavioral task (e.g., Nicoleilis *et al.*, 1995). Similarly, optical imaging methods allow visualization of activity in relatively large regions of the cortex (Grinvald *et al.*, 1988). This approach has been used to obtain clearer concepts of cortical organization, such as columns in visual cortex, but has been limited because the technique has a low signal-to-noise ratio, so images are usually averaged over multiple presentations of a stimulus, leading to a loss of information on temporal dynamics. However, this limitation is being overcome by methods that allow the membrane potentials of populations of neurons to be monitored in real time (Arieli *et al.*, 1995; Senseman, 1996).

3. Hodgkin and Huxley through Network Models

3.1. Hodgkin and Huxley

One influence of the work of Hodgkin and Huxley on characterizing the generation of action potentials in the giant axons of squid has already been mentioned. The immediate reaction was that the mathematical description they provided of the ionic conductances underlying action potential generation was too complex to be used easily in modeling cortical neurons, and integrate-and-fire models were developed to represent neurons in models. However, two aspects of their work formed the basis for later efforts that led to effective models of cortical neurons, circuits, and networks which incorporate varying levels of biophysical detail. The first set of studies concerned current flow in axons. This work, by Hodgkin (1937), used frog sciatic nerve. The approach was to view the axon as analogous to an electrical cable and use the differential equations that describe current flow in wires to explain current flow in axons. The second set of studies took advantage of electrical circuits that enable the investigator to control, or clamp, the membrane voltage of a neuron to characterize the sodium and potassium currents involved in generating the sodium-dependent action potentials (Hodgkin and Huxley, 1952).

3.2. Wilfrid Rall

These two features of work on axons were extended to models of entire neurons by Wilfrid Rall. The major papers representing this body of work have been recently brought together in a single volume (Segev *et al.*, 1995). One contribution was to develop a mathematical theory of dendritic function based on cable theory (e.g., Rall, 1959, 1960). Rall showed that simple, unbranched dendrites could be modeled as cables in the same way that axons can be described. He also showed that a larger class of dendrites (those conforming to the so-called 3/2-power law) were mathematically equivalent to unbranched cables (Fig. 6A). Synapses on dendrites could be modeled as synaptic currents, and Rall framed a series of ideas about how the spatial distribution of excitatory and inhibitory synapses upon dendrites was important in understanding the integrative features of dendrites.

Rall (e.g., Rall *et al.*, 1967) also developed a more powerful approach, which was to conceptualize neurons as composed of many interconnected compartments (Fig. 6B). He borrowed the approach from the field of chemical engineering in which chemical reactions take place in a sequence of interconnected vats. Each vat is a compartment with specific concentrations of different chemical species and chemicals are allowed to flow from one compartment to another. Differential equations are then written to describe the rates at which concentrations in each vat vary as a function of time. Rall described dendrites as a series of anatomical segments and wrote differential equations that describe the membrane potential of each segment or compartment as a function of the currents flowing into and out of the compartment. This formulation duplicated the results of cable theory, but permit-

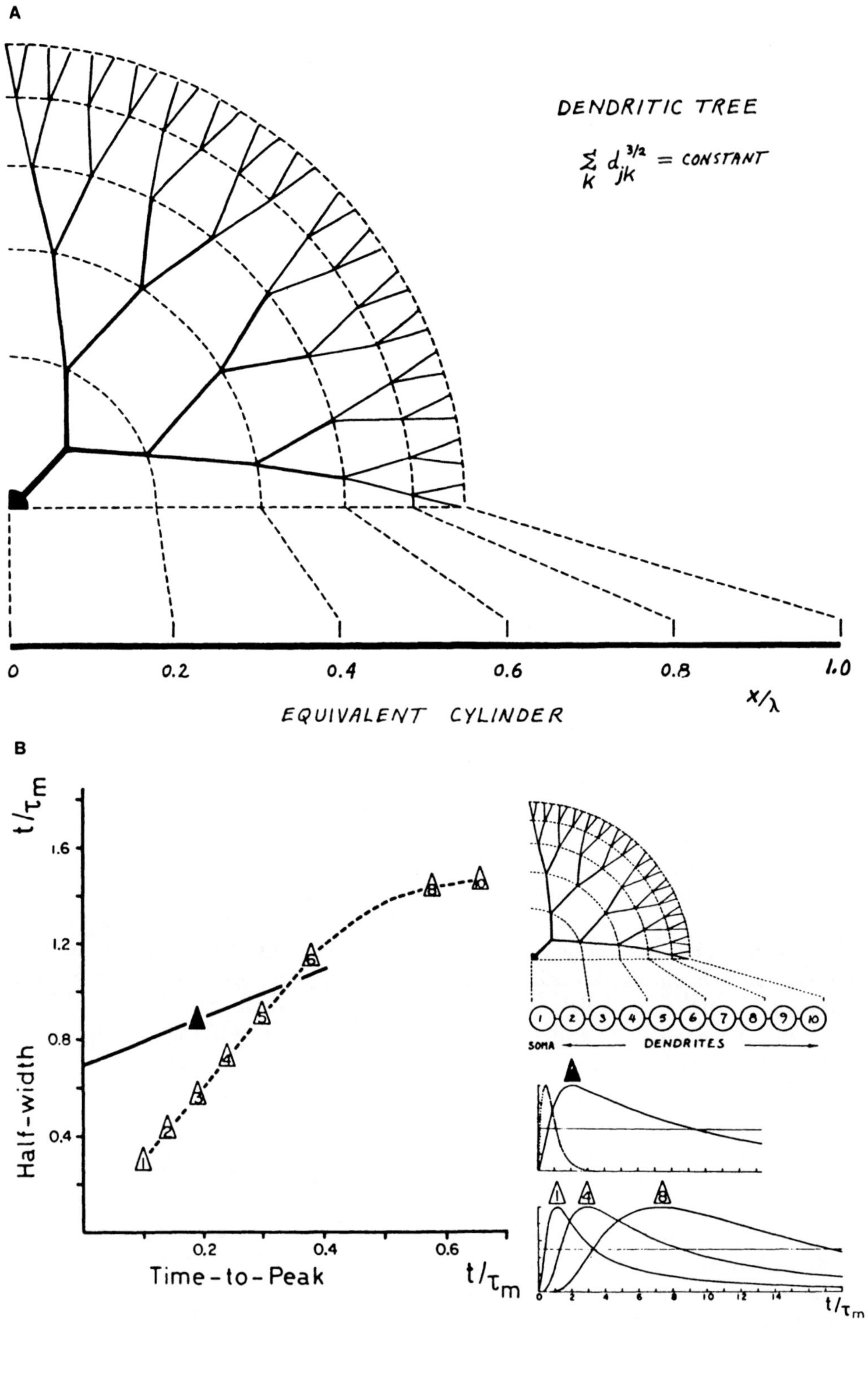
A
DENDRITIC TREE
$\sum_K d_{jK}^{3/2}$ = CONSTANT
0
0.2
0.4
0.6
0.8
1.0
x/λ
EQUIVALENT CYLINDER
B
t/τm
1.6
1.2
0.8
0.4
Half-width
0.2
0.4
0.6
Time-to-Peak
t/τm
SOMA
DENDRITES
t/τm

ted the introduction of voltage-gated conductances into the membranes of the various compartments. The use of compartmental models was limited for a time because the mathematical description of the neuron produced a large number of differential equations and solution of these equations exceeded the capability of most computers. Also, the detailed characterizations of voltage- and ligand-gated conductances that eventually resulted from the application of voltage-clamp methods, especially with patch-clamp electrodes, were not then available. However, contemporary compartmental models are providing very detailed analyses of the biophysical properties of neurons. These have proven useful in understanding how various voltage-gated channels interact with the passive properties of the neuron's membrane to produce the firing patterns characteristic of neurons and in beginning to understand the contribution of voltage-gated channels present on dendrites to the integrative physiology of cortical neurons. Like Hebb's ideas about the plasticity of synapses, these model are changing the relatively static view of neurons embodied in the McCulloch–Pitts idea of neurons as doing algebraic sums into a more complex view of neurons as computing devices.

3.3. Software Packages

A key step in establishing compartmental models as effective research tools was the development of software packages for modeling individual neurons and groups of neurons. The first packages to be used in neuronal modeling included SPICE, which was designed for applications in electrical engineering that involved complex networks of resistors and capacitors (e.g., Segev *et al.*, 1989). Since neuronal membranes can be represented as circuits consisting of resistors and capacitors, it was possible to use these packages for neuronal models. However, they were not particularly convenient when it came to representing individual segments of dendrites and somata. Several software packages designed specifically for neuronal modeling overcome this difficulty. These include GENESIS (Bower and Beeman, 1995), NEURON (Hines, 1993), and NODUS (DeSchutter, 1989), each of which contains menus that represent parts of neurons, voltage-gated conductances, etc., in ways that facilitate the construction of realistic models of neurons. There is no longer any reason for individuals interested in neuronal modeling to design and implement their own programs because of the ready availability of user-friendly software packages.

←

Figure 6. Cable and compartmental models of neurons. These are figures from Rall's work on dendrites. All modern biophysically realistic models are based on these formulations. (A) An equivalent cylinder model. The upper part of the diagram shows a neuron with a cell body and a dendritic arbor. If the branches of the dendrites follow the 3/2-power law, they can be collapsed into a single cable shown at the bottom of the diagram. Current flows within this cable can be modeled using cable theory. (B) Compartmental model. The same neuron is reduced to a series of compartments symbolized by a series of numbered circles. The soma is represented by compartment 1 and dendritic segments are represented by higher number compartments. Waveforms below the compartmental model show simulated synaptic potentials. The graph to the left is a plot of shape parametres of the waveforms. From Rall (1962) and Rall *et al.* (1967), by permission of the publisher.

3.4. Realistic Models of Cortical Networks

An important feature of compartmental models of neurons is that they can be used to construct biophysically realistic models of cortical networks. Roger Traub and colleagues have used this approach to study the bursting properties of neurons in the hippocampal formation (Traub and Miles, 1991). Several groups have now constructed models of the geniculocortical pathway in cats (Somers *et al.*, 1994; Suarez *et al.*, 1995; Wörgötter and Koch, 1991). Retinal ganglion cells and geniculate neurons are typically represented in these models as linear or nonlinear filters, but neurons in the cortex are represented by compartmental models that often include specific ligand-gated and voltage-gated conductances. These models are currently being used to understand the origin of orientation and direction selectivity in cortical neurons. There are also models of pyriform cortex that address the mechanisms of associative learning in the olfactory system (Hasselmo and Barkai, 1995).

4. Prospectus

Models of cortical neurons have become accepted in attempts to understand cortical organization and it is now commonplace for authors to combine modeling and experimental approaches. It is likely that this marriage of approaches will be increasingly important in attempts to resolve a variety of issues in cortical function, but three particularly promising areas can be highlighted here.

The first is the use of compartmental models to understand the interactions that occur in cortical neurons between synapses and voltage-gated conductances. It has often been asserted that such attempts were doomed to sterility because of the impossibility of characterizing the biophysical properties of dendrites and axons. However, the past few years have seen dramatic advances in our ability to record from visually identified dendrites using infrared microscopy (e.g., Stuart and Spruston, 1995), so this particular form of pessimism seems unwarranted. It is also now becoming possible to fill individual neurons with voltage-sensitive dyes and calcium indicators so that detailed descriptions of voltage and calcium transients even within individual dendritic spines are possible (Magee and Johnston, 1995; Spruston *et al.*, 1995; Svoboda *et al.*, 1997). Modeling approaches will become critical as we try to place these detailed findings into a functional context.

The second area in which modeling approaches are likely to be valuable is in understanding the spatiotemporal dynamics of populations of cortical neurons. The now classic era of cortical physiology founded by Mountcastle (1957) and Hubel and Wiesel (1962) focused on the firing patterns of individual neurons. However, most behavioral tasks involve the coordinated activity of thousands of neurons both within and between cortical areas. We lack a conceptual framework for interpreting, or even storing, reducing, and analyzing, data sets involving hundreds of thousands of neurons. Simply looking at firing patterns or calculating firings rates of individual neurons will obviously not suffice, so some mathematical approach will be necessary.

Finally, computational approaches to understanding cortical function are beginning to find engineering applications in attempts to build artifical devices that can duplicate the functions carried out by the cortex. This area is called neuromorphic engineering and involves attempts to duplicate the essential features of cortical circuits. An important aspect of this form of engineering is the use of circuits fabricated from silicon chips. Many complicated circuits can be constructed using very large scale integrated (VLSI) technologies. VLSI circuits are being fabricated to carry out several basic tasks that are presumably done in cerebral cortex by circuits composed of different types of neurons (Mahowald, 1992; Koch and Mathur, 1996).

5. References

Anderson, J. A., and Rosenfeld, E., 1988, *Neurocomputing. Foundations of Research,* MIT Press, Cambridge, MA.

Anderson, J. A., Pellionisz, A., and Rosenfeld, E., 1990, *Neurocomputing 2: Direction for Research,* MIT Press, Cambridge, MA.

Arbib, M., 1995, *The Handbook of Brain Theory and Neural Networks,* MIT Press, Cambridge, MA.

Arieli, A., Shoham, D., Hildesheim, R., and Grinvald, A., 1995, Coherent spatiotemporal patterns of ongoing activity revealed by real-time optical imaging with single-unit recording in cat visual cortex, *J. Neurophysiol.* **73**:2072–2093.

Bower, J. M., and Beeman, D., 1994, *The Book of Genesis,* Springer-Verlag, New York.

Carnap, R., 1938, *The Logical Syntax of Language,* Harcourt, New York.

Churchland, P. S., and Sejnowski, T. J., 1992, *The Computational Brain,* MIT Press, Cambridge, MA.

Cotterill, R. M. J., 1989, *Models of Brain Function,* Cambridge University Press, Cambridge.

DeSchutter, E., 1989, Computer software for development and simulation of compartmental models of neurons, *Comput. Biol. Med.* **19**:71–81.

Fitzhugh, R., 1960, Thresholds and plateaus in the Hodgkin–Huxley nerve equations, *J. Gen. Physiol.* **43**:867–896.

Freeman, W. J., 1975, *Mass Action in the Nervous System,* Academic Press, New York.

Fulton, J. F., 1938, *Physiology of the Nervous System,* Oxford, University Press, Oxford.

Gardner, D., 1993, *The Neurobiology of Neural Networks,* MIT Press, Cambridge, MA.

Gluck, M. A., and Rumelhart, D. E., 1990, *Neuroscience and Connectionist Theory,* Erlbaum, Hillsdale, NJ.

Grinvald, A., Frostig, R. D., Lieke, E., and Hildesheim, R., 1988, Optical imaging of neuronal activity, *Physiol. Rev.* **68**:1285–1366.

Grossberg, S., 1976, Adaptive pattern classification and universal recoding: I. Parallel development and coding of neural features detectors, *Biol. Cybernet.* **23**:121–134.

Hasselmo, M., and Barkai, E., 1995, Cholinergic modulation of activity-dependent synaptic plasticity in the pyriform cortex: Brain slice physiology and computational modeling, *J. Neurosci.* **15**:6529–6604.

Haykin, S., 1994, *Neural Networks,* Macmillan, New York.

Hebb, D. O., 1949, *The Organization of Behavior,* Wiley, New York.

Hines, M., 1993, NEURON—A program for simulation of nerve equations, in: *Neural Systems: Analysis and Modeling* (F. H. Eeckman, ed.), Kluwer, Boston, pp. 127–136.

Hodgkin, A. L., 1937, Evidence for electrical transmission in nerve. Part I, *J. Physiol.* (Lond.) **90**:183–210.

Hodgkin, A. L., and Huxley, A. F., 1952, A quantitative description of membrane current and its application to conduction and excitation in nerve, *J. Physiol.* (Lond.) **117**:500–544.

Hubel, D. H., and Wiesel, T. N., 1962, Receptive fields, binocular interaction and functional architecture in the cat's visual cortex, *J. Physiol.* (Lond.) **160**:106–154.

Koch, C., and Davis, J. L., 1994, *Large-Scale Neuronal Theories of the Brain,* MIT Press, Cambridge, MA.

Koch, C., and Mathur, B., 1996, Neuromorphic vision chips, *IEEE Spectrum* **33**(5):38–46.

Koch, C., and Segev, I., 1989, *Methods in Neuronal Modeling,* MIT Press, Cambridge, MA.

Lashley, K. S., 1950, In search of the engram, in: *Psychological Mechanisms in Animal Behavior,* Cambridge University Press, Cambridge, pp. 454–482.

Lehky, S. R., and Sejnowski, T. J., 1988, Network model of shape-from-shading: Neural function arises from both receptive and projective fields, *Nature* **333**:452–454.

Lorente de Nó, R., 1934, Studies on the structure of the cerebral cortex. II. Continuation of the study of the Ammonic system, *J. Psychol. Neurol.* (Lpz.) **46**:117–177.

Lorente de Nó, R., 1938, Analysis of the activity of the chains of internuncial neurons, *J. Neurophysiol.* **1**:207–244.

Magee, J. C., and Johnston, D., 1995, Synaptic activation of voltage-gated channels in dendrites of hippocampal pyramidal neurons, *Science* **268**:301–304.

Mahowald, M. A., 1992, Evolving analog VLSI, in: *Single Neuron Computation* (T. McKenna, J. Davis, and S. F. Zornetzer, eds.), Academic Press, San Diego, CA, pp. 413–436.

McCulloch, W. S., and Pitts, W., 1943, A logical calculus of the ideas immanent in nervous activity, *Bull. Math. Biophys.* **5**:115–133.

McKenna, T., Davis, J., and Zornetzer, S. F., 1992, *Single Neuron Computation,* Academic Press, San Diego, CA.

Minsky, M., and Papert, S., 1969, *Perceptrons,* MIT Press, Cambridge, MA.

Mountcastle, V. B., 1957, Modality and topographic properties of single neurons of cat's somatic sensory cortex, *J. Neurophysiol.* **20**:408–434.

Nagumo, J. S., Arimoto, S., and Yoshizawa, S., 1962, An active pulse transmission line simulating a nerve axon, *Proc. IRE* **50**:2061–2070.

Nicolelis, M. A., Baccala, L. A., Lin, R. C., and Chapin, J. K., 1995, Sensorimotor encoding by synchronous neural ensemble activity at multiple levels of the somatosensory system, *Science* **268**:1353–1358.

Pitts, W., and McCulloch, W. S., 1947, How we know universals: The perception of auditory and visual forms, *Bull. Math. Biophys.* **9**:127–147.

Rall, W., 1959, Branching dendritic trees and motoneuron membrane resistivity, *Exp. Neurol.* **1**:491–527.

Rall, W., 1960, Membrane potential transients and membrane time constant of motoneurons, *Exp. Neurol.* **2**:503–532.

Rall, W., 1962, Theory of physiological properties of dendrites, *Ann. N. Y. Acad. Sci.* **96**:1071–1092.

Rall, W., Burke, R. E., Smith, T. G., Nelson, P. G., and Frank, K., 1967, Dendritic location of synapses and possible mechanisms for the monosynaptic EPSP in motoneurons, *J. Neurophysiol.* **30**:1169–1193.

Ramón y Cajal, S., 1995, *Histology of the Nervous System of Man and Vertebrates,* Oxford University Press, Oxford.

Rieke, F., Warland, D., de Ruyter van Steveninck, R., and Bialek, W., 1997, *Spikes: Exploring the Neural Code,* MIT Press, Cambridge, MA.

Rose, D., and Dobson, V. G., 1985, *Models of the Visual Cortex,* Wiley, Chichester, U.K.

Rosenblatt, F., 1958, The perceptron: A probabilistic model for information storage and organization in the brain, *Psyschol. Rev.* **65**:386–408.

Rumelhart, D. E., Hinton, G. E., and Williams, R. J., 1986, Learning internal representations by error propagation, in: *Parallel Distributed Processing: Explorations in the Microstructure of Cognition* (D. E. Rummelhart and J. L. McClelland, eds.), MIT Press, Cambridge, MA, Vol. 1, pp. 318–362.

Schwartz, E. L., 1990, *Computational Neuroscience,* MIT Press, Cambridge, MA.

Segev, I., Fleshman, J. W., and Burke, R. E., 1989, Compartmental models of complex neurons, in: *Methods in Neuronal Modeling,* (C. Koch and I. Segev, eds.), MIT Press, Cambridge, MA, pp. 63–96.

Segev, I., Rinzel, J., and Shepherd, G. M., 1995, *The Theoretical Foundation of Dendritic Function: Selected Papers of Wilfrid Rall with Commentaries,* MIT Press, Cambridge, MA.

Senseman, D. M., 1996, Correspondence between visually evoked voltage-sensitive dye signals and synaptic activity recorded in cortical pyramidal cells with intracellular microelectrodes, *Vis. Neurosci.* **13**:963–977.

Somers, D. C., Nelson, S. B., and Sur, M., 1995, An emergent model of orientation selectivity in cat visual cortical simple cells, *J. Neurosci.* **15**:5448–5465.

Spruston, N., Schiller, Y., Stuart, G., and Sakmann, 1995, Activity-dependent action potential invasion and calcium influx into hippocampal CA1 dendrites, *Science* **268**:297–300.

Strogatz, S. H., 1994, *Nonlinear Dynamics and Chaos,* Addison-Wesley, Reading, MA.

Stuart, G., and Spruston, N., 1995, Probing dendritic function with patch pipettes, *Curr. Opin. Neurobiol.* **5**:389–394.

Suarez, H., Koch, C., and Douglas, R., 1995, Modeling direction selectivity of simple cells in striate visual cortex within the framework of the canonical microcircuit, *J. Neurosci.* **15**:6700–6719.

Svoboda, K., Denk, W., Kleinfeld, D., and Tank, D. W., 1997, *In vivo* dendritic calcium dynamics in neocortical pyramidal neurons, *Nature* **385:**161–165.

Traub, R. D., and Miles, R., 1991, *Neuronal Networks of the Hippocampus,* Cambridge University Press, Cambridge.

Ulinski, P. S., 1993, Computational approaches to comparative neurobiology, *Am. Zool.* **33:**1–103.

Wilson, H. R., and Cowan, J. D., 1972, Excitatory and inhibitory interactions in localized populations of model neurons, *Biophys. J.* **12:**1–24.

Wörgötter, F., and Koch, C., 1991, A detailed model of the primary visual pathway in the cat: Comparison of afferent excitatory and intracortical inhibitory connection schemes for orientation selectivity, *J. Neurosci.* **11:**1959–1979.

Zipser, D., and Andersen, R. A., 1988, A back-propagation programmed network that simulates response properties of a subset of posterior parietal neurons, *Nature* **331:**679–684.

2

Interpretations of Data and Mechanisms for Hippocampal Pyramidal Cell Models

LYLE J. BORG-GRAHAM

1. Introduction

A neuron model organizes, collates, and directs our expanding electrophysiological and biophysical knowledge. This chapter reviews single-neuron models, with hippocampal pyramidal cells (HPCs) as the reference, emphasizing the integration of electrophysiological data and biophysical theory into a cohesive hypothesis. We discuss the process of translating data into specific mechanistic hypotheses. We critically review the parameters and mechanisms that have been used in several published models, comparing their predictions with the data and against each other. A final goal is the description of an updated HPC model that may be a reference for future refinement and elaboration.

1.1. Neuron Model Evolution—following Electrophysiology

Single-neuron models, inspired by the remarkable success of Hodgkin and Huxley (1952b), have evolved in fits and starts in parallel with advances in experi-

LYLE J. BORG-GRAHAM • Equipe Cognisciences, Institut Alfred Fessard-CNRS, 91198 Gif-sur-Yvette, France.

mentally based (wet) understanding. Happily, the tools for both approaches have become more sophisticated and easier to use—consider the widespread use of patch-clamp setups and sophisticated neuronal simulator programs (cf. sharp electrode protocols and SPICE).

Unhappily, while in the experimental realm natural constraints can be relied upon—barring gross mishandling of the data, brain tissue sets the rules—the "garbage in, garbage out" maxim in computer science is a ubiquitous gnat in simulations. Thus, for wet experiments biology sets the "equations," and the job of the electrophysiologist is to be clever and careful enough to read the results. With dry experiments, the experimenter-programmer defines both the input and the black box, exposing the result to a much greater possibility of error.

The success of the Hodgkin–Huxley model of the squid axon and the associated Na^+ and K^+ channels was due to at least two reasons. First, there was a scientific design in which experiment and theory were directly intertwined (with good fortune; Hodgkin 1976). Second, the system under scrutiny was relatively simple and well measured, with two dominant and separable nonlinear components and very good space clamp of a large, accessible structure. All the model parameters were taken directly from data (more or less, in the case of gating particle exponents) and, in retrospect, it is not completely surprising that such a model reproduces "itself" so well.

On the other hand, applying the Hodgkin–Huxley approach or extensions of it to neurons in which many, or in fact most, of the parameters are not directly measurable has been more problematic. This difficulty stems from at least the following factors. First, it is difficult to constrain the free parameters of nonlinear dynamical systems in order to reconstruct uniquely some data set—a minimal data set is hard to specify and there is the possibility of multiple formal solutions. Second, most nonlinear modeling is accomplished with fairly easy to apply, robust numerical methods; thus it is not too hard to get something that "looks" reasonable, even when bugs linger in the code or understanding.

That being said, detailed, carefully constructed, and qualified single-cell models can yield useful insights, suggest testable predictions, and assist in quantitative comparison of data from different sources that describe the same phenomena (Borg-Graham, 1987c; Bower and Koch, 1992). The intent of this chapter is to point out some of the pitfalls as well as useful strategies in this process.

The ultimate utility of these models is for understanding the role of individual cells, particular channel types, or other mechanisms in brain physiology and pathology. For example, the identification of several membrane currents as the principal effector of various components of the voltage response of a cell—e.g., I_{CT} and the fast after-hyperpolarization (fAHP), I_{AHP} and the slow after-hypoerpolarization (sAHP)—is well established. Correlations between intrinsic, low-level phenomena and disease states have been reported, e.g., alteration of various phases of spike repolarization and repetitive firing characteristics in scrapie-infected mouse HPCs (A. R. Johnston *et al.*, 1997). More ambitious proposals include assigning specific membrane potentials, and by implication their causitive membrane currents, to cognitive phenomena, e.g., the proposal by Lisman and colleagues as to the role of the slow after-depolarization in short-term memory in hippocampus (O. Jensen *et al.*, 1996).

Although much work remains to be done for the predictive success of this level of model [see, for example, Goldstein and Colatsky (1996) in relation to channel models], given the increasing confluence of experiment and theory, this is indeed an exciting time for this fundamental component in the evolving field of computational neuroscience.

1.2. Neuron Model Evaluation—following the Parameters

In this chapter we shall review both the experimental basis for model parameters and the interpretation of these parameters in various published models. In the process of analyzing the modeling reports, it became clear that a significant obstacle was the difficulty of reproduction, independent of any issues associated with specific interpretations of the electrophysiological data—essentially every model analyzed within this review (including the author's previous work!) presented various degrees of internal inconsistencies at the parameter or equation level. In some cases, consultations with the authors resulted in the necessary corrections (e.g., typographical errors); in others, careful analysis and adjustment of model parameters bore fruit. We believe that the essential problem is that of reliable communication of the large parameter sets characteristic of these sorts of models. To this end, the parameters presented here are available on-line in the Surf-Hippo neuron simulator distribution (Borg-Graham, 1995; see Section 3.5). Readers wishing to use the mechanisms described in this chapter thus have convenient access to the relevant parameters, with the benefit that updates to the on-line parameters will be made as warranted by future work or contributions from other workers.

A recent collection of papers describing single-neuron models, including hippocampus (Claiborne *et al.*, 1992; T. H. Brown *et al.*, 1992), is found in McKenna *et al.* (1992). Segev (1992) discusses various issues regarding the level of detail appropriate for single-cell models, and a review of models focusing especially on computations in dendritic trees is given by Mel (1994).

1.3. Why Hippocampus?

Models of many cell types, and just about every brain region, have appeared in the last decade. Models of hippocampal cells are of particular interest because the electrophysiological literature on these cells and the hippocampal region is especially rich—HPCs are among the most-studied cell types of the mammalian brain, and the circuitry of the hippocampus is known to a large degree. While there are certainly functional physiological differences between cortical cells of different areas, and between cells within the same area (see Section 2.7 for how this applies to the hippocampus), choosing the HPC as a representative, canonical cortical pyramidal cell is not unreasonable.

The data and models in this chapter derive mainly from studies of HPCs in the CA1–CA3 regions in rat and guinea pig. Data from both sharp-electrode and patch-electrode recordings will be considered and compared, with the majority of cases being from the *in vitro* brain-slice preparation. Occasionally, specific references will be made to data and models of non-pyramidal hippocampal neurons (interneurons, cells of the dentate gyrus), cortical pyramidal neurons, and other neurons—such references are intended to be illustrative, not definitive.

1.4. Organization of This Chapter

There are several interrelated themes to this chapter:

- Model components—anatomy, linear properties, channel models, concentration dynamics
- Review of HPC electrophysiology
- Review of HPC models
- New model structure and parameters

We focus on a single-cell model that reproduces the intrinsic behavior of a neuron, whose elements include the linear membrane properties and the variety of voltage and second messenger gated channels which shape a cell's response. The discussion is primarily on properties observed from responses to intracellular current or voltage stimulation, as opposed to synaptic stimulation; occasionally we will mention single-neuron models in the context of network models. Likewise, our view of the neuron here is in the context of relatively short time scale dynamics, under the assumption that the basic architecture is stable. The resulting model could serve as a starting point for looking at longer time scale phenomena, such as plasticity.

We start with a review of the electrophysiological methods from which model data are derived, since parameters relevant to a given mechanism or phenomenon can strongly depend on the experimental protocol. We then present a discussion of neuron model methodology, the components that go into single-cell models, and what these models may be used for. This is followed by a discussion of the morphological component of cell models, and estimates of the linear properties of HPCs. Next, the phenomenological templates that constrain model mechanisms in HPCs are presented, including the important issue as to what makes some cells burst and others not. Next, we review some of the published models of HPCs, with emphasis on single-cell models. This is followed by descriptions of various models of channel biophysics and concentration dynamics, with a review of the data regarding activity-driven changes in intracellular Ca^{2+}. We then describe electrophysiological data and models concerning sodium channels, calcium channels, potassium channels, and other channels of HPCs. We close the chapter with selected examples of model simulations.

At various points in this chapter we will propose parameters and show results for a relatively simple HPC model, referred to as the Working model. This model is intended to illustrate some of the issues discussed in the chapter, and sample simulations will be shown of some key behaviors—all traces in this chapter are from either this model or examples of more elaborate passive anatomical models.

We assume that the reader has an introductory exposure to cable theory, compartmental modeling techniques, and electrophysiology—throughout the text we shall mention appropriate background references.

A reconstruction of the neuron's electrical properties is a test of the electrophysiologist's basic, reductionist starting position: the total response of a neuron arises from an ensemble of channel types, each of which is characterized by protocols designed to describe it individually using some combination of electrical (voltage clamp or current clamp), pharmacological, and kinetic isolation. Many mechanisms contribute to particular characteristics, and it is of considerable interest to understand which of these mechanisms are physiologically modulated (Kaczmarek and Levitan, 1987). Observed variability of the cell response may have profound implications on the computational role of the neuron—one goal of a model may be to determine what variability is "real" and what is a function of the experimental setup.

In this section we describe some of the characteristics and limitations of the experimental procedures from which model parameters are derived. Different experimental techniques can give different values of model parameters; therefore it is quite important to understand the variety of procedures in order to better integrate data for different sources. Useful introductions to practical electrophysiological technique can be found in Kettenmann and Grantyn (1992), *The Axon Guide* (Sherman-Gold, 1993), and in Sakmann and Neher (1995).

In reality, isolation of specific mechanisms is rarely, if ever, completely possible, notwithstanding the issue of the potential lability of channel properties under various recording contexts (*in vivo,* slice, culture, electrode type, etc.). An integrative model, therefore, must incorporate the data with a certain flexibility—all the models referenced here base their channel models with varying degrees of fidelity to the literature. The model parameters that we will develop in this chapter are molded in this context, that is, as a hypothesis of specific mechanisms which may suggest additional tests, and are constructed so that parameter evolution is as incremental as possible.

2.1. Voltage Clamp versus Current Clamp

In an ideal electrophysiology world voltage-clamp data would supply parameters of voltage-dependent parameters directly, and the reconstruction of current-clamp protocols would follow accordingly. But despite the necessary working assumption by electrophysiologists that voltage clamp will yield a "purer" picture of channel behavior, the direct insertion of voltage clamp data plus assumptions of kinetic mechanisms (in particular the Hodgkin–Huxley model; Section 8.3) into current-clamp simulations typically fails. The reality is that for all but a very limited set of circumstances, the available voltage-clamp data can only provide a framework for parameter estimation.

There are two reasons for this. The first has to do with limitations inherent in the various types of voltage-clamp preparations, which include intact cells or reduced preparations (cultured or isolated cells). The major problem with intact cells is that accurate voltage clamp is limited to membrane near the point of electrode penetration, so that currents arising from distant, poorly clamped membrane may distort the results. Model-based discussions of these sorts of errors are given by Müller and Lux (1993) in the context of voltage-dependent membrane, and Spruston *et al.* (1993), looking at synaptic events (see also D. Johnston and Brown, 1983). On the other hand, in the reduced preparations the control of membrane voltage is better, but the isolation or culturing methods may alter the cellular properties. For example, occasionally reports using the reduced preparations include representative current-clamp behaviours, and the spike characteristics and firing patterns typically are quite different from what is considered "normal" (e.g., Sah *et al.*, 1988b).

The second reason derives from the fundamental challenge for any experiment for elucidating phenomenological mechanisms: the decision as to which are the most important dimensions of the experimental parameter space. In other words, since usually it is impossible to measure everything, the primary question in experimental design is how best to place the bet. When electrophysiological mechanisms are analyzed with voltage-clamp methods, the protocols are designed assuming a kinetic model, even implicitly; thus, measurements are specifically chosen in order to characterize that model. For measuring membrane currents, the model is usually the Hodgkin–Huxley model. However, there is increasing evidence from both single-channel data and macroscopic channel modeling that this kinetic model is inadequate for certain currents (see Section 8). For this reason, and since the system is nonlinear, it is quite possible to miss important characteristics if they are not measured explicitely (the great advantage of linear systems analysis, of course, lies in the comprehensiveness of the impulse response).

In summary, since our aim is to describe a model that reproduces functional characteristics, we place an emphasis on the reproduction of properties seen under current clamp. Voltage-clamp data are used to estimate parameters, but not necessarily for direct reproduction, unless those data were obtained under more or less "normal" conditions.

2.2. Single-Channel versus Macroscopic Currents

In a similar vein, in principle one should be able to record single-channel properties, including kinetics and density, and then apply the results to whole-cell models. However, there have been few reports of this approach to date. One difficulty is that there may be significant alterations in channel behavior when going from the single-channel recording environment to the "natural" state. We have tried this approach for some descriptions of Na^+ channels (Section 10.2.3) with limited success at reproducing expected macroscopic current behavior (see also Destexhe *et al.*, 1994). In particular, the limitations of whole-cell models based on measured channel densities (see Section 6.9) may reflect, among other things, an inadequate kinetic description.

2.3. Type of Preparation

Electrophysiological protocols for measuring electrical properties of neurons *in toto* and specific currents use a wide variety of preparations, each with its own advantages and disadvantages. Interpretation of data must take into account how the relevant protocol compares to that corresponding to reproduced behavior. As stated at the beginning of the chapter, the assumed standard condition is the slice preparation, using either sharp or whole-cell patch (Sakmann and Neher, 1995) electrodes. Note of course that this standard, dictated by the primacy of this preparation in the literature, is rather arbitrary if we are eventually interested in understanding the "true" state of the cell. Consequently, we can assume that if the slice (Alger *et al.*, 1984) is the benchmark preparation, then data from, progressively, acutely isolated cells, cultured slice (Thompson and Gähwiler, 1992), and finally cultured cells may require more flexibility in the interpretation. Likewise, perforated patch* and on-cell (cell-attached) patch recordings (Marty and Neher, 1995) are incrementally further removed from the control case.† Similar considerations apply to the age of the animal, species, and the hippocampal subfield in which the HPCs are found (for example, the degree in which properties of CA3 cells may be definitive of CA1 cells, or vice versa, may depend on the specific mechanism under consideration). Additional considerations for interpreting sharp versus patch electrode recordings are discussed in Section 2.8. A review of the variables implicit in the techniques for slice electrophysiology is given by Reid *et al.* (1988).

From the point of view of parameter estimation of specific channel properties, the slice preparation allows recordings of cells with minimum disruption of the membrane properties and morphology, with the disadvantage of limited access for pharmacological manipulations and poor control of membrane voltage. Isolated cells, on the other hand, allow better control of the cell's extracellular environment, and better voltage clamp due to truncation of the dendritic tree. It follows then that recordings from isolated cells will tend to emphasize somatic membrane properties, rather than dendrites, an important consideration when postulating the existence of a given dendritic channel based on isolated cell studies. While cultured preparations may offer better access and control than intact or isolated preparations, these techniques may significantly alter the expression of biophysical mechanisms. A further caveat with the culture technique is that direct identification of pyramidal cells per se is not possible.

2.4. Kinetic and Pharmacological Dissection

Pharmacological and kinetic techniques are often used to isolate individual nonlinear components of the cell response. Neither method is perfect, since both the sensitivities to various drugs or chemicals and the time scales for activation/inactivation may overlap. In the case of pharmacological studies, it is also quite important to consider the side effects of a given "cocktail" in the interpretation of published results.

For example, the choice of the electrode solution is a critical factor in deter-

*In principle the perforated patch technique may give a better picture of the "true" physiological condition; however, it as yet cannot be considered the standard since the literature using this technique is relatively sparse.

†On the other hand, in theory single-channel measurements from the cell-attached configuration should provide the most unambiguous picture of detailed channel kinetics.

mining which currents will be observed. Certainly the standard procedure is to include or exclude specific ingredients in the pipette with this in mind. This situation is more critical in the case of whole-cell patch recordings, because of dialysis of the intracellular medium by the pipette solution. For example, Zhang *et al.* (1994) have done a recent systematic study on the effect of various whole-cell pipette solutions on one hallmark of the HPC response, the slow after-hyperpolarization (sAHP; Section 6.4.4). The authors found that the sAHP (and associated accommodation) was very similar to classical results when the patch electrodes were filled with $KMeSO_4$, whereas other commonly used patch pipette solutions attenuated this response. The $KMeSO_4$ patch recordings were also very similar to $KMeSO_4$-filled sharp-electrode recordings presented in the study.

While adding some pharmacological agent or blocking ion, either to the bath or pipette solution, is the more obvious way of isolating specific currents, manipulation of the effective concentration of a permeant ion can also be an important part of a given protocol. For antagonist actions, replacement of an ion with an impermeant one may be used. For example, while extracellular Cs^+ is known to be a specific blocker of I_Q/I_H (Section 14.1), recording with an electrode solution whose principal cation is Cs^+ ("cesium loading") has a general K^+ channel blocking action, simply because there are no K^+ ions to flow out, and K^+ channels typically have a low permeability to Cs^+.

Increasing the gradient of K^+ or Na^+ is often used to enhance the respective currents, but the already high gradient of Ca^{2+} requires other strategies. One approach is to replace Ca^{2+} with Ba^{2+}, which has a higher permeability than Ca^{2+} in many Ca^{2+} channels—in the later discussion on Ca^{2+} channels there will be references to both measurements made of Ca^{2+} currents and Ba^{2+} currents (via Ca^{2+} channels) for this reason.

Another factor in the expression of various neuronal properties is the pH of the external environment, and the buffer system used in the preparation. While a modulation in pH per se is linked to various pathological states in brain tissue and concomitant changes in firing behavior, the dependence on buffer system is of particular relevance to the *in vitro* situation and the subsequent interpretation of reported experiments. Church (1992) has examined this for rat CA1 pyramidal cells in the slice, and found several properties which depend on whether an HCO_3–CO_2- or HEPES-buffered medium is used. He found that a switch from the former to the latter system caused a fall in E_{rest}, a rise in spike threshold, a reduction in R_{in}, and a reduction in rectifying responses around E_{rest}. All major phases of after-hyperpolarizations following spikes (fAHP, mAHP, and sAHP; see Section 6.4) were also reduced, and consistent with these reductions was a reduction in spike adaptation. The HEPES buffer also blocked current-evoked bursting activity in those cells that burst under the HCO_3–CO_2 buffer.

2.5. Temperature Dependence

In general the kinetics of biophysical mechanisms are temperature dependent, and this dependence has been characterized in a variety of theoretical descriptions (Tinoco *et al.*, 1978; Hille, 1992). A major, and difficult, consideration in integrating data from different sources is taking this dependence into account, since it is often the case that a model's database may include experimental conditions with temperatures ranging over 10°C.

While a few studies of channel kinetics have included temperature as an experimental variable, in general the Q_{10} of a mechanism must be postulated ($Q_{10} \approx 3$ is a typical assumption). All Working model parameters presented in this chapter will be nominally adjusted for room temperature, but additional refinement to account for temperature will be necessary.

One of the few explicit studies on temperature and HPC responses were reported in Thompson *et al.* (1985). They describe a strong temperature dependence of R_{in} (see Table I), various spike characteristics (see Table II), afterpotentials (see Section 6.4.4), and spike adaptation, for temperatures ranging from 27 to 37°C. Briefly, they found that E_{rest} and the amplitude of the initial spike in response to a current step were not significantly changed by the temperature, but that reducing the temperature increased both the sAHP and the adaptation of spike trains, and reduced the spike amplitude for spikes after the first spike in a train.

2.6. Age Dependence

Various cell properties depend on the age of the preparation, in particular at early developmental stages [e.g., see work on the characterization of K^+ currents with early development by Spigelman *et al.* (1992)] or in very old animals (e.g., Landfield and Pitler, 1984; Thibault and Landfield, 1996; Campbell *et al.*, 1996).

There is a somewhat implicit correlation between properties recorded with patch electrodes and the age of the animal, since in general patch recordings are done on younger animals (before P25 in rat), whereas sharp-electrode recordings are done at essentially any age.

2.7. Hippocampal Subfield Dependence

Are there fundamental differences in the electrophysiology between pyramidal cells in the various hippocampal subfields? The traditional view is that certain basic firing types correlate with subfield: regular-spiking cells predominate in CA1 (Schwartzkroin, 1977), and bursting cells are mainly in CA3 (Bilkey and Schwartzkroin, 1990). A more refined view from Masukawa *et al.* (1982) was that bursters are found in both subfields, but primarily on their lateral borders (CA1a, CA1c, CA3a, and CA3c), while the medial parts (CA1b, CA3b) showed regular spiking responses.

Clearly there are (sub)regional specializations, but whether these imply fundamental differences in the cells' intrinsic mechanisms is still open to question. The recent result of Jensen and colleagues (M. S. Jensen *et al.*, 1994; see Section 6.7) that an apparently normal-looking regular-spiking cell can be induced to look like a bursting cell by (among other things) lowering $[K^+]_o$ is an indication that the differences are more quantitative than qualitative. Along these lines are the demonstrations by models of Traub *et al.* (1991) and Migliore *et al.* (1995) showing how nondrastic quantitative modifications in the distributions of nonlinearities in the same basic structure could account for the various electrophysiological types.

2.8. Differences in Firing Properties between Sharp versus Patch Recordings

In Section 5.2.1 we will discuss the different "passive" properties of HPCs recorded with sharp microelectrodes versus whole-cell patch electrodes. It is also important to consider whether or not there is a difference in firing properties between the two techniques. This question has not been specifically addressed in the literature, other than the paper by Zhang *et al.* (1994) discussed above.

A very basic and perhaps obvious point to keep in mind is that the size of stimulus currents used for whole-cell patch recordings is typically on the order of one fifth of that used in sharp electrodes, because input resistances with patch electrodes tend to be higher than those for sharp-electrode recordings (Section 5.2.1). Thus, for example, current-clamp recordings using sharp electrodes will typically use stimuli on the order of 0.5–2.0 nA to elicit spike trains, while with patch electrodes similar responses are evoked with typically 0.1–0.5 nA.

In the perforated patch recordings by Spruston and Johnston (1992) the examples shown of firing patterns for the pyramidal cells deviate from the classical pattern. Spikes were followed by strong hyperpolarizations that are more reminiscent of interneurons (see discussion of pre-hyperpolarizations in Section 6.1.2) and thus look very "nonpyramidal." Since these trajectories reflect activation of significant conductances, and can have a strong effect on various firing properties, this discrepancy may be more than anecdotal. For this reason it may be of interest to examine in more detail particular characteristics of perforated patch recordings (this pattern can be seen in sharp recordings, however; e.g. Deuchars and Thomson, 1996).

A careful analysis of current-clamp artifacts intrinsic to the design of several popular patch-clamp amplifiers has been presented recently by Magistretti *et al.* (1996). They show that the additional feedback loop in these circuits introduces significant distortions of fast events, in particular action potentials, including overestimation of spike amplitude and after-hyperpolarizations. Interestingly, due to the simpler input stage design of the classical "bridge-mode" amplifiers of the sort typically used with sharp microelectrodes, a practical result is that these signals are more faithfully recorded using the "older" amplifier design [note that the study of Spruston and Johnston (1992) used an AxoClamp amplifier, which is a "bridge-mode" design].

Magistretti *et al.* (1996) proposed a modification to the patch-clamp amplifier circuit to circumvent this problem. In summary, quantitative analysis of current-clamp patch-electrode recordings must take into account the hardware used to a greater extent, considering that this issue was probably somewhat taken for granted previously. For the purposes of this chapter, with respect to quantitative parameters of spikes we shall place more emphasis on data made with sharp recordings.

2.9. The Measured Voltage

A final consideration in reviewing electrophysiological data is the accuracy of measured voltages in light of two main types of artifacts. The first is the presence of a tip potential which derives from the different ionic composition of the intracellular milieu and the electrode solution, which may or may not be correctly estimated. The second artifact arises from the resistance of the electrode generating a voltage drop when current is passed. This property may be compensated for by either bridge circuits or by switched clamp protocols, but perfect compensation is difficult, especially when the currents are large. In sum, these artifacts can generate errors on the order of a few millivolts, depending on the experiment, and should be considered in the quantitative evaluation of measured cell properties.

3. Strategies for Single-Neuron Models

Single-cell behaviors that we may choose to model form a large-dimensional space, considering the possible variables which define the context of the model (form of the input, chemical/pharmacological milieu, temperature, species, cell type, etc.). Basic subsets of the model behavior include the linear or quasilinear properties of the cell and the basic firing patterns observed in the slice preparation in response to intracellular current injection or activation of extrinsic (afferent or orthodromic) pathways. As mentioned earlier, in this chapter we shall primarily, but not exclusively, consider intrinsic stimulation.

3.1. Building the Model

If we assume that the mechanisms that determine single-spike characteristics are shared to some degree with those that determine repetitive firing, then a reasonable strategy is to start small and work up. Thus, in constructing a single-cell model a first step can be to reproduce a single, isolated spike and then extend the description to match various repetitive firing characteristics. Tractability encourages this initial hierarchy, especially since it is quite common for the set of model parameters that satisfies one behavior to be disjoint from that which reproduces another. In particular, since parameters that work well at one time scale may fail on another time scale, the process of parameter fitting is profoundly iterative.

Concentrating on reproducing responses to current stimulation assumes that derived mechanisms will respond correctly to properly characterized synaptic input. Thus we assume that dividing the modeling effort into two distinct parts—synaptic versus nonsynaptic mechanisms—is a reasonable strategy, despite the obvious direct overlaps between the two (e.g., synaptic channels with voltage dependences).

Often an additional assumption for single-cell models is that many channel properties can be reproduced to a good first approximation by neglecting dendritic channel distributions and placing all nonlinearities at the soma (see Section 4). A related issue, when considering synaptic input, is the nonlinear effect of the dendritic tree on synaptic integration (see Section 5.5).

The Relevance of *Accurate* Conductance Models

It is important to emphasize that these considerations imply that models built with reduced representations of input/output mechanisms—e.g., at one extreme integrate-and-fire models, with perhaps simple adaptive mechanisms—may neglect crucial elements. This is a difficult question, since large network models require some simplification in order to be testable in any practical way.

A challenge of formulating a neuron model is that such models are often underconstrained: at one extreme, a wide variety of kinetic descriptions can give rise to a stereotypical action potential. As will be pointed out later, various published HPC models use quite different kinetic parameters for the membrane channels, for example in characterizing the Na^+ current (Section 10).

However, the detailed dynamics of a neuron's conductance state that arise from intrinsic properties may be quite important for understanding the interaction between synaptic (that is, conductance-based) input and the action potential output. Simplifications that decouple input integration and firing properties may neglect or misconstrue important effects (Holt, 1998), especially since the conductance changes of functional synaptic input can be several times the cell's resting conductance (Borg-Graham *et al.*, 1998). It is not sufficient to assume that a model that generates a realistic-looking voltage trace under some limited stimulus conditions can adequately predict neuron output in response to synaptic activation.

In summary, the development of reduced models, appropriate for network models, should derive from a careful analysis of the more complete type of model described here, with as full an understanding as possible of what may be lost by the simplification.

3.2. Parameter Searching

As alluded to earlier, a nonlinear system of many parameters is likely to have many local error minima with respect to the reproduction of a given behavior. At some point it becomes important to decide (usually implicitly) which minimum is closet to the "best" solution. To my knowledge no formal method for avoiding this problem exists [but there have been attempts; see Bhalla and Bower (1993) and Foster *et al.* (1993)]. Therefore a specific choice of parameters for these models should be considered as a possible solution, open to the possibility of further refinement. Indeed, as will be demonstrated in this chapter, the parameters of published models tend to be quite distinct from one another, and tend to work only in a limited testing domain.

Of course, the very notion of an "error" in the output of a simulation implies that there is some specific data that we are trying to reproduce. One of the goals of this chapter is to present the quantitative variation of a few aspects of HPC output, at least to provide a neighborhood of behaviors for which single-cell models may be designed to match. Certainly this list is incomplete, and it is difficult to say *a priori* whether or not these aspects are crucial for what a neuron is *really* doing.

3.3. Model Robustness and Validation—as Robust as Nature?

Assuming that we have chosen something to reproduce, quantitative fitting of cell models to data is hampered by the fact that, first, biology is probably quite variable and, second, biological measurements are definitely variable. This situation is often exacerbated by the experimental design and/or presentation of data in the literature, which may not be well matched for the derivation of model parameters (Bower and Koch, 1992; Foster *et al.*, 1993).

For example, quantitative parameters are almost always presented as averages (±SD) over a group of cells. However, these statistics say nothing about the intrinsic variability of *each* cell. This information may be even more important when more than one parameter, say A and B, is measured: model parameter estimation may be very different if there is a significant covariance between A and B.

A related issue is parameter sensitivity, or fragility, of a given model. In some respect the biology of neurons is not very fragile, in that many basic behaviors at the cellular level are maintained over some nonzero range of parameters (e.g., temperature, extracellular constituents, measurement techniques). The problem is to predict the relationship between the sensitivity of neuronal phenomena to that of experimental parameters, and the sensitivity of model behavior to that of model parameters.

In our work we use the following ad hoc criteria with respect to parameter precision (and by implication set a criterion for robustness): every model parameter is constrained to at most three significant figures, with the majority having only one or two significant figures. This is a bit more conservative than allowing the unit step of a given parameter to correspond to the likely range of the parameter divided by 30–300 (Sokal and Rohlf, 1981).*

Another test of model design is the consistency of parameter precision for a particular mechanism. Different mechanisms may have different intrinsic precisions, thus it may be reasonable for a cell model to have components defined with single-digit precision (e.g., R_m or C_m), with others defined with higher precision, especially sensitive nonlinear mechanisms. This is related to the fact that neuron models represent stiff systems, that is, they display dynamics on a very wide range of time scales (Mascagni and Sherman, 1998). In any event, the free parameters that define a specific mechanism should have more or less the same precision. For example, some of the reviewed models include mechanisms in which the precision of the parameters of a single mechanism ranges from one to five significant digits [e.g., "10" and "18.975" in the definition of I_C kinetics in Traub *et al.* (1991); see Section 7.1.1]. This range suggests a parameter sensitivity that may be unrealistic, and perhaps a choice of an unlikely mechanism.

*For example, if we are specifying the reversal potential of a current, then the range of values from first principles is roughly 100 mV. This implies that the precision of this parameter should be on the order of 0.3–3 mV; e.g., a value of −72 is reasonable, but −72.25 is probably overspecified. If the test behavior is significantly affected by a change from −72 to −72.2, this may indicate that some part of the model (hopefully related to this parameter in an obvious way) is poorly constructed. The relevant precision of a given model parameter is also indicated by the error in any reported measurements; more specifically, the measurement gives a lower bound on the likely biological, true precision. For example, if the reversal potential in the example above was reported as −72.3 ± 0.1 mV, we might consider that this parameter is quite fixed in the real system. On the other hand, a reported value of −72.3 ± 4.2 mV could mean either that there is a lot of variation in the fundamental mechanism or there is a large variation inherent in the measurement technique.

3.4. The Purpose of Modeling—Testing Speculations

At minimum, a neuron model should reproduce the database from which it was built, that is, the templates used for determining model parameters. It is at this step that the original model of Hodgkin and Huxley succeeded brilliantly, and in doing so showed the plausibility of their speculations on channel kinetics. Next, the generalizability of the model should be tested against other, independent sets of data. Such tests have led to various and sometimes drastic revisions to the Hodgkin–Huxley squid axon model (e.g., Patlak, 1991; see also Pallotta and Wagoner, 1992); the success of the published HPC models at this stage is spotty. If a model does not pass this stage, then the raison d'être of models, generating testable predictions, becomes barely tenable—at best the flawed theory encapsulated in such models may still yield intriguing predictions.

3.5. Tools for Compartmental Models

Both analytical and numerical techniques have been applied in the analysis of neuron models. Analytical approaches are more commonly used for the analysis of linear mechanisms (discussed in Section 5.3 and elsewhere), although some work has been done for nonlinear mechanisms (Jack *et al.*, 1983). Numerical techniques are much more common, due to their applicability to a variety of nonlinear mechanisms and their overall ease of use. These techniques are exemplified by compartmental models of neurons (for review see Segev *et al.*, 1998), in which the branching cable structure of the cell is approximated by a series of isopotential compartments interconnected with resistors. All of the single-cell models that we discuss in this chapter are of this type.

The descriptions by Hines (1984) of $O(n)$ integration of tree circuit topologies, as naturally found in individual neurons, and mid-step solving of time-dependent nonlinear elements (e.g., voltage-dependent gating particles) are exploited by most modern simulators, and have been vital contributions in making simulations of detailed anatomies practical. Hines and Carnevale (1994) have written a short review of numerical methods appropriate for neuronal modeling; a more detailed discussion may be found in Mascagni and Sherman (1998).

There are now several compartmental model simulation packages in the public domain designed for the computational neuroscience community, including NEURON (Hines, 1992), GENESIS (Bower and Beeman, 1994), and the program used for the simulations presented in this chapter, Surf-Hippo (Borg-Graham, 1995). Each of these programs, which can handle both single-neuron models and networks, runs on Unix workstations and x86 boxes (PCs). A recent survey of tools is given by De Schutter (1992), although current availability of public domain tools is better found using the WWW.

As mentioned earlier, the necessity for numerical methods has one major drawback: formal verification of even moderately complicated neuron models is not possible. One approach to verification is by the cross-validation of models using independent tools, possibly written in different languages and with major or minor differences in algorithms. This becomes practical with the increasing availability of sophisticated public domain packages.

For example, GENESIS allows various numerical integration methods, which may point out particular instabilities in a given formulation. Surf-Hippo is one of the few complete packages written in Lisp, as opposed to C. Lisp has the advantage that the user is communicating directly with the Lisp interpreter environment, and thus has complete access to all components of a simulation. Simulation scripts, a necessity for serious parameter searching, are also written in Lisp, whose flexible and powerful syntax is arguably more transparent than most other languages. Surf-Hippo allows both fixed and adaptive time step integration—using an adaptive time step can give much faster run times for typical simulations, with the option of verifying selected results using the more conservative fixed time integration. Surf-Hippo can simulate both nonideal and ideal voltage clamp, in the latter case by effectively dividing the circuit into two parts around the clamped node. The ideal voltage-clamp algorithm not only allow analysis of voltage-clamp protocols without the complications of a finite source series resistance, but, because the method is very stable, it also makes for much faster simulations. Both GENESIS and Surf-Hippo can simulate recordings from extracellular electrodes, that is, measurement of the field potential, assuming a purely resistive, homogeneous extracellular medium.

Recently, a set of benchmarks ("Rallpacks," after Wilfrid Rall) have been proposed that are specifically tailored for neuron simulations (Bhalla *et al.,* 1992). These include benchmarks for which analytical solutions are available (signal propagation in linear dendritic trees), and nonlinear structures based on the "canonical" Hodgkin–Huxley Na^+ and K^+ channel equations. This last test is an example of the technique of cross-validation mentioned above, where solutions (waveforms) have been published for comparison. The results of the Rallpack benchmarks for NEURON, GENESIS, and Surf-Hippo are comparable.

The graphical user interfaces for NEURON, GENESIS, and Surf-Hippo are quite sophisticated, an important point for practical work. For example, Surf-Hippo allows automatic saving of edited model parameters into files which are both loadable and human readable. This feature helps to avoid errors in saving results at any time during a simulation session. "Point-and-click" histology rendering in these packages (with 3D anatomical representations in NEURON and Surf-Hippo) allow easy modification of cell models.

4. Anatomy and the Model: Data and Methods

The circuit topology of the single-cell model—the mapping of the cell anatomy to a compartmental model—can have a range of complexity, depending on the question being addressed. As illustrated in Fig. 1, this complexity may range from a few compartments to a circuit with several thousand compartments, corresponding to the level of detail typically found in anatomical studies (for a general discussion see Turner and Schwartzkroin, 1984).

There are essentially two stages of the mapping. The first concerns the representation of the branching structure of the dendritic (and possibly axonal) tree, and may be described in terms of cable sections (regular or tapering cylinders). The second stage translates the connected cables into connected isopotential compartments.

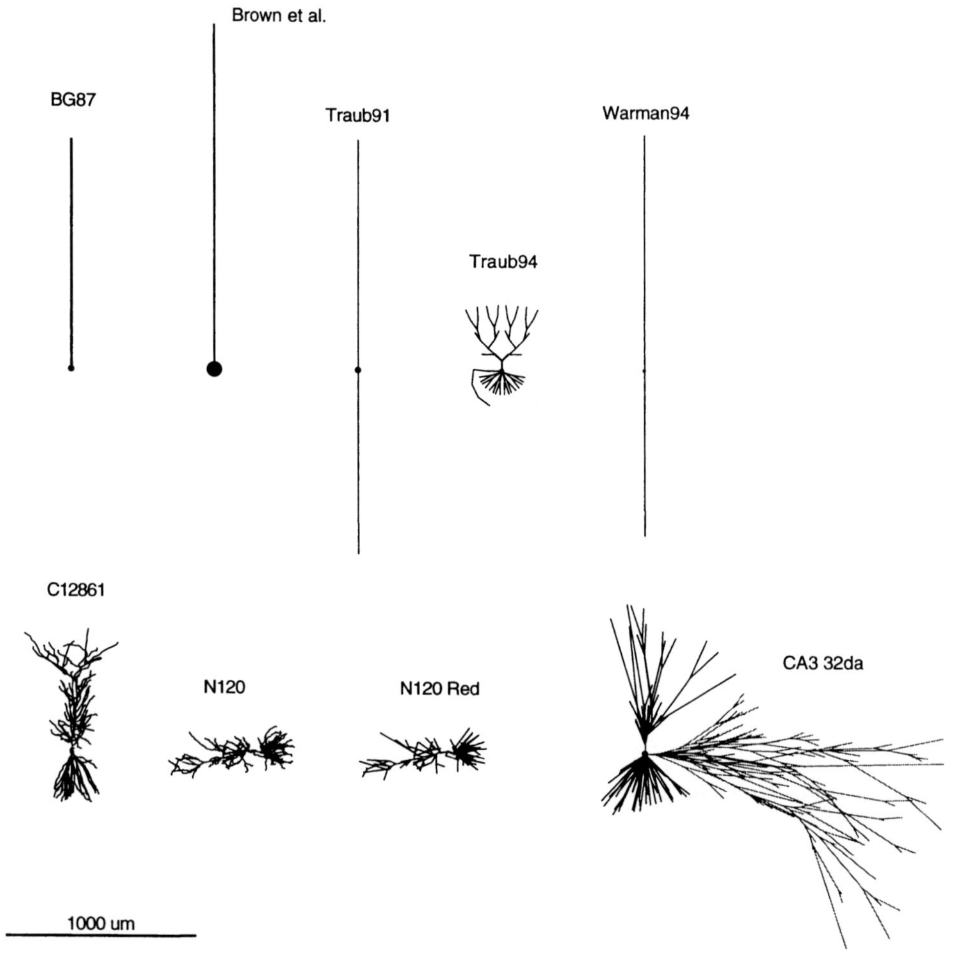

Figure 1. CA1 and CA3 pyramidal cell model anatomies illustrating a range from simplified abstractions to those derived from fully detailed anatomical data. Models are drawn to scale as represented by the Surf-Hippo neuron simulator. Diameters of nonvertically or horizontally oriented processes are distorted in this figure because of the printing resolution, such that they are drawn as if the diameter was 5 μm. Abstracted model anatomies include the BG87 (used in the Working model, six compartments), Traub91 (19 compartments), Traub94 (66 compartments), and Warman94 (16 compartments) models (Section 7.1). The process curving left and down from the soma in the Traub94 anatomy is the axon. The Brown *et al.* anatomy is a reverse-engineered soma/short-cable model derived from parameters given in T. H. Brown *et al.* (1981), as described in Section 5.3.2. The remaining representations are taken directly or derived from anatomical files. C12861 (941 compartments) is a CA1 cell (supplied by David Amaral, in the ntscable program distribution; Wathey, 1989). A 400-compartment version of this cell, with an axon, was used for the Wathey92 model (Section 7.1). The N120 cell (3205 compartments) is also from CA1, and was supplied by Dennis Turner [appears in Pyapali and Turner (1994, Fig. 3A); although this paper describes dendritic alterations subsequent to kainic acid lesions, this particular cell is from a normal animal]. The N120 Red cell (193 compartments) is derived from N120 (see Section 4.2 and Fig. 3). The CA3 32da anatomy (367 compartments) was supplied by Guy Major; the original cell is drawn in Fig. 9 of Major *et al.* (1994). The anatomy file did not include 3D locations, so in this rendering an ad hoc procedure was applied to branch points to give a reasonable overall shape. Also, the axon was traced in this cell, and is the structure branching off to the right.

4.1. Reducing Branching Cable Structures and the Placement of Channels

The classical work of Rall (for review see Rall *et al.,* 1992) on the analysis of linear dendritic trees included the derivation of his "3/2" rule for the collapsing of a set of daughter branches and their parent branch into a single cylinder with the same proximal input impedance.* Unfortunately, HPC anatomy does not appear to conform to these restrictions: for example, Turner *et al.* (1995) quote a range of average values for $d_r^{3/2}$ in CA3 HPCs between 1.25 to 3.36 [Turner and Schwartzkroin (1980) reported similar results for CA1]. It must also be emphasized that this technique is based on linear structures.

Nonetheless, it is often reasonable to reconstruct basic cell properties using soma-centric models with simplified dendritic trees, for example, a soma/short-cable model consisting of a single-compartment soma with one or two unbranched dendritic cables. Under some conditions it may even be reasonable to estimate parameters using a single-compartment model.

On a practical level, estimating parameters of distributed nonlinearities is much more difficult than when a point distribution is assumed—a first approximation of somatic responses to somatic stimuli restricts nonlinearities to the soma. Here the dendritic tree is a linear load for the soma, and as such a simple cable (with on the order of five compartments; Borg-Graham, 1987b) can be constructed whose linear input impedance approximates the full tree very well (Turner, 1984b).

In addition, until recently almost all quantitative measurements of electrical properties were derived from somatic impalements [early exceptions in hippocampus include Wong *et al.* (1979), Benardo *et al.* (1982), and Masukawa and Prince (1984)]. Such measurements suggest a first interpretation with a soma/short-cable model.

However, simulated responses to dendritic input using models with vastly simplified dendritic trees must be interpreted very carefully. Simulating nonsomatic input by stimulating the "dendrite" in these sorts of structures (e.g., Traub *et al.,* 1991) in the best case assumes a linear response to simultaneous stimulation at a set of locations with a fixed electrotonic distance from the soma, an unrealistic experimental situation. In the more realistic case, the contribution of the local dendritic nonlinearities would be highly dependent on the local input impedance at the site or sites of stimulation. Thus, even if experimental simultaneous stimulation at equidistant sites could be assumed, their composite response could not be lumped together in a model by stimulation of a single location in a short-cable geometry. A compromise approach is to model a local patch of dendrite in fine detail, while representing the rest of the tree in a much more schematic fashion. An example of this approach is given by Holmes and Levy (1990) in a model of synaptic interaction with internal Ca^{2+} in a dentate granule cell.

In summary, the short-cable representation of the dendritic tree can at least represent the dendritic linear impedance load on the soma, and at most, with careful interpretation, may provide a substrate for a first approximation of the

*An equivalent cylinder is possible if the electrotonic lengths of all the daughter branches are the same and they have identical terminations, and if the ratio $d_r^{3/2}$ of the sum of the daughter diameters each raised to the 3/2 power to the parent branch diameter raised to the 3/2 power is equal to one.

somatic view of intrinsic dendritic nonlinear mechanisms, and possibly the integration of synaptic input.

4.2. Dendritic Tree Reduction from the Raw Anatomy

Assuming a defined branching structure, various algorithms may be employed to transform anatomical coordinate data into a compartmental model (e.g., Stockley *et al.*, 1993; Henze *et al.*, 1996). The simplest approach, of course, is a one-to-one mapping, where there is one compartment (dendritic segment) for each anatomical location, with the location and dimensions of that compartment taken as some function of the associated anatomical point and either one or two adjacent points.

For most practical simulations this produces an overdetailed representation since typical anatomical studies include on the order of thousands of dendritic locations and associated diameters. Here we describe one method for reducing such a description to fewer compartments, with the minimum number given by twice the number of branch points in the dendritic tree, plus the number of dendritic trunks. The motivation for this method is to preserve both the three-dimensional shape of the neuron and the electrotonic properties of the tree as best as possible.

The reduction is iterative and operates on a pair of segments at a time. Starting from the soma and working down each proximal trunk, two consecutive segments, call them *A* and *B*, are combined into one if there is no other segment common to both (i.e., the connection is not a branch point) *and* if the electrotonic length of the candidate replacement segment is less than that set by some *a priori* criterion (i.e., a parameter L_{max}). The process continues distally—if *A* and *B* were combined in the previous step into a segment *AB*, the next pair consists of *AB* and the segment distal to the original segment *B* (call it *C*). Otherwise, the next pair to be considered is *B* and *C*. Figure 2 illustrates the procedure on a short chain of dendritic segments.

The parameters of the replacement segment are derived according to the following constraints:

1. The total axial series resistivity of the two original segments is conserved.
2. The total membrane impedance (area) of the two original segments is conserved.
3. The endpoints of the new segment correspond to the noncommon endpoints of the original two segments.

In order to meet these constraints, an additional parameter for each segment must be introduced, which we are choosing to be a coefficient, a, of the cytoplasmic resistivity that is used when determining the segment intracellular axial resistance (the default value for a is 1). This coefficient and the dimensions of the new segment are calculated as follows. Let l_i be the segment length, d_i the segment diameter, and a_i the coefficient for the segment axial resistivity, where i is p, d, or n for the proximal, distal, or new segment, respectively. As stated above, the length of the new segment l_n is fixed by the far endpoints of the original proximal and distal

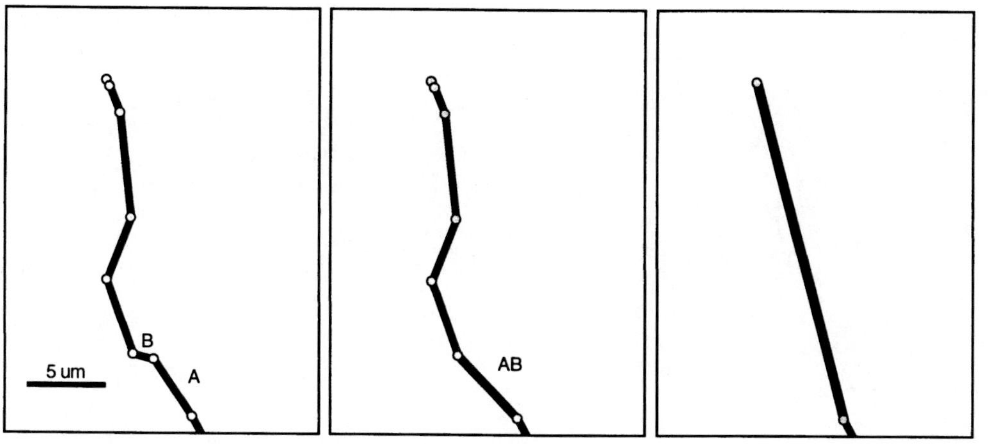

Figure 2. Demonstration of the consolidation of dendritic compartments. The algorithm described in the text works on successive pairs of dendritic segments, working distally. In the panel on the left, segments *A* and *B* are chosen to be consolidated into segment *AB*, shown in the middle panel. The proximal and distal endpoints of *AB* correspond to the proximal endpoint of *A* and the distal endpoint of *B*, respectively. The panel on the right shows the final result after continuing this procedure down the dendrite, assuming that the electrotonic length of a segment resulting from the consolidation of all the original segments is less than some *a priori* criterion. Notice that the diameter of the final replacement segment is slightly larger than that of the original segments, reflecting the conservation of membrane area in the algorithm and the shorter final path length of the branch.

segments. Conservation of membrane area (that is, conservation of membrane capacitance and resistance) gives

$$d_n = \frac{(l_p \times d_p) + (l_d \times d_d)}{l_n}$$

Conservation of axial resistance gives

$$a_n = \frac{(l_p \times a_p)/d_p^2 + (l_d \times a_d)/d_d^2}{l_n / d_n^2}$$

This technique yields a new segment whose cable parameters are similar to the concatenation of the two parents. However, the actual proximal input impedance of the new segment will always be greater than or equal to the input impedance of the former segment pair, since the membrane impedance is now farther away electrically from the proximal end of the new segment relative to the membrane impedances of the original pair. Thus, the input impedance of a transformed (consolidated) tree will always be greater than or equal to the original tree, and the difference in the impedances will be greater, the larger the consolidation (equivalently, the larger the L_{max}).

This algorithm is implemented in the Surf-Hippo simulator (Borg-Graham, 1995). An example is illustrated in Figs. 1 and 3, where a CA1 cell anatomy description with 3205 points is reduced to a 193-compartment model. Somatic and dendritic responses for the two models to current stimulation at each soma and similar positions in the dendritic tree are compared in Fig. 4. This figure shows that both

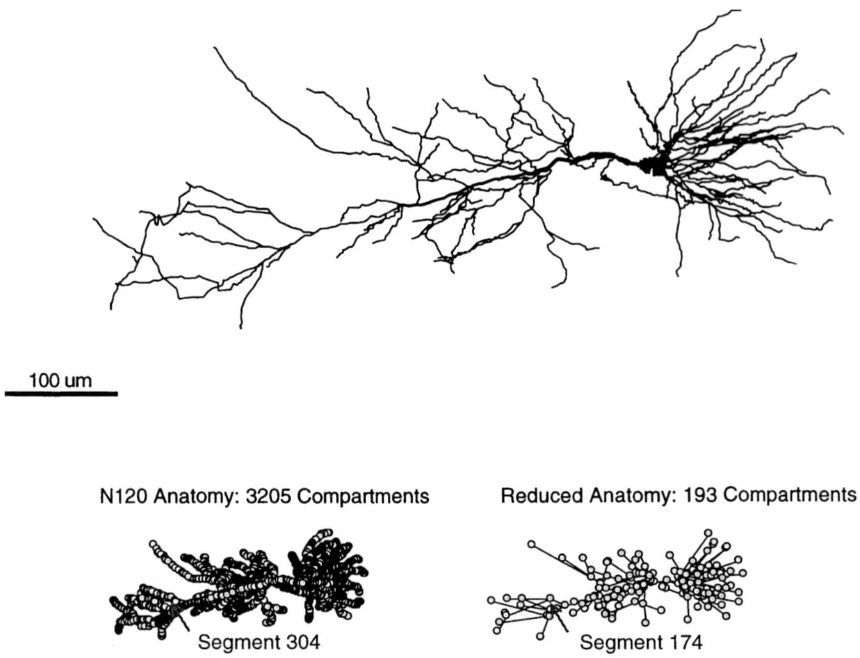

Figure 3. Top: Larger view of the cell model derived from original anatomy file of CA1 N120 cell (Pyapali and Turner, 1994, Fig. 3A). Somatic "notches" result from approximating the soma with a series of right cylinders. Bottom: Comparison of compartment density for original anatomy model and reduced model (the N120 Red model in Fig. 1) derived by the algorithm described in Section 4.2. Current sources and voltages associated with segments 304 and 174 are referenced in simulations shown in Fig. 4. Scale bar applies to the cell model at the top.

centripetal and centrifugal passive propagation and local input impedance are very similar for the two models.

4.3. Accounting for Dendritic Spines

HPCs are covered with thousands of dendritic spines. Depending on the question being addressed, spines may be incorporated in a compartmental model with either explicit representations (e.g., a few compartments per spine; Zador *et al.*, 1990; Schiegg *et al.*, 1995), mixed implicit and explicit representations, where a subset of spines (e.g., for which synaptic input is applied) have detailed representations, with the membrane properties of the remainder of the tree adjusted for spine membrane area (Holmes and Levy, 1990), or purely implicit spine representation via membrane parameter adjustment (Major *et al.*, 1994). On the other hand, it is not uncommon for single-cell models with partial or full anatomical representations to omit spines all together [e.g., the model of Jaffe *et al.* (1994); see Section 7.1], even when synaptic input is an explicit part of the model [e.g., the models of Traub *et al.* (1994), Wathey *et al.* (1992), and Migliore *et al.* (1995); see Section 7.1].

As mentioned, a simple method for incorporating the centrifugal view of dendritic spines, while maintaining the overall geometry of the dendritic tree, is to adjust R_m and C_m to account for the membrane area of the spines (Holmes and

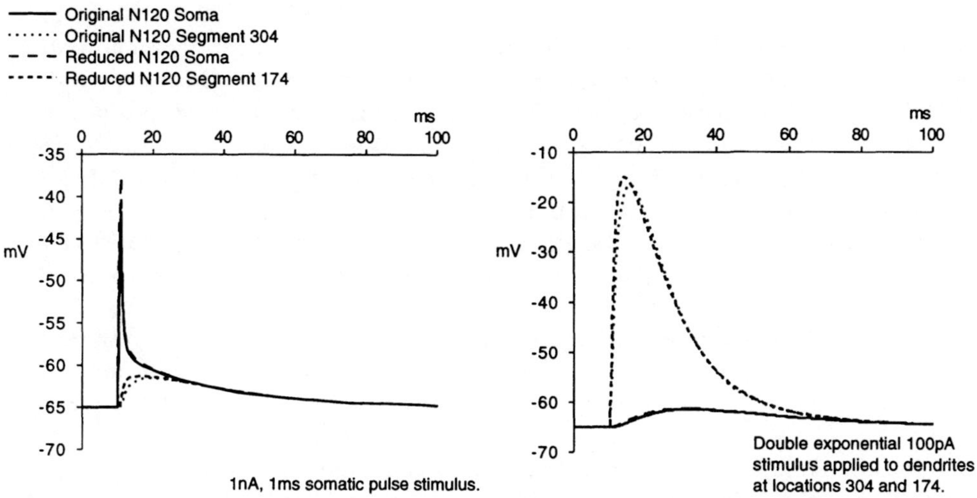

Figure 4. Responses to dendritic and somatic stimulation for the original N120 anatomy and the reduced N120 anatomy described in Section 4.2, demonstrating the close equivalence of these passive models. Dendrite segments 304 and 174 correspond to locations shown in Fig. 3. Left: Somatic and dendritic response to 1-nA, 1-msec current pulses applied at the somas. Right: Somatic and dendritic response to double-exponential current waveforms (difference of 1-msec τ and 10-msec τ exponentials, with waveform amplitude of 100 pA) applied at dendrite locations 304 and 174. The linear parameters for both cells are $R_m = 40$ kΩ cm², $R_i = 200$ Ω cm, and $C_m = 0.7$ μF/cm². The R_{in} of the original N120 cell is 191 MΩ, and of the reduced N120 cell is 199 MΩ.

Levy, 1990; see also Holmes and Rall, 1992b). This method assumes that the voltage drop across the spine neck is negligible (for nonstimulated spines).

For example, in a study of the electrotonic properties of rat CA1 HPCs, Mainen *et al.* (1996) estimated an average cumulative spine area of 2.85 μm²/μm of dendritic length (Harris *et al.*, 1992). For the cells analyzed in this study, this constant factor resulted in an average 52% increase in the total membrane area of the cell.* Similar parameters for spine incorporation were used in the analysis of CA3 HPCs by Major *et al.* (1994).

4.4. Modeling the Axon

Axon modeling is all the more difficult since data on their nonlinear structure are scant, probably less than that for dendrites. Of course, the original Hodgkin–Huxley model was of an axon, and the first application of nonlinear compartmental models was to the Hodgkin–Huxley squid axon model (Cooley and Dodge, 1966). Unfortunately, there are much fewer quantitative data for mammalian axons for which to build similar models (Segev *et al.*, 1992).

Another obstacle to realistic axon modeling in the case of HPCs is that axon morphologies are often not included in the anatomical data. One exception is the study by Major *et al.* (1994), where the axon arborization was traced with the same fidelity as the dendritic tree (the CA3 32da cell in Fig. 1).

*However, some of the interpretations in this paper may be problematic since the authors do not take into account the effect of a soma shunt from sharp microelectrodes, as described in Section 5.2.1.

We can use this anatomy to estimate the contribution of the axonal arbor to a linear model of the cell. As given in this study, assuming a homogeneous membrane resistivity R_m of 200 kΩ-cm^2 and an intracellular resitivity R_i of 270 Ω-cm, one finds a somatic input resistance of 267 MΩ. The total dendritic/axonal tree resistance is 277 MΩ, and the input resistance of the axon alone is 1555 MΩ. Thus, in this particular cell if the axon is ignored, the somatic input resistance becomes 321 MΩ, a 20% increase. Note that this difference may be an upper bound, since any myelination of the axon would be expected to increase greatly the appropriate R_m (this possibility may also mean that the estimated "overall" R_m in this study may be an overestimate).

Most published single-cell models ignore the axon completely; a few reproduce the contribution to the cell response by a simplified version of an axon, including I_{Na} and I_K for somewhat schematic action potentials [e.g., the models of Traub *et al.* (1994) and Pongracz *et al.* (1992); see Section 7.1].

In summary, *for the linear model,* neglecting the axon may not be wholly unreasonable. For the full nonlinear model, on the other hand, the validity of this assumption remains an open question. For example, it has been proposed that inhomogenous densities of Na^+ channels and/or different types of Na^+ channels with respect to the soma, dendrites, and axon can account for details of the location of spike initiation and threshold; we will return to this point in Section 6.3. The default assumption is that the contribution of axon nonlinearities to the somatic response can be approximated by "moving" those nonlinearities to the soma, essentially the same approximation that may be made with respect to dendritic nonlinearities as above.

4.5. Anatomy of the Working Model

The anatomy of the Working model is that used in the earlier soma/short-cable model by the author (Borg-Graham, 1987b; see Section 7.1.2; see also Fig. 1, the BG87 model). The parameters of this structure are a soma diameter of 35 μm, a dendrite length of 1200 μm, and a dendrite diameter of 12 μm.

5. The Linear Model: Data and Methods

In this section we summarize the basic data on the linear properties of HPCs, and discuss various reports and interpretations of these properties.

5.1. Reproduction of Linear Properties

Perhaps the most basic properties of neurons are those that characterize the resting state of the cell, that is, the somatic input resistance R_{in}, the main time constant τ_0, and the resting potential E_{rest}. The first two parameters implicitly assume linear characterizations of what is, of course, a profoundly nonlinear system (see Section 4). For practical purposes, unless specifically defined otherwise, R_{in} is taken from the small-signal response around rest (within about 5 mV), and τ_0 is the slowest time constant characterizing the small-signal impulse response.

The R_{in}, τ_0, and E_{rest} of a cell model are determined by the model geometry, the specific membrane resistivity R_m, the reversal potential E_{leak} associated with R_m, the cytoplasmic resistivity R_i, the specific membrane capacitance C_m, and membrane conductances activated at rest which can change the *effective* membrane resistance and reversal potential. For a given model, R_i and C_m are typically time invariant and assumed constant over the entire cell (for exceptions, see model discussions in Sections 7.1.1, 7.1.4, and 7.1.5). On the other hand the effective R_m and E_{leak} of the linear model may deviate from the intrinsic values of these parameters by consideration of an electrode shunt, which we will discuss shortly, and activated channels or synapses. Thus the modulation of one of the many identified non- or weakly voltage-dependent channels found in neurons, in particular, those modulated via second messenger pathways over time scales that are long (tens of seconds) compared to intrinsic firing patterns, can change the effective R_m and E_{leak}. Although we will not model these channels explicitly, examples of this type of channel in HPCs will be mentioned in this chapter. Synaptic effects, more relevant when reproducing the *in vivo* state, can include that of tonic, or "background" synaptic activity [recent *in vivo* data on this point in visual cortex are given in Borg-Graham *et al.* (1996); a model-based study of this question for neocortical cells may be found in Bernander *et al.* (1991)]. For all of these mechanisms—shunts, non-voltage-dependent channels, and background synapses—the functional sequelae may be captured by a static modulation of R_m and E_{leak}. In other words, these factors need not be considered in the dynamic structure of the model circuit. The participation of explicitly modeled channels that are open around the resting potential will also certainly influence the final value of E_{rest}, but these effects are not lumped into the linear model parameters.

A linear system approximation of the dendritic tree has been quite productive: linear cable theory has been applied in great detail to neurons (for review see Rall *et al.*, 1992). For example, as we discussed in Section 4, a first approximation to the distribution of nonlinearities in a single-neuron model places them on the soma, with the dendrites being linear (and ignoring the axon). This is an especially attractive strategy for evaluating dendritic tree properties, given the elegance of linear systems theory. Thus in a practical sense it is useful to assume that the electrical properties of dendritic membrane can be thought of as a set of nonlinearities on a linear substrate, where the linear portion is the dominant current path near rest (but see discussion in Section 5.5).

5.2. Measured Properties of the Resting State

In Table I measured values of E_{rest}, R_{in}, and τ_0 are listed from various reports, including data from both CA1 and CA3 HPCs, as well as dentate granule cells for comparison. This table illustrates a key difference between recordings made with fine-tipped, or "sharp" microelectrodes, and those made with whole-cell patch electrodes, specifically that the latter give significantly higher estimates for R_{in} and τ_0 than the former. On the other hand, the resting potential does not seem to depend much on the recording technique.

Table I. Values of Measured Linear Parameters from Various Sources, along with Values for the Working Model[a]

Reference	Temp. (°C)	Cell type/ method	E_{rest} (mV)	R_{in} (MΩ)	τ_0 (msec)
Turner and Schwartzkroin (1980)	—	CA1/S	—	28±7	11±4
T. H. Brown *et al.* (1981)	37.5	CA1/S	—	43±4	15±1
		CA3/S	—	39±3	19±2
		DGC/S	—	38±3	12±1
Thompson *et al.* (1985)	37	CA1/S	−64±9	30±9	—
	27		−60±7	52±13	—
Staley *et al.* (1992)	32	DGC/S	−74±2	54±3	14±2
		DGC/P	−84±1	228±14	27±1
Spruston and Johnston (1992)	32	CA1/PP	−64±2	104±10	28±2
		CA3/PP	−66±1	135±8	66±4
		DGC/PP	−74±2	446±87	43±4
Buckmaster *et al.* (1993)	35	CA3c, burst?/S	−62±0.9	52±3.5	16±1.2
Scharfman (1993)	—	CA3c, burst/S	−66±2	63±5	23±2
Major *et al.* (1994)	20–24	CA3/P	—	190[b]	93[b]
Zhang *et al.* (1994)	33	CA1/S	−66±1	38±2	—
		CA1/P[c]	−60±1	59±2	—
		CA1/P[d]	−63±1	51±3	—
M. S. Jensen *et al.* (1996)	33.5	CA1, nonburst/S	−63±3	25±8	8±2
		CA1, burst/S	−63±3	26±7	14±5
		CA1, all/S	−63±3	26±7	13±5
Fraser and MacVicar (1996)[e]	34–35	CA1/P	−65±0.1	149±3	20.7±0.4
Working model	—	—	−65.7[f]	39	14

[a]S, Sharp microelectrode; P, whole-cell patch electrode; PP, perforated patch electrode. For this and subsequent tables, most values are rounded to two significant figures, and are followed by the reported standard deviation when available; a blank field means parameter not given or not applicable; DGC, dentate granule cell. Measurements from CA1 and CA3 are all from pyramidal cells. Patch pipettes in both Major *et al.* and Fraser and MacVicar were filled with K-gluconate solutions. For the different temperatures tested by Thompson *et al.*, and the different patch pipettes filling solutions tested by Zhang *et al.*, there was a statistically significant difference in the R_{in} values, but not the E_{rest}. In the different firing classifications used by Jensen *et al.*, there was a significant difference in the value of τ_0 between bursting and nonbursting CA1 HPCs.
[b]Average of four cells.
[c]$KMeSO_4$-filled patch pipettes.
[d]K-gluconate-filled patch pipettes.
[e]Data from 188 cells.
[f]"True" E_{rest} obtained after 20-sec relaxation to steady state from a holding potential of −65 mV.

Considering a Somatic Shunt

A parsimonious explanation for the difference in R_{in} and τ_0 is that the sharp electrode introduces a shunt, or leak, at the recording site [see analysis in Staley *et al.* (1992)]. The whole-cell patch technique is thought to give a more accurate picture of the cell impedance because the gigaseal prevents an electrical shunt. While patch electrodes may cause certain membrane properties to vary due to fast dialysis of the cell interior with the electrode solution, the high resistance is stable from the beginning of the recording (L. J. Borg-Graham, unpublished data from *in vitro* turtle retina and *in vivo* cat visual cortex). Similar high impedances are also seen with perforated patch electrodes, which preclude dialysis (Spruston and Johnston, 1992; Marty and Neher, 1995).

When appropriate, a somatic shunt must be considered in deriving various linear parameters from impulse or step responses, that is, by including it in the somatic membrane resistance. An example of how the assumption of a homogeneous R_m (i.e., neglecting the local shunt) can result in incorrect parameter estimation is given below in Section 5.3.2.

The problem remains, however, of how to account biophysically for the cell's apparent resistance to the assumed nonspecific permeability of a sharp electrode shunt. Typical values of the shunt are estimated on the order of 50 MΩ, which is in the neighborhood of the cell's "intrinsic" impedance. A simple voltage divider analysis would predict that a nonspecific shunt would reduce the cell resting potential by about half, yet this is not observed. Various candidate mechanisms have been proposed for the apparent paradox, including the upregulation of either membrane-bound ion pumps or K^+ channels with reversal potentials below rest (Jones, 1989; Staley *et al.*, 1992).

5.3. Analytical Methods and Results

The extensive literature of Rall laid the foundation for methodologies of cable analysis which are specifically appropriate for linear properties of dendritic trees (for review, see Rall, 1989; Rall *et al.*, 1992; also Jack *et al.*, 1983). Turner and colleagues have presented extensive studies on the electrotonic analysis specifically on hippocampal dendritic trees, both under normal conditions (Turner and Schwartzkroin, 1980, 1984; Turner, 1984a,b; Turner *et al.*, 1995) and pathological conditions (Pyapali and Turner, 1994).

5.3.1. Analysis of Dendritic Transients

Classically, closed-form solutions for the behavior of transient signals in linear dendrites, for example, the speed of a signal, have been difficult to obtain. Recognizing that the major obstacle arose from the *definition* of the quantity of interest, Agmon-Snir and colleagues (Agmon-Snir and Segev, 1993; Agmon-Snir, 1995; Zador *et al.*, 1995) observed that when a given parameter was described as a linear measure (specifically, in terms of moments of the waveform), the analysis was greatly facilitated. For example, the usual definition of "speed" is the time it takes for the *peak* of signal to get from point *A* to point *B*. However, the peak is a nonlinear measurement; in these reports the landmark for measuring speed was the center of gravity of the waveform (the ratio of the first and zeroth moments), a linear measure.

5.3.2. Pitfalls in Evaluating Linear Parameters Derived from Intracellular Measurements

Assumption of a homogeneous R_m, for example when using the Rall model of equivalent cylinders, has given erroneous estimates of cell properties. This can be illustrated by examining the often-cited report of T. H. Brown *et al.* (1981). In this paper, analysis of the somatic current step response was based on the critical assumption that the cell could be approximated by the soma/short-cable model with a homogeneous membrane time constant. Essentially three parameters were derived from hyperpolarizing current steps (responses assumed linear) of hippocampal pyramidal cells: a homogeneous R_m, L (the electrotonic length of the short-cable approximation), and ρ (ratio of dendritic and somatic conductance). The response to a current step at the soma of this structure can be represented by a linear combination of exponential terms:

$$V_f - V = \sum_{i=0,\infty} C_i \exp(-t / \tau_i)$$

where V is the response at the soma relative to rest, V_f is the steady-state soma voltage, τ_0 is the membrane time constant ($\tau_0 = R_m C_m$), and the remaining τ_i are shorter time constants due to charge redistribution down the dendrite cable. Standard exponential peeling techniques were used to estimate the first few terms of this expression, which were then used in three different methods to estimate the cable model parameters. Measured somatic R_{in} averaged about 39 MΩ, and τ_0 about 19 msec (means from CA3 cells). R_m was then derived from the measured τ_0: since C_m was taken to be 1.0 μF/cm², this immediately gives R_m equal to 19 kΩ-cm². With R_i assumed to be 75 Ω-cm, these techniques resulted in average values of ρ as 1.2 and L as 0.93.

To see how well this procedure works, it is straightforward to "reverse-engineer" a soma/short-cable geometry based on the above parameters. For this we need to derive the radius of the soma, a_{soma}, the radius of the dendrite, a, and the length of the dendrite, l. The first step is to derive a_{soma}. Now, G_{soma} is given by ρ and R_{in}—since

$$\rho = \frac{G_{dendrite}}{G_{soma}}$$

and

$$G_{dendrite} + G_{soma} = \frac{1}{R_{in}}$$

then

$$G_{soma} = \frac{1}{R_{in}(1+\rho)}$$

This gives G_{soma} = 11.7nS. The radius of the soma is then calculated from G_{soma} and R_m:

$$a_{soma} = \sqrt{\frac{R_m G_{soma}}{4\pi}}$$

This results in a_{soma} = 42 μm. With λ as the electrotonic space constant of the cable, R_{in} is a function of G_{soma}, l, and a, given by

$$R_{in} = \left(\frac{1}{r_a \lambda} \tanh(L) + G_{soma} \right)^{-1} \qquad (1)$$

where the first term on the right is the input resistance of a finite cable with an open circuit termination [e.g., Equation (4.11) in Jack *et al.* (1983)] and where

$$L = l/\lambda$$

and

$$\lambda = \sqrt{\frac{R_m a}{2R_i}}$$

Estimates for l and a may be obtained by calculating R_{in}, L, and ρ, over a range of l and a with Equation (1), and finding the best fit to the experimentally derived values for R_{in}, L, and ρ. This procedure gives $l \approx 1800$ μm and $a \approx 3$ μm. The resulting geometry is shown in Fig. 1, compared against other model geometries and actual cell anatomies. The most striking feature of the geometry derived from this report is the estimated soma radius of 42 μm, whereas a more realistic value is on the order of 15 μm (this difference gives about an eightfold difference in soma surface area). Clearly the parameters reported in this paper cannot correspond to the actual cell properties, stemming from the nonconsideration of the somatic shunt.

Holmes and colleagues present a more theoretical analysis of the sorts of errors illustrated by the above example (Holmes *et al.*, 1992; Holmes and Rall, 1992a). In another paper of this series, Holmes and Rall (1992c) develop a general approach for estimating electrotonic dendritic structure that does not rely on equivalent cylinder approximations.

5.3.3. Combined Experimental and Modeling Results

Using different analytical approaches, two recent studies (Spruston and Johnston, 1992; Major *et al.*, 1994) carefully estimated linear membrane properties of hippocampal cells from patch-clamp recordings.

Spruston and Johnston (1992) used the perforated path technique on CA1 HPCs, CA3 HPCs, and dentate gyrus cells. In this paper it was assumed that the perforated patch recordings precluded a somatic shunt, and therefore R_m could be estimated directly from τ_0. Assuming a value of 1.0 μF/cm² for C_m, this gave an R_m of 28 and 66 kΩ-cm² for CA1 and CA3 HPCs, respectively.

One interesting finding from this study concerned the prediction that with a somatic shunt the longest time constant of a step response is less than the intrinsic membrane time constant. Estimates for parameters of a soma/short-cable model were evaluated from the data using analytical techniques similar to those of T. H. Brown *et al.* (1981), as discussed earlier, but with an explicit consideration of a somatic shunt term. These results were checked against independent estimates derived from numerical simulations of a soma/short-cable model. Spruston and Johnston showed that realistic values for the soma/short-cable model parameters, constrained by their experimental measurements, quantitatively supported the role of a soma shunt in the apparent discrepancy in the relationships between τ_m and R_N in sharp versus whole-cell recordings. The estimates presented in this paper were consistent with predictions shown in a linear model by Borg-Graham (1987b). In this study the benchmark τ_0 (from T. H. Brown *et al.*, 1981) was taken to be 19 msec—allowing an inhomogeneous R_m gave a (nonunique) estimate of 40 and 2.5 kΩ-cm^2 for the dendritic and somatic regions, respectively; thus the measured τ_0 of 19 msec underestimated the 40 msec τ_m of the dendrites (C_m was taken as 1.0 μF/cm^2).

Major and colleagues have derived analytical expressions for the analysis of arbitrary characterized linear dendritic trees (Evans *et al.*, 1992).* Major *et al.* (1994) applied these methods to their data from CA3 HPCs, and used an optimization procedure to match responses to short current pulses from specific cells to their anatomies by varying R_m, C_m, and R_i, taking into account estimates for the pipette contribution to the response. This resulted in estimates for R_m of 150–200 kΩ-cm^2, R_i of 170–340 Ω-cm, and C_m of 0.7–0.8 μF/cm^2.

Since this study was constrained by quantitative anatomy, as opposed to an assumption of the equivalent cylinder implicit in the soma/short-cable model, and since the recordings were made with whole-cell patch electrodes, thus minimizing the somatic shunt, these estimates may represent the best picture to date of the resting properties of the HPC electrotonic structure. It is intersting, then, to note that the estimates for R_m and R_i are significantly higher, and for C_m significantly lower, then those used in most previous studies. One caveat may be that the techniques used in this paper required fairly long recording times (around 30 min, for reduction of noise) and thus washout effects, as mentioned in the paper, may have biased the cell properties.

5.4. Numerical Methods: Morphoelectrotonic Analysis of Dendritic Trees

Zador and Brown and their colleagues (Zador 1993; Zador *et al.*, 1995; Tsai *et al.*, 1994; Carnevale *et al.*, 1995) have developed a numerical technique known as the morphoelectrotonic transform. This results in a vivid anatomical rendering of the dendritic tree based on the frequency-dependent transfer impedance between each point in the tree and a fixed reference point. This visualization emphasizes the point that steady-state electrotonic signal transfer is much larger than that for signal frequencies typical of synaptic events. The major limitation of this linear transform is that it is applicable for current, not conductance inputs, and thus is difficult to apply to synaptic integration under more realistic conditions.

5.5. How Far Can We Take Linear Dendritic Models?

As the ubiquity of voltage-dependent properties in dendrites becomes more apparent (D. Johnston *et al.*, 1996), it is not obvious how much we can safely extrapolate from purely linear representations of dendritic structure. For example, even a small-signal linear approximation must consider the evidence for voltage-dependent mechanisms that are activated near rest, serving to "clamp" the resting potential (see Section 6.1, and later discussions of I_D, I_M, I_{AHP}, I_Q/I_H, and $I_{Cl(V)}$).

*Note that in formulations of the simple soma/short-cable model, the dimensions of model are some of the parameters estimated from analysis the voltage response (refer to the reverse-engineered model presented earlier).

In addition, even if the dendritic membrane is reasonably voltage independent for small signals (on the order of several millivolts), analyzing the integration and transmission of synaptic signals requires consideration of the nonlinear interactions *between* synapses since they are modulated conductances (Koch *et al.*, 1983), and it is hard to imagine synapses being activated alone in a functional context. Specifically, the expected interactions between synapses put a severe restriction on the transmission properties predicted by purely linear methods (e.g., with the use of linear moments and the morphoelectrotonic models discussed earlier). In summary, it is quite important to determine whether functionally relevant synaptic input *ever* operates in a linear regime.

Nonetheless, despite its limitations, the linear model forms the necessary backbone for the estimation of nonlinear properties.

5.6. Model Linear Parameters

The following summarizes typical default values for linear parameters, both measured and derived, with the parameters of the Working model in parentheses:

- Resting potential: −65 mV (E_{leak} = −65 mV, E_{rest} = −65.7 mV)
- Input resistance R_N: 40 MΩ (sharp recordings) to 200 MΩ (patch recordings) (39 MΩ)
- Time constant τ_0: 20 msec (sharp recordings) to 100 msec (patch recordings) (14.4 msec)
- Estimated membrane capacitance C_m: 0.7 μF/cm² (0.7 μF/cm²)
- Estimated membrane resistivity R_m: 150–200 kΩ-cm² (dendrite: 40 kΩ-cm², soma: 2500 Ω-cm²)
- Estimated cytoplasmic resistivity R_i: 150–250 Ω-cm (200 Ω-cm)

The small-signal step response of the Working model is shown in Fig. 5. The value for the dendritic R_m in the Working model is probably an underestimate, and the low somatic R_m reflects a somatic shunt from a sharp recording. This value also results in a slightly low τ_0.

In all the simulations presented in this chapter, simulations start from a holding potential of −65 mV, corresponding to the value assigned to E_{leak} (an exception is the demonstration of I_D shown in Fig. 16). Simulation of the "true" E_{rest} in the Working model by letting the system relax (over about 20 sec) to steady state from the holding potential of −65 mV gives E_{rest} equal to −65.7 mV, primarily due to a small activation of I_{CaT}, which in turn raises the resting level of $[Ca^{2+}]_{in}$ enough to slightly activate I_{AHP}.

6. Phenomenological Templates

The shapes of action potentials, the trajectories of the membrane voltage prior to and after the spike, and the responses of a neuron to long-lasting imputs are complex and variable, and vary significantly between different types of cells. In contrast to the modulation of non- or weakly voltage-dependent channels, or as-

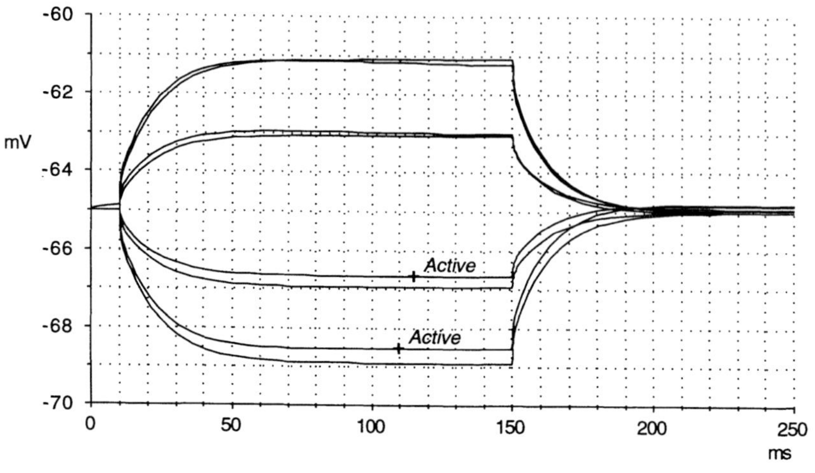

Figure 5. Response of the Working model to small current steps (±0.05 and 0.1 nA), both with all the channels active and blocked. The value for τ_0 in the passive case is 14.4 msec, and in the active case the τ_0 for the response to a 0.1-nA step is 14.1 msec; the passive R_{in} is 39 MΩ. Note the small inward rectification in the two hyperpolarizing active responses, due mainly to I_H. The depolarizing transient in the active case at the beginning of the trace is due to the fact that the simulation starts with an imposed resting potential of −65 mV; after settling of the small activated currents around rest the final resting potential is −65.7 mV.

sumptions of tonic synaptic activation, mechanisms with a strong voltage or Ca^{2+} dependence and fast kinetics need to be explicitly considered in the model.

This section describes the main properties of HPC responses, generally under control conditions, that are relevant to single-cell models, and issues concerning dendritic channels. Experimental manipulations of these properties, in particular as used to isolate specific membrane currents, will be discussed here and in later sections detailing the different currents.

We start with a quantitative review of the various potentials that precede and follow the individual spike, illustrated in Fig. 6. We then discuss the site of spike initiation in the cell, the differences in firing patterns of various cell types, the characteristics of repetitive firing in regular spiking (nonbursting) cells, the characteristics and mechanisms for bursting, and so-called "D" spikes. We conclude this section with a short discussion on dendritic channel properties.

6.1. Morphology of the Potentials Preceding the Spike

In this section we discuss the responses to current steps that precede spikes, both starting from the resting state and the voltage trajectories between spikes during repetitive firing.

6.1.1. Rectification around the Resting Potential

Subthreshold inputs which change the membrane potential by more than plus or minus several millivolts begin to activate a variety of time-dependent nonlinearities. For hyperpolarizing current steps, this is manifested as a depolarizing

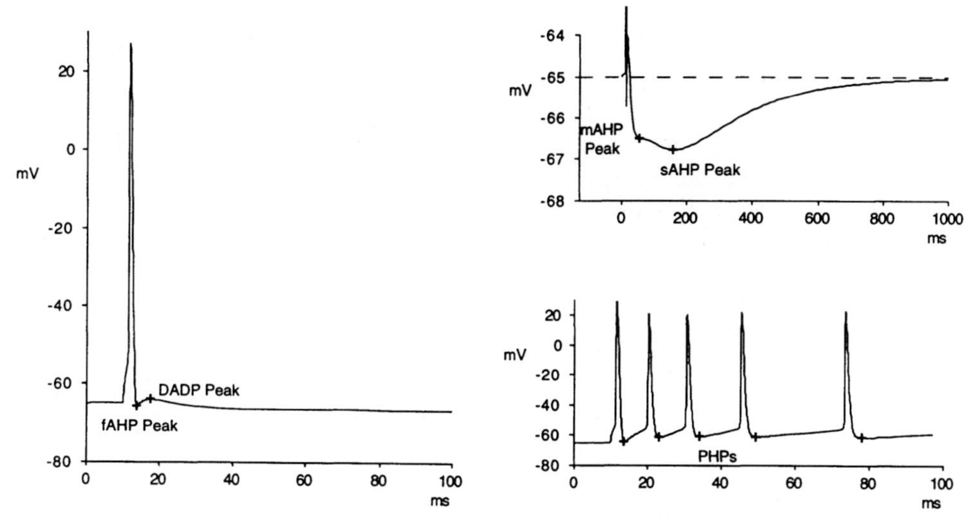

Figure 6. Single-spike and spike-train responses of the Working model under control conditions illustrating the various landmarks of firing. These include the fast after-hyperpolarization (fAHP), the after-depolarization (ADP), which in this case is a depolarizing ADP (thus DADP), the medium after-hyperpolarization (mAHP) and the slow after-hyperpolarization (sAHP) after a single spike, and the pre-hyperpolarization peak between spikes during repetitive firing (PHP). Note the change of scale in the single-spike responses illustrated in the left and upper right panels. In this model, the mAHP moves directly into the sAHP; real cells may show a much more distinct transition. The spike train is in response to a 1.0-nA current step; single spikes are in response to 2.5-msec, 1.0-nA pulses. The dashed line in the upper right figure shows the resting potential.

sag due to a slowly developing (typically about 100 msec time constant) inward rectification. For depolarizing current steps, depending on the conditions, there may be either an inward rectification (so that the apparent input resistance becomes larger; e.g., Hotson *et al.,* 1979) or an outward rectification (e.g., Storm, 1988b). Hyperpolarization-activated inward rectification (see Section 14.1) and depolarization-activated outward rectification (see Sections 13.1, 13.3, and 13.7) serve to stabilize the resting potential; in addition, time-dependent inactivation of depolarization-activated outward rectification may act to integrate excitatory synaptic input over time. For depolarizing current steps in particular, the holding potential (or equivalently, the effective resting potential) of the neuron will to some degree determine which membrane currents will be available for shaping the subthreshold response; however, it is not always clear from the literature what determines the dominant mechanism (e.g., inward or outward rectification) in a given experiment.

6.1.2. The Pre-Hyperpolarization between Repetitive Spikes

For isolated single spikes, the next characteristic of interest is the spike threshold, discussed in the next section. For repetitive firing another important characteristic is the lowest voltage that precedes a spike, which we will refer to as the pre-hyperpolarization (PHP). PHPs are usually referred to as after-hyperpolarizations; we use the term PHP to emphasize the constraints imposed by this potential on the firing mechanisms for subsequent spikes, in particular the removal of I_{Na} inactivation, which we discuss in Section 11.1.7. In essence, it may be better to view this potential not so much as the spike's aftermath, but rather as its overture.

A conspicuous difference between the firing of HPCs (and typically cortical pyramidal cells) and other neurons is that in pyramidal cells the difference between PHPs and spike threshold is typically less than 10 mV, while in other cells this difference is usually at least 10–20 mV, a classical example being the spike train of the squid axon. Hippocampal and cortical interneurons exhibit similar large differences between the PHP and threshold. During pyramidal cell bursts, the PHP may be almost eliminated, with little difference between the apex of the repolarization and the threshold for the subsequent spike, but for our purposes we will concentrate on the PHP characteristics during the repetitive spikes of regular firing cells.

A further characteristic of pyramidal cells is that the threshold of the *first* spike of a spike train is often below subsequent PHPs, whereas in nonpyramidal cells the combination of a generally deeper PHP and a generally smaller threshold difference between initial and later spikes in a train result in PHP profiles that are more hyperpolarized than the thresholds of all the spikes in the train. We also note that there are exceptions to these characteristics of PHPs in HPCs (see Section 2.8), but otherwise this means that recovery of spike current inactivation can occur at potentials above the lowest spike threshold. A general discussion regarding differences in firing properties between pyramidal (specifically, spiny) cells and interneurons (specifically nonspiny) cells in the hippocampus may be found in Scharfman (1993).

In the study by Lanthorn *et al.* (1984), various characteristics of repetitive firing in CA1 HPCs were reported. This study described "slow potentials," a depolarizing ramp between spikes whose starting point is the PHP. Interestingly, the difference between the PHP and the spike threshold during steady-state repetitive firing was quite stable (about 6.7 mV), whereas the difference between the peak of the fAHP (see below) and the spike threshold became larger with higher spike rates (roughly 5 mV with 400 msec interspike interval, and about 10 mV for a 100 msec interspike interval; refer to Fig. 4 in Lanthorn *et al.*, 1984). If the peak of the fAHP is relatively fixed in absolute terms (M. S. Jensen *et al.*, 1996; discussed later), then this suggests that there is a fixed relationship between the PHP and the spike threshold over some voltage range. Such a relationship is consistent with a Na^+ channel whose kinetics provides a variable threshold which is a function of the previous repolarization (see Section 11.1.7).

The PHPs of faster spike trains can be compared in the paper by Thompson *et al.* (1985). In Fig. 6 of this report, spike trains with steady-state interspike intervals of 19 msec (elicited at 37°C) and 27 msec (elicited at 27°C) had steady-state PHP to spike threshold differences of 7.5 mV. Again, the difference between PHP and subsequent spike threshold appears stable in a given cell, in this case with respect to both temperature and firing rate.

6.2. Morphology of the Action Potential

Here we shall discuss quantitative parameters for action potentials including voltage threshold, rate of rise and fall, and the spike amplitude and duration. Table II summarizes these parameters from various references.

Table II. Values of Single-Spike Parameters of Regular Spiking Cells[a] Compared with Values for the Working Model[b]

Reference	Temp. (°C)	Threshold (HP) (mV)	Amplitude (mV)	dV/dt Rise (mV/msec)	dV/dt Fall (mV/msec)	Duration (msec)
Thompson *et al.* (1985)	37	−50 (−55)	78±10	394±90	105±21	0.78±0.1
	27	−45 (−55)	79±10	265±79	43±7	1.5±0.3
Andersen *et al.* (1986)	34	−49 (−65)	—	—	—	—
Storm (1987a)	28–33	−55/−45 (−60/−70)	98±8	238±16	85±19	
Spruston and Johnston (1992)	32	−55 (−66)	—	—	—	—
Buckmaster *et al.* (1993)	35	—	66±1	—	—	2.6±.1
Scharfman (1993)	—	−47 (−63)	85±2	410±25	87±5	1.7±0.2
Jensen *et al.* (1996)	33.5	−56±3 (−63±3)	89±5	—	—	—
Working model	—	−54 (−65)	78	467	98	.83

[a]Except for those from Scharfman, which were quasibursting cells.
[b]Cell types are either CA1 pyramidal cells or CA3c cells (Buckmaster *et al.* and Scharfman), and recordings are either with sharp electrodes or perforated patch (Spruston and Johnston). In all cases parameters were taken from either the single spike in response to a brief current pulse or the intial spike in response to a sustained, near-threshold stimulus. Threshold (HP) refers to the spike threshold, with the holding potential in parentheses; note that this may or may not be the "true" resting potential. The values given for Storm are ranges. Spike threshold in Andersen *et al.* was defined as the point at which dV/dt was greater than 50 mV/msec, for the Working model this point was taken as when dV/dt was greater than 12 mV/msec, and the thresholds reported in Jensen *et al.* were apparently determined by eye. Similarly, the remainder of the values were estimated by the author from figures in the cited reference. Spike amplitude was measured from threshold in all cases, except for Storm, where amplitude was referenced from the holding potential, and Buckmaster *et al.* who defined spike onset as the point at which $dV/dt > 12$ mV/msec. The maximum rates of rise and fall were reported by Thompson *et al.*, and used for the Working model. Storm used the 10–80% and 80–20% points (relative to spike threshold and peak), Scharfman the slope of lines fitted from the threshold to the peak and the peak back to the threshold voltage, as references for the rates of rise and fall, respectively. The spike duration was taken in Thompson *et al.* and the Working model as the width at half amplitude, for Buckmaster *et al.* as the time from spike onset to the recrossing of the onset voltage, and for Scharfman and Jensen *et al.* as the time from spike threshold to the recrossing of the threshold voltage.

6.2.1. Action Potential Threshold: Absolute and Relative to Holding Potential

The reporting of threshold values is problematic since there is rarely a precise definition given in the literature; indeed, there is no such thing as a definite threshold that can be interpreted out of context, for example, without specifying the current stimulus level or firing mode. Unambiguous definitions of the threshold include the point at which the slope of the voltage trace exceeds some ad hoc value (e.g., 50 mV/msec; Andersen *et al.*, 1986), or the point corresonding to the maximum of the second derivative (e.g., Mainen *et al.*, 1995). The thresholds listed in Table II, measured from either isolated spikes or the initial spikes in spike trains, range from about −55 to −45 mV. Most of these values are estimated by eye. From this table we can see that the difference between the holding potential and single-spike threshold is also variable, ranging from about 5 to 15 mV.

The threshold for firing in HPCs tends to be more abrupt relative to non-pyramidal neurons. Quantitative data on this characteristic, such as the time course of the second derivative, would be useful for the determination of model parameters (in particular I_{Na} kinetics), but these data are lacking in the literature. In some reports (e.g., Storm 1987a), the sharpness of the threshold can be appreciated by the almost passive character of just-subthreshold current step responses; however, this phenomenon is not always observed (see Section 6.4.2).

Another observation that may be made concerning spike threshold is that typically the threshold for the initial spike of a train from a regular-firing cell is several millivolts less than later spikes, with a total span of 5–10 mV (Andersen *et al.*, 1986). Of course, the thresholds for spikes within a burst span even a greater range (on the order of 20 mV). All of these characteristics are most directly related to the voltage dependence for the activation of fast Na^+ channels, and to a lesser (but important) extent to the inactivation properties of these channels.

Action potential threshold is a function of where in the cell the excitatory stimulus occurs. Andersen *et al.* (1986) reported that the thresholds of somatic spikes evoked by either somatic current injection or stimulation of synaptic inputs on the proximal apical dendrite were about 3 mV higher (around 15 mV above the resting potential) as compared to those evoked by stimulation of inputs to the distal apical dendrite (around 12 mV above the resting potential). These results were interpreted as evidence for regenerative properties of the HPC dendrite.

6.2.2. Action Potential Rate of Rise and Fall

Rates of rise and fall may be taken in terms of the maxima or the average slope between the relative positions (e.g., 10–90%) of the spike; of course, the latter measure depends on the definition of threshold. The dV/dt for the spike upstroke is a good indicator of the minimum magnitude of the inward currents (i.e., without taking into account opposing outward currents), estimated from the *effective* capacitive load that must be charged. The effective capacitance arises from the somatic capacitance plus the capacitive component of the dendritic tree's input impedance as seen from the soma. On the other hand, the rate of fall for the spike gives a somewhat weaker estimate of the magnitude of the outward currents during repolarization—this is more complicated than the estimation of depolarizing current since there is more likely to be residual inward currents active, and the slower rate of fall means that the cell conductance has to be considered more carefully (see discussion in Storm, 1987a). Unfortunately, the *dV/dt* for the spike is not readily available in the literature for later spikes during repetitive firing, information which would be useful in reconstructing the kinetics of I_{Na} for repetitive firing.

In Table II the values for the rate of rise range from about 200 to 400 mV/msec and for the rate of fall from about 40 to 100 mV/msec.

6.2.3. Action Potential Amplitude and Duration

Spike amplitude is usually measured from the spike threshold, and sometimes from the resting potential—obviously the former measure can be more problematic due the ambiguity of threshold. As seen in Table II, the peak of the spike under control conditions can reach from about +20 to +35 mV; assuming a typical value of E_{Na} as around +60 mV (Hille, 1992), we see that this large range suggests that spike amplitude is determined mainly by inactivation properties of I_{Na}, in addition to the activation properties of the repolarizing currents, rather than E_{Na}.

Spike duration, as reported in the literature, may be taken from either the threshold crossing to when the voltage recrosses the threshold (the "base" of the spike), or the width at half-amplitude. Using the latter measure, which is less sensitive to firing mode, we find that spike duration varies from about 0.5 to 1.5 msec.

Figenschou *et al.* (1996) describe Ca^{2+}-dependent cholinergic overall depression of the spike waveform (including increase of threshold, slower rise, slower repolarization, spike broadening, and amplitude reduction) which is not readily attributed to modulation of Ca^{2+} currents or the slow Ca^{2+}-dependent K^+ current I_{AHP} The authors postulate that the effect may be due to activation of a pathway involving protein kinase C, which in turn modulates either Ca^{2+} channels, Na^+ channels, or both.

6.3. Where Does the Spike Start?

The precise location of spike initiation is classically taken as the axon hillock, or initial segment. There has been increasing interest in the spike-generating properties of dendrites (e.g., Benardo *et al.*, 1982; Stuart and Sakmann, 1994; Andreason and Lambert, 1995; see review in Stuart *et al.*, 1997). Under various conditions "ectopic" spikes (those arising from other locations) may occur, for example, in certain *in vitro* models of epileptogenesis (Stasheff *et al.*, 1993), or with certain strengths of synaptic input to the apical dendritic tree (Turner *et al.*, 1991). The report by Stasheff *et al.* in particular combined electrophysiological measurements with simulations to show that under their conditions the ectopic spikes were arising from axon terminals. Other modeling studies have examined this question [Moore *et al.* (1983) for motorneurons; Mainen *et al.* (1995) and Rapp *et al.* (1996) for neocortical pyramidal cells] with the general conclusion that excitable dendrites under certain conditions can start the spike.

It may be that the relationships between the PHP and spike threshold, among others, can be explained in part by the spiking behavior of structures with distributed channels, for example, the consequence of a higher density of Na^+ channels in the initial segment that is electrically decoupled from the soma (Mainen *et al.*,

1995; Rapp *et al.*, 1996). Experimental data to date for HPCs, through, does not support such a hot spot (Colbert and Johnston, 1996). In the Working model we shall interpret the data in the context of a somatic channel model, and therefore apply these constraints solely to the determination of channel kinetics.

6.4. Morphology of the Afterpotentials

The trajectory of the voltage after the spike has several prominent features, all of which vary under different conditions, in particular whether the spike is isolated or part of a repetitive train.

6.4.1. The Fast After-Hyperpolarization

The first 1 or 2 msec of spike repolarization under control conditions has a distinct fast phase followed by an abrupt transition that can precede either a slower repolarization or a depolarization. Classically, when the transition leads to a depolarization, the transition is referred to as the fast after-hyperpolarization (fAHP), with the peak of the fAHP assigned to the voltage of the transition. This peak can vary in relation to other landmarks such as the holding potential and the spike threshold. Although there is some variability among references, it is important to consider quantitative measurements of the fAHP since the speed of the event (indicative of a large current) suggests that the underlying mechanism has a significant role in shaping the cell's overall response.

Two of the earlier studies focusing on the fAHP were that of Storm (1987a) and Lancaster and Nicoll (1987), both done on rat CA1 cells in the slice, demonstrating the Ca^{2+} dependence of the fAHP. In an example trace under control conditions Storm showed a peak of the fAHP (−67 mV) which was quite a bit below both the spike threshold (−45 mV) and the resting potential (−62 mV); this paper reports an overall range of −60 to −65 mV for the fAHP peak. Similar values are seen in an example shown in Lancaster and Nicoll (fAHP peak of −67 mV, threshold −55 mV, and resting potential −63 mV).

The values in these studies are in contrast with pooled data for regular-firing cells in rat CA1 slice in a more recent study (M. S. Jensen *et al.*, 1996). Here, with a resting potential of about −63 mV, the peak of the fAHP (about −53 mV) was higher than the spike threshold (about −56 mV). This paper also reported the point of maximum repolarization after the initial spike in bursting cells and called this an fAHP. Although this term is more typically used in the context of regular-firing cells, it is interesting that Jensen *et al.* found no significant difference between this measure and the regular firing cell's fAHP.

This variance in the peak fAHP relative to the spike threshold may be seen in traces of many papers, and applies to both single spikes and the initial spikes during repetitive firing [e.g., see Deuchars and Thompson (1996) for both types]. A more consistent characteristic is that during repetitive firing the fAHP is only seen in the first one or two spikes, with occasional smaller fAHPs showing up later in the train. This diminution of the fAHP during repetitive firing is not simply a result of the higher baseline voltage for the later spikes, since M. S. Jensen *et al.* (1996) show that the peak of the fAHP of single spikes is remarkably insensitive to the holding potential (up to ±8 mV around the true resting potential). Spigelman *et al.* (1992), however, show such a dependence in rat P26 CA1 HPCs.

We may note that although not mentioned in the paper, a figure in Thompson *et al.* (1985, Fig. 5) showed a possible strong dependence on fAHP inactivation with temperature—a prominent fAHP was evident throughout a spike train at 37°C, whereas the more typical diminution of the fAHP was seen in a spike train recorded at 29°C. Although the experiments of Storm (1987a) were done over a wide range of temperatures (28–33°C), there was no mention of an fAHP temperature dependence. The experiments of Lancaster and Nicoll (1987) were done at 29–31°C.

6.4.2. The After-Depolarization

After single evoked spikes, the shift in trajectory after the fAHP has been designated the after-depolarization (ADP) or the depolarizing afterpotential (DAP). Probably the first term is more useful since this phase may be either a continuation of the spike repolarization consisting of a simple "passive" decay down to the resting potential, or a distinct "active" positive-going depolarization of several millivolts, which may then decay to rest or be immediately followed by either additional spikes or a full-fledged burst. We will make the distinction that the ADP is that portion of the voltage trajectory after a single spike which is simply above resting potential, and that a *depolarizing* ADP (DADP) includes a positive-going component.

The issue of whether the ADP is passive or active has been the subject of some interest, in particular as the ADP leads to further spikes for bursting cells. Several studies [e.g., Traub (1982) and Storm *et al.* (1987) in hippocampus] have pointed out that the distributed capacitance of the dendritic tree would contribute a small passive inward current to the soma after a fast and local spike repolarization. The parameters of the Working model predict such a current for at least a small, fast DADP (Fig. 7). However, the time course of the passive contribution is on the order of a few milliseconds, appropriate for some depolarizing effect at the beginning phase of the DADP, but too fast for the longer DADPs (10–20 msec) often seen in the literature. More prominent DADPs probably reflect the activation of various somatic conductances during this phase, which would further limit the effect of the passive dendritic current. Finally, the concurrent activation of channels present in the proximal dendrite (D. Johnston *et al.*, 1996) would argue against a significant capacitive (passive or linear) component. This possibility is not considered in the Working model prediction since here the dendrite is linear.

An active ADP in some CA1 HPCs (those which tend to burst; see below) has been identified as due to a TTX-sensitive Na^+ current (M. S. Jensen *et al.*, 1996; Azouz *et al.*, 1996). This ADP was seen both after single spikes and as a prolonged decay after subthreshold current pulses.

6.4.3. The Medium After-Hyperpolarization

At the end of a stimulus that elicits one or more spikes there is a medium after-hyperpolarization, or mAHP, which increases with the number of spikes and lasts on the order of 100 msec. Storm (1989) described the properties of the mAHP in detail, showing that the size of the mAHP grows roughly linearly with the number of

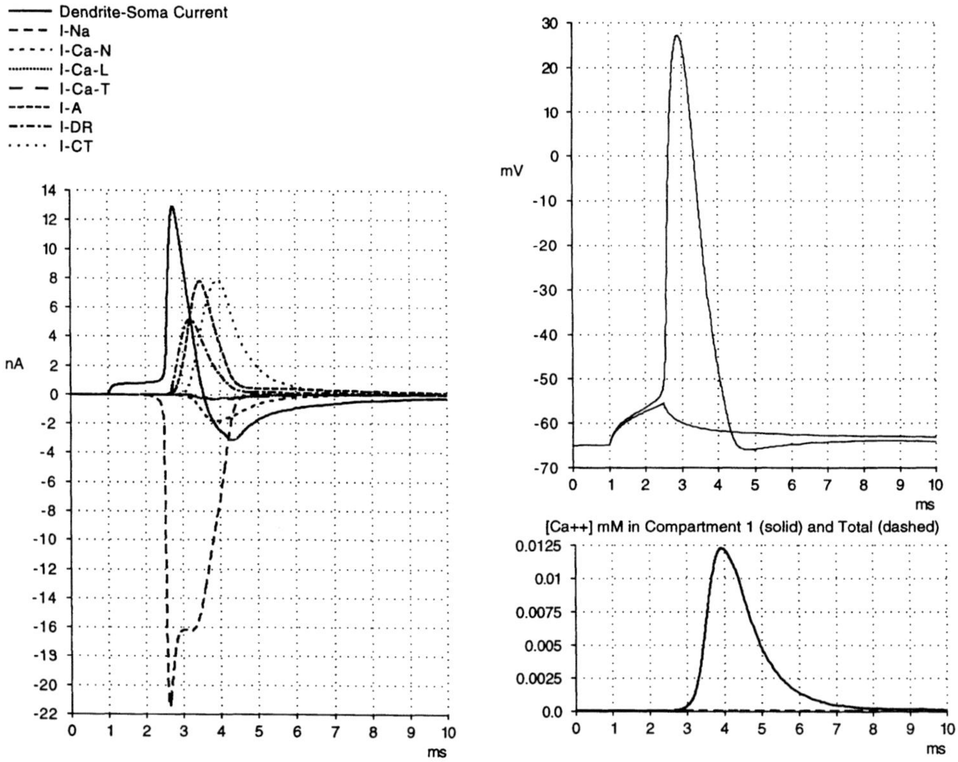

Figure 7. Single spike of the Working model under control conditions, in response to 1.5-msec, 1.0-nA pulse, the major currents active during the spike, and the transient change in $[Ca^{2+}]_{in}$ for the small Ca^{2+} compartment assigned to the Ca^{2+} channels and I_{CT}. Also shown is the subthreshold response to a 1.5-msec, 0.9-nA pulse, showing the sharp threshold in this model. In this version of the model the kinetics of the inward currents are very transient for single spikes, thus the DADP, whose peak occurs between the times of 8 and 9 msec, is mediated almost completely by the capacitive redistribution current from the dendrite cable. Note, however, that the DADP under these conditions is rather short, and that for other HPC cells with more prominent ADPs the role of active currents is more likely. Also note that the concentration "spike" in compartment 1 is much larger and more transient than the concentration change averaged over the entire soma (compare with Fig. 15).

spikes in a train, increasing from about 1 mV with a single spike to about 5 mV with seven spikes.

The mAHP is attenuated by cholinergic agents (Storm, 1989), and both Storm (1989) and a later study by Williamson and Alger (1990) showed that when a train of action potentials was elicited from a cell depolarized from rest, the mAHP had a substantial Ca^{2+}-dependent component. For example, when cells were held between -49 and -57 mV, the amplitude of the mAHP with control saline was about 4 mV. When Ca^{2+}-blockers (nifedipine or Cd^{2+}) were added, or external $[Ca^{2+}]$ was reduced, the amplitude decreased to around 2 mV. Manipulations that increase Ca^{2+} influx (addition of BAY K 8644, raising external $[Ca^{2+}]$) increased the amplitude of the mAHP to about 6 mV. Conversely, spike trains evoked from hyperpolarized levels ended in an mAHP that was antagonized by Cs^{+}.

It is not unusual for the peak of the fAHP to be within millivolts of the peak of the mAHP for single evoked spikes. For repetitive firing, a similar relation is observed between the fAHP and the minimum voltage between spikes—that is the PHP described earlier, which might be considered the analogue to the mAHP for repetitive firing (see Deuchars and Thomson, 1996, Fig. 2B).

6.4.4. The Slow After-Hyperpolarization

The slow after-hyperpolarization (sAHP) is a Ca^{2+}-dependent afterpotential classically seen after a train of current-induced action potentials. As described by Hotson and Prince (1980), the sAHP has a magnitude of up to several millivolts below the resting potential, may peak with a delay up to 500 msec after the spike train, and can last over 1 sec. These authors also showed that the magnitude of the sAHP saturates abruptly when measured as a function of either current stimulus duration or amplitude (e.g., sAHP saturation with a 125-msec, 0.5-nA current stimulus). The magnitude of the sAHP is also strongly dependent on the resting potential—in this paper the amplitude of the sAHP of cells with a resting potential between −65 and −70 mV was increased by up to 5 mV when the cells were deplolarized 5–10 mV. In some cells with a low resting potential of around −70 mV no sAHP was observed unless a steady depolarizing current was applied.

Although the sAHP is typically discussed in terms of firing accommodation after several spikes, the sAHP is present after a single action potential. For example, Storm (1987a) showed a single-spike-evoked sAHP which peaked at about 1.5 mV below an imposed resting potential of −65 mV. Another example from Storm (1987b) showed a 5-mV sAHP after five spikes (resting potential not given). Another quantitative analysis of the sAHP by Storm (1989) included the dependence of the sAHP amplitude on the number of spikes, showing, as with the mAHP, a roughly linear relationship from about 0.5 mV after one spike to a little over 3 mV with seven spikes (resting potential not given). In this paper the sAHP was also studied with a hybrid clamp protocol, in which voltage clamp was imposed immediately after current clamp evoked spikes (Lancaster and Adams, 1986). This technique showed a rectification of the sAHP current, possibly consistent with the GHK permeation model (Section 8.2), as the voltage-clamp potential approached the reversal potential of about −100 mV. The linear portion of the I–V characteristic intersected the voltage axis at an effective reversal potential of about −85 mV. Similar results were reported by Lancaster and Adams (1986).

As mentioned earlier, Thompson *et al.* (1985) reported that a reduction in temperature increased the sAHP; in one example the sAHP peak amplitude after a train of six spikes increased from about 2 mV at 36°C to about 4 mV at 32°C (resting potential apparently −57 mV). The authors suggested that a facilitation of a Ca^{2+}-dependent K^+ current could account for the changes, caused by both an increase of Ca^{2+} entry per spike and a slowing down of the removal of intracellular Ca^{2+}.

An increase in Ca^{2+} entry could be a result of a broader spike (giving more time for Ca^{2+} channel activation), while a decrease in Ca^{2+} removal may be a function of a temperature dependence on a Ca^{2+} pump. However, if a slowing down of channel kinetics (e.g., I_{Na} inactivation and/or repolarizing K^+ channel activation) can account for the broadening of the spike, then it remains to be demonstrated whether any expected concomitant slowing of the Ca^{2+} channel kinetics would not offset an increase in total Ca^{2+} influx caused by the longer depolarization. It will be interesting in the future to examine such tradeoffs with simulations, also taking into account the different Q_{10} values of channel conduction and channel kinetics.

6.4.5. The Slow After-Depolarization and Plateau Potentials

In the presence of cholinergic agents a Ca^{2+}-dependent slow after-depolarization (sADP), roughly 7 mV and lasting on the order of 100 msec or longer, may be observed in CA3 HPCs after a spike train instead of the sAHP (Caeser *et al.*, 1993). In CA1 HPCs application of cholinergic agonists can reveal either an sADP or a long-lasting (seconds) plateau potential of about −20 mV after a spike train (Storm, 1989; Fraser and MacVicar, 1996). Both potentials require Ca^{2+} and Na^{+} in the bath solution; they may be related to hippocampal seizure activity, also known to be facilitated by cholinergic agents. Contrary to the conclusions from Figenschou *et al.* (1996), data in the paper by Fraser and MacVicar (1996, Fig. 1A, not commented in the discussion) suggest that 20 μM of carbachol irreversibly blocks the fAHP.

6.5. Differences in Firing Properties between HPCs and Other Cell Types

Physiologically, HPCs (and spiny cortical cells) are divided into two major classes, regular-firing (or regular-spiking) and bursting cells (for a review of the firing patterns in neocortical cells see Connors and Gutnick, 1990). These classes refer to the pattern of action potentials in response to both short and sustained current injections. Regular-firing cells respond with spikes for only as long as the stimulus, in a regular, adapting manner. Bursting neurons, on the other hand, typically respond to a short stimulus (e.g., suprathreshold for several milliseconds) with several spikes riding on a slow depolarizing "hump" lasting on the order of 100 msec (e.g., Wong and Prince 1978). Sustained stimuli tend to elicit an initial burst followed by either an adapting "regular" firing pattern or additional bursts. The morphology of individual bursts is variable, but usually show a diminution of height and an increase in duration of the fast spikes riding the hump. Quite often, the repolarization of the somewhat fat and slow final spike is continuous with the repolarization of the slow hump. The mechanisms that determine these firing patterns may or may not overlap those that shape individual spike waveforms, and we will mention several theories as to the crucial burst mechanism (see Section 6.7).

The most conspicuous difference between current-evoked spike trains of HPCs (as well as spiny, presumably excitatory cortical neurons) and hippocampal interneurons (as well as nonspiny, presumably inhibitory cortical neurons) is that the former show either adapting regular firing or bursting behavior, while the latter tend to show nonadapting ("fast-spiking") regular firing. Note that the term fast-spiking also refers to the fact that the spikes of interneurons tend have a shorter duration than that of pyramidal cells. There are also various subtle differences between the responses of HPCs and other cell types, for example, in the details of the morphology of the action potential and the after-hyperpolarizations, which are beyond the scope of this chapter. However, it is worthwhile to note that these differences may reflect differences in channel distribution and electrotonic structure, as well as fundamental difference in channel types. This latter point should be kept in mind in the discussion later on specific channel properties of HPCs, in that the details of a given current identified in HPC may not be the same for a similar current in another cell, even if the current in question is referred to by the same name. Finally, it may very well be that a particular channel type described in hippocampus may be unique to this structure.

6.6. Regular Firing Characteristics—Adaptation and *f*/*I* Characteristics

Adaptation (*deaccelerando* over the spike train) in regular-firing HPCs varies, but in general is mediated by two main mechanisms, one primarily Ca^{2+}-dependent (via I_{AHP}) and the second primarily voltage dependent (via I_M). A common way to represent the pattern of repetitive firing in response to steady current injection is with the so-called "*f*/*I*" plot, which plots the instantaneous firing frequency (the reciprocal of the interspike interval) at some point in the spike train against stimulus intensity. The *f*/*I* plots for the first few interspike intervals tend to show three components as the stimulus is increased, corresponding to the primary (relatively flat), secondary (relatively steep), and tertiary (relatively flat) ranges, similar to motorneurons (Lanthorn *et al.,* 1984). The *f*/*I* plots for steady-state firing, on the other hand, tend to lack an intermediate steep portion of the firing curve as a function of stimulus strength.

The adaption in regular-spiking HPCs can be extreme, such that firing can be momentarily suppressed for some intermediate range of stimulus strengths. The report, by Lanthorn *et al.* (1984) and others (e.g., Madison and Nicoll, 1984; Lancaster and Nicoll, 1987) show that the interruption of firing occurs after about 200 msec of repetitive firing for stimuli which typically generate about 5–10 spikes during this initial period. This suppression is a relatively robust phenomenon, occurring over a range of stimulus strengths from about 0.5 to 1.5 nA (outside of this range Lanthorn *et al.* show no pause). The pause in firing is transient, lasting a few hundred milliseconds (Lanthorn *et al.,* 1984), after which there is typically a nonadapting spike train.

6.7. Mechanisms for Bursting

The characteristics of bursting in central neurons and the underlying mechanisms have historically attracted much interest. Computational roles for bursting have been proposed (e.g., Lisman, 1996), and bursting is associated with pathological states. In particular, bursting in the hippocampal slice under various conditions has served as a model system for studying epilepsy (for a review see Alger, 1984).

Despite certain differences in the proportion of cells that burst in CA1 and CA3, it is not clear whether there is a fundamentally different bursting mechanism between HPCs in these two regions, or for that matter for other bursting cells in the central nervous system. Most likely there is some overlap in the responsible mechanisms, given that there are several plausible and mutually compatible candidates. Although bursting in epilepsy models is typically associated with synaptic activation (especially under protocols in which GABAergic inhibition is blocked), the association with current-evoked bursts and other evidence indicate that the fundamental burst mechanisms are intrinsic. Key mechanisms or conditions for bursting in HPCs that have been proposed from experimental work include the following:

- Ca^{2+} entry/spikes (Wong and Prince, 1978, 1981; Konnerth *et al.*, 1986)
- Dendritic Ca^{2+} spikes (Wong *et al.*, 1979)
- Cholinergic agonists (Benardo and Prince, 1982a)
- Low extracellular Ca^{2+} (e.g., Haas *et al.*, 1984; Azouz *et al.*, 1996)
- Persistent tetrodotoxin (TTX)-sensitive Na^+ current (M. S. Jensen *et al.*, 1996; Azouz *et al.*, 1996)

Bursting may also depend on the holding potential: in cells which normally respond with bursts, long current steps applied from hyperpolarized levels tend to result in regular firing patterns (Wong and Prince, 1981; see also examples in Scharfman, 1993). Interestingly, thalamic neurons show the opposite behavior, in that they tend to burst only after being hyperpolarized (see, for example, McCormick and Pape, 1990). This points out the fact that different cell types could very well rely on different mechanisms for bursting.

Several cell models have examined the role of dendritic Ca^{2+} spikes as the mechanism underlying bursting, some of which are described later in the chapter (Traub and Llinas, 1979; Traub, 1982; Traub *et al.*, 1991, 1994; Rhodes and Gray, 1994, for neocortical pyramidal cells). The HPC model by Migliore *et al.* (1995), described later, focused on the relative balance of Ca^{2+}-independent K^+ currents as determining whether a given structure would burst or not.

6.7.1. Is the Dendritic Tree Important for Bursting?

Various attempts have been made to relate firing pattern with morphology [for a recent model-based analysis, see Mainen and Sejnowski (1996)]. For example, Bilkey and Schwartzkroin (1990) reported a positive correlation between the incidence of bursting and the length of the apical dendrite in CA3 cells.

However, bursting in CA3 cells does not require an extensive dendritic tree, since bursts are observed in acutely isolated cells with just apical dendrites [data in Wong *et al.* (1986), referenced in Traub and Miles (1991, Fig. 2.4); also see Traub *et al.* (1991)]. Also, any conclusions from modeling studies of the role of the electrotonic structure and channel distribution depend critically on the choice of channel models. For example, in the study by Mainen and Sejnowski (1996) it is not obvious that the channel models are directly relevant to HPC mechanisms.

6.7.2. Constraints for the Slow Depolarization Mechanism underlying Bursts

If the mechanisms for the spikes during the burst are roughly the same as that for regular spikes, then the inward current underlying the slow depolarizing potential must be strong enough to drive an input impedance that is significantly lower than the resting state because of the activation of the various repolarizing currents. Since these currents that are active during the later stages of spike repolarization are probably on the order of hundreds of picoamperes or more, this sets a lower bound on the size of the depolarizing current.

Under these assumptions, the major problem with attributing the slow depolarization to the classical persistent Na^+ current I_{NaP} (French *et al.*, 1990; see Section 11.1.3) is that the size of this current, on the order of a few hundred picoamperes at around −53 mV (the voltage at the end of the first burst spike repolarization; e.g., M. S. Jensen *et al.*, 1996), is too small, even if we assume that this current activates instantaneously. Similar conclusions apply to Ca^{2+} channel-mediated slow depolarizations, using estimates of Ca^{2+} channel kinetics that reproduce other firing patterns (e.g., Ca^{2+} spikes), as discussed below.

6.7.3. Is Ca^{2+} Entry the Key?

The classical explanation of a Ca^{2+} current generating the slow depolarization underlying bursts, at least in CA3 cells, was supported by studies such as Wong and Prince (1978, 1981). In these reports, either extracellular Mn^{2+} or low extracellular Ca^{2+} arrested burst firing, and focal application of Ba^{2+} restored a burstlike component. However, the effects of extracellular Mn^{2+}/low Ca^{2+} often developed over 1 hr after application, and from the figures in the papers apparently were accompanied by increases of more than 10 mV in spike threshold. These results suggest that the effect of the Ca^{2+} manipulation may be mediated by local membrane interactions, i.e., shifts in the voltage "seen" by the Na^+ channels via charge screening by divalent ions [the effect of divalent ions on neuronal membranes is discussed in Hille (1992)]. Since the Na^+-dependent component of the active ADP is quite sensitive to the holding potential (M. S. Jensen *et al.*, 1996), such shifts might serve to inactivate the slow Na^+ component. The slow depolarization with Ba^{2+} [whose time course shown in Wong and Prince (1981) was at least twice as long as the control burst depolarization] may be due to the higher permeability of Ba^{2+} through Ca^{2+} channels providing enough current to support a regenerative event.

Also, estimates of Ca^{2+} channel density (see Section 12) constrained by Ca^{2+} spike data do not predict I_{Ca}-mediated slow depolarizations—this phenomenon in models discussed later (Traub *et al.*, 1991, 1994) included Ca^{2+}-channel dendritic densities which were probably unrealistically high, as well as Ca^{2+} channel biophysics that possibly overestimated I_{Ca}. Thus, the evidence seems to be against a role for Ca^{2+} influx supplying the necessary *charge* for the burst depolarization.

Madison and Nicoll (1984) showed the contemporaneous development of initial bursts and reduction of accommodation to long pulses in CA1 cells with ethyleneglycoltetraacetic acid (EGTA)-filled electrodes. They proposed the appearance of a burst resulted from a Ca^{2+} spike facilitated by an increased driving force for Ca^{2+} (EGTA → lower $[Ca^{2+}]_{in}$). However, the reduction in current predicted by a Goldman–Hodgkin–Katz (GHK) permeation model (as probably appropriate for Ca^{2+} currents; Section 8.2) due to Ca^{2+} accumulation at the potentials in the range of this hump is small, and probably cannot account for suppression of a regenerative Ca^{2+} potential. On the other hand, in this paper application of the Ca^{2+} channel blocker Cd^{2+} only shows a reduction in accommodation, without a *de novo* burst.

One differential effect of EGTA versus external Ca^{2+} blockers (e.g., Mn^{2+}, Co^{2+}, Ni^{2+}, or Cd^{2+}) with respect to Ca^{2+}-dependent mechanisms is that the former blocks I_{AHP} and not I_C (because of the high affinity of I_C to Ca^{2+} influx), whereas the latter blocks both because of global Ca^{2+} current block (e.g., Storm,

1987a). It could be that the mechanism underlying the burst has a similar sensitivity to Ca^{2+} entry as I_C.

Finally, the most conspicuous evidence against the hypothesis that Ca^{2+} is necessary for bursting is found in the recent report by Azouz *et al.* (1996). This paper shows that burst firing in CA1 HPCs is resistant to inorganic Ca^{2+} blockers, and in some cases was even enhanced. CA1 bursts also persisted in 0 Ca^{2+}, 2 mM Mn^{2+} bath solutions. On the other hand, application of TTX in 0 Ca^{2+}, 2mM Mn^{2+} bath solution progressively blocked bursts in response to sustained current steps, before blocking the remaining full-amplitude spikes. This latter finding was cited by the authors as evidence for a distinct persistent Na^+ current underlying the burst slow depolarization.

6.7.4. Bursting in the Hippocampal Network

With respect to the hippocampal slice as a model for epilepsy, the low-Ca^{2+} preparation is used to study nonsynaptic epileptiform activity (e.g., Haas *et al.*, 1984; Roper *et al.*, 1993), providing further evidence against the hypothesis that substantial Ca^{2+} entry is a necessary component of intracellular bursts.

Synchronous bursting, e.g., that associated with epileptogenesis, is a function of both intrinsic bursting properties and patterns of connectivity. Although normally associated with CA3, such patterns of activity can be seen in CA1 under various conditions. In particular, and relevant to the present discussion, when $GABA_A$ inhibition has been blocked with bicuculline, synchronized bursts in CA1 can be seen when this region has been completely isolated from the rest of the hippocampus, as demonstrated by Meier and Dudek (1996). Such behavior is further evidence for the role of intrinsic CA1 bursting mechanisms, particularly because the intracellular records show bursts with a classical morphology.

6.7.5. So What Causes Bursting?

Recent reports (M. S. Jensen *et al.*, 1994, 1996; Azouz *et al.*, 1996) show that CA1 HPCs with stereotypical regular firing patterns may be reversibly transformed into bursting cells under various ionic manipulations. For example, there is an upregulation of bursting with increases in extracellular $[K^+]$ (M. S. Jensen *et al.*, 1994), which is qualitatively different than simply an increase in excitability that would be expected in general with raised $[K^+]_o$. A similar result was reported by Andreasen and Lambert (1995) for dendritic recordings of CA1 HPCs: raising $[K^+]_o$ from 3.25 to 5.25 mM converted a "regular" train of dendritic action potentials to bursts. Also, in this paper both dendritic bursts and slow Ca^{2+}-mediated dendritic spikes were found to be resistant to blockers of T- and L-type Ca^{2+} channels, leaving open the possibility for a necessary role of other types of Ca^{2+} channels in dendritic bursts (see Section 12).

The recently described Ca^{2+}-dependent nonspecific cation currents in CA1 and CA3 cells may offer another explanation behind the apparent Ca^{2+} dependence of bursting (Caeser *et al.*, 1993; Fraser and MacVicar, 1996; see Section 14.3). The pH of the extracellular medium is also important: Church (1992) showed that bursting CA1 rat HPCs reversibly switched to regular firing when the buffer was changed from HCO_3-CO_2 to HEPES (pH 7.4 throughout). In the same experiment, lowering the pH (under the HCO_3-CO_2 buffer) to 6.9 also reversibly changed the cell to regular firing.

Taken together, these findings support the view that the apparent dichotomy between regular and bursting patterns is at base a reflection of a delicate, and quite possibly modulatable balance of nonlinear mechanisms. We also note that explaining bursts requires not only identifying the mechanism or mechanisms underlying the slow depolarization, but also an adequate description of the fast I_{Na} kinetics that actually generates the spikes and the repolarizing mechanisms, which may be different than the regular-firing case. All cell models that we have examined which generate some sort of burst utilize a Hodgkin–Huxley-based description of I_{Na} with properties essential for the bursts, specifically, that the recovery of inactivation may occur at relatively depolarized potentials. However, as we shall discuss in Section 11, the use of this type of model overestimates steady-state Na^+ rectification, precisely because of the relatively depolarized inactivation curve.

It may be that bursting involves an interplay of I_{Na} inactivation that is less complete and K^+ current inactivation that is faster than that occurring during regular firing. This would be consistent with the proposal by M. S. Jensen *et al.* (1996) and Azouz *et al.* (1996) for a crucial role of a persistent TTX-sensitive Na^+ current, where the current may reflect a qausipersistent kinetic state of I_{Na} distinct from the reported I_{NaP}.

The channels that we include in the Working model do not generate bursting responses. For example, the preliminary model of I_{Na} that we propose in Section 11.3 is not able to generate spikes originating from the more depolarized portion of the depolarizing burst envelope, and even though it has a persistent component that is similar to the classical I_{NaP}, the kinetics of this current is not sufficient to drive the hump. On the other hand, this I_{Na} model can be readily extended to provide for higher thresholds, and possibly more complicated kinetics of inactivation. Finally, it will be important to reevaluate channel kinetics in the Working model after the addition of dendritic and axonal channels in order to see how the behavior of the distributed channels can contribute to bursting properties not evident with the somatic channel model.

In summary, bursting mechanisms in the cell models reviewed in this chapter do not seem to be consistent with all the available data. We feel that the bursting question is still open, and perhaps a refinement of the channel models described in this chapter will burst properly.

6.8. D Spikes

D ("dendritic") spikes are fast 5 to 15 mV depolarizing events in HPCs that are robust to changes in the holding potential (Kandel and Spencer, 1961). D spikes may occur singly or in clusters (with a timing reminiscent of bursts), they may be either subthreshold or capable of eliciting full spikes, and they may be spontaneous or triggered by current injection (in which case they appear as later events of a burst).

In the context of single-cell models, the important issue regarding D spikes is their relationship to intrinsic mechanisms; specifically, should our model be able to reproduce these waveforms in isolation? Although early explanations centered on the idea that D spikes were due to intrinsic dendritic mechanisms (Kandel and Spencer, 1961; Schwartzkroin, 1977), a more likely explanation is that these events are echos of single spikes or bursts from electrotonically coupled cells, i.e., via gap junctions (Valiante *et al.*, 1995; M. S. Jensen *et al.*, 1996) [for a discussion of electrotonic interactions in hippocampus, see Alger (1984)]. We may note that none of the models discussed in Section 7.1 describes D spikes; also, attempts at reproducing these "miniregenerative" waveforms in our more recent studies have argued against an intrinsic basis for them. Thus, in summary, it would appear that network interactions are crucial for this phenomenon.

6.9. Dendritic Channels and "Hotspots"

The data on the distribution of specific channels and their kinetics in the dendritic tree are increasing at an impressive rate (see the reviews by Johnston *et al.*, 1996; Magee *et al.*, 1998; Chapter 3, this volume). As mentioned earlier (Sections 4.1 and 4.5), voltage-dependent dendritic membrane properties can have profound implications for synaptic integration (e.g., Magee and Johnston, 1995b; Andreason and Lambert, 1998). In our discussion of specific channel properties in Sections 11–13 we will summarize some of the more recent key findings.

Certainly models can help focus this question, at least by proposing plausible density patterns. The practical problem is that without knowing precisely the kinetics of a specific channel, it is all the more difficult to establish reasonable predictions for its contribution in a distributed structure (Section 4.1). The more conservative models that deal with this question have tested uniform or fairly uniform dendritic channel distributions, and probably such constraints are consistent with the present state of knowledge regarding channel kinetics [e.g., the HPC model of Migliore *et al.* (1995) described in Section 7.1.6; see also the Purkinje cell model in De Schutter and Bower (1994)]. Other HPC models (discussed in Section 7.1) have used much more ad hoc distributions of various channels in the dendrites, for example, local concentrations in so-called "hotspots." Such distributions are equivalent to adding many more degrees of freedom to a model's parameter space, and as such make it more difficult to extrapolate general conclusions from the model behavior.

Direct electrophysiological evidence for hotspots in either the dendrites or in a perisomatic location (e.g., the axon hillock) of HPCs is lacking. Channel densities determined with cell-attached patch recordings show roughly homogenous distributions of Na^+ channels, sustained K^+ channels (e.g. I_{DR}), and the total number of all Ca^{2+} channel types throughout the soma and dendritic tree (Magee and Johnston, 1995; Magee *et al.*, 1998). The density of transient K^+ channels (e.g., I_A), on the other hand, increases with distance from the soma (Hoffman *et al.*, 1997). Simulations of somatic and dendritic spikes made under these constraints, however, appear to require channel densities at least two times larger than shown in the data (Hoffman *et al.*, 1997). It may be that on-cell patch measurements underestimate the true channel density (D. Johnston, personal communication), the kinetic models for the channels are inadequate, or that highly localized hotspots have so far escaped detection.

We now present a brief review of HPC models, emphasizing those models oriented toward understanding intrinsic properties, and summarize the major results.

7.1. Single-Cell Models Emphasizing Intrinsic Nonlinear Properties

Several of the single-cell models in this section have been referenced earlier in discussions of specific mechanisms, and later sections will do likewise (for convenience, these models will be labeled as described). Characteristics of different models will be taken both from the original reports and, in the case of the Traub91, Traub94, Warman94, Jaffe94, and Migliore95 models (see references below), from additional simulations run using Surf-Hippo.

7.1.1. Models of Traub and Colleagues

The work of Traub and colleagues (Traub and Llinas, 1979; Traub, 1982; Traub *et al.*, 1991, 1994) comprises a progressive evolution of HPC models that have been particularly influential not only in describing hippocampal function, but in application to models of other cortical neurons. The more recent models (Traub *et al.*, 1991, referred to as the Traub91 model; Traub *et al.*, 1994, referred to as the Traub94 model; see Fig. 1) are primarily based on CA3 HPC physiology. They include descriptions of six currents (I_{Na}, I_{Ca}, I_{DR}, I_C, I_{AHP}, and I_A), based to varying degrees on voltage-clamp data. These models also include strongly active dendrites. The Traub94 model included a moderately complex dendritic tree and axon, with significant variations of the kinetic descriptions and a different value of E_K (essentially making them models of *different* channels) for axonal I_{Na} and I_{DR} (Westenbroek *et al.*, 1989). This model was used mainly to examine synaptic-evoked firing patterns under various conditions.

There was no emphasis in either model on the reproduction of single-spike morphologies from the data. In particular, the morphology of the bursts generated by these models, which showed the fast spikes riding on the slow burst depolarization initially getting smaller and then getting larger, differ substantially from that which is typically seen in the data (Section 6.5).

A rather high value of C_m (3 μF/cm^2) and a rather low value of R_m (10,000 Ω-cm^2) were used for the entire Traub91 model. In the Traub94 model, more conventional values for R_m and C_m were used for the soma and dendrites (50,000 Ω-cm^2 and 0.75 μF/cm^2, respectively), but the axon membrane had a significantly lower R_m of 1000 Ω-cm^2. The R_i was also different in the soma/dendrites and axon in this model (200 and 100 Ω-cm, respectively). An essentially ad hoc pattern of dendritic channel densities was used in the models, with hotspots of I_{Ca} and associated I_{DR} and I_C, in order to reproduce Ca^{2+} spikes. A minor difficulty in both models for quantitative comparison with data is that all potentials are given relative to the resting potential; in our analysis we have assumed an E_{rest} of -60 mV.

Both the Traub91 and Traub94 models predicted bursts driven by a Ca^{2+} current and differential repetitive firing characteristics that depended on somatic versus dendritic stimulation. The Traub94 model showed bursting evoked by antidromic stimulation, which was blocked by appropriately timed dendritic inhibitory postsynaptic currents (IPSCs). Reduction of dendritic I_{Ca} and I_C, and an increase in somatic I_{DR}, caused the Traub91 model to switch from burst firing, reminiscent of classical CA3 HPC properties, to adaptive repetitive firing, more like CA1 HPCs. Neither model has a stable resting potential, in that they require constant holding currents to avoid spontaneous firing. For example, the Traub94 model shows spontaneous bursting even if the axon is removed (simulations not shown). When Ca^{2+} channels are blocked, this model fires spontaneous single spikes (simulations not shown).

The role of dendritic Ca^{2+} spikes was crucial in determining the behavior of both of these models. However, the dendritic spikes predicted by the Traub91 model, in particular under control conditions, are larger than either the early dendritic recordings of Wong *et al.* (1979) or Benardo *et al.* (1982), or recent recordings such as those of Spruston *et al.* (1995). The peak of these spikes in the model was about +40 mV, as compared to about +10 mV for the somatic (Na^+) spikes. In the Traub94 model dendritic Ca^{2+} spikes were somewhat smaller (about +20 mV) and the somatic spikes somewhat larger (about +30 mV).

It is likely that the magnitude of I_{Ca} in the models was overestimated, especially due to overestimation of the driving force (see Sections 12.2.3 and 12.3). We can consider the functional consequence of the predictions, namely that the excitability of the dendrites in these models predicted all-or-nothing responses which propagated throughout the tree, thereby greatly limiting the capability of the tree to act as a substrate for any subtlety in local synaptic interactions. These consequences are especially important in the critical evaluation of the many network models that have been built using these cell models (e.g., Traub and Miles, 1991; Traub *et al.*, 1996).

7.1.2. Borg-Graham (1987) Model

An earlier single-cell model (Borg-Graham, 1987a,b; referred to as the BG87 model), using a soma/short-cable model (Fig. 1) with all channels at the soma, was developed by the author to explore a variety of channel properties in a single context. Analysis was restricted to somatic responses to somatic stimulation. The model, based on both CA1 and CA3 data, included 11 channels: 3 Na^+ channels (2 fast transient channels, $I_{Na(Trig)}$ and $I_{Na(Rep)}$, and a persistent channel, $I_{Na(Tail)}$), 6 K^+ channels (I_{DR}, I_A, I_C, I_{AHP}, I_M, and I_Q), and 2 Ca^{2+} channels (I_{Ca} and I_{CaS}). Linear parameters were chosen under the assumption that measurements of HPCs [from the sharp-electrode recordings of T. H. Brown *et al.* (1981)] could be accounted for by incorporating a somatic shunt. The two putative fast transient Na^+ channels were included to account for the available voltage and current clamp data in the context of the Hodgkin–Huxley model, which required a mechanism that allowed for both distinct thresholds of activation and removal of inactivation above the lowest spike threshold without significant steady-state rectification. The model included a description of the dynamics of $[Ca^{2+}]_{in}$ based on a colocalization of I_{Ca} and I_C channels. This structure was proposed to account for the fast kinetics of I_C as opposed to the slower, more integrative kinetics of I_{AHP}. Finally, the model used a reformulation of the Hodgkin–Huxley model that casts the Hodgkin–Huxley parameters in a more biophysical and intuitive way, an updated version of which is discussed in Section 8.4.

The inclusion of specific mechanisms—two fast Na^+ channels with distinct kinetics—for different spike thresholds suggested that selective regulation of those mechanisms could have useful computational consequences (specifically, variation of failure point, or lockup, for large sustained inputs; Borg-Graham, 1987c).

While the BG87 model reproduced single-spike morphology and many of the characteristics of single-spike modulation in response to various pharmacological procedures, it was less successful at reproducing patterns of repetitive firing. For example, while there was a strong adaptation and an sAHP mediated by I_{AHP} in the model, I_M played a minor role in the cell response, in contrast to its more likely important contribution to regulation of repetitive firing (see Section 13.7). Also, there is no experimental evidence to support two distinct fast Na^+ channels: the Markovian I_{Na} model described in Section 11.3 collapses the functional characteristics of $I_{\mathrm{Na}(Trig)}$, $I_{\mathrm{Na}(Rep)}$, and $I_{\mathrm{Na}(Tail)}$ into a single channel.

7.1.3. Wathey, Lytton, Jester, and Sejnowski (1992) Model

When the increase in excitatory postsynaptic potential (EPSP) slope associated with long-term potentiation (LTP) is compared to the amplitude of the extracellularly measured population spike, there is often a larger increase in the latter relative to the former measurement. This relationship (E–S potentiation) was investigated by Wathey *et al.* (1992; referred to as the Wathey92 model), with a detailed anatomical model of a CA1 cell (Fig. 1) that included simple models, that is not necessarily matched to hippocampal properties, of somatic/axonal I_{Na} and I_{K}, with I_{Ca} distributed in hotspots in the dendritic tree. This paper described how the electrotonic relationship between the I_{Ca} locations and the locations of synaptic enhancement predicted whether the E–S potentiation was specific or nonspecific to the potentiated synapses.

7.1.4. Pongracz, Poolos, Kocsis, and Shepherd (1992) Model

Pongracz *et al.* (1992) examined the interaction of voltage-dependent NMDA receptors with a soma/short-cable structure cell model (apical and basal cables with five compartments each, and a two-compartment axon). Somatic channels included I_{Na}, I_{DR}, I_{Ca}, I_{AHP}, and I_C; all dendritic compartments included I_{Ca}, I_{AHP}, and I_C; and dendritic hotspots included I_{Na} and I_{DR}. The proximal axonal compartment (the "initial segment") included I_{Na} and I_{DR} channels; the I_{Na} was of a slightly higher density than the soma. In order to speed up the response of the axon, the passive properties of the initial segment and distal axonal compartment were significantly different than the rest of the cell: R_m of 1000 and 500 $\Omega\text{-cm}^2$ as opposed to 80,000 $\Omega\text{-cm}^2$, respectively, and C_m of 0 and 1 $\mu F/cm^2$ as opposed to 1 $\mu F/cm^2$, respectively. Notably, this model took into account the accumulation and removal of extracellular $[K^+]$ in a thin (200 Å) shell around each compartment—this was used to update dynamically the reversal potential of the K^+ currents and $GABA_B$ synaptic currents, but apparently not of the K^+ contribution to the leak battery. Intracellular Ca^{2+} accumulation in *very* thin (2 Å) compartmental shells was also considered, in a somewhat schematic way—a normalized $[Ca^{2+}]_{in}$ figured in the activation of the Ca^{2+}-dependent K^+ currents, but not in the driving force for the Ca^{2+} currents.

A key finding with the parameters used in this paper was that synaptic responses increased after repetitive input [concurrent activation of AMPA (α-amino-3-hydroxy-5-methyl-4-isoxazole proprionic acid), NMDA (*N*-methyl-D-aspartate), $GABA_A$, and $GABA_B$ (α-aminobutyric acid) synapses], in part due to external $[K^+]$ accumulation. Although model results were directly compared quantitatively with experimental data, the omission of any reference for the data made detailed comparison difficult.

7.1.5. Warman, Durand, and Yuen (1994) Model

Warman *et al.* (1994; referred to as the Warman94 model) constructed a CA1 HPC model that was similar in structure and intent to the earlier models, i.e., analysis of interactions between multiple somatic channels (I_{Na}, I_{Ca}, I_{CT}, I_{AHP}, I_M, I_A, and I_{DR}) within a simple electrotonic structure (Fig. 1). As in the BG87 model, the value of R_m for the soma was significantly lower than for the dendrites to account for the somatic shunt. In addition, the somatic C_m was taken as 1 μF/cm², whereas the dendritic C_m was taken as 1.85 μF/cm², although the reason was not specified. This model further refined the channel models, with better success in reproducing accommodation characteristics. An earlier model from this group with similar intent was made for a dentate granule cell (Yuen and Durand, 1991).

As with the BG87 model, this model reproduces several aspects of single-spike morphologies in response to the modulation of various currents. Also, in contrast with the likely underestimation of I_M in the BG87 model, a notable prediction of the model was that I_M prevents spontaneous repetitive bursting. This was a result of tonic activation of both I_M and I_{Na} at rest, suggesting in part that I_M played a major role in maintaining the resting potential. The resting activation of I_{Na} was related to a large "window current" for this channel, which also was the main current causing the slow depolarization underlying bursts (as compared to a Ca^{2+} current). While indeed there is some evidence for I_M activation at the resting potential (see Section 13.7), cholinergic suppression of this current does not necessarily lead to spontaneous activation. This prediction of the model, though, may reflect more an overestimation of I_{Na} around rest, rather than an overestimation of I_M (see Section 11.1.4).

7.1.6. Models of Jaffe *et al.* (1994) and Migliore *et al.* (1995)

Two related models with complicated dendritic structure from Jaffe *et al.* (1994; referred to as the Jaffe94 model) and Migliore *et al.* (1995; referred to as the Migliore95 model) focused mainly on reproducing the pattern of Ca^{2+} accumulation in the cell and patterns of repetitive firing.

The Jaffe94 model included I_{Na} limited to within 150 μm of the soma, and I_{DR}, I_C, I_{CaT}, I_{CaN}, and I_{CaL} in every cell compartment. In this model less attention was paid to the descriptions of channel kinetics, excepting those for the Ca^{2+} channels, since the analysis was constrained to the dynamics of $[Ca^{2+}]_{in}$ driven by uncomplicated spike trains. Although the morphology of the model was taken from a CA3 cell (Fig. 1), model parameters were chosen to represent both CA1 and CA3 HPCs; the author claimed that in their experience $[Ca^{2+}]_{in}$ profiles were essentially the same for the two types. This model was able to reproduce characteristics of $[Ca^{2+}]_{in}$ profiles consistent with data generated by this group, including a slower decay for somatic versus dendritic $[Ca^{2+}]_{in}$ that could be accounted for by the difference in surface-to-volume ratios.

The Migliore95 model, on the other hand, was specific to CA3 HPC properties. This model included a more complete repertoire of channels (I_{Na}, I_{CaT}, I_{CaN}, I_{CaL}, I_A, I_M, I_{AHP}, I_{DR}, and I_C), and considered how various patterns of dendritic channel density in several detailed anatomical reconstructions of HPCs could generate regular versus bursting behavior. Patterns of channel distributions were carefully constrained by placing fixed densities of Na^+ and Ca^{2+}-independent K^+ channels on the soma and the proximal apical dendrites (within 100 μm from the soma), and by varying a fixed density of the remaining channels throughout the dendritic tree and the soma. A main result of the Migliore95 model was that given the model channel kinetics, similar rules for distributions of all channel types would predict regular or bursting behavior for all the structures, despite some nontrivial variations in the dendritic morphology.

7.1.7. Other Single-Cell Models

As mentioned in Section 2.1, Spruston *et al.* (1993) presented a model-based analysis of somatic measurement of events in the dendritic tree. This paper looked at both a detailed cell anatomy (in fact the same cell that was used in the Jaffe94 model) and soma/short-cable equivalents. The cell structures were completely passive. The main result was that under a wide range of reasonable model cable parameters, under somatic voltage clamp distant synaptic events will be significantly distorted.

Two recent studies (Sah and Bekkers, 1996; Lipowsky *et al.*, 1996) directly combined experimental and modeling analyses of nonlinear dendritic properties, specifically with respect to the interaction between the propagation of synaptic potentials and dendritic channels. Both studies used simple soma/short-cable structures (including a basal and an apical dendrite), with various dendritic channels.

Sah and Bekkers (1996) examined the interaction between somatic and dendritic GABAergic IPSCs with I_{AHP}, and concluded that I_{AHP} is present in the proximal 200 μm of apical dendrite. Since the I_{AHP} was taken to be voltage independent (see Section 13.5), their I_{AHP} model was simply a fixed-conductance waveform. No other channels nor Ca^{2+} dynamics were included in the cell model. The proximal distribution of I_{AHP} was inferred by the response of this current to changes in voltage-clamp holding potentials that was similar to that of IPSCs of known dendritic origin. The restriction of I_{AHP} to the apical tree was suggested since EPSPs evoked in stratum radiatum (thus arising from the apical tree) were shunted more by the sAHP than EPSPs evoked in stratum oriens (thus arising from the basal tree).

Lipowsky *et al.* (1996) investigated the role of dendritic persistent Na^+ current (I_{NaP}) on the amplification of EPSPs in CA1 pyramidal cells. The model included I_{Na}, I_{NaP}, I_{Ca}, I_C, I_{AHP}, I_M, I_{DR}, and I_A, with parameters of the $[Ca^{2+}]$ dynamics and all channels other than I_{NaP} adapted from the Warman94 model. The description of I_{NaP} was taken directly from the experimental report of French *et al.* (1990). The key issue addressed by the model was the effect on subthreshold distal apical input by the inclusion of either somatic or dendritic I_{NaP}, and in fact the relevance of the model channels other than I_{NaP} and I_A was minimal to the results reported in this paper. Addition of dendritic I_{NaP} amplified and slowed the EPSP; subsequent incorporation of dendritic I_A resulted in EPSP waveforms that were well matched to the data. Although a simple dendritic structure was used (single cable), the fact that nonlinearities were included only in the proximal (<200 μm of the soma) apical dendrite, and simulated synaptic input was quite distal (600 μm from the soma), means that the issue of nonlinearities local to the input was avoided.

7.1.8. Recapitulation of the Working Model

The Working model that we demonstrate in this chapter has the soma/short-cable geometry of the BG87 model, again with all nonlinear mechanisms located at the soma. The model includes the following channels: I_{Na}, I_{CaN}, I_{CaT}, I_{CaL}, I_D, I_{DR}, I_A, I_M, I_{CT}, I_{AHP}, and I_H. A multiple-compartment Ca^{2+} concentration system is included in the soma, incorporating a Ca^{2+} pump and an instantaneous buffer.

7.1.9. On-Line Parameters of Reviewed Models

Complete "turnkey" implementations of the Traub91, Traub94, Warman94, and Working models, and specific mechanisms of the Jaffe94 and Migliore95 models are included in the Surf-Hippo distribution. Also available at the Surf-Hippo site are NEURON files for some of these mechanisms (written by Ragnhild Halvorsrud).

7.2. Models Emphasizing Interactions between Synaptic Processes and Intracellular Ion Dynamics

Although not specific to hippocampus, Qian and Sejnowski (1989, 1990) described a one-dimensional model of ionic flow in dendritic structures which takes into account coupled electromotive and diffusional flow, and which they used to predict voltage and concentration responses to synaptic input. These results were compared with that of a model based on electromotive flow only (the standard cable model, which is the type of circuit used by all other models described in this chapter), and that of a modified cable model which calculates, at each time step, the contribution of diffusional flow after electromotive flow has been evaluated. They showed that for large structures the contribution of diffusional processes is, as expected, negligible due to the small changes in concentration, while for small structures, such as spines, the incorporation of diffusional processes gave potentially significant differences from the cable model with respect to predicted concentration profiles and saturation characteristics in response to varying strengths of synaptic input. They also showed that the modified cable model accounts for much of these differences, providing that the synaptic input is not too great, with the benefit of being computationally much more efficient than the full diffusion formulation.

The models of Zador *et al.* 91990) and Schiegg *et al.* (1995; see Section 9.3) combined a detailed representation of a single spine with a simplified HPC dendritic structure to analyze Ca^{2+} dynamics in the spine microenvironment. Holmes and Levy (1990) presented a detailed compartmental model of a dentate gyrus neuron which focused on the interaction between Ca^{2+} influx via NMDA-mediated synaptic channels on explicit spine representations and detailed $[Ca^{2+}]_{in}$ dynamics relevant to conditions consistent with the expression of LTP. They found that transient saturation of local buffer in the spine head was necessary for $[Ca^{2+}]_{in}$ increases that were sufficient to account for LTP.

8. Channel Models

In this section we will describe various models for nonlinear channel properties, including both voltage and Ca^{2+}-dependent channels. These models are at an intermediate level of biophysical detail, appropriate for describing whole-cell currents.

For a given current I_X, the most common models for the underlying channel are based on the assumption of a gated, membrane-spanning ionophore, where the mechanism of conduction $f(V, \Delta[X])$ (flow of ions down a voltage gradient) and that of gating $h(V, t, \ldots)$ (voltage- and/or ligand-dependent modulation of conduction) are separable:

$$I_X = h(V, t, \ldots)f(V, \Delta[X])$$

Here V is the membrane voltage, t is time, and $\Delta[X]$ represents the concentration gradients for the permeable ions of the channel. The ellipsis in the argument of $h(\cdot)$ stands for the various ligand-dependent processes (e.g., Ca^{2+} dependence).

Here we review two models of the conduction term $f(V, \Delta[X])$: Ohmic (thermodynamic equilibrium conduction) and constant-field permeation (nonequilibrium conduction). We shall also describe two related models of the gating term $h(V, t, \ldots)$: the Hodgkin–Huxley model (and a reformulation of the original equation parameters), and Markov models (which include the Hodgkin–Huxley model as a special case). We shall also review models of Ca^{2+}-dependent gating. We shall see that in principle both conduction models for $f(V, \Delta[X])$ have a nonlinear negative feedback characteristic which in certain cases has important functional consequences; therefore the nonlinearity of neuronal ionic currents can arise from *both* the conduction term and the channel gating term.

8.1. Linear Conductance: Driving Force from Equilibrium Thermodynamics

In most cell models channel conduction is taken as ohmic (linear), with the driving force determined by the membrane voltage, $\Delta[X]$, and the relative permeabilities of the ions that pass through the channel. The general form of the ohmic model is

$$f(V, \Delta[X]) = \bar{g}_X(V - E_X)$$

where, assuming that the units of $f(\cdot)$ are in nanoamperes, $\bar{g}_X$ is the absolute conductance in μS and E_X is the reversal potential for the channel, the latter given by either the Nernst equation if only one ion is involved, or the Goldman–Hodgkin–Katz (GHK) voltage equation (e.g., Hille, 1992) for more than one ion (voltages in millivolts). In this model, the nonlinearity of the conductance term arises implicitly from the Nernst or GHK equation-based estimate of the driving force E_X coupled with a nonzero change in $[X]$ with respect to I_X. Thus, the more current, the more the concentration gradient is reduced, resulting in negative feedback. In practice, many cell models that use the ohmic description neglect changes in concentration, under the assumptions that these activity-driven changes are small.

The use of the Nernst equation for the driving force in the ohmic model is based on the assumption of thermodynamic equilibrium; strictly speaking, this corresponds to the situation in which no current is flowing. Thus this description is more appropriate for those ions in which the concentration gradient is not very high and the integrated currents are small compared to the relevant concentrations. These conditions are approximated for K^+ and Na^+, which is similar to saying that these ions are closer to equilibrium. Although widely used (as in all the non-Ca^{2+}-channel descriptions we shall present here), the validity of this approximation for describing nonequilibrium conditions (that is, when the neuron is *doing* something electrically) has not been systematically tested.

Setting E_{rev} for Ohmic Channel Models

The most basic way to determine the ions associated with a particular channel experimentally is by estimating the reversal potential and relating that value to that expected from the Nernst or GHK voltage equations. Manipulating the concentrations of specific ions and comparing the change in the reversal potential with that predicted by these equations may also be done. Especially for the case of K^+ channels, there is a dichotomy between the experimental values for E_{rev} and those used by many models, in that in the former case the measured E_{rev} is rarely that predicted for a pure K^+ conductance, whereas in the latter case models often assume that all K^+ channels share the same $E_{rev} = E_K$. A more flexible approach is to allow the E_{rev} for nominally K^+ channels to be set to within a maximum of 20 mV or so above the Nernst E_K (typically this is −90 to −95 mV), a range typically seen in the data. We have found that this additional free parameter allows for some useful fine tuning in the shape of the afterpotentials.

We note as well that the value of E_K (and sometimes E_{Na}) in some models is not given explicitly, despite the fact that there are no reliable canonical values, and variations in these parameters can have a significant effect on cell behavior.

8.2. Intrinsic Nonlinearity of Pore Conduction

As a particular ion moves farther from equilibrium (specifically the case for Ca^{2+}), the ohmic model becomes less accurate. One of the most widely used nonequilibrium models of pore conduction permeation is the constant-field model, described by the GHK current equation. In this equation (Jack *et al.*, 1983; Hille, 1992), the nonlinearity of the permeation term is explicit:

$$f(V,\Delta[X]) = \bar{p}_X \frac{Vz^2F^2}{RT} \frac{[X]_{in} - [X]_{out}\exp(-zFV/RT)}{1-\exp(-zFV/RT)}$$

where, again assuming that the units of $f(\cdot)$ are in nanoamperes, R is the gas constant, F is Faraday's constant, T is temperature in degrees Kelvin, $\bar{p}_X$ is the permeability (*not* the conductance) of the channel in cm^3/sec, $[X]_{in}$ and $[X]_{out}$ are the intracellular and extracellular concentrations in millimolar, and z is the valence of the permeant ion; here we assume that channel conduction is mediated by only one type of ion. The nonlinear behavior of this term manifests itself as negative feedback for inward currents, particularly Ca^{2+} (e.g., Hess and Tsien, 1984), both because of the sublinear characteristic at typical values of $[Ca^{2+}]$ and because the driving force is reduced as the gradient of $[X]$ is reduced with current flow. We note, however, that the dynamic change of this characteristic due to typical concentration changes is not very large.

Neuron models which have used this formulation, in all cases for Ca^{2+} currents, include the Wathey92, Jaffe94, and Migliore95 models, and models of thalamic relay neurons (Huguenard and McCormick, 1992) and cortical pyramidal cells (Lytton and Sejnowski, 1991). At membrane potentials far from the equilibrium point (e.g., $V_m < -20$ mV for Ca^{2+} channels) the GHK current equation becomes linear, and if the voltages in the model under consideration stay in this range it may be reasonable to substitute an ohmic equivalent (e.g., De Schutter and Bower, 1994). This conversion also has to be considered when interpreting many of the reports on Ca^{2+} channels, where conduction values at some holding potential are reported instead of permeabilities.

8.3. Channel Gating à la Hodgkin and Huxley: Independent Voltage-Dependent Gating Particles

Hodgkin and Huxley (1952b) described channel gating as an interaction between different, independent two-state (open and closed) elements or "particles," all of which must be in the open state for channel conduction. The state dynamics of each particle is described with first-order kinetics:

$$x_C \overset{\alpha_x(V),\beta_x(V)}{\rightleftharpoons} x_O \tag{2}$$

where x_C and x_O represent the closed and open states of gating particle x, respectively. Here $\alpha_x(V)$ and $\beta_x(V)$ are the forward and backward rate constants of the particle as a function of voltage, respectively. Taking x to be the probability of the open state, we obtain the associated differential equation

$$\frac{dx}{dt} = \frac{x_\infty(V) - x}{\tau_x(V)}$$

with the steady-state value of x, $x_\infty(V)$, and the time constant, $\tau_x(V)$, given by $\alpha_x(V)$ and $\beta_x(V)$ (see below). A useful generalization of this kinetic scheme is to assume that the rate constants can be functions of other signals (e.g., $[Ca^{2+}]$); the general form of the gating term in this scheme is the product of the open probability of the channel's gating particles:

$$h(V, t, \ldots) = \Pi\, x_i^{n_i}(V, t, \ldots)$$

where x_i is the open probability of a given type of gating particle, and n_i is the number of particles of a given type associated with the channel. This system is a specific type of Markovian model (see Section 8.7), constrained by the independence condition; the general Markovian model allows arbitrary transitions between the states describing the channel kinetics.

The success of the Hodgkin and Huxley theory is in the utility of a kinetic scheme which is described by ensembles of independent two-state gating particles, each of which has transition rates which are functions of the membrane voltage. While Hodgkin and Huxley hypothesized that the steady-state behavior of each particle fits a Boltzman distribution, the underlying rate equations were fit to the data with various functions that included both linear and exponential voltage terms. Although taken as a canonical form by countless cell models, a consequence of the specific form of the equations is that there is not an obvious relationship between the equations' parameters and the standard kinetic description, i.e., in terms of the voltage dependence of the steady-state value x_∞ and time constant τ_x.

8.4. An Extended Hodgkin and Huxley Model

For gating particles that are dependent only on voltage, this theory can be recast in more explicit form by considering parameters of a single-barrier kinetic model for each particle (Jack *et al.*, 1983; Borg-Graham, 1991). The advantages of this form are parameters which are both more directly related to the functional behavior of the channel, and which have specific biophysical interpretations. Various neuron models have used this formalism [the BG87, Migliore95, and Jaffe94 models; also Lytton and Sejnowski (1991) and Bush and Sejnowski (1994) for cortical pyramidal cells, Quadroni and Knöpfel (1994) for medial vestibular nuclei neurons, and Burgi and Grzywacz (1994) for retina].

8.4.1. Comparison to the Original Hodgkin–Huxley Equations

A final advantage is that in its basic form, this formulation has five parameters for each particle ($V_{1/2}$, z, γ, K, and τ_0; see below), whereas the Hodgkin–Huxley model has six parameters, in part because because the dependence between $\alpha(V)$ and $\beta(V)$ is not explicit. Nevertheless, this formulation may be readily fitted to the original Hodgkin–Huxley equations of squid axon I_{Na} and I_K; the error is comparable to the error between the original equations and the data to which they were fit (cf. Figs. 4, 7, and 9 in Hodgkin and Huxley, 1952b).*

*We have fitted this formulation to all Hodgkin–Huxley mechanisms in the channels used in the models of Section 7.1—in every case model behaviors using the fitted versions were qualitatively the same as using the published versions.

The formulation which describes the first-order kinetics represented in Equation (2) is defined as follows [the following derivation is somewhat modified from the description given in Borg-Graham (1991)].

We first derive the expressions of $\alpha'_x(V)$ and $\beta'_x(V)$, the forward and backward rate constants of the single-barrier transition. For a given particle x, z (dimensionless) is the *effective* valence of the gating particle, where a positive value means that the particle tends to the open position with depolarization—i.e., it is an activation particle. Likewise, a negative value of the valence means that the particle tends to the open position with hyperpolarization—i.e., it is an inactivation particle.* The effective valence is the product of the actual valence of the particle (an integer) and the proportion of the membrane thickness that the particle moves through during state transitions; thus this parameter is not restricted to integers. γ (dimensionless, between 0 and 1) is the asymmetry of the gating particle voltage sensor within the membrane. K (1/msec) is the leading coefficient of both $\alpha'_x(V)$ and $\beta'_x(V)$. This term can be described in terms of Eyring rate theory, an explicit version of which may include an additional temperature term (Borg-Graham, 1991). We shall not consider this component explicitly here, and just take K as a constant. $V_{1/2}$ is the voltage for which $\alpha'_x(V)$ and $\beta'_x(V)$ are equal. The final equations for $\alpha'_x(V)$ and $\beta'_x(V)$ are then

$$\alpha'_x(V) = K \exp\left(\frac{z\gamma(V - V_{1/2})F}{RT}\right)$$

$$\beta'_x(V) = K \exp\left(\frac{-z(1-\gamma)(V - V_{1/2})F}{RT}\right)$$

An additional parameter, τ_0 (not the same as the linear membrane time constant), represents a rate-limiting step in the state transition, for example, "drag" on the particle conformation change.† This parameter is crucial for fitting the expressions to the original Hodgkin–Huxley equations. There are two equivalent ways of incorporating τ_0 into the kinetics. The more intuitive approach is to include τ_0 explicitly in the expression for the time constant $\tau_x(V)$ of the particle state differential equation:

$$\tau_x(V) = \frac{1}{\alpha'_x(V) + \beta'_x(V)} + \tau_0$$

We can also derive the appropriate forms for $\alpha_x(V)$ and $\beta_x(V)$ that characterize the kinetics of Equation (2):

*This definition for the sign convention of z is opposite to that found in Borg-Graham (1991), for no profound reason. The convention used here is the same as that used in Jack *et al.* (1983) [Equations (8.33) and (8.34), p. 242] and depends only on the (arbitrary) definition of polarity of the gating particle with respect to the membrane inner and outer surfaces.

†The formulation in Borg-Graham (1991) used this parameter as a lower bound for $\tau_x(V)$. The form defined above not only avoids discontinuities in the characteristic, but also seems more plausible. Similar considerations have been explored for other, more general, kinetic schemes (e.g., Patlak, 1991).

$$\alpha_x(V) = \frac{\alpha'_x(V)}{\tau_0(\alpha'_x(V) + \beta'_x(V)) + 1}$$

$$\beta_x(V) = \frac{\beta'_x(V)}{\tau_0(\alpha'_x(V) + \beta'_x(V)) + 1}$$

Note that when τ_0 is 0, $\alpha_x(V)$ is equal to $\alpha'_x(V)$ and $\beta_x(V)$ is equal to $\beta'_x(V)$. Finally, the expression for $x_\infty(V)$ is

$$x_\infty(V) = \frac{\alpha'_x(V)}{\alpha'_x(V) + \beta'_x(V)} = \frac{\alpha_x(V)}{\alpha_x(V) + \beta_x(V)} \quad (3)$$

Two additional parameters, α_0 and β_0, may be considered in some cases, though they are not necessary in reproducing the original Hodgkin–Huxley equations. These parameters are forward and backward rate constants, respectively, of parallel state transitions which are voltage independent, but nonetheless figure into the total kinetics. If considered, α_0 and β_0 alter the expressions for $\alpha_x(V)$, $\beta_x(V)$, and thus $\tau_x(V)$ and $x_\infty(V)$, which we will not consider here.

8.4.3. Relationships between the Extended Hodgkin–Huxley Model Parameters and $x_\infty(V)$ and $\tau_x(V)$

As stated, an advantage of this formulation is a clearer connection between the parameters and the macroscopic characteristics $x_\infty(V)$ and $\tau_x(V)$. In particular, $V_{1/2}$ gives the midpoint and z sets the steepness of the $x_\infty(V)$ sigmoid. The symmetry parameter γ determines the skew of $\tau_x(V)$: $\gamma = 0.5$ gives a symmetrical bell-shaped curve for $\tau_x(V)$, which otherwise bends to one side or the other as γ approaches 0 or 1. The value of z sets the width of $\tau_x(V)$, unless γ is equal to either 0 or 1, in which case $\tau_x(V)$ becomes sigmoidal and thus z sets the steepness as for $x_\infty(V)$.

With this scheme a particle with a voltage-independent rate constant can be represented by setting $1/K \ll \tau_0$, and both α_0 and β_0 to zero, thus making τ_0 the effective time constant. Likewise, both the time constant and steady-state become voltage independent by setting $K = 0$ and choosing the appropriate α_0 and β_0. We should emphasize that the α_0 and β_0 parameters are included for flexibility; in all of the Hodgkin–Huxley-type parameters that we shall discuss in this chapter, and in the discussion below, both parameters are taken as zero.

8.4.4. Determining the Number of Particles in Hodgkin–Huxley Models

The exponent n of a particular gating particle in the Hodgkin–Huxley description implies the number of that type of particle in the channel. The motivation of Hodgkin and Huxley for introducing this parameter was to account for sigmoidal characteristics of current activation under voltage clamp (the power for the inactivation particle h in the Hodgkin–Huxley I_{Na} model is one—however, this protocol did not directly measure the kinetics for the removal of inactivation). Some experimental papers report fitting integer powers of hypothetical gating particles according to the observed kinetics, but more typically steady-state activation or inactivation data are simply the observed macroscopic behavior (that is, reflecting the steady state of the ensemble of particles). Thus, for channel models, gating particle powers can often be considered as another free parameter.

There are three immediate consequences of assuming more than one particle of a given type. First, while the transition from the closed to open state under voltage clamp of a single particle is exponential, as noted above (and key to Hodgkin and Huxley's choice of parameters), powers greater than one result in a sigmoidal transition (arising from the time course of $[1 - \exp(-t)]^n$ when n is greater than one). This trajectory implements a kind of delay in the closed-to-open state transition, which is often a major motivation for hypothesizing $n > 1$ in Hodgkin–Huxley models. Second, although the collective steady-state behavior of an ensemble of particles of a given type as a function of voltage is sigmoidal, the expression $[1 - \exp(V)]^{-n}$ goes to zero faster than $[1 - \exp(V)]^{-1}$ for $n > 1$ [these functions in V are the basic form of Equation (3)]. Thus, if the parameters of the extended Hodgkin–Huxley model are fitted to the measured steady-state behavior, assuming n greater than one can attenuate the zero end of this characteristic. In some cases, we have found that using n greater than one has restricted this tail activation which otherwise could complicate the contribution of the channel in question at certain voltages. Finally, in this model when $n = 1$ the peak (when $0 < \gamma < 1$) or inflection point (when $\gamma = 0$ or 1) of $\tau_x(V)$ is roughly around $V_{1/2}$. Higher powers of n tend to shift these points in the direction of voltage for which $x_\infty(V)$ tends to 0. This shift may have a noticeable functional effect between different values of n.

8.5. Activation versus Deactivation; Inactivation versus Deinactivation

It is useful to briefly review the terminology of activation and inactivation. Traditionally, in the context of the Hodgkin–Huxley model, the difference between activation and inactivation is simply that they have an opposite dependence on membrane voltage. On the other hand, in more general kinetic descriptions inactivation is often used to describe the transition into closed states after activated or open states. In the Hodgkin–Huxley context, we can refer to a channel becoming open either by activation (assuming no concurrent inactivation) or by deinactivation (assuming concurrent activation). Likewise, a channel can become closed by either by inactivation or by deactivation.

Finally, there are channels that become open with hyperpolarization, e.g., I_H. Assuming that they are persistent channels (no inactivation), it is clearer to say that these channels are gated by activation particles, even though this is inconsistent with the definitions given above for the voltage dependence of gating particles.

8.6. The Hodgkin–Huxley Model: Some Conclusions

The independence of gating particles in the Hodgkin–Huxley model has at least two advantages: presented in the proper form, the kinetics of the model can be predicted in an intuitive way, especially when considering experimental data, and as a practical issue the numerical integration may be accomplished efficiently (Hines, 1984). On the other hand, this independence constrains the equivalent state space description (vis-à-vis the relative transition rates; Destexhe *et al.*, 1994) of the more general Markovian model.

8.7. Channel Gating as Dynamical Systems à la Markov Models

The behavior of a dynamical system is characterized by some trajectory in state space, where each "state" is defined fundamentally in terms of observed kinetics. For models of ion channels, including the Hodgkin–Huxley model, the general assumption is that each state corresponds to a specific conformation of the membrane protein, thus an individual channel can be in only one state at a time. The description of an ensemble of channels (e.g., as appropriate for a cell model) assigns a probability at any given time for the occupancy of each state. The observable states of a channel are those that conduct current (the open states), but of course the nonopen states (called closed or inactivated, depending on the particular model) are as important in understanding channel behavior.

When the transitions between states in a dynamical system are stochastic with probabilities that are independent of the *history* of the system, the resulting model is a Markov model. This classification does not mean that the transitions cannot be functions of time per se; however, most models of this sort do not use functions in which time is an explicit independent variable. State transitions may be functions of other quantities, certainly—for channel models the most common dependence is that of voltage.

Markov kinetic models are the standard description for detailed biophysical analysis of channel kinetics, but there have been relatively few applications in the neural modeling literature (however, see Bush and Sejnowski, 1991). One practical limitation for these models is that they are typically much more computationally expensive than two-state models such as the Hodgkin–Huxley model (for a hybrid of Markov and Hodgkin–Huxley, see below). The general methodology of Markov models as applied to channel mechanisms is beyond the scope of this chapter (see Colquhoun and Hawkes, 1995). Destexhe *et al.* (1994) have written a concise discussion on Markov models which is aimed in particular at cell-level modeling of channels and synapses.

Although the Markovian framework puts no restrictions on the functions that define state transitions (other than the no-memory condition), the form presented above of the $\alpha_x(V)$ and $\beta_x(V)$ functions for the extended Hodgkin–Huxley model is convenient and very general. The voltage-dependent-transitions Markov models that we shall use in the Working model for I_{Na} and I_{CT} (Sections 11.3 and 13.4.1) may be defined in such terms; however, we shall use the following squeezed-exponential formula for the transition rate of $\alpha_{ij}(V)$ from state i to state j:

$$\alpha_{ij}(V) = \left\{ \tau_{min} + \left[(\tau_{max} - \tau_{min})^{-1} + \exp\left(\frac{V - V_{1/2}}{k} \right) \right]^{-1} \right\}^{-1} \tag{4}$$

where the inverses of τ_{min} (analogous to τ_0 in the extended Hodgkin–Huxley model) and τ_{max} put upper and lower bounds, respectively, on the rate constant

$\alpha_{ij}(V)$. Note that there is an implicit coefficient of the exponential term of 1/msec (same units as either $1/\tau_{min}$ or $1/\tau_{max}$) in this equation.

8.7.1. Markov Mechanisms Can Decouple the Time Course of Activation/Inactivation Signals and Functional Channel State Transitions

There are several cases in which there is an apparent delay between a modulating signal S, e.g., voltage or Ca^{2+}, and the functional state transition, say $A \rightarrow B$. For example, the activation of the I_{AHP} channel is dependent on a rise in $[Ca^{2+}]$, but the actual channel opening may lag by a few hundred milliseconds, by which time the $[Ca^{2+}]$ may have fallen below the activation threshold (Section 13.5). In a general Hodgkin–Huxley scheme of a collection of independent particles (where the modulating signal S is not necessarily voltage), the time course of S is directly coupled to the associated state transition, such that the subsequent removal of S would arrest a slow transition. The sigmoidal activation when $n > 1$ in this model effects a kind of delay, but the result is somewhat gradual, since the underlying time constant for the transition dependent on S is the same as that determining the delay. In the Markov description, a more independent delay may be accommodated by adding an additional state, say A', from which the transition to state B is defined by some constant slow rate. Thus, a change in S could move the channel from A to A', from which it would then move more slowly to B, assuming that the reverse transition $A' \rightarrow A$ was even slower and relatively independent of S.

We may note that the Moczydlowski and Latorre (1983) model described below has some degree of this sort of decoupling, in that the full closed-to-open transition is a sequence comprised of a Ca^{2+}-dependent, a Ca^{2+}-independent, and finally a Ca^{2+}-dependent step.

8.7.2. Linear Markov Models Based on the Hodgkin–Huxley Model

The standard interpretation of the Hodgkin–Huxley gating model is that all particles associated with a channel must be in the open state for channel conduction. A generalization of this theme is to assume that for a channel with N Hodgkin–Huxley gating particles of the same type, the channel conducts when any M ($\leq N$) of the particles are in the open state [e.g., the model of retinal I_H of Barnes and Hille (1989)]. We call this a linear Markov model since the state diagram has no loops. Since all the particles are of the same type, the computation of the channel gating is much more efficient than that for the general Markov case. Specifically, given N particles of type x, each with the same open probability p (since they are assumed to be identical and independent), then the contribution $h_x(V, t, \ldots)$ of the ensemble of particles to the channel gating term $h(V, t, \ldots)$ is given by a summation of binomial terms:

$$h_x(V,t, \ldots) = \sum_{i=M}^{N} \binom{N}{i} p^i (1-p)^{(N-i)}$$

8.8. Ca^{2+}-Dependent Gating

In this section we review various data and models for the Ca^{2+} dependence of K^+ channel activation.

8.8.1. The Barrett, Magleby, and Pallotta (1982) Model

Using single-channel recordings of the Ca^{2+}-dependent K^+ channel (BK channel) in cultured rat muscle cells, Barrett *et al.* (1982) found an approximate third-power relationship between channel open times and $[Ca^{2+}]$ that was strongly facilitated by depolarization. For example, at a membrane voltage of 10 mV, channels became open with a $[Ca^{2+}]$ threshold of about 1 μM. The steady-state characteristic of this data was adapted in the Warman94 model for the kinetics of I_{CT}.

8.8.2. The Moczydlowski and Latorre (1983) Model

Moczydlowski and Latorre (1983) recorded from similar channels, reconstituted in planar lipid bilayers, and proposed a detailed Markovian kinetic scheme for the Ca^{2+}- and voltage-dependent gating of these channels. Other work analyzing the BK channel in rat muscle supports the Markov interpretation (McManus and Magleby, 1989).

The Moczydlowski and Latorre model has been applied in several neuron models (e.g., the Jaffe94 and Migliore95 models), in part, presumably, because this model is included in the NEURON package. We found it possible to apply the Moczydlowski and Latorre model to the Ca^{2+} dependence of both I_{CT} and I_{AHP}, after appropriate adjustments of the model parameters for intracellular Ca^{2+} dynamics (Section 9.7). However, considering that the six parameters for the Moczydlowski and Latorre model do not have an obvious relationship with properties readily observed from whole-cell currents, and that our particular fit required parameter values quite different from the original report, we chose to use a simpler Ca^{2+} dependence formulation (see below). Also, in the case of I_{AHP} the data do not indicate a voltage dependence (but see discussion in Section 13.5). We note as well that the parameters used in the cell models mentioned above are also quite different from the parameters in the original paper (and from each other), which puts into question the utility of this description for neuron models without further data specific to neuronal Ca^{2+}-dependent K^+ channels. We also note that the specific Markovian kinetic model chosen by these authors was not unique, but rather was the simplest (least number of states) given their data; this strategy is typical for Markovian analysis of single-channel kinetics. A perhaps more robust general conclusion from the Moczydlowski and Latorre analysis was that channel kinetics were voltage as well as Ca^{2+} dependent, and that the fully activated state required at minimum the binding of two Ca^{2+} ions.

8.8.3. A Simple Kinetic Model

A simple model for Ca^{2+}-dependent kinetics of a two-state gating particle can be described by the following reaction; a similar scheme was used in the BG87 and Migliore95 models [also Quadroni and Knöpfel (1994) for medial vestibular nuclei neurons and Burgi and Grzywacz (1994) for retina]. Assume that w represents the open probability of a Ca^{2+}-dependent gating particle:

$$w + n\mathrm{Ca}^{++} \overset{\alpha,\beta}{\rightleftharpoons} w^*$$

where the asterisk denotes the open state bound to n Ca^{2+} ions.* In the more general Markovian framework, w and w^* refer to two adjacent states out of the entire state space. The forward and backward rate constants are α and β, respectively. This scheme assumes that binding with Ca^{2+} ions is cooperative—either all binding sites are occupied or none are. We may also consider a τ_0 parameter as in the extended Hodgkin–Huxley model. If we assume that the binding of Ca^{2+} in this reaction does not appreciably change $[Ca^{2+}]$, then the steady-state value for w^*, w_∞, and the time constant for the kinetics τ_w are given by

$$w_\infty = \frac{\alpha[Ca^{2+}]^n}{\alpha[Ca^{2+}]^n + \beta} \tag{5}$$

$$\tau_w = \frac{1}{\alpha[Ca^{2+}]^n + \beta} + \tau_0 \tag{6}$$

Voltage dependence of this scheme can be added in a number of ways. For example, in the Markov model of I_{CT} in the Working model the forward rate constant is defined by

$$\alpha_w(V,[Ca^{2+}]) = \alpha_V(V) \times \alpha[Ca^{2+}]^n \tag{7}$$

where $\alpha_V(V)$ is the squeezed-exponential function of voltage in Equation (4), such that the forward reaction speeds up with depolarization.

9. Ionic Concentration Dynamics

The ionic fluxes which underly neuronal activity necessarily change the concentrations of those ions on either side of the membrane. Whether or not these changes have any functional consequences depends on the relative size of the changes and on any secondary effects (e.g., those not directly determining the driving force for the ionic current) of the concentrations. The necessary role of internal Ca^{2+} on both short and fast time scales in the control of various membrane properties and processes such as LTP is well established, and makes this ion the most important candidate for considering the concentration dynamics in models.

For our purposes, at least two major K^+ currents have been clearly demonstrated to depend on the influx of Ca^{2+}: the large and fast I_{CT} and the small and much slower I_{AHP}. Thus for the Working model we develop a relatively simple model for $[Ca^{2+}]_{in}$ in order to account for the activation of these currents. Other Ca^{2+}-dependent currents that may be relevant to the reproduction of basic firing properties, but will not be included in the Working model, include the "slow" I_C, discussed in Section 13.4, the Ca^{2+} dependence of I_M, discussed in Section 13.7, and the Ca^{2+}-dependent nonspecific cation currents, discussed in Section 14.3.

Other modeling studies have considered the interaction between cell re-

*One can also imagine a similar but reverse reaction for the description of Ca^{2+}-dependent inactivation, for example, as reported for Ca^{2+} channels (Section 12.2.3).

sponses and changes in $[K^+]$ [e.g., Yamada *et al.* (1989) for bullfrog ganglion cells, and Pongracz *et al.* (1992) for HPCs; Section 7.1.4], and/or $[Cl^-]$ [e.g., Qian and Sejnowski (1989, 1990) for a nonspecific dendritic spine model; Section 7.2].

In this section we shall discuss recent data on measurements of intracellular $[Ca^{2+}]$ during synaptic activity and current-evoked spikes. We review the factors which set the resting level of $[Ca^{2+}]_{in}$ and the role of intracellular Ca^{2+} stores. We discuss membrane pumps both for Na^+ and, more relevant to the model, for Ca^{2+}. Models for intracellular Ca^{2+} buffers are described, and we then describe a fairly simple model which captures the basic characteristics of $[Ca^{2+}]$ dynamics, comparing this system with previous hippocampal models. We note that constructing an adequate model of $[Ca^{2+}]$ dynamics is very difficult since the quantitative data are somewhat inconsistent (see below) and the mechanisms for regulation of intracellular Ca^{2+} are poorly understood. Also, estimating Ca^{2+} flux in a model is of course quite dependent on the model for the Ca^{2+} channel or channels.

9.1. The Data on Ca^{2+} Concentration Dynamics

As we shall see, estimates on the absolute change in $[Ca^{2+}]$ in response to either intracellular current or synaptic stimulus vary among quantitative studies, as summarized in Table III. The key figures that we must extract from the data are the incremental change of $[Ca^{2+}]$ per spike, which, since the sAHP amplitude increases incrementally with the number of spikes, is directly related to the Ca^{2+} sensitivity of the current underlying the sAHP, I_{AHP} (Section 13.5), and the time constant for decay, which is related to the duration of the sAHP and thus I_{AHP}. These figures, combined with direct and indirect estimations of crucial I_{Ca} parameters and a buffering parameter (Section 9.5.1), form the backbone of our model for $[Ca^{2+}]$. Note that the structure of the Ca^{2+} system that we shall use (Section 9.7), and its relationship to the Ca^{2+} dependence of I_{CT}, implies that the large transient localized $[Ca^{2+}]_{in}$ relevant for I_{CT} will not be directly evident from, and thus not directly constrained by, $[Ca^{2+}]$ changes integrated over many micrometers of a cell process, as given by the available data.

We also note that the size of changes in proximal dendrite $[Ca^{2+}]$, as opposed to somatic $[Ca^{2+}]$, may be more relevant for determining parameters of the I_{AHP}, since the functional location of this current may be primarily the proximal apical dendrite (Sah and Bekkers, 1996; see Section 13.5). This factor will be neglected in the Working model, however, since all channels are located at the soma.

9.1.1. Ca^{2+} Influx via Voltage-Gated Ca^{2+} and Synaptic Channels

First, it is important to recall the major sources of intracellular Ca^{2+} in general, that is, via Ca^{2+} channels, synaptic Ca^{2+} entry, and release from Ca^{2+} stores. For example, Regehr and Tank (1992) describe synaptically evoked $[Ca^{2+}]$ changes in CA1 in the slice, distinguishing between those arising from voltage-gated Ca^{2+} channels versus NMDA channels (see also Miyakawa *et al.*, 1992). Although a large fraction of the rise in $[Ca^{2+}]$ was blocked by intracellular application of QX-314 (indicating a dependence on I_{Na}-mediated depolarization), they also characterized a significant AP5- (NMDA-receptor blocker) sensitive component that was insensitive to Na^+ channel block. However, since we intend to model responses to intrinsic stimulation, we will focus only on Ca^{2+} influx via voltage-gated Ca^{2+} channels. Future development of the model will include evaluation of Ca^{2+} release from Ca^{2+} stores.

9.1.2. $\Delta[Ca^{2+}]$ from Current-Evoked Spikes

Jaffe *et al.* (1992) studied the dynamics of $[Ca^{2+}]_{in}$ and its relationship with Na^+ entry in hippocampal neurons using Ca^{2+}- and Na^+-sensitive dyes. They reported an increase of about 3 nM in $[Ca^{2+}]_{in}$ 70 μm from the soma with a single current-evoked spike. A spike train of 29 spikes in response to a 500-msec current step evoked a steady increase in $[Ca^{2+}]_{in}$ both at the soma and 60 μm away, with a total change of about 60 and 100 nM, corresponding to 2.1 and 3.5 nM/spike, respectively. The correlation of Na^+ concentration changes in the dendrites with $[Ca^{2+}]_{in}$ changes indicated that dendritic Na^+ spikes were necessary for the major component of $[Ca^{2+}]$ changes in the tree. Little change in $[Ca^{2+}]$ was seen at locations greater than 250 μm from the soma.

A similar study by Spruston *et al.* (1995), but using dual soma/dendritic recordings instead of Na^+-sensitive dye, also concluded that $[Ca^{2+}]$ changes in the dendritic tree were strongly correlated with dendritic action potentials. Single action potentials evoked an average 60 nM increase in $[Ca^{2+}]$, roughly independent of location up to 400 μm from the soma. Trains of action potentials, however, resulted in a peak $[Ca^{2+}]$ that fell roughly exponentially moving away from the soma, with a space constant of 200 μm. This falloff was mirrored by the progressive attenuation of dendritic spike trains. The particular $[Ca^{2+}]$ profile during sustained activity was spatially heterogeneous, in that adjacent dendritic branches could have quite different profiles.

In experiments with neocortical pyramidal cells, Schiller *et al.* (1995) reported qualitatively similar results to Spruston *et al.* (1995) with respect to the maximum changes of $[Ca^{2+}]$ per spike. In this case, the involvement of dendritic Na^+ spikes was inferred by the marked reduction of dendritic $[Ca^{2+}]$ changes when somatic spike voltage-clamp waveforms were imposed under TTX. The primary difference

Table III. Values for $[Ca^{2+}]_{in}$ Changes in Response to Current-Evoked Spikes from the Literature and the Working Model, Expressed as $\Delta[Ca^{2+}]$/Spike[a]

Reference	Cell	Temp. (°C)	Location	$\Delta[Ca^{2+}]$/spike (nM)	Decay τ (msec)
Jaffe *et al.* (1992)	HPC	30	Soma	2.1	>500[est]
			60 μm from soma	3.5	>500[est]
			>250 μm from soma	No change	
Spruston *et al.* (1995)	HPC CA1	22–24	Within 400 μm of soma	60	500[est]
Schiller *et al.* (1995)	LV CPC	24	Proximal dendrite	110	560
		34		110	300
Christie *et al.* (1995)	HPC CA1	30	Soma	11	2000[est]
			1–50 μm from soma	16	600[est]
			50–200 μm from soma	11–16	400[est]
Working model	—	—	Soma	12	1000[est]

[a]Values marked with *est* are measured directly from figures in the cited reference. LV CPC, Layer V cortical pyramidal cell.

between the results for cortical pyramidal cells and HPCs is in the pattern of [Ca^{2+}]: in Schiller *et al.* (1995) as a first approximation [Ca^{2+}] increased with electrotonic distance from the soma, whereas in HPCs there is either little dependence on location (Spruston *et al.*, 1995) or peak concentrations within 100 μm from the soma (Jaffe *et al.*, 1992). Possibly these differences could be most simply explained by differences in the dendritic distribution of voltage-gated Ca^{2+} channels of the each cell type.

The paper by Christie *et al.* (1995) described the contribution of specific Ca^{2+} channel types in HPCs on [Ca^{2+}] as a function of location, by measuring the effect of specific Ca^{2+}-channel blockers on spike-evoked [Ca^{2+}]. Transients were measured in response to trains of 15 spikes. Under control conditions the peak concentrations ranged from 170 nM at the soma (11 nM/spike) to 235 nM in the proximal 50 μm of the apical dendrite (16 nM/spike), with intermediate values up to 200 μm along the apical dendrite. The time constant of decay of [Ca^{2+}] in the figures from this paper shows values ranging from about 2 sec at the soma down to about 500 msec 200 μm from the soma. This study implied that high-threshold Ca^{2+} channels (N, P, Q, and L type; see Section 12.1) were concentrated at the soma, with the low-threshold T types distributed on the apical dendrite.

9.2. Determinants of the Resting $[Ca^{2+}]_{in}$

The resting level for $[Ca^{2+}]_{in}$ is generally estimated at around 50 nM, and in all models that we have examined this value is taken as a constant. An implicit justification for this may be that the resting value reflects an equilibrium state of the various buffering, extrusion, and internal Ca^{2+} store mechanisms that regulate $[Ca^{2+}]_{in}$, under the assumption that there is no current through voltage-dependent Ca^{2+} channels at rest.

However, a recent report by Magee *et al.* (1996) showed that dihydropyridine blocked a hyperpolarization-induced reduction in $[Ca^{2+}]_{in}$, suggesting that Ca^{2+} channels may be a major factor influencing the resting $[Ca^{2+}]_{in}$. They propose that under physiological conditions, the L-type Ca^{2+} channel (normally high-voltage activated; Section 12.1), which is blocked by dihydropyridines, can start activating at potentials as low as −70 mV, and thus contribute to the resting $[Ca^{2+}]_{in}$. The paper pointed out that this would provide a direct link between past voltage activity in the cell and the resting $[Ca^{2+}]_{in}$.

In the Working model, simulations are run with a starting value of 50 nM in all the Ca^{2+} compartments. However, the "true" resting value for $[Ca^{2+}]_{in}$ when the system is allowed to settle over about 20 sec is slightly higher (about 65 nM), due to a small activation of I_{CaT}, as mentioned earlier.

9.3. Considering Intracellular Ca^{2+} Stores

The model of Schiegg *et al.* (1995) focused on how $[Ca^{2+}]_{in}$ dynamics may be affected by Ca^{2+} release from intracellular Ca^{2+} stores. Specifically, they proposed that such release was necessary to account for a 1- to 2-sec elevation of $[Ca^{2+}]_{in}$ that is crucial for the induction of LTP (Malenka *et al.*, 1992). In the structure of their model the Ca^{2+} store had apparently little effect on overall cellular $[Ca^{2+}]_{in}$ profile, since the store was only in a single spine head compartment, and the $[Ca^{2+}]_{in}$ "tail" in the spine head due to release from the store was less than 10% of the peak $[Ca^{2+}]_{in}$ in this compartment. Although we shall not consider this further in this chapter, it may be of interest to examine the effect of similar models of Ca^{2+} stores on $[Ca^{2+}]_{in}$ dynamics in the soma and dendrites.

9.4. Membrane Pumps

Important components for the maintenance of ion gradients include electrogenic membrane-bound pumps that transport Ca^{2+}, K^+, and Na^+ against their respective concentration gradients (Läuger, 1991).

9.4.1. The Sodium Pump

Sodium and potassium gradients are maintained by a pump driven by Na, K-ATPase that is typically referred to as the "sodium pump" (Rose and Ransom, 1997; for a review see Glynn, 1993). The sodium pump extrudes three Na^+ ions for every two K^+ ions that are brought inside the cell, thus generating a net outward current.

Thompson and Prince (1986) showed that after prolonged depolarizations induced by focal application of glutamate (which results in high intracellular Na^+ accumulation), the current generated by this pump can be large enough (on the order of a few hundred pA) to significantly hyperpolarize the cell. Thus in principle the sodium pump contributes to the electrical stability of the cell both by maintaining ionic gradients and by direct electrical action. The functional importance of the Na, K-ATPase pump has been suggested, for example, with respect to abnormal firing characteristics in epileptic hippocampus (Haglund and Schwartzkroin, 1990; Fukuda and Prince, 1992; Anderson *et al.*, 1994) and bipolar illness (el-Mallakh and Wyatt, 1995).

A literature search failed to locate neuron models with explicit descriptions of the sodium pump; modeling this mechanism appears to be limited to isolated membrane systems, myocytes, or oocytes (e.g., Sagar and Rakowski, 1994). In the Working model we neglect the sodium pump since the focus is on current-evoked responses, and we assume that changes in $[Na^+]_i$ are small under these conditions.

9.4.2. Ca^{2+} Pumps and Models of Ca^{2+} Removal

Any model of $[Ca^{2+}]$ must take into account not only the influx of Ca^{2+}, but also some mechanism for the removal of Ca^{2+}. The simplest method is to include a steady-state term in the differential equation(s) describing $[Ca^{2+}]$. In the general case this value is associated with a second parameter corresponding to the time constant for concentration decay. The BG87, Traub91, and Traub94, and Warman94 models used this type of description [see also Huguenard and McCormick (1992) for thalamic relay neurons].

More explicit models of Ca^{2+} removal include mechanisms such as membrane-bound pumps. The spine model by Zador *et al.* (1990) included two pumps with

Michaelis–Menton kinetics: one high-affinity, low-capacity, corresponding to a CaA-TPase-driven mechanism, and the other low-affinity, high capacity, corresponding to a Ca^{2+}/Na^{+} exchange mechanism (see also Yamada *et al.*, 1989). These pumps were treated as a separate currents in the $[Ca^{2+}]$ differential equation. The parameters of this model were adapted in the model of Schiegg *et al.* (1995). The Jaffe94 model included a similar mechanism, but with a single pump. The major constraints for the pump parameters in this model were a $[Ca^{2+}]_{in}$ decay time constant of about 600 msec in the proximal apical dendrites and the parameters for the buffer (see below).

The Migliore95 model incorporated a pump which bound to intra- and extracellular Ca^{2+} with various rate constants—these reactions were solved simultaneously with another binding reaction between $[Ca^{2+}]_{in}$ and a buffer.

9.4.3. Pump Model for the Working Model

For the removal of intracellular Ca^{2+} in the Working model, we shall adapt the Michaelis–Menton mechanism used by the Jaffe94 model:

$$J_{Ca^{2+}} = V_{max} \frac{[Ca^{2+}]_{in}}{K_d + [Ca^{2+}]_{in}} - J_{leak}$$

where $J_{Ca^{2+}}$ is the removal rate of Ca^{2+} per unit area, V_{max} is the maximum flux rate per unit area, K_d is the half-maximal $[Ca^{2+}]_{in}$, and J_{leak} compensates for the resting pump rate. The parameters for the Working model are

$$V_{max} = 6 \times 10^{-11} \text{ mmole msec}^{-1} \text{ cm}^{-2}$$

$$K_d = 0.01 \text{ mM}$$

with J_{leak} adjusted so that there is no net pump current at rest, assuming no resting activation of Ca^{2+} channels (however, there is evidence for such resting activation—as mentioned in Section 9.2—and in the Working model there is a very slight activation of I_{CaT} at the usual holding potential of −65 mV). This pump is associated with the membrane assigned to the second juxtamembrane compartment described in Section 9.7—this is essentially equivalent to the entire surface area of an anatomical compartment. We ignore any electrogenic component of the Ca^{2+} pump.

9.5. Buffer Models

Endogenous intracellular Ca^{2+} buffers have a strong effect on the free $[Ca^{2+}]_{in}$. Cell models that consider Ca^{2+} dynamics have incorporated buffer mechanisms of varying complexities, including solving the dynamical equations for the buffer–Ca^{2+} reaction during the course of the simulation [e.g., the Migliore95 model and the models of Zador *et al.* (1990) and Schiegg *et al.* (1995); also see Yamada *et al.* (1989)].

Neither the Traub91 and Traub94 models, which treated $[Ca^{2+}]_{in}$ as a dimensionless variable, nor the BG87 and Warman94 models, which were in principle quantitative with respect to $[Ca^{2+}]_{in}$, include buffer mechanisms. None of these four models attempted to replicate the data for $[Ca^{2+}]_{in}$; the key functional parameters that "replaced" a buffering mechanism are found in the Ca^{2+} sensitivities of their Ca^{2+}-dependent K^+ channels.

Instantaneous Buffer Model

A simple way to treat intracellular buffering of Ca^{2+} is to assume a nonsaturated buffer (i.e., $[Bu] \gg [Ca^{2+}]_{in}$, where $[Bu]$ is the concentration of buffer binding sites) with instantaneous kinetics [e.g., the Jaffe94 model; also see the Stockbridge and Moore (1984) model of frog neuromuscular junction].

The key parameters, β_{Bu}, in this mechanism equals the ratio of the concentration of bound Ca^{2+} and free Ca^{2+}:

$$\beta_{Bu} = \frac{[Ca^{2+}]^{bound}}{[Ca^{2+}]^{free}}$$

Thus, this mechanism implies that the predicted *measured* $[Ca^{2+}]$ is equal to the total $[Ca^{2+}]$ divided by $(\beta_{Bu} + 1)$. For nondiffusional models of $[Ca^{2+}]$, e.g., where there is one Ca^{2+} compartment per electrical compartment (see below in Section 9.6), this is the only role of the instantaneous buffer. For multiple-compartment systems, the effective diffusion constant D' applied to the difference in total $[Ca^{2+}]$ between compartments must also be adjusted to take into account the instantaneous buffer, by setting D' equal to $D/(\beta_{Bu} + 1)$, where D is the diffusion constant for the free ion. A variation on this scheme would be to assume that β_{Bu} is a function of each compartment; in this case the diffusion equation between compartments would reference the original D, with the concentration difference between any two compartments determined by the difference of the total concentrations, weighted by the appropriate β_{Bu}'s.

What are reasonable values for β_{Bu}? We can start with an estimate of total buffer that is primarily calmodulin, with a concentration of 30 μM (Schiegg *et al.*, 1995). Calmodulin binds four Ca^{2+} ions independently; thus the effective buffer concentration $[Bu]$ becomes 120 μM. The dissociation constant of each calmodulin-binding site is estimated at 1 μM (Schiegg *et al.*, 1995). Thus

$$[Ca^{2+}]^{free} + [Bu]^{free} \rightleftharpoons [Ca^{2+}]^{bound}, \quad Kd = 1\ \mu M$$

and, assuming that the change in $[Bu]$ from the binding reaction is negligible,

$$K_d = \frac{[Ca^{2+}]^{free}[Bu]}{[Ca^{2+}]^{bound}}$$

$$\beta_{Bu} = \frac{[Ca^{2+}]^{bound}}{[Ca^{2+}]^{free}} = \frac{[Bu]}{K_d}$$

$$\beta_{Bu} = \frac{120\ \mu M}{1\ \mu M} = 120$$

We note that the Jaffe94 model tested a range of 100–1000 for the values of β_{Bu}, with 1000 used for the simulations shown in the paper. These values can be compared to the range of 20–600 for β_{Bu} in a model of the frog neuromuscular junction (Stockbridge and Moore, 1984). In the Working model we use a somewhat low value of $\beta_B{}^u$ equal to 20; this was necessary to generate reasonable (about 10 nM) incremental changes of free $[Ca^{2+}]_{in}$ per spike.

9.6. Models of Ca^{2+} Diffusion

At a minimum, models for Ca^{2+} concentration dynamics assign one compartment for integrating $[Ca^{2+}]$ per electrical compartment (e.g., the Traub91, Traub94, and Jaffe94 models). However, studying the implications of nonuniform local $[Ca^{2+}]_{in}$ as well as better approximations for diffusional equilibrium requires more than a single compartment. The BG87 model used two juxtamembrane shells (one of which was subject to Ca^{2+} current influx), coupled with a constant concentration core compartment. Diffusional flow was considered between all three compartments; a similar model is discussed below in Section 9.7. Two uncoupled "partial" shells were used in the Warman94 model as an approximation of the BG87 model structure. In the Migliore95 model, each somatic/dendritic electrical compartment was divided into four concentric subshell diffusion compartments, as supplied in NEURON. Finally, separate compartments for the spine neck and head were used by Zador *et al.* (1990); six diffusion compartments were assigned to the spine neck and head in Schiegg *et al.* (1995).

It remains to be seen if accounting for any *lateral* or longitudinal diffusion of Ca^{2+} along a dendritic process is important. For example, this could be especially relevant in understanding mechanisms for heterosynaptic plasticity. Preliminary indications that such diffusion is in fact not significant to account for overall changes in internal Ca^{2+} has been mentioned in some studies (e.g., Migliore *et al.*, 1995), but systematic analysis of this question is lacking.

9.7. The $[Ca^{2+}]_{in}$ System in the Working Model

The primary behavior that we wish to capture in this model includes the following:

1. Profiles of locally averaged $[Ca^{2+}]_{in}$ in response to somatic current stimulation and at rest.
2. Differential and localized $[Ca^{2+}]_{in}$ profiles appropriate for the activation of I_{CT} and I_{AHP}
3. Changes in $[Ca^{2+}]_{in}$ as seen by Ca^{2+} channels consistent with estimated currents from these channels.

9.7.1. A Multicompartment Ca^{2+} System with Instantaneous Buffer

We shall use a three-compartment model to represent the internal Ca^{2+} dynamics in any anatomical (electrical) compartment (i.e., soma or dendritic segment) that includes Ca^{2+} channels, although in the version presented here only the soma has any channels. There is no Ca^{2+} diffusion between anatomical compartments. This model is designed to be intermediate in detail, based on quantitative estimates of the various parameters in a relatively simple geometry, and can be readily extended for more sophisticated schemes.

This model assumes that a juxtamembrane shell with thickness r_{shell} is split up into two diffusional compartments, with compartment 1 corresponding to a collection of "islands" or domains associated with colocalized Ca^{2+} channels and I_{CT}, which occupy some fraction α_1 of the cell element surface (Borg-Graham, 1987b). The remaining fraction $1 - \alpha_1$ of the juxtamembrane shell is assigned to compartment 2. If the surface area of the cell element is A_{total} (this definition approximates the actual area of the inside surface of the shell), then the volumes of compartments 1 and 2 (V_1 and V_2, respectively) are given by

$$V_1 = \alpha_1 \times A_{total} \times r_{shell}$$
$$V_2 = (1 - \alpha_1) \times A_{total} \times r_{shell}$$

The volume of the inner compartment 3 is taken as the remainder of the cell element volume.

For simplicity, diffusional flow is considered only between compartments 1 and 2, and between compartments 2 and 3. A relatively simple way to assign the diffusional parameters describing Ca^{2+} flux between the islands assigned to compartment 1 and the surrounding shell compartment 2 is by using a free parameter $\alpha_{1:2}$, which is an interdigitation coefficient (in units of 1/distance). Thus the effective diffusional area $A_{1:2}$ between compartments 1 and 2 is given by

$$A_{1:2} = \alpha_{1:2} \times A_{total} \times r_{shell}$$

In the present version of the Working model, $\alpha_{1:2}$ is set to 1. The diffusional area between compartments 2 and 3 is given by

$$A_{2:3} = (1 - \alpha_1) \times A_{total}$$

Diffusional flux between the compartments is then given by the standard Fick relationship, using a diffusion coefficient for Ca^{2+} of 8×10^{-6} cm^2 sec^{-1} (25°C; Hille, 1992). As stated, instantaneous buffering with a $\beta_{Bu} = 20$ is assumed for all compartments.

The thickness r_{shell} of the juxtamembrane shell is 1μm. The value for α_1 is 0.01%: this represents a limiting value since smaller values give essentially the same peak transients of [Ca^{2+}] in compartment 1. The profile of the concentration transient in this compartment is shown in Fig. 7, and the change in concentration averaged over the entire soma compartment during a spike train is seen in Fig. 15. The motivation for the two juxtamembrane compartments, which is solely concerned with the Ca^{2+} dependence of I_{CT} and I_{AHP}, is discussed in Section 13.6.

Finally, as mentioned previously, the Ca^{2+} pump efflux is computed according to the surface area of compartment 2. Since α_1 is very small, this area represents practically the entire membrane of the anatomical (in this case somatic) compartment.

An important limitation with the present form of the model is in the computation of the Ca^{2+} gradients. Thus the spatial concentration gradient between compartments i and j is given as:

$$d[Ca^{2+}]_{ij}/dx = D_{ij}A_{ij}\frac{[Ca^{2+}]_i - [Ca^{2+}]_j}{\Delta x_{ij}}$$

where Δx_{ij} is the diffusional distance between the compartments. In the present version, Δx_{ij} is taken to be 1 cm (to match the units of the diffusion coefficient), whereas the true diffusional distances are on the order of 1 micron. Given that the present hypothetical system has several rather weakly constrained parameters (including α_1, $\alpha_{1:2}$, r_{shell} and β_{Bu}, as well as the Ca^{2+} sensitivities of I_{AHP} and I_{CT}, and the

Ca^{2+} channel δ parameter described below), the model should allow future refinement, which will yield similar quantitative transients for the compartment [Ca^{2+}].

9.7.2. Consequences of the Soma/Short-Cable Model: Adjusting Ca^{2+} Channel Current Influx for [Ca^{2+}] Integration

Placing all Ca^{2+} (or other type) channels at the soma in a soma/short-cable approximation may give a reasonable first estimation as to the *electrical* image (more so under current clamp) from the soma of channels that are in reality distributed throughout the dendritic tree. However, the concentration profile elicited by the "unnatural" concentration of Ca^{2+} channels could be overestimated, all other factors being equal (e.g., distribution of buffer and pump mechanisms). Thus, we introduce an additional parameter δ, which is a scaling coefficient, less than 1, for the Ca^{2+} influx from the integrated Ca^{2+} current. As a first approximation, δ is estimated by the ratio of the actual somatic conductance of a channel divided by the total conductance of the channel for the entire cell.

Here we will make the following approximation: we assume that in reality all Ca^{2+} channels have a constant membrane density over the soma and the proximal 200 μm of apical dendrite (referring to the soma/short-cable geometry). The coefficient δ is then given by the ratio of somatic area divided by the sum of the total Ca^{2+} channel membrane area. If the soma diameter is 35 μm and the dendrite diameter is 12 μm, this sets δ to be about 1/3. Note that this value is probably a high estimate, since Ca^{2+} channels may be more spread out throughout the dendritic tree.

10. Nonsynaptic Channels of Hippocampal Pyramidal Cells

We shall now review the nonsynaptic channels that have been characterized in HPCs, with the goal of building a link between the main experimental findings and model descriptions for the more prominent channels. The discussions will also mention some of the major pharmacological and neuromodulator susceptibilities for each channel type. These presentations are not exhaustive—more complete reviews of membrane currents and their functional roles include Rudy (1988), Adams and Galvan (1986), Storm (1990), and D. A. Brown *et al.* (1990), with the latter three focusing on hippocampus.

The sections on sodium and calcium channels will describe properties of the various types as a group, whereas potassium and the nonspecific cation channels will be described individually by type. The channels that are included in the Working model are summarized in Table IV. The model parameters for these channels and their associated gating particles are listed in Tables V–VIII.

11. HPC Sodium Channels

Probably the choice of kinetic description for the sodium current generating the spike is the most problematic component of a neuron model, essentially because these channels underly the fundamental instability of the membrane. Much progress has occurred in the measurement of microscopic and macroscopic sodi-

Table IV. Summary of the Channels Used in the Working Model[a]

Channel	Major ion(s)	Gating equation	E_{rev} (mV)	$\bar{g}$ (μS)	$\bar{p}$ (cm^3/sec)	Major functional role
I_{Na}	Na^+	x	+65	1.2	—	Spike upstroke
I_{CaN}	Ca^{2+}	m^2h	+135[b]	—	1e-8	Trigger I_{CT}, I_{AHP} Ca^{2+} spikes
I_{CaL}	Ca^{2+}	m^2	+135[b]	—	3e-9	Trigger I_{CT}, I_{AHP} Ca^{2+} spikes
I_{CaT}	Ca^{2+}	m^2h	+135[b]	—	3e-9	Trigger I_{CT}, I_{AHP} Ca^{2+} spikes
I_D	K^+	x^4y^4	−95	0.05	—	Subthreshold integration
I_{DR}	K^+	xy	−70	0.4	—	Spike repolarization
I_A	K^+	x^4y^3	−70	2.3	—	Spike repolarization
I_{CT}	K^+	x	−80	0.4	—	Repolarization of first spikes in train
I_M	K^+	x^2	−80	0.4	—	Medium adaptation
I_{AHP}	K^+	w^2	−85	0.02	—	Slow adaptation
I_H	K^+, Na^+	y	−17	0.003	—	Stability of E_{rest}

[a]The x particles in the gating expressions for I_{Na} and I_{CT} are the Markov gating particles shown in Fig. 9, with parameters given in Tables VI and VIII. The parameters for the remaining particles referenced in the gating expressions, defined with the extended Hodgkin–Huxley model, are given in Table V. The parameters $\bar{g}$ and $\bar{p}$ characterize ohmic and GHK currrent equation-based constant-field conductance models, respectively.
[b]Based on the Nernst potential equation with resting $[Ca^{2+}]_{in}$ of 50 nM and $[Ca^{2+}]_{out}$ of 1.8 mM; this overestimates the true driving force since the current for these channels is calculated using the GHK current equation.

um currents and in the theory of individual channel biophysics. However, the interpretation of these data for macroscopic models remains at an early stage.

The question remains as to what kinetic model is appropriate for a macroscopic model, e.g., the traditional Hodgkin–Huxley type or the more biophysically accurate Markovian description. Kinetic subtleties which may not be functionally

Table V. Parameters of the Hodgkin–Huxley Model Particles in the Working Model[a]

Channel	Particle	Order	z	γ	K (1/msec)	$V_{1/2}$ (mV)	τ_0 (msec)
I_{CaL}	m	2	4.6	—	—	−1	1.5
I_{CaN}	m	2	3.4	—	—	−21	1.5
	h	1	−2	—	—	−40	1.5
I_{CaT}	m	2	3	—	—	−36	75
	h	1	−5.2	—	—	−68	10
I_D	x	4	3	—	—	−63	1
	y	4	−2.5	0	2e-4	−73	0
I_{DR}	x	1	3	0.8	1.7e-1	−5	0.8
	y	1	−1	—	—	−68	300
I_A	x	4	2.8	0.85	8e-2	−41	1
	y	3	−3	1	4e-2	−49	2
I_M	x	2	6	0.6	3e-3	−45	8
I_H	y	1	−2	—	—	−98	180

[a]The number of particles of a given type assigned to a channel is given by the "order" parameter, as listed in the "Gating equation" column in Table IV. An empty entry for K means that $1/K << \tau_0$, and thus the time constant for the particle is essentially voltage independent and equal to τ_0. In this case the entry for γ is blank since this parameter only affects the voltage-dependent time constant. The parameters α_0 and β_0 described in Section 8.4.2 are zero for all these models.

Table VI. Parameters of the I_{Na} Markov Gating in the Working Model, According to Fig. 9 and Equation (4)[a]

Transition	$1/\tau_{max}$ (1/msec)	k (mV)	$V_{1/2}$ (mV)	$1/\tau_{min}$ (1/msec)
$C_2 \rightarrow O$	0	1	−51	3
$O \rightarrow C_2$	0	−2	−57	3
$C_1 \rightarrow O$	0	1	−42	3
$O \rightarrow C_1$	0	−2	−51	3
$O \rightarrow I$	—	—	—	3
$I \rightarrow C_1$	0.01	−1	−53	1
$C_1 \rightarrow C_2$	0.01	−1	−60	1

[a]The $O \rightarrow I$ transition is voltage independent and is equal to 3/*msec*. The τ_{min} parameter represents a rate-limiting step: its inverse gives the maximum rate for a state transition. Likewise, the inverse of τ_{max} acts as a voltage-independent additive rate for a transition. In this version of the I_{Na} model, the rates for the transitions $I \rightarrow O$, $C_1 \rightarrow I$, and $C_2 \rightarrow C_1$ are assumed to be very small and thus are ignored.

important for other channels may have an important effect in the case of descriptions I_{Na}, simply because I_{Na} is by far the largest and fastest membrane current. Thus, conflicts arise when trying to simultaneously construct a model for I_{Na} that accounts for the more dramatic roles, but at the same time does not intrude *sotto voce* during other phases of firing behavior.

The original parameters of Hodgkin–Huxley squid I_{Na} have been applied to recent models of mammalian neurons (e.g, the Jaffe94 model), which at best can be justified in that these parameters are well known and certainly will generate basic spiking behavior. For more detailed replication of firing properties, however, these parameters are likely to be inappropriate, and I_{Na} channel data from more relevant preparations must be considered.

11.1. Electrophysiology of Sodium Channels in Mammalian Central Neurons

Over the past several years, there have been various reports documenting voltage-clamp measurements of sodium currents and single channels from cortical

Table VII. Parameters of the I_{CT} Markov Gating in the Working Model According to Fig. 9, Equation (4) (τ_{max}, k, $V_{1/2}$, τ_{min}), and Equations (5) and (6) (α, β, n)

Transition	$1/\tau_{max}$ (1/msec)	k (mV)	$V_{1/2}$ (mV)	$1/\tau_{min}$ (1/msec)	α (mM^{-n})	β (1/msec)	n
$O \rightarrow I$	0	−3.5	−64	10	—	—	—
$I \rightarrow C$	0	−10	−120	0.1	—	—	—
$C \rightarrow O$	1	7	−20	1000	1e6	—	3
$O \rightarrow C$	—	—	—	—	—	0.05	—

[a]Note that the $C \rightarrow O$ transition is defined [Equation (7)] as the product of α (V) [Equation (4)] and $\alpha[Ca^{2+}]^n$. The τ_0 parameter in Equation (6) is taken as zero. Note that the units of α in this table (mM^{-n}) lacks 1/msec as implied by Equation (6) to account for the units of Equation (4). The exponent n in the units for α is the number of bound Ca^{2+} ions, as given in the table. Similar to the model of I_{Na}, for I_{CT} the rates for the transitions $I \rightarrow O$ and $C \rightarrow I$ are assumed to be very small and are thus ignored.

Table VIII. Parameters of I_{AHP} Ca^{2+}-Dependent Gating Particle w in the Working Model, as Defined by Equations (5) and (6)[a]

Particle	Order	α ($msec^{-1}$ mM^{-n})	β (1/msec)	n	τ_0 (msec)
w	2	2e14	0.01	4	100

[a]The exponent n in the units for α is the number of bound Ca^{2+} ions, as given in the table.

and noncortical mammalian cells (e.g., Sah *et al.*, 1988b; French *et al.*, 1990; Steinhäuser *et al.*, 1990; Ogata and Tatebayashi, 1990; Cummins *et al.*, 1994). Evidence for the functional role of dendritic I_{Na} was reported as early as Wong *et al.* (1979). Magee and Johnston (1995) measured single TTX-sensitive Na^+ channels in the apical dendrites of HPC CA1 neurons that had an activation threshold around −60 mV. On the other hand, Masukawa *et al.* (1991) reported that a fast, inactivating TTX-sensitive inward current was consistently localized at the soma as compared to the dendrites in cultured hippocampal neurons.

11.1.1. Interpretations Based on the Sah *et al.* (1988b) Data

One of the more influential papers is the one by Sah *et al.* (1988b), both because it was one of the first to do a quantitative voltage-clamp study on I_{Na} in hippocampal (dissociated) neurons, and, indirectly, because of the reference to this paper by various cell models. However, the interpretations of Sah *et al.* have often been inconsistent with or strongly modified from the original paper.

For example, D. A. Brown *et al.* (1990) describe the voltage dependence of I_{Na} by (implicitly) referencing data from a single cell shown in a figure from Sah *et al.* (Fig. 10 of Sah *et al.*, 1988b) whose characteristics are significantly different from the pooled analysis in Sah *et al.*, in particular with respect to the slope and position of the steady-state activation and inactivation curves. That the properties suggested by the figure are more consistent with that expected from current clamp behavior of I_{Na} may have contributed to the discrepancy being overlooked by these and other authors.

The interpretation of the Sah *et al.* data by Traub *et al.* (1991, 1994) in turn has been used in models such as the Wathey92 and Migliore95 models, the HPC model of Pongracz *et al.* (1992), and, for example, in models of cortical pyramidal cells (Rhodes and Gray, 1994) and of thalamic reticular neurons (Destexhe *et al.*, 1995). In their model of I_{Na}, Traub *et al.* increased the rate constants given in the Sah *et al.* data by a factor of about five to generate a spike upstroke dV/dt that was at least 100 mV/msec. This model also included a rightward shift in inactivation kinetics (moving the steady-state curve to more depolarized levels) that was justified assuming a temperature of 32°C for the model, and considering the temperature of the experimental data (22°C). Although not discussed in the text, the inactivation characteristic was also much steeper than that shown in Sah *et al.* (1988b). Taken together, these changes result in a large overlap between activation and inactivation in the model. Also, this shift would suggest that at the lower temperatures the spike would have a higher threshold, which is consistent with the report of Thompson *et al.* (1985; see Table II of this chapter). The Warman94 model also used modified Sah *et al.* (1988b) parameters, compensating for the higher temperature of the model simulations as compared to the data by a similar shift in kinetics to Traub *et al.*, and speeding the time constants by a factor of three on the basis of an assumed Q_{10} of 2. Unlike Traub *et al.*, the precise adaptation of the Sah *et al.* data in the Warman94 model was constrained by matching spike threshold and reduction of spike amplitude during spike trains to data and, significantly, avoiding intrinsic bursting by an overexcitable I_{Na}.

11.1.2. I_{Na}—Fast, Slow, Persistent Components?

A persistent, TTX-sensitive Na^+ current, I_{NaP}, has been characterized in HPCs (French and Gage, 1985; French *et al.*, 1990). Other reports of slow or persistent I_{Na} currents include TTX-insensitive Na^+ currents [Hoehn *et al.* (1993) for striatum and HPCs; Deisz (1996) for neocortex]. In addition, MacVicar (1985) reported a subthreshold TTX-sensitive Na^+ current that contributed to the prepotentials described by Lanthorn *et al.* (1984).

As discussed in the recent review by Crill (1996), it is not altogether clear if in fact there is either a distinct channel underlying I_{NaP}, if I_{NaP} is a window current of the transient I_{Na} (see below), or finally if the pesistent current arises from a non-inactivating state of normally transient I_{Na} channels (Alzheimer *et al.*, 1993). Azouz *et al.* (1996) argued for a distinct I_{NaP} versus an I_{Na} window current since they found a different sensitivity of TTX between the active, Na^+-mediated ADP and the spike.

For the present discussion we will assume that whatever the actual source for the persistent Na^+ current, the data of French *et al.* (1990) provide an upper bound for the size and range of activation for this component. We suggest that this report may be a more reliable quantitative reference than the reports of the fast I_{Na}, such as the Sah *et al.* (1988b) data, since the French *et al.* (1990) data were taken from both the intact slice and disassociated cell preparation (with comparable results), and the slow kinetics of the current makes for a more accurate measurement. On the other hand, the reason that this data may be considered as an upper bound is that persistent components of transient currents may be overestimated if there are significant space-clamp errors (White *et al.*, 1995).

11.1.3. Slow Inactivation of I_{Na}

The Na^+ channels underlying the action potential can show a slow inactivation that is dependent on past activity (Colbert *et al.*, 1997; Jung *et al.*, 1997; Toib *et al.*, 1998). This kinetic component appears to have a functional role at least with respect to dendritic Na^+ channels, causing a progressive decrease in dendritic spike amplitude during repetitive firing (Colbert and Johnston, 1998).

11.1.4. I_{Na} Modulatic

I_{Na} is susceptible to a muscarinic downregulation of the peak current and a slowing of inactivation (Cantrell *et al.*, 1996), acting via a protein kinase C pathway (Colbert and Johnston, 1998).

Cultured fetal hippocampal neurons in a mouse model of Down syndrome (Galdzicki *et al.*, 1993) have about 25% less sodium current (and a 20% slower action potential rise time) than control. It remains to be seen if this subtle difference in basic action potential properties can be linked to the cognitive deficit in Down syndrome.

11.1.5. Stability and the "Window Current"

Steady-state rectification of nominally transient currents, in the context of the Hodgkin–Huxley description, is due to an overlap of the activation and inactivation steady-state curves. This rectification, also known as the window current, is kinetically indistinguishable from an independent persistent, or noninactivating current. In the case of I_{Na} (and possibly other transient channels), the window current may have an influence on the stability of the resting potential.

The data for a persistent Na^+ current set an upper bound on any true window current of a transient I_{Na}, which, in the context of the Hodgkin–Huxley model, sets a limit as to the closest possible positions of the activation and inactivation curves. However, postulating a window current based on voltage-clamp measurements of steady-state activation and inactivation curves is based on the assumption of the simple Hodgkin–Huxley model. Since evidence for this model has been increasingly put into doubt by detailed analysis of I_{Na} gating kinetics (Armstrong, 1981), this means that the interpretation of the Hodgkin–Huxley window current is problematic. As we shall see below (Section 11.2), most descriptions of I_{Na} in published cell models include a very large window current that is at variance with whole-cell voltage-clamp data (Sah *et al.*, 1988b). Since in these models the window current has a profound effect on firing properties, it is necessary to consider kinetic schemes which follow the experimental data more faithfully. Apropos of this, a recent detailed model of Purkinje cells included a Hodgkin–Huxley form of the fast Na^+ current, and the potential error associated with the window current in their formulation was discussed (De Schutter and Bower, 1994).

The report on Na^+-dependent plateau potentials by Fraser and MacVicar (1996; Section 6.4.5) would seem to be evidence for a substantial I_{Na} window current, at least under cholinergic stimulation. On the other hand, this phenomenon was not blocked by TTX, and therefore probably does not arise from the classical Na^+ spike channel.

11.1.6. Na^+-Only Spikes

The dynamic contribution of a putative Na^+ window current can be assessed in experiments in which all Ca^{2+} and K^+ currents are blocked [CsCl-filled sharp electrode with extracellular tetraethylammonium, 4-aminopyridine (4-AP), and Mn^{2+}]. In this situation single action potentials may be elicited by current pulses in HPCs which repolarize slowly back to the resting potential (J. F. Storm, unpublished data, shown in Borg-Graham, 1987a).* This behavior is further evidence against a large steady-state rectification. Long current steps under the same conditions can generate Na^+-only repetitive firing (J. F. Storm, unpublished data, shown in Borg-Graham, 1987a), which adds additional constraints on the kinetics of recov-

*In Fig. 12 of Storm (1987a) this condition is approximated, showing that Na^+ spikes can eventually repolarize without active K^+ currents.

ery from inactivation. The two Hodgkin–Huxley-type I_{Na} models in the BG87 model were developed specifically to reproduce this behavior—in this model repetitive firing of Na^{+}-only spikes required a very delicate adjustment of the inactivation kinetics of the higher threshold current.

11.1.7. The Subthreshold/Threshold Regime

Another important functional characteristic of I_{Na} kinetics concerns the sharpness of the threshold. For example, Storm (1987a) shows several protocols in which short pulses were adjusted to elicit just-threshold responses—in particular, the associated subthreshold responses were almost passive [though this is not always the case; for example, see the quasi-active subthreshold "ADPs" of M. S. Jensen *et al.* (1994)].

11.1.8. Relating I_{Na} Kinetics to Spike Properties and the PHP

The relationship between the resting potential, the initial and later spike thresholds during repetitive firing, and the depth of the interspike repolarizations (the PHP) also provides constraints on the activation and inactivation curves of the Hodgkin–Huxley model.

The assumption of a nominal value of the resting potential of −65 mV and an initial spike threshold of −55 mV puts a lower bound (corresponding to unchanging inactivation during the spike initiation phase) on the product of the inactivation curve at rest and the maximum conductance of the channel, since this value is directly related to the slope of the spike upstroke. Likewise, the spike threshold places the foot of the activation curve within a few millivolts of −55 mV, and the change in voltage trajectory during the approach to threshold and beyond depends strongly on the steepness of the voltage dependence of activation.

The PHP constrains the parameters of the inactivation process during repetitive firing in that inactivation must be fairly well removed by the PHP, essentially because spike amplitude is well maintained throughout the train, at least at 37°C (Thompson *et al.*, 1985). As mentioned earlier, a more sensitive measure of the I_{Na} magnitude would be the dV/dt of the later spikes, but these data are not readily available. In other words, spikes during a train require a substantial deinactivation, which, in the context of the preceding spike, can result from a combination of incomplete inactivation and/or sufficient recovery from inactivation during the repolarizing phase. Thus the relationship between the PHP and the threshold is directly related to the kinetics of repetitive firing, and reflects limitations imposed on the firing by the interelationship of the repolarizing currents and the recovery kinetics of I_{Na}.

While the large PHP in nonpyramidal cells (or more generally, nonspiny hippocampal cells; Scharfman, 1993) may seem to relax the restrictions on the relative positions of inactivation/activation voltage dependences; in fact, in many of these cells the final approach to the sharp threshold, lasting tens of milliseconds, can be quite shallow. If we assume a Hodgkin–Huxley mechanism, then this loitering near threshold also puts a constraint on the steady-state voltage dependence of inactivation, in that if there is little overlap in the activation and inactivation steady-state curves, then this time course would tend to inactivate the I_{Na} channels before the spike. This supports the coupling of inactivation and activation, contrary to the Hodgkin–Huxley model (Armstrong, 1981; Patlak, 1991).

Finally, simulations (not shown) suggest that a model with stationary Boltzmann-type kinetics that has a voltage dependence which is not arbitrarily steep must show a nonzero depolarizing phase prior to threshold. On the other hand, the trajectories of some intraburst spikes (which essentially have a zero PHP) suggest that spikes can arise abruptly from a *downward*-moving voltage. At the moment we shall leave this as an unaddressed issue, other than to mention that this characteristic might be evidence for a nonsomatic source of the fast spike during bursting, e.g., from the axon.

11.2. Comparison of I_{Na} Models from the Literature

We shall now discuss the various forms of I_{Na} models that have been proposed, both in the context of cell models and in the context of single-channel kinetic models.

11.2.1. Hodgkin–Huxley Type I_{Na} Models

We have mentioned several cell models that adapt the Hodgkin–Huxley squid axon formula for I_{Na} to match the data from more relevant preparations. In Fig. 8 representative voltage-clamp protocols are used to compare the kinetics and steady-state properties of some of these models with each other and the canonical Hodgkin–Huxley I_{Na} parameters. We also include the model of I_{Na} used by Huguenard and McCormick (1992) in their model of thalamic relay cells, which was adapted from I_{Na} measurements in cortical pyramidal cells (Huguenard *et al.*, 1988). As described in the figure, the maximum conductances of each model were adjusted so that they would give approximately the same spike current. Although this results in a relatively crude quantitative comparison (since there was not a detailed fitting of this parameter), it is clear from the figure that the various I_{Na} models, although all adjusted to do more or less the same thing, have a wide range of kinetic and steady-state properties. The implications of these differences are seen in close examination of not only the individual spike properties of the various models, but also the variation in repetitive or bursting characteristics.

11.2.2. A Model with Multiple Hodgkin–Huxley Type I_{Na} Channels

As mentioned, another approach is to reconstruct the observed I_{Na} behavior with more than one Hodgkin–Huxley-type model (Borg-Graham, 1987b), each with relatively nonoverlapping activation and inactivation curves in order to limit Na^+ steady-state rectification. Note that this latter approach can be reduced to an equivalent single-Markov model (Destexhe *et al.*, 1994); an advantage with the separate channels is that it may be easier to do an initial map of desired kinetic characteristics to this form of an I_{Na} model than to determine an appropriate Markov form. On the other hand, as mentioned previously, there are no data to support these distinct channels.

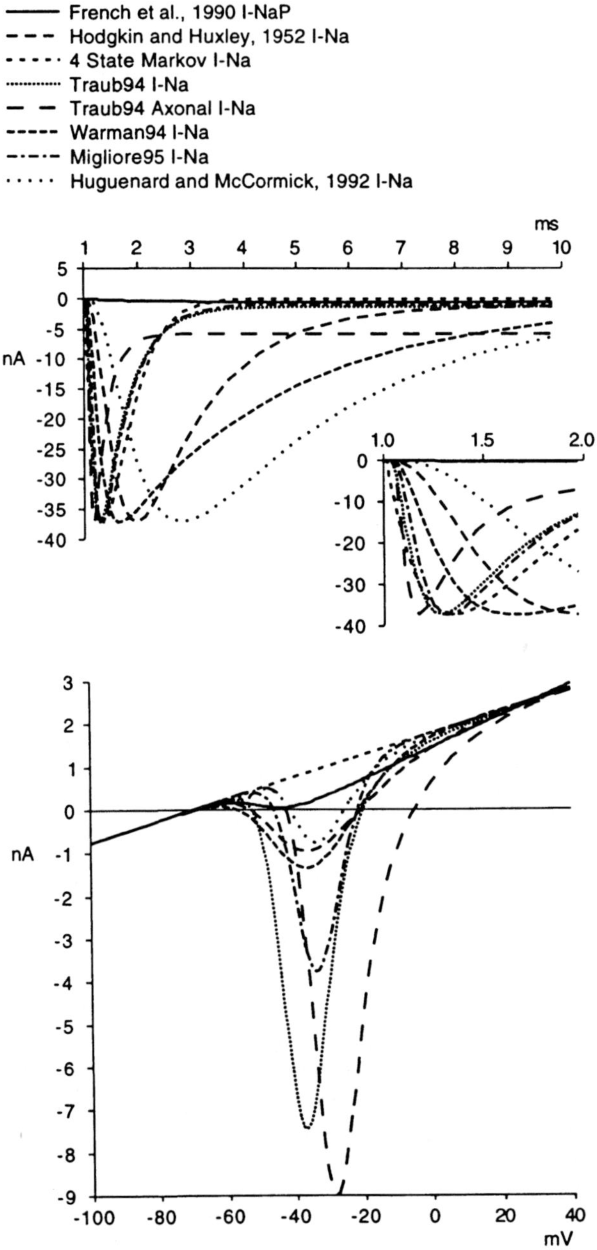

Figure 8. Representative ideal voltage-clamp currents and steady-state rectification for various models of I_{Na} and I_{NaP}, corresponding to experiments in which all other voltage-dependent channels are blocked. Thus, for each trace the associated I_{Na} or I_{NaP} is the only activated nonlinear current. All I_{NaP} or I_{Na} of a given type are at the soma, and the dendrite is passive. Cell geometry is the BG87 soma/short-cable model, with an input impedance of 39 MΩ, and a reversal potential of the leak conductance of −70 mV. For this comparison, the maximum conductances for each I_{Na} model was adjusted by normalizing the peak current of each under the voltage clamp protocol shown here to that of the four-state Markov model. The maximimum conductance of the latter model was adjusted to reproduce the spike characteristics discussed in the text. Thus, each adjusted I_{Na} model would give spike currents of similar (though not exactly the same) magnitude. Also, the four-state Markov I_{Na} shown here lacks the small steady-state rectification term discussed in the text; the steady state of the full model is shown in Fig. 10. The parameters for the I_{NaP} model are those described by French *et al.* (1990), with a maximum conductance of 1.2 nS (corresponding to measurements in the hippocampal slice preparation). The reversal potentials for the currents are: I_{NaP}, 30 mV; Hodgkin and Huxley (1952b) I_{Na}, 55 mV; four-state Markov I_{Na}, 65 mV; both Traub94 I_{Na}, 55 mV; Warman94 I_{Na}, 65 mV; Migliore95 I_{Na}, 50 mV; Huguenard and McCormick (1992) I_{Na}, 45 mV. Top (main and inset): Voltage clamp currents in response to clamp voltage step to −20 mV, from a holding potential of −70mV. Bottom: Steady-state rectification of I_{Na}, or window current, compared with I_{NaP}, demonstrated by steady-state I–V characteristics.

11.2.3. Evaluating Published Markovian Kinetic Models

Markov-type models with complex state descriptions may offer a solution to the constraints that we have discussed which the standard Hodgkin–Huxley approach cannot match [Kuo and Bean (1994) from hippocampus; Vandenberg and Bezanilla (1991) and Patlak (1991) from squid axon].

However, simulations (not shown) of the model described by Kuo and Bean (1994), with modifications limited to a shifting of the voltage-dependent rate constants by 20 mV in the hyperpolarizing direction (consistent with the authors' comment regarding the expected shift in voltage dependence for their whole-cell recordings), generated a spike with a shallow threshold, slow inactivation, and a significant steady-state rectification starting around the resting potential and peaking on the order of 0.5–1 nA. It may be expected that the kinetic details for suprathreshold voltages may not be well characterized by this report, since the model predictions were based on quantitative measurements at relatively hyperpolarized levels.

11.3. An ad Hoc Markovian Model

In this section we propose a new Markovian model for I_{Na} in the Working model. The state diagram for the general framework of the model and the specific simple version used for simulations are shown in Fig. 9, where the arrowheads correspond to the predominant transitions. The basic constraints for the model are the absolute spike thresholds taken from the relationships between spike threshold and PHPs, the generation of spikes of similar amplitude to the initial spike during spike trains, and the stability of resting potential. In particular, the version of the model that we demonstrate shows a sharp threshold over a 5- to 10-mV range, depending on the recent history of the channel, and the ability to recover from inactivation during repolarizations positive to earlier thresholds, without a large window current. This model is less constrained from the single-channel data other than that inactivation is voltage independent and coupled to activation (Armstrong, 1981; Patlak, 1991; Kuo and Bean, 1994); it remains to be shown that the kinetic scheme described here can be reconciled with the experiments.

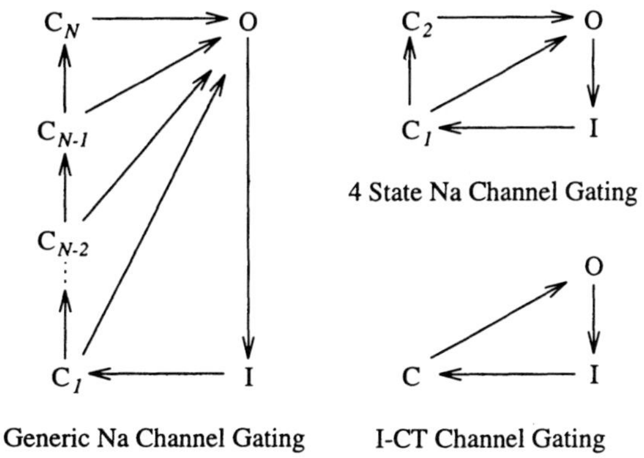

Figure 9. Left: General state diagram for a hypothetical Markov gating model used for I_{Na} in the Working model. From the single inactivated state I, the N closed states C_i are reached with increasing hyperpolarization. The $C_i \rightarrow O$ transitions implement in effect distinct thresholds, occurring at progressively lower potentials with increasing i. Likewise, the $I \rightarrow C_1$ and $C_i \rightarrow C_{i+1}$ transitions occur at voltages hyperpolarized to the associated $C_1 \rightarrow O$ and $C_{i+1} \rightarrow O$ transitions, respectively, somewhat like a rachet mechanism. The $O \rightarrow I$ transition is voltage independent. Top right: The four-state version of the general framework, used in the current version of I_{Na}. Bottom right: A three-state Markov model for I_{CT} gating. The $C \rightarrow O$ transition is Ca^{2+} and voltage dependent. The $O \rightarrow I$ transition is strongly voltage dependent, accounting for the fast turnoff of I_{CT} during spike repolarization, giving rise to the fAHP. The much slower $I \rightarrow C$ transition is also voltage dependent, such that removal of inactivation is slow while the cell is depolarized. In all these schemes the arrows denote the dominant transitions during spike depolarization/repolarization.

11.3.1. A Variable Threshold without a Large Window Current

The most important idea of this model is that the available threshold depends on the history of firing, shifting down during the course of spike repolarization *without* giving a strong window current.

Crucially, inactivation occurs after the open state—this process is not decoupled from activation as in the Hodgkin–Huxley model. After reaching the open state (i.e., the spike) and inactivating, the channel passes through a series of closed states each of which offers a path to the open state at progressively more hyperpolarized thresholds. The effective removal of inactivation is determined by the voltage dependence of the path from the inactivated state to the chain of closed states. Steady-state rectification is avoided by requiring that the transition *into* a given closed state occurs at a voltage range that is hyperpolarized relative to the closed-to-open transition for that state.

Thus the threshold of the closed-to-open transition ($C_1 \rightarrow O$) from the first closed state reached after inactivation determines the most depolarized threshold of the channel, and the necessary PHP for this firing transition is determined by the voltage dependence of the inactivated-to-first closed state transition ($I \rightarrow C_1$). Likewise, the threshold to the open state from the final, most hyperpolarized closed state ($C_N \rightarrow O$) determines the lowest threshold of the channel. The specific closed-to-open transition that is dominant for a subsequent spike depends on the strength of the repolarizing mechanisms (e.g., K^+ currents) and the stimulus current.

Figure 10 illustrates the voltage dependences of the major state transitions in the simple four-state version of the model, as set by the parameters listed in Table VI. Note that the two closed state-to-open state transitions correspond to the range of thresholds in the model, and the lack of overlap for the respective rates into the closed states avoids significant steady-state activation. In contrast, measurements of steady-state activation and inactivation kinetics using standard voltage-clamp protocols (activations measured as peak currents at various potentials from a holding potential of -100 mV and inactivations measured as peak currents after a step to -20 mV from various holding potentials) predict a large (and probably overestimated) overlap in the "Hodgkin–Huxley" interpretation.

11.3.2. Accounting for the Persistent Na^+ Current

This model can account in part for a persistent component of I_{Na} simply by adding a small voltage-independent rate term to the appropriate paths into the closed states [the α_0 term in Equation (4) and given in Table VI, analogous to the α_0 and β_0 terms described in Section 8.4.2]. The effect of these terms is kinetically similar to the overlap of the inactivation and activation curves of the Hodgkin–Huxley model, e.g., a window current, with the important difference that the rectification is not limited for voltages higher than the threshold. The resulting rectification, the amplitude of which is related to the relative values of α_0 and τ_0 used for the transition pairs $C_1 \rightarrow C_2$ and $C_2 \rightarrow O$, and $I \rightarrow C_1$ and $C_1 \rightarrow O$, respectively, is shown in Fig. 10, and can be compared with the properties of the persistent Na^+ current described by French *et al.* (1990), shown in Fig. 8. The major difference is in the threshold of the persistent component of I_{Na}, around -55 mV, which is determined in this scheme by the threshold of the $C_2 \rightarrow O$ transition. In

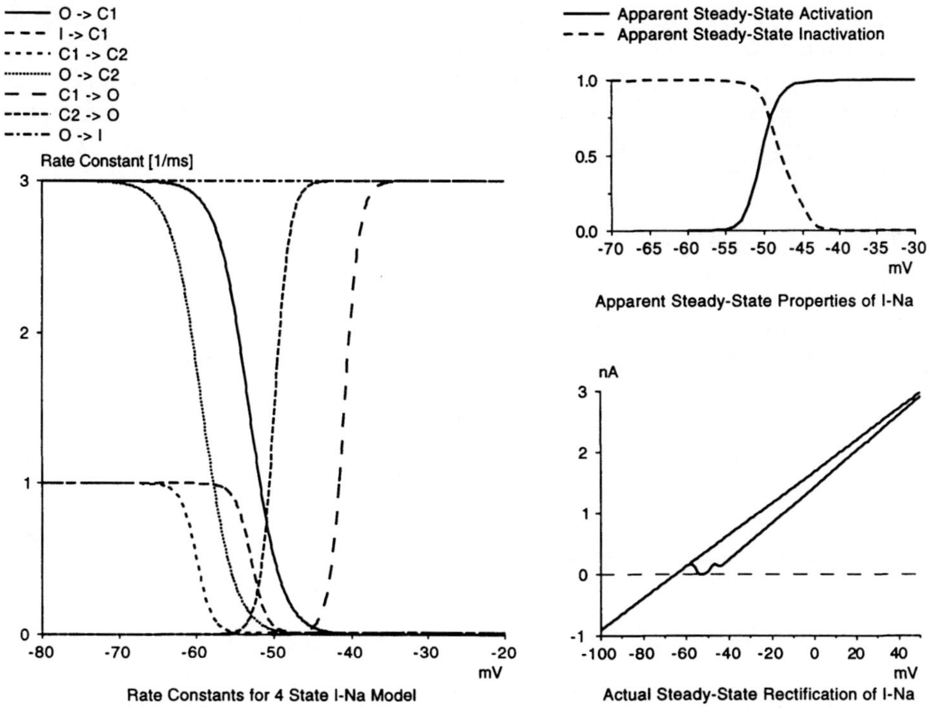

Figure 10. Rate functions for the major state transitions in a four-state Markov I_{Na} model and voltage clamp simulations. The parameters for the rate functions plotted in the left panel are given in Table VI, according to Equation four. Note that the "steady-state" activation and inactivation curves derived with standard voltage-clamp protocols (upper right panel) predict a substantial overlap, and thus a large window current. However, the actual steady-state rectification is minimal (lower right panel), and compares well with the measurements of the persistent Na$^+$ current I_{NaP} (French *et al.*, 1990), as illustrated in Fig. 8.

the data of French *et al.*, I_{NaP} turns on at about −70 mV, thus implicating a small, but possibly significant, substhreshold activation of this current. This characteristic will be considered in future refinements of the I_{Na} Markov model used here.

11.3.3. Shortcomings of the I_{Na} Model

As can be seen, the range of thresholds and associated necessary PHPs of this model is determined by the different closed-to-open state paths. In the version illustrated here, there are two such paths, spanning a voltage range of about 5 mV. The parameters for these transitions are chosen so that the lowest threshold is about −55 mV, and the highest threshold is about −50 mV. The associated paths *into* the closed states have thresholds of about −60 and −55 mV, respectively; the higher value corresponds to the minimum PHP required to reset the channel. This value is clearly too low to account for the resetting of I_{Na} between the spikes during a burst if we assume that the somatic voltage directly determines the kinetics of the fast spikes during the burst. In the context of this kinetic scheme the solution is to add more closed states that may be entered from progressively higher membrane potentials, as shown in Fig. 9.

Also, in its present form this model has a steep voltage dependence of steady-state inactivation and a very large overlap in the measured activation and inactivation characteristics, as shown in Fig. 9, both of which are inconsistent with published data. It may be possible to adjust these characteristics with some additional hypoerpolarized closed states.

12. HPC Calcium Channels

In this section we review the various Ca^{2+} channels and related phenomenology for HPCs. As we have mentioned in earlier sections, our description of Ca^{2+} channels has to consider not only the charge contribution of Ca^{2+} currents, but also their effect on $[Ca^{2+}]$.

In the literature Ca^{2+} currents are often described in terms of conductance, e.g., an equivalent conductance density per unit area of membrane in a given preparation. Some references cite a permeability model explicitly (Section 8.2); for the model description that we use here all conductances are converted to permeability units, assuming some value for the voltage at which the conductance value was estimated.

12.1. T, N, R, P, Q, and/or L Type: How Many Ca^{2+} Channels Are There in HPCs?

Several subtypes of Ca^{2+} channels have been characterized in neurons, with the T (LVA, small conductance, rapidly inactivating), N, R, and P or Q (HVA, rapidly inactivating), and L (HVA, large conductance, slowly inactivating) types, as well as nonspecified types, variously identified in HPCs from different studies. The nomenclature of LVA (for low-voltage-activated) and HVA (for high-voltage-activated) is commonly used in the Ca^{2+} channel literature; an early review of the variety of neuronal calcium channels can be found in Hagiwara and Byerly (1981). With respect to HPCs, the existence of specific Ca^{2+} channel types in these cells is somewhat controversial (see discussion in D. A. Brown *et al.*, 1990), with different reports claiming from one to five channel types.

12.2. Electrophysiology of Calcium Channels

In general, under control conditions Ca^{2+} currents are too small to affect significantly the voltage waveform, so that the functional significance for these currents is primarily via their effect on $[Ca^{2+}]_{in}$. Under noncontrol conditions, however, we can consider at least two electrical signatures of Ca^{2+} currents.

First, when Na^+ channels are blocked, calcium channels in HPCs (as with other neurons) have the necessary voltage-dependent kinetics and are sufficiently concentrated to generate Ca^{2+} spikes (e.g., Schwartzkroin and Slawsky, 1977). The amplitude of these spikes is greatly enhanced by blocking K^+ repolarizing channels. As we have mentioned earlier, although the involvement of Ca^{2+} spikes has been classically implicated for burst generation (e.g., Wong and Prince, 1978), this assumption is open to question. Rather, the severely nonphysiological conditions necessary for the observation of Ca^{2+} spikes suggest that this phenomenon is restricted to the experimenter's dish, and may be used at least to help constrain the parameters for a model of Ca^{2+} channels.

The second electrical signature of Ca^{2+} channels occurs under conditions closer to control. Concentrations of TEA sufficient to block the large outward current I_C (see Section 13.4) may reveal a small Ca^{2+}-mediated shoulder on the repolarizing phase of single Na^+ spikes (e.g., Storm, 1987b). We may note that it is not clear from the literature whether, under control conditions, a similar shoulder component of the repolarizing phase exists for later spikes in a train, when I_C is probably much smaller than during the initial spike. On the other hand, the kinetics of other repolarizing currents may shift during repetitive firing, so the comparison may not be appropriate.

12.2.1. I_{Ca} Modulation

Doerner *et al.* (1990) investigated second messenger pathways that modulate Ca^{2+} currents in HPCs via protein kinase C (PKC). Ca^{2+} channels may be modulated by adenosine (Madison *et al.*, 1987a); the various types of these channels have different pharmacological profiles (Nowycky *et al.*, 1985).

12.2.2. Characteristics of Calcium Spikes

Segal and Barker (1986) looked at putative Ca^{2+}-mediated responses in cultured rat hippocampal cells under TTX and TEA (25 mM), including recordings with and without 4-AP. Ca^{2+} spikes elicited with current steps had a threshold of about −30 mV, and peaked around 10–20 mV. Under voltage clamp I_{Ca} activated with a time constant of about 20 msec at −20 mV, with an extrapolated reversal potential of 10 mV, similar to what is predicted by the GHK model. This study also identified a Ca^{2+}-dependent K^+ current as "I_C," but since I_C is blocked by TEA, and the kinetics of the reported current was slow, it is more likely that this current was I_{AHP}.

The levels of TEA used in this study would also be expected to block I_M and I_{DR} as well as I_C (Storm 1990), thus the estimation of the size of the Ca^{2+} current during spikes needs to take into account the involvement of I_A. In this regard, Segal and Barker (1986) showed that Ca^{2+} spikes without 4-AP had a duration of about 50–100 msec; under 1 mM 4-AP (which blocks I_A) the spikes developed a long, slowly decaying plateau that started at about 0 mV, decaying to about −10 mV over about 200 msec before a much more rapid return to rest. In addition, a direct facilitation of the Ca^{2+} current by 4-AP was also determined. This study did not find evidence for more than one Ca^{2+} current, e.g., a low-threshold transient current or persistent current.

Recordings of Ca^{2+}-mediated activity under Na^+ and K^+ blockers may also be seen in the paper by Fraser and MacVicar (1996)—amplitudes of slow Ca^{2+} spikes were similar to those of Na^+ spikes prior to addition of blockers. The Ca^{2+} spikes under these conditions repolarized faster than those shown in Segal and Barker; however, the 10 mM TEA used in this paper may not have blocked repolarizing K^+ currents as efficiently as the 25 mM TEA used by Segal and Barker.

12.2.3. Kinetics of Activation and Inactivation of Ca^{2+} Currents

A persistent slow inward calcium current was described by J. Johnston *et al.* (1980) and D. Brown and Griffith (1983b), with characteristics of an L-type Ca^{2+} current.

Kay and Wong (1987) described the activation of an Ca^{2+} current in isolated CA1 HPCs, fitting the current to Hodgkin–Huxley-type m^2 kinetics. In a later study using the same protocol, Kay (1991) described the voltage and Ca^{2+}-dependent inactivation of this current. He argued for a single type of Ca^{2+} current in HPCs since the three kinetic components of inactivation (roughly a fast τ of about 200 msec, a slower τ of about 2 sec, and a constant component) varied in concert under various conditions, and activation kinetics had but a single component.*

Kay argued that the T current is not functionally present in CA1 cells.† A complex, and unspecified, interaction between voltage and Ca^{2+} entry through individual channels (a "domain" model) was proposed for the inactivation process [a theoretical analysis of this type of process was described by Sherman *et al.* (1990) for mouse pancreatic β cells].

On the other hand, Fisher *et al.* (1990) proposed that there are three distinct types of Ca^{2+} channels (T, L, and N type) in both CA1 and CA3 HPCs (and dentate gyrus granule cells), based on cell-attached recordings in the exposed slice preparation, a preparation somewhat between slice and acutely isolated cells. This conclusion was reached primarily because the single-channel conductances fell into three groups and there was a strong correlation between the channel conductances and voltage-dependent kinetics. The proposed T-type channel had a conductance of about 8 pS, activated above -40 mV (half-maximal activation at -18 mV), and inactivated within 50 msec at holding potentials of -10 mV or more. Inactivation was much slower at smaller depolarizations. The proposed N-type channel had a conductance of about 14 pS, activated above -25 mV (half-maximal activation at 0 mV), with inactivation described as "variable." Finally, an approximately 25-pS channel was identified as an L-type Ca^{2+} channel. It activated at potentials greater than -10 mV, and did not show inactivation.

Recently, Avery and Johnson (1996) reported that the LVA Ca^{2+} current in acutely isolated CA3 HPCs derives from both a transient T-type channel and a sustained Ca^{2+} channel. This second component was tentively identified as an L-type channel, despite its activation at potentials more negative than that typically associated with HVA Ca^{2+} channels.

Fisher *et al.* (1990) also reported that T-, L-, and N-type calcium channels were in CA1 and CA3 HPC dendrites, as well as on the soma (see below). Magee and Johnston (1995) examined single channels in the apical dendrites of CA1 HPCs, and described L, T, and R (or possibly Q) channel types, with a smaller density of N-type channels. On the other hand, as mentioned earlier, in CA1 HPCs Christie *et al.* (1995) reported that N-, P-, Q-, and L-type channels were concentrated at the soma, with T-type channels on the apical dendrite.

*However, considering similar results from CA3 HPCs in slice, Zbicz and Weight (1986) interpreted multiple inactivation components as indicative of more than one Ca^{2+} current.

†Note that in the Jaffe94 and Migliore95 models I_{CaT} had a relatively minor contribution to the $[Ca^{2+}]_{in}$ profiles.

Various HPC models have included from one to three types of Ca^{2+} channels, in every case with a Hodgkin–Huxley-type kinetic description. Some models use an ohmic conduction scheme and others a constant-field model; the former group includes channels for which the driving force is dynamically calculated based on the Nernst potential from explicit $[Ca^{2+}]_{in}$ calculation, while others assume a constant E_{Ca}.

12.3.1. Models with a Single Type of Ca^{2+} Channel

Variations on the single-Ca^{2+}-channel description were used in the Traub91, Traub94, and Warman94 models, the differences being mainly in the inactivation kinetics. Although the BG87 model used two Ca^{2+} channels, a fast I_{Ca} and a slow I_{CaS}, the latter current [based on the report of D. Johnston *et al.* (1980)] had little functional effect. In the Traub91 model steady-state inactivation as a function of voltage was set to 1 below the resting potential, and to a decreasing exponential function of voltage above resting potential, with a voltage-independent time constant of 200 msec. The Traub94 model eliminated the inactivation term entirely. In the Warman94 model I_{Ca} steady-state inactivation was defined with a Boltzmann expression of similar shape to the function used in the Traub91 model; the primary difference was that inactivation was quite rapid (bell-shaped function of voltage peaking at 8 msec). A result of this characteristic is that Ca^{2+} influx in the Warman94 model is significantly reduced during subsequent spikes in a train (i.e., it is frequency dependent), in contrast with the rather constant increment in $[Ca^{2+}]_{in}$ per spike seen in imaging experiments during spike trains (see Section 9).

Because of the different inactivation characteristics, these models have very different Ca^{2+} spikes when all channels other than I_{Ca} are blocked. The Traub91 model generates a slow spike with a threshold at about −40 mV, a peak at 80 mV (E_{Ca} in the model), and a slow decay that halts at about 20 mV because of a strong Ca^{2+} steady-state rectification. Since there is no I_{Ca} inactivation, under the same conditions the Traub94 model "locks up" at E_{Ca}. At the other extreme, the fast inactivation of I_{Ca} in the Warman94 model prevents any regenerative Ca^{2+}-only response. The description of I_{Ca} in the BG87 model reproduced the threshold and shape of the spikes shown in Segal and Barker (1986); this was not surprising since the Segal and Barker data were a major template in determining the I_{Ca} parameters for this model.

12.3.2. Models Incorporating Three Types of Ca^{2+} Channels

In contrast, the Jaffe94 and Migliore95 HPC models adopted the three-Ca^{2+} channel position (T, N, and L-type), adapting the kinetics from both HPC and non-HPC data. The three Ca^{2+} channel types were included over the entire cell, with a uniform density. These models assume a total Ca^{2+} channel density of about 50 pS/μm^2, derived from an estimate of 5–20 channels per μm^2 and 10 pS per channel, and adjusted to match $[Ca^{2+}]_{in}$ profiles. A T:N:L channel ratio of 1:10:10 was cited from the work of Fisher *et al.* (1990), with the density of the N and L channels set to 25 pS/μm^2 and that of the T channels set to 2.5 pS/μm^2.

The driving force for the calcium channels was derived with the GHK rectification formula (Hagiwara and Byerly, 1981), modified to accommodate units of conductance (as above) instead of permeability. It should be noted that strictly speaking this translation of units is incorrect, since the permeability appropriate for the constant-field model is quite different than the conductance of the ohmic model (Section 8).

These authors used a least-squares fitting of referenced Boltzmann equation parameters, adjusted to m^2h or m^2 kinetics (T- and N-type channels for the former, L-type channels for the latter), to derive parameters for Hodgkin–Huxley-type rate equations. No justification was given for the use of m^2 (as opposed to simply m).* The kinetic data referenced in the paper included voltage-independent time constants that were also used in the fitting to the Hodgkin–Huxley parameters. Because the inactivation kinetics reported by Fisher *et al.* (1990) were quite variable, inactivation parameters in this model were derived from nonhippocampal cells. The final result was a good fit to the steady-state data, but improbable time constant functions (in particular for inactivation), since the canonical Hodgkin–Huxley form has a poor fit to a time constant expression with a large voltage-independent term. In particular, the reference data for the τ's of inactivation for I_{CaT} and I_{CaN} were 10 and 50 msec, respectively: the fitted τ functions were bell-shaped, peaking at 8000 and 1600 msec, respectively. A consequence may be an overestimation of the Ca^{2+} influx, all other factors being equal.†

The experiments of Fisher *et al.* were done with 95 mM Ba^{2+} in patch pipette in order to accentuate the single-channel currents—thus, the voltage parameters of activation in the model were shifted −10 mV to account for the charge-screening effect of the Ba^{2+}.

12.4. Major Properties for the Working Model Ca^{2+} Channels

For the Ca^{2+} channels in the Working model, we shall adapt the three-channel models from the Jaffe94 and Migliore95 HPC models (T, N, and L type), fitting the gating particles to the extended Hodgkin–Huxley scheme. To summarize, the major goals for our description of the Ca^{2+} currents are as follows:

1. Generate Ca^{2+} spikes when Na^+ and K^+ channels are blocked.
2. Sufficient Ca^{2+} entry during single spikes for the activation of Ca^{2+}-dependent K^+ channels.
3. Ca^{2+} entry consistent with measurements of $[Ca^{2+}]$ changes.
4. Current/voltage relationship consistent with the large concentration gradients seen by Ca^{2+} channels.

The associated goals for the description of internal Ca^{2+} dynamics were described in Section 9. Absolute permeabilities were adjusted in order to reproduce regenerative Ca^{2+}-only spikes (Segal and Barker, 1986).

*Although m^2 kinetics is consistent with that proposed by Kay, he was fitting one, not three, channel models to the data.

†However, the slow inactivation of I_{CaT} mentioned earlier (Fisher *et al.*, 1990) means that the fit may not be completely unreasonable.

This section describes the variety of K^+ channels that have been identified in HPCs. These channels cover a wide range of kinetic scales, and thus have a variety of distinct roles in determining the overall cell response.

13.1. The Slow Subthreshold K^+ Current: I_D

I_D is a transient K^+ current that has been implicated in the control of the onset of firing under certain conditions (Storm, 1988b), primarily those in which the cell has been at rest, or, especially, below rest, for at least several seconds. In earlier reports of transient K^+ currents (e.g., Segal and Barker, 1984), it is possible that I_D was lumped together with I_A, since both are sensitive to 4-AP. However, Storm was able to dissect out I_D from I_A taking advantage of the fact that I_D is about 100 times more sensitive to 4-AP, and I_D is much slower than I_A.

The kinetics of I_D from Storm (1988b) shows slow inactivation, ranging from 4.7 sec at −110 mV (measured in terms of recovery from inactivation) to about 1 sec at −26 mV (these values measured as current decays). Activation is apparently much faster, essentially instantaneous on the time base of seconds of Fig. 2 in Storm (1988b). Storm has suggested that the slow inactivation of I_D makes it particularly suitable for making the responsiveness of the neuron dependent on the past several seconds of activity. Taking into account the voltage dependence of inactivation, a maximum conductance of I_D on the order of 50 nS can be inferred from this paper.

Lüthi *et al.* (1996) reported a slow K^+ current which may complicate the interpretation of I_D derived from Storm (1988b) (Section 13.8; this current corresponds to a slow component of the 4-AP-sensitive current described by Storm (1986)]. For our present purposes, since this current is about one tenth the size of I_D, we neglect it in the current Working model.

A Model of I_D

None of the models in Section 7.1 includes an explicit I_D, although the kinetics of I_A in the BG87 model was adjusted to account for a 4-AP-sensitive delay in initial firing. In the Warman94 model the role of I_D was also assumed to be mediated by I_A, although the kinetics of I_A was not adjusted accordingly. In the BG87 model the description of I_A resulted in a fairly delicate, nonrobust mechanism; in the Working model, accounting for the delay and the 4-AP-sensitive component of spike repolarization with two channels (I_D and I_A, respectively) gives a much more robust result.

Our model for I_D uses fourth powers of both an activation and inactivation particle, primarily to restrict the voltage range of steady-state activation (window current) while still providing a good fit to the data in Storm (1988b) (see Section 8.4.4). The justification for this restriction was to avoid contribution of I_D to the resting state and thus simplify the model evaluation. The sigmoidal time course of activation resulting from setting $n = 4$ is not important for the model. However, it will be important to consider the possible contribution of I_D to resting potential mechanism in future versions of the model.

Figure 16 illustrates the effect of I_D on repetitive firing with conditions similar to that used in Storm (1988b). Note that the inclusion of I_D has an effect mainly on the timing of the first spike. Apropos to this point, an interesting finding by Storm was the apparent relationship between I_D and accommodation. Under control conditions in HPCs, for low-current stimulus strengths there is little accommodation; in particular, the initial interspike interval during repetitive firing is close to the steady-state value. As the data in the paper by Storm (1988b) show, this is correlated with the long delay mediated by I_D, since this delay is only seen if the stimulus is sufficiently small (larger current steps appear to overwhelm the I_D rectification, giving the more classical immediate first spike in response to the current). When the cell is held at hyperpolarized holding potentials, which enhances I_D, the lack of accommodation is seen for a much larger range of current stimulus strengths. Conversely, when the cell is held at depolarized holding potentials, which inactivates I_D, accommodation is evident over the entire range of current strengths.

In the model, the time constant for inactivation is relatively fast at the peak of the spike, so that a substantial portion of this current is inactivated by the first few spikes. It may be that the connection between I_D and accommodation is indirect, in that it is not I_D per se that influences the interspike interval, but rather that the delay to first spike mediated by I_D may tend to inactivate more classical accommodation mechanisms (specifically, I_{AHP} or I_M). In our simulations, we have found that a kinetic description for I_D which mediates a delay to first spike and is consistent with the estimate for the magnitude of I_D (Storm, 1988b) results in a rather feeble I_D between spikes, which contributes little in and of itself to the determination of the interspike interval.

13.2. The Delayed-Rectifier K$^+$ Current: I_{DR}

There is a variety of data in hippocampus concerning a slowly inactivating, TEA-sensitive, classical "delayed rectifier" K$^+$ current I_{DR}, analogous to the I_K described in the squid axon (Hodgkin and Huxley, 1952a). Although I_{DR} has been thought not to play a substantial part in the repolarization of the spike in cortical neurons, primarily due to the slower activation seen in some reports (see below), the observation (Storm, 1987a) that a TEA-sensitive component remains for repolarization after apparent block of I_A, I_C, and I_M suggests that I_{DR} is available for this role. Similarily, the limited degree of spike broadening reported in Storm (1988b) when both I_A and I_C are blocked pharmacologically suggests that I_{DR} must be large. In fact, during repetitive firing there is a downregulation of both I_A and I_C, suggesting that I_{DR} might become more predominant during the repolarization of later spikes in a train. Finally, the analysis of Williamson and Alger (1990) suggests that I_{DR} may play a small role in the mAHP. In summary, although I_{DR} in HPCs certainly does not take the center stage for spike repolarization as it does in squid axon, it probably does make a significant contribution.

I_{DR} may also be present in dendrites—Masukawa and Hansen (1987) and Masukawa *et al.* (1991) reported a fast "outward delayed-rectifier" in patches from both the cell body and dendrites in cultured hippocampal neurons. Hoffman *et al.*

(1997) reported that a sustained component of dendritic K^+ channels, with properties similar to I_{DR}, was homogeneously distributed in the tree and the soma.

A Model of I_{DR}

Most of the previous HPC models use some variation of the Hodgkin–Huxley I_{DR}, with kinetics derived from, for example, either the original Hodgkin and Huxley (1952a,b) data (e.g., the Jaffe94 model), Segal and Barker (1984; the BG87 model), Sah *et al.* (1988a; e.g., the Traub91, Traub94, and Warman94 models), or Ficker and Heinemann (1992; the Migliore95 model). The role of I_{DR} in the models varies from that of a primary repolarizing current (e.g., the Traub91 and Traub94 models) to being negligible (e.g., the Warman94 model). In the BG87 model, I_{DR} was strong enough to repolarize the spike if all other K^+ channels were blocked. Also, in this model I_{DR} was adjusted under control conditions to generate the mAHP; in the present version of the Working model the mAHP is mediated solely by I_M.

The voltage sensitivity of activation for the I_{DR} that we shall use in the Working model is based on measurements of the TEA-sensitive outward K^+ current in Storm (1988a), Segal and Barker (1984), Numann *et al.* (1987), and Sah *et al.* (1988a), all of which describe this current as activating at around −60 to −40 mV. The reported kinetics of activation, however, varies considerably, ranging from over 100 msec (e.g., Segal and Barker, 1984) to 1 msec (Sah *et al.*) at depolarized levels. Furthermore, Sah *et al.* report relatively fast kinetics of activation/deactivation over a broad voltage range, described as a bell-shaped curve with a peak of 7 msec at about −60 mV (23°C). The time constant for inactivation has been reported as ranging from 450 msec (Sah *et al.*) to over seconds (Segal and Barker, 1984).

Adjusting the kinetics of I_{DR} so that it can substantially repolarize the spike requires a reasonably fast turn on (within milliseconds) at depolarized levels. Since inactivation is slow, this activation gives rise to a substantial mAHP-like component by I_{DR} unless deactivation is also fairly rapid (we assume that the majority of the mAHP is mediated by I_M; Section 12.7). For this reason, the activation kinetics for I_{DR} was chosen to match more closely the data of Sah *et al.* (1988a). Additionally, these workers fit a single exponential to the time course of activation, suggesting a single activation particle with respect to the Hodgkin–Huxley model framework (as compared to the four activation particles of the Hodgkin–Huxley squid I_{DR} model).

The reason for the large difference in kinetics for the putative I_{DR} (TEA-sensitive, non-Ca^{2+}-dependent current) in the literature remains to be determined. In the context of the model, it does not seem plausible to postulate a single mechanism whose kinetics could be sufficiently complex to account for the various findings. One obstacle is the lack of data implicating a specific role of I_{DR} independent of I_A and I_C (e.g., as might be observed when I_{DR} is blocked in isolation), in part because the concentrations of TEA needed to block I_{DR} in hippocampus (at least 20 mM; Storm, 1988a; Sah *et al.*, 1988a) also block I_C.

The final version of I_{DR} used in the model is the first repolarizing current to be activated during a spike (but not the largest), mainly because the single activation particle caused an immediate activation with depolarization (cf. I_A below).

13.3. The Fast Transient Repolarizing K^+ Current: I_A

A fast, transient, 4-AP-sensitive K^+ current I_A is found in many neuronal types. In hippocampus there are conflicting data concerning the repolarization of the spike with agents that would be expected to block both I_C and I_A. For example, Fig. 3 of Segal *et al.* (1984) shows recordings with 5 mM Co^{2+} (to block Ca^{2+} currents, and thus I_C), with and without 5mM 4-AP. The addition of 4-AP acts primarily to decrease the latency of the first spike (suggesting action of a suppressed I_D) with ". . . [no] detectable changes in . . . [spike] duration," although this cannot be easily ascertained by the scale of the figure. More importantly, the recordings both with and without 4-AP show a prominent AHP that peaks about 15 mV below spike threshold, contrary to the recordings of Storm (1988b), and stronger than would be expected by contributions of I_M or I_{DR}. The depth of the AHP is more reminiscent of aspiny nonpyramidal hippocampal neurons (e.g., Scharfman, 1993): although the recordings in this study were made from "pyramidal-shaped" cultured cells, it is possible that some cells were nonpyramidal in origin.

I_A in cultured hippocampal cells was also described in Nakajima *et al.* (1986), showing that this current is antagonized by acetylcholine. I_A may also be modulated by the concentration gradient of K^+: lowering extracellular $[K^+]$ shifts the kinetics of I_A in the hyperpolarizing direction, and results in a Mg^{2+}-dependent *reduction* of I_A at potentials above −20 mV (the voltage range at which the spike repolarizing role of I_A is initiated) (Eder *et al.*, 1996). Since modulation of $[K^+]_o$ is associated with various pathological states (e.g., seizure-like events), this finding suggests counterintuitive effects on the repolarization of the spike as mediated by I_A, in particular when $[K^+]_o$ is reduced, which would be expected to *increase* a K^+ current via an increase in the driving force. I_A is also susceptible to dendrotoxin at nanomolar concentrations, which may explain the mechanism of this toxin in the induction of epileptiform activity (Halliwell *et al.*, 1986).

Voltage-clamp data references for I_A include Halliwell *et al.* (1986) and Storm (1988b) for slice preparations and Ficker and Heinemann (1992) for cultured cells. The voltage-clamp data in Storm (1988b) shed some light on the kinetics of I_A, for example, showing that I_A inactivation has a time constant of about 20 msec for voltage steps above −50 mV. The steady-state inactivation was shallow, with a midpoint at about −50 mV.

Under current clamp Storm (1987a) demonstrated that the action of 4-AP is seen immediately at the peak of the spike, and acts to broaden the spike by about 10–20% until the peak of the fAHP.

Masukawa *et al.* (1991) reported an I_A-like channel in the dendrites of cultured hippocampal neurons. Andreasen and Lambert (1995) reported that 4-AP had quite strong effects on the firing patterns of CA1 HPC dendrites. For example, under concentrations (0.5 mM) that uniformly broaden the somatic spike by 10% or 20% (as above), 4-AP converted current-evoked fast dendritic spikes into incompletely repolarized fast spikes that subsequently led directly into complex, Ca^{2+}-mediated slow spikes. These results suggest that I_A may play a greater role in maintaining stability in the dendrites rather than the soma. More recently, Hoffman *et al.* (1997) report I_A channels in the dendrites whose density increases markedly with distance from the soma.

Functionally I_A may gate action potential transmission in either direction from the soma—both down the axonal tree (Debanne *et al.*, 1997) or in the dendrites (Hoffman *et al.*, 1997). In the dendrites at least the voltage-dependency of I_A kinetics shifts in the depolarizing direction by activation of either PKA or PKC (Hoffman and Johnston, 1998). The resulting functional downregulation of these channels is reflected in a stronger backpropagation of dendritic action potentials. I_A may also participate in the shaping of subthreshold dendritic EPSPs (Hoffman *et al.*, 1997; see also Lipowsky *et al.*, 1996).

A Model of I_A

For our Working model the primary determinant for the functional properties of I_A is the report of Storm (1987a). In particular, the uniform broadening of the entire action potential when I_A is blocked implies that this current must activate quickly and strongly during the spike under control conditions.

Williamson and Alger (1990) note that 4-AP only broadens the early spikes in a spike train, somewhat consistent with the steady-state inactivation curve shown in Storm (1988b), but perhaps suggesting another component of inactivation (leading to stronger inactivation of I_A with prolonged depolarization).

The time course of activation of I_A is probably voltage dependent, since this current must activate quickly at very depolarized levels, but at the same time must not activate too quickly at subthreshold potentials, lest its effect be seen on the spike upstroke. A delay in activation can also be achieved by postulating more than one activation particle. Here, we define the kinetics of the activation particle with both characteristics. We note that the data in Storm (1988b) do not show an obvious delay in activation, but it may be that the time resolution of the presentation is inadequate to determine this. In the model, under control conditions I_A is the second repolarizing current to be activated, after I_{DR}.

The model that we use for I_A has several drawbacks. While the activation particle definition is quite reminiscent of the Hodgkin–Huxley squid I_{DR} model (activation raised to the fourth power and voltage-dependent time constant, needed here to account for the apparent delay in activation under current clamp), it does not seem to match the kinetics in the HPC voltage-clamp data. Second, the time course of I_A during the repolarization, as inferred from the blocking experiments (which show a broadening of the action potential under 4-AP all the way until the peak of repolarization), is suggestive of a current source, which implies a steady increase of conductance throughout the repolarization to compensate for the reduction in driving force. This behavior is difficult to address with the Hodgkin–Huxley kinetics. One possibility is that a full Markov formulation is more consistent with the data.

13.4. The Fast Ca^{2+}-Dependent Repolarizing K^+ Currents: I_C and I_{CT}

A fast, TEA-sensitive component of the Ca^{2+}-activated K^+ current in hippocampus, referred to generally as I_C (also known as $I_{K(C)}$), was first described as distinct from a slow, non-TEA-sensitive component (I_{AHP}, see below) in hippocampus by Lancaster and Adams (1986). This current was found to be responsible for the fAHP during the repolarization of single spikes (Lancaster and Nicoll, 1987; Storm, 1987a).

However, there is substantial evidence that there is more than one type of channel underlying this current. Voltage-clamp studies suggest that I_C may represent more than one current (for review see Storm, 1990), in particular a transient I_{CT} (also known as $I_{K(CT)}$) and a noninactivating I_C (e.g., Zbicz and Weight, 1985).

The evidence demonstrating a partial dependence of the mAHP on Ca^{2+} influx, as mentioned earlier, has been attributed to participation of I_C in the mAHP, especially when the cell was depolarized. This suggests that although I_C must inactivate quickly in order to account for the fAHP, residual I_C must be present for about 100 msec after a spike or spike train, during the mAHP. Likewise, Alger and Williamson (1988; see also Alger and Nicoll, 1980) reported a Ca^{2+}-dependent K^+ current, independent of the I_{AHP} (see below), that generated an after-hyperpolarization of about 150 msec duration after epileptiform bursts.

Thus, the kinetics of the Ca^{2+}-dependent K^+ current must either be rather complex, and cover a relatively wide range, or in fact represent more than one current. For the Working model we shall focus on kinetics which mediate the fAHP, therefore we will refer to this current as I_{CT} (note that the Warman94 model refers to I_{CT}, not I_C). Most other cell models do not make this distinction, and just refer to I_C (e.g., although the description of this current in the BG87 model also was focused on reproducing the fAHP, the current was called I_C).

The main references that we will use regarding the role of I_{CT} during the repolarization of the action potential, and the subsequent fAHP, are Lancaster and Nicoll (1987) and Storm (1987a). We note that voltage-clamp dissection of Ca^{2+}-dependent currents is even more difficult than other currents due to the difficulty of simultaneous voltage and $[Ca^{2+}]_{in}$ control [for one approach to this problem, see Ikemoto *et al.* (1989)]. Accordingly, the parameters for the model I_{CT} will be primarily determined by matching behavior under current clamp.

I_C has been tentatively identified with the large conductance voltage and Ca^{2+}-dependent BK channel recorded in patchs from hippocampal isolated and cultured cells (Sah, 1996). This channel has a threshold/half-activation point of 1/4 μM $[Ca^{2+}]$ at 0 mV (Storm, 1990). On the other hand, the kinetics of the BK channel is probably too slow to account for the fAHP-generating role of I_{CT} (Ikemoto *et al.*, 1989).

The description of I_C in the BG87 model included four voltage-dependent fast activation particles, so that the abrupt phase of the fAHP was caused by the deactivation of these particles. A separate Ca^{2+}-dependent activation particle was also part of the model. The kinetics of the fAHP was replicated in the Warman94 model by an activation particle with both voltage and Ca^{2+} dependence, based on the data of Barrett *et al.* (1982). Kinetics for I_C in the Traub91 and Traub94 models were adjusted from bullfrog sympathetic neuron data. As mentioned earlier, the description of I_C in the Jaffe94 and Migliore95 models was adapted from the Moczydlowski and Latorre model of the rat muscle BK channel.

The finding that dendritic fast spikes show a slight broadening under concentrations of TEA (1–2 mM) that would be expected to block I_C (Andreasen and Lambert, 1995) allows the possibility that this current, or at least I_{CT}, is present in HPC dendrites.

A Three-State Markov Model for I_{CT} with Voltage- and Ca^{2+}-Dependent Transitions

Most reports on the kinetics of I_C describe it as a noninactivating current, but this is at odds with the overall transient nature of the fAHP during repetitive firing (Section 6.4.1). This is further support for attributing a separate I_{CT} to the fAHP. For the slow inactivation during spike trains the I_C in the BG87 model and the I_{CT} in the Warman94 model included a distinct slow voltage-dependent inactivation particle.

Close examination of the behavior of the fAHP suggests that inactivation occurs *after* activation: the preferential appearance of the fAHP on the *first* few spikes of a train, even though subsequent spikes traverse similar voltage ranges, argues against independent activation and inactivation voltage-dependent mechanisms. In particular, it is difficult to adjust the parameters of an independent voltage-dependent inactivation component such that the first spike causes much less inactivation (thus sparing the fAHP) than the second and subsequent spikes. For this reason we will adopt a simple three-state Markov model to describe I_{CT} gating in the Working model, including an inactivated state that is reachable only through the open state. In this model, shown in Fig. 9, the disappearance of the fAHP during repetitive firing is a result of a voltage-dependent transition between the inactivated state and the closed state that is slow at depolarized potentials and speeds up with hyperpolarization. It may be that this transition accelerates with temperature, which could account for the apparent persistence of the fAHP during spike trains seen in Thompson *et al.* (1985), as mentioned earlier.

The transition from the closed state to the open state is Ca^{2+} dependent (following the $[Ca^{2+}]$ in compartment 1 of the associated $[Ca^{2+}]_{in}$ integration system, being fast at $[Ca^{2+}]$ above 0.01 mM) and voltage dependent. It is not clear if this codependece is absolutely necessary for activation for the I_{CT} model since the large transient in the local $[Ca^{2+}]$ seen by the channel is strictly correlated with the spike peak, and further adjustment of the parameters defining the Ca^{2+}-mediated kinetics may allow the dropping of the voltage term. The fast shutoff of I_{CT}, necessary in order to mediate the fAHP, is accomplished by a strong voltage dependence in the transition from the open state to the inactivated state.

13.5. The Slow Ca^{2+}-Dependent K^+ Current: I_{AHP}

The slow Ca^{2+}-dependent after-hyperpolarization (sAHP) is attributed to a Ca^{2+}-dependent and apparently voltage-independent I_{AHP} potassium current (Lancaster and Adams, 1986). The major functional role assigned to this current is that of regulating repetitive firing (Madison and Nicoll, 1984).

The sAHP is insensitive to TEA and charybdotoxin, unlike the fAHP, whereas in hippocampal interneuron cells the sAHP is sensitive to apamin (in HPCs apamin does not block the sAHP) (Storm, 1990). Both noradrenaline and cholinergic agonists, among other neurotransmitters, block I_{AHP}. Second messengers in addition to Ca^{2+} are involved in the regulation of I_{AHP}. Results from Pedarzani and Storm (1993, 1995a, 1996) indicate that PKA mediates the suppression of I_{AHP} by all the monoamine transmitters: noradrenaline, serotonine, histamine, and dopamine; and there is a synergistic interaction between α- and β-receptor agonists. Zhang *et al.* (1996) showed that activation of muscarinic afferents blocks the I_{AHP} only if this occurs before the relevant rise in $[Ca^{2+}]_{in}$.

In single-channel recordings from cultured rat E17 hippocampal neurons, Lancaster *et al.* (1991) identified a Ca^{2+}-dependent 19-pS K^+ channel that they associate with I_{AHP}. In these recordings they noted a rise time of Ca^{2+} activation of approximately 70 msec at −70 mV. Noting the longer delay in the peak I_{AHP} of a few hundred milliseconds (consistent with the delay of the sAHP peak), they discount the possibility of Ca^{2+}-mediated Ca^{2+} release in hippocampus, and suggest that there must be some mechanism which restricts access of the channel to intracellular Ca^{2+}.

The I_{AHP} has been attributed to the SK class of channels, distinct from the BK channels mentioned earlier (Sah, 1996). Generally it is found that 100–400 nM $[Ca^{2+}]$ is necessary for the activation of the SK channel; this may be contrasted to the activation of hippocampal I_{AHP} by a rise from 30 to 60 nM $[Ca^{2+}]$ (Knöpfel *et al.*, 1989).

Recently, a candidate channel for I_{AHP}, has been cloned in mammalian brain (Köhler *et al.*, 1996), which has the general attributes that would be expected from the macroscopic behavior of I_{AHP}. This paper suggested that the Ca^{2+} dependence of the channel fit a Hill relationship with a coefficient of about four, in other words implying an involvement of four Ca^{2+} ions in channel gating, and a $K_{0.5}$ between 0.5 and 1.0 μM.

Despite the lack of direct supporting data, the model of I_{AHP} in the BG87 model included explicit voltage dependence via two types of voltage-dependent inactivation gating particles. The rationale for these particles was to account for the delay to peak of the sAHP and to avoid interaction between I_{AHP} and the shape and time course of either Na^+ or Ca^{2+} spikes. Although the latter motivation may be somewhat problematic, the former will be reconsidered below.

Lancaster and Adams (1986) found no voltage dependence for I_{AHP} using either voltage clamp or hybrid clamp protocols. However, the maximum holding potential used was −47 mV; thus it is conceivable that voltage-dependent inactivation, fast and relatively complete at spiking potentials, and removed slowly (hundreds of milliseconds) around rest, could account for the observed delay in the sAHP peak.

Steady-state activation based on the log of $[Ca^{2+}]_{in}$ was used for I_{AHP} in the Warman94 model, adapted from the work of Blatz and Magleby (1987), with a time constant for activation of 48 msec.

A Model of I_{AHP}

In our Working model of I_{AHP}, we will use the simple Ca^{2+}-dependent gating model described in Section 8.8.3. This is similar to that used for the Ca^{2+} dependence of both I_C and I_{AHP} in the BG87 model. On the other hand, for this model of I_{AHP} we will neglect the voltage-dependent mechanisms used in the BG87 model. We shall assume cooperative binding of four Ca^{2+} ions in order to reproduce the distinct threshold in Ca^{2+} dependence that is necessary for this channel, assuming that activation of I_{AHP} depends on slow, accumulative changes in $[Ca^{2+}]_{in}$, and that the quiescent activation of I_{AHP} is negligible. The Ca^{2+} signal is taken from compartment 2 of the associated (somatic) $[Ca^{2+}]_{in}$ integration system.

In order to reconcile the small incremental change in $[Ca^{2+}]_{in}$ per spike (about 10 nM) relative to the resting level (50 nM), and to avoid a substantial resting activation of I_{AHP}, it was necessary to include two Ca^{2+}-dependent gating particles. This, plus the requirement of the binding of four Ca^{2+} ions for activation and the particular values used for α and β results in a distinct activation threshold around 50 nM.

An important question is that of the mechanism underlying the long delay to peak of the sAHP, and thus I_{AHP}. We can consider at least two candidates, delay arising from either the intracellular diffusion of Ca^{2+} or from the intrinsic channel gating kinetics. Arguments for the latter cause include the strong temperature dependence of the sAHP peak (Thompson *et al.*, 1985; Lancaster and Adams, 1986), and the relatively fast rise time of $[Ca^{2+}]$ signals in response to single spikes (e.g., Jaffe *et al.*, 1992), which presumably reflect a global or average concentration over many micrometers. Accordingly, in the Working model I_{AHP} a value for τ_0 of 100 msec for the Ca^{2+}-binding kinetics contributes to a delay in the I_{AHP} peak after the rise in $[Ca^{2+}]_{in}$ in compartment 2 of the associated Ca^{2+} concentration integrator. A similar rate-limiting term was used in the I_{AHP} in the Traub91 and Traub94 models.

On the other hand, the value of 100 msec does not account for the somewhat longer delays in the peak I_{AHP} reported in the literature [e.g., 400–700 msec in Lancaster and Adams (1986)]. Using a longer value for τ_0 causes the I_{AHP} gating to "miss" the peak of the $[Ca^{2+}]_{in}$. It may be that either a voltage-dependent inactivation term should be retained, as in the BG87 model, or a Markov description that decouples the time course of the relevant $[Ca^{2+}]_{in}$ and the activation of I_{AHP} should be considered (see discussion in Section 8.7.1).

More importantly, recent data from Lancaster and Zucker (1994) show that photolytically released intracellular Ca^{2+} can trigger I_{AHP} with a delay on the order of 50 msec, suggesting that the delay of the sAHP is due to factors extrinsic to the I_{AHP} channel. These workers suggested that delays due to diffusion, taking into account immobile buffers (see Section 9.5.1), would reach on the order of 400 msec for distances of about 3 μm. They suggest that the strong temperature dependence on the delay to sAHP peak is related to the temperature dependence of the Ca^{2+} pump and buffer systems. Future refinements of the diffusion models described in Section 9 may support this scheme.

We note that choosing model parameters in order to avoid a resting contribution of I_{AHP} was done primarily for convenience, and it will be important to check if in fact the assumed 50 nM level for resting $[Ca^{2+}]_{in}$ does not slightly activate I_{AHP}. If so, then the numerous modulatory influences on this current may also be expected to affect the neuron's resting state. In fact, as pointed out elsewhere, the true resting state of the Working model is reached after 10–20 sec, during which time a small Ca^{2+} influx from a small resting activation of the Ca^{2+} channels leads to a buildup of the resting $[Ca^{2+}]_{in}$, which in turn slightly activates I_{AHP}, hyperpolarizing the cell.

13.6. Is the Variance in Kinetics between I_{CT} and I_{AHP} Due to Unique Pools of Ca^{2+}?

The difference in the kinetics of I_{CT} and I_{AHP} can be accounted for in part by a difference in the Ca^{2+} "pools" that each are exposed to. This can be inferred by the following reasoning.

The time course of the rise of $[Ca^{2+}]$ averaged over the entire volume of the soma or a section of dendrite is consistent with the buildup of the sAHP and thus by implication I_{AHP}. On the other hand, the I_{CT}-mediated fAHP is present at the first spike from the quiescent state. One way to account for this difference is for I_{CT} to be exposed to a Ca^{2+} pool with much faster rise time, such that the concentration in this pool rises significantly above the resting level with a single spike. This could be accomplished if I_{CT} channels were colocalized with one or more of the Ca^{2+} channel types, so that the I_{CT} channels would be triggered by very local and large increases of $[Ca^{2+}]$ that would occur adjacent to Ca^{2+} channels with every spike (Lancaster and Nicoll, 1987; the BG87 and Warman94 models).

This hypothesis is supported by the observation that a fast intracellular Ca^{2+} buffer [*bis*-(*o*-aminophenoxy)-ethane-*N,N,N′,N′*-tetracetic acid, BAPTA] will block the repolarization by I_{CT}, whereas injection of a slower buffer (EGTA) blocks only the I_{AHP} component, as shown by Storm (1987b). Note that BAPTA does not prevent putative Ca^{2+}-mediated inactivation of Ca^{2+} channels (Kay, 1991; see Section 12.2.3), which suggests that I_{CT} channels may be close, but not immediately adjacent to Ca^{2+} channels.

The functional significance of microdomains of large and fast $[Ca^{2+}]$ changes in the vicinity of Ca^{2+} channels has been considered previously in the context of synaptic release, and recent studies confirm this phenomenon in the squid giant synapse (for references see Llinás *et al.*, 1995). In addition, recent evidence for the functional colocalization of Ca^{2+} and Ca^{2+}-dependent K^+ channels has been reported in presynaptic membranes of hair cells and the neuromuscular junction (Roberts *et al.*, 1990; Robitaille *et al.*, 1993). Finally, patch recordings of both somatic and dendritic membrane in cultured hippocampal cells (Masukawa *et al.*, 1991) support the general idea of colocalized channels: they showed a high proportion of patches with both a "delayed-rectifier" type of outward current and either a noninactivating or inactivating inward current. However, the evidence presented in this paper suggests that the inward currents are mediated by Na^+, rather than by Ca^{2+}.

The domain model, mentioned earlier, for Ca^{2+}-dependent inactivation of Ca^{2+} channels has been proposed based on the assumption of functionally relevant local high concentrations of Ca^{2+}. The model analyzed by Sherman *et al.* (1990), although investigated in the context of Ca^{2+} channel kinetics, is of the same structure as the model for I_{CT} Ca^{2+} dependence described above.

13.7. The Muscarinic K^+ Current: I_M

Cholinergic agonists regulate a variety of channels, but the classical target (though not the most sensitive) is the noninactivating K^+ current, I_M (Halliwell and Adams, 1982)—the "*M*" stands for "muscarinic." The major hallmark of I_M is the mAHP; the major functional role assigned to this current is for regulating repetitive firing (Benardo and Prince, 1982b; Madison and Nicoll, 1984).

I_M is activated over a narrow range, from about −60 to −40 mV (Halliwell and Adams, 1982). This threshold suggests that the contribution of I_M to the resting potential is minimal. Time constants of I_M relaxations under voltage clamp (and, by implication, the time constant of activation) at 25°C are about 210 msec at −45 mV and 95 msec at −55 mV. Measurements by these workers at 30°C suggest a Q_{10} of about 5.

The voltage-dependent activation of I_M may also be inferred from the data of Williamson and Alger (1990). In one experiment, an mAHP-like potential was elicited under TTX and low Ca^{2+} by long current steps to various potentials. The amplitude of this potential increased somewhat linearly with the plateau potential, beginning when the step reached about −45 mV. The amplitude of the mAHP-like potential increased for plateau potentials up to at least −32 mV, but this may reflect a voltage-dependent increase in the rate of activation for I_M rather than a steady-state characteristic that is still rising at this voltage.

The reversal potential of I_M has been estimated at between −78 mV (Halliwell and Adams, 1982) and −73 mV (Williamson and Alger, 1990), suggesting that this is not a pure K^+ conductance.

As mentioned earlier, I_M in the models of Section 7.1 had different functional effects, including an important role in stabilizing the resting potential, playing a role in accommodation, and terminating bursts. The Traub91 and Traub94 models did not include I_M, and the I_M used in the BG87 model was underestimated. On the other hand, I_M in the Warman94 model was adjusted to account for both the mAHP and the early regulation of repetitive firing. The question of whether or not I_M contributes to the resting potential is complicated by the many effects of muscarinic agonists, particularly the modulation of mechanisms that are active around rest.

A Model of I_M

The Working model description of I_M is based primarily on the voltage-clamp data of Halliwell and Adams (1982) (and D. Brown and Griffith, 1983a), and the more functionally oriented analysis in Storm (1989) and Williamson and Alger (1990), such that I_M parameters were adjusted in order to account for the major properties of the mAHP.

In this description, I_M is completely shut at rest. The observation that there is a roughly constant increment per spike in the mAHP (Storm, 1989) can be accounted for in part if we use a second order in the activation term (that is, two activation particles), with a fast time constant at depolarized levels and a much slower value around threshold. The second power tends to equalize the incremental increase of I_M from spike to spike (the Warman94 model used a second-order description of the I_M gating particle to implement a temporal delay in activation).

13.8. Anomalous Inward Rectification and Other K^+ Currents

Here we summarize the mechanism of anomalous inward rectification, and discuss additional K^+ currents that have been reported for HPCs. These currents will not be considered in the Working model, but are left for future work.

13.8.1. Anomalous Inward Rectification

Anomalous inward rectification is classically defined as an increase in a K^+ conductance with hyperpolarization (Katz, 1949; Hotson *et al.*, 1979; Hille, 1992). The term "anomalous" is to highlight the difference from the classical delayed rectifier, or in general the fact that most K^+ conductances open with depolarization. However, as described below, anomalous inward rectifiers may also be mixed conductances.

There are two classes of inward rectifying currents. The first class consists of mixed cation inward rectifiers whose nonlinearities are characterized in terms of voltage-dependent gating mechanisms (Section 12.1). The second class of inward rectifiers, which are pure K^+ currents, demonstrate an instantaneous voltage dependence that shifts with the log of $[K^+]_o$, in other words, relative to the difference between the membrane potential and E_K, and have an inflection in the I–V curve around E_K. These channels may also show a time-dependent component of the rectification, as well as a $[K^+]_o$ dependence on the steepness of the steady-state I–V characteristic. Rectification of these channels arises directly from the conduction properties of the channel ionophore, but they do so in a sense opposite to that which would be predicted by the constant-field permeation model (Section 8.2). Hille (1992) discusses a multi-ion pore model with a monovalent intracellular blocking ion that seems to account for many of the pure K^+ inward rectifier properties.

As a side note, since under normal conditions the ion with the lowest equilibrium potential is K^+, there can be no channel-mediated current (synaptic or otherwise) that can drive the membrane potential below E_K. This would suggest that K^+-mediated inward rectifiers are functionally relevant only for limiting below-E_K hyperpolarizations due to membrane pumps. With respect to HPCs, there is limited evidence for this type of channel under control conditions [but see Owen (1987), Section 14.1, and discussion in D. A. Brown *et al.* (1990)]. A serotonin-activated inward rectifier has been reported, however, and this current will be discussed next.

13.8.2. A Serotonin-Activated Inward Rectifier

Serotonin (5-HT) has both inhibitory and excitatory effects on HPCs causing an increase in one or more K^+ currents, which may include a true inward rectifier, and causing a decrease in a resting K^+ conductance and the Ca^{2+}-dependent sAHP (Colino and Halliwell, 1987). The inward-rectifying characteristics of the serotonin-activated current include a dependence on the slope conductance with $[K^+]_o$ (Andrade and Nicoll, 1987). Rectification is relatively shallow, with approximately 3 nS of activated conductance at −120 mV compared to about 1 nS at −30 mV ($[K^+]_o$ of 3 mM and 60 μM 5-HT; Okuhara and Beck, 1994). This current may be preferentially expressed in the dendrites and axons, while another serotonin-activated K^+ current, which has less obvious inward-rectifying properties, may be localized to the soma (Uneyma *et al.*, 1993).

13.8.3. $I_{K,rest}$, $I_{T,slow}$, $I_{K(slow)}$

One candidate for the leak current is a K^+ current, $I_{K,rest}$ (see discussion by Storm, 1990), that is linear for tens of millivolts around rest, and persists with blockers of the major K^+ currents. There may also be a muscarinic-sensitive component to this current (Madison *et al.*, 1987b).

Ficker and Heinemann (1992) have described a slowly inactivating, TEA-sensitive K^+ current, $I_{T,slow}$, in cultured hippocampal cells. This current is not unlike I_{DR} other than that $I_{T,slow}$ has a substantially higher threshold (between −40 and −30 mV), and it is sensitive to 4-AP.

Lüthi *et al.* (1996) describe a $I_{K(slow)}$ which contributes particularly to the latency of the first action potential, along with I_D and perhaps I_A (see also Storm, 1986). This current had a reversal potential of about −93 mV that shifted with changing $[K^+]_o$ as predicted by the Nernst equation given a pure K^+ current. The fully activated conductance was about 6 nS (estimated from data presented in the paper). Several characteristics of $I_{K(slow)}$ are similar to I_D, a major difference being a slower inactivation.

14. Nonspecific Cation and Chloride Currents

We now mention a variety of other channels that have been reported in hippocampal pyramidal neurons. Other than I_H, these currents will not be demonstrated in the Working model. As before with the nonmodeled K^+ currents, it remains to be seen how including these mechanism will influence estimated parameters or functional consequences of the channel models described earlier.

14.1. The Hyperpolarization-Activated Sag Current: I_Q/I_H

Various descriptions of a hyperpolarizaton-activated, mixed-cation inward current have been reported for hippocampal neurons. Although earlier reports (e.g., Halliwell and Adams, 1982) called this current I_Q (for "queer," presumably in the same vein as "anomalous"), more recently this current has been identified as I_H due to the similarity to the I_H identified in several other cells [e.g., first in photoreceptors and heart cells, then in neurons, e.g., thalamic cells (McCormick and Pape, 1990); for a thorough review see Pape (1996)]. This current, blocked by Cs^+, is mainly responsible for a depolarizing sag in response to long (100 msec or more) hyperpolarizing current steps, and may contribute to resting properties. The biophysical details and, as suggested above, the name of this current have been somewhat inconsistent in the literature.

I_Q was characterized by Halliwell and Adams (1982) as a persistent current which was probably carried by both K^+ and Na^+ ions (they estimated the reversal potential for I_Q as around −55 mV). In this report the activation of I_Q occurred at minimum in the range between −120 and −80 mV, and activation was well fitted to a single exponential with a time constant that decreased monotonically with hyperpolarization (184 msec at −82 mV to 100 msec at −130 mV, 25°C; experiments at 30°C suggested at $Q_{10} = 5$). The maximum conductance of this current can be estimated from the paper at about 4–5 nS. Later papers also referred to either I_{AR} (for anomalous rectifier; Segal and Barker, 1984) or I_Q (D. A. Brown *et al.*, 1990). Grove and Halliwell (1990) estimated E_{rev} for I_Q as -49 ± 11 mV. Colino and

Halliwell (1993) showed that I_Q is potentiated by muscarinic agonists. A Cs^+-dependent component of the mAHP when the cell was hyperpolarized was demonstrated by Storm (1989), and later by Williamson and Alger (1990), suggesting a contribution of I_Q in the mAHP under these conditions.

I_H is enhanced by noradrenaline via beta receptors and cAMP (Pedarzani and Storm, 1995b), and by serotonin, histamine, and dopamine, also by cAMP (Storm *et al.*, 1996).

Maccaferri *et al.* (1993) described an anomalous rectifier current, I_H, in rat CA1 cells that contributes significantly to the resting potential and is involved in after-hyperpolarizations: 2 mM extracellular Cs^+ lowered the resting potential about 4 mV lower than control, and blocked the mAHP. This paper reported a reversal potential for I_H of about −17 mV, with a shallow activation starting at −50 mV and saturating at −140 mV, and a maximum conductance of around 2.6 nS. The major difference between this and earlier reports of I_Q seems to be the more depolarized activation range of I_H (which gives it an important role under non-hyperpolarized conditions) and a significantly higher E_{rev}. Although these authors report a time constant of activation of about 80 msec at −70 mV, this seems to be too slow to account for the hyperpolarization-induced sag shown in the paper under current clamp, thus suggesting that the time constant becomes faster with hyperpolarization. Although the rectification by I_H seems to be due to voltage-dependent gating, the slope conductance was shown to increase with $[K^+]_o$, reminiscent of a true inward rectifier.

Perkins and Wong (1995) demonstrated intracellular blocking by QX-314 of what they refer to as I_Q; interestingly, examination of data shown in this paper suggests that average resting potentials were 5–10 mV *higher* when QX-314 was present in the whole-cell patch electrodes. In general the biophysical characteristics of this current differed slightly from that reported by Maccaferri *et al.* (1993). Perkins and Wong showed a steeper activation that saturated at about −120 mV, and estimations made from the data shown in this paper suggest a total conductance of about 10 nS with an E_{rev} of about −40 mV. Although the activation kinetics for current relaxations at −120 mV was well fit to two exponentials, an approximate fit for modeling purposes may be made to a single exponential with a 60 msec time constant.

It may be that there is more than one current contributing to non-transmitter-activated inward rectification in HPCs. In a short abstract, Owen (1987) described three types of inward-rectifier currents: a current that he identified as I_Q, an inactivating fast current that he associated with the I_{fir} seen in other cell types, and a third, low-threshold current (−45 mV, complete activation at −115 mV), activating with a 100-msec time constant at −80 mV, that he called I_{uIR}. Grove and Halliwell (1990) demonstrated an additional inward-rectifying current other than I_Q, which is activated at hyperpolarizing potentials (below −80 mV), with an E_{rev} close to E_K.

None of the models in Section 7.1 (other than a cursory nonfunctional description in the BG87 model) includes I_Q.

A Model of I_H

For the description of I_H in the Working model we primarily refer to the data of Maccaferri *et al.* (1993). A single activation particle is used, and the effect at this stage of the model is mainly to generate a depolarizing sag in response to hyperpolarizing current steps (see Figs. 5 and 12). Blocking this current in the model hyperpolarizes the resting potential by only about 300 μV, an effect about ten times less than that reported above. For future refinement it will be instructive to see if the linear-Markov model of I_h in retina proposed by Barnes and Hille (1989) is relevant for hippocampus.

14.2. The Persistent Cl^- Current: $I_{Cl(V)}$

Madison *et al.* (1986; rat HPCs in slice) and D. A. Brown *et al.* (1990; rat CA3 HPCs in cultured slice) describe a slowly activating (hundreds of milliseconds), persistent, voltage-dependent Cl^- current, $I_{Cl(V)}$. Estimates of the physiological value vary for the associated reversal potential E_{Cl}, but is normally around the resting potential (−70 mV). In the first report this current has a shallow steady-state inactivation with a midpoint about −50 mV, and thus may contribute to the resting potential under control conditions. In the second report, however, inactivation was complete below −60 mV, so its contribution to the maintenance of the resting potential in this case is less obvious. In either case, $I_{Cl(V)}$ is blocked by either activation of PKC (demonstrated with phorbol esters) or 100 μM Cd^{2+}. The susceptibility to PKC activation implies that this current may have a modulatory role, particularly with respect to synaptic integration. A primarily dendritic location for $I_{Cl(V)}$ was inferred by difficulty of detection when K^+ channel blockers were not used.

The conductance of $I_{Cl(V)}$ of about 8 nS at −70 mV, as reported by Madison's group, would be expected to contribute on the order of 30% of the resting input impedance for sharp-electrode recordings; for whole-cell recordings this value could account for the *entire* input impedance (8 nS corresponds to 125 MΩ). This clearly presents a problem, for at least the following reasons: first, the voltage range for the current inactivation depends on the preparation (see above), thus this is not a "reliable" mechanism; second, recordings with KCl electrodes would considerably raise the reversal potential of this current, and thus have a strong depolarizing effect on the resting potential.

14.3. The Ca^{2+}-Dependent Nonspecific Cation Currents: I_{ADP}, I_{CAN}, I_{ACPD}

Nonspecific cation (CAN) currents have been described for HPCs, and may be a candidate for the crucial slow depolarizing mechanism underlying bursts (Crépel *et al.*, 1994).

In cultured slice Caeser *et al.* (1993) describe a Ca^{2+}-dependent ADP that is carried by at least a mix of K^+ and Na^+ ions, observed after either cholinergic or metabotropic glutamate receptor (mGluR)-mediated blockade of the sAHP. The magnitude (about 250 pA at holding potentials between −65 and −50 mV) and timing of the underlying I_{ADP} current (also referred to as I_{CAN}) suggest that the neurotransmitter agonists enhance this current, rather than the ADP being more or less "hidden" during the sAHP under control conditions. This interpretation, also reported by Colino and Halliwell (1993), suggests that the delay of the I_{AHP} peak is not due to an early interval of activated I_{ADP}. The decay of I_{ADP} under voltage clamp was significantly slower than the time course of simultaneous $[Ca^{2+}]_{in}$ measurements. This current apparently has an intrinsic voltage dependence, activating at potentials higher than −90 mV and becoming maximal at about −60 mV. The reversal potential of this current has been estimated as about −20 mV (Colino and Halliwell, 1993).

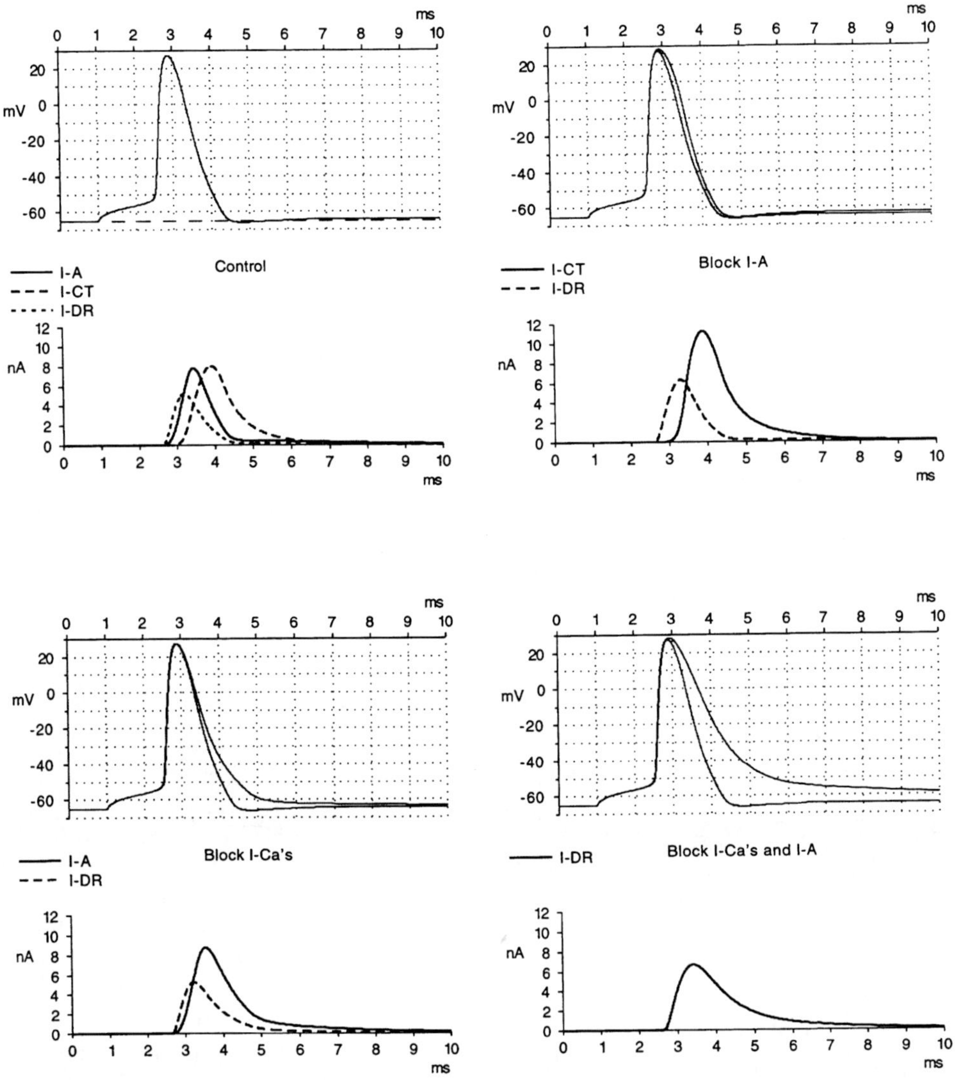

Figure 11. Simulations of the Working model of single spikes and the major K^+ currents evoked with just-threshold short current pulses (1.0 nA, 1.5 msec), comparing control spike with spikes after the blocking of various channels (cf. Figs. 2 and 10 in Storm, 1987a). In the simulations, I_{CT} is effectively blocked when the Ca^{2+} currents are blocked. Note that various combinations of I_{DR}, I_A, and I_{CT} can combine to repolarize the spike, with each combination having a distinct spike signature.

A similar (but voltage-independent) Ca^{2+}-dependent nonspecific cation current, upregulated by activation of mGluRs, was reported by Crépel *et al.* (1994), who named it I_{ACPD} in reference to the mGluR agonist used in the experiment. This current, which was about 100 pA at a holding potential of −60 mV, had a reversal potential around −20 mV. The contribution to the cell's conductance by the activation of I_{ACPD} by mGluRs may be partially offset by the reduction of K^+ currents by mGluRs (I_M and I_{AHP}; Charpak *et al.*, 1990), and may generate a long-lasting depolarization in response to rises in $[Ca^{2+}]_{in}$.

15. Simulations of HPC Properties with the Working Model

We conclude this chapter with some additional simulations illustrating various characteristics of the Working model.

The role of the major repolarizing currents in shaping the spike is seen in Fig. 11, where single evoked spikes are shown under various conditions. Note that the subthreshold time course and spike upstroke are identical under all conditions,

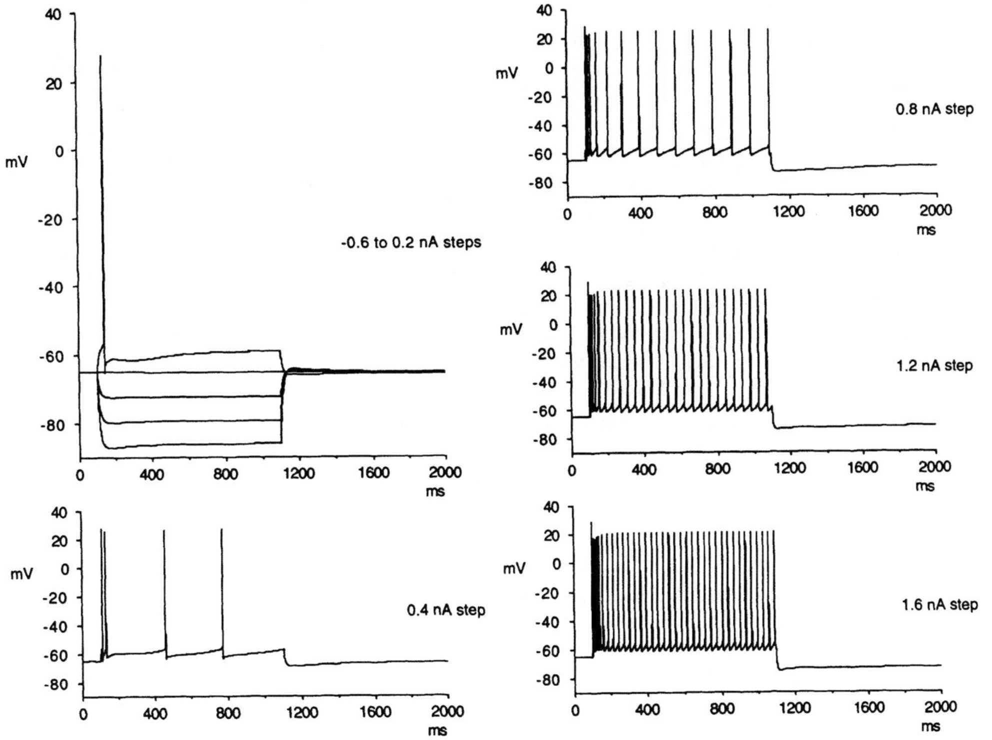

Figure 12. Basic responses to long current steps of the Working model under control conditions. Characteristics of these responses include depolarizing sags to hyperpolarizing current steps, fast initial spiking, followed by strong accommodation, and long-lasting hyperpolarizations after spike trains. Also, the prominent fAHP after the spike (more evident in the 0.2-nA and 0.4-nA responses) disappears during the spike train as the firing frequency increases.

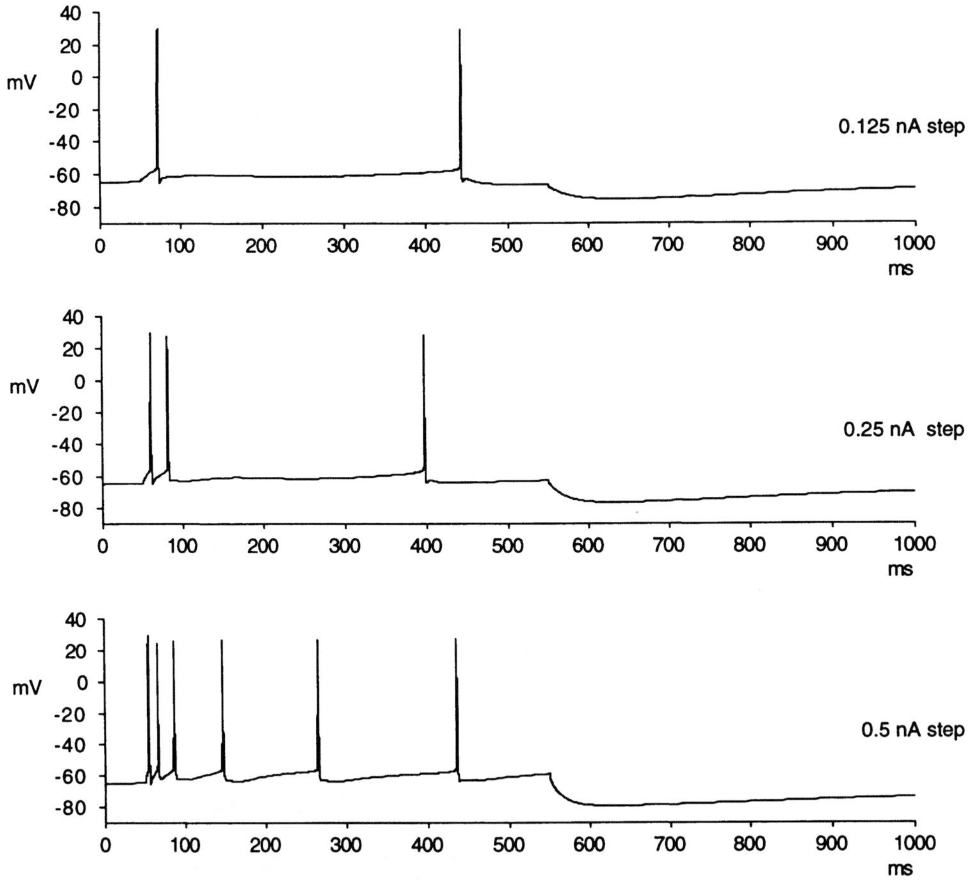

Figure 13. Simulations of the Working model of spike trains after setting R_m for both the soma and dendrites to 100 kΩ-cm², giving an R_{in} of about 210 MΩ, thus corresponding more to parameters from whole-cell patch recordings. The major differences are the higher gain for the cell with respect to the firing frequency as a function of stimulus intensity, and deeper sAHPs. Otherwise, the cell response is quite similar to that seen with the default linear parameters.

Figure 14. Responses of the Working model to long 0.5-nA current steps, illustrating the roles of I_M and I_{AHP} in mediating the early and late phases of accommodation. The associated mAHP and sAHP, or lack thereof, can also be appreciated in these simulations by comparing the after-hyperpolarizations with the resting potential of −65 mV, marked by the dashed line. Note the reappearance of the fAHP in the later spike (at around 800 msec) for the response when I_M is blocked.

Figure 15. Evolution of the mAHP, sAHP, and the average somatic $[Ca^{2+}]_{in}$ of the Working model under control conditions, in response to 0.85-nA current steps. Durations of the steps were adjusted to evoke from one to ten spikes. For these simulations the "true" steady-state $[Ca^{2+}]_{in}$ of about 66 nM was used as an initial condition (obtained after allowing the system to relax for about 20 sec). The mAHP peak can be readily distinguished, preceding the sAHP peak after each spike train. The peak of the sAHP for the shorter trains occurs about 200 msec after the stimulus end. Note that the PHPs during the spike train are about 6 mV, starting from about −62 mV until spike threshold between −56 and −55 mV.

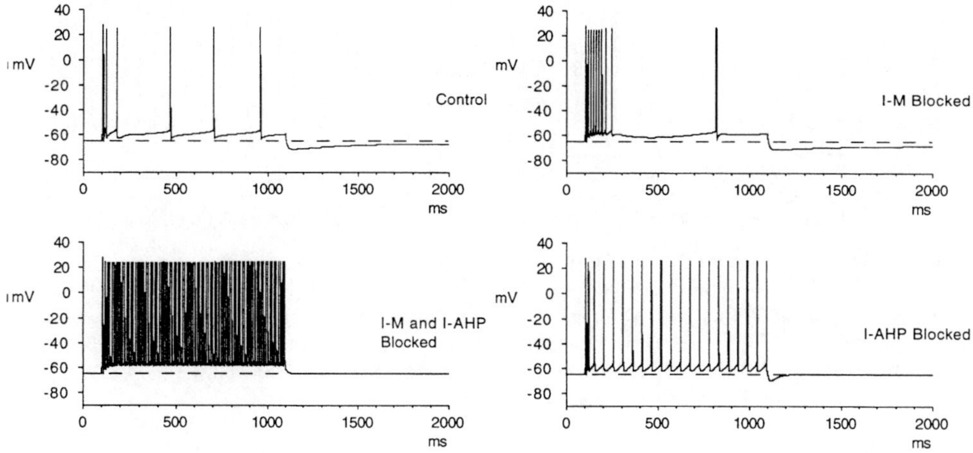

Figure 14.

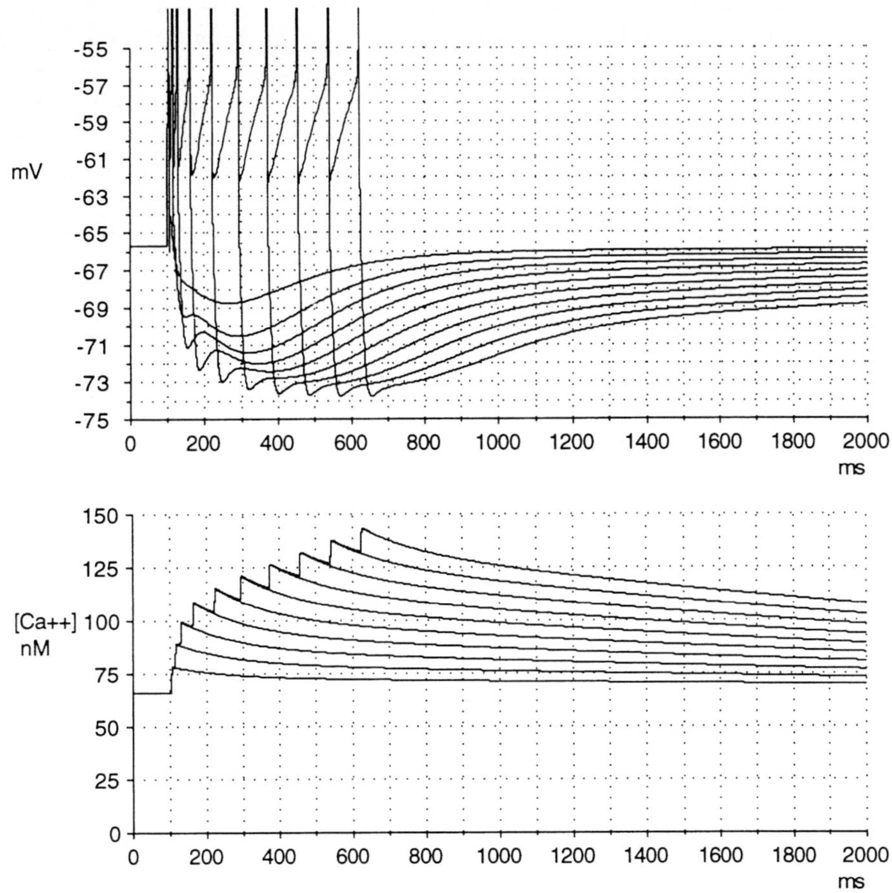

Figure 15.

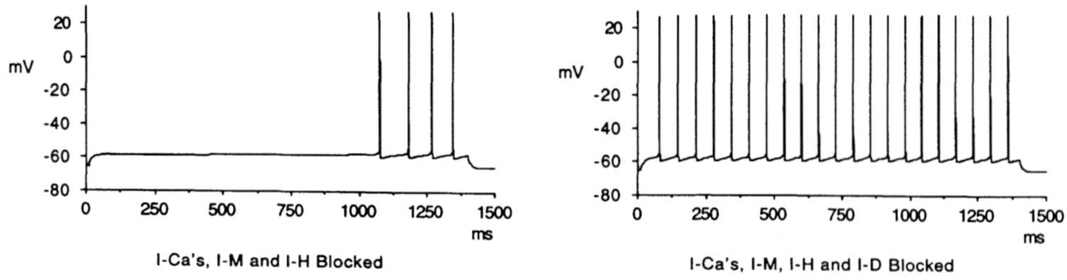

Figure 16. Repetitive firing in response to 0.19-nA current step showing the delay in the initial spike mediated by I_D (cf. Fig. 2c in Storm, 1988b). In both traces Ca^{2+} currents I_M and I_H were blocked, and I_D is blocked in the trace on the right. Although the resting potential of this model was −65 mV, a virtual holding potential of −75 mV was used to enhance I_D by partial deinactivation. This corresponds to the figure mentioned above in Storm (1988b), where the resting potential of the cell was −74 mV.

showing that for the initiation of firing, modulation of either the Ca^{2+} currents (and, by implication, I_{CT} and I_{AHP}) or I_A has no effect on these characteristics. The same holds for I_{DR} (simulations not shown).

The classical basic computational characteristic of a neuron is the transformation of stimulus intensity into the frequency of repetitive spikes. The response of the Working model to long current steps is shown in Fig. 12, which among other things illustrates the time dependence of the output due to various adaptation mechanisms (Figs. 14 and 15). Increasing R_m to bring the linear properties of the model closer to that seen with whole-cell patch recordings changes the response to long current steps quantitatively, but not qualitatively, as seen in Fig. 13. The f/I plot, as discussed in Section 6.6, encapsulates these characteristics in a more quantitative form. The validity of this measure in a functional context remains to be

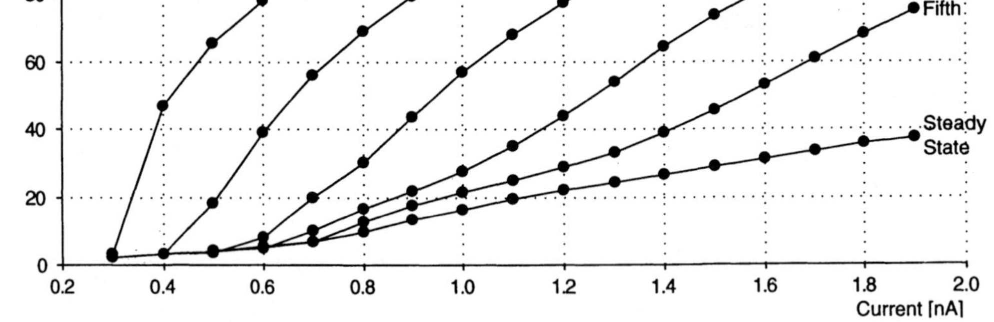

Figure 17. The f/I plot for the Working model under control conditions (compare with Lanthorn *et al.*, 1984). Each trace is the instantaneous firing frequency for the spike interval indicated on the right for different intensities of long (1000 msec) current steps.

demonstrated, primarily since *in vivo* input is never as stereotypically constant as the current steps used in experimental protocols. Nonetheless, comparing the model *f*/*I* plot with those reported in the literature is another test of the collective properties of the model mechanisms (Fig. 17).

ACKNOWLEDGMENTS. Part of this work was covered by HFSP grant RG69/93, and grants from the Fondation Fyssen, Fondation Phillipe, and a CNRS ATIPE Fellowship. I would like to acknowledge the very useful discussions, both past and present, with Johan Storm and John Lisman, and the initial impetus for this project from Tommy Poggio, Christof Koch, and Paul Adams. I would also like to acknowledge Yves Frégnac for his comments and support during the writing of this chapter, and Nicolas Gazeres, Cyril Monier, and Ragnhild Halvorsrud for very thorough feedback including the testing of model properties using Surf-Hippo, GENESIS, and NEURON.

16. References

Adams, P. R., and Galvan, M., 1986, Voltage-dependent currents of vertebrate neurons and their role in mmembrane excitability, in: *Advances in Neurology, Vol. 4, Basic Mechanisms of the Epilepsies* (A. Delgado-Escueta, A. A. Ward, D. M. Woodbury, and R. Porter, eds.), Raven Press, New York, pp. 137–170.

Agmon-Snir, H., 1995, A novel theoretical approach to the analysis of dendritic transients, *Biophys. J.* **69**:1633–1656.

Agmon-Snir, H., and Segev, I., 1993, Signal delay and input synchronization in passive dendritic structures, *J. Neurophsyiol.* **70**:2066–2085.

Alger, B. E., 1984, Hippocampus: Electrophysiological studies of epileptiform activity *in vitro*, in: *Brain Slices* (R. Dingledine, ed.), Plenum Press, New York, pp. 157–199.

Alger, B. E., and Nicoll, R. A., 1980, Epileptiform burst afterhyperpolarization: Calcium-dependent potassium potential in hippocampal CA1 pyramidal cells, *Science* **210**:1122–1124.

Alger, B. E., and Williamson, A., 1988, A transient calcium-dependent potassium component of the epileptiform burst after-hyperpolarization in rat hippocampus, *J. Physiol.* **399**:191–205.

Alger, B. E., Dhanjal, S. S., Dingledine, R., Garthwaite, J., Henderson, G., King, G. L., Lipton, P., North, A., Schwartzkroin, P. A., Sears, T. A., Segal, M., Whittingham, T. S., and Williams, J., 1984, Brain slice methods, in: *Brain Slices* (R. Dingledine, ed.), Plenum Press, New York, pp. 381–437.

Alzheimer, C., Schwindt, P. C., and Crill, W. E., 1993, Modal gating of Na^+ channels as a mechanism of persistent Na^+ current in pyramidal neurons from rat and cat sensorimotor cortex, *J. Neurosci.* **13**(2):660–673.

Andersen, P., Storm, J., and Wheal, H. V., 1986, Thresholds of action potentials evoked by synapses on the dendrites of pyramidal cells in the rat hippocampus *in vitro*, *J. Physiol.* **383**:509–526.

Anderson, W. R., Franck, J. E., Stahl, W. L., and Maki, A. A., 1994, Na, K-ATPase is decreased in hippocampus of kainate-lesioned rats, *Epilepsy Res.* **17**(3):221–231.

Andrade, R., and Nicoll, R. A., 1987, Pharmacologically distinct actions of serotonin on single pyramidal neurones of the rat hippocampus recorded *in vitro*, *J. Physiol.* **394**:99–124.

Andreasen, M., and Lambert, J. D. C., 1995, Regenerative properties of pyramidal cell dendrites in area CA1 of the rat hippocampus, *J. Physiol.* **483**:421–441.

Andreasen, M., and Lambert, J. D. C., 1998, Factors determining the efficacy of distal excitatory synapses in rat hippocampal CA1 pyramidal neurones, *J. Physiol.* **507**:441–462.

Armstrong, C. M., 1981, Sodium channels and gating currents, *Physiol. Rev.* **61**:644–683.

Avery, R. B., and Johnston, D., 1996, Multiple channel types contribute to the low-voltage-activated calcium current in hippocampal CA3 pyramidal neurons, *J. Neurosci.* **16**(18):5567–5582.

Azouz, R., Jensen, M. S., and Yaari, Y., 1996, Ionic basis of spike after-depolarization and burst generation in adult rat hippocampal CA1 pyramidal cells, *J. Physiol.* **492**:211–223.

Barnes, S., and Hille, B., 1989, Ionic channels of the inner segment of tiger salamander cone photoreceptors, *J. Gen. Physiol.* **94**:719–743.

Barrett, J. N., Magleby, K. L., and Pallotta, B. S., 1982, Properties of single calcium-activated potassium channels in cultured rat muscle, *J. Physiol.* **331**:211–230.

Benardo, L., Masukawa, L., and Prince, D., 1982, Electrophysiology of isolated hippocampal pyramidal dendrites, *J. Neurosci.* **2**(11):1614–1622.

Benardo, L. S., and Prince, D. A., 1982a, Cholinergic excitation of mammalian hippocampal pyramidal cells, *Brain Res.* **249**:315–331.

Benardo, L. S., and Prince, D. A., 1982b, Ionic mechanisms of cholinergic excitation in mammalian hippocampal pyramidal cells, *Brain Res.* **249**:333–344.

Bernander, O., Douglas, R. J., Martin, K. A. C., and Koch, C., 1991, Synaptic background activity influences spatiotemporal integration in single pyramidal cells, *Proc. Natl. Acad. Sci. USA* **88**:11569–11573.

Bhalla, U. S., and Bower, J. M., 1993, Exploring parameter space in detailed single neuron models: Simulations of the mitral and granule cells of the olfactory bulb, *J. Neurophysiol.* **69**(6):1948–1968.

Bhalla, U. S., Bilitch, D. H., and Bower, J. M., 1992, Rallpacks: A set of benchmarks for neuronal simulators, *Trends Neurosci.* **15**(11)453–458 [Available from ftp://genesis.bbb.caltech.edu/pub/genesis].

Bilkey, D. K., and Schwartzkroin, P. A., 1940, Variation in electrophysiology and morphology of hippocampal CA3 pyramidal cells, *Brain Res.* **514**:77–83.

Blatz, A. L., and Magleby, K. L., 1987, Calcium-activated potassium channels, *Trends Neurosci.* 463–467.

Borg-Graham, L., 1987a, Modelling the electrical behavior of cortical neurons—Simulations of hippocampal pyramidal cells, in: *Computer Simulation in Brain Science* (R. M. J. Cotterill, ed.), Cambridge University Press, Cambridge, pp. 384–404.

Borg-Graham, L., 1987b, Modelling the somatic electrical behavior of hippocampal pyramidal neurons, Master's thesis, Massachusetts Institute of Technology, Cambridge, MA [Available as MIT AI Laboratory Technical Report 1161 (1989)].

Borg-Graham, L., 1987c, Simulations suggest information processing roles for the diverse currents in hippocampal neurons, in: *Neural Information Processing Systems* (D. Z. Anderson, ed.), American Institute of Physics, New York, pp. 82–94.

Borg-Graham, L., 1991, Modelling the non-linear conductances of excitable membranes, in: *Cellular Neurobiology: A Practical Approach* (J. Chad and H. Wheal, eds.), IRL/Oxford University Press, Oxford, pp. 247–275.

Borg-Graham, L., 1995, The Surf-Hippo neuron simulation program [http://www.cnrs-gif.fr./iaf/iaf9/surf-hippo.html, v. 2.5].

Borg-Graham, L., Monier, C., and Frégnac, Y., 1996, Voltage-clamp measurements of visually-evoked conductances with whole-cell patch recordings in primary visual cortex, *J. Physiol.* (Paris) **90**:185–188.

Borg-Graham, L., Monier, C., and Frégnac, Y., 1998, Visual input evokes transient and strong shunting inhibition in visual cortical neurons, *Nature* **393**:369–373.

Bower, J. M., and Beeman, D., 1994, *The Book of GENESIS: Exploring Realistic Neural Models with the GEneral NEural SImulation System,* TELOS/Springer-Verlag, New York.

Bower, J. M., and Koch, C., 1992, Experimentalists and modelers: Can we all just get along? *Trends Neurosci.* **15**(11):458–461.

Brown, D., and Griffith, W., 1983a, Calcium activated outward current in voltage clamped hippocampal neurones of the guinea pig, *J. Physiol.* **337**:287–301.

Brown, D., and Griffith, W., 1983b, Persistent slow inward calcium current in voltage clamped hippocampal neurones of the guinea pig, *J. Physiol.* **337**:303–320.

Brown, D. A., Gähwiler, B. H., Griffith, W. H., and Halliwell, J. V., 1990, Membrane currents in hippocampal neurons, in: *Progress in Brain Research,* Vol. 83 (J. Storm-Mathisen, J. Zimmer, and O. P. Ottersen, eds.), Elsevier, Amsterdam, pp. 141–160.

Brown, T. H., Fricke, R. A., and Perkel, D. H., 1981, Passive electrical constants in three classes of hippocampal neurons, *J. Neurophysiol.* **46**(4):812–827.

Brown, T. H., Zador, A. M., Mainen, Z. F., and Claiborne, B. J., 1992, Hebbian computations in hippocampal dendrites and spines, in: *Single Neuron Computation* (T. McKenna, J. Davis, and S. F. Zornetzer, eds.), Academic Press, Boston, pp. 81–116.

Buckmaster, P. S., Strowbridge, B. W., and Schwartzkroin, P. A., 1993, A comparison of rat hippocampal mossy cells and CA3c pyramidal cells, *J. Neorophysiol.* **70**(4):1281–1299.

Burgi, P. -Y., and Gryzwacz, N. M., 1994, Model based on extracellular potassium for spontaneous synchronous activity in developing retina, *Neural Computation* **6**:983–1004.

Bush, P. C., and Sejnowski, T. J., 1991, Simulations of a reconstructed cerebellar purkinje cells based on simplified channel kinetics, *Neural Computation* **3**:321–332.

Bush, P. C., and Sejnowski, T. J., 1994, Effects of inhibition and dendritic saturation in simulated neocortical pyramidal cells, *J. Neurophysiol.* **71**(6):2183–2193.

Caeser, M., Brown, D. A., Gähwiler, B. H., and Knöpfel, T., 1993, Characterization of a calcium-dependent current generating a slow afterdepolarization of CA3 pyramidal cells in rat hippocampal slice cultures, *Eur. J. Neurosci.* **5**:560–569.

Campbell, L. W., Hao, S., Thibault, O., Blalock, E. M., and Landfield, P. W., 1996, Aging changes in voltage-gated calcium currents in hippocampal CA1 neurons, *J. Neurosci.* **16**:6286–6295.

Cantrell, A. R., Ma, J. Y., Scheuer, T., and Catterall, W. A., 1996, Muscarinic modulation of sodium current by activation of protein kinase C in rat hippocampal neurons, *Neuron* **16**:1019–1026.

Carnevale, N. T., Tsai, K. Y., and Claiborne, B. J., 1995, Qualitative electrotonic comparison of three classes of hippocampal neurons in the rat, in: *Neurobiology of Computation* (J. Bower, ed.), Kluwer, Boston, Vol. 4, pp. 67–72.

Charpak, S., Gähwiler, B. H., Do, K. Q., and Knöpfel, T., 1990, Potassium conductances in hippocampal neurons blocked by excitatory amino-acid transmitters, *Nature* **347**:765–767.

Christie, B. R., Eliot, L. S., Ito, K., Miyakawa, H., and Johnston, D., 1995, Different Ca^{2+} channels in soma and dendrites of hippocampal pyramidal neurons mediate spike-induced Ca^{2+} influx, *J. Neurophysiol.* **73**(6):2553–2557.

Church, J., 1992, A change from HCO_3^--CO_2- to HEPES-buffered medium modifies membrane properties of rat CA1 pyramdical neurones *in vitro*, *J. Physiol.* **455**:51–71.

Claiborne, B. J., Zador, A. M., Mainen, Z. F., and Brown, T. H., 1992, Computational models of hippocampal neurons, in *Single Neuron Computation* (T. McKenna, J. Davis, and S. F. Zornetzer, eds.), Academic Press, Boston, pp. 61–80.

Colbert, C. M., and Johnston, D., 1996, Axonal action-potential initiation and Na+ channel densities in the soma and axon initial segment of subicular pyramidal neurons, *J. Neurosci.* **16**(21):6676–6686.

Colbert, C. M., and Johnston, D., 1998, Protein kinase C activation decreases activity-dependent attenuation of dendritic Na+ current in hippocampal CA1 pyramidal neurons, *J. Neurosci.* **79**(1):491–495.

Colbert, C. M., Magee, J. C., Hoffman, D. A., and Johnston, D., 1997, Slow recovery from inactivation of Na^+ channels underlies the activity-dependent attentuation of dendritic action potentials in hippocampal CA1 pyramidal neurons, *J. Neurosci.* **17**(17):6512–6521.

Colino, A., and Halliwell, J. V., 1987, Differential modulation of three separate K-conductances in hippocampal CA1 neurons by serotonin, *Nature* **328**:73–77.

Colino, A., and Halliwell, J. V., 1993, Carbachol potentiates Q current and activates a calcium-dependent non-specific conductance in rat hippocampus *in-vitro*, *Eur. J. Neurosci.* **5**:1198–1209.

Colino, A., Garcia-Seoane, J. J., and Valentin, A., 1998, Action potential braodening induced by lithium may cause a presynaptic enhancement of excitatory synaptic transmission in neonatal rat hippocampus, *Eur. J. Neurosci.* **10**:2433–2443.

Colquhoun, D., and Hawkes, A. G., 1995, The principles of the stochastic interpretation of ion-channel mechanism, in: *Single-Channel Recording*, 2nd ed. (B. Sakmann and E. Neher, eds.), Plenum Press, New York, pp. 397–482.

Connors, B. W., and Gutnick, M. J., 1990, Intrinsic firing patterns of diverse neocortical neurons, *Trends Neurosci.* **13**(3):99–104.

Cooley, J., and Dodge, F., 1966, Digital computer solutions for excitation and propagation of the nerve impulse, *Biophys. J.* **6**:583–589.

Crépel, V., Aniksztejn, L., Ben-Ari, Y., and Hammond, C., 1994, Glutamate metabotropic receptors increase a Ca^{2+}-activated nonspecific cationic current in CA1 hippocampal neurons, *J. Neurophysiol.* **72**(4):1561–1569.

Crill, W. E., 1996, Persistent sodium current in mammalian central neruons, *Annu. Rev. Physiol.* **58**:349–362.

Cummins, T. R., Xia, Y., and Haddad, G. G., 1994, Functional properties of rat and human neocortical voltage-sensitive sodium currents, *J. Neurophysiol.* **71**(3):1052–1064.

Debanne, D., Guérineau, N. C., Gähwiler, B. H., and Thompson, S. M., 1997, Action potential propagation gated by an axonal I_A-like K^+ conductance in hippocampus, *Nature* **389**:286–289.

De Schutter, E. D., 1992, A consumer guide to neuronal modeling software, *Trends Neurosci.* **15**(11):462–464.

De Schutter, E., and Bower, J. M., 1994, An active membrane model of the cerebellar Purkinje cell I. Simulation of current clamps in slice, *J. Neurophysiol.* **71**(3):375–400.

Deisz, R. A., 1996, A tetrodotoxin-insensitive sodium current initiates burst firing of neocortical neurons, *Neuroscience* **70**(2):341–351.

Destexhe, A., Mainen, Z. F., and Sejnowski, T., 1994, Synthesis of models for excitable membranes, synaptic transmission and neuromodulation using a common kinetic formalism, *J. Comput. Neurosci.* **1**:195–230.

Destexhe, A., Contreras, D., Steriade, M. D., Sejnowski, T., and Huguenard, J., 1995, *In vivo, in vitro* and computational analysis of dendritic calcium currents in thalamic reticular neurons, *J. Neurosci.* **16**(1):169–185.

Deuchars, J., and Thomson, A. M., 1996, CA1 pyramid–pyramid connections in rat hippocampus *in-vitro:* Dual intracellular recordings with biocytin filling, *Neuroscience* **74**(4):1009–1018.

Doerner, D., Abdel-Latif, M., Rogers, T. B., and Alger, B. E., 1990, Protein kinase C-dependent and -independent effects of phorbol esters on hippocampal calcium channel current, *J. Neurosci.* **10**(5):1699–1706.

Eder, C., Klee, R., and Heinemann, U., 1996, Modulation of A-currents by $[K^+]_o$ in acutely isolated pyramidal neurones of juvenile rat entorhinal cortex and hippocampus, *Neuroreport* **7**:1565–1568.

el-Mallakh, R. S., and Wyatt, R. J., 1995, The Na, K-ATPase hypothesis for bipolar illness, *Biol. Psychiatry* **37**(4):235–244.

Evans, J. D., Kember, G. C., and Major, G., 1992, Techniques for obtaining analytical solutions to the multicylinder somatic shunt cable model for passive neurones, *Biophys. J.* **63**:350–365.

Ficker, E., and Heinemann, U., 1992, Slow and fast transient potassium currents in cultured rat hippocampal cells, *J. Physiol.* **445**:431–445.

Figenschou, A., Hu, G.-Y., and Storm, J. F., 1996, Cholinergic modulation of the action potential in rat hippocampal neurons, *Eur. J.Neurosci.* **8**:211–219.

Fisher, R. E., Gray, R., and Johnston, D., 1990, Properties and distribution of single voltage-gated calcium channels in adult hippocampal neurons, *J. Neurophysiol.* **64**(1):91–104.

Foster, W. R., Ungar, L. H., and Schwaber, J. S., 1993, Significance of conductances in Hodgkin–Huxley models, *J. Neurophysiol.* **70**(6):2502–2518.

Fraser, D. D., and MacVicar, B. A., 1996, Cholinergic-dependent plateau potential in hippocampal CA1 pyramidal neurons, *J. Neurosci.* **16**(13):4113–4128.

French, C., and Gage, P., 1985, A threshold sodium current in pyramidal cells in rat hippocampus, *Neurosci. Lett.* **56**:289–293.

French, C., Sah, P., Buckett, K. J., and Gage, P., 1990, A voltage-dependent persistent sodium current in mammalian hippocampal neurons, *J. Gen. Physiol.* **95**:1139–1157.

Fukuda, A., and Prince, D. A., 1992, Postnatal development of electrogenic sodium pump activity in rat hippocampal pyramidal neurons, *Dev. Brain Res.* **65**:101–114.

Galdzicki, Z., Coan, E., and Rapoport, S. I., 1993, Cultured hippocampal neurons from trisomy 16 mouse, a model for Down's syndrome, have an abnormal action potential due to a reduced inward sodium current, *Brain Res.* **604**:69–78.

Glynn, I. M., 1993, All hands to the sodium pump, *J. Physiol.* **462**:1–30.

Goldstein, S. A. N., and Colatsky, T. J., 1996, Ion channels: Too complex for rational drug design? *Neuron* **16**:913–919.

Grove, E. A., and Halliwell, J. V., 1990, Two inward rectifier currents revealed in CA1 neurones of the rat hippocampus *in vitro, J. Physiol.* **424**:46P.

Haas, H. L., Jefferys, J. G. R., Slater, N. T., and Carpenter, D. O., 1984, Modulation of low calcium induced field bursts in the hippocampus by monoamines and cholinomimetics, *Eur. J. Physiol.* **400**:28–33.

Hagiwara, S., and Byerly, L., 1981, Calcium channel, *Annu. Rev. Neurosci.* **4**:69–125.

Haglund, M. M., and Schwartzkroin, P. A., 1990, Role of Na–K pump potassium regulation and IPSPs in seizures and spreading depression in immature rabbit hippocampal slices, *J. Neurophysiol.* **63**(2):225–239.

Halliwell, J., and Adams, P., 1982, Voltage clamp analysis of muscarinic excitation in hippocampal neurons, *Brain Res.* **250**:71–92.

Halliwell, J. V., Othman, I. B., Pelchen-Matthews, A., and Dolly, J. O., 1986, Central action of dendrotoxin: Selective reduction of a transient K conductance in hippocampus and binding to localized receptors, *Proc. Natl. Acad. Sci. USA* **83**:493–497.

Harris, K. M., Jensen, F. E., and Tsao, B., 1992, Three-dimensional structure of dendritic spines and synapses in rat hippocampus (CA1) at postnatal day 15 and adult ages: Implications for the maturation of synaptic physiology and long-term potentiation, *J. Neurosci.* **12**:2685–2705.

Henze, D. A., Cameron, W. E., and Barrionuevo, G., 1996, Dendritic morphology and its effects on the amplitude and rise-time of synaptic signals in hippocampal CA3 pyramidal cells, *J. Comp. Neurol.* **369**:331–344.

Hess, P., and Tsien, R., 1984, Mechanism of ion permeation through calcium channels, *Nature* **3**:453–456.

Hille, B., 1992, *Ionic Channels of Excitable Membranes*, 2nd ed., Sinauer, Sunderland, MA.

Hines, M., 1984, Efficient computation of branched nerve equations, *Int. J. Bio-Med. Computing* **15**:69–76.

Hines, M., 1992, NEURON—A program for simulation of nerve equations, in: *Neural Systems: Analysis and Modeling* (F. Eeckman, ed.), Kluwer, Boston, Vol. 2, pp. 127–136 [http://www.nnc.yale.edu/HTML/YALE/NNC/neuron/neuron.html].

Hines, M., and Carnevale, N. T., 1994, Computer simulation methods for neurons, in: *The Handbook of Brain Theory and Neural Networks* (M. Arbib, ed.), MIT Press, Cambridge, pp. 226–230.

Hodgkin, A. L., 1976, Chance and design in electrophysiology: An informal account of certain experiments on nerve carried out between 1934 and 1952, *J. Physiol.* **263**:1–21.

Hodgkin, A. L., and Huxley, A. F., 1952a, Currents carried by sodium and potassium ions through the membrane of the giant axon of *Loligo*, *J. Physiol.* **116**:449–472.

Hodgkin, A. L., and Huxley, A. F., 1952b, A quantitative description of membrane current and its application to conduction and excitation in nerve, *J. Physiol.* **117**:500–544.

Hoehn, K., Watson, T. W. J., and MacVicar, B. A., 1993, A novel tetrodotoxin-insensitive, slow sodium current in striatal and hippocampal neurons, *Neuron* **10**:543–552.

Hoffman, D. A., and Johnston, D., 1998, Downregulation of transient K^+ channels in dendrites of hippocampal CA1 pyramidal neurons by activation of PKA and PKC, *J. Neurosci.* **18**(10):3521–3528.

Hoffman, D. A., Magee, J. C., Colbert, C. M., and Johnston, D., 1997, K^+ channel regulation of signal propagation in dendrites of hippocampal pyrmidal neurons, *Nature* **87**:869–875.

Holmes, W. R., and Levy, W. B., 1990, Insights into associative long-term potentiation from computational models of NMDA receptor-mediated calcium influx and intracellular calcium concentration changes, *J. Neurophysiol.* **63**(5):1148–1168.

Holmes, W. R., and Rall, W., 1992a, Electrotonic length estimates in neurons with dendritic tapering or somatic shunt, *J. Neurophysiol.* **68**(4):1421–1437.

Holmes, W. R., and Rall, W., 1992b, Electrotonic models of neuronal dendrites and single neuron computation: in: *Single Neuron Computation* (T. McKenna, J. Davis, and S. F. Zornetzer, eds.), Academic Press, Boston, pp. 7–25.

Holmes, W. R., and Rall, 1992c, Estimating the electrotonic structure of neurons with compartmental models, *J. Neurophysiol.* **68**(4):1438–1452.

Holmes, W. R., Segev, I., and Rall, W., 1992, Interpretation of time constant and electrotonic length estimates in multicylinder or branched neuronal structures, *J. Neurophysiol.* **68**(4):1401–1420.

Hotson, J. R., and Prince, D. A., 1980, A calcium-activated hyperpolarization follows repetitive firing in hippocampal neurons, *J. Neurophysiol.* **43**(2):409–419.

Hotson, J. R., Prince, D., and Schwartzkroin, P., 1979, Anomalous inward rectification in hippocampal neurons, *J. Neurophysiol.* **42**:889–895.

Huguenard, J. R., and McCormick, D. A., 1992, Simulation of the currents involved in rhythmic oscillations in thalamic relay neurons, *J. Neurophysiol.* **68**(4):1373–1400.

Huguenard, J. R., Hamill, O. P., and Prince, D. A., 1988, Developmental changes in Na^+ conductances in rat neocortical neurons: Appearance of a slowly inactivating component, *J. Neurophysiol.* **59**(3):778–795.

Ikemoto, Y., Ono, K., Yoshida, A., and Akaike, N., 1989, Delayed activation of large-conductance Ca^{2+}-activated K channels in hippocampal neurons of the rat, *Biophys. J.* **56**:207–212.

Jack, J. J. B., Noble, D., and Tsien, R. W., 1983, *Electric Current Flow in Excitable Cells*, Clarendon Press, Oxford.

Jaffe, D. B., Johnston, D., Lasser-Ross, N., Lisman, J. E., Miyakawa, H., and Ross, W. N., 1992, The spread of Na^+ spikes determines the pattern of dendritic Ca^{++} entry into hippocampal neurons, *Nature* **357**:244–246.

Jaffe, D. B., Ross, W. N., Lisman, J. E., Lasser-Ross, N., Miyakawa, H., and Johnston, D., 1994, A model for dendritic Ca^{2+} accumulation in hippocampal pyramidal neurons based on fluorescence imaging measurements, *J. Neurophysiol.* **71**(3):1065–1077.

Jensen, M. S., Azouz, R., and Yaari, Y., 1994, Variant firing patterns in rat hippocampal pyramidal cells modulated by extracelluar potassium, *J. Neurophysiol.* **71**(3):831–839.

Jensen, M. S., Azouz, R., and Yaari, Y., 1996, Spike after-depolarization and burst generation in adult rat hippocamplal CA1 pyramidal cells, *J. Physiol.* **492**:199–210.

Jensen, O, Idiart, M. A., and Lisman, J. E., 1996, Physiologically realistic formation of autoassociative memory in networks with theta/gamma oscillations: Role of fast NMDA channels, *Learning Memory* **3**:243–256.

Johnston, A. R., Black, C., Fraser, J., and MacLeod, N., 1997, Scrapie infection alters the membrane and synaptic properties of mouse hippocampal CA1 pyramidal neurones, *J. Physiol.* **500**(1):1–15.

Johnston, D., and Brown, T., 1983, Interpretation of voltage-clamp measurements in hippocampal neurons, *J. Neurophysiol.* **50**(2)**2**:464–486.

Johnston, D., Hablitz, J., and Wilson, W., 1980, Voltage clamp discloses slow inward current in hippocampal burst-firing neurones, *Nature* **286**:391–393.

Johnston, D., Magee, J. C., Colbert, C. M., and Christie, B. R., 1996, Active properties of neuronal dendrites, *Annu. Rev. Neurosci.* **19**:165–186.

Jones, S. W., 1989, On the resting potential of isolated frog sympathetic neurons, *Neuron* **3**:153–161.

Jung, H. -Y., Mickus, T., and Spruston, N., 1997, Prolonged sodium channel inactivation contributes to dendritic action potential attenuation in hippocampal pyramidal neurons, *J. Neurosci.* **17**(17):6639–6646.

Kaczmarek, L. K., and Levitan, I. B., eds., 1987, *Neuromodulation: The Biochemical Control of Neuronal Excitability*, Oxford University Press, Oxford.

Kandel, E. R., and Spencer, W. A., 1961, Electrophysiology of hippocampal neurons: IV. Fast prepotentials, *J. Neurophysiol.* **24**:272–285.

Katz, B., 1949, Les constantes electriques de la membrane du muscle, *Arch. Sci. Physiol.* **3**:285–299.

Kay, A. R., 1991, Inactivation kinetics of calcium current of acutely dissociated CA1 pyramidal cells of the mature guinea-pig hippocampus, *J. Physiol.* **437**:27–48.

Kay, A. R., and Wong, R. K. S., 1987, Calcium current activation kinetics in isolated pyramidal neurones of the CA1 region of the mature guinea-pig hippocampus, *J. Physiol.* **392**:603–616.

Kettenmann, H., and Grantyn, R., eds., 1992, *Practical Electrophysiological Methods*, Wiley-Liss, New York.

Knöpfel, T., Brown, D. A., Vranesic, I., and Gähwiler, B. H., 1989, Depression of Ca^{2+} activated potassium conductance by muscarine and isoproternenol without alteration of depolarization-induced transient rise in cytosolic free Ca^{2+} in hippocampal CA3 pyramidal cells, *Eur. J. Neurosci. Suppl.* **2**.

Koch, C., Poggio, T., and Torre, V., 1983, Nonlinear interactions in a dendritic tree: Localization, timing, and role in information processing, *Proc. Natl. Acad. Sci. USA* **80**:2799–2802.

Köhler, M., Hirschberg, B., Bond, C. T., Kinzie, J. M., Marrion, N. V., Maylie, J., and Adelman, J. P., 1996, Small-conductance, calcium-activated potassium channels from mammalian brain, *Science* **273**:1709–1714.

Konnerth, A., Lux, H. D., and Heinemann, U., 1986, Ionic properties of burst generation in hippocampal pyramidal cells, in: *Calcium Electrogenesis and Neuronal Functioning* (U. Heinemann, M. Klee, E. Neher, and W. Singer, eds.), Springer-Verlag, Berlin, pp. 368–374.

Kuo, C. C., and Bean, P. B., 1994, Na^+ channels must deactivate to recover from inactivation, *Neuron* **12**:819–829.

Lancaster, B., and Adams, P., 1986, Calcium dependent current generating the afterhyperpolarization of hippocampal neurons, *J. Neurophysiol.* **55**:1268–1282.

Lancaster, B., and Nicoll, R. A., 1987, Properties of two calcium-activated hyperpolarizations in rat hippocampal neurones, *J. Physiol.* **389**:187–203.

Lancaster, B., and Zucker, R. S., 1994, Photolytic manipulation of Ca^{2+} and the time course of slow, Ca^{2+}-activated K^+ current in rat hippocampal neurones, *J. Physiol.* **475**(2):229–239.

Lancaster, B., Nicoll, R. A., and Perkel, D. J., 1991, Calcium activates two types of potassium channels in rat hippocampal neurons in culture, *J. Neurosci.* **11**(23):23–30.

Landfield, P. W., and Pitler, T. A., 1984, Prolongation of calcium-dependent afterhyperpolarizations in hippocampal neurons in aged rats, *Science* **226**:1089–1091.

Lanthorn, T., Storm, J., and Anderson, P., 1984, Current to frequency transduction in CA1 hippocampal pyramidal cells: Slow prepotentials dominate the primary range firing, *Exp. Brain Res.* **53**:431–443.

Läuger, P., 1991, *Electrogenic Ion Pumps*, Sinauer, Sunderland, MA.

Lipowsky, R., Gillessen, T., and Alzheimer, C., 1996, Dendritic Na^+ channels amplify EPSPs in hippocampal CA1 pyramidal cells, *J. Neurophysiol.* **76**(4):2181–2191.

Lisman, J., 1996, Bursts as a unit of neural information: Making unreliable synapses reliable, *Trends Neurosci.* **20**(1):38–43.

Llinás, R., Sugimori, M., and Silver, R. B., 1995, Time resolved calcium microdomains and synaptic transmission, *J. Physiol.* (Paris) **89**:77–81.

Lüthi, A., Gähwiler, B. H., and Gerber, U., 1996, A slowly inactivating potassium current in CA3 pyramidal cells of rat hippocampus *in-vitro*, *J. Neurosci.* **16**(2):586–594.

Lytton, W. W., and Sejnowski, T. J., 1991, Simulations of cortical pyramidal neurons synchronized by inhibitory interneurons, *J. Neurosci.* **66**(3):1059–1079.

Maccaferri, G., Mangoni, M., Lazzari, A., and DiFrancesco, D., 1993, Properties of the hyperpolarization-activated current in rat hippocampal CA1 cells, *J. Neurophysiol.* **69**(6):2129–2136.

MacVicar, B. A., 1985, Depolarizing prepotential are Na^+ dependent in CA1 pyramidal neurons, *Brain Res.* **333**:378–381.

Madison, D. V., and Nicoll, R. A., 1984, Control of repetitive discharge of rat CA1 pyramidal neurones *in vitro*, *J. Physiol.* **354**:319–331.

Madison, D. V., Malenka, R. C., and Nicoll, R. A., 1986, Phorbol esters block a voltage-sensitive chloride current in hippocampal pyramidal cells, *Nature* **321**:695–697.

Madison, D. V., Fox, A. P., and Tsien, R. W., 1987a, Adenosine reduces an inactivating component of calcium current in hippocampal CA3 neurons, *Biophys. J.* **51**:30a.

Madison, D. V., Lancaster, B., and Nicoll, R. A., 1987b, Voltage clamp analysis of cholinergic action in the hippocampus, *J. Neurosci.* **7**(3):733–741.

Magee, J. C., and Johnson, D., 1995, Characterization of single voltage-gated Na^+ and Ca^{2+} channels in apical dendrites of rat CA1 pyramidal neurons, *J. Physiol.* **487**(1):67–90.

Magee, J. C., and Johnston, D., 1995b, Synaptic activation of voltage-gated channels in the dendrites of hippocampal pyramidal neurons, *Science*, **268**:301–304.

Magee, J. C., Avery, R. B., Christie, B. R., and Johnston, D., 1996, Dihydropyridine-sensitive, voltage-gated Ca^{2+} channels contribute to the resting intracellular Ca^{2+} concentration of hippocampal CA1 pyramidal neurons, *J. Neurophysiol.* **76**(5):3460–3470.

Magee, J., Hoffman, D., Colbert, C., and Johnston, D, 1998, Electrical and calcium signaling in dendrites of hippocampal pyramidal neurons, in: *Ann. Rev. Physiol.* **60**:327–346.

Mainen, Z. F., and Sejnowski, T. J., 1996, Influence of dendritic structure on firing patterns in model neocortical neurons, *Nature* **382**:363–366.

Mainen, Z. F., Joerges, J., Huguenard, J. R., and Sejnowski, 1995, A model of spike initiation in neocortical pyramidal neurons, *Neuron* **15**:1427–1439.

Mainen, Z. F., Carnevale, N. T., Zador, A. M., Claiborne, B. J., and Brown, T. H., 1996, Electrotonic architecure of hippocampal CA1 pyramidal neurons based on three-dimensional reconstructions, *J. Neurophysiol.* **76**(3):1904–1923.

Major, G., Larkman, A. U., Jonas, P., Sakmann, B., and Jack, J. J. B., 1994, Detailed passive cable models of whole-cell recorded CA3 pyramidal neurons in rat hippocampal slices, *J. Neurosci.* **14**(8):4613–4638.

Malenka, R. C., Lancaster, B., and Zucker, R. S., 1992, Temporal limits on the rise in postsynaptic calcium required for the induction of long-term potentiation, *Neuron* **9**:121–128.

Marty, A., and Neher, E., 1995, Tight-seal whole-cell recording, in: *Single-Channel Recording*, 2nd ed. (B. Sakmann and E. Neher, eds.), Plenum Press, New York, pp. 31–51.

Mascagni, M., and Sherman, A. S., 1998, Numerical methods for neuronal modeling, in: *Methods in Neuronal Modeling* (C. Koch and I. Segev, eds.), MIT Press/Bradford Books, Cambridge, MA, Vol. 2.

Masukawa, L. M., and Hansen, A., 1987, Regional distribution of voltage-dependent single channel conductances in cultured hippocampal neurons from the rat, *Neurosci. Abstr.* **13**:1442.

Masukawa, L., and Prince, D., 1984, Synaptic control of excitability in isolated dendrites of hippocampal neurons, *J. Neurosci.* **4**(1):217–227.

Masukawa, L. M., Benardo, L. S., and Prince, D. A., 1982, Variations in electrophysiological properties of hippocampal neurons in different subfields, *Brain Res.* **242**:341–344.

Masukawa, L. M., Hansen, A. J., and Shepard, G., 1991, Distribution of single-channel conductances in cultured rat hippocampal neurons, *Cell. Mol. Neurobiol.* **11**(2):231–243.

McCormick, D. A., and Pape, H. C., 1990, Properties of a hyperpolarization-activated cation current and its role in rhythmic oscillation in thalamic relay neurones, *J. Physiol.* **431**:291–318.

McKenna, T., Davis, J., and Zornetzer, S. F., eds., 1992, *Single Neuron Computation,* Academic Press, Boston.

McManus, O. B., and Magleby, K. L., 1989, Kinetic time constants independent of previous single-channel activity suggest Markov gating for a large-conductance Ca-activated K channel, *J. Gen. Physiol.* **94**:1037–1070.

Meier, C. L., and Dudek, F. E., 1996, Spontaneous and stimulation-induced synchronized burst after-discharges in the isolated CA1 of kainate-treated rats, *J. Neurophysiol.* **76**(4):2231–2239.

Mel, B., 1994, Information processing in dendritic trees, *Neural Computation* **6**:1031–1085.

Migliore, M., Cook, E. P., Jaffe, D. B., Turner, D. A., and Johnston, D., 1995, Computer simulations of morphologically reconstructed CA3 hippocampal neurons, *J. Neurophysiol.* **73**(3):1157–1168.

Miyakawa, H., Ross, W. N., Jaffe, D., Callaway, J. C., Lasser-Ross, N., Lisman, J. E., and Johnston, D., 1992, Synaptically activated increases in Ca^{2+} concentration in hippocampal CA1 pyramidal cells are primarily due to voltage-gated Ca^{2+} channels, *Neuron* **9**:1163–1173.

Moczydlowski, E., and Latorre, R., 1983, Gating kinetics of Ca^{2+}-activated K^{+} channels from rat muscle incorporated into planar lipid bilayers, *J. Gen. Physiol.* **82**:511–542.

Moore, J., Stockbridge, N., and Westerfield, M., 1983, On the site of impulse initiation in a neurone, *J. Physiol.* **336**:301–311.

Müller, W., and Lux, H. D., 1993, Analysis of voltage-dependent membrane currents in spatially extended neurons from point-clamp data, *J. Neurophysiol.* **69**(1):241–247.

Nakajima, Y., Nakajima, S., Leonard, R. J., and Yamaguchi, K., 1986, Acetylcholine raises excitability by inhibiting the fast transient potassium current in cultured hippocampal neurons, *Proc. Natl. Acad. Sci. USA* **83**:3022–3026.

Nowycky, M., Fox, A., and Tsien, R., 1985, Three types of neuronal calcium channel with different calcium agonist sensitivity, *Nature* **316**:440–443.

Numann, R. E., Wadman, W. J., and Wong, R. K. S., 1987, Outward currents in single hippocampal cells obtained from the adult guinea-pig, *J. Physiol.* **393**:331–353.

Ogata, N., and Tatebayashi, H., 1990, Sodium current kinetics in freshly isolated neostriatal neurones of the adult guinea pig, *Eur. J. Physiol.* **416**:594–603.

Okuhara, D. Y., and Beck, S. G., 1994, 5-HT_{1A} receptor linked to inward-rectifying potassium current in hippocampal CA3 pyramidal cells, *J. Neurophysiol.* **71**(6):2161–2167.

Owen, D. G., 1987, Three types of inward rectifier currents in cultured hippocampal neurones, *Neurosci. Lett. Suppl.* **29**:S18.

Pallotta, B. S., and Wagoner, P. K., 1992, Voltage-dependent potassium channels since Hodgkin and Huxley, *Physiol. Rev. Suppl.* **72**(4):S49–S67.

Pape, H. C., 1996, Queer current and pacemaker: The hyperpolarization-activated cation current in neurons, *Annu. Rev. of Physiol.* **58**:299–327.

Patlak, J., 1991, Molecular kinetics of voltage-dependent Na^{+} channels, *Physiol. Rev.* **71**(4):299–327.

Pedarzani, P., and Storm, J., 1993, PKA mediates the effects of monomamine transmitters on the K^{+} current underlying the slow spike frequency adaptation in hippocampal neurons, *Neuron* **11**:1–20.

Pedarzani, P., and Storm, J., 1995a, Dopamine modulates the slow Ca^{2+}-activated current I_{AHP} via cyclic AMP-dependent protein kinase in hippocampal neurons, *J. Neurophysiol.* **74**(6):2749–2753.

Pedarzani, P., and Storm, J., 1995b, Protein kinase A-independent modulation of ion channels in the brain by cyclic AMP, *Proc. Natl. Acad. Sci. USA* **92**:11716–11720.

Pedarzani, P., and Storm, J. F., 1996, Interaction between α- and β-adrenergic receptor agonists modulating the slow Ca^{2+}-activated K^{+} current I_{AHP} in hippocampal neurons, *Eur. J. Neurosci.* **8**:2098–2110.

Perkins, K. L., and Wong, R. K. S., 1995, Intracellular QX-314 blocks the hyperpolarization-activated inward current I_q in hippocampal CA1 pyramidal cells, *J. Neurophysiol.* **73**(2):911–915.

Pongracz, F., Poolos, N. P., Kocsis, J. D., and Shepherd, G. M., 1992, A model of NMDA receptor-mediated activity in dendrites of hippocampal CA1 pyramidal neurons, *J. Neurophysiol.* **68**(6):2248–2259.

Pyapali, G. K., and Turner, D. A., 1994, Denervation-induced dendritic alterations in CA1 pyramidal cells following kainic acid hippocampal lesions in rats, *Brain Res.* **652**:279–290.

Qian, N., and Sejnowski, T. J., 1989, An electro-diffusion model for branching dendrites, spines and axons, *Biol. Cybernet.* **62**:1–15.

Qian, N., and Sejnowski, T. J., 1990, When is an inhibitory synapse effective? *Proc. Natl. Acad. Sci.* **87**:8145–8149.

Quadroni, R., and Knöpfel, T., 1994, Compartmental models of Type A and Type B guinea pig medial vestibular neurons, *J. Neurophysiol.* **72**(4):1911–1924.

Rall, W., 1989, Cable theory for dendritic neurons, in: *Methods in Neuronal Modeling* (C. Koch and I. Segev, eds.), MIT Press/Bradford Books, Cambridge, MA, pp. 9–62.

Rall, W., Burke, R. E., Holmes, W. R., Jack, J. J. B., Redman, S. J., and Segev, I., 1992, Matching dendritic neuron models to experimental data, *Physiol. Rev. Suppl.* **72**(4):S159–S186.

Rapp, M., Yarom, Y., and Segev, I., 1996, Modeling back propagating action potential in weakly excitable dendrites of neocortical pyramidal cells, *proc. Natl. Acad. Sci. USA* **93**:11985–11990.

Reid, K. H., Edmonds, H. L., Schurr, A., Tseng, M. T., and West, C. A., 1988, Pitfalls in the use of brain slices, *Prog. Neurobiol.* **31**(1):1–18.

Regehr, W. G., and Tank, D. W., 1992, Calcium concentration dynamics produced by synaptic activation of CA1 hippocampal pyramidal cells, *J. Neurosci.* **12**(11):4202–4223.

Rhodes, P. A., and Gray, C. M., 1994, Simulations of intrinsically bursting neocortical pyramidal neurons, *Neural Computation* **6**:1086–1110.

Roberts, W. M., Jacobs, R. A., and Hudspeth, A. J., 1990, Colocalization of ion channels involved in frequency selectivity and synaptic transmission at presynaptic active zones of hair cells, *J. Neurosci.* **10**(11):3664–3684.

Robitaille, R., Garcia, M. L., Kaczorowski, G. J., and Charlton, M. P., 1993, Functional colocalization of calcium and calcium-gated potassium channels in control of transmitter release, *Neuron* **11**:645–655.

Roper, S. N., Obenaus, A., and Dudek, F. E., 1993, Increased propensity for nonsynaptic epileptiform activity in immature rat hippocampus and dentate gyrus, *J. Neurophysiol.* **70**(2):857–862.

Rose, C. R., and Ransom, B. R., 1997, Regulation of intracellular sodium in cultured rat hippocampal neurones, *J. Physiol.* **499**(3):573–587.

Rudy, B., 1988, Diversity and ubiquity of K channels, *Neuroscience* **25**(3):729–749.

Sagar, A., and Rakowski, R. F., 1994, Access channel model for the voltage dependence of the forward-running Na^{+}/K^{+} pump, *J. Gen. Physiol.* **103**(5):869–893.

Sah, P., 1996, Ca^{2+}-activated K^{+} currents in neurones: Types, physiological roles and modulation, *Trends Neurosci.* **19**(4):150–154.

Sah, P., and Bekkers, J. M., 1996, Apical dendritic location of slow afterhyperpolarization current in hippocampal pyramidal neurons: Implications for the integration of long-term potentiation, *J. Neurosci.* **16**(15):4537–4542.

Sah, P., Gibb, A. J., and French, C., 1988a, Potassium current activated by depolarization of dissociated neurons from adult guinea pig hippocampus, *J. Gen. Physiol.* **92**:263–278.

Sah, P., Gibb, A. J., and Gage, P. W., 1988b, The sodium current underlying action potentials in guinea pig hippocampal CA1 neurons, *J. Gen. Physiol.* **91**:373–398.

Sakmann, B., and Neher, E., eds., 1995, *Single-Channel Recording,* 2nd ed., Plenum Press, New York.

Scharfman, H. E., 1993, Spiny neurons of area CA3c in rat hippocampal slices have similar electrophysiological characteristics and synaptic responses despite morphological variation, *Hippocampus* **3**(1):9–28.

Schiegg, A., Gerstner, W., Ritz, R., and Leo van Hammen, J., 1995, Intracellular Ca^{2+} stores can account for the time course of LTP induction: A model of Ca^{2+} dynamics in dendritic spines, *J. Neurophysiol.* **74**(3):1046–1055.

Schiller, J., Helmchen, F., and Sakmann, B., 1995, Spatial profile of dendritic calcium transients evoked by action potentials in rat neocortical pyramidal neurones, *J. Physiol.* **487**(3):583–600.

Schwartzkroin, P., 1977, Further characteristics of hippocampal CA1 cells *in vitro, Brain Res.* **128**:53–68.

Schwartzkroin, P. A., and M. Slawsky, 1977, Probable calcium spikes in hippocampal neurons, *Brain Res.* **135**:157–161.

Segal, M., and Barker, J., 1984, Rat hippocampal neurons in culture: Potassium conductances, *J. Neurophysiol.* **51**(6):1409–1433.

Segal, M., and Barker, J., 1986, Rat hippocampal neurons in culture: Calcium and calcium-dependent potassium conductances, *J. Neurophysiol.* **55**(4):751–766.

Segal, M., Rogawski, M., and Barker, J., 1984, A transient potassium conductance regulates the excitability of cultured hippocampal and spinal neurons, *J. Neurosci.* **4**(2):604–609.

Segev, I., 1992, Single neuron models: Oversimple, complex and reduced, *Trends Neurosci.* **15**(11):414–421.

Segev, I., Rapp, M., Manor, Y., and Yarom, Y., 1992, Analog and digital processing in single nerve cells: Dendritic integration and axonal propagation, in: *Single Neuron Computation* (T. McKenna, J. Davis, and S. F. Zornetzer, eds.), Academic Press, Boston, pp. 173–198.

Segev, I., Burke, R. E., and Hines, M., 1998, Compartmental models of complex neurons, in: *Methods in Neuronal Modeling* (C. Koch and I. Segev, eds.), MIT Press/Bradford Books, Cambridge, MA, pp. 63–96.

Sherman, A., Keizer, J., and Rinzel, J., 1990, Domain model for Ca^{2+}-inactivation of Ca^{2+} channels at low channel density, *Biophys. J.* **58**:985–995.

Sherman-Gold, R., ed., 1993, *The Axon Guide,* Axon Instruments, Foster City, CA.

Sokal, R. R., and Rohlf, F. J., 1981, *Biometry: The Principles and Practice of Statistics in Biological Research,* Freeman, San Francisco.

Spigelman, I., Zhang, L., and Carlen, P. L., 1992, Patch-clamp study of postnatal development of CA1 neurons in rat hippocampal slices: Membrane excitability and K^+ currents, *J. Neurophysiol.* **68**(1):55–69.

Spruston, N., and Johnston, D., 1992, Perforated patch-clamp analysis of the passive membrane properties of three classes of hipppocampal neurons, *J. Neurophysiol.* **67**(3):508–529.

Spruston, N., Jaffe, D. B., Williams, S. H., and Johnston, D., 1993, Voltage- and space-clamp errors associated with the measurement of electrotonically remote synaptic events, *J. Neurophysiol.* **70**(2):781–802.

Spruston, N., Schiller, Y., Stuart, G., and Sakmann, B., 1995, Activity-dependent action potential invasion and calcium influx into hippocampal CA1 dendrites, *Science* **268**:297–300.

Staley, K. J., Otis, T. S., and Mody, I., 1992, Membrane properties of dentate gyrus granule cells: Comparison of sharp microelectrode and whole-cell recordings, *J. Neurophysiol.* **67**(5):1346–1358.

Stasheff, S. F., Hines, M., and Wilson, W. A., 1993, Axon terminal hyperexcitability associated with eplieptogenesis *in vitro.* I. Origin of ectopic spikes, *J. Neurophysiol.* **70**(3):961–975.

Steinhüauser, C., Tennigkeit, M., Matthies, H., and Gündel, J., 1990, Properties of the fast sodium channels in pyramidal neurones isolated from the CA1 and CA3 areas of the hippocampus of postnatal rats, *Pflügers Arch. Eur. J. Physiol.* **415**:756–761.

Stockbridge, N., and Moore, J. W., 1984, Dynamics of intracellular calcium and its possible relationship to phasic transmitter release and facilitation at the frog neuromuscular junction, *J. Neurosci.* **4**(3):803–811.

Stockley, E. W., Cole, H. M., Brown, A. D., and Wheal, H. V., 1993, A system for quantitative morphological measurement and electrotonic modelling of neuron: Three-dimensional reconstruction, *J. Neurosci. Meth.* **47**:39–51.

Storm, J. F., 1986, A-current and Ca-dependent transient outward current control the initial repetitive firing in hippocampal neurons, *Biophys. J.* **49**:369a.

Storm, J. F., 1987a, Action potential repolarization and a fast after-hyperpolarization in rat hippocampal pyramidal cells, *J. Physiol.* **385**:733–759.

Storm, J. F., 1987b, Intracellular injection of a Ca^{2+} chelator inhibits spike repolarization in hippocampal neurons, *Brain Res.* **435**:387–392.

Storm, J. F., 1988a. Four voltage-dependent potassium currents in adult hippocampal pyramidal cells (HPCs), *Biophys. J.* **53**:148a.

Storm, J. F., 1988b, Temporal integration by a slowly inactivating K^+ current in hippocampal neurons, *Nature* **336**:379–381.

Storm, J. F., 1989, An after-hyperpolarization of medium duration in rat hippocampal pyramidal cells, *J. Physiol.* **409**:171–190.

Storm, J. F., 1990, Potassium currents in hippocampal pyramidal cells, in: *Progress in Brain Research,* Vol. 83 (J. Storm-Mathisen, J. Zimmer, and O. P. Otterson, eds.), Elsevier, Amsterdam, pp. 161–187.

Storm, J. F., Borg-Graham L., and Adams, P. R., 1987, A passive component of the afterdepolarization (ADP) in rat hippocampal pyramidal cells, *Biophys. J.,* **51**:65a.

Storm, J. F., Winther, T., and Pedarzani, P., 1996, h-current modulation by norepinephrine, serotonin, dopamine, histamine and cyclic-AMP analogues in rat hippocampal neurons, *Neurosci. Abstr.* **22**:1444.

Stuart, G. J., and Sakmann, B., 1994, Active propagation of somatic action potentials into neocortical pyramidal cell dendrites, *Nature* **367**:69–72.

Stuart, G., Spruston, N., Sakmann, B., and Häusser, M., 1997, Action potential initiation and back-propagation in neurons of the mammalian CNS, *Trends Neurosci.* **20**(3):125–131.

Thibault, O., and Landfield, P. W., 1996, Increase in single L-type calcium channels in hippocampal neurons during aging, *Science* **272**:1017–1020.

Thompson, S. M., and Gähwiler, B. H., 1992, Organotypic hippocampal slice cultures, in: *Practical Electrophysiological Methods* (H. Kettenmann and R. Grantyn, eds.), Wiley-Liss, New York, pp. 58–61.

Thompson, S. M., and Prince, D. A., 1986, Activation of electrogenic sodium pump in hippocampal CA1 neurons following glutamate-induced depolarization, *J. Neurophysiol.* **56**(2):507–522.

Thompson, S. M., Masukawa, L. M., and Prince, D. A., 1985, Temperature dependence of intrinsic membrane properties and synaptic potentials in hippocampal CA1 neurons *in vitro*, *J. Neurosci.* **5**(3):817–824.

Tinoco, Jr., I., Sauer, K., and Wang, J. C., 1978, *Physical Chemistry: Principles and Applications in Bioloical Sciences.* Prentice-Hall, Englewood Cliffs, NJ.

Toib, A., Lyakhov, V., and Marom, S., 1998, Interaction between duration of activity and time course of recovery from slow inactivation in mammalian brain Na^+ channels, *J. Neurosci.* **18**(5):1893–1903.

Traub, R., 1982, Simulation of intrinsic bursting in CA3 hippocampal neurons, *Neuroscience* **7**(5):1233–1242.

Traub, R., and Llinas, R., 1979, Hippocampal pyramidal cells: Significance of dendritic ionic conductances for neuronal function and epileptogenesis, *J. Neurophysiol.* **42**(2):467–496.

Traub, R. D., and Miles, R., 1991, *Neuronal Networks of the Hippocampus,* Cambridge University Press, Cambridge.

Traub, R. D., Wong, R. K. S., Miles, R., and Michelson, H., 1991, A model of a CA3 hippocampal pyramidal neuron incorporating voltage-clamp data on intrinsic conductances, *J. Neurophysiol.* **66**(2):635–650.

Traub, R. D., Jefferys, J. G. R., Miles, R., Whittington, M. A., and Tóth, K., 1994, A branching dendritic model of a rodent CA3 pyramidal neurone, *J. Physiol.* **481**(1):79–95.

Traub, R. D., Whittington, M. A., Colling, S. B., Buzsáki, G., and Jefferys, J. G. R., 1996, Analysis of gamma rhythms in the rat hippocampus *in vitro* and *in vivo*, *J. Physiol.* **493**(2):471–484.

Tsai, K. Y., Carnevale, N. T., Claiborne, B. J., and Brown, T. H., 1994, Efficient mapping from neuroanatomical to electrotonic space, *Network* **5**:21–46.

Turner, D., 1984a, Conductance transients onto dendritic spines in a segmental cable model of hippocampal neurons, *Biophys. J.* **46**:85–96.

Turner, D., 1984b, Segmental cable evaluation of somatic transients in hippocampal neurons (CA1, CA3, and dentate), *Biophys. J.* **46**:73–84.

Turner, D., and Schwartzkroin, P. A., 1980, Steady-state electrotonic analysis of intracellularly stained hippocampal neurons, *J. Neurophysiol.* **44**(1):184–199.

Turner, D., and Schwartzkroin, P. A., 1984, Passive electrotonic structure and dendritic properties of hippocampal neurons, in: *Brain Slices* (R. Dingledine, ed.), Plenum Press, New York, pp. 25–50.

Turner, R. W., Meyers, D. E. R., Richardson, T. L., and Barker, J. L., 1991, The site for initiation of action potential discharge over the somatodendritic axis of rat hippocampal CA1 pyramidal neurons, *J. Neurosci.* **11**(7):2270–2280.

Turner, D. A., Li, X.-G., Pyapali, G. K., Ylinen, A., and Buzsaki, G., 1995, Morphometric and electrical properties of reconstructed hippocampal CA3 neurons recorded *in vivo*, *J. Comp. Neurol.* **356**:580–594.

Uneyma, H., Ueno, S., and Akaike, N., 1993, Serotonin-operated potassium current in CA1 neurons dissociated from rat hippocampus, *J. Neurophysiol.* **69**(4):1044–1052.

Valiante, T. A., Velazquez, J. L., Jahromi, S. S., and Carlen, P. L., 1995, Coupling potentials in CA1 neurons during calcium-free-induced field burst activity, *J. Neurosci.* **15**(10):6946–6956.

Vandenberg, C. A., and Bezanilla, F., 1991, A sodium channel gating model based on single-channel, macroscopic ionic, and gating currents in the squid giant axon, *Biophys. J.* **60**:1511–1533.

Warman, E., Durand, D., and Yuen, G., 1994, Reconstruction of hippocompal CA1 pyramidal cell electrophysiology by computer simulation, *J. Neurophysiol.* **71**(6):2033–2045.

Wathey, J. C., 1989, ntscable—Neuron anatomy file conversion program [Modified version available from http://www.cnrs-gif.fr/iaf/iaf9/Surf-Hippo.html].

Wathey, J. C., Lytton, W. W., Jester, J. M., and Sejnowski, T. J., 1992, Computer simulations of EPSP–spike (E–S) potentiation in hippocampal CA1 pyramidal cells, *J. Neurosci.* **12**(2):607–618.

Westenbroek, R. E., Merrick, D. K., and Catterall, W. A., 1989, Differential subcellular localization of the R_I and R_{II} Na^+ channel subtypes in central neurons, *Neuron* **3**:695–704.

White, J. A., Sekar, N. S., and Kay, A. R., 1995, Errors in persistent inward currents generated by space-clamp errors: A modeling study, *J. Neurophysiol.* **73**(6):2369–2377.

Williamson, A., and Alger, B. E., 1990. Characterization of an early afterhyperpolarization after a brief train of action potentials in rat hippocampal neurons *in vitro*, *J. Neurophysiol.* **63**(1):72–81.

Wong, R. K. S., and Prince, D. A., 1978, Participation of calcium spikes during intrinsic burst firing in hippocampal neurons, *Brain Res.* **159**:385–390.

Wong, R. K. S., and Prince, D. A., 1981, Afterpotential generation in hippocampal pyramidal cells, *J. Neurophysiol.* **45**(1):86–97.

Wong, R. K. S., Prince, D. A., and Basbaum, A. I., 1979, Intradendritic recordings from hippocampal neurons, *Proc. Nat. Acad. Sci.* **76**(2):986–990.

Wong, R. K. S., Traub, R. D., and Miles, R., 1986, Cellular basis of neuronal synchrony in epilepsy, in *Basic Mechanisms of the Epilepsies. Molecular and Cellular Approaches* (A. V. Delgado-Esueta, A. A. Ward, Jr., D. M. Woodbury, and R. J. Porter, eds.), Raven Press, New York, pp. 583–592.

Yamada, W. M., Koch, C., and Adams, P. R., 1989, Multiple channels and calcium dynamics, in: *Methods in Neuronal Modeling* (C. Koch and I. Segev, eds.), MIT Press/Bradford Books, Cambridge, MA, pp. 97–134.

Yuen, G. L. F., and Durand, D., 1991, Reconstruction of hippocampal granule cell electrophysiology by computer simulation, *Neuroscience* **41**(2/3):411–423.

Zador, A. M., 1993, Biophysics of computation in single hippocampal neurons, Ph.D. thesis, Yale University, New Haven, Connecticut.

Zador, A., Koch, C., and Brown, T. H., 1990, Biophysical model of a hebbian synapse, *Proc. Nat. Acad. Sci. USA* **87**:6718–6722.

Zador, A. M., Agmon-Snir, H., and Segev, I., 1995, The morphoelectrotonic transform: A graphical approach to dendritic function, *J. Neurosci.* **15**(3):1669–1682.

Zbicz, K., and Weight, F., 1985, Transient voltage and calcium dependent outward currents in hippocampal CA3 pyramidal neurons, *J. Neurophysiol.* **53**(4):1038–1058.

Zbicz, K., and Weight, F., 1986, Voltage-clamp analysis of inward calcium current in hippocampal CA3 pyramidal neurons, *Neurosci. Abstr.* **11**:520.

Zhang, L., Weiner, J. L., Valiante, T. A., Velumian, A. A., Watson, P. L., Jahromi, S. S., Schertzer, S., Pennefather, P., and Carlen, P. L., 1994, Whole-cell recording of the Ca^{2+}-dependent slow afterhyperpolarization in hippocampal neurones: Effects of internally applied anions, *Pflügers Arch. Eur. J. Physiol.* **426**:247–253.

Zhang, L., Han, D., and Carlen, P. L., 1996, Temporal specificity of muscarinic synaptic modulation of the Ca^{2+}-dependent K^+ current (I_{sAHP}) in rat hippocampal neurones, *J. Physiol.* **496**(2):395–405.

3

Functional Implications of Active Currents in the Dendrites of Pyramidal Neurons

PAUL A. RHODES

1. Introduction

The original analyses of the transformation of synaptic inputs into voltage deflections at the soma were based largely on studies of motoneurons (Coombs *et al.,* 1957a, b) and suggested that motoneuron dendrites were electrically passive. The view was that synaptic inputs summed algebraically, with excitatory postsynaptic potentials (EPSPs) and inhibitory postsynaptic potentials (IPSPs) offsetting each other, and an output signal emitted if and only if the net summed voltage reached at a spike initiation zone at the axon hillock exceeded a threshold. The notion that the input–output transform accomplished by neurons was based upon the simple summation of inputs at a single integration point was so strongly embedded that when Eccles *et al.* (1958) found evidence for active dendritic spikes in axotomized motoneurons, they regarded the nonlinear boosting and multiple regions of integration implied by these events to be undesirable (further discussion in Sections 2 and 4 below).

Motoneuron responses were analyzed by Rall, who developed analytical methods for modeling dendrites as capacitative cables (Rall, 1959, 1962, 1964). His

PAUL A. RHODES • Mathematical Research Branch, NIDDK, National Institutes of Health, Bethesda, Maryland 20814.

studies were the first to explicitly treat dendrites in neuronal modeling (see collected papers and commentary in Segev *et al.*, 1995), and were able to reproduce motoneuron experimental results with models incorporating passive dendrites. Even with an electrically passive dendritic membrane, Rall predicted a number of nonlinear characteristics of synaptic integration, such as the sublinear summation of synaptic inputs, which results from the reduction of efficacy as synaptic current increases local dendritic voltage toward the synaptic reversal potential. He also predicted that the somatic EPSP caused by distal synaptic input would be greatly attenuated (Rall, 1964; Rinzel and Rall, 1974; Stratford *et al.*, 1989) for a wide range of values for the electrotonic parameters R_m (membrane resistance), C_m (membrane capacitance), and R_i (cytoplasmic resistivity). These compartment model results suggested that with electrically passive dendrites the efficacy of inputs to the apical tuft would be limited, underpinning the commonly held view that these distal inputs by themselves do not trigger activity, and act instead in a modulatory or biasing role, with the inputs to the basal dendrites and proximal obliques primarily responsible for driving firing (e.g., Vogt, 1990).

The Presence of Active Dendritic Membrane in Pyramidal Neurons

When this review was first proposed by the editor in July 1994 the subject of active dendrites and their effects had recently begun a period of renewed excitement, with new fundamental experimental results about voltage-gated channels in dendritic membrane appearing monthly. While there has been an accumulation of evidence for decades that pyramidal cells have active dendrites, the new-found recognition and interest within the neuroscience community was fueled by a crystal-clear experimental demonstration in which the soma and apical trunk of a layer 5 pyramidal cell were simultaneously impaled to demonstrate the extent of Na^+ channels in the apical trunk as revealed by the impact of tetrodotoxin (TTX) on the backpropagation of somatic firing into the apical trunk (Stuart and Sakmann, 1994). Previous experimental studies, following a rich history of studies of Purkinje cells by Llinás and colleagues (Llinás *et al.*, 1968; Llinás and Nicholson, 1971; Llinás and Sugimori, 1980), had reported clear evidence for active dendrites in direct recordings in the apical dendrites of hippocampal (Wong and Prince, 1978; Wong *et al.*, 1979; Benardo *et al.*, 1982, Poolos and Kocsis, 1990) and neocortical (Huguenard *et al.*, 1989; Amitai *et al.*, 1993) pyramidal cells, current source density analyses indicating active outward propagation through the dendrites of pyramidal neurons following somatic firing (Richardson *et al.*, 1987; Turner *et al.*, 1991), optically imaged Ca^{2+} and Na^+ influx into the dendrites of hippocampal CA1 (Jaffe *et al.*, 1992) and neocortical (Yuste *et al.*, 1994) pyramidal neurons, and antibody staining of Ca^{2+} and Na^+ channels in pyramidal dendrites (Westenbröek *et al.*, 1992) (for recent reviews see Johnston *et al.*, 1996; Yuste and Tank, 1996).

Compartment models of burst generation in CA3 (Traub and Llinás, 1979) and neocortical (Rhodes and Gray, 1994) intrinsically bursting (IB) pyramidal cells supported the view that not only dendritic Ca^{2+}, but also Na^+ channels were integrally required for burst generation in this cell type, along with voltage and $[Ca^{2+}]$-gated (Friedman and Gutnick, 1987) K^+ currents. It now seems likely that not only intrinsically bursting, but also regular-spiking (RS) cortical pyramidal cells have dendritic Na^+ and Ca^{2+} currents (Rhodes *et al.*, 1995), since an array of evidence suggests that RS cells can be shifted toward bursting behavior (Friedman and Gutnick, 1989; Jensen *et al.*, 1996; Y. Amitai, personal communication), and Ca^{2+} influx has been observed throughout RS pyramidal dendrites (R. Yuste and D. W. Tank, personal communication).

In sum, there has been an overwhelming accumulation of evidence establishing the ubiquity of a variety of active channels in pyramidal dendrites. It is self-evident that the presence of voltage-gated electrical currents in the dendrites may change the effects of synaptic inputs, and therefore the realization that dendritic membrane is electrically active compels a reanalysis of dendritic function.

2. Historical Perspective

2.1. Partial Responses in Axotomized Motoneurons

In studies of chromatolyzed (axotomized) motoneurons with intracellular electrodes, Eccles *et al.* (1958) noted what they termed "partial responses," small (1–5 mV) fast (4–6 msec time course) voltage deflections at the soma (Fig. 1) not seen in normal intact motoneurons. These events could be triggered synaptically, but did not appear to be associated with an outgoing spike on the truncated remainder of the axon (Figs. 1a–1c), suggesting they did not originate in the axon hillock. Eccles *et al.* inferred that axotomization caused channels to be inserted in dendritic membrane, where "exceptionally excitable" patches generated the partial responses. They suggested that these partial responses sum and thus amplify inputs (Figs. 1d–1f), one of the very early suggestions for a function of active dendrites. Interestingly, Eccles *et al.* viewed the nonlinear boosted mode of integration in motoneuron dendritic trees as imperfect when compared with the classical picture (of which Eccles was an author) of passive collection of inputs at a single point of

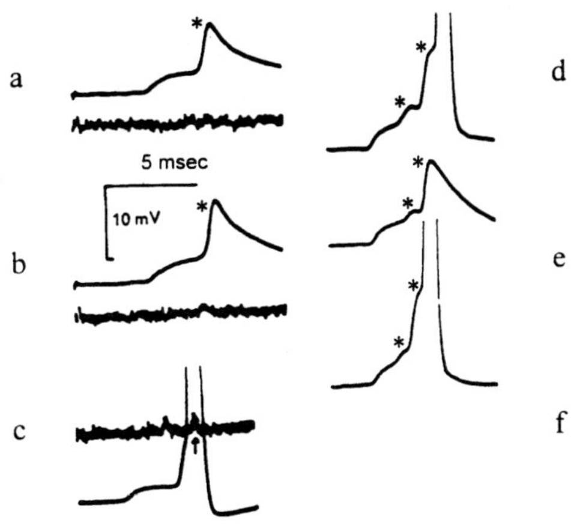

Figure 1. The first evidence for the presence of active dendrites came from the observation of small (1–5 mV), quick (4–6 msec) depolarizations, termed "partial responses" (marked by asterisks), triggered by synaptic input to motoneurons following the severing of their axon. Activity in the truncated axon was monitored extracellularly (lower traces in panels a and b and upper trace in panel c), and it was noted that the deflection in the extracellular record produced when the cell fired (arrow in panel c) was not associated with partial responses (a, b), suggesting that they were not produced by firing of the initial segment. It was hypothesized that the axotomization caused channels to be deployed in the dendrites, and the partial responses were due to firings of the dendritic membrane. Synaptic stimulation was observed to elicit partial responses of variable amplitude and latency which contributed to triggering of firing (d–f). Eccles *et al.*, wedded to the classical notion of arithmetic summation of excitatory and inhibitory inputs at the axon hillock, viewed the ability of subregions of the dendritic tree to separately integrate and fire as a "defective" mode of integration. Modified from Eccles *et al.* (1958).

summation at the axon hillock, writing that "the chromatolysed motoneurone is a defective integrator, for a partial response can now be generated in a dendrite by a special strategic grouping of excitatory synapses" (Eccles *et al.*, 1958, p. 36). It is ironic that this view of excitation of dendrites is both suggested by more recent studies of the effects of physiological levels of dendritic currents and is thought to confer considerable functional power (see Section 4 below).

Kuno and Llinás (1970a, b) also noted that afferent input more sensitively excited chromatolyzed motoneurons. They elaborated on the view that partial responses were all-or-none and were produced at a variety of electrotonically distinct dendritic sites, with new partial responses appearing abruptly rather than with graded increase as input stimulus was enhanced. From their variation in size and latency from synaptic input Kuno and Llinás inferred electronically separate dendritic spike trigger patches, concluding that these all-or-none dendritic events implied segregated "hotspots" of active membrane in the dendrites. They also made the prescient observation that propagation of pulses could, by analogy with axons, be bidirectional in dendrites.

2.2. Spikelets in Hippocampal Pyramidal Neurons

In what has been a very influential study, Spencer and Kandel (1961) applied many of the techniques previously used to analyze motoneuron firing (Coombs *et al.*, 1957a, b) to the components of excitation of pyramidal cells intracellularly recorded *in vivo*. They noted that an inflection in the voltage trace immediately preceded action potential firing, and by applying hyperpolarization were able to isolate what they termed a "fast prepotential" (Fig. 2), a small, fast deflection averaging 5 mV or so in height, with an average time to peak of just 0.7 msec. They noted that the rate of fall of these events ($\tau \approx 3$ msec) was much quicker than the 8 msec membrane time constant they measured would suggest. Following Eccles *et al.*, they suggested that spikelets were the somatic effect of larger spikes occurring at distal excitable hotspots of dendritic membrane in the apical trunk (Fig. 2c). The small size of spikelets in comparison to somatic action potentials was assumed to be a result of electrotonic attenuation between the distal dendritic site of generation and the site of recording at the soma. They inferred further that the absence of propagation of a full action potential in the dendrites implied a tract of passive membrane between the soma and a hot patch of membrane.* Rather than consider boosting an integration defect, as had Eccles *et al.*, Spencer and Kandel suggested that amplification was a normal functional effect of active dendrites. Spikelets, described variously as D-spikes, fast prepotentials, partial responses, etc., have been subsequently reported in both hippocampal (Schwartzkroin, 1977; Turner *et al.*, 1989; Wong and Stewart, 1992; Turner *et al.*, 1993) and neocortical (Purpura, 1965; Deschênes, 1981) pyramidal neurons.

*Compartment model simulations do not support this inference; in dendritic trees with uniform Na^+ channel density, an active spike may occur in distal apical trunk, but fail to propagate inward, or alternatively a side branch may fire, but fail to activate the abutting apical trunk (P. Rhodes, unpublished observations).

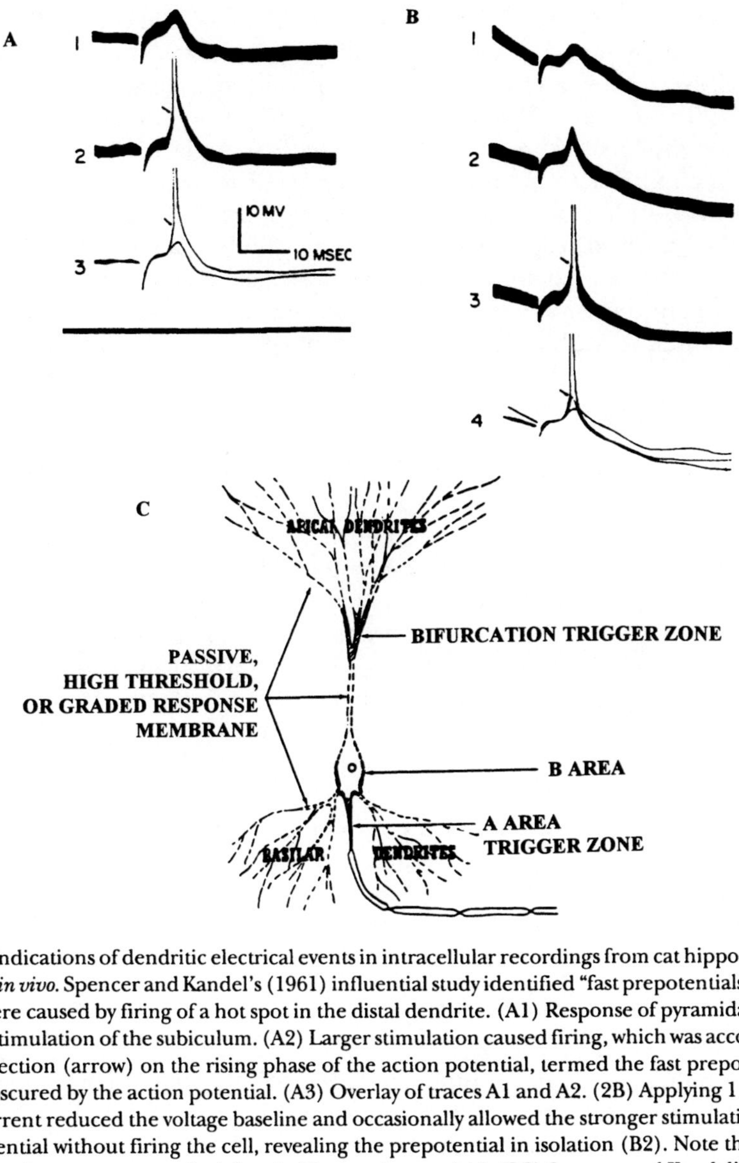

Figure 2. Indications of dendritic electrical events in intracellular recordings from cat hippocampal pyramidal cells *in vivo*. Spencer and Kandel's (1961) influential study identified "fast prepotentials" which they believed were caused by firing of a hot spot in the distal dendrite. (A1) Response of pyramidal cell to subthreshold stimulation of the subiculum. (A2) Larger stimulation caused firing, which was accompanied by a small inflection (arrow) on the rising phase of the action potential, termed the fast prepotential, here partially obscured by the action potential. (A3) Overlay of traces A1 and A2. (2B) Applying 1 nA hyperpolarizing current reduced the voltage baseline and occasionally allowed the stronger stimulation to trigger the prepotential without firing the cell, revealing the prepotential in isolation (B2). Note that it is much sharper than the components of subthreshold events in panel A1. (2C) Spencer and Kandel's conjecture, following Eccles *et al.* (1958), that fast prepotentials are generated by the firing of a distant patch of isolated active membrane, a "hotspot," which can boost distal input. Spencer and Kandel placed the hotspot in this location because they assumed that the small amplitude of the spikelet (5 mV) required the active dendrite to be far from the soma, and that an intervening tract of passive apical trunk was needed in order to prevent full propagation of firing inward. Compartment simulations, which were not available when this study was published, do not support either requirement. From Spencer and Kandel (1961).

2.3. Analysis of the Effects of Active Dendrites Using Compartment Models

The development of compartment simulation methods (Rall, 1964), in which dendritic cable is subdivided into equipotential segments (compartments) which interact via Ohmic current flow, allowed for the incorporation of arbitrarily complex dendritic geometries and distributions of voltage-gated membrane currents (Rall, 1964; Cooley and Dodge, 1966; Rall and Shepherd, 1968; Dodge and Cooley, 1973; Traub, 1975; Traub and Llinás, 1977, 1979; Pellionisz and Llinás, 1977; Perkel *et al.*, 1981; Segev *et al.*, 1989). The first application of compartment models to the simulation of neurons with active dendrites was a remarkable study of mitral cells in the olfactory bulb (Rall and Shepherd, 1968). In attempting to reproduce the observed response of the olfactory bulb to antidromic stimulation, Rall and Sheperd extended compartment models from single neurons to the construction of an entire neural circuit, including a mitral cell with axon, along with a connected granule cell. They computed the extracellular field potential generated by the simulated bulb circuitry following antidromic input with and without action potential-generating currents in the mitral dendrites, comparing simulated with experimentally measured extracellular potentials to infer whether dendritic action potentials were present. They reached the conclusion that the measured field potentials were consistent with either passive or weakly active dendrites. This study pioneered the incorporation of active dendrites in compartment model neurons, was the first to simulate a full circuit with compartment methods, and was the first in which simulated and measured extracellular potentials were compared to assess the likelihood of active dendritic membrane. Simulations of neurons with active dendrites and the effects of dendritic voltage events were not actively pursued again until years later (Dodge and Cooley, 1973; Traub, 1975; Traub and Llinás, 1977, 1979), and have only recently become vogue following the recent dendritic recordings mentioned previously. It is perhaps surprising that, despite the availability of compartment model techniques just a few years (Rall, 1964) subsequent to the first suggestion of active dendritic events in pyramidal neurons (Spencer and Kandel, 1961), compartment models have rarely been utilized to test the hypothesis that spikelets are generated by the firing of hotspots, or the conjecture that dendritic spikes cause amplification of synaptic inputs.

3. Amplification of Synaptic Inputs

The amplification of distal inputs was the first functional role suggested for active dendrites (Eccles *et al.*, 1958; Spencer and Kandel, 1961). There has been surprisingly little direct evidence of boosting, in part because experiments which test it are hard to execute. Some of the experimental literature concerning the amplification of distal inputs will be briefly reviewed before turning to compartment model studies of amplification. Experimental and compartment model results concerning effects of active currents on EPSP shape, a special case of amplification, will be reviewed in a later section.

3.1. Current Source Density Measurements Suggest Inwardly Propagating Dendritic Action Potentials in the Proximal Apical Trunk of CA1 Cells

Using current source density analysis of hippocampal area CA1 pyramidal cells, Herreras (1990) reported that following threshold stimulation of stratum radiatum fibers, a large depolarization arose in the proximal apical trunk about 100–150 μm from the soma, and appeared to propagate inward toward the soma. The apparent rate of travel of the peak dendritic inward current (0.2 m/sec) is consistent with simulated propagation velocity (P. Rhodes, unpublished observations) in dendrite with physiological levels of Na^+ channel density (see Section 4.1). Drawing an analogy with axons, where Na^+ and K^+ currents enable transmission of voltage signals over long distances, Herreras suggested that excitable membrane in the apical trunk enabled distal synaptic inputs to ignite a propagating dendritic action potential which, in contrast to incremental boosting of synaptic inputs via the graded summation of spikelets, always fired the soma.

3.2. Synaptic Inputs Restricted to Layer 1 Can Fire the Soma

Cauller and Connors (1992, 1994) reported that synaptic input to the apical tuft is capable of firing layer 5 pyramidal cells in the absence of proximal input. They selectively activated synaptic input restricted to layer 1 and upper layer 2 using a slice of sensorimotor rat neocortex in which a cut had been made from layers 2–6, electrically stimulating in layer 1 on one side of the cut while recording in somata of pyramidal neurons on the other side of the cut (Fig. 3a). Synaptic inputs to the apical tuft caused EPSPs that increased in size in a graded fashion, reaching 8 mV or more before firing the postsynaptic cell as stimulus strength was increased (Fig. 3c). Remarkably, they found that almost all (19/21; Cauller and Connors, 1994) cells in which synaptic input had any effect at all (i.e., at least some EPSP was produced) could be fired as stimulus strength increased.

Cauller and Connors (1994) constructed morphologically accurate compartment models of their neurons, setting ranges for R_m, R_i, and somatic impalement leak to be consistent with their measurements of decay time constant and of chargeup from rest (Cauller and Connors, 1992) and found that with passive dendrites responses to simulated layer 1 synaptic inputs were small and did not satisfactorily reproduce their measurements. They therefore inferred that the layer 1 responses they measured were likely due in part to amplification via activation of membrane currents. They also found that EPSP amplitude was extraordinarily insensitive to somatic potential in layer 5 cells, and argued that if active currents boosted the EPSP observed in layer 5 cells following apical tuft input, and if manipulation of the somatic voltage did not affect the amplitude of these EPSPs, then the active currents providing the boosting could not be at the soma (Cauller and Connors, 1992, 1994).

Cauller and Connors' experimental results contradicted the classical view (e.g., Vogt, 1991) that distal layer 1 inputs are incapable of driving firing in the absence of proximal inputs, and thus heighten the potential functional importance of inputs to the layer 1 apical tuft. Because corticocortical backprojections have been found to send fibers into layer 1 (Rockland and Pandya, 1979), where they produce *en passant* synapses along several millimeters of travel (Rockland and Virga, 1989; Cauller and Connors, 1992), the surprising efficacy of layer 1 input enables a functional role for the backprojections from higher to lower cortical areas in triggering firing in the absence of feedforward input (Rhodes, 1990).

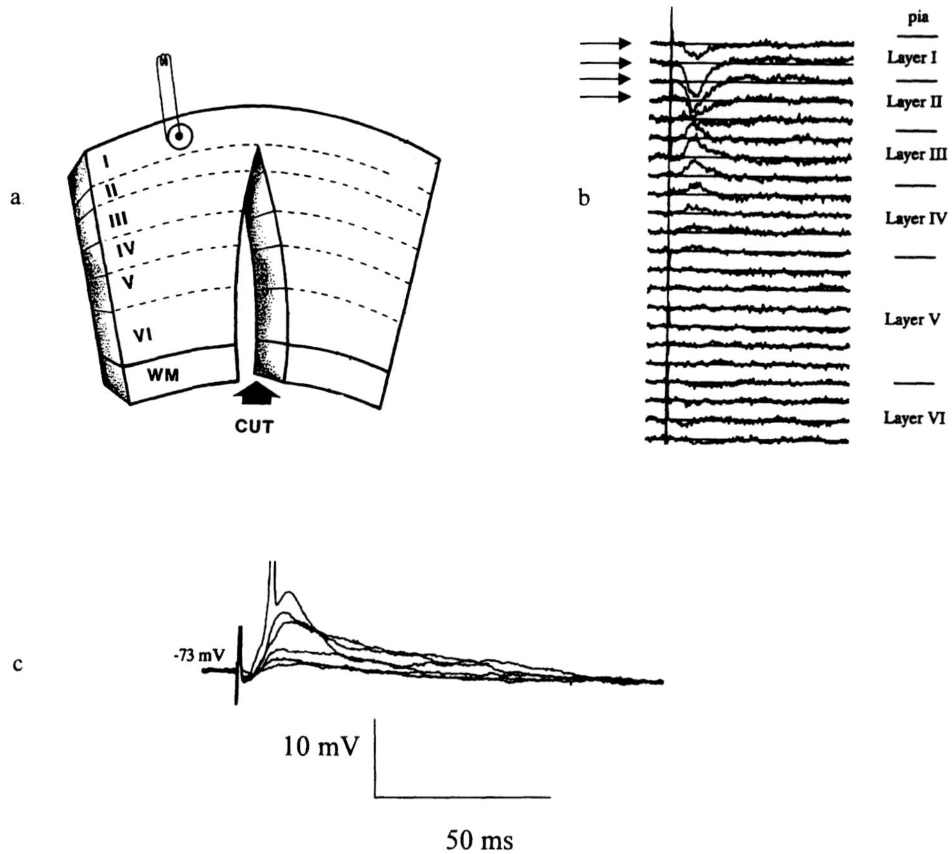

Figure 3. Synaptic inputs to layer 1 *in vitro* elicit large EPSPs and somatic firing. (a) Diagram from Cauller and Connors (1992) showing the slice preparation (left) used to isolate synaptic inputs to layer 1. A cut was made from layer 2 through layer 6, with electrical stimulation applied in layer 1 on one side of the cut while recording in somata of pyramidal neurons on the other side of the cut. (b) The substantial isolation of synaptic input to layer 1 was verified by current source density analysis along the vertical axis of the layer 5 pyramidal cell, with initial sinks (arrows) following synaptic input restricted to layer 1 and the top 50 μm of layer 2. (c) Responses of a layer 5 intrinsically bursting pyramidal cell to a series of layer 1 inputs elicited by extracellular shocks of increasing magnitude. The resulting EPSPs increased in size in a graded fashion, in this case reaching 8 mV before firing the postsynaptic cell. Adapted from Cauller and Connors (1992).

3.3. Tetrodotoxin-Sensitive Effect of Glutamate Iontophoresis onto the Distal Apical Dendrite

In order to test directly whether dendritic currents boost synaptic input, it is necessary to be able to compare responses to the same input with and without the contribution of dendritic currents. This requires the ability to repeatedly deliver an unchanging input to the same cell while manipulating dendritic current activity, which is inherently difficult using axonal stimulation protocols because bath blockade of currents such as Na^+ and Ca^{2+}, which are needed for axonal propagation and synaptic transmission, cannot be employed (though presumably intracellular agents such as the channel blocker QX-314 could).

Schwindt and Crill (1995) tested the involvement of dendritic currents in amplification by stimulating the distal apical dendrite with focally applied glutamate using a pipette positioned along the apical dendrite between 145 and 555 μm from the soma of layer 5 neocortical pyramidal cells. They gauged the effect of dendritic Na^+ currents by comparing the current flowing into the soma during iontophoretic stimulation in control and TTX-containing bath. As an indication of how focal this stimulation was, they reported that moving the tip of the stimulating pipette 2 μm significantly reduced the postsynaptic response, and 10 μm eliminated it. In order to be certain that somatic voltage-sensitive channels were not involved in the response, the soma was voltage clamped, ensuring that current flowing through Na^+ and other voltage-activated channels in and near the soma was an unchanging baseline at each holding voltage. They found that with somatic voltage held at resting level (−76 mV) bath TTX did not diminish the current flowing into the soma from the dendrites following stimulation. However, when the soma was held at −61 mV the addition of bath TTX reduced the current caused by application of glutamate to the dendrite (Fig. 4). The increase in current observed when stimulating with the soma clamped at a depolarized voltage was attributed to a stimulation-triggered increase in current at Na^+ channels in the dendrites. Because the TTX-sensitive component of depolarizing current endured through a 1-sec glutamate application, it was suggested that the persistent (noninactivating) Na^+ current Na_P (Connors *et al.*, 1982; Deisz *et al.*, 1991; Alzheimer *et al.*, 1993a, b) was involved, since the transient fast Na^+ current inactivates with a time constant on the order of milliseconds. They also noted a TTX-resistant component of inward

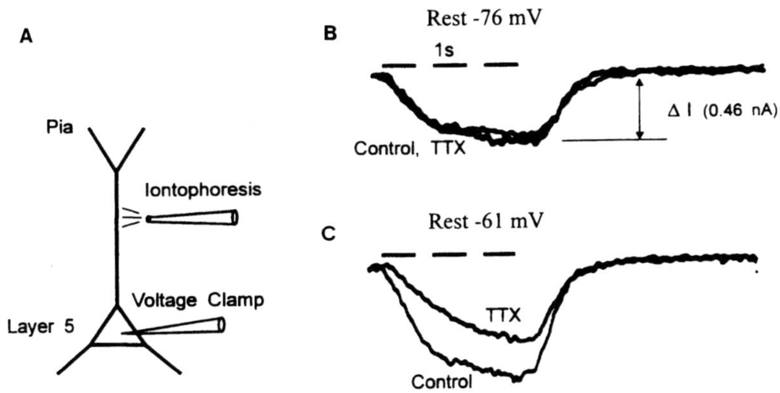

Figure 4. The importance of dendritic sodium currents in amplification of distal inputs was tested by focal iontophoresis of glutamate while holding somatic voltage fixed. While glutamate was applied by a pipette positioned at the apical dendrite of a layer 5 pyramidal cell (A), the current necessary to maintain somatic voltage clamp in control was compared with that in bath TTX. (B) With somatic voltage clamped at −76 mV, bath TTX did not affect the clamp current during stimulation, indicating the glutamate did not activate sodium channels in that condition. (C) When clamp voltage was increased to −61 mV, however, TTX reduced the clamp current, suggesting the stimulation activated TTX-sensitive voltage-activated currents. The iontophoretic glutamate also invoked a residual voltage-sensitive current resistant to TTX which was eliminated with APV (not shown). In this experiment the voltage-gated currents turned on by the stimulus were not on or near the soma, since the somatic voltage was clamped. From Schwindt and Crill (1995).

current which increased with depolarization and which was essentially abolished by 5-phosphonovaleric acid (APV), indicating that *N*-methyl-D-aspartate (NMDA) channels contributed to the enhanced current reaching the soma in the depolarized case (Thomson *et al.*, 1988, 1993). Blockers of Ca^{2+} currents had little effect (Schwindt and Crill, 1995), indicating that in this experiment noninactivating dendritic Ca^{2+} currents did not participate in boosting. It should be noted that the role of the transient low threshold current, which presumably would have inactivated in the first 100 msec during the long (1 sec) continuous glutamate application, would not be probed by this experimental protocol.

4. Compartment Model Simulations of Amplification

While experiments testing the effect of dendritic channels on boosting are difficult, it is comparatively straightforward to examine this issue with compartment simulations. It is therefore surprising that though boosting of distal input has long been believed to be a function of dendritic active currents (Eccles *et al.*, 1958; Spencer and Kandel, 1961), there have been few attempts (but see Shepherd *et al.*, 1991) to apply compartment model tools to ask some of the natural questions: Are spikelets produced by dendritic firings in compartment models, do they imply hotspots, and do they boost synaptic input? If compartment models predict EPSPs to be enhanced by dendritic currents, which currents might provide that function? In what way does amplification depend upon variables such as dendritic channel density and synaptic input amplitude? Is boosting uniform, with inputs to any location in the dendritic tree amplified a fixed percentage, or rather do inputs arriving at different locations in the dendritic tree get differentially boosted?

4.1. Estimation of Realistic Dendritic Na^+ Conductance Densities

It may be that compartment models have only recently been used to study boosting in part because data on dendritic ion channel conductance densities have been lacking. Information from patch electrode experiments has, however, begun to allow estimates of dendritic channel densities, beginning to establish a foundation for modeling studies which examine their functional implications.

4.1.1. Channel Density Estimates from Acutely Dissociated Cells

One approach utilizes whole-cell recordings in acutely dissociated rat neocortical pyramidal cells (Hamill *et al.*, 1991; Cummins *et al.*, 1994; Klee *et al.*, 1995), where densities are estimated by blocking all but one channel and then dividing whole-cell peak current during a voltage clamp step by total membrane area and driving force to obtain an average peak conductance density. This value then must be divided by peak open probability (Rhodes and Gray, 1994, Methods section) to yield the conductance density utilized in compartment models. These densities pertain to the entire dissociated cell (which consists of a soma and a limited amount of truncated axon and primary dendritic branches), and so blend somatic and dendritic conductance densities. The values suggested for Na^+ conductance density inferred from these measurements range up to 25 mS/cm^2 for mature (P30) rat pyramidal cells, with a three- to sixfold increase in density observed between postnatal days 7 and 30. The sensitivity of many aspects of pyramidal cell function to Na^+ current and its rapid change in density during the first postnatal weeks suggest that experiments addressing the implications of dendritic currents in immature rat pyramidal cells will produce different results at different ages.

4.1.2. Channel Density Estimates in Patches of Dendritic, Somatic, and Axonal Membrane

The recent advent of direct patch recording from the soma, apical trunk (Huguenard *et al.*, 1989; Stuart and Sakmann, 1994; Magee and Johnston, 1995b; Christie *et al.*, 1995) and axon (Colbert and Johnston, 1996a) has enabled conductance density measurements to be made with heretofore unprecedented location specificity. These studies either measure peak current during a voltage clamp step, this time applied to the patch itself instead of to a whole dissociated cell, or count the maximum number of channels per patch in single-channel records. Somewhat surprisingly, the results all suggest that Na^+ channel density in the apical trunk and the soma is uniform, with estimates of 6 mS/cm^2 for 2- to 8-week hippocampal CA1 pyramidal cells (Magee and Johnston, 1995b), and 8–11 mS/cm^2 in young (P14) neocortical layer 5 cells (Huguenard *et al.*, 1989; Stuart and Sakmann, 1994). Stuart and Sakmann (1994) reported 4.4 mS/cm^2, but overlooked dividing this peak conductance density by the peak open probability during the voltage clamp step (which is just 40%), an adjustment necessary for estimation of conductance density corresponding to 100% open conductance, the parameter utilized in compartment equations [i.e., $g_{max,Na}$ in the equation $g_{Na} = g_{max,Na} m^3 h$, with m and h the Hodgkin–Huxley (1952) activation and inactivation variables, respectively]. Finally, Colbert and Johnston (1996a) recently found similar densities fairly uniformly distributed in the first 50 μm of the axon. While these individual patch estimates depend upon a visual measurement of the membrane area per patch, about which there is some uncertainty (D. Johnston, C. Colbert, personal communication), with that caveat, the picture which emerges is that the entire membrane surface of the cell, including soma and dendrites and even proximal axon, incorporates Na^+ channels with an average conductance density of 6–11 mS/cm^2.

The profile of Ca^{2+} channels, which have only been studied with this method in CA1 pyramidal cells (Magee and Johnston, 1995b; Christie *et al.*, 1995), suggests a systematic variation in the blend of L-, N-, T-, and R-type conductance density with distance from the soma with the proportion of L diminishing and T-channel density increasing with distance from the soma. Interestingly, the aggregate density of all Ca^{2+} channels appeared to be fairly constant. The functional implications of the apparent predominance of T channels in the distal reaches of the apical trunk are intriguing, and not yet understood. Density estimates of the K^+ channels have just begun to be examined in CA1 cells, and these early results suggest remarkably high values (Hoffman *et al.*, 1996). One hopes that this work will be soon extended to the specification of Ca^{2+} and K^+ channel distributions in cortical pyramidal cells.

4.1.3. Conductance Density Estimates Inferred from Comparison of Modeled with Measured Backpropagated Amplitude

A very different approach to estimation of dendritic Na^+ conductance density is to compare the amplitude of a backpropagated voltage wave into the dendrites following somatic firing in compartment model simulations (Rhodes and Gray, 1993; Mainen *et al.*, 1995; Rapp *et al.*, 1996; Rhodes, n.d.) with measurements from dendritic impalement data (Stuart and Sakmann, 1994). This comparison informs an estimate of Na^+ channel density because, for a given capacitance and temperature, the amplitude of the voltage wave which backpropagates through the apical dendrite following somatic firing is particularly sensitive to Na^+ channel density (Rhodes, n.d.). In this approach, Na^+ conductance along the apical trunk is varied in the model cell until simulated and experimentally measured backpropagated amplitudes match, yielding estimates of 3–5 mS/cm^2 for immature (P14) layer 5 pyramidal cells, and 8–10 mS/cm^2 in mature pyramidal dendrites (Mainen *et al.*, 1995; Rapp *et al.*, 1996; Rhodes, n.d.).*

Thus the dendritic Na^+ channel density inferred from comparing modeled and measured backpropagation is in the same range, 5–10 mS/cm^2, as that determined from measurements of currents in dendritic patches. The convergence of values estimated for dendritic Na^+ channel density using a variety of disparate approaches encourages the study of boosting in compartment models.

4.2. Models of Layer 1 Input

Cauller and Connors' (1994) experiments provide a special data set with which to study the effects of dendritic membrane on synaptic integration because, in contrast to the input caused by electrical stimulation of cortex or white matter in uncut slices, their protocol assures a relatively well localized site of synaptic input, as they showed with current source density (CSD) analysis of current sinks along the apical dendrite (Fig. 3b). Further, the location of this input in the most distal apical tuft ensures a large stretch of dendritic cable intervening between input and soma, so that effects of dendritic currents upon integration of synaptic input will likely be elicited.

Development of compartment model neurons which similarly respond to apical tuft input (P. Rhodes, unpublished observations) can be constrained by two of Cauller and Connors' observations: they found that sufficiently large layer 1 synaptic inputs almost always succeeded in firing the target layer 5 soma some 800–1000 μm away from the synaptic inputs, yet they found that excitation was not all-or-none, but rather that subthreshold stimuli produced a graded range of subthreshold EPSPs, reaching >8 mV in amplitude (Fig. 3c). To mimic their experiments, simulated synaptic conductance was applied to compartments in the apical tuft and the most distal 50 μm of apical trunk of the model layer 5 pyramidal neuron (further simulation details in Fig. 5 caption) and the set of EPSPs measured by Cauller and Connors over a range of input magnitudes was compared to simulated responses. In this approach Na^+ conductance was uniformly distributed through-

*It should be noted that these estimates are sensitive to temperature, membrane capacitance, spine density, and especially the voltage dependence of Na^+ channel inactivation (P. Rhodes, unpublished observations).

out the apical dendrite and was iteratively modified, with the goal of obtaining a good match between simulated and measured EPSPs over a wide range of subthreshold levels all the way through firing. A summary of the results is as follows (Fig. 5):

1. With a passive apical dendrite, synaptic input to the apical tuft was incapable of firing the simulated layer 5 pyramidal cell (as noted by Cauller and Connors, 1994), with EPSPs reaching a maximal amplitude of about 4 mV (Fig. 5a). As they observed in their simulations of apical tuft input to passive dendrite (Cauller and Connors, 1994), increasing synaptic conductance had diminishing effect due to saturation of synaptic driving force (Rall, 1962).

2. With dendritic Na^+ conductance* at densities up to about 4 mS/cm^2 (a value sensitive to the voltage dependence of Na^+ channel inactivation; see Section 4.3.1) uniformly distributed throughout the apical dendrite including obliques and the tuft, the synaptic input failed to trigger the firing observed *in vitro.* At 4 mS/cm^2, input readily activated Na^+ currents in the tuft, but inward propagation of the distal activity failed, and firing did not occur, though this activity boosted the EPSP somewhat, to about 6 mV (Fig. 5b).

3. When Na^+ density was increased to 5–6 mS/cm^2, however, a threshold of excitability was crossed, with a small synaptic input (peak conductance 9.6 nS) to the apical tuft readily firing the cell body 800 μm away. Examination of these simulations (not shown) shows that the synaptic input triggered branch firings which readily spread throughout the tuft. These together ignited the firing of the most distal trunk, which then propagated inward at about 0.2 m/sec (at 35°C), firing the soma. This is the same velocity as the inwardly traveling CSD peak measured by Herreras (1991) in CA1 pyramid apical dendrites (see Section 3.1 above).† However, with Na^+ and K_{DR} currents alone the large subthreshold EPSPs observed *in vitro* were never produced, with the largest subthreshold EPSP about 3 mV (Fig. 5c).

4. To address the observation (Cauller and Connors, 1992) that large subthreshold EPSPs are produced by layer 1 input to layer 5 cells *in vitro,* the transient potassium current I_A known to exist in CA1 pyramidal cell dendrites (Andreason and Lambert, 1995) was incorporated into the simulated apical dendrite. The A current is well suited for modulation of subthreshold EPSPs because, unlike the delayed rectifier, it activates in the same voltage range as the Na^+ current, allowing it to become involved in opposing subthreshold dendritic Na^+-driven depolarization. Adding 0.5 mS/cm^2 dendritic K_A density‡ along with 6 mS/cm^2 Na^+ enabled subthreshold EPSPs to reach 8 mV in amplitude while still permitting larger input to fire the cell, with firing threshold about 14.4 nS (Fig. 5d, which may be compared with Fig. 3c). The possibility that I_A plays a critical role in enabling larger subthreshold EPSPs leads to a testable experimental prediction: large subthreshold EPSPs will not be obtainable in layer 5 pyramids following layer 1 stimulus in bath 4-aminopyridine (4-AP), though the ability to produce small (2–3 mV) EPSPs and full firing will both be preserved.

*K_{DR} was incorporated at one half the conductance density of Na^+; the conclusions were relatively insensitive to the density of K_{DR}.

†The axon, which contained the same Na^+ channel density as the soma (100 mS/cm^2), generally fired just ahead of the cell body in these simulations, due to its higher local input impedance (P. Rhodes, unpublished observations) in a manner similar to that seen in orthodromic activation of motoneurons (Coombs *et al.,* 1957a, b).

‡Note: The channel used here was a relatively slow A-current, often termed the Δ-current because of its sensitivity to dendrodotoxin.

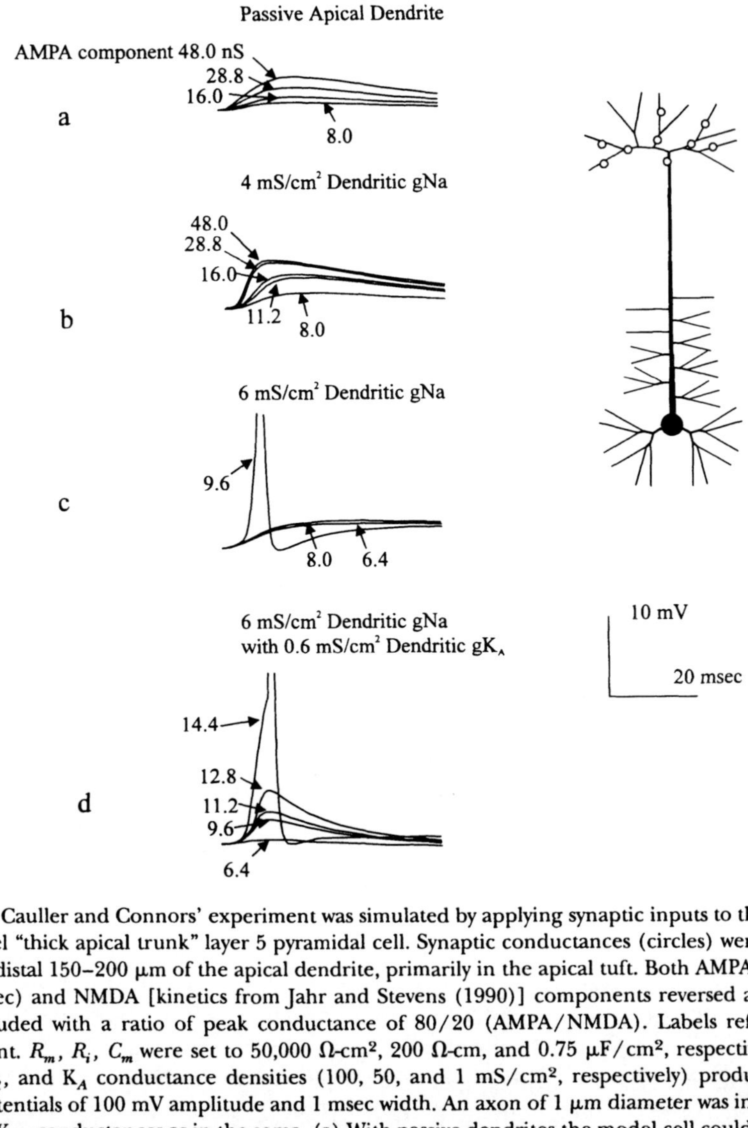

Figure 5. Cauller and Connors' experiment was simulated by applying synaptic inputs to the apical tuft of a model "thick apical trunk" layer 5 pyramidal cell. Synaptic conductances (circles) were applied to the most distal 150–200 μm of the apical dendrite, primarily in the apical tuft. Both AMPA (α function = 0.5 msec) and NMDA [kinetics from Jahr and Stevens (1990)] components reversed at 0 mV, and were included with a ratio of peak conductance of 80/20 (AMPA/NMDA). Labels refer to AMPA component. R_m, R_i, C_m were set to 50,000 Ω-cm², 200 Ω-cm, and 0.75 μF/cm², respectively. Somatic Na^+, K_{DR}, and K_A conductance densities (100, 50, and 1 mS/cm², respectively) produced somatic action potentials of 100 mV amplitude and 1 msec width. An axon of 1 μm diameter was included, with Na^+ and K_{DR} conductances as in the soma. (a) With passive dendrites the model cell could not be fired by apical tuft input, no matter how strong, supporting Cauller and Connors' conclusion that active conductances in the dendrites are needed to reproduce their experimental observations. (b) Adding uniform dendritic Na^+ conductance at 4 mS/cm² enabled a larger and faster-rising subthreshold EPSP, but not firing. The EPSP peak amplitude reached 6 mV before saturating. The jump in EPSP amplitude when input increased from 8 to 11 nS reflected firing of a subbranch of the tuft. As input was increased from 11 to 16 nS little further increase in amplitude occurred. Thus, in this case, Na^+ channels activated in the distal tuft and trunk, but regenerative inward propagation failed. (c) With dendritic Na^+ density increased to 6 mS/cm² a threshold was crossed, and the model cell became extremely sensitive to layer 1 synaptic input, with just 9.6 nS triggering firing. An action potential initiated in the distal apical trunk and propagated inward, at about 0.2 m/sec. Unlike *in vitro* observations, however, large subthreshold EPSPs were not produced, with the largest only reaching 3 mV before firing was triggered. (d) Adding dendritic I_A conductance at 0.5 mS/cm² (kinetics utilized were like those of the D-current) extended the amplitude range of subthreshold EPSPs to 7.5 mV, while larger synaptic input still fired the cell. Comparison of panel d with Fig. 3c suggests that the range of subthreshold and superthreshold responses seen *in vitro* could be reproduced in the simulation. From P. Rhodes (unpublished observations).

It might be thought that one could further increase Na^+ with compensating increases in I_A, but when 10 mS/cm² Na^+ density was tried, increasing I_A did not allow the experimentally observed transition from large subthreshold EPSPs to firing in these models; rather, the dendrite became so excitable that behavior as in Fig. 5b, with either firing or a small EPSP, but no large subthreshold EPSP, was produced (not shown). These simulations of Cauller and Connors' observations therefore suggested that if Na^+ channel density is uniformly distributed, its value is between 4 and 10 mS/cm².* Thus the dendritic Na^+ conductance density which allowed simulated layer 1 input to produce both the large EPSPs and full firing observed *in vitro,* about 6 mS/cm² in the distal apical trunk, is very close to that inferred by requiring the amplitude of simulated action potentials backpropagating into the dendrites match that measured in dual patch studies (see Section 4.1 above), and further matches the range reported from patches of CA1 and layer 5 apical trunks. Such accord among several different means of comparing simulated with measured results provides evidence that compartment models of dendrites with active currents are sound, and that processes as different as distal synaptic input and outward propagation of firing can be simulated and compared with experimental measurement to converge on consistent estimations of current densities.

4.3. Inputs to Basal Branches

Most inputs to pyramidal cells arrive not on the apical trunk or apical tuft, but rather on the basal and the apical oblique dendrites (Larkman, 1991b), and any picture of pyramidal cell synaptic integration must consider inputs to those areas, which because of their relative proximity to the cell body have classically been assumed to be most effective locations for synaptic input. These fine-caliber branches are not yet accessible to patch recordings or other means to measure density, though optical recordings of influx of Ca^{2+} in these regions following somatic firing (Yuste *et al.*, 1994; Schiller *et al.*, 1995) coupled with the simulation result that Na^+ channels are required for propagation of voltage into dendrites (see Section 8) implies that the basal dendrites incorporate Na^+ channels.

To make a preliminary inquiry into effects of active dendrites on synaptic integration in these regions, the density of Na^+ conductance inferred for the apical trunk was extended into the basal dendrites. In these simulations (P. Rhodes, in preparation), responses to a range of levels of synaptic input to a basal branch were examined with dendritic Na^+, K_{DR}, and I_A conductance densities uniformly distributed at 6, 3, and 0.5 mS/cm² respectively, the densities used in the layer 1 input simulation of Fig. 5d. Synaptic input was applied to a single basal subbranch (inset to Fig. 6), and peak gAMPA/gNMDA was maintained at 80/20. The model neuron was stimulated from rest, without background input, and the simulations did not examine the effects of I_h, Ca_T, or Na_p, all of which activate or inactivate around rest and so may play a role in synaptic integration, which further work will have to better

*These quantitative values are very sensitive to the voltage dependence of Na^+ channel inactivation (P. Rhodes, unpublished results). Further, the addition of low threshold T-type Ca^{2+} channels, a persistent component of the Na^+ channel current (Alzheimer *et al.*, 1993; Lipowsky and Alzheimer, 1995; Schwindt and Crill, 1995), and I_h (see Section 7.6) were not systematically considered in this study, though all are activated by subthreshold input. Future work simulating layer 1 synaptic physiology should take into account their presence in the apical dendrite.

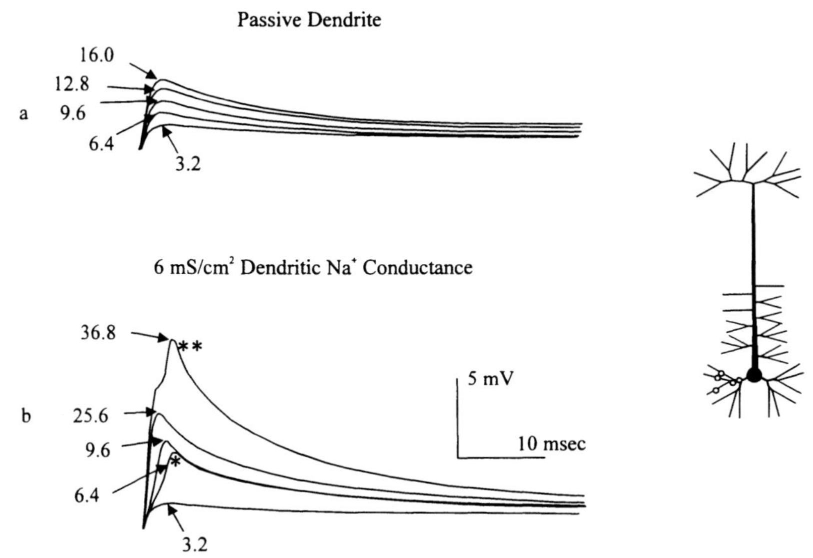

Figure 6. Simulated responses to synaptic input to the basal branches with passive and with active dendrites. Thick-apical-trunk layer 5 cell-compartment morphology, with two of six basal branches shown; synaptic current kinetics and somatic conductances as in Fig. 5. Labels refer to peak AMPA component in nS. (a) Responses to synaptic inputs to basal branches with passive dendrites. Increasing synaptic input causes sublinearly increasing EPSP with dendrites passive, as expected, since increasingly depolarized dendrite reduces driving force and synaptic current (Rall, 1964). (b) Responses to synaptic inputs to basal branches with dendrites incorporating dendritic Na^+, K_{DR}, and K_A conductance densities of 6, 3, and 0.5 mS/cm^2, respectively, as in Fig. 5d. At synaptic conductance up to 4.8 nS, the resulting dendritic depolarization does not activate enough membrane current to affect the amplitude of the EPSP. With input increased to 6.4 nS, however, an Na^+ current-driven firing of the dendritic subbranch receiving input was elicited, with voltage rapidly rising 70 mV in amplitude within the subbranch (see Fig. 7). The somatic EPSP abruptly increased (single asterisk) as a result of this "subbranch spike." Further increasing synaptic conductance beyond branch excitation threshold caused comparatively little increment in EPSP magnitude. Thus, once enough synaptic input was applied to fire the subbranch, further input had diminishing efficacy. Finally, at 36.8 nS, enough input was applied to recruit a neighboring subbranch of the same basal branch, causing another abrupt augmentation of the EPSP (double asterisks). From P. Rhodes (in preparation).

address. With these caveats, the results of synaptic stimulation of a basal branch with increasing levels of inputs were as follows (Fig. 6):

1. Inputs too small to activate significant dendritic currents produced similar EPSPs in both model cells. The threshold synaptic conductance required to activate dendritic Na^+ currents, which depended on input resistance at the synapse, was 4.8 nS in this example.

2. Increasing synaptic conductance just a little beyond this threshold, to 6.4 nS, triggered a dramatic event: the regenerative activation of the dendritic Na^+ currents and the firing of the subbranch receiving input. Voltage in the dendrite climbed to 70 mV above rest in a couple of milliseconds (Fig. 7b). The firing was isolated to the subbranch receiving input even though dendritic conductances were uniform. The failure of propagation and the isolation of firing to the basal sub-

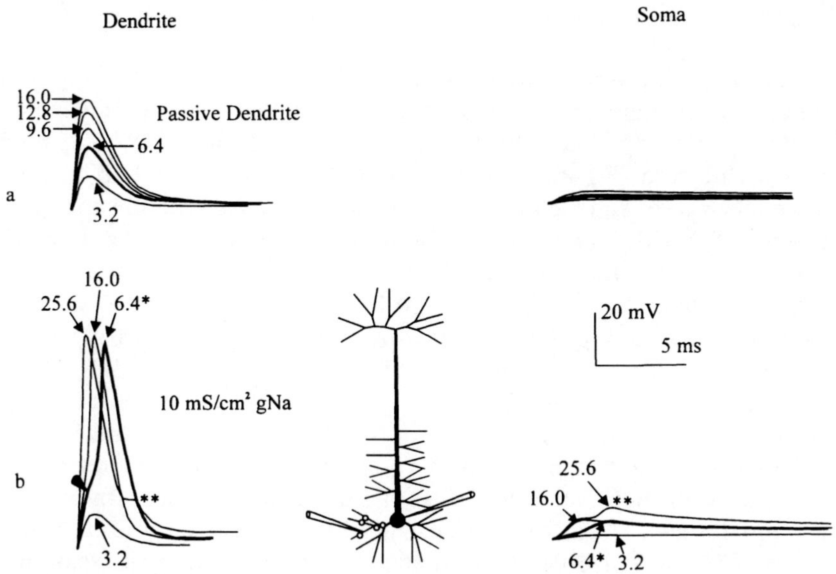

Figure 7. With dendritic Na^+ conductance, simulated input may elicit an all-or-none "subbranch" firing. Synaptic inputs (80/20 AMPA/NMDA; labels refer to peak *g*AMPA) were applied to a portion of one basal branch of a model layer 5 cell. Morphology, passive parameters, and somatic conductances were as in Figs. 5 and 6. Dendritic voltage is shown from the segment indicated in the diagram inset. (a) Passive dendrites. The dendritic voltage response (left, top) to a 6.4-nS input (thick trace) was 25 mV and the corresponding somatic EPSP (right, top) 2.5 mV. Increasing inputs led to sublinearly increasing dendritic and somatic response, with the EPSP saturating at 5–6 mV with very large inputs (not shown). (b) With 10 mS/cm² dendritic *g*Na (K_{DR}, K_A at 3, 0.3 mS/cm²) (Note voltage scale the same for both dendritic and somatic traces), the response to 3.2 nS was the same as the passive case, but 6.4 nS (thick trace) triggered regenerative activation (inflection marked by arrowhead) of the Na^+ currents in the dendritic membrane, powering a full-blown dendritic spike reaching >70 mV above rest. This firing causes an abrupt jump to 7.5 mV in the somatic EPSP (thick line in bottom right panel). Further increasing the synaptic input applied to the subbranch had little effect on either dendritic or somatic voltage response other than a small reduction in latency to the subbranch spike, reflecting the all-or-none nature of the dendritic event. After a fourfold increase in input, to 25.6 nS, an abutting subbranch is finally recruited to fire, causing a second partial response component in the somatic EPSP (double asterisks in right bottom panel).* Firing of the second subbranch also causes a small shoulder in the declining phase of the dendritic voltage in the first (double asterisks in left bottom panel). The relative isolation of subbranches evident in the small size of this bump is due to load mismatch at branch points in the tree, which contributes to the failure of regenerative inward propagation, notwithstanding that dendritic Na^+ conductance density was uniform. From P. Rhodes (in preparation).

branch or apical oblique branch receiving input was a typical response over a wide range of inputs and dendritic current densities (P. Rhodes, unpublished observations). Though it did not propagate inward to the soma, the isolated dendritic branch firing caused an abrupt increase in the somatic EPSP amplitude to 5 mV (Fig. 6b, asterisk), versus just 2 mV with the same synaptic input applied to the passive model (Figs. 6b). Subtracting the somatic voltage traces with and without active dendritic membrane (not shown) reveals a component which can be considered the incremental effect of the firing of a secondary branch and which is reminiscent of the partial responses reported for axotomized motoneurons (Eccles

*Compare with Figure 2A1.

et al., 1958) (Fig. 1), with 3 mV amplitude, 2 msec time to peak, and about 10 msec half-width at half-height.

3. Once enough input was applied to fire the subbranch, further input to that subbranch had surprisingly little additional effect. EPSP amplitude only increased 40% (5 to 7 mV) as synaptic conductance increased 400%, from 6.4 to 25.6 nS, in the active dendrite case (Fig. 6b; see also Fig. 7b, right panel). While a sublinear response is expected for passive dendrites (Fig. 7a), because synaptic current driving force diminishes with increased input (Rall, 1964), once a subbranch has fired, the saturation efficacy of further coincident synaptic input is even more severe in an active branch, since the voltage climbs so high during the several milliseconds that branch firing is ongoing that further inputs have almost no driving force.

4. Once branch threshold is reached, the voltage time course in the branch was almost identical as synaptic input is doubled, with this elemental event initiating 1 msec sooner, but lasting no longer with inputs three- or fourfold above threshold (Fig. 7b). Thus, with these dendritic Na^+ conductances, a subbranch firing is an elemental event, with their somewhat isolated nature and their all-or-none character (Kuno and Llinás, 1970a, b) imposing a very different organization on the boosting conferred by active dendrites than a simple percentage increase in efficacy.

5. Finally, with synaptic input further increased, an adjacent subbranch was recruited to fire, marked by another abrupt augmentation of the EPSP, from 8 up to 11 mV (double asterisk in Fig. 6b and right panel in Fig. 7b). Even though in this example the entire basal branch receiving input fired (until the soma was finally driven to fire with further increases to 55.6 nS), the active currents in the other five basal branches on the model neuron were insignificant; they were isolated from the activated branch by the large capacitance of the soma.

A 10-mV Shift in the Voltage Dependence of Inactivation of the Na^+ Channel Alters Synaptic Integration

The results of every modeling study noted in this review depend in part on the properties of the voltage-gated currents incorporated in the dendritic membrane. What is often not made clear is that the voltage dependence of both the steady-state values of activation and inactivation and the rates with which these values are approached often differ among studies. This complicates assessment of conclusions reported in the modeling literature, because it turns out that many of the functional effects of active dendrites, including synaptic integration, are profoundly altered with seemingly modest changes in current kinetics. In particular, many results seem to be very sensitive to the voltage dependence of inactivation of the Na^+ channel (P. Rhodes, unpublished observations).

As an example, the response of a layer 5 cell to synaptic input to one of its basal branches in two very similar cases was simulated (Fig. 8). The two model cells incorporated identical cell morphology, dendritic current distributions, and passive parameters, and were subjected to the same sequence of increasing synaptic inputs, with identical synaptic properties, applied to the same segments of a single basal branch. The only difference between them was the voltage at which 50% inactivation of the Na^+ channel was reached in steady state, a parameter denoted by $V_{1/2,h}$ in the expression $h = 1/[1 + \exp(V - V_{1/2,h})/k]$, where h is the Hodgkin–Huxley

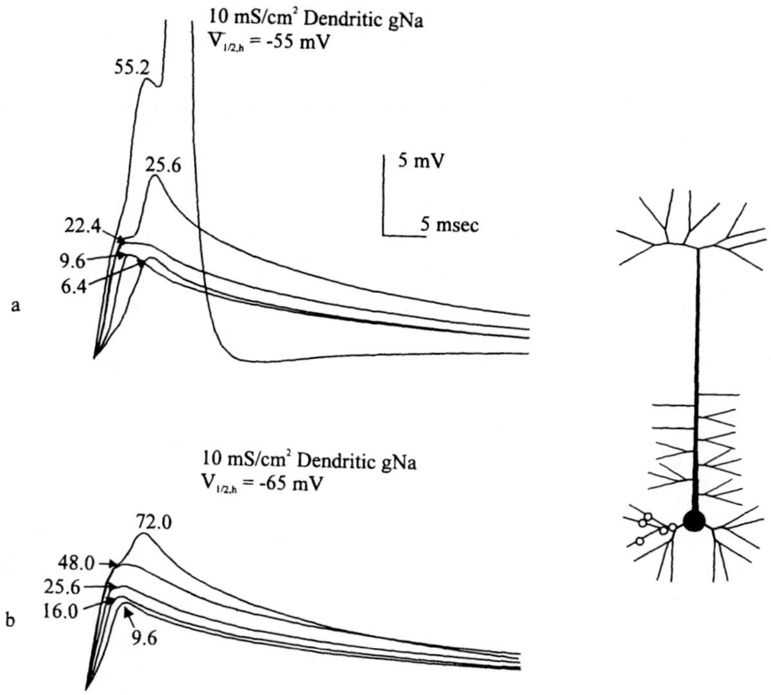

Figure 8. A 10-mV shift in the voltage dependence of inactivation of the Na^+ channel alters synaptic integration. The response of a layer 5 cell to synaptic input to one of its basal branches with cell morphology, dendritic current distributions, and passive parameters as in Figs. 5 and 7. (a, b) The same sequence of increasing synaptic inputs, with identical synaptic properties, is applied to the same segments of a single basal branch (shown in inset). The only difference between them is that the voltage at which 50% inactivation of the Na^+ channel is reached in steady state, $V_{1/2,h}$ (with h the Hodgkin–Huxley inactivation variable) is −55 mV in panel a and −65 mV in panel b, with the time constant of inactivation correspondingly 10 mV shifted along the voltage axis. (a) 6.4 nS fires a first branch spike, 25.6 nS triggers a second, and 55 nS fires the cell. (b) With $V_{1/2,h} = -65$ mV, 9.6-nS input fires the recipient subbranch, but further increases in input do not excite further branch firings, and 72 nS input fails to fire the soma. Thus in this instance shifting the voltage dependence of inactivation by just 10 mV profoundly altered synaptic integration and excitability. P. Rhodes, unpublished observations.

inactivation variable. In the two model cells $V_{1/2,h}$ was −55 mV in one case and −65 mV in the other, with the time constant of inactivation correspondingly shifted by 10 mV along the voltage axis.

With $V_{1/2,h} = -55$ mV, a 6.4-nS input triggered an Na^+ spike in the recipient branch, with 25.6 nS recruiting a second branch to fire, and 55 nS firing the cell (Fig. 8a). When $V_{1/2,h}$ was shifted just 10 mV on the voltage axis, to −65 mV, however, the pyramidal cell became much less excitable. In this case, after an initial branch firing with 9.6 nS synaptic conductance, further increasing input to the target branch to 72 nS caused relatively little change in response, and failed to fire the cell (Fig. 8b).

A review of the literature on the voltage dependence of inactivation of the Na^+ channel reveals that experimentally measured values seem to vary much more than 10 mV, with reported values for $V_{1/2,h}$ from −72 mV (Fleidervish *et al.*, 1996) to −46 mV (Cantrell *et al.*, 1996) in pyramidal cell preparations. It appears that the voltage dependence of inactivation can be shifted by 10–15 mV by the action of protein kinases (A. Cantrell, personal communication), and even by inclusion of CsF in the slice bath (Isom *et al.*, 1995), which is occasionally used in *in vitro* intracellular recordings. One is struck by the powerful opportunity for regulation of excitability that is conferred by mechanisms to shift voltage dependence of Na^+ inactivation.

Because of the measured variability in channel kinetics and because of the evident impact that a 10-mV shift in voltage dependence can have on neuronal function, the clarity of the modeling literature (and perhaps its credibility with electrophysiologists) will be benefited by the adoption of some standard format with which to specify the properties of the ion channels, perhaps a standardized table to be routinely offered in Methods sections, so that the reader can appraise the details of the channel models used and their degree of difference from other modeling and experimental values, and in that light assess the simulated neuronal behavior reported.

4.4. Functional Consequences of the Characteristics of Amplification of Synaptic Inputs into Layer 5 Pyramidal Cell Dendrites

4.4.1. Excitability in Response to Apical Tuft Input

A comparison can be made between the relative efficacy of input to the distal apical tuft versus that to the proximal dendrites with active dendritic conductance densities as described in the foregoing. The results in Sections 4.2 and 4.3 show that 36.8-nS synaptic input to a basal branch caused an 11-mV EPSP, whereas just 14.4 nS applied to the apical tuft, 800 μm from the soma, fired the model cell. Thus, in this example, in which currents were uniformly distributed throughout the dendritic tree, the layer 5 pyramidal cell was much more readily fired from the apical tuft than from a basal dendrite.

Why would the cell be so sensitive to apical tuft input despite a uniform distribution of dendritic conductances and its comparatively great distance from the cell body? First, its extremely high input impedance facilitates the excitation regenerative events in its branches. Second, the 1-μm-diameter most-proximal segments of the tuft (morphology adapted from Larkman, 1991a) abut the ~1.50-μm-diameter most-distal end of the tapered apical trunk, a comparatively modest load mismatch, with further propagation facilitated by gradual widening of trunk diameter and load as the soma is approached. In addition, when both tuft "daughter" branches are excited by layer 1 input, the "parent" segment, the most distal trunk, is driven by both. In contrast, there is a greater load mismatch between each apical oblique first segment and the abutting trunk, with typical diameters of 1.5 and 3 μm, respectively (Larkman, 1991a), and a much greater mismatch between a basal branch and the cell body. As a result, with uniform dendritic current distributions, input to the tuft can be more capable of firing cells than equivalent input to individual apical oblique or basal branches.

While the effect of synaptic input to the basal dendrites is dependent upon how the inputs are distributed (see below), clearly the notion that firing can be very readily triggered by apical tuft input inverts the classical picture of synaptic integration (Vogt, 1990) in electrotonically extended dendrites. Since serial reconstruction of individual axons projecting from higher to lower cortical areas suggests many of their terminations are conveyed by fibers in layer 1 (Rockland and Virga, 1989), the ability of layer 1 inputs to fire pyramidal cells (Cauller and Connors, 1992) has implications for the function of topdown inputs in a hierarchy of cortical areas (Rhodes, 1990).

4.4.2. With Uniform Dendritic Na^+ Conductance, Inputs to Basal Branches May Sum Most Effectively When Segregated into Clusters on Different Branches

The observation in Section 4.3 that applying sufficient synaptic input to fire a subbranch boosted the EPSP, but that further input had diminished effect, suggests that applying larger inputs to distinct subbranches on different branches would have greater relative efficacy. An example is shown in Fig. 9, in which the division of 21.6 nS of synaptic input into three clusters targeted to separate subbranches triggers a branch spike in each, which together fired the model cell (Fig. 9c). When the same total input was applied to one or to two separate subbranches, EPSPs of about 6 and 12 mV, respectively, were generated, suggesting that in this case the subbranch firings had an elemental all-or-none effect (Kuno and Llinás, 1970a, b), and when occurring on separate branches sum in a simple fashion. Millonas and Ulinski (1996) reached the same result in simulations of pyramidal cells with hotspots isolated on different branches. As with many other issues in this field, Rall and coworkers (Miller *et al.*, 1985; Rall and Segev, 1988) got there early, though they addressed the effect of amplification of synaptic input to subbranches and cluster sensitivity in the context of Na^+ channels on spines* with passive dendrites, rather than with active dendrites.

Mel (1992, 1993) demonstrated that a variety of local voltage threshold mechanisms can cause clustered distributions of input to have increased efficacy in simulations of a layer 5 pyramidal cell. He distributed a fixed number (100) of synapses into clusters of varying size randomly, with the clusters segregated on 40-μm dendritic compartments which were themselves situated throughout the dendritic tree. With each synapse independently and asychronously firing at 100 Hz, he considered the effect on output spike rate of variation in cluster size. Somewhat complicating interpretation of his results, in these simulations Mel set peak conductance of each synapse to be inversely proportional to the local input resistance to insure similar local dendritic voltage response for any individual input, so distal synapses had 17-fold greater conductance than those near the soma.

As a benchmark, output rate of the target cell was computed as a function of cluster size for passive dendrites with pure AMPA synapses (Fig. 10aA), confirming the expected sublinearity due to saturation of driving force for the synaptic current (Rall, 1964). When all synapses carried predominantly NMDA current (gNMDA = 10× gAMPA), however, increasing cluster size steadily increased input efficacy (Fig. 10aB), revealing a net synergy between nearby synapses due to the activation of voltage-sensitive NMDA current. Alternatively, including a very high density

*Na^+ channels have not been observed on spines, but recently Ca^{2+} channels have (Jaffe *et al.*, 1994; Yuste and Denk, 1995).

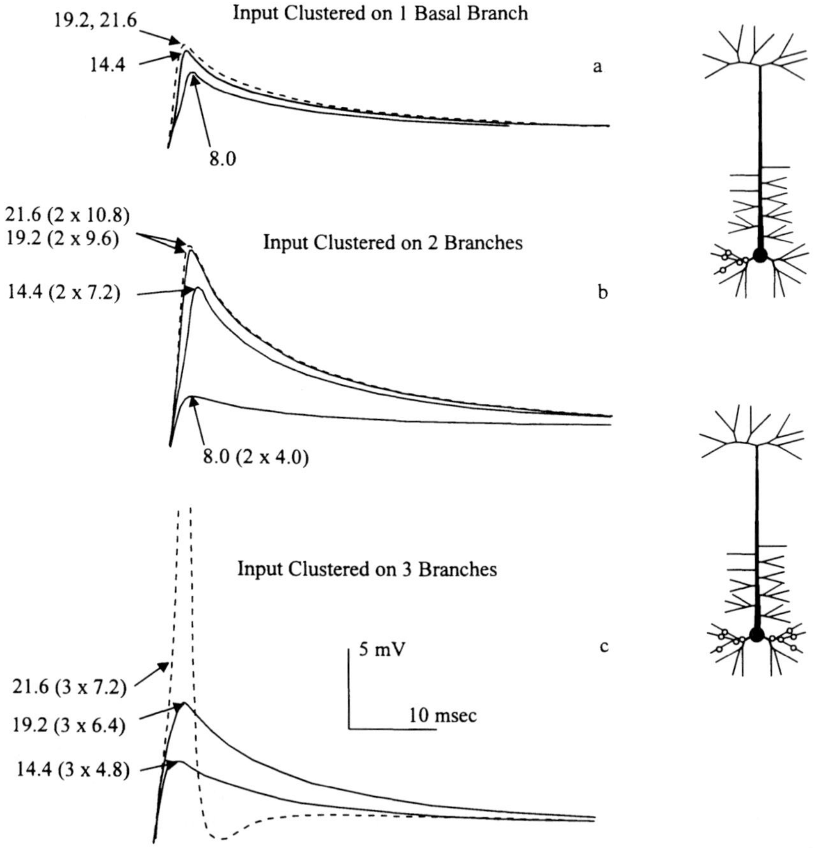

Figure 9. The effect of the same total synaptic input when distributed to one, two, or three basal branches. Compartment model layer 5 cell parameters as in Fig. 6, with dendritic Na^+, K_{DR}, and K_A conductance at 6, 3, and 0.5 mS/cm^2, respectively. For clarity, only two of six basal branches are shown. (a) Synaptic input clustered on a single basal subbranch. An 8.0-nS (peak *g*AMPA) input fired the branch, but further input had little additional effect. (b) The same total synaptic inputs divided equally into two identical subbranches from two basal branches. An 8.0-nS input (2 × 4 nS to each) was inadequate to trigger firing of either subbranch, and the resulting EPSP was smaller than caused by 8 nS when concentrated as in panel a. However, the response to 14.4 nS when divided into two clusters was much greater (about 11 vs. 6.5 mV), since each cluster of 7.2 nS was adequate to activate dendritic Na^+ currents, and both subbranches fired. (c) The same total synaptic input now divided into three clusters on three subbranches branches (only two shown). A 14.4-nS input had much less effect when divided into three, since the local input was inadequate to fire any of the three subbranches, but 21.6 nS (7.2 nS to each subbranch) had much more efficacy, since each of the three subbranches fired, in turn firing the cell. Thus the pattern of distribution of synaptic input to basal branches affects efficacy, with the all-or-none firing of isolated dendritic subbranches an elemental unit of excitation. From P. Rhodes (in preparation).

(50 mS/cm^2) of a high-threshold Ca^{2+}-like current in the distal dendritic segments only (all channels were excluded from the trunk) caused a sensitivity to clustering, even without any NMDA conductance (Fig. 10aC). Mel observed a decrement with further increases in cluster size, reflecting the balance between the payoff in activating the dendritic depolarizing current and the diminishing effect of adding further synapses.

Mel (1992) described this form of input summation as "Sigma-Pi," to refer to the sum (Σ) of local products (Π) of clusters of input, and showed it has great power in discriminating afferent patterns. The confounding between overlapping patterns which occurs in linear association matrix memories is greatly reduced—allowing linearly inseparable patterns to be discriminated by the cell output—when there are a large number of afferent fibers, images are represented by sparsely

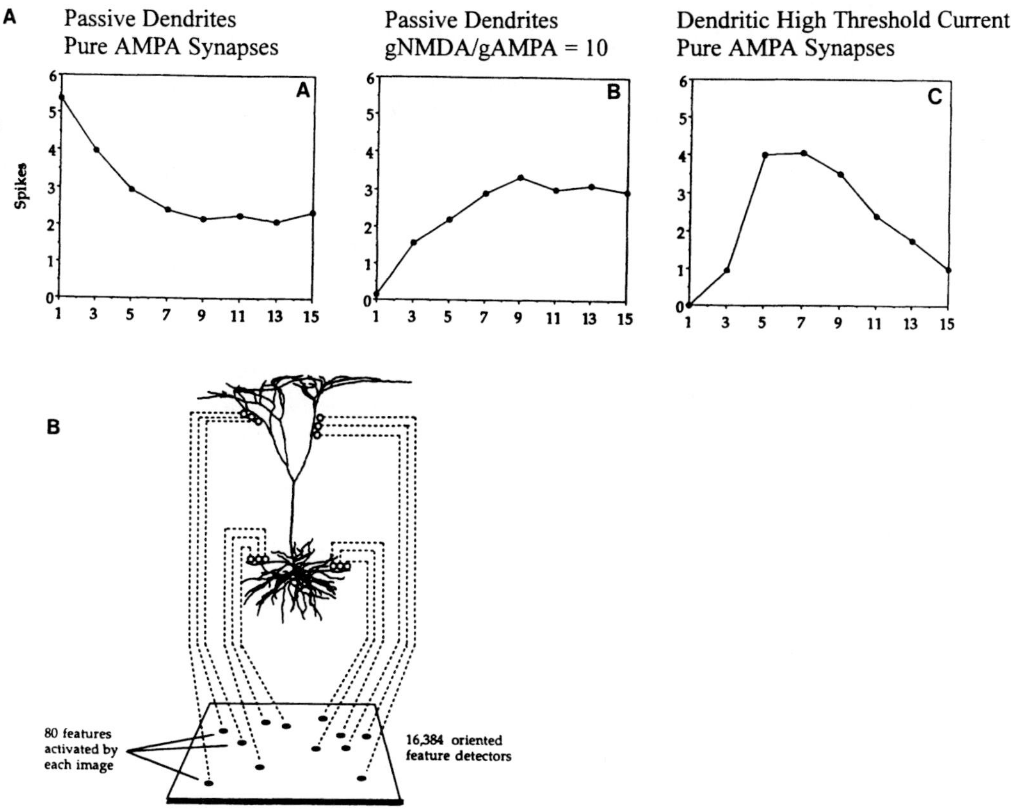

Figure 10. Clustering of synaptic inputs can increase efficacy when voltage-gated synaptic or membrane currents are present. Mel (1993) assessed the effect of clustering of inputs to a simulated layer 5 pyramidal cell. One hundred synaptic inputs were grouped in clusters, each segregated onto a 40-μm compartment, placed at random locations throughout the dendritic tree. Each synapse was asynchronously active at 100 Hz. Synapse conductance magnitudes varied inversely with local input resistance, and so were 17-fold greater near the soma than in distal segments. Output rate of the target cell in spikes/100 msec is plotted versus number of synapses per cluster. (A-A) With input to passive dendrite transmitted by purely AMPA synapses, the expected sublinearity was observed as clustering increased, reflecting mutual saturation of driving force between nearby inputs. (A-B) With synaptic current primarily NMDA, unclustered input did not produce enough dendritic depolarization to activate NMDA current, and so did not drive any firing, but increasing cluster size increased the effect of the 100 inputs. (A-C) Alternatively, with very high density (50 mS/cm^2), high-threshold Ca^{2+}-like current in the distal dendritic segments (channels were excluded from the trunk), clusters of inputs enhanced efficacy. In this case, however, beyond seven synapses per cluster further increases began to reduce efficacy, presumably reflecting the balance between turning on the dendritic current and the diminished effect of adding synapses after this had been achieved (see also Figs. 6 and 7). (B) Mel suggested that cluster sensitivity in a pyramidal cell with many synaptic inputs provides a computational substrate enabling a selective response to an enormous set of distinct input patterns. From Mel (1992, 1993).

distributed sets of these afferents, and the local voltage threshold imposes a cluster requirement.

4.4.3. "Coincidence Detection" on a 15- to 20-msec Time Scale

Whether the notion of "coincidence detection" can be sensibly related to the triggering of a branch depends on the time window within which inputs must "coincidently" arrive. While Softkey (1993) has argued for a very small time window (1 msec) of coincidence discrimination in cells with active dendrites, his simulations depended on dendritic currents tenfold more powerful than those suggested by data, and further utilized modifications of kinetics for the Na^+ and K_{DR} currents which are unconventional and thus hard to assess. He observed, however, that extremely strong K^+ currents could sharpen temporal discrimination by quickly curtailing dendritic depolarizations, so that if two inputs are to sum, they must arrive coincidently, a conjecture which bears further study in light of a recent report of extremely high K^+ channel densities in CA1 apical trunks (Hoffman *et al.*, 1996).

Mel *et al.* (1995) found that in simulated pyramidal cells with either active or passive dendrites, the mutual synchrony of 100 randomly placed inputs was rather *unimportant* in determining the output rate of the target neuron. Indeed, increasing the regularity of the spacing between arrivals at individual synapses increased their effect, presumably because saturation of driving force at each synapse was reduced when its presynaptic arrivals were spaced apart.

An alternative way to examine the effect of coincidence among synaptic inputs and its interaction with dendritic currents is to consider its effect on the triggering of a branch firing. To test this, three synaptic inputs were applied to a subbranch of a compartment-model layer 5 cell, with dendritic current densities and other parameters as in Sections 4.2 and 4.3. The effect of the time interval between inputs, a measure of their degree of synchrony, on the somatic EPSP was compared with passive versus active dendrites (P. Rhodes, unpublished observations). Surprisingly, depending upon the total synaptic conductance applied, the effect caused by synaptic inputs arriving synchronously was often no greater and sometimes *less* than when they arrived spread over a 20-msec interval (10 msec between each of the three inputs), presumably because the later inputs in a staggered train arrive when local dendritic voltage has repolarized, increasing their synaptic current. When the spacing of inputs was further increased, so that they arrived over 30 rather than 20 msec, the EPSP diminished. This issue awaits fuller examination taking into account the effects of background activity, passive parameters, membrane current densities, and other relevant variables; with that caveat, preliminary results seem to suggest that synchronous arrival within a 15- to 20-msec window is the temporal scale of meaningful coincidence for triggering branch firings in pyramidal cells with dendrites incorporating active channels at the density suggested by recent data, and hence in this context the precise time of spikes is not meaningful at shorter time scales.

4.4.4. Boolean Logic Circuitry in Coordinated Excitatory and Inhibitory Connections

The effect of temporally coincident versus staggered inputs on the efficacy of excitatory synaptic conductances is somewhat related to the possibility that the effect of ordering the temporal and spatial relationship of synaptic excitatory and inhibitory inputs might give rise to Boolean logic computations. For example, when an EPSP is followed immediately by a more proximal IPSP, its effect is quenched, implementing the AND-NOT logical element, in which input effect is dependent upon arrival of the EPSP AND-NOT an immediately subsequent arrival of an appropriately located IPSP. Alternatively, a local threshold for excitation of active current is viewed as implementing an AND operation. A considerable literature exists on this subject (e.g., Shepherd and Brayton, 1987; Rall and Segev, 1988; Shepherd *et al.*, 1991), which tends to focus on the implications of active spines rather than active dendrites* and so is outside the scope of this review.

4.5. Hotspots?

Hotspots, hot patches of excitable dendrite, have been conjectured to be responsible for spikelets in axotomized motoneurons (Eccles *et al.*, 1958; Kuno and Llinás, 1970a, b) and intact hippocampal cells (Figs. 1 and 2; Spencer and Kandel, 1961). While the notion of hotspots continues to animate ideas about the distribution of active currents in the dendrites (Reuveni *et al.*, 1993; Yuste *et al.*, 1994) and their function in boosting distal inputs, and compartment simulations have been developed (M. M. Millonas and P. S. Ulinski, n.d.) in which hotspots cause spikelets,† wherever a direct survey of channel density has been made by sampling along the apical trunk with patch electrodes, evidence has supported fairly uniform densities of channels in the apical trunk (Huguenard *et al.*, 1989; Magee and Johnston, 1995b; Stuart and Sakmann, 1994) of pyramidal cells. Spencer and Kandel argued that the small size of the fast prepotentials they observed suggested the firing of a distal dendritic region, and asserted that the absence of full-blown inward propagation required the intervening dendritic membrane to be passive, but examination of this conjecture with compartment simulations of pyramidals with active dendrites does not support it. First, an apical dendrite of uniform conductance can fire distally, but propagation may still fail, as occurred in the example in Fig. 5b, so only an attenuated EPSP reaches the soma. In this case the EPSP resulting was broader than a spikelet, an example of the familiar dispersion of voltage transients as they travel through cable (Rinzel and Rall, 1974). This suggests that sharp spikelets are more likely to be generated close to the soma. Second, with the entire dendritic tree endowed with uniform conductance density, basal or apical branches with uniform conductance distributions can themselves readily fire, but fail to further propagate (Section 4.3), a result of the smaller branch being unable to sufficiently charge the capacitatively heavier abutting apical trunk or soma. However, in these instances the firings of proximal subbranches trigger small, relatively brief responses (Fig. 6b). The observations of Kuno and Llinás (1970a, b), who reported that synaptic input to chromatolyzed motoneurons elicited all-or-none partial responses of various size and latency and inferred electronically separate dendritic

*But see Shepherd *et al.* (1991), who considered the effects of small patches of active dendrite under passive spines along a short, isolated 1-μm dendritic branch.

†As the authors noted, the dendritic Na^+ conductance density incorporated into hotspots in these simulations (≈300 mS/cm^2) was comparatively high (Millonas and Ulinski, n.d.).

spike trigger patches, can be reinterpreted as the firing of separate motoneuron branches in a motoneuron dendritic tree of uniform conductance properties, rather than of hotspots.

5. Effects of Dendritic Active Currents on EPSP Shape

Amplification of synaptic inputs is often quantified by an increase in amplitude of the EPSP resulting from the input, but the effects of dendritic currents on EPSPs also include broadening of the EPSP, which can enhance summation among inputs even when individual EPSP peak amplitude is not itself enhanced. The effects of dendritic currents on EPSP shape have been the subject of a number of recent experimental and modeling studies.

5.1. Experimental Results on the Effect of Active Currents on EPSP Shape

5.1.1. Experiments with Minimal Extracellular Stimulation Found QX-314-Sensitive Broadening of the EPSP with Postsynaptic Depolarization

Sutor and Hablitz (1989b) stimulated intracellularly recorded layer 2/3 pyramidal cells with extracellular shock in layer 4 in the same "column" as the recorded cell, producing EPSPs which broadened with depolarization of the postsynaptic cell. This effect was eliminated with application of intracellular QX-314, but very little curtailed with 10 μM APV, and these authors proposed that active membrane currents were responsible for the broadening of the EPSP. Hirsch and Gilbert (1991) also studied synaptic inputs to layer 2/3 pyramids, utilizing "minimal" extracellular stimulation horizontally displaced 800 μm in layer 2/3. They observed EPSPs following threshold stimuli, with fast and slow IPSPs appearing at higher stimulation amplitudes, and found that depolarizing the postsynaptic cell increased EPSP magnitude and width (Fig. 11). Again, incorporating QX-314 in the recording pipette eliminated the broadening. The effect of QX-314 was assumed to suggest that Na^+ channels were responsible for much of the broadening, but recent work (Talbot and Sayer, 1996) has suggested that QX-314 suppresses not only Na^+, but also Ca_T (as well as some K^+ channels), so these experiments may be reinterpreted as suggesting the involvement of Na^+ and/or Ca_T channels in boosting. These studies did not address whether the responsible channels were in the soma, dendrites, or both.

5.1.2. Effects of Focal TTX on "Synthetic" EPSPs Initiated by Direct Current Injection into the Apical Trunk Suggest Somatic but Not Dendritic Na^+ Current Is Involved in Boosting

Stuart and Sakmann (1995) reached somewhat different conclusions from the other studies cited in this and the preceding section, using an elegant protocol with which to examine boosting of distal input to "thick"-trunked (Larkman and Mason, 1990) layer 5 pyramidal cells *in vitro.* They simultaneously impaled the soma and distal apical trunk, and injected artificial excitatory post-synaptic currents (EPSCs) with the dendritic electrode, using a current waveform of the same time course (0.3 msec, 20–80% rise time, 3 msec time constant of decay) as the fastest miniature synaptic currents they measured with the dendritic electrode. This mode of stimulation has the disadvantage of being restricted to the apical trunk, which has much lower input impedance than smaller branches. Further, current injection does not activate NMDA channels at all and so does not address their contribution to EPSP broadening (Thomson, 1986). Direct current injection does, however, allow for localization of input with certainty while permitting application of either focal or bath TTX, which could interfere with axonal stimulation protocols, and it also mitigates concern about polysynaptic stimulation that might arise (Sutor and Hablitz, 1989a) with axonal stimulation.

They found that EPSP amplitude measured at the cell body increased nonlinearly with peak input current either when the somatic resting voltage was depolarized from rest (Fig. 12a) or when EPSPs were quite large (most examples shown were >10 mV). Both findings suggest the participation of a depolarization-gated inward current. This enhancement of EPSP amplitude was abolished in bath TTX (Fig. 12b), indicating depolarization-gated amplification of inputs was due to Na^+ channels.

The authors presented three pieces of evidence suggesting the Na^+ channels involved were only at the soma, and not in the dendrites. First, the relationship between injected current magnitude and peak voltage at the dendritic electrode was linear for input magnitudes as high as several nA and dendritic voltage excursions up to 40–50 mV, suggesting that dendritic active currents did not enhance the local peak voltage attained when current was injected there. Second, the incremental enhancement of the EPSP waveform had the same time course whether measured by the stimulating electrode in the apical trunk or by the somatic electrode, consistent with the hypothesis that an active broadening at the soma passively propagated into the dendrites. Finally, when TTX was focally applied with a pipette, dendritic TTX application has no effect on EPSP boosting, whereas somatic application did (Fig. 12b, right panel).

Some questions arise in reflecting on these observations: First, Na^+ channels are known to be present in the apical trunk of these cells (Huguenard *et al.*, 1989;

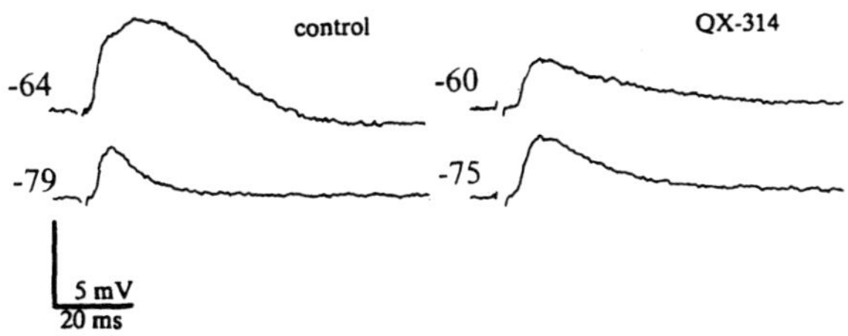

Figure 11. The effect of QX-314 on extracellularly stimulated EPSPs in layer 2/3 pyramidal cells in cat visual cortex *in vitro.* Synaptic inputs to layer 2 pyramids were elicited utilizing "minimal" extracellular electric shock 800 μm away in layer 2/3. Depolarizing the postsynaptic cell 15 mV greatly increased EPSP magnitude and width, an enhancement entirely absent when QX-314 was present in the recording pipette. From Hirsch and Gilbert (1991).

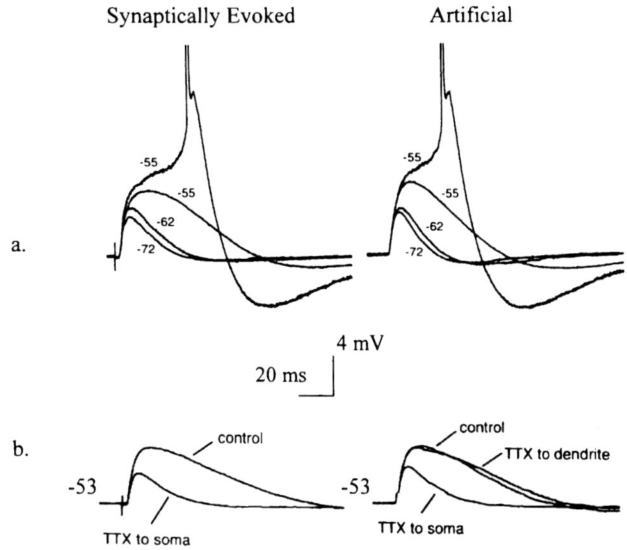

Figure 12. Effects of focal TTX on artificial EPSPs suggest somatic, but not dendritic Na^+ currents boost input. Artificial input (right) was directly applied to the apical trunk, (a) 380 and (b) 240 μm from the soma, by passing current in the waveform of a fast EPSC through a patch electrode while recording the response in the soma. Resting voltages stated should be adjusted downward about 7 mV to account for electrode junction potential. (a) Both synaptically evoked and artificial inputs broadened considerably with depolarization of the soma, suggesting the participation of depolarization-gated inward currents. (b) The location of Na^+ channels contributing to depolarization-dependent broadening was investigated by focally applying TTX to the soma or to the dendrite during stimulation. TTX focally applied to the apical dendrite had no effect on boosting, whereas somatic application did, both for synaptic and artificial stimulation, suggesting somatic Na^+ channels play a primary role in boosting EPSPs. The original figures were stretched so that the same scale applied to panels a and b. Experiments were done in bath APV, so voltage-dependent effects of NMDA receptor channels were not involved. Adapted from Stuart and Sakmann (1995).

Stuart and Sakmann, 1994), with a speed of activation at least comparable to the time to peak of the dendritic current waveform, so it is not clear why the authors did not find that peak voltage in the dendrites would be augmented as these channels activated. Second, the conclusion that a somatic active current source is implied by the similarity of the slow broadening of the EPSP they measured in the dendrites in the soma and dendrites (Fig. 5a, b of Stuart and Sakmann, 1995) seems uncertain. The very slow (half-width >50 msec) time course of the broadening and the low-pass nature of passive voltage propagation through dendrites suggests that activation of relatively sustained currents such as Ca_T or Na_P or a window current of the transient Na^+ channel (Rhodes and Thomson, 1996) in *either* dendrites and/or soma could result in similar broadening of the EPSP in the dendrites and in the soma. There also may be some uncertainty about the degree to which artificial EPSPs driven by injecting a waveform into the trunk mimic synaptic EPSPs. One indication of some difference in this regard is that despite the author's efforts to match waveforms, the rise time of evoked EPSPs was *shorter* (2.3 vs. 2.7 msec) and yet the half-width *longer* (16.5 vs. 14.6 msec) than obtainable with the EPSPs with a ratio between these EPSP shape parameters of 7.2 for evoked versus just 5.4 for artificial EPSPs. EPSP shape parameters are affected by location of input, but generally covary, and so the observation that artificial EPSP rise time was longer, but its width shorter, suggests that differences in input location and passive dendritic properties cannot reconcile the discrepancy in EPSP shape. There might have been some difference in the activation of nonlinear currents with synaptically delivered versus artificial input.

Further experiments with this elegant technique and others (see below) may shed light on these issues. In that regard, a similar study by the same group in CA1 cells found evidence for the role of Cs^+-sensitive currents in shaping EPSPs. Spruston *et al.* (1995b) dually impaled a single CA1 pyramidal cell, and used direct stimulation of the apical trunk with an EPSP voltage waveform to examine the factors affecting EPSP shape. They found, like Stuart and Sakmann (1995), that Na^+ channel amplification occurred when the artificial EPSPs were >5 mV, but only when Cs^+ was used to block K^+ currents activated by the EPSP, and suggested that in CA1 pyramidal cells K^+ currents ordinarily provide a counterveiling influence by curtailing the EPSP amplification, opposing the amplification caused by depolarizing currents.

5.1.3. Single-Axon EPSPs Broaden with Depolarization

Thomson and coworkers (Thomson *et al.*, 1988; Thomson and West, 1993; Thomson and Deuchars, 1995) have observed depolarization-dependent broadening in the EPSP triggered by a single presynaptic axon. In these pioneering studies of cortical circuitry, simultaneous impalements of two synaptically connected pyramidal cells were made, enabling control of the firing of one while recording EPSPs in the other. This technically difficult protocol allows certainty that successive stimulations activate the same presynaptic axon, overcoming the problem inherent in experiments where the presynaptic input is activated with extracellular shock. They found that in some, but not all connections EPSPs broaden, sometimes considerably (Fig. 13), at depolarized voltages. In some cases the broadening was curtailed when APV was bath applied, providing evidence that in these instances depolarization-gated enhancement of synaptic NMDA conductance contributed to amplification (Thomson, 1986; Thomson *et al.*, 1988; see also Nicoll *et al.*, 1993). Consistent with the results from extracellular stimulation experiments (Sutor and Hablitz, 1989b; Hirsch and Gilbert, 1991) with QX-314 in the recording electrode the depolarization-dependent broadening was not seen (A. Thomson, personal communication).

5.1.4. Focal Application of TTX and Ni^{2+} Suggests the Involvement of Dendritic Currents in Amplification of Extracellularly Stimulated EPSPs in CA1 Pyramidal Cells

Alzheimer and colleagues (Lipowsky *et al.*, 1996; Gillessen and Alzheimer, 1997) have reported evidence that both dendritic Na^+ (Fig. 14) and Ca_T channels boost EPSP amplitude in rat CA1 pyramidal cells. Working in young (P12–24) rats in slices with NMDA, $GABA_A$, $GABA_B$, and K^+ channels all blocked, they triggered synaptic stimulation by electric shock of stratum radiatum fibers, and applied either TTX (to block Na^+ channels) or amiloride or Ni^{2+} (to block Ca_T) with a pipette positioned several hundred μm from the soma.* They found that dendritic application of either Na^+ or Ca_T channel blockers reduce EPSP amplitude by about

*The interpretation of these measurements is somewhat uncertain because concentration of Ni^{2+} in the drug pipette was 5 μM, which has been reported in other studies (Avery and Johnston, 1996) to be below the concentration required to have substantial effect on T currents (C. Colbert, personal communication).

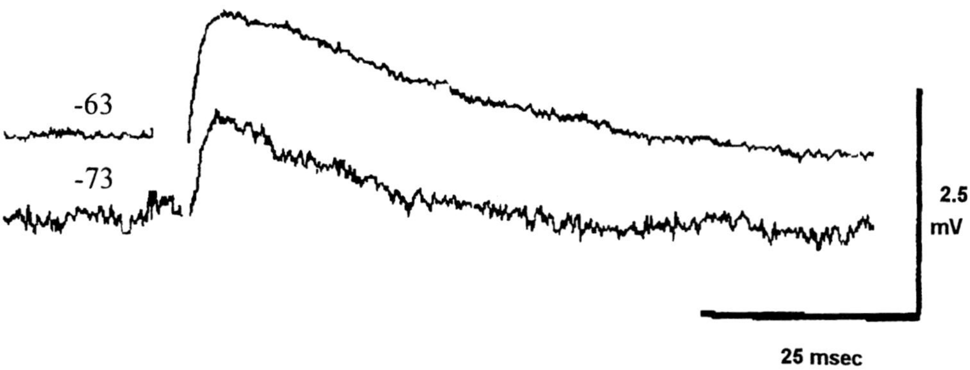

Figure 13. Single-axon EPSPs broaden with depolarization. Simultaneous impalements of two synaptically connected layer 5 neocortical pyramidal cells enable control of the firing of one while recording EPSPs in the other, and allowing certainty that successive stimulations activate the same presynaptic axon. In this case the presynaptic pyramid was in layer 5 and its postsynaptic target was an intrinsically bursting layer 5 cell. In this case (though not for all connections between pyramidal cells), the EPSP is enhanced by depolarization rather than diminished, suggesting the presence of voltage-gated synaptic and/or membrane currents. Data supplied by A. Thomson.

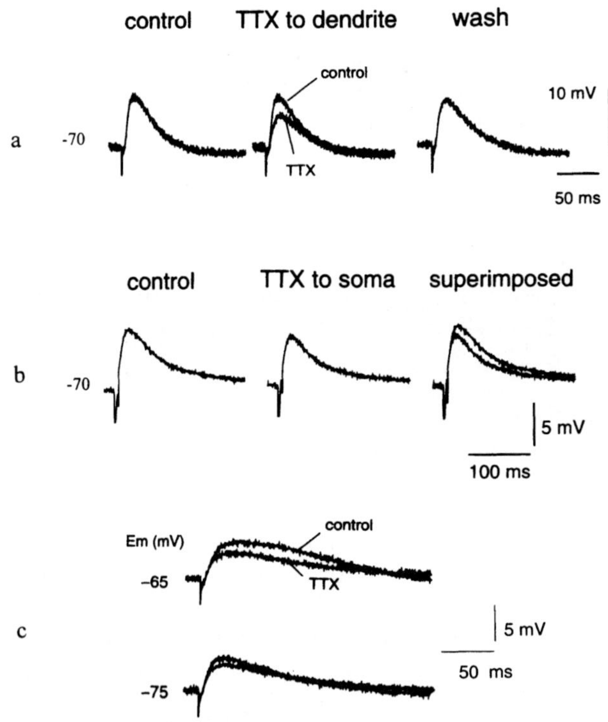

Figure 14. The effects of focal application of TTX on extracellularly evoked EPSPs in CA1 cells *in vitro.* (a) Application of TTX via extracellular pipette to the proximal apical dendrite of P12-24 CA1 pyramidal cells curtail the amplitude of EPSPs elicited by extracellular shock in slice. (b) When the TTX pipette was positioned at the soma, the effect on amplitude lessened, suggesting that somatic and perisomatic Na^+ channels were not the dominant component of the TTX-sensitive boosting of the EPSP. It should be noted that the blocker was estimated to diffuse into a sphere at least 100 μm in radius based on dispersion of food coloring, so the "dendritic" or "somatic" localization was of necessity imprecise. In panels a and b the soma was at rest (−70 mV). (c) Somatic depolarization enhanced the effect of dendritic TTX, a control suggesting that much of the TTX effect was not related to Na^+ channels in the presynaptic axon or terminal. Comparison of control traces at −65 and −75 mV indicates depolarization-dependent broadening. From Lipowsky *et al.* (1996).

25–30%, whereas with the antagonist pipette positioned at the soma the reduction in amplitude decreased to around 10% in both cases. Because food color in the pipette solution spread over a sphere of influence 100 μm in radius (Gillessen and Alzheimer, 1997), somatic application would be expected to affect channels in the proximal 100 μm of dendrite, so their results could be interpreted to indicate that almost all the boosting conferred by Na^+ and by Ca_T currents is caused by dendritic channels. As they noted, it may be that the effects of these two dendritic inward currents synergize, since depolarization caused by each current will enhance activation of the other.

In a somewhat similar study of hippocampal CA3 pyramidal cells, Urban and Barrionuevo (1996) stimulated with perforant path input from the entorhinal cortex, which targets the most distal portion of the apical dendrites, or with extracellular current in the immediate proximity of the distal apical dendrite of the recorded cell in the presence of transmission blockers. They found that depolarization of the postsynaptic cell consistently broadened the EPSP, while more rarely boosting its amplitude. Citing controls which establish no effect on transmission, they noted that bath application of 30 μm Ni^{2+} caused an approximate 20% reduction in amplitude for EPSPs above 2–3 mV, at odds with Magee *et al.* (1995). They also found that TTX focally applied to the apical dendrite blocked action potential generation by distal perforant path inputs, suggesting that dendritic Na^+ boosts the amplitude of distal EPSPs.

Finally, Nettleton and Spain (1996) reported evidence that active Na^+ and Ca_T conductances in the apical dendrite cause nonlinear summation of AMPA-mediated EPSPs in rat neocortical layer V pyramidal neurons. They used extracellular shock to elicit two synaptic inputs, with one stimulating electrode in layer 1 and a second in layers 2–5, with the two inputs separated by a cut from layer 2 through layer 6 as in Fig. 3a. In bath APV, the effect of the two inputs, separated by 5–10 msec, sometimes summed superlinearly, as measured by comparing the area of the separate versus combined EPSPs (in half the cases the integral resulting when the inputs were activated together was >110% the sum of them separately). Consistent with other studies (Sutor and Hablitz, 1989b; Hirsch and Gilbert, 1991; A. Thomson, personal communication), the superlinearity was abolished in all but 10% of the cells with QX-314 in the postsynaptic electrode, suggesting that Na^+ and/or Ca_T currents (Talbot and Sayer, 1996) were responsible for the amplification. They noted that when, instead of a second dendritic EPSP, the layer 1 input was accompanied by somatic current injection with the waveform of the second EPSP, the combined effect of both inputs together no longer summed nonlinearly, indicating the nonlinear interaction of EPSPs was due to dendritic and not to somatic voltage-gated currents.

5.2. Compartment Models

The effects of dendritic currents on EPSP shape are difficult to disentangle experimentally, because of the need to block selectively dendritic channels while maintaining synaptic input. Compartment models of pyramidal cells are a natural tool with which to examine how Na^+, Ca^{2+}, and K^+ currents in the soma and dendrites might alter EPSP shape and thus the efficacy of synaptic inputs.

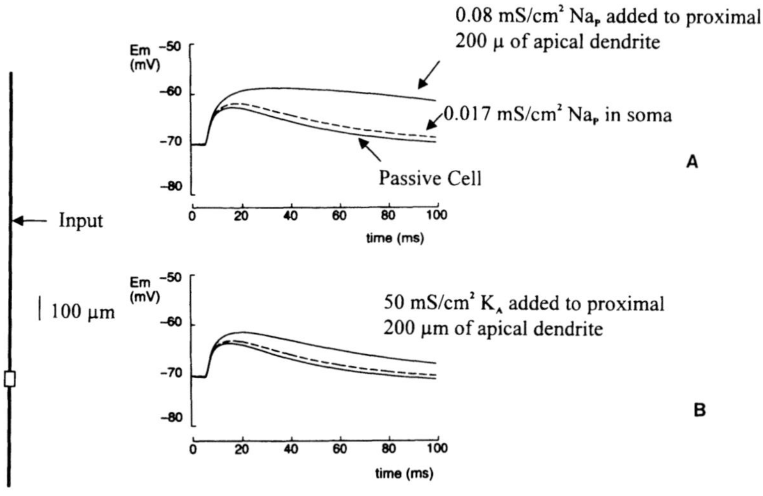

Figure 15. Compartment simulation of EPSPs onto an equivalent cylinder suggest dendritic Na_P current could broaden EPSPs. An equivalent cylinder reduction of a CA1 cell was utilized to examine the effects of dendritic Na_P and K_A currents on EPSPs. The model cell (inset) consisted of a 1200-μm-long, 3-μm-diameter apical dendrite and a 400-μm-long, 4-μm-diameter basal dendrite (both broken into 100-μm compartments) with R_m, C_m, and R_i values of 50,000 Ω-cm, 1 μF/cm², and 200 Ω-cm, respectively. A 50-nS synaptic conductance was applied to a point (arrow) on the apical dendrite 600 μm from the soma, and the resulting EPSPs compared with several channel distributions. All EPSPs were initiated from normal resting voltage. (A) Starting with a passive neuron (lowest trace), addition of 0.17 mS/cm² Na_P to the soma (dashed line) had little effect, but incorporating 0.08 mS/cm² Na_P in the proximal 200 μm of the apical trunk broadened the EPSP. (B) Adding 100 and 50 mS/cm² I_A conductance density to the soma (dashed line) and proximal 200 μm of dendrite curtailed the EPSP. From Lipowsky *et al.* (1996).

5.2.1. Equivalent Cylinder Modeling of a CA1 Pyramidal Cell

Lipowsky *et al.* (1996) utilized an equivalent cylinder model (details in Fig. 15) of a CA1 pyramidal cell to test the effect of Na_P in the soma and dendrites. They collapsed the apical and basal dendritic trees into cylinders 1200 and 400 μm long and 3 and 4 μm in diameter, respectively,* applied a 50-nS synaptic conductance at a point 600 μm distal to the soma, and examined the EPSPs produced at resting voltage with several conductance distributions. Incorporating a small (0.17 mS/cm²) density of Na_P into the soma had little effect on the EPSP (Fig. 15a), whereas they found that alternatively less density (0.08 mS/cm²) in the proximal 200 μm of the apical dendrite caused broadening. Adding K_A conductance (at a very high density; 100 and 50 mS/cm² in the soma and dendrite, respectively)

*The validity of using equivalent cylinder models (Rall, 1962) to analyze inputs requires that identical inputs be applied to the actual neuron at all comparable levels of the dendritic tree, a condition unlikely to hold during experimental simulation. Any nonlinearity in the synaptic response, whether due to the effect of dendritic voltage change on driving force or on the activation of dendritic membrane currents, invalidates the equivalent cylinder model. For example, when synaptic input arrives at a single apical oblique branch, enough depolarization may be elicited to activate dendritic currents, resulting in a branch firing (Section 4.3 and Fig. 7b), which affects the EPSP. However, if the dendritic tree is collapsed to a 3-μm-diameter equivalent cylinder, input impedance is much lower and the same synaptic input may not depolarize the cylinder sufficiently to activate either the dendritic currents or NMDA conductance. For this reason the equivalent cylinder simplification is not well suited to simulate synaptic integration where dendritic or synaptic voltage-gated currents have an effect.

curtailed EPSP width, providing a counterveiling influence (Fig. 15b), consistent with the experimental effect of Cs^+ in these cells (Spruston *et al.*, 1995b).

5.2.2. Compartment Model Simulations of Synaptic Responses in Layer 5 Regular Spiking Cells Suggest Dendritic Na^+, K_A, and Ca_T Conductances All Affect EPSP Shape

To examine the factors which might affect EPSP shape, Rhodes and Thomson (1996) compared compartment model simulations of a reconstructed neocortical layer 5 cell of the "thin" regular-spiking type with the EPSPs observed in single-axon pyramid-to-pyramid connections. The somatic membrane incorporated sufficient Na^+ and K_{DR} conductance density to allow the production of action potentials with physiological shape (100 mV height, 500 V/sec rise and 100 V/sec fall, and 0.6 msec width at half-height), and a 1-μm-diameter axon was included incorporating conductance densities identical to the soma. Two synaptic contacts were applied to two different apical oblique branches, a pattern of synaptic innervation typical of that revealed by serial section reconstruction of recorded pairs (Deuchars *et al.*, 1994). Synaptic peak *g*AMPA (with α function = 0.5 msec) was 1.33 nS, resulting in a 1.1-mV EPSP, consistent with the range measured in single-axon connections. NMDA conductance was colocalized with AMPA and examined at a number of amplitudes from 0.33 to 1.33 nS. NMDA kinetics, adjusted to 35°C, was that of Jahr and Stevens (1990), with external $[Mg^{2+}]$ fixed at 1 mM.

A variety of ion currents (Na^+, K_{DR}, K_A, low- and high-threshold Ca^{2+} conductance, at densities of 6, 3, 0.5, 6, and 0.5 mS/cm^2, respectively) were systematically introduced (individually and in combination) into the somatic and/or dendritic membrane, and the effect of postsynaptic depolarization on EPSP shape was compared with that observed in single-axon paired recordings. Though Na_P was not included, the voltage dependence of inactivation of the transient current was such ($V_{1/2} = -50$ mV) that a small (0.1% of transient peak) steady-state (window) current occurred around −40 mV. In each case, the EPSP was computed at rest (−71 mV) and at a depolarized holding potential (−63.5 mV). The effects of removal of the APV, Ni^{2+}, 4-AP, focal somatic and dendritic Na^+, and 0 Mg^{2+} were simulated.

The results included the following (Fig. 16):

1. Somatic Na^+, dendritic Na^+, and dendritic Ca_T currents all importantly contributed to EPSP broadening, whereas high-threshold Ca^{2+} channels did not. Somatic Na^+ channels, however, contributed to the broadening even though no persistent component was included. While the persistent mode (Alzheimer *et al.*, 1993) may well be involved, in these simulations the comparatively massive aggregate Na^+ conductance at and around the soma enabled a 1-mV EPSP to elicit enough activation of window current in the transient channel to boost the EPSP without a separate persistent current. Dendritic and somatic K_A channels both curtailed EPSP width.

2. With peak NMDA conductance (measured without Mg^{2+} blockade) at a ratio of 20/80 with colocalized AMPA, synaptic NMDA current was not integral to EPSP broadening, and simulated APV had little effect.

3. The experimentally observed broadening could alternatively be reproduced with passive dendrites if peak NMDA conductance was increased to the level of peak AMPA. In this case, APV had a marked effect.

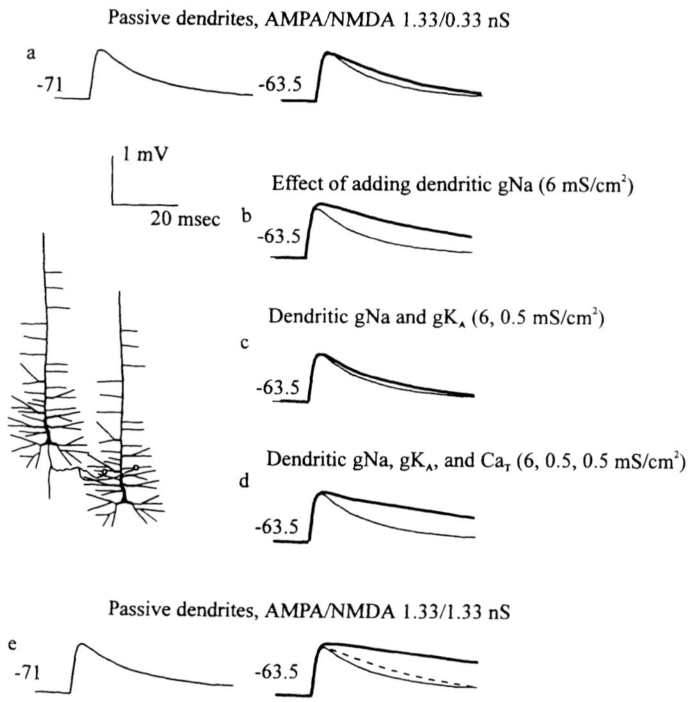

Figure 16. Effects of dendritic currents on simulated EPSPs. Synaptic input (peak *g*AMPA 1.33 nS, *g*NMDA 0.33 nS, kinetics as in Fig. 5, 35°C, 1 mM [Mg^{2+}]) was applied to two apical oblique branches (inset) of a regular-spiking layer 5 cell from rat visual cortex (morphology kindly supplied by G. Major). R_m, C_m, and R_i values of 50,000 Ω-cm², 0.75 μF/cm², and 200 Ω-cm, respectively. The soma included sufficient Na^+ density to produce 100-mV, 1-msec action potentials (100 mS/cm²), and an axon with the same conductance was included. The Na^+ channel model did not include a persistent component, but the midpoint of steady-state inactivation was −50 mV, allowing a small window current. (a) Passive dendrites. The modest broadening caused with depolarization to −63.5 mV was due to somatic Na^+ current (see text). The EPSP at −71 mV is overlayed (thin line). In panels b–d uniformly distributed dendritic conductances are cumulatively added. (b) Dendritic *g*Na added at 6 mS/cm² increased and broadened the EPSP. (c) Dendritic gK_A (0.5 mS/cm²) (kinetics similar to a D-channel) curtailed the EPSP at the depolarized voltage, counterveiling the effect of dendritic gNa. (d) Dendritic gCa_T, at 0.5 mS/cm², restored the broadening of the EPSP. Panels c and d imply predictions about the effects of dendritic application of K_A or Ca_T channel blockers. In panel 3 there is little APV sensitivity (not shown), indicating that dendritic Na^+ and Ca_T along with somatic Na^+ currents were the cause of the EPSP boosting observed with depolarization in this case. (e) Alternatively, a comparable broadening was achieved with passive dendrites, but an increased (1.33 nS) NMDA component. In this case, the effect of APV (dashed line) was considerable. From Rhodes and Thomson (1996).

4. The results therefore suggest that depolarization-gated broadening of single EPSPs may be caused either by membrane currents of the soma (Na^+) and dendrites (Na^+ and Ca_T), by a large synaptic NMDA conductance, or by a combination of these factors. The experimental observation that EPSP broadening is sensitive to APV in some but not all cases may therefore imply that a variety of ratio's between NMDA and AMPA conductance are exhibited in synapses, with at least on some occasions their peaks reaching comparable magnitude.

These simulation results account for the experimentally observed sensitivity of broadening to QX-314, and further predict that Ni^{2+} will reduce depolarization-dependent EPSP broadening, while 4-AP will broaden EPSPs (Rhodes and Thomson,

1996). Because I_A and Ca_T are transient, the time history of membrane voltage prior to arrival of the EPSP affects the degree to which Ca_T enhances or K_A curtails EPSP width, imposing a complex active temporal filtering on input efficacy, with dendritic hyperpolarization prior to input enhancing broadening due to dendritic Ca_T.

5.2.3. K_{sAHP} in the Proximal Apical Dendrite Curtails EPSP Amplitude and Width

Sah and Bekkers (1996) reported evidence for the presence of the slow afterhyperpolarizing [Ca^{2+}]-gated K^+ current (I_{sAHP}) (Schwindt *et al.*, 1988a) in the apical dendrite of CA1 pyramidal cells, and further that dendritic I_{sAHP} current can sculpt EPSPs. To establish the location of this current, they first measured the time constant with which two types of IPSP, one known to synapse onto the soma and the other onto the apical dendrite, "shut down" after clamped somatic voltage was stepped from a poststimulus command of −50 mV to −65 mV, near IPSP reversal potential. The somatic effect of the more distally located inhibitory synaptic input faded more slowly than did the somatic IPSP, with time constants of 8 and 3 msec, respectively, presumably because of the time it takes for voltage transients to travel through cable. The same measurement of shutdown time constant (Fig. 17a) was made (during other trials) for I_{sAHP} following its activation with a short current step to 0 mV, finding a similar value ($\tau \approx 8$ msec) to that for the distally located IPSP. They thus inferred a primarily dendritic location for the K_{sAHP} channels. They found that when I_{sAHP} was turned on following a current step, subsequent EPSPs due to synaptic input (via stratum radiatum) to the apical dendrites were curtailed in amplitude and width, while that to the basal dendrites (via stratum oriens) was reduced in width only (Fig. 17c), suggesting I_{sAHP} was present in apical, but not the basal dendrites.

Sah and Bekkers supported this inference with simulations incorporating three alternative distributions of I_{sAHP}, somatic, proximal, and entire apical dendrite, in an equivalent cylinder compartment model of the CA1 pyramid ($C_m = 0.7$ $\mu F/cm^2$, $R_i = 100$ Ω-cm, $R_m = 50{,}000$ Ω-cm^2). With these parameters, the rate at which the current shut off *in vitro* during a voltage clamp step from −50 mV to −65 mV was best reproduced with I_{sAHP} distributed within the proximal 150 μm or so of the apical dendrite (Fig. 17b). Thus, from this line of inference, simulation, and measurement they conclude that I_{sAHP} channels in the proximal 100–200 μm of the apical dendrite diminish EPSPs arriving at the distal apical dendrite. They further noted the intriguing possibility that modulators such as norepinephrine (NE) which reduce I_{sAHP} might selectively enhance the effect of those synaptic inputs which arrive on the CA1 distal apical dendrite and might selectively gate LTP through its action on dendritic I_{sAHP} channels.

6. The Effect of Dendritic Active Currents in Shaping the Intrinsic Firing Properties of Pyramidal Cells

The previous section was concerned with the amplification of subthreshold synaptic inputs, whereby active dendrites have functional effect by altering the efficacy of input in bringing the postsynaptic cell to firing threshold. Another means by which dendritic currents may alter the efficacy of input is to alter the

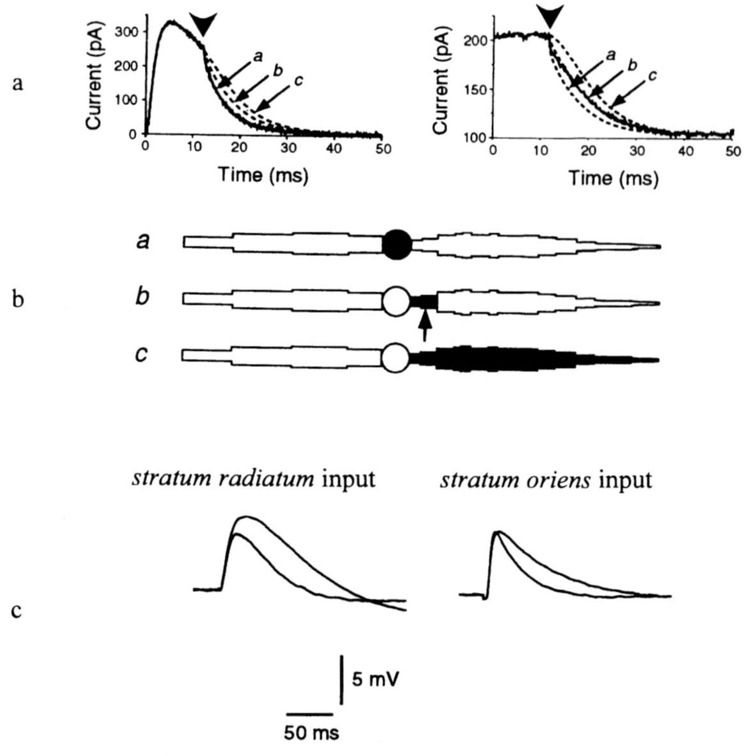

Figure 17. Indirect evidence for the presence of the slow after-hyperpolarizing [Ca^{2+}]-gated K^+ current (I_{sAHP}) in the proximal apical dendrite of CA1 pyramidal cells. Sah and Bekkers (1996) measured the time constant for IPSPs of known location to shut off when during activation of the IPSP, voltage was clamped to the reversal potential (−65 mV) and compared the same measurement with I_{sAHP} to infer the location of K_{sAHP} channels. (a) After the somatic inhibitory postsynaptic current (IPSC) (left trace) is shut down (arrowhead), the current decays to rest with $\tau \approx 3$ msec. Following activation by a current step to 0 mV, I_{sAHP} shut down (arrowhead) with $\tau \approx 8$ msec, closer to the rate ($\tau \approx 10$ msec) for the distally located IPSP (not shown) suggesting a primarily dendritic rather than somatic location for the I_{sAHP} channels. (b) Sah and Bekkers supported this inference with simulations incorporating three alternative distributions of I_{sAHP}, somatic (*a*), proximal (*b*), and entire apical (*c*) dendrite in an equivalent cylinder compartment model of the CA1 pyramid ($C_m = 0.7\ \mu F/cm^2$, $R_i = 100\ \Omega$-cm, $R_m = 50{,}000\ \Omega$-cm^2). The simulated shutdown of an IPSC or I_{sAHP} localized with these three alternative distributions is overlayed (dashed lines in panel a) upon the experimental measurement. The *in vitro* measurement matched simulated shutdown for I_{sAHP} best with the channels distributed to the proximal apical dendrite. (c) Finally, they measured the effect of I_{sAHP} on EPSPs *in vitro*. Following activation of I_{sAHP} with a brief current step, EPSPs elicited (lower traces after I_{sAHP} activated) from excitation in *stratum radiatum* were curtailed more than inputs elicited from *stratum oriens*. From Sah and Bekkers (1996).

output, the train of action potentials in the axon, produced when somatic firing threshold is reached. Active currents in the dendrites of neurons are thought to control these so-called intrinsic firing properties in many neuronal types (Llinás, 1988), including neocortical pyramidal cells. There is a considerable literature on the intrinsic properties of cortical cells (e.g., Connors *et al.*, 1982; McCormick, *et al.*, 1985; Friedman and Gutnick, 1987, 1989; Connors and Gutnick, 1990; Mason and Larkman, 1990) and the role of many different currents in their production and modulation. Here we concentrate on the generation of fast bursts of spikes in pyramidal neurons, which is both thought to require dendritic active currents (Traub and Llinás, 1979; Rhodes and Gray, 1994; Pinsky and Rinzel, 1994) and to constitute an output signal of special effectiveness (Gamble and Koch, 1987).

6.1. Bursts

Bursts in cortical cells have been defined as a set of two or more spikes fired at a rapid rate (>200 Hz), riding atop a depolarization of the soma (Connors *et al.*, 1982). Presuming that all the spikes in a burst are successfully transmitted along the axon (a nontrivial assumption for high firing rates; P. Rhodes, unpublished observations), they could exert a powerful effect on their target cells for several reasons. First, a rapid barrage of presynaptic spikes could drive a large depolarization in the recipient dendrites and thus be well suited to activate depolarization-gated synaptic (NMDA) and membrane channels and to cause a buildup of Ca^{2+} in the spine (Holmes and Levy, 1990). The resulting enhancement of Ca^{2+} buildup is further magnified by the superlinear (fourth order) dependence on $[Ca^{2+}]$ of processes leading to long-term potentiation (LTP) induction (e.g., Ca^{2+}-calmodulin activation) (Gamble and Koch, 1987; Zador *et al.*, 1990). Indeed, the experimental induction of LTP was first observed (Bliss and Lomo, 1973) in response to high-rate inputs. The potential impact of burst output highlights the relevance of the possibility that dendritic currents may be responsible for transforming a threshold input into a burst output in pyramidal cells.

Models of Burst Generation Rely on Dendritic Active Current

An *intrinsic* burst is a firing pattern that is not dependent upon input, except insofar as that is needed to reach firing threshold for an initial spike. Thus, once triggered by a brief (several msec) threshold current injection, an intrinsically bursting cell can produce a full extended (≈15 msec) burst, including a set of spikes and depolarizing envelope, irrespective of whether the current input is sustained. This implies that the cell's intrinsic membrane currents are dominant in the generation of the burst itself. Models of burst generation both for hippocampal CA3 (Traub and Llinás, 1979; Traub *et al.*, 1991) and for neocortical layer 5 intrinsically bursting pyramidal cells (Rhodes and Gray, 1994) have been developed which take into account the intrinsic nature of bursting and which reproduce the range of output to different magnitudes of current step seen *in vitro* (Fig. 18a). The models combine high-threshold (L- or N-type) Ca^{2+} channels in the dendrites with a powerful Na^+-driven fast spike mechanism in the cell body. A lower density of Na^+ channels is required to be colocalized with the high-threshold Ca^{2+} channels in the dendrites in order for somatic firing to activate the dendritic Ca^{2+} current (Rhodes and Gray, 1994), and dendritic K^+ channels including I_C and slower [Ca]-gated K^+ currents all seem to be required to match measured with simulated patterns of response to current step. The burst of spikes is driven by a sustained Ca^{2+} channel-driven depolarization of the dendrites, which in turn drive the soma, creating a depolarizing envelope and the rapid burst of somatic Na^+ spikes (Rhodes and Gray, 1994). These compartment models produce "intrinsically" bursting model cells (Fig. 18b) because once the initial spike has propa-

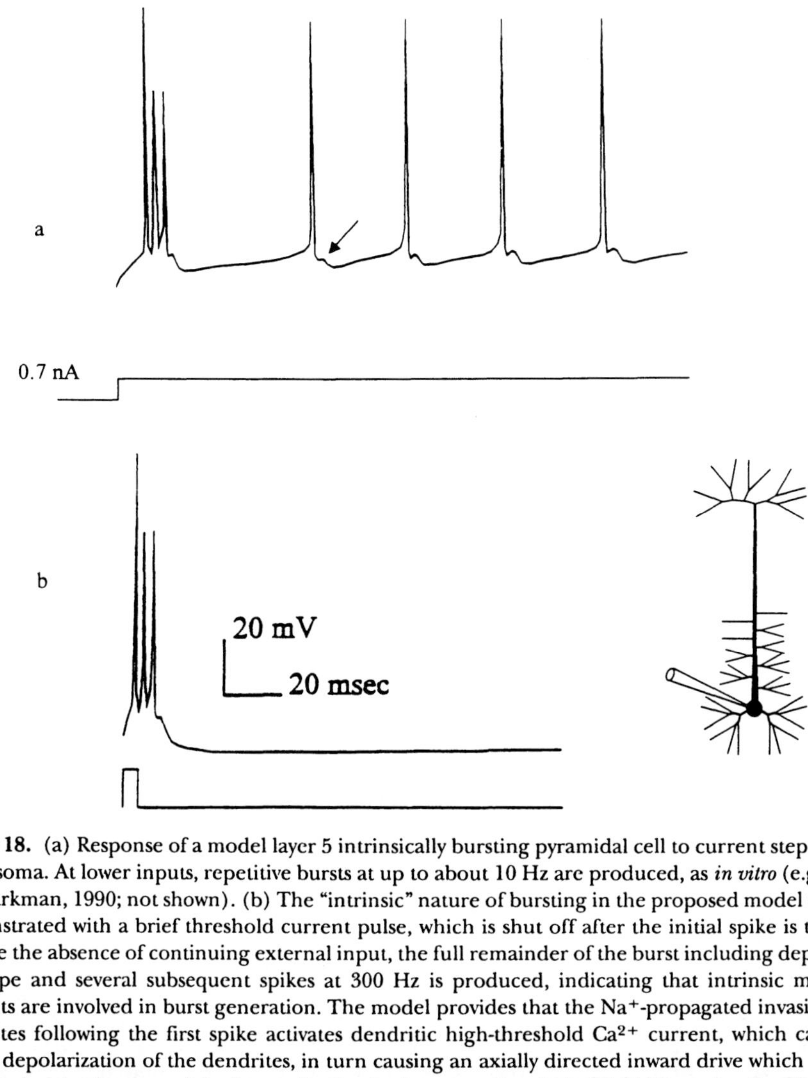

Figure 18. (a) Response of a model layer 5 intrinsically bursting pyramidal cell to current step injection at the soma. At lower inputs, repetitive bursts at up to about 10 Hz are produced, as *in vitro* (e.g., Mason and Larkman, 1990; not shown). (b) The "intrinsic" nature of bursting in the proposed model neuron is demonstrated with a brief threshold current pulse, which is shut off after the initial spike is triggered. Despite the absence of continuing external input, the full remainder of the burst including depolarizing envelope and several subsequent spikes at 300 Hz is produced, indicating that intrinsic membrane currents are involved in burst generation. The model provides that the Na^+-propagated invasion of the dendrites following the first spike activates dendritic high-threshold Ca^{2+} current, which causes sustained depolarization of the dendrites, in turn causing an axially directed inward drive which produces the depolarizing hump and the burst of spikes in the soma. Dendritic $[Ca^{2+}]$ lingering for 100 msec or so following influx during the initial burst activates dendritic $[Ca^{2+}]$-gated K^+ current, which quenches subsequent dendritic Ca^{2+} spikes, preventing repetitive bursting at rates faster than 10 Hz. Instead, subsequent somatic spikes are only accompanied by depolarizing afterpotentials (arrow) caused by the abbreviated dendritic Ca^{2+} spike. Further details of currents and simulation in Rhodes and Gray (1994).

gated into the dendrites and activated a dendritic high-threshold Ca^{2+} spike, the sustained Ca^{2+} current depolarizes the dendrites for the remaining duration of the burst, driving the soma, without any necessity for continuing current injection. Manipulations of this compartment-model neuron in which the dendrites are eliminated in a sequential fashion show that the apical dendrite is not required for burst generation, but the basal dendrites are (Rhodes and Gray, 1994; P. Rhodes, unpublished observations), consistent with the fact that layer 5 intrinsic bursting responses had been reported in slices where a horizontal cut in layer 4 truncated the apical dendrite (A. Telfeian and B. Connors, unpublished observations). Thus active currents in dendrites and the interplay between the dendritic Ca^{2+} spike and the somatic fast Na^+-driven spike are essential elements of burst generation in models of intrinsically bursting pyramidal cells (Traub and Llinás, 1979; Rhodes and Gray, 1994; Pinsky and Rinzel, 1994).

6.2. Ca^{2+} and Na^+ Channels in the Dendrites of Regular-Spiking Cells and the Transformation of Regular-Spiking toward Intrinsically Bursting Properties

Regular-spiking (RS) pyramidal neurons (Connors *et al.*, 1982) are the most plentiful neuronal type in the cortex, appearing in all cell layers. Their characteristic physiological response following somatic current step injection is a train of single spikes with increasing interspike interval (Connors *et al.*, 1982; McCormick *et al.*, 1985; Mason and Larkman, 1990), which can be modeled without incorporating active currents in dendrites (Durand and Yuan, 1990). Indeed, Kim and Connors (1993) classified two subtypes of dendritic responses in direct dendritic records in the apical trunk of layer 5 pyramidal cells. One class exhibited large Ca^{2+}-dependent dendritic depolarizations, correlated with thicker apical trunks, and may correspond to the intrinsically bursting cell type, while the second subtype had a thinner apical trunk, fewer oblique branches, and smaller and much narrower dendritic spikes, and may represent regular-spiking cells, both because of their respective morphologies (Mason and Larkman, 1990; Chagnac-Amitai *et al.*, 1990) and because of the proposed association of a dendritic Ca^{2+} spike with burst generation (previous section). In the first (putative bursting) class, they noted that spike amplitude did not decrease with distance from the soma, clearly indicating local electrogenesis in the dendrites, while in the second (putative regular-spiking) class, dendritic spike amplitude reduced steadily with distance along the trunk. From comparisons of spike decrement with distance along the trunk in compartment simulations they suggested that the rate of decrement in the second (putative regular-spiking) subtype could be consistent with passive dendrites.

Other evidence, however, points to the presence of both Ca^{2+} and Na^+ channels in the dendrites of RS pyramids. First, it has become increasingly clear that when K^+ currents are suppressed by increase in external $[K^+]$ (Morton *et al.*, 1994), bath application of the I_A blocker 4-aminopyridine (Andreason and Lambert, 1995), or injection of $[Ca^{2+}]$ chelators (Friedman and Gutnick, 1989), the intrinsic output pattern of regular spiking pyramidal cells transforms toward a bursting mode (see Section 7.2).* Most models for burst generation in pyramidal cells (previous section) depend upon dendritic Na^+ and Ca^{2+} currents to initiate and sustain a dendritic depolarization driving a burst of somatic spikes, and so the observation of bursting in RS cells implies that those depolarizing currents exist in RS dendrites (Rhodes *et al.*, 1995), but that under normal conditions both K_A and

*These data argue against the view that the output mode of a pyramidal cell is primarily determined by its morphology, since in these studies the morphology is fixed, but the output mode changes as currents are modified. Rather, a blend of factors, including dendritic morphology, the absolute and relative densities of dendritic Na^+, Ca^{2+}, and K^+ currents, and even the particulars of Ca^{2+} chelation all determine the voltage events in the dendrites and their consequent impact on intrinsic firing properties.

[Ca^{2+}]-gated K^+ currents in the dendrites of regular-spiking cells prevent expression of a dendritic Ca^{2+} spike.

Physiologically accurate models of a regular spiking must be capable of shifting to bursting mode following suppression of K^+ currents in the ways reported, while at the same time reproducing the familiar graded series of regular-spiking responses (Connors *et al.*, 1982; McCormick *et al.*, 1985; Mason and Larkman, 1990) under normal conditions. Model layer 5 regular spiking morphology pyramidal cells which satisfy these constraints have been developed (Rhodes *et al.*, 1995) (Fig. 19). Both dendritic Na^+ and high-threshold Ca^{2+} channels were required, along with a set of dendritic [Ca^{2+}]-gated and other K^+ currents. To avoid production of bursts under normal conditions, dendritic Ca^{2+} conductance density was lower for RS than for IB dendrites (3 versus 6–8 mS/cm^2). Simulated RS pyramidal cells with these properties produced the regular-spiking train of fully repolarizing accommodating single spikes under ordinary conditions, but when an [ethylene glycol-bis N,N,N^{1}N^1-tetraacetic acid] (EGTA)-like buffer (dissociation constant 1 μM, 1.5 μM sec^{-1} and 1.5 sec^{-1} forward and backward rate constants) with the internal concentration used by Friedman and Gutnick (1989; estimated experimentally at 1.3 mM; see Friedman and Gutnick 1989) was added in the simulation, suppression of [Ca^{2+}]-gated K^+ currents in the dendrites permitted dendritic Ca^{2+} current to sustain a dendritic Ca^{2+} spike, driving bursts (Rhodes *et al.*, 1995). The ability of dendritic K^+ currents to shift the firing mode of RS cells points to a mechanism

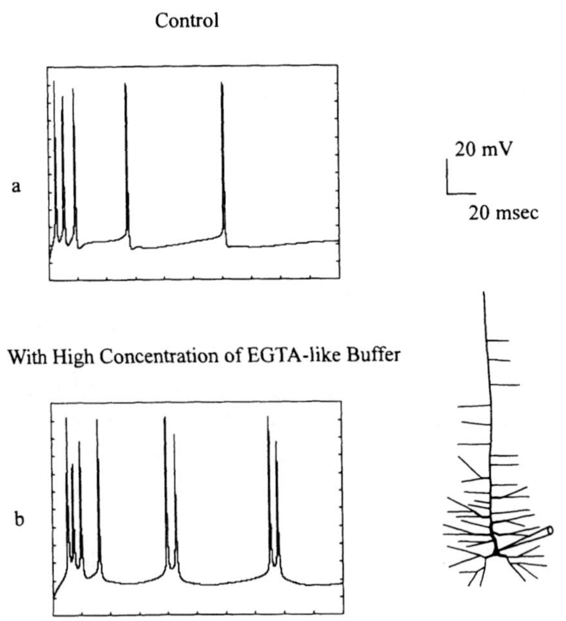

Figure 19. Suppression of dendritic K^+ currents shifts the output mode of regular-spiking cells from single spiking toward bursting. Injection of high concentrations of the [Ca^{2+}] chelator EGTA has been reported (Friedman and Gutnick, 1989) to shift the output mode of regular-spiking neurons toward bursting. A model cell with morphology (inset) taken from a reconstructed regular-spiking layer 5 pyramid (courtesy of G. Major) was developed which produced the familiar accommodating train of single spikes in response to current step in control conditions (a), but which also began to burst (b) when a high (1.3 mM) concentration of an EGTA-like buffer (K_d1 μM and forward rate constant = 1.5 μM/sec) was introduced, as reported *in vitro* (Friedman and Gutnick, 1989). To obtain these results, it was necessary to include both Na^+ and high-threshold Ca^{2+} currents in the dendrites (somatic Ca^{2+} was not necessary) of the model RS cell (6 and 3 mS/cm^2, respectively) along with a set of [Ca^{2+}]-gated and other K^+ currents. The balance between the dendritic Ca^{2+} and K^+ currents was such that under normal conditions dendritic Ca^{2+}-gated K^+ currents suppressed the activation of dendritic Ca^{2+} current; when suppressed by high chelator concentration dendritic calcium currents were revealed, and the cell produced bursts (see Fig. 18). This example illustrates that for a fixed dendritic morphology either regular spiking or bursting output may be produced, and these alternative intrinsic firing properties can be controlled through modulation of dendritic currents. From Rhodes *et al.* (1995).

by which endogenous modulation of dendritic currents may switch the output mode of cortical pyramidal cells *in vivo* (Wang and McCormick, 1993).

7. Effects of Potassium Currents

7.1. Potassium Currents in Pyramidal Cells

Pyramidal neurons incorporate a remarkable number of distinct classes of potassium currents including the fast delayed rectifier (I_{DR}), the fast transient current known as the *A*-current (I_A), at least three K^+ currents gated by internal [Ca^{2+}] (Schwindt *et al.*, 1988a) differentiated in part by their speeds of activation, the hyperpolarization-activated I_h (McCormick and Pape, 1990), and the acetylcholine-blocked *M*-current I_M. The fastest [Ca^{2+}]-gated current, the *C*-current I_C, activates with a time constant in milliseconds, as do I_{DR} and I_A, and all three contribute to repolarization of the somatic action potential. A slower [Ca^{2+}]-gated K^+ current termed K_{mAHP} by Schwindt *et al.* (1988a) and I_M both activate over tens of milliseconds and are gated in part by depolarization, whereas the slow K_{sAHP} activates over hundreds of milliseconds and may be gated by [Ca^{2+}] alone (Schwindt *et al.*, 1988a). K_{sAHP} is likely to contribute to the adaptation of spike rate in regular-spiking neocortical and hippocampal CA1 pyramidal cells (Madison and Nicoll, 1982, 1984), with its blockade curtailing adaptation.

There is very little direct knowledge about whether and in which types and densities these K^+ channels are present in dendrites (but see Hoffman *et al.*, 1996). I_A has recently been identified in the apical trunk of CA1 pyramidal cells (Andreason and Lambert, 1995) and it seems reasonable to suppose that [Ca^{2+}]-gated K^+ currents are present where Ca^{2+} channels are colocalized, since Ca^{2+} does not diffuse for long distances in dendritic cytoplasm (e.g., Hille, 1984), and Ca^{2+} channels are present throughout the dendritic tree (Regehr *et al.*, 1989; Jaffe *et al.*, 1992; Magee and Johnston, 1995b; Christie *et al.*, 1995; Markram and Sakmann, 1994; Yuste *et al.*, 1994; Schiller *et al.*, 1995). The following pages will argue that dendritic K^+ currents have profound and varied effects on pyramidal cell function, and ascertaining their densities and distribution in the dendrites is an underexplored area of great value.

7.2. Dendritic K^+ Currents Suppress Dendritic Ca^{2+} Spikes and May Thereby Regulate Firing Mode between Bursting and Single Spiking

The control exerted by K^+ currents over burst generation is revealed by several manipulations which suppress K^+ currents and which transform regular-spiking pyramidal cells toward burst production. Elevating external [K^+], which reduces the driving force for all the K^+ currents, elicits bursting in regular-spiking CA1 (Andreason and Lambert, 1995; Jensen *et al.*, 1994) and cortical (Y. Amitai, personal communication) pyramidal cells. Injection of large concentrations of the [Ca^{2+}]-chelator EGTA, which presumably reduces [Ca^{2+}]-gated K^+ currents, has been reported (Friedman and Gutnick, 1989) to convert regular-spiking pyramidal cells to repetitively burst in cortical slice.

These experiments did not address whether suppression of somatic or dendritic K^+ currents was responsible for shifting regular-spiking cells toward bursting mode, but models of bursting (Traub and Llinás, 1979; Rhodes and Gray, 1994; Rhodes *et al.*, 1995; Pinsky and Rinzel, 1994) suggest dendritic K^+ currents are central. In these models the initial burst is often followed by a train of single spikes because dendritic Ca^{2+}, elevated during the dendritic Ca^{2+} spike associated with the initial burst, enables increased activation of dendritic $[Ca^{2+}]$-gated K^+ current, which then prevents the development of successive dendritic Ca^{2+} spikes (see Discussion in Rhodes and Gray, 1994). A causal role of I_C in quenching bursts was suggested in these simulations by increasing dendritic I_C in compartment models of these cells in a graded manner, which led to a graded transition from full bursting to smaller bursts with fewer spikes, to single spiking (P. Rhodes, unpublished observations).

Finally, the involvement of I_A in quenching dendritic Ca^{2+} events has been suggested by experiments in which dendritic impalements were made in the distal apical trunk of mature CA1 pyramidal cells 350–400 μm distal from the soma (Andreason and Lambert, 1995). The authors found that after blocking I_A with 4-aminopyridine (4-AP), the fast-spiking dendrites in which dendritic current injection produced an accommodating train of single spikes began to produce an initial fast spike followed by a broader, Ca^{2+}-dependent spike complex (Fig. 20), suggesting that under normal conditions I_A curtailed Ca^{2+} spiking in the dendrites (Andreason and Lambert, 1995). This finding suggests that in regular-spiking CA1 cells, as in cortical RS cells (Friedman and Gutnick, 1989; Rhodes *et al.*, 1995), sufficient depolarizing dendritic currents are present to drive bursting, but that under normal conditions in CA1 cells activation of I_A quenches the production of dendritic Ca^{2+} events.

Thus it appears plausible that at least dendritic I_C and I_A are involved in control of bursting in cortical pyramidal cells. The involvement of dendritic K^+ currents in controlling the firing mode suggests that agents which act on these currents may be capable of shifting pyramidal cell output mode and functional state *in vivo* (Wang and McCormick, 1992), with dendritic Ca^{2+}-driven spikes and the bursts they drive suppressed as dendritic I_A and I_C activation increases.

7.3. Dendritic K^+ Currents Can Extinguish Backpropagation into Branches

Firing at the cell body initiates a voltage wave which invades the dendritic tree in both hippocampal (Richardson *et al.*, 1987) and neocortical (Rhodes and Gray, 1993; Stuart and Sakmann, 1994) pyramidal cells, with propagation actively sustained by dendritic Na^+ channels (Rhodes and Gray, 1993, 1994; Stuart and Sakmann, 1994; Mainen *et al.*, 1995). Potential consequences of the backpropagation of a wave of voltage into the dendrites include implementation of the Hebbian linkage between somatic firing and strengthening of synapses active just before firing (Rhodes and Gray, 1993; Section 8), along with a variety of processes (e.g.,

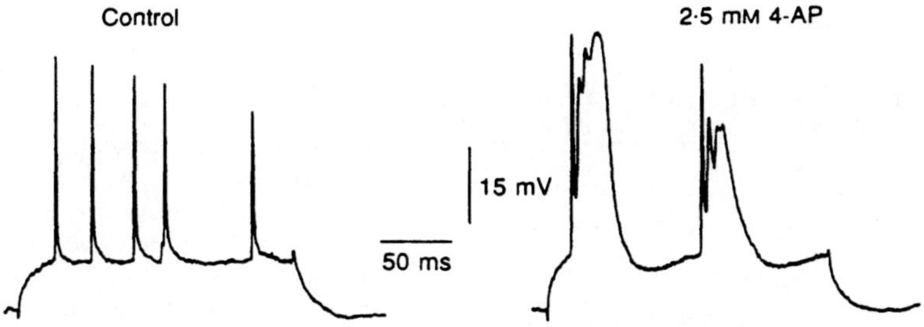

Figure 20. Suppression of I_A enhances production of dendritic Ca^{2+} spikes. Dendritic impalements were made in the distal apical trunk of mature CA1 pyramidal cells, 350–400 μm from the soma. After blocking I_A with 4-aminopyridine (4-AP), dendrites in which current injection normally produced an accommodating train of single spikes became more excitable, with an initial fast spike followed by a broader Ca^{2+}-dependent spike complex. These observations suggested that under normal conditions I_A controls Ca^{2+} spiking in the dendrites (Andreason and Lambert, 1995).

Llinás and Sugimori, 1979) gated by an influx of Ca^{2+} into all the dendrites where the backpropagation reaches.

Jaffe *et al.* (1992) observed both Ca^{2+} and Na^+ inflow in CA1 pyramidal cells with optical methods, and found that while initial firing of the cell was accompanied by Ca^{2+} and Na^+ entry throughout, with repeated firing the pattern of distribution of entry became increasingly proximal, suggesting that the backpropagated invasion of the apical dendrite was diminishing as the train of firing proceeded. Spruston *et al.* (1995a) confirmed this finding by measuring the backpropagated voltage wave directly using a dual impalement of the soma and distal apical trunk, finding activity-dependent action potential invasion which began to fail as a pulse train continued. Jaffe *et al.* (1992) postulated activation of Ca^{2+}-gated K^+ currents as a candidate mechanism for the suppression of invasion of the dendrite. This mechanism would depend on the buildup of Ca^{2+} with each successive backpropagating wave, as was subsequently observed in neocortical layer 5 cells (Markram and Sakmann, 1994).

Experimental data indicating that dendritic K^+ currents can block LTP induction comes from the already noted study (Section 5.2.2) of Sah and Bekkers (1996), who found that adrenergic suppression of K_{sAHP} channels, which they localized to the proximal apical dendrite, diminished the strength of tetanizing input required to induce LTP (see Section 8 for further discussion).

7.4. K^+ Currents Can Dynamically Modify the Spatial Range of Synaptic Interaction

In one of the most systematic compartment model studies of the functional effects of dendritic currents, Wilson (1995) examined the effects of I_{DR} and I_A-like currents on dendritic cable properties, synaptic efficacy, and the interaction between coactive synapses.

Using a simplifying geometry of a long cable attached to a somatic compartment, he charted the effects of dendritic I_{DR} and I_A on the interaction between synaptic inputs as a function of their separation in time and distance along the

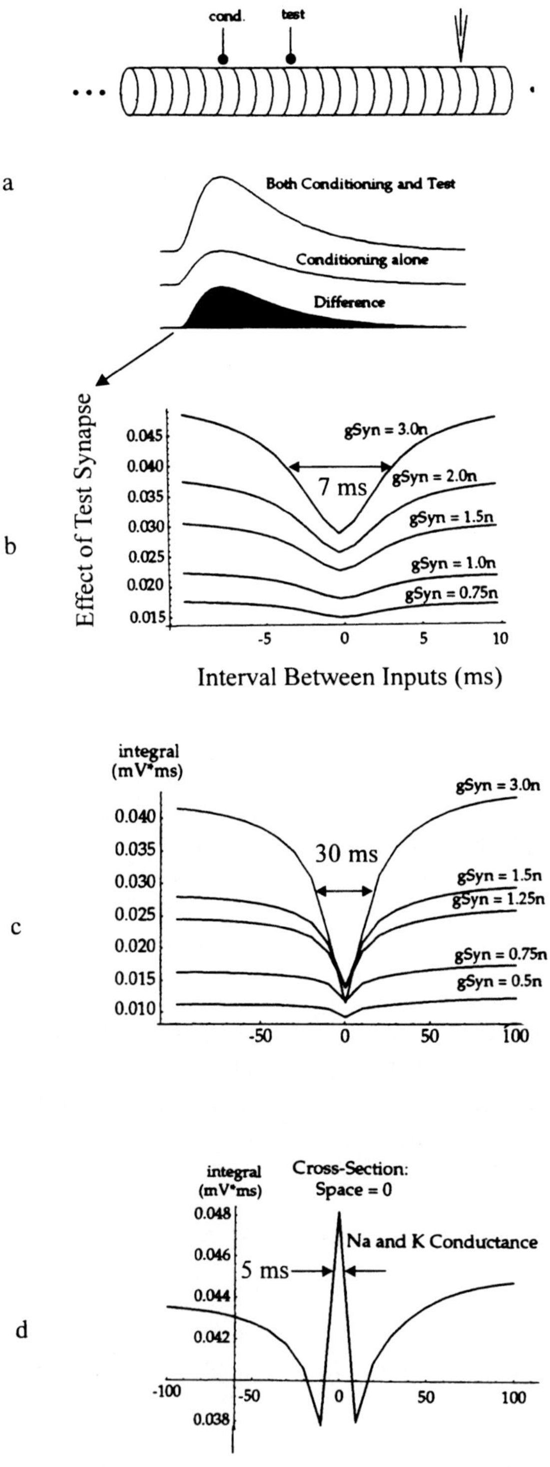

Figure 21. The spatiotemporal interaction between synaptic inputs is altered by the presence of dendritic currents. Wilson (1995) simulated the interaction between synaptic inputs to a long cable, comparing the results for passive and active membrane. (a) The interaction between synaptic inputs was characterized by computing the efficacy of a test synapse as a function of the position and relative arrival time of a second "conditioning" EPSP. Efficacy was quantified to be the increase in voltage time integral at a distant (1.2λ away) somatic compartment caused by the arrival of the test synapse. (b) In passive dendrite the closer the conditioning and test synapse, both in space and in time, the lower the efficacy of the test input, the classic sublinearity due to saturation of driving force (Rall, 1964). (c) With a K_{DR} conductance density of 2 mS/cm^2 added to the cable, the window in time over which the conditioning input reduced a test input expanded (note difference in *x*-axis scale). One possible explanation is that the first EPSP causes a sustained depolarization which, when increased by a later test input, further activates dendritic K^+ currents, in turn effectively reducing its effect. The cooperative interaction between inputs required the synapses to be spatially close, within about 0.25λ, and to arrive within a time window of about 10 msec. (d) With 10 mS/cm^2 dendritic Na^+ along with K_A at 0.5 mS/cm^2 a more complex interaction between inputs is revealed, with inputs coincident within 5 msec causing a small (10%) increase in efficacy (inputs also needed to by within 0.2λ, not shown), whereas abutting inputs arriving more than 10 msec apart, or alternatively which arrive synchronously, but are separated spatially by >0.2λ, actually *reduced* each other's efficacy. These results, which are dependent upon both the absolute and the relative densities of the currents, their kinetics, and the magnitude of the inputs, illustrate the interplay between dendritic currents and spatiotemporal patterns of input. From Wilson (1995).

cable by computing the efficacy of a test synapse as a function of the position and relative arrival time of a second "conditioning" EPSP. Efficacy was quantified to be the increase in voltage time integral at a fixed distant measuring point caused by the arrival of the test synapse (Fig. 21a).

With this measure of interaction between synapses, in passive dendrite the conditioning synapse always reduced the effect of the test, a result of the reduction in driving force for the synaptic current as successive EPSPs depolarize the dendrite (Rall, 1964). As expected, the saturation and resulting sublinear summation is greater with input magnitude, spatial proximity, and temporal proximity, with about a 30% reduction* in efficacy suffered by the second of two adjacent and temporally coincident 2-nS synapses (Fig. 21b).

The functional effect of adding K_{DR} was assessed by comparison with the passive case. With a K_{DR} conductance density of 2 mS/cm^2, the efficacy (as defined above) of the test input was reduced by about 45% (again for adjacent and coincident 2-nS inputs). A more significant effect was that addition of dendritic K_{DR} expanded the region over which the conditioning input reduced a test input (Fig. 21c), presumably because the first EPSP caused depolarization over an extended dendritic region, enabling a second, test EPSP arriving some distance away to further activate dendritic K currents, thereby reducing its incremental effect on the integrated EPSP.

Dendritic K_{DR} not only extended the distance between which synapses could mutually suppress, but also extended the window in time within which two inputs negatively interacted, from about 5 to 15 msec (at half maximum effect) (see legend to Fig. 21c).† Thus in the absence of other currents dendritic K_{DR} conductance of just 2 mS/cm^2 dynamically altered the manner in which inputs are integrated in cable. Wilson described "a dynamic adjustment of the electrotonic structure of dendrites. In addition to altering the strength of individual synapses, this . . . mechanism allows for very complex interaction between synapses depending on their location in the dendritic tree relative to each other and also relative to the distribution of ion channels" (Wilson, 1995, p. 112). Wilson's study and commentary emphasize that the input–output function of pyramidal cells is itself activity dependent.

Turning briefly to the complementary effect of excitatory dendritic currents, Wilson applied the same test protocol with 10 mS/cm^2 Na^+ conductance density in the dendritic cable. Now, when two 2-nS inputs arrived within 10 msec of each other, the net effect of the test input *increased* about 40% due to activation of Na^+ conductance (not shown). For these 2-nS inputs, the cooperative interaction between inputs required the synapses to be located within about (0–0.2)λ (space constants). Presumably, larger inputs would interact over a larger region, and of course their interaction could cause a dendritic regenerative event which could have effects not captured by a measure devised for comparing subthreshold EPSPs.

Of course, examining the effect of any one current in isolation must be viewed as a technical building block which assists in understanding the real system, where

*The quantitative effects also depend upon cable parameters, were estimated by eye from Wilson's published figures, and are not precise.

†It seems plausible that this is because the interaction time window is governed by the synaptic conductance time course alone in passive dendrites, whereas adding K_{DR} involves a convolution of the synaptic time course with that of activation of K_{DR}. In this simulation, Wilson assumed the time constant for K_{DR} activation was 10 msec.

all the actual dendritic currents are interacting. One might have expected a negative mutual interaction with only K^+ current, and cooperativity between inputs with dendritic Na^+, but we cannot as easily intuit what the effect will be on subthreshold EPSPs when both K^+ and Na^+ conductances are present. Wilson begins addressing this question by adding* K_A at 0.5 mS/cm^2 along with the 10 mS/cm^2 Na^+, and finds that the cooperative enhancement of the second EPSP is now much diminished, with only a 10% increase in efficacy for a test synapse arriving within 10 msec and 0.2λ (Fig. 21d). Further, inputs arriving at the same location but more than 10–15 msec apart, or alternatively which arrive synchronously but are separated spatially by >0.2λ, actually *reduce* each other's efficacy by about 15% notwithstanding the presence of Na^+ conductance, presumably because the K^+ current is slower to deactivate and inactivate, whereas the very fast kinetics of the Na^+ current requires inputs to arrive within 10 msec for a cooperative effect to occur. The complexity of the interactions between synaptic input and dendritic K^+, Na^+, and Ca^{2+} currents, and the nonlinear dependence of the interactions on current density, channel kinetics, and the inputs themselves, suggest a vast space which must be characterized in order to comprehend how pyramidal cells process information.

7.5. Dendritic I_A, which Activates at Subthreshold Voltages, Extends the Range of Amplitudes for Subthreshold EPSPs

The transient K^+ current I_A begins to activate near rest, and so is well suited to dampen the efficacy of excitatory synaptic input, much as it was classically shown to be suited to the regulation of firing rate (Connor and Stevens, 1971). The examples shown in Section 5.2 illustrate that dendritic I_A can curtail the duration of individual EPSPs (Lipowsky *et al.*, 1996; Rhodes and Thomson, 1996). Further, the addition of 0.5 mS/cm^2 I_A† to the apical dendrite of a simulated layer 5 cell already made excitable with 6 mS/cm^2 Na^+ conductance density was shown to allow layer 1 synaptic inputs to produce the large range of subthreshold EPSP magnitudes before triggering firing (Fig. 5d; P. Rhodes, unpublished observations) which are observed *in vitro* (Fig. 3; Cauller and Connors 1992).

In a study related to the role of K^+ currents in extending the dynamic range of pyramidal cell input/output functions, Bernander *et al.* (1994) addressed the inverse problem of presuming a linear relationship between synaptic input and model cell output, then deriving the kinetics of dendritic currents which would linearize the relationship using a highly simplified mathematical representation of the cell. They came up with a K^+ current which deactivates at high voltage, along with a variation of high-threshold depolarizing current like a Ca^{2+} channel. They then tested this analytical result by incorporating these currents into a morphologically realistic layer 5 pyramid compartment model, finding the linear relationship they sought with a distribution restricting the high-threshold depolarizing conductance to a very small (40 μm) piece of dendrite, while placing the counterveiling hyperpolarizing conductance throughout the distal half of the dendritic tree. It is hard to

*In this work the kinetic model for I_A inactivated with τ = 100 msec, but it should be noted that, for example, in CA1 cells I_A inactivates with τ in the range of 10–30 msec at 20–24°C (Klee *et al.*, 1995), presumably severalfold faster at body temperature.

†This I_A, like Wilson's, had the slower kinetics associated with the subtype of A-current known as the D-current.

assess the applicability of these results to physiological cortical cells because the K^+ current they derived in order to linearize input response is not similar to observed currents.

7.6. Dendritic Cs^+-Sensitive Current May Dampen the Effects of Distal Excitement at the Cell Body

There are several reported observations of the effects of distal synaptic input which hint at an unexpectedly great attenuation for small subthreshold inputs between distal and somatic voltage. Cauller and Connors (1994) found that the amplitude of layer 1 EPSPs was remarkably unaffected by variation in somatic voltage over the range of resting potentials between −90 and −65 mV, implying that the voltage at layer 1 where the synaptic and relevant membrane currents were generated was very little altered by a 25-mV shift in somatic voltage. Nettleton and Spain (1996) reported layer 1 synaptic input combined with a second, more proximal input summed superlinearly, suggesting the interaction of inputs with voltage-activated currents in the membrane, but that the effect was not altered by shifts in the voltage of the soma over a wide range, implying both that dendritic rather than somatic voltage-gated membrane currents were responsible for the nonlinear summation, and also that variation of somatic voltage did not affect voltage in the dendritic location.

Spruston and Stuart (1996) examined this issue directly in layer 5 pyramidal cells of the thick morphology (Mason and Larkman, 1990), by dually impaling soma and trunk with patch electrodes. They noted that the deflection caused by a small (1–2 mV) current step at the soma was enormously attenuated in the apical trunk (Fig. 22), with a 2-mV response in the soma reduced by 80% 500 μm away. This falloff in voltage, which was linear with distance from the soma, is considerably greater than can be accounted for with passive dendrite with R_m and R_i at levels (e.g., Cauller and Connors, 1992) which allow simulations of these cells to reproduce measured input resistance, particularly in view of the wide-diameter (2–4 μm) apical trunks in these cells, suggesting therefore that the attenuation is enhanced by an active current. Indeed, Spruston and Stuart (1996) found the attenuation was greatly diminished with 5 mM CsCl, a blocker of K^+ currents, added to the bath (Fig. 22), and at the same time the inward-rectifying "sag" following a sustained current step was abolished. Since the Cs^+-sensitive current did not inactivate over the several hundred-microsecond current-step durations they used, Spruston and Stuart suggested that the hyperpolarization-activated depolarizing mixed-ion current I_h (Budde *et al.*, 1994) which deactivates with and thereby opposes depolarization, was the current contributing an active component to the attenuation along the dendritic trunk.

Spruston and Stuart examined these findings with compartment models of layer 5 cells, first setting R_m and R_i by matching input resistance and attenuation in Cs^+, then adding I_h to account for the observed sag. They found that either a uniform density of 0.07 mS/cm² throughout soma and dendrites, or alternatively, a graded density of I_h starting at 0.02 mS/cm² in the soma and increasing to 4.5 mS/cm² in the distal apical trunk, reproduced much of their results (N. Spruston, personal communication). If these preliminary findings hold up, then future mod-

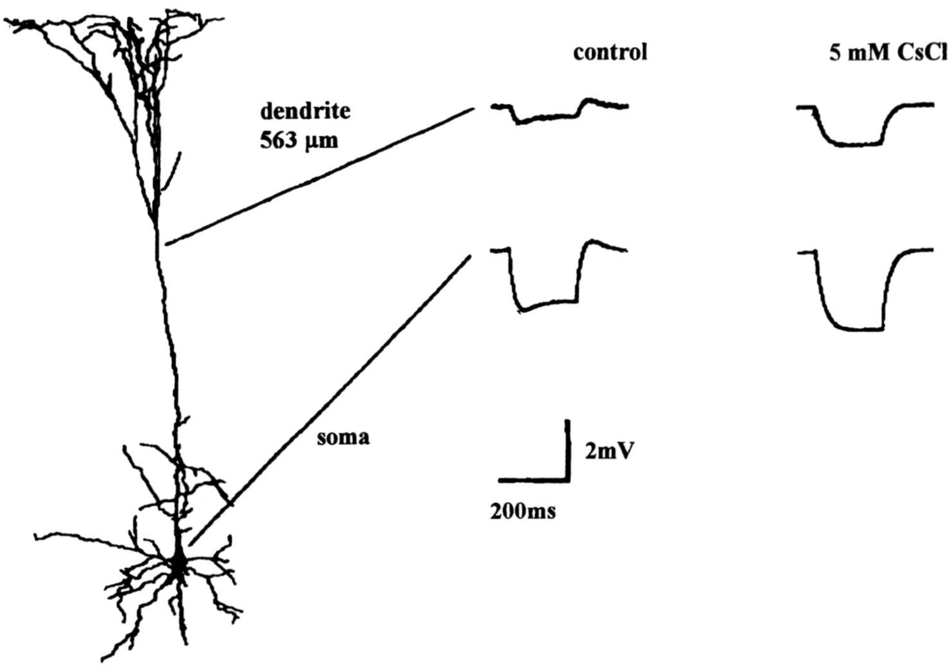

Figure 22. Sustained dendritic K^+ currents cause an enormous attenuation of subthreshold inputs in the apical dendrite of thick layer 5 cells. The attenuation of current step response along the apical trunk was directly examined with dual patch impalement of distal apical trunk and soma of layer 5 cells in 21- to 28-day-old rat. Small somatic inputs (1- to 2-mV deflections) were attenuated 80% over a 550-µm length of apical trunk in the example shown here, much more attenuation than expected from passive cable. Note also the active "sag" in the response to current step, which has been suggested to be due to the hyperpolarization-activated depolarizing current I_h. The attenuation was much reduced and the sag abolished in 5 mM CsCl, suggesting a sustained K^+ current such as I_h contributed to both. From Spruston and Stuart (1996).

els of synaptic integration, at least in thick layer 5 cortical cells, will clearly need to incorporate the presence of a dendritic Cs^+-sensitive noninactivating current which is involved in subthreshold voltages.

8. Linking Firing of the Soma to Depolarization at Distal Synapses and the Implementation of Hebb's Hypothesis

8.1. Somatic Firing Does Not Travel Far into Passive Dendrites

The proposition that correlation between presynaptic and postsynaptic firing leads to a strengthening of neuronal connections (Hebb, 1949) is a fundamental underpinning of models of associative memory and of the self-organization of ordered maps and connection patterns during development (e.g., Miller *et al.*, 1989). The properties of the NMDA channel, which requires both presynaptic activity and postsynaptic membrane depolarization in order to conduct, are thought to provide a basis for implementing the conjunctive requirement in Hebbian synapse modification. It is still necessary, however, to establish how the firing of the postsynaptic neuron causes depolarization in the vicinity of distal synapses.

While it might be assumed that somatic firing leads to a general depolarization While it might be assumed that somatic firing leads to a general depolarization throughout the neuron, simulations of pyramidal neurons with passive dendrites have suggested that for much of the dendritic tree this is not the case (Rall, 1964; Stratford *et al.,* 1989), since fast transients such as somatic spikes decrement rapidly in dendritic cable (Rinzel and Rall, 1974) (Fig. 23a). Even in comparatively short basal dendrites, with passive membrane, peak voltage following a somatic firing quickly drops from +30 mV at the soma to −35 mV at the midpoint of the basal branches (Fig. 23b), well below what is needed to relieve blockade at physiological levels of [Mg^{2+}] (Jahr and Stevens, 1990). There is therefore a problem in implementing the Hebbian notion of linkage between correlated pre- and post-synaptic firing and synaptic modification when the postsynaptic cell has electrotonically extended passive dendrites.

8.2. The Spread of the Somatic Spike into Apical Dendrite Requires Na^+ Channels

Which dendritic currents might be capable of sustaining the dendritic invasion of somatic firing? Examining this question with compartment models of layer 5 cells, it has been shown (Rhodes and Gray, 1993; Rhodes, n.d.) that dendritic high-threshold Ca^{2+} channels, which are present in pyramidal dendrites (Westenbroek *et al.,* 1992), do not sustain propagation. Moreover, at conductance densities in excess of 10 mS/cm^2, higher than reported in patch studies of CA1 apical trunks (Magee and Johnston, 1995b), neither L- nor N-type high-threshold Ca^{2+} channels in apical dendritic tree had any significant effect on the voltage reached in the apical trunk following somatic firing (Fig. 23a, dashed line). Further increasing their levels did not cause propagation of a voltage transient, but instead eventually caused a plateau depolarization (P. Rhodes, unpublished observations). Incorporation instead of low-threshold Ca^{2+} channels also does not seem to sustain propagation of a voltage transient, but can cause plateau depolarizations (P. Rhodes, unpublished observations), which may have been the phenomenon underlying the mention by Zador *et al.* (1990) that low-threshold Ca^{2+} channels enabled invasion of somatic firing into a linear cable.*

Voltage propagation into the dendrites is transformed, however, when dendritic Na^+ conductance is added to passive dendrites (Rhodes and Gray, 1993), with the somatic spike spreading throughout a dendritic tree incorporating physiological levels (e.g., 5–10 mS/cm^2) of Na^+ conductance density (Fig. 23a). This finding echoed the observation which had arisen in the development of compartment models of pyramidal cell bursting (Traub and Llinás, 1979; Rhodes and Gray, 1993, 1994; Pinskey and Rinzel, 1994) that dendritic Na^+ current is necessary to propagate firing into the dendrites, in turn activiting dendritic high-threshold

*Whether large quantities of counterveiling K^+ currents along with high densities of various Ca^{2+} channels might be an alternative means of propagating a voltage signal from somatic firing into the dendrites remains a possibility which has not been systematically explored.

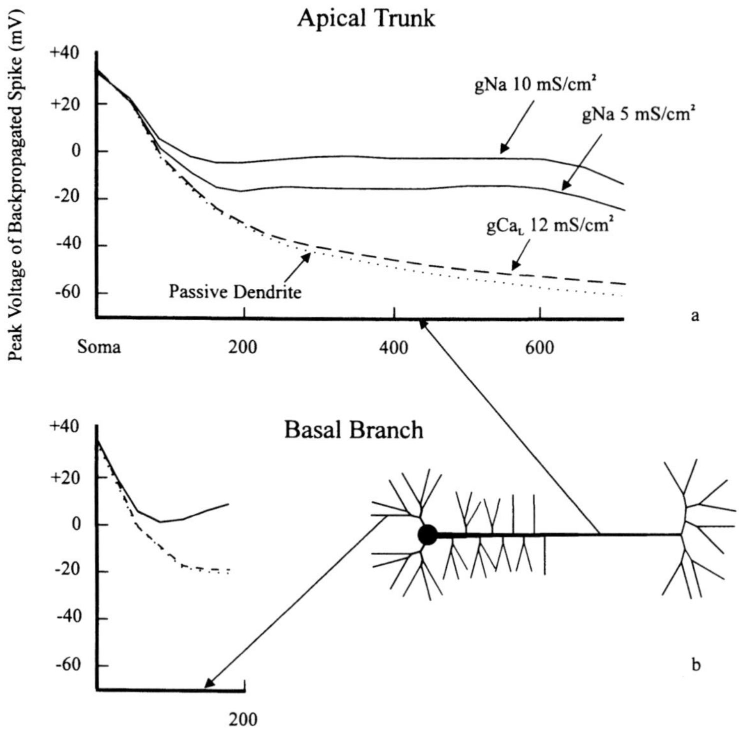

Figure 23. Dendritic Na^+ channels are essential for the backpropagation of somatic firing into the dendrites. A compartment-model layer 5 cell of "thick apical trunk" morphology was used to examine the currents capable of sustaining backpropagation of somatic firing into the dendrites. (a) Peak voltage reached in the apical trunk following somatic firing. With passive dendrites the action potential faded within 150 μm (dotted line), and high-threshold Ca^{2+} conductance did not support backpropagation (dashed line), notwithstanding a full-height action potential being initiated in the soma. Voltage propagation into the dendrites was transformed, however, when modest (5 mS/cm²) levels of dendritic Na^+ conductance were present. The amplitude of the backpropagated wave was also sensitive to C_m, spine density, and temperature, with amplitude at 22°C considerably greater than at 35°C (not shown). (b) Somatic firing rapidly attenuated in passive basal dendrites as well, indicating that dendritic Na^+ current is necessary for the effective invasion of even comparatively proximal regions of the dendritic tree. From Rhodes and Gray (1993) and Rhodes (n.d.).

Ca^{2+} channels which cause the sustained dendritic depolarization driving the burst. These results suggest that absence of outward propagation of somatic firing in Purkinje dendrites (Llinás *et al.*, 1968) is a direct consequence of the absence of dendritic Na^+ channels in Purkinje cell dendrites (Lasser-Ross and Ross, 1992).

Stuart and Sakmann (1994) confirmed the dependence of dendritic invasion on Na^+ channels in an elegant dual impalement experiment in which, in bath TTX, they replicated somatic firing by applying the action potential voltage waveform in the soma with voltage clamp. They found that the amplitude of depolarization at the dendritic electrode was greatly diminished in comparison to that measured in control bath, verifying that Na^+ channel function is essential for the propagation of voltage into the dendrites.

8.3. The Na^+-Dependent Invasion of Somatic Firing into the Dendrite Gates NMDA Conductance at Active Synapses

What is the functional effect of backpropagation into the dendrites? In particular, is it of sufficient amplitude and duration to open NMDA conductances at active synapses? This issue was addressed in a compartment simulation of a layer 5 pyramidal cell (Rhodes and Gray, 1993; Rhodes, n.d.) by applying a 1-nS synaptic conductance (0.8 nS AMPA, 0.2 nS NMDA, with 2 mM [Mg^{2+}]) to explicitly simulated spines. These synaptic test conductances were initiated 10 msec prior to the firing of a single spike (driven by brief somatic current injection) in the model neuron, and the resulting NMDA conductance was computed and compared in two cases, one with passive dendrites and the other with the model cell incorporating a uniform dendritic Na^+ conductance density of 5 mS/cm², at the low end of the range of densities implied by *in vitro* measurement. The results (Fig. 24) showed that with the test synapse located on a proximal apical oblique branch the voltage propagation afforded by this dendritic Na^+ conductance enabled peak NMDA conductance fivefold greater than in the passive case, with identical presynaptic input. At more distal synapses, such as those on the apical tuft, where somatic firing causes almost no effect in passive dendrites, the ability to implement correlation-based gating of NMDA influx becomes entirely dependent (Fig. 24b) upon the invasion of the dendrites enabled by dendritic Na^+ channels (Rhodes and Gray, 1993; Rhodes, n.d.).

The results just reviewed suggest that dendritic Na^+ channels are essential for somatic firing to cause depolarization in the dendrites, and thus for the linkage between postsynaptic firing and enhanced NMDA current at distal synapses, and that they are therefore a critical element of the mechanism by which Hebbian synaptic modification is implemented in electrotonically extended neurons.

In this regard, it is intriguing that the backpropagation through the dendrites has been shown to be subject to fatigue with repeated somatic firing (Jaffe *et al.*, 1992; Spruston *et al.*, 1995a; Colbert and Johnston, 1996b), disconnecting the voltage signal proposed to gate Hebbian plasticity, and further that the acetylcholine agonist carbachol has recently been shown to relieve the failure of Na^+-dependent backpropagation (Tsubokawa and Ross, 1996). Taken with the rest of the results in this section, this suggests a means by which basal forebrain cholinergic input might regulate backpropagation and thus Hebbian synaptic modification, either by altering the inactivation of dendritic Na^+ channels, to which backpropagation is very sensitive (P. Rhodes, unpublished observations) or by downregulating dendritic K^+ channels, which can quench the backpropagating voltage signal (see Section 7.2). Another means might be available through the NE-dependent reduction of K_{sAHP} channels proposed for the apical dendrite of CA1 cells (Sah and Bekkers, 1996; Section 5.2.2). One is led to envision a beautiful system for control of the ability of somatic firing to gate NMDA influx and thus memory formation exerted by acetyl-choline and NE through their effect on the dendritic currents involved in sustaining or curtailing backpropagation.

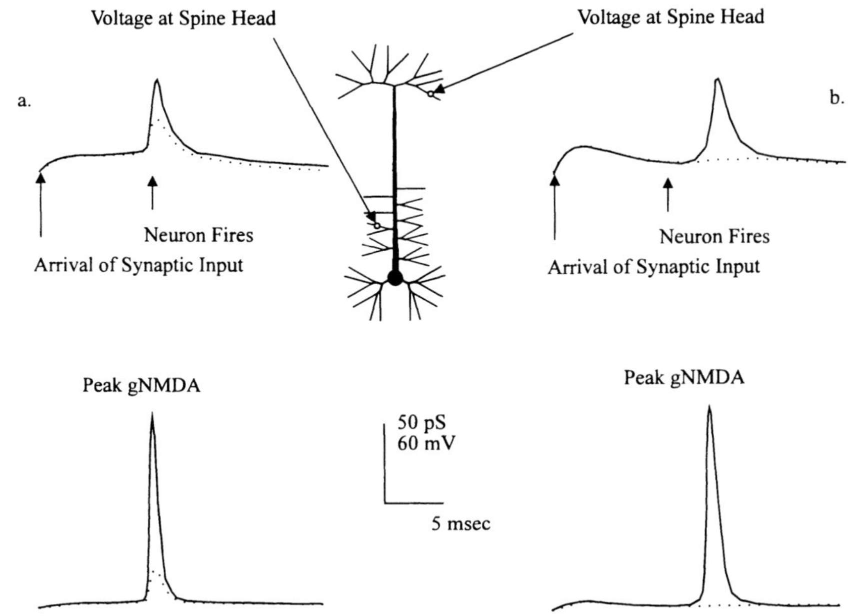

Figure 24. Dendritic Na^+ current-dependent propagation of somatic firing into the apical dendrite gates NMDA conductance at active synapses. In a compartment simulation of a layer 5 pyramidal cell (Rhodes and Gray, 1993; Rhodes, n.d.) a 1-nS synaptic conductance (0.8 nS AMPA, 0.2 nS NMDA, kinetics as in Jahr and Stevens, 1990; 2 mM [Mg^{2+}]) was applied to explicitly simulated spines (shaft 0.1 × 1 μm; head 0.6 × 0.6 μm) on the dendritic tree of two model layer 5 pyramidal neurons, which differed only in that one had a passive apical dendrite and the other had uniform dendritic Na^+ conductance density of 5 mS/cm^2, at the low end of the range of densities implied by *in vitro* measurement. The synaptic test conductance was initiated (arrow) 10 msec prior to the firing (arrow) of a single spike (driven by brief somatic current injection) in the postsynaptic neuron. (a) Peak gNMDA reached at a test synapse located on a proximal apical oblique branch was fivefold higher, with identical presynaptic input, when the somatic spike actively propagated outward. Because of the nonlinear voltage dependence of relief of Mg^{2+} blockade, the less than twofold increase in depolarization magnified into a much greater increase in peak gNMDA. (b) At more-distal synapses, such as those on the apical tuft, the linkage between somatic firing and gating of NMDA conductance was almost completely dependent upon the invasion of the dendrites enabled by dendritic Na^+ channels. From Rhodes and Gray (1993) and Rhodes (n.d.).

8.4. Effects of Order of Arrival and Latency between Pre- and Postsynaptic Activity: An Asymmetric 100-msec Time Window for Associative Synaptic Plasticity

In the simulations reported in the preceding section, synaptic input arrived a fixed 10 msec prior to postsynaptic firing. To determine the manner in which the enhancement of NMDA conductance fades as these events are separated in time, the relative times of the pre- and postsynaptic activity were manipulated by altering the start time of the brief somatic current injection driving the postsynaptic neuron. The results, and similar observations by Zador *et al.* (1990) in modeling Ca^{2+} accumulation in a spine, help explain the experimental finding (Levy and Steward, 1983) that it is not necessary for pre- and postsynaptic activity to be synchronous for a spike in the postsynaptic neuron to gate increased NMDA conductance at an active synapse. In simulation, a spike occurring 50 msec after arrival of the presynaptic input still gated 58.5% of the peak NMDA conductance reached when the postsynaptic neuron fired shortly after synaptic input, suggesting a long interval following synaptic input during which postsynaptic firing can gate synaptic NMDA conductance. Further, the temporal window in simulation was asymmetric: Though somatic firing could *follow* presynaptic input by 50–100 msec and still gate NMDA current, if somatic firing occurred even a few msec *before* synaptic input, there was no enhancement of NMDA conductance (Fig. 25, asterisk).

This interaction between the backpropagated dendritic voltage wave and synaptic NMDA conductance suggests a straightforward explanation for the experimental observation (Levy and Steward, 1983; Gustafsson *et al.*, 1987) that LTP induction requires presynaptic input (the conditioned or weak stimulus) to precede postsynaptic firing (the unconditioned or strong stimulus): There is no enhancement of synaptic NMDA conductance when the postsynaptic firing caused by the strong stimulus occurs more than a few milliseconds prior to arrival of presynaptic input, because in this case the voltage waves triggered by spiking in the soma have already passed the synaptic locus when the presynaptic input arrives, and so have no effect on synaptic NMDA conductance. It is satisfying to realize that this mechanism incorporates into the conditions for synaptic enhancement the desired causal relationship (Hebb, 1949), whereby those inputs which may be inferred to contribute to firing are the ones strengthened.

9. Apologies

This work has dwelt upon the active properties of pyramidal cell dendrites and their functional consequences. It does not address the large literature concerning

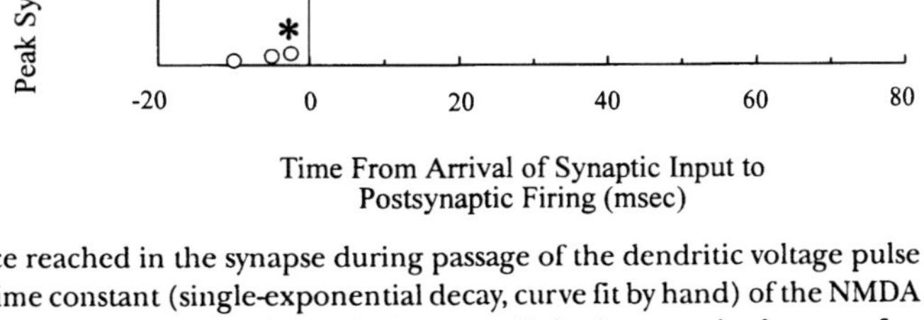

Figure 25. The time window for Hebbian synaptic modification. To determine the effects of temporal order and degree of coincidence between synaptic input and the spike in the postsynaptic neuron, the timing between these two events was varied by changing the onset time of postsynaptic current injection. Na^+ conductance density in the apical dendrite was at 5 nS/cm^2. As the time between synaptic input and postsynaptic spike increased, peak NMDA conductance reached in the synapse during passage of the dendritic voltage pulse decayed slowly, with the 75-msec time constant (single-exponential decay, curve fit by hand) of the NMDA conductance. Peak NMDA conductance dropped off sharply, however, if the input arrived even a few milliseconds after the postsynaptic spike (asterisk), since the voltage wave triggered by postsynaptic firing had already swept past when the synaptic input arrived, and thus had no effect. The results define a 50- to 100-msec period after (but not before) synaptic input during which NMDA conductance can be gated to an large fraction of its maximum level if postsynaptic firing occurs, establishing a relatively long asymmetric time window for the Hebbian strengthening of synapses connecting coactive neurons. From Rhodes and Gray (1993) and Rhodes (n.d.).

the properties and effects of active dendrites in cerebellar Purkinje cells. Llinás's studies of Purkinje cells in the late 1960s (Llinás *et al.*, 1968, 1969) were pivotal in opening up the idea of active dendrites and their consequences, and his group's dendritic impalements of Purkinje dendrites in 1980 (Llinás and Sugimori, 1980) complemented the pyramidal dendritic impalements of Prince and coworkers (Wong and Prince, 1979; Benardo *et al.*, 1982) in for the first time going beyond inference from CSD and somatic spikelets to direct observation of dendritic currents.

The modeling work examining the functional implications of active currents in Purkinje cells (Pellionisz and Llinás, 1977; De Schutter and Bower, 1994a, b) warrants a review chapter in itself. Since many aspects of both synaptic integration and outward invasion of the dendritic tree depend upon the presence of dendritic Na^+ channels (see Sections 4 and 8 above), it seems likely that the absence of dendritic Na^+ currents in Purkinje cells (Lasser-Ross and Ross, 1992) implies profound functional differences between pyramidal and Purkinje cells. It therefore seems likely that the principal neurons of the cortex and cerebellum may play deeply different computation roles, which other, more complete reviews might consider.

There are many issues arising in pyramidal cells which should have been more fully addressed here, but were not. The consequences of Ca^{2+} influx into the dendrites are beyond the scope of this review (see Llinás and Sugimori 1979), but are clearly part of the functional implications of active dendritic currents. The recent observation that voltage-activated Ca^{2+} channels apparently exist in the spines themselves (Jaffe *et al.*, 1994; Yuste and Denk, 1995), gating Ca^{2+} entry into all (not just active) spines wherever there is dendritic depolarization, suggests a mechanism for heterosynaptic depression. These and even more complicated interactions among varied dendritic currents and dendritic (Magee and Johnston, 1995a) and spine Ca^{2+} influx, and all the resulting functional implications, have scarcely been addressed.

In discussing amplification of synaptic inputs, the effects of addition of dendritic Ca_T has been largely neglected. This is an egregious omission both because this channel has recently been determined not only to exist in pyramidal dendrites (Magee and Johnston, 1995a, b), but to be more particularly distributed there than the other Ca^{2+} channels (Christie *et al.*, 1995), and because, like Na^+, Ca_T activates at voltages which come into play with subthreshold synaptic input. While Magee *et al.* (1995) found that Ni^{2+} did *not* reduce EPSP amplitude in CA1 pyramidal cells, simulation results find dendritic Ca_T *does* amplify subthreshold inputs (Rhodes and Thomson, 1996) by broadening EPSPs, a result supported by recent experiments on CA1 cells in slice (Gillessen and Alzheimer, 1997). Further work will emerge in both experimental and modeling literature on the role of dendritic Ca_T channels which will clarify their functional role.

Much of the analysis of amplification presented here concerning the functional effects of dendrites on synaptic integration addressed only the effect of dendritic currents on a single barrage of synaptic input. The amplification of patterned trains of inputs, which are of course more physiologically relevant, will be much more complex, both because of the stochastic and dynamic nature of synaptic transmission in pyramidal neurons (Thomson *et al.*, Deuchars *et al.*, 1994; Thomson and Deuchars, 1995) and because of the presence of inactivating currents, including Na^+, Ca_T, and K_A. Since these currents all turn off with sustained input, the temporal organization of input plays a vital part in determining the relationship between patterns of input and their effect upon consequent outputs, a complex subject matter about which there is again little modeling or experimental work to review. The effect of a background of excitatory and inhibitory synaptic input to the neuron in the *in vivo* environment, which both alters membrane voltage and reduces effective R_m (Holmes and Woody, 1989; Bernander *et al.*, 1994), on many of the functional effects reviewed above must also be systematically addressed. One hopes that some of the results presented here might help build a foundation of intuition to bring to more complex cases.

Finally, the functional effects addressed in this review all refer to events within a single pyramidal neuron. This begs the question of what consequences these effects will themselves have when pyramids with active dendrites are synaptically interconnected into cortical circuits. Traub's work in particular (e.g., Traub and Jeffrys, 1994, and references therein) has gone beyond study of single neurons to the properties of networks of neurons with active dendrites, but a consideration of the role of dendritic active currents in the network phenomena he describes is beyond the scope of this work. Furthermore, because the induction of plasticity can be gated by the voltage-sensitive NMDA receptor current, the electrical activity of dendrites will clearly alter the rules by which circuit connections self-organize during neuronal activity (Rhodes, 1994). The effects of active dendrites which arise in this setting are very rich and have scarcely begun to be considered.

The above enumeration of the functional effects of active dendrites will not prove inclusive; it is certain that as data continue to accumulate rapidly and analyses of the functional implications follow, the set of functional consequences of active dendrites will become extended and refined, rendering this review incomplete shortly after it appears. Nevertheless, it is hoped that it will encourage the systematic consideration of the presence, properties, and effects of dendritic currents, without which, if the preceding pages are any indication, one cannot develop accurate models and intuitions about the functioning of pyramidal cells and cortical circuitry.

10. Concluding Observations

As of this writing, at least 14 distinct currents have been found to be present in pyramidal cells, including two modes of Na^+, Ca_T, a high- and low-threshold Ca_L, Ca_N, Ca_R, K_{DR}, K_A, K_D, K_C, K_{mAHP}, K_{sAHP}, K_M, and I_H, with many demonstrated to exist in dendrites (references in first section). While modeling studies have begun to be systematically applied to the effect of individual currents (e.g., Wilson, 1995), the real challenge is to catalog how these currents interact. Many of these currents have been shown to be modified in their kinetics by phosphorylation of the Ca^{2+}-gated and cAMP-gated protein kinases C and A (Catterall, 1997), causing shifts in the voltage dependence of channels which themselves have profound consequences (e.g., Sections 4.3.1 and 5.2.3). For each channel there is a remarkable degree of further variation with the multiplicity of splice and expression variants. While we have not yet begun to differentiate the functional roles of the variants of each current, or to comprehend the effects of the ongoing modification of their activity by protein kinases and other second messengers, the evolution of such an enormous pallette of distinct currents is striking. One suspects that the array of 10–20 or more distinct ion channels, many apparently concurrently present and most subject to dynamic and reversible alteration, coordinates subtleties of information flow in dendrites which we are far from grasping.

The presence of so many coexisting voltage and Ca^{2+}-gated ion currents in central neurons, and the apparently enormous range of modifications of kinetics for each type, reveals a picture of enormous complexity. Yet all these varieties of currents evolved, and are maintained in various combinations in pyramidal dendrites, compelling us to move from appreciation of their presence to the obvious subsequent question: *what are they there for?*

While this review has attempted to gather what little is known about their combined effects on single-neuron firing and integration of inputs, we will not have adequate perspective on these issues until we know how these individual pyramidal cells function within circuits, and how these circuits operate on incoming information. How are patterns of input from one cortical area to another organized, and what further filtering and reorganization occurs as activity flows in turn to the higher and lower areas, and to and from thalamus? How is encoding of sensory information built up, and used to decode subsequent sensory input, and to produce both internal and motor output? The very presence of all the currents identified in cortical dendrites, and just the present partial summary of their profound effects on the input/output properties of cells, suggests dendritic currents are integrally involved in pyramidal cell function, but until we understand the role of pyramidal neurons in the larger setting of the operation of cortical circuitry, we lack the context with which to appraise the potential functions that dendritic currents provide.

ACKNOWLEDGMENTS. I thank Drs. Ariel Agmon, Christian Alzheimer, Yael Amitai, Angela Cantrell, Larry Cauller, Costa Colbert, Barry Connors, Alon Friedman, John Huguenard, Dan Johnston, Jeff Magee, Tom Richardson, Jeremy Seamans, Nelson Spruston, Greg Stuart, Alex Thomson, Ray Turner, Phil Ulinski, and Rafael Yuste for discussing their work and in some instances sharing unpublished data; I thank Drs. Charles Gray and John Rinzel for providing me environments in which to work; and finally I thank, the editor, Dr. Philip Ulinski, for his kind support and patience.

NOTE ADDED IN PROOF. This review was completed around January 1997, and a great many relevant results have been thereafter reported in this quickly developing area of research. The author has, as a matter of necessity, not attempted to incorporate newer results into this review, but rather hopes its deficiencies will motivate preparation of a superceding analysis of this subject matter in the future.

11. References

Alzheimer, C., Schwindt, P. C., and Crill, W. E., 1993, Postnatal development of a persistent Na^+ current in pyramidal neurons from rat sensorimotor cortex, *J. Neurophysiol.* **69**:290–292.

Amitai, Y., Friedman, A., Connors, B. W., and Gutnick, M. J., 1993, Regenerative electrical activity in apical dendrites of pyramidal cells in neocortex, *Cerebral Cortex* **3**:26–38.

Anderson, P., Storm, J., and Wheal, H., 1987, Thresholds of action potentials evoked by synapses on the dendrites of pyramidal cells in the rat hippocampus *in vitro, J. Physiol.* **383**:509–526.

Andreason, M., and Lambert, J. D. C., 1995, Regenerative properties of pyramidal cell dendrites in area CA1 of the rat hippocampus, *J. Physiol.* **483:**421–441.

Bekkers, J. M., and Stevens, C. F., 1989, Dual modes of excitatory synaptic transmission in the brain, in: *Excitatory Synaptic Transmission in Brain* pp. 39–50.

Benardo, L. S., Masukawa, L. M., and Prince, D. A., 1982, Electrophysiology of isolated pyramidal dendrites, *J. Neurosci.* **2:**1614–1622.

Bernander, O., Koch, C., and Douglas, R. J., 1994, Amplification and linearization of distal synaptic input to cortical pyramidal cells, *J. Neurophys.* **72:**2743–2753.

Bliss, T. V., and Lomo, T., 1973, Long-lasting potentiation of synaptic transmission in the dentate area of the anaesthetized rabbit following stimulation of the perforant path, *J. Physiol.* **233:**331–356.

Budde, T., White, J. A., and Kay, A. R., 1994, Hyperpolarization-activated Na^+-Ja^+ current (Ih) in neocortical neurons is blocked by external proteolysis and internal TEA, *J. Neurophysiol.* **72:**2737–2742.

Cantrell, A. R., Ma, J. Y., Scheuer, T., and Catterall, W. A., 1996, Muscarinic modulation of sodium current by activation of protein kinase C in rat hippocampal neurons, *Neuron* **16:**1019–1026.

Catterall, W. A., 1997, Modulation of sodium and calcium channels by protein phosphorylation and G proteins, in: *Signal Transduction in Health and Disease* (J. Corbin and S. Francis, eds.), Lippincott-Raven, Philadelphia, pp. 159–181.

Cauller, L. J., and Connors, B. W., 1992, Functions of very distal dendrites: Experimental and computational studies of layer I synapses on neocortical pyramidal cells, in: *Single Neuron Computation* (T. McKenna, J. Davis, and S. F. Zornetzer, eds.), Academic Press, Boston, pp. 199–230.

Cauller, L. J., and Connors, B. W., 1994, Synaptic physiology of horizontal afferents to layer I in slices of rat S1 neocortex, *J. Neurosci.* **14:**751–762.

Chagnac-Amitai, Y., Luhmann, H. J., and Prince, D. A., 1990, Burst generating and regular spiking layer 5 pyramidal neurons of rat neocortex have different morphological features, *J. Comp. Neurol.* **296:**598–613.

Christie, B. R., Eliot, L. S., Ito, K., Miyakawa, H., and Johnston, D., 1995, Different Ca channels in the soma and dendrites of hippocampal pyramidal neurons mediate spike-induced Ca influx, *J. Neurophysiol.* **73:**2553–2557.

Colbert, C. M., and Johnston, D., 1996a, Axonal action-potential initiation and Na^+ channel densities in the soma and axon initial segment of subicular pyramidal neurons, *J. Neurosci.* **16:**6676–6686.

Colbert, C. M., and Johnston, D., 1996b, A decrease in Na^+ current contributes to loss of action potential amplitude in dendritic spike trains, *Soc. Neurosci. Abstr.* **22:**791.

Connor, J. A., and Stevens, C. F., 1971, Prediction of repetitive firing behaviour from voltage clamp data on an isolated neurone somata, *J. Physiol.* **213:**31–53.

Connors, B. W., Gutnick, M. J., and Prince, D. A., 1982, Electrophysiological properties of neocortical neurons *in vitro*, *J. Neurophysiol.* **48:**1301–1320.

Cooley, J. W., and Dodge, F. A., 1966, Digital computer solutions for excitation and propagation of the nerve impulse, *Biophys. J.* **6:**583–600.

Coombs, J. S., Curtis, D. R., and Eccles, J. C., 1957a, The interpretation of spike potentials of motoneurones, *J. Physiol.* **139:**198–231.

Coombs, J. S., Curtis, D. R., and Eccles, J. C., 1957b, The generation of impulses in motoneurones, *J. Physiol.* **139:**232–249.

Cummins, T. R., Xia, Y., and Haddad, G. G., 1994, Functional properties of rat and human neocortical voltage-sensitive sodium currents, *J. Neurophysiol.* **71:**1052–1064.

Deisz, R. A., Fortin, G., and Zieglgansberger, W., 1991, Voltage dependence of excitatory postsynaptic potentials of rat neocortical neurons, *J. Neurophysiol.* **65:**371–382.

Deschênes, M., 1981, Dendritic spikes induced in fast pyramidal tract neurons by thalamic stimulation, *Exp. Brain Res.* **43:**304–308.

De Schutter, E., and Bower, J. M., 1994a, An active membrane model of the cerebellar Purkinje cell I. Simulation of current clamps in slice, *J. Neurophysiol.* **71:**375–400.

De Schutter, E., and Bower, J. M., 1994b, An active membrane model of the cerebellar Purkinje cell II. Simulation of synaptic responses, *J. Neurophysiol.* **71:**401–419.

Deuchars, J., West, D. C., and Thomson, A. M., 1994, Relationships between morphology and physiology of pyramid–pyramid single axon connections in rat neocortex *in vitro*, *J. Physiol.* **478:**423–435.

Dodge, F. A., and Cooley, J. W., 1973, Action potential of the motoneuron, *IBM J. Res. Dev.* 219–229.

Eccles, J. C., Libet, B., and Young, R. R., 1958, The behaviour of chromatolysed motoneurons studied by intracellular recording, *J. Physiol.* **143:**11–40.

Fleidervish, I. A., Friedman, A., and Gutnick, M. J., 1996, Slow inactivation of Na^+ current and slow cumulative spike adaptation in mouse and guinea-pig neocortical neurones in slices, *J. Physiol.* **493:**83–97.

Friedman, A., and Gutnick, M. J., 1989, Intracellular calcium and control of burst generation in neurons of guinea pig neocortex *in vitro, Eur. J. Neurosci.* **1:**374–381.

Gamble, E., and Koch, C., 1987, The dynamics of free calcium in dendritic spines in response to repetitive input, *Science* **236:**1311–1315.

Gillessen, T., and Alzheimer, C., 1997, Amplification of EPSPs by low Ni^{2+}- and amiloride-sensitive Ca^{2+} channels in apical dendrites of rat CA1 pyramidal neurons, *J. Neurophysiol.* **77:**1639–1643.

Gustafsson, B., Wigstrom, H., Abraham, W. C., and Huang, Y.-Y., 1987, Long-term potentiation in the hippocampus using depolarizing current pulses as the conditioning stimulus to single volley synaptic potentials, *J. Neurosci.* **7:**774–780.

Hamill, O. P., Huguenard, J. R., and Prince, D. A., 1991, Patch clamp studies of voltage-gated currents in identified neurons of the rat cerebral cortex, *Cerebral Cortex* **1:**48–61.

Hebb, D. O., 1949, *The Organization of Behavior,* Wiley, New York.

Herreras, O., 1990, Propagating dendritic action potential mediates synaptic transmission in CA1 pyramidal cells *in situ, J. Neurophys.* **64:**1429–1441.

Hille, B., 1984, Ionic Channels of Excitable Membranes, Sinauer, Sunderland, MA.

Hirsch, J. A., and Gilbert, C. D., 1991, Synaptic physiology of horizontal connections in the cat's visual cortex, *J. Neurosci.* **11:**1800–1809.

Hoffman, D., Magee, J., and Johnston, D., 1996, Characterization of voltage-gated K^+ channels in the soma and dendrites of hippocampal CA1 pyramidal neurons, *Soc. Neurosci. Abstr.* **22:**793.

Holmes, W. R., and Levy, W. B., 1990, Insights into associative long-term potentiation from computational models of NMDA receptor-mediated calcium influx and intracellular calcium concentration changes, *J. Neurophys.* **63:**1148–1168.

Holmes, W. R., and Woody, C. D., 1989, Effects of uniform and non-uniform synaptic 'activation distributions' on the cable properties of modeled cortical pyramidal neurons, *Brain Res.* **505:**12–22.

Huguenard, J. R., Hamill, O. P., and Prince, D. A., 1989, Sodium channels in dendrites of rat cortical pyramidal neurons, *Proc. Natl. Acad. Sci. USA* **86:**2473–2477.

Isom, L. L., Scheuer, T., Brownstein, A. B., Ragsdale, D. S., Murphy, B. J., and Catterall, W. A., 1995, Functional co-expression of the $\beta 1$ and type IIA α subunits of sodium channels in a mammalian cell line, *J. Biol. Chem.* **7:**3306–3312.

Jaffe, D. B., Johnston, D., Lasser-Ross, N., Lisman, J. E., Miyakawa, H., and Ross, W. N., 1992, The spread of Na^+ spikes determines the pattern of dendritic Ca^{2+} entry into hippocampal neurons, *Nature* **357:**244–246.

Jaffe, D. B., Fisher, S. A., and Brown, T. H., 1994, Confocal laser scanning microscopy reveals voltage-gated calcium signals within hippocampal dendritic spines, *J. Neurobiol.* **25:**220–233.

Jahr, C. E., and Stevens, C. F., 1990, Voltage dependence of NMDA-activated macroscopic conductances predicted by single-channel kinetics, *J. Neurosci.* **10:**3178–3182.

Jenson, M. S., Azouz, R., and Yaari, Y., 1996, Spike after-depolarization and burst generation in adult rat hippocampal CA1 pyramidal cells, *J. Physiol. (London)* **492:**199–210.

Johnston, D., Magee, J. C., Colbert, C. M., and Christie, B. R., 1996, Active properties of neuronal dendrites, *Annu. Rev. Neurosci.* **19:**165–186.

Kim, H. G., and Connors, B. W., 1993, Apical dendrites of the neocortex: Correlation between sodium- and calcium-dependent spiking and pyramidal cell morphology, *J. Neurosci.* **13:**5301–5311.

Klee, R., Ficker, R., and Heinemann, U., 1995, Comparison of voltage-gated potassium currents in rat pyramidal neurons acutely isolated from hippocampal regions CA1 and CA3, *J. Neurophysiol.* **74:**1982–1995.

Kuno, M., and Llinás, R., 1970a, Enhancement of synaptic transmission by dendritic potentials in chromatolysed motoneurones of the cat, *J. Physiol.* **210:**807–821.

Kuno, M., and Llinás, R., 1970b, Alterations of synaptic action in chromatolysed motoneurones of the cat, *J. Physiol.* **210:**823–838.

Larkman, A. U., 1991a, Dendritic morphology of pyramidal neurones of the visual cortex of the rat: I. Branching patterns, *J. Comp. Neurol.* **306:**332–343.

Larkman, A. U., 1991b, Dendritic morphology of pyramidal neurones of the visual cortex of the rat: III. Spine distributions, *J. Comp. Neurol.* **306:**307–319.

Larkman, A. U., and Mason, A. J. R., 1990, Correlations between morphology and electrophysiology of pyramidal neurones in slices of rat visual cortex. I. Establishment of cell classes, *J. Neurosci.* **10:**1407–1414.

Lasser-Ross, N., and Ross, W. N., 1992, Imaging voltage and synaptically activated sodium transients in cerebellar Purkinje cells, *Proc. R. Soc. Lond. B* **247:**35–39.

Levy, W. B., and Steward, O., 1983, Temporal contiguity requirements for long-term associative potentiation/depression in the hippocampus, *Neuroscience* **8:**791–797.

Lipowsky, R., Gilleson, T., and Alzheimer, C., 1996, Dendritic Na^+ channels amplify EPSPs in hippocampal CA1 pyramidal cells, *J. Neurophysiol.* **76:**2181–2191.

Llinás, R., 1975, Electroresponsive properties of dendrites in central neurons, in: *Advances in Neurology,* Vol. 12 (G. W. Kreutzberg, ed.), Raven Press, New York, pp. 1–13.

Llinás, R., 1988, The intrinsic electrophysiological properties of mammalian neurons: Insights into central nervous system function, *Science* **242:**1654–1664.

Llinás, R., and Nicholson, C., 1971, Electrophysiological properties of dendrites and somata in alligator Purkinje cells, *J. Neurophysiol.* **34:**532–551.

Llinás, R., and Sugimori, M., 1979, Calcium conductances in Purkinje cell dendrites: Their role in development and integration, *Prog. Brain Res.* **51:**323–334.

Llinás, R., and Sugimori, M., 1980, Electrophysiological properties of *in vitro* Purkinje cell dendrites in mammalian cerebellar slices, *J. Physiol.* **305:**197–213.

Llinás, R., Nicholson, C., Freeman, J. A., and Hillman, D. E., 1968, Dendritic spikes and their inhibition in alligator Purkinje cells, *Science* **160:**1133–1135.

Llinás, R., Nicholson, C., and Precht, W., 1969, Preferred centripetal conduction of dendritic spikes in alligator Purkinje cells, *Science* **163:**184–187.

Madison, D. V., and Nicoll, R. A., 1982, Noradrenaline blocks accommodation of pyramidal cell discharge in the hippocampus, *Nature* **299:**636–638.

Madison, D. V., and Nicoll, R. A., 1984, Control of the repetitive discharge of rat CA1 pyramidal neurones *in vitro, J. Physiol.* **354:**319–331.

Magee, J. C., and Johnston, D., 1995a, Synaptic activation of voltage-gated channels in the dendrites of hippocampal pyramidal neurons, *Science* **2688:**301–304.

Magee, J. C., and Johnston, D., 1995b, Characterization of single voltage-gated Na^+ and Ca^{2+} channels in apical dendrites of rat CA1 pyramidal neurons, *J. Physiol.* **487:**67–90.

Magee, J. C., Christofi, G., Miyakawa, H., Christie, B., Lasser-Ross, N., and Johnston, D., 1995, Subthreshold synaptic activation of voltage-gated Ca^{2+} channels mediates a localized Ca^{2+} influx into the dendrites of hippocampal pyramidal neurons, *J. Neurophysiol.* **4:**1335–1342.

Mainen, Z. F., Joerges, J., Huguenard, J. R., and Sejnowski, T. J., 1995, A model of spike initiation in neocortical pyramidal neurons, *Neuron* **15:**1427–1439.

Malenka, R. C., Kauer, J. A., Zucker, R. S., and Nicoll, R. A., 1988, Postsynaptic calcium is sufficient for potentiation of hippocampal synaptic transmission, *Science* **242:**81–84.

Markram, H., and Sakmann, B., 1994, Calcium transients in dendrites of neocortical neurons evoked by single subthreshold excitatory postsynaptic potentials via low-voltage activated channels, *Proc. Natl. Acad. Sci. USA* **91:**5207–5277.

Markram, H., Helm, P. J., and Sakmann, B., 1995, Dendritic calcium transients evoked by single back-propagating action potentials in rat neocortical pyramidal neurons, *J. Physiol.* **45:**1–20.

Mason, A. J. R., and Larkman, A. U., 1990, Correlations between morphology and electrophysiology of pyramidal neurons in slices of rat visual cortex, *J. Neurosci.* **10:**1415–1428.

McCormick, D. A., and Pape, H.-C., 1990, Properties of a hyperpolarization-activated cation current and its role in rhythmic oscillation in thalamic relay neurones, *J. Physiol.* **431:**291–318.

McCormick, D. A., Connors, B. W., Lighthall, J. W., and Prince, D. A., 1985, Comparative electrophysiology of pyramidal and sparsely spiny stellate neurons of the neocortex, *J. Neurophysiol.* **54:**782–806.

Mel, B. W., 1992, NMDA-based pattern discrimination in a modeled cortical neuron, *Neural Computation* **4:**502–517.

Mel, B. W., 1993, Synaptic integration in an excitable dendritic tree, *J. Neurophysiol.* **70:**1086–1101.

Mel, B. W., 1994, Information processing in dendritic trees, *Neural Computation* **6:**1031–1085.

Mel, B. W., Niebuhr, E., Croft, D. W., and Gabbiani, F., 1995, Why neurons make bad coincidence detectors but good periodicity detectors, *Soc. Neurosci. Abstr.* **21:**1506.

Miller, J. P., Rall, W., and Rinzel, J., 1985, Synaptic amplification by active membrane in spines, *Brain Res.* **325:**325–330.

Miller, K. D., Keller, J. P., and Stryker, M. P., 1989, Ocular dominance column development: Analysis and simulation, *Science* **245**:605–615.

Millonas, M. M., and Ulinski, P. S., 1996, Integrative properties of voltage-gated conductances on the dendrites of cortical pyramidal cells, *Soc. Neurosci. Abstr.* **22**:1214.

Millonas, M. M., and Ulinski, P. S., n.d., Dendritic origin of fast prepotentials in pyramidal cells from turtle visual cortex.

Nettleton, J. S., and Spain, W. J., 1996, Post-synaptic conductances cause non-linear summation of AMPA-mediated EPSP's in rat neocortical layer V pyramidal neurons, *Soc. Neurosci. Abstr.* **22**:796.

Nicoll, A., Larkman, A., and Blakemore, C., 1993, Modulation of EPSP shape and efficacy by intrinsic membrane conductances in rat neocortical pyramidal neurons *in vitro*, *J. Physiol.* **468**:693–710.

Pellionisz, A., and Llinás, R., 1977, A computer model of cerebellar Purkinje cells, *Neuroscience*, **2**:37–48.

Perkel, D. H., Mulloney, B., and Budelli, R. W., 1981, Quantitative methods for predicting neuronal behavior, *Neuroscience* **6**:823–837.

Pinsky, P. F., and Rinzel, J., 1994, Intrinsic and network rhythmogenesis in a reduced Traub model for CA3 neurons, *J. Comp. Neurosci.* **1**:39–60.

Poolos, N. P., and Kocsis, J. D., 1990, Dendritic action potentials activated by NMDA receptor-mediated EPSP's in CA1 hippocampal pyramidal cells, *Brain Res.* **524**:342–346.

Purpura, D. P., Shofer, R. J., and Scarff, T., 1965, Properties of synaptic activities and spike potentials of neurons in immature neocortex, *J. Neurophysiol.* **28**:925–942.

Rall, W., 1959, Branching dendritic trees and motoneuron membrane resistivity, *Exp. Neurol.* **1**:497–527.

Rall, W., 1962, Theory of physiological properties of dendrites, *Ann. N.Y. Acad. Sci.* **96**:1071–1092.

Rall, W., 1964, Theoretical significance of dendritic trees for neuronal input-output relations, in: *Neural Theory and Modeling* (R. F. Reiss, ed.), Stanford University Press, Stanford, CA, pp. 73–97.

Rall, W., and Segev, I., 1988, Dendritic spine synapses, excitable spine clusters, and plasticity, in: *Cellular Mechanisms of Conditioning and Behavioral Plasticity* (C. D. Woody, D. L. Alkon, and J. L. McGaugh, eds.), Plenum Press, New York, pp. 221–236.

Rapp, M., Yarom, Y., and Segev, I., 1996, Modeling back propagating action potentials in weakly excitable dendrites of neocortical pyramidal cells, *Proc. Natl. Acad. Sci. USA* **93**:11985–11990.

Regehr, W. G., Connor, J. A., and Tank, D. W., 1989, Optical imaging of calcium accumulation in hippocampal pyramidal cells during synaptic activation, *Nature* **341**:533–536.

Reuveni, I., Friedman, A., Amitai, Y., and Gutrick, M. J., 1993, Stepwise repolarization from Ca^{2+} plateaus in neocortical pyramidal cells: Evidence for nonhomogeneous distribution of HVA Ca^{2+} channels in dendrites, *J. Neurosci.* **13**:4609–4621.

Rhodes, P. A., 1990, A computational study indicates cross-modal association arises naturally in neocortex via feedback projections, *Soc. Neurosci. Abstr.* **16**:286.

Rhodes, P. A., 1992, The long open time of the NMDA channel facilitates the self-organization of invariant object responses in cortex, *Soc. Neurosci. Abstr.* **18**:740.

Rhodes, P. A., 1994, Dendritic Na^+ channels interact with somatic firing to shape neocortical circuit formation: A compartment model study, *Soc. Neurosci. Abstr.* **20**:715.

Rhodes, P. A., n.d., Dendritic sodium channels are required for somatic firing to invade the dendrites: Effects on NMDA conductance at distal synapses.

Rhodes, P. A., and Gray, C. M., 1993, Effects of electrically active dendrites and NMDA-type synaptic conductances, *Soc. Neurosci. Abstr.* **19**:242.

Rhodes, P. A., and Gray, C. M., 1994, Simulations of intrinsically bursting neocortical pyramidal neurons, *Neural Computation* **6**:1086–1110.

Rhodes, P. A., and Thomson, A. M., 1996, Effects of dendritic membrane currents on the shape of EPSP's in pyramidal neurons, *Soc. Neurosci. Abstr.* **22**:791.

Rhodes, P. A., Yuste, R., and Tank, D. W., 1995, Ca^{2+} and Na^+ channels throughout the basal dendrites of regular spiking neocortical pyramidal neurons: Implications of bursting with EGTA, *Soc. Neurosci. Abstr.* **21**:1996.

Richardson, T. L., Turner, R. W., and Miller, J. J., 1987, Action potential discharge in hippocampal CA1 pyramidal neurons: Current source-density analysis, *J. Neurophysiol.* **58**:981–996.

Rinzel, J., and Rall, W., 1974, Transient response in a dendritic neuron model for current injected in one branch, *Biophys. J.* **14**:759–790.

Rockland, K. S., and Pandya, D., 1979, Laminar origins and terminations of cortical connections of the occipital lobe in the rhesus monkey, *Brain Res.* **179**:3–20.

Rockland, K., and Virga, A., 1989, Terminal arbors of individual "feedback" axons projecting from area

V2 to V1 in the macaque monkey: A study using immunohistochemistry of anterogradely transported *Phaseolus vulgaris*-leucoagglutinin, *J. Comp. Neurol.* **285**:54–72.

Sah, P., and Bekkers, J. M., 1996, Apical dendritic location of slow afterhyperpolarization current in hippocampal pyramidal neurons: Implications for the integration of long-term potentiation, *J. Neurosci.* **16**:4537–4542.

Schiller, J., Helmchen, F., and Sakmann, B., 1995, Spatial profile of dendritic calcium transients evoked by action potentials in rat neocortical pyramidal neurons, *J. Physiol.* **487**:583–600.

Schwartz, E. I., Desimone, R., Albright, T. D., and Gross, C. G., 1983, Shape recognition and inferior temporal neurons, *Proc. Natl. Acad. Sci. USA* **80**:5776–5778.

Schwartzkroin, P. A., 1997, Further characteristics of hippocampal CA1 cells *in vitro*, *Br. Research* **128**:53–68.

Schwindt, P. C., and Crill, W. E., 1995, Amplification of synaptic current by persistent sodium conductance in apical dendrite of neocortical neurons, *J. Neurophysiol.* **74**:2220–2224.

Schwindt, P. C., Spain, W. J., Foehring, R. C., Stafstrom, C. E., Chubb, M. C., and Crill, W. E., 1988a, Multiple potassium conductances and their functions in neurons from cat sensorimotor cortex *in vitro*, *J. Neurophysiol.* **59**:424–449.

Schwindt, P. C., Spain, W. J., Foehring, R. C., Chubb, M. C., and Crill, W. E., 1988b, Slow conductances in neurons from cat sensorimotor cortex *in vitro* and their role in slow excitability changes, *J. Neurophysiol.* **59**:450–467.

Segev, I., Fleshman, J. W., and Burke, R. E., 1989, Compartment models of complex neurons, in: *Methods in Computational Neuroscience* (I. Segev and C. Kock, eds.), MIT Press, Cambridge, MA, pp. 63–96.

Segev, I., Rinzel, J., and Sheperd, G. M., 1995, *The Theoretical Foundation of Dendritic Function*, MIT Press, Cambridge, MA.

Shepherd, G. M., and Brayton, R. K., 1987, Logic operations are properties of computer-simulated interactions between excitable dendritic spines, *Neuroscience* **21**:151–165.

Shepherd, G. M., Brayton, R. K., Miller, J. P., Segev, I., Rinzel, J., and Rall, W., 1985, Signal enhancement in distal cortical dendrites by means of interactions between active dendritic spines, *Proc. Natl. Acad. Sci. USA* **82**:2192–2195.

Shepherd, G. M., Woolf, T. B., and Carnevale, N. T., 1991, Comparisons between active properties of distal dendritic branches and spines: Implications for neuronal computations, *J. Cognit. Neurosci.* **1**:273–286.

Softkey, W., 1993, Sub-millisecond coincidence detection in active dendritic trees, *Neuroscience* **58**:13–41.

Spencer, W. A., and Kandel, E. R., 1961, Electrophysiology of hippocampal neurons. IV. Fast prepotentials, *J. Neurophysiol.* **24**:272–285.

Spruston, N., and Stuart, G. 1996, Voltage attenuation and intracellular resistivity in neocortical pyramidal neurons, *Soc. Neurosci. Abstr.* **22**:792.

Spruston, N., Schiller, Y., Stuart, G., and Sakmann, B., 1995a, Activity-dependent action potential invasion and calcium influx into hippocampal CA1 dendrites, *Science* **268**:297–300.

Spruston, N., Stuart, G., and Sakmann, B., 1995b, How do voltage-activated channels shape EPSP's in hippocampal CA1 neurons, *Soc. Neurosci. Abstr.* **21**:584.

Stratford, K. J., Mason, A. J. R., Larkman, A. U., Major, G. M., and Jack, J. J. B., 1989, The modeling of pyramidal neurones in the visual cortex, in: *The Computing Neuron* (R. M. Durbin, R. C. Miall, and G. J. Mitchison, eds.), Addison-Wesley, Reading, MA, pp. 296–321.

Stuart, G. J., and Sakmann, B., 1994, Active propagation of somatic action potentials into neocortical pyramidal cell dendrites, *Nature* **367**:69–72.

Stuart, G. J., and Sakmann, B., 1995, Amplification of EPSP's by axosomatic sodium channels in neocortical pyramidal neurons, *Neuron* **15**:1065–1076.

Sutor, B., and Hablitz, J. J., 1989a, EPSP's in rat neocortical neurons *in vitro*. I. Electrophysiological evidence for two distinct EPSP's, *J. Neurophysiol.* **61**:621–634.

Sutor, B., and Hablitz, J. J., 1989b, EPSP's in rat neocortical neurons *in vitro*. II. Involvement of *N*-methyl-D-aspartate receptors in the generation of EPSP's, *J. Neurophysiol.* **61**:607–620.

Talbot, M. J., and Sayer, R. J., 1996, QX-314 inhibits calcium currents in hippocampal CA1 pyramidal neurons, *J. Neurophysiol.* **76**:2120–2124.

Thomson, A. M., 1986, A magnesium sensitive postsynaptic potential in rat cerebral cortex resembles neuronal responses to *N*-methylaspartate, *J. Physiol.* **370**:531–549.

Thomson, A. M., Girdlestone, D., and West, D. C., 1988, Voltage-dependent currents prolong single-axon postsynaptic potentials in layer III pyramidal neurons in rat neocortical slices, *J. Neurophysiol.* **60**:1896–1907.

Thomson, A. M., and Deuchars, J., 1995, Divers pre- and post-synaptic properties of fast excitatory synapses, in: *Excitatory Amino Acids and Synaptic Function* (H. V. Wheal, and A. M. Thomson, eds.), Academic Press, London, pp. 146–170.

Thomson, A. M., and West, D. C., 1993, Fluctuations in pyramid–pyrammid excitatory postsynaptic potentials modified by presynaptic firing pattern and postsynaptic membrane potential using paired intracellular recordings in rat neocortex, *Neuroscience* **54:**329–346.

Thomson, A. M., Deuchars, J., and West, D. C., 1993, Large, deep layer pyramid-pyrammid single exon EPSPs in slices of rat motor cortex display paired pulse and frequency-dependent depression, mediated presynaptically and self-facilitation, mediated postsynaptically, *J. Neurophysiol.* **70:**2354–2369.

Traub, R. D., and Jeffrys, J. G., 1994, Simulations of epileptiform activity in the hippocampal CA3 region *in vitro, Hippocampus* **4:**281–285.

Traub, R. D., and Llinás, R., 1977, The spatial distribution of ionic conductances in normal and axotomized motoneurons, *Neuroscience* **2:**829–849.

Traub, R. D., and Llinás, R., 1979, Hippocampal pyramidal cells: Significance of dendritic ionic conductances for neuronal function and epileptogenesis, *J. Neurophysiol.* **42:**476–496.

Traub, R. D., Wong, R. K. S., Miles, R., and Michelson, H., 1991, Model of a CA3 hippocampal pyramidal neuron incorporating voltage-clamp data on intrinsic conductances, *J. Neurophysiol.* **66:**635–650.

Tsubokawa, H., and Ross, W. N., 1996, Pharmacological modulation of spike propagation in the apical dendrites of hippocampal pyramidal cells, *Soc. Neurosci. Abstr.* **22:**791.

Turner, R. W., Meyers, D. E. R., and Barker, J. L., 1989, Localization of tetrodotoxin-sensitive field potentials of CA1 pyramidal cells in rat hippocampus, *J. Neurophysiol.* **62:**1375–1387.

Turner, R. W., Meyers, D. E. R., Richardson, T. L., and Barker, J. L., 1991, The site for initiation of action potential discharge over the somatodendritic axis of rat hippocampal CA1 pyramidal neurons, *J. Neurosci.* **11:**2270–2280.

Turner, R. W., Meyers, D. E. R., and Barker, J. L., 1993, Fast pre-potential generation in rat hippocampal CA1 pyramidal neurons, *Neuroscience* **53:**949–959.

Urban, N. N., and Barrionuevo, G., 1996, The monosynaptic perforant path input to CA3 pyramidal neurons: II. The role of voltage-dependent conductances in propagation of EPSP's, *Soc. Neurosci. Abstr.* **22:**791.

Vogt, B. A., 1990, The role of layer 1 in cortical function, in: *Cerebral Cortex,* Vol. 8 (A. Peters and E. G. Jones, eds.), Plenum Press, New York, pp. 49–80.

Wang, Z., and McCormick, D. A., 1993, Control of firing mode of corticotectal and corticopontine layer V burst-generating neurons by norepinephrine, acetylcholine, and 1S,3R-ACPD, *J. Neurosci.* **13:**2199–2216.

Westenbröek, R. E., Hell, J. W., Warner, C., Dubel, S. J., Snutch, T. P., and Catterall, W. A., 1992, Biochemical properties and subcellular distribution of an N-type calcium channel $\alpha 1$ subunit, *Neuron* **9:**1099–1115.

Wilson, C. J., 1995, Dynamic modification of dendritic cable properties and synaptic transmission by voltage-gated potassium channels, *J. Comp. Neurosci.* **2:**91–115.

Wong, R. K. S., and Prince, D. A., 1978, Participation of calcium spikes during intrinsic burst firing in hippocampal neurons, *Brain Res.* **159:**385–390.

Wong, R. K. S., and Stewart, M., 1992, Different firing patterns generated in dendrites and somata of CA1 pyramidal neurones in guinea-pig hippocampus, *J. Physiol.* **457:**675–687.

Wong, R. K. S., Prince, D. A., and Basbaum, A. I., 1979, Intradendritic recordings from hippocampal neurons, *Proc. Natl. Acad. Sci. USA* **76:**986–990.

Yuste, R., and Denk, W., 1995, Dendritic spines as basic functional units of neuronal integration, *Nature* **375:**682–684.

Yuste, R., and Tank, D. W., 1996, Dendritic integration in mammalian neurons, a century after Cajal, *Neuron* **16:**701–716.

Yuste, R., Gutnick, M. J., Saar, D., Delaney, K. D., and Tank, D. W., 1994, Calcium accumulations in dendrites from neocortical neurons: An apical band and evidence for functional compartments, *Neuron* **13:**23–43.

Zador, A., Koch, C., and Brown, T. H., 1990, Biophysical model of a Hebbian synapse, *Proc. Natl. Acad. Sci. USA* **87:**6718–6722.

4

Comparing Different Modeling Approaches of Visual Cortical Cell Characteristics

FLORENTIN WÖRGÖTTER

1. Introduction

The goal of a modeling study should be to arrive at a better analytical and intuitive understanding of the underlying system. To achieve this, every model needs to represent an abstract reflection of reality, and the level of abstraction largely depends on the complexity of the system which is to be modeled. As soon as this complexity exceeds a certain degree, a multitude of possible model levels can be designed and these models can coexist without mutually contradicting each other. A hopeful assumption at this point would be to believe that this diversity could be reduced if all models were required to represent only one common aspect of the system. The complexity of the visual cortex of vertebrates, however, thwarts this hope, because even a single aspect of cortical cell behavior can lead to quite

FLORENTIN WÖRGÖTTER • Institut für Physiologie, Ruhr Universität Bochum, D-44780 Bochum, Germany.

different successful modeling approaches. In this chapter *visual cortical orientation specificity* shall be the common aspect on which a comparison of different models will be based. The goal shall be to try to develop cross-links between quite different approaches in order to arrive at a "comparative anatomy" of those cortex models. Such a comparison shall subserve the process of establishing a common ground on which the different aspects of visual cortical cell behavior including orientation specificity might be based.

Soon after the first studies about single-cell recordings in visual cortex were published about 35 years ago (Hubel and Wiesel, 1959, 1962), it became clear that those cells behave in a rather peculiar way. Unlike their precursors in the retina or the lateral geniculate nucleus (LGN), already cortical cells in Area 17 (visual area 1, V1) respond optimally only to a stimulus with relatively complex features. Some cells prefer only one color, or a certain velocity of a moving stimulus, others need binocular stimulation with a well-defined disparity and so on [for a review of cortical specificities see Orban (1984)]. The most prominent feature, however, for which almost all cortical cells in the lower cortical areas (V1, V2) share a certain degree of specificity is the orientation of the stimulus. In particular, many visual cortical simple cells only respond to an elongated stimulus presented at one orientation. Even small deviations from the optimal orientation (e.g., $<\pm 10°$) can totally abolish the response in these cells. More commonly, deviations of about $\pm(20–40°)$ are tolerated, but will lead to a significant response reduction (Henry *et al.* 1973, 1974a, b).

This so-called *orientation specificity* has given rise not only to a large number of experimental studies, but has also raised interest in the community of theoreticians and modelers of brain function. The number of theoretical papers published on this topic in the last 35 years is counted by several hundreds. The reason for such a strong effort lies in the fact that cortical orientation specificity has a strong impact on more general models of computer vision and image analysis. The interest in this type of specialized cell behavior therefore extends beyond the field of neurophysiology and reaches into computer science and engineering.

Yet it is rather intriguing that after all this effort there is still quite some disagreement about the origin of orientation specificity. Partly this disagreement is fueled by the lack of clear-cut experimental evidence in favor or against certain models; partly, however, it might also be caused by the strongly differing viewpoints introduced as the consequence of different modeling approaches.

In this chapter I will therefore attempt to outline the conceptual differences among modeling approaches ranging from biophysically realistic models to models of self-organization of orientation specificity. To this end, I shall compare the models with each other, highlighting the virtues and drawbacks of each approach. In addition, I will try to embed the ideas underlying the different approaches in the broader context of other aspects of visual cortical cell behavior. Finally, the comparison of models of cortical orientation specificity should also subserve the more general purpose of pointing out the complexity of problems in the design of models of brain function. In general, only cortical simple cells shall be considered. Before I discuss the models in greater detail, however, the next two shorter sections are devoted to a brief review of the "why" and "how" of modeling orientation specificity.

2.1. Reasons for Modeling Orientation Specificity

Orientation specificity is part of a more generalized description of cortical receptive fields and cannot be considered just by itself. From a computational point of view at least one more aspect in the spatial domain needs to be taken into account, which is the spatial frequency selectivity resulting from the ON–OFF substructure of the receptive fields. The elongated ON or OFF subfields of a cortical cell are arranged in a side-by-side order forming a receptive field that responds to alternating contrasts.

Marr (1982) defined several steps that need to be performed for image analysis. One of the first steps is the detection of contrast discontinuities, or its equivalents, the zero crossings of the second derivative of the image intensity function. He argued that the $\nabla^2 G$ operator is optimally suited for this purpose, where ∇^2 describes the Laplacian operator and G is a two-dimensional Gaussian distribution. This operator has an ON–OFF center–surround organization and is isotropic. It will therefore respond to all contrast edges only limited by the bandwidth σ of the Gaussian.

A cortical receptive field with is ON–OFF substructure basically detects contrast changes in the same way. As an additional feature it will, however, respond to elongated contrast edges, but only if they are oriented appropriately.

Therefore, an image processing system with a reasonably complete set of oriented filters similar to the receptive fields in the visual cortex will be able to compute the complete outline of an image preserving the orientation information of the edges. In fact, a vast majority of artificial image processing systems utilize the principle of such oriented filters in one way or another.

There are many computational problems associated with Marr's approach (e.g., low noise robustness because of the second derivative) which are beyond the scope of this chapter, and many alternative approaches exist. The general applicability of oriented edge detectors in computer vision, however, strongly emphasizes the necessity to try to understand the underlying neuronal principles, and modeling approaches can subserve this purpose.

2.2. Basic Physiological Findings

This section gives a short reintroduction to the phenomenon of cortical orientation specificity. It shall be considered as a "keyword list" in order to allow faster access to the models described below. A highly detailed review about the cortical physiology which underlies orientation specificity is given by Henry *et al.* (1994); other aspects of cortical cell behavior are also covered by Orban (1984) and Martin (1988).

In the first step, methods for the quantification of orientation specificity shall be briefly discussed, before considering the physiological origin of cortical orientation specificity.

The optimal stimulus for most cells in the lower visual cortical areas is a moving light (or dark) bar, and mostly orientation specificity is measured using such a stimulus (Fig. 1A). Consequently, a motion-dependent, direction-sensitive part of the response is superimposed onto the orientation-specific response. In order to quantify orientation specificity, one should realize that a response curve, measured with moving bar stimuli covering 360° directions, is in fact a circular probability distribution (Fig. 1D). Thus, the first and second trigonometric moments (Fig. 1E) of this distribution carry the information about the direction and the orientation specificities of a cell (Batschelet, 1981; Wörgötter and Eysel, 1987) [see Zhang (1990) and Li *et al.* (1994) for a more elaborate treatment]. The trigonometric moments come in pairs a_i, b_i, respectively, the real and imaginary (cosine and sine) coefficients of a Fourier series. Therefore, the normalized amplitude $[A_i = (1/A_0)\sqrt{a_i^2 + b_i^2}]$ and the phase values $[\phi_i = \arctan(b_i/a_i)]$ encode the relevant variables of the cell behavior. For the orientation specificity (i.e., $i = 2$) we define two variables: (1) the angle of the preferred orientation (in short; the *preferred orientation* $PO = \phi_2 < 180°$) and (2) the bandwidth of the response curve (the *orientation tuning strength* $O = A_2$). In many studies O is measured as half-width at half-height of the response curve, but one should realize that for a moving stimulus this measure confounds direction and orientation tuning. A flashing stimulus would only elicit an orientation-specific response, but it is often less efficient in driving the cell.

In the second step, the question of the origin of orientation specificity shall be considered. In general there are three possible points of origin: subcortical, afferent convergence, and intracortical.

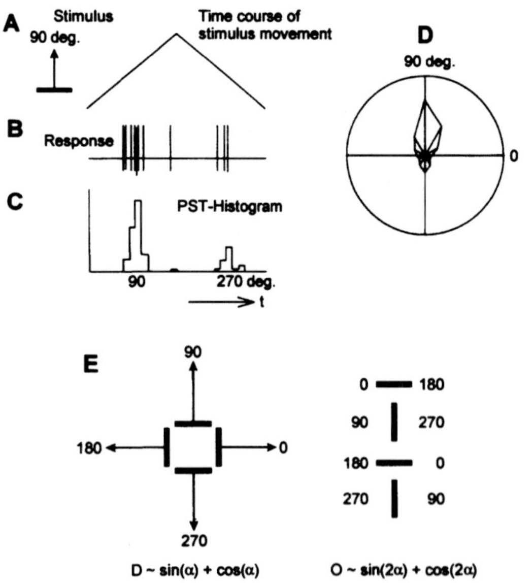

Figure 1. Schematic of the most commonly used stimulus protocol for the assessment of orientation specificity. (A) Stimulus moving up and down and crossing the receptive field of the recorded cell. (B) Cell response (C) Peristimulus–time histogram (PSTH) for up (90°) and down (270°) motion. Note that the PSTH of *one* bar orientation contains the responses for *two* motion directions. This cell is directionally tuned for upward motion. The same stimulus protocol is repeated for all orientations of the bar in small steps (e.g., 30° between orientations). The orientation sequence is randomized. PSTHs for each orientation are obtained this way and (D) the orientation tuning curve is plotted in a polar coordinate diagram (polar plot). In the polar plot only the maximal response for each motion direction is plotted against the angle of movement. The shape of the polar plot gives a first estimate of the direction and orientation tuning of the cell. (E) A more quantitative analysis can be obtained if one associates the directional component D of the response with the first harmonic and the orientation component O with the second harmonic of a Fourier series, as explained in more detail in the text.

Most influential in this context was Hubel and Wiesel's qualitative model of a cortical simple cell (Hubel and Wiesel, 1962). Relying on the fact that cells in the LGN were believed to have totally isotropic receptive fields, they suggested that the elongated receptive field of a simple cell could be generated by structured convergence of several tens of LGN cells. If the converging LGN cells were aligned like pearls on a row, the cortical target cell would most strongly respond to an elongated stimulus oriented in the same way as the long axis of the converging cell row. The simplicity of this model accounted for its success, even though more direct experimental evidence for it could not be obtained until 1991, when Chapman *et al.* actually described such an aligned ordering of afferent convergence. The Hubel and Wiesel model also plays a major role in the remainder of the chapter because many feedforward models of cortical orientation specificity are directly based on it.

Despite its success, there are several factors which are in disagreement with the basic version of the Hubel and Wiesel model. Already several early reports described inhibitory influences acting on the cortical cells (Benevento *et al.*, 1972; Blakemore and Tobin, 1972; Bishop *et al.*, 1973). In particular, the so-called sideband inhibition could play a role in shaping the orientation specificity, because it acts on the flanks of the receptive field (Bishop *et al.*, 1973; Sherman *et al.*, 1976; Kato *et al.*, 1978), making it "narrower." In addition, it has been shown that a receptive field formed by purely excitatory convergence from the LGN would perform nonlinear spatial summation when subjected to a drifting sinusoidal grating stimulus (Tolhurst and Dean, 1987). To further prove the dominant influence of inhibition, Sillito used micro-iontophoretical application of bicuculline to block the action of the inhibitory transmitter γ-aminobutyric acid (GABA) directly at the recorded cell (Sillito, 1975, 1977, 1979). He showed that orientation specificity could be formed entirely by inhibitory processing. The topographical arrangement of such inhibitory connections was assessed in other reports (Wörgötter and Eysel, 1991; Kisvárday and Eysel, 1993; Kisvárday *et al.*, 1993, 1994).

While these studies focused on inhibitory intracortical interactions, in particular some long-ranging excitatory connections between cortical cells could also be of relevance in shaping the orientation specificity. In order to understand the significance of long-range connections first one needs to look at the three-dimensional arrangement of orientation specificity in the visual cortex, which follows certain ordering principles.

The preferred orientation varies smoothly in all but a few *XY* locations, but in a complicated way on the cortical surface. On the other hand, it is almost constant throughout the whole depth of the cortex ("cortical columns"; Hubel and Wiesel, 1963; Albus, 1975; Murphy and Sillito, 1986; but see Bauer, 1982; Bauer *et al.*, 1983). More recently complete maps of the preferred orientation column structure have been obtained using optical imaging methods (Blasdel and Salama, 1986; Grinvald *et al.*, 1986). Two-dimensional autocorrelation functions obtained from these maps confirmed the smooth transition of the preferred orientation at short distances (local correlation) already observed with electrode mappings. In addition, a secondary positive correlation peak was found at distances between 0.5 and 1.0 mm, pointing to a repetitive occurrence of similar preferred orientations. Already in 1972, Hubel and Wiesel suggested that the visual cortex should be arranged in repeating slabs, each of which contains a complete set of feature detector cells. Such a slab was called a hypercolumn, and the underlying concept of a modular cortex is directly supported by the repetitive peaks in the autocorrelation function.

In the context of single-cell orientation specificity these findings are of particular relevance together with reports on a patchy and clustered axon distribution found predominantly in cortical pyramidal cells (Gilbert and Wiesel, 1979; 1989; Martin and Whitteridge, 1984; Gilbert, 1985; Ts'o *et al.*, 1986). The spacing of the patches roughly corresponds to a distance where cells with an orientation preference similar to each other should be located. Excitatory connections between these patches would be ideally suited to enhance the response in the preferred orientation, thereby strengthening the orientation specificity of cells in both patches.

Other reports emphasize a possible subcortical origin of the cortical orientation specificity. An orientation bias already exists at the level of the retina (Levick and Thibos, 1980, 1982; Leventhal and Schall, 1983) and the preferred orientation of retinal cells is mostly radial, i.e., parallel to the line connecting their receptive fields to the area centralis. A similar tendency is found for LGN cells (Vidyasagar and Urbas, 1982; Vidyasagar and Heide, 1984; Shou and Leventhal, 1989; Smith *et al.*, 1990). The orientation bias in the retina and the LGN is spatial frequency dependent and usually increases for very narrow gratings close to the upper cutoff frequency of the response (Soodak *et al.*, 1987; Smith *et al.*, 1990). For such gratings the tuning strength can be similar to that of cortical cells; for wider gratings, however, the orientation tuning of retinal or LGN cells drops sharply and the remaining bias is rather weak. The bias does not seem to be introduced by the corticofugal feedback connections (Thompson *et al.*, 1994; but see Vidyasagar and Urbas, 1982), while it might play a role in the generation of cortical orientation specificity. Vidyasagar (1992) showed that a removal of the orientation bias in the LGN also reduced the orientation specificity in the corresponding cortical region.

Consequently, the orientation bias in retina and LGN can also be applied in models of cortical orientation specificity.

A novice modeler of cortical orientation specificity should at this stage start to become rather frustrated, because apparently there is no clear evidence in favor or against any possible origin of cortical orientation specificity. But things will get worse. . . .

2.3. Choosing the Model Level

Particularly for very complex systems like the vertebrate brain it does not make sense to try to incorporate each and every known aspect of the anatomy or physiology of the cells, because one would arrive at models of a complexity similar to the original system. This complexity has prevented a detailed analysis of the brain and would prevent the analysis of any such overly detailed model. Realizing this, however, the question arises of how to restrict the approach of modeling cortical orientation specificity.

Most obviously all modeling approaches are restricted in a natural way by the limited data available. Given the large number of publications on cortical orientation specificity, often containing conflicting data, from a practical viewpoint this restriction is still not very helpful for the actual design of a model.

Therefore, the ends should define the means. In other words, any modeling approach must be guided by the desired level of abstraction. For example, in order to implement orientation-specific behavior in a robot usually one would not try to implement realistic cells incorporating membrane characteristics, etc., rather one should try to use some sort of orientation-specific operator (e.g., Gabor function; see below).

The following discussion of models of orientation specificity will mostly concentrate on a few examples taken from four classes of models with different levels of abstraction, which shall be briefly introduced now:

1. Filter models.
2. Developmental models.
3. Structurally defined models.
4. Biophysically realistic models.

Without trying to be complete, Table I gives a brief summary of the differences among these model classes. It cannot be avoided that this brief introduction of the different model classes contains some overlap, and other subdivisions could also be defined if desired. For the practical purpose of reviewing this divergent field of research, however, they will be appropriate.

Filter models subsume all approaches in which orientation specificity is modeled by the convolution of the image intensity function with an oriented filter. Usually only single cells are modeled and the level of abstraction is rather high, disregarding all biophysical characteristics. The output of these models is continuously varying and can be interpreted in most cases as a neuronal impulse rate. In some of the models negative impulse rates are permitted and interpreted as an inhibitory signal; in many others the output is clipped below zero. To account for a contrast reversal, in these cases cell pairs are modeled with antagonistic filter profiles and their output is combined for further processing. In addition, filter models are often extended by the temporal domain applying spatiotemporal filter functions, and their scope almost always lies beyond modeling of the mere orientation-specific behavior. All these approaches can be treated analytically to a large degree, which is one of their central advantages. The convolutions can often be solved in the Fourier domain, which saves computational power, and in many cases the derived equations can be directly implemented as "operators" in technical systems. This way an orientation-filtered version of the original image can be generated much in the sense of Marr's *raw primal sketch* (Marr, 1982, Chapter 2.2). Membrane characteristics cannot be modeled with filter models and the interactive behavior of larger networks of cells is usually also excluded because of the increasing computational complexity.

Developmental models concentrate on the arrangement of orientation specificity on the cortical surface. Large arrays of cells (e.g., $>10^3$) are modeled in most cases and the orientation preference is often the only variable considered. The output is therefore a map of angular values at each simulated cortical *XY* coordinate yielding the so-called *orientation preference map*. Some of these models try to combine the orientation preference map with other cortical maps (mostly the ocular dominance map). Others include more features of a cortical receptive field (e.g., the ON–OFF substructure). Developmental models utilize self-organizing

Table I. Comparison of the Different Modeling Approaches for Cortical Orientation Specificity[a]

	Filter models	Developmental models	Structurally defined models	Biophysically realistic models
Approach	Gabor/DOG filters (other similar filters ⇒ Motion-energy/directional/push–pull models Single cell only	Self-organization Artificial neural networks Maps of the preferred orientation and orientation tuning strength	"Graphical"/one-step computation or mathematical description of the map Final map of the preferred orientations	Multicell models Membrane characteristics Convergence/divergence, topography, excitation, inhibition, cells, layers, areas
Method	Convolutions (in Fourier domain) Sometimes analytically tractable, else computer simulations	Hebbian learning, many other learning rules Kohonen maps, radial basis function maps Correlation techniques Computer simulation	"Hand-drawn" or analytical models Sometimes computer simulations	Differential equations of the membrane characteristics Spike propagation, synaptic transmission Computer simulation
Goal	Reproduce abstractions from the data (e.g., DI) Often direct predictions possible Technical implementation (operators)	Reproduce the maps Predictions about development (e.g., developmental distortions) Determine the connectivity pattern Provide "energy" estimates of the final maps	Understand the structure of the final map Understand cortical design principles	Reproduce membrane behavior Mimic electrophysiologic experiments Rule out qualitative models Assess robustness, control experimentally inaccessible parameters
Comment	In many ways very successful model class, some models are also rather intuitive Multicell models mostly impossible	Relations exist to applied and theoretical physics (e.g., spin-glass models)	Conceptional questions arise from this model class Recent mathematical models provide a solid theory of the final map structure	Often very complex/opaque, complete assessment of the parameter space impossible Have to be tested much in the same way as the real system

[a]DOG, Differences of Gaussians; DI, Direction Index.

algorithms for the formation of the map. They apply all kinds of different learning rules, which can range from physiologically realistic rules to abstract rules for self organization in any maplike structure. Therefore, a strong relationship of some of these models exists to artificial neural network models and also to self-organizing processes in other physical or biological systems (e.g., spin-glasses, fur patterns in animals, etc.).

In most of the **structurally defined models** also the orientation preference map also is simulated. Unlike in the developmental models, however, here the goal is to give an explanation for the mechanisms which govern the structure of the final map. Some of them, for instance, define the map by interactions between pointlike "centers" on the cortical surface, from which the distribution of the preferred orientations is initialized. The advantage of many structural models is their high level of abstractions, which makes them very intuitive. In many cases one can "see" how the model behaves. This feature can be used in trying to understand how certain connection patterns would influence orientation specificity. Many of the older models are graphical and only semiquantitative. More recent approaches, however, are rather mathematical and provide solid theories of the structure of cortical maps. These theories are advanced and it seems that we are approaching a high level of understanding of the final map structure. In addition, several conceptually important questions can be asked with the help of this model class. The answers to these questions are of particular relevance for the comparison with the other model classes (i.e., developmental and biophysical models) because they give indications about the possible connectivity in the mature cortex. For this reason a larger section shall be devoted to this model class (see Appendix).

Biophysical realistic models are by far the most complex model class. They can range from single-cell models to models of single columns and finally to those of more than 10,000 cells including several substructures of the visual pathway. All of them incorporate membrane mechanisms, which usually are modeled by differential equations, more or less related to the Hodgkin and Huxley equation [for an overview see, e.g., the collection of articles in Koch and Segev (1989)]. Thus, the output of these models is the membrane potential of the cells, which can lead to a spiking behavior. The degree of complexity of the implementation of each single cell depends here crucially on the number of modeled cells. Some single-cell models, for instance, try to model cortical specificities as the result of intradendritic computations and therefore require a highly realistic implementation of the cell's dendrite and membrane channels. In models of large networks, on the other hand, cells only consist of very few compartments (e.g., soma and axon) and the differential equation which describes the single-cell behavior is often reduced to an integrate-and-fire characteristic. Network models usually include topography and realistic numbers of convergence and divergence. In most of them all parameters are subject to statistical fluctuations similar to those observed in the real system.

An advantage of biophysically realistic models is that they allow access to aspects of the cell behavior that cannot be measured directly in an experiment, for example, the intracellular behavior of a single spine or the extracellular characteristics of several hundreds of simultaneously "recorded" cells. While this can lead to interesting predictions and observations, it poses the problem of how to test the model experimentally. In addition, one often observes that biophysically realistic models are almost as opaque as the underlying real system and their central disadvantage is the vast complexity, with parameter spaces of a dimension which can easily exceed 10^3. A modeler is therefore tempted to try to achieve each and every kind of behavior from the model components by just raising the level of complexity and/or by twisting the parameters. In particular, the "second temptation" is facilitated by the fact that our knowledge about the visual system is still spotty. Missing information can thus be easily filled in by arbitrary parameter settings in a model. Obviously both temptations very efficiently lead to models that do not prove anything except for the fact that "they work."

3. Models of Cortical Orientation Specificity

In the following sections I will describe the different models and will limit the discussion of these models to basic aspects concerning orientation specificity. In most cases I will not attempt to describe the different models in great detail; instead the reader is referred to the original articles. Rather, the goal shall be to summarize central features and/or questions which arise from the individual model classes and which can be used to compare them (see Table I and last section).

3.1. Filter Models

Filter models for orientation specificity are relevant in the context of the well-known tradeoff between the resolution of a visual filter in the spatial frequency and the spatial (viz. orientation) domain ("joint resolution problem"). This problem was introduced by the discussion between experimentalists of whether cortical simple cells would perform a spatial feature analysis (Hubel and Wiesel, 1962, 1965) or decompose an image in the Fourier domain rather acting as spatial frequency analyzers (Campbell and Robson, 1968; Pollen *et al.*, 1971; Maffei and Fiorentini, 1973). About 1980 theoretical considerations were published which suggested that the dichotomy between the feature detector and the frequency analyzer hypotheses is in fact nonexistent, because selectivities in the one domain imply complementary ones in the other (Marcelja, 1980; Kulikowski and Bishop, 1981).

A similar joint resolution problem was discussed by Gabor (1946), who noted that a fundamental uncertainty principle limits the degree of specificity that a signal can have simultaneously in time and temporal frequency. He derived a set of filters—called Gabor functions—which optimize this tradeoff in the sense that they occupy a minimal surface area in the X/Y = *time/frequency* plane. The analogy of temporal filters to visual filters becomes obvious if one replaces the words *time* and *temporal frequency* with *space* and *spatial frequency*, and Marcelja (1980) established a parallel concept for the visual system. He derived a similar set of Gabor filters by which the response profile of a simple cell can be approximated (Marcelja, 1980; Kulikowski and Bishop, 1981). While these studies analyzed one-dimensional cross sections of cell responses, it was soon realized that Gabor functions can be extended in the second dimension, thereby more realistically describing the receptive

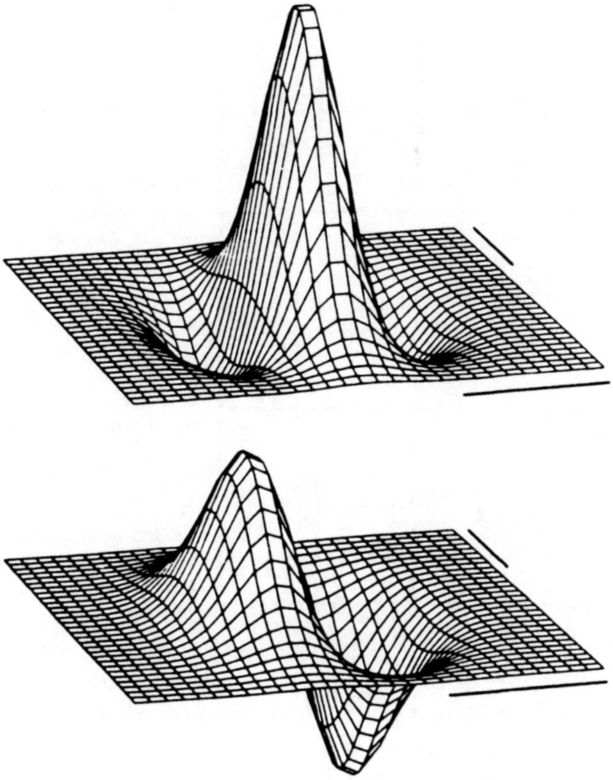

Figure 2. Typical even (top) and odd (bottom) symmetrical Gabor functions.

field of a simple cell (Daugman, 1980; Kulikowski *et al.*, 1982). The corresponding Gabor functions are given by

$$g(x, y) = \exp\{-\pi[(x - x_0)^2a^2 + (y - y_0)^2b^2]\} \exp\{-2\pi i[u_0(x - x_0) + v_0(y - y_0)]\} \quad (1)$$

where x_0, y_0 and u_0, v_0 are the coordinates in the spatial and the spatial frequency domains (Fig. 2). Equation (1) describes a two-dimensional sinewave of a certain phase undulating orthogonally to the preferred orientation and multiplied by a two-dimensional Gaussian envelope centered at the receptive field center. The Gaussian envelope has an elliptical shape with aspect ratio b/a, the long axis is not necessarily aligned to the preferred orientation, which is given by arctan (v_0/u_0). The center of the receptive field is located at (x_0, y_0) and it has width and length dimensions of $1/a$ and $1/b$, respectively. The optimal spatial frequency is given by $\sqrt{u_0^2 + v_0^2}$ (Daugman, 1985). Regardless of the choice of the parameters, the product of the two surfaces that are covered by these filter functions in the spatial and the spatial frequency domains (Fig. 3) is given as

$$(\Delta x \Delta y)(\Delta u \Delta v) = \frac{1}{16\pi^2} \quad (2)$$

where Δ is a standard measure of the linear dimension (width, length) of a filter function defined as the square root of the variance of its energy distribution. It has

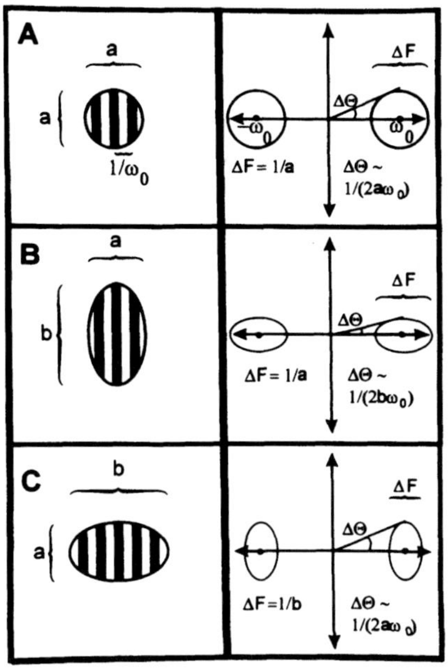

Figure 3. Schematic diagram explaining the tradeoff between spatial and spatial frequency resolution. (A–C) A surface view of three different even symmetrical Gabor functions in the spatial domain (left) and the spatial frequency domain (right). Stripes indicate the polarity change of the Gabor functions. The preferred spatial frequency is given by ω_0. Note that a distortion of the Gabor function along one axis of either domain will lead to an equivalent, but inverse distortion in the other domain, leaving the total surface area constant. $\Delta\Theta$ is the resolution in the orientation domain and is indicated in the figure by the angle underneath the $\Delta\Theta$ symbol. Thus, the resolution of the orientation specificity ($\Delta\Theta$) is inversely related to the resolution in the spatial frequency domain (ΔF). Note also that this inverse relation holds only for Gabor filters of equal size. Recompiled from Daugman (1985).

been shown that this product is the minimum joint surface area that can be obtained for all possible two-dimensional filter functions (Daugman, 1985). Thus, two-dimensional Gabor functions optimally resolve the tradeoff between space and spatial frequency resolution.

With the introduction of the second dimension, orientation specificity was automatically also included in the filter models. Orientation specificity can be regarded as an aspect of the spatial domain. Therefore, the tradeoff between space and spatial frequency resolution also applies to orientation specificity. Let us consider a set of Gabor filters of different shape with vertical preferred orientation, which occupy an identical spatial surface area (Daugman, 1985). Thus, any dilation of the filter area along one of the axes of the Gaussian envelope (a or b) must be accompanied by a contraction along the other axis (Fig. 3, left side). Due to Equation (2) also the surface area in the spatial frequency domain must be the same for the whole set. Therefore, continuous deformations in the spatial domain of the kind described above are followed by equally strong, but oppositely directed deformations in the spatial frequency domain (Fig. 3, right side). Thus, spatial frequency resolution ΔF and orientation resolution $\Delta\Theta$ (orientation tuning strength) are inversely related.

This tradeoff, however, only holds for Gabor filters of exactly the same spatial (and hence spatial frequency) area, which is usually not the case for receptive fields in the visual cortex. If we take a different set of Gabor filters for which we give up the constraint of constant area and instead assume a constant aspect ratio $\Delta x/\Delta y$ (and hence constant $\Delta v/\Delta u$), then we will get not an inverse relation between spatial frequency and orientation resolution, but a positive correlation between both variables. Let $\Delta x/\Delta y = \lambda$; then

$$\Delta\Theta = \arcsin\left(\lambda \frac{2^{\Delta\omega} - 1}{2^{\Delta\omega} + 1}\right) \tag{3}$$

where

$$\Delta\omega = \log_2\left(\frac{\omega_0 + \Delta u/2}{\omega_0 - \Delta u/2}\right) \tag{4}$$

Such a positive correlation between spatial frequency and orientation resolution has been observed experimentally (DeValois *et al.,* 1982; Hammond and Pomfrett, 1990).

The Gabor function model for cortical receptive fields benefits from the fact that only four parameters are needed to define its shape, while at the same time it provides a compact description of the characteristics of the field in the spatial and the spatial frequency domains. In addition, very good fits of two-dimensional receptive field profiles measured extracellularly in the visual cortex of the cat were obtained with this model (Jones *et al.,* 1987; Jones and Palmer, 1987a, b).

Hawken and Parker (1987), however, using a different experimental setup, found significant mismatches of the Gabor model with spatial frequency tuning curves recorded in monkey striate cortex. To avoid these differences they introduced a class of models based on combinations of spatially displaced differences of Gaussians (DOG), which yielded a better approximation of their data. Unfortunately, the best fit was obtained with a model containing nine free parameters. It is conceivable that the Gabor model does not capture all aspects of the spatial/spatial frequency behavior of cortical receptive fields, in particular, if one considers different species. On the other hand, it seems that it is a very good compromise which produces a reasonably accurate behavior, while it does not violate the requirement of trying to give the minimal description length of the underlying system. In this respect it is not clear if models with more free parameters or more complex approaches (e.g., Atick and Redlich, 1990) which try to embed the Gabor model into a broader theoretical context will necessarily lead to a better understanding of simple cell behavior.

Since the first report on the possible importance of Gabor functions in 1980 by Marcelja these models have been extended in the temporal domain, yielding the class of motion energy models (Adelson and Bergen, 1985; Heeger, 1987) and have been very successfully used to explain the motion and direction sensitivity of cortical cells (Heeger, 1993; see also Carandini *et al.,* this volume).

There is, however, one more consideration which needs to be discussed in the context of image and orientation processing by sets of oriented filters. From the viewpoint of machine vision, Gabor functions have in fact one significant disadvantage. Functions included in this class are not mutually orthogonal. Thus, a simple orthogonal projection of an image onto a set of Gabor "basis" functions will produce a set of coefficients from which the image is only poorly reconstructed. This problem can be solved by means of so-called "relaxation networks" which essentially perform an orthogonalization of the representation without orthogonalizing the visual code primitives, thus retaining their redundancy (Daugman, 1988, 1990; Pattison, 1992).

3.2. Developmental Models

Filter models usually provide an abstract description of several features of a cortical receptive field (orientation specificity, spatial frequency tuning, ON–OFF substructure, etc.), which are parametrized with respect to the two-dimensional visual field. They do not require an iterative process, because those features are explicitly defined by the set of equations which govern the particular model. All developmental models (but see *structurally defined models* in section 3.3), on the other hand perform a stepwise and iterative process for the generation of the desired features, which usually are descriptions of one or more cortical maps. The developmental aspect makes these models always "more realistic" than the filter models, because some connectivity pattern (connections, synaptic weights) between the simulated cells must be specified in order to allow for iterations.

The goal for a developmental model is usually to gain insight into the processes which form the preferred orientation map (or other maps) rather than to understand the final orientation specificity of single cells by itself. In this respect these models are beyond the scope of this chapter. On the other hand, the underlying connectivity pattern between the simulated cells makes this model class also relevant for the discussion of single-cell orientation specificity.

Concerning the number of output parameters, developmental models can be divided into two groups, of either a higher or a lower level of abstraction as compared to the filter models. Cells are described in the more abstract group of developmental models by a few basic variables (e.g., preferred orientation angle, orientation specificity, ocular dominance) without specifying the receptive field of the cell. Examples of this kind are given by Swindale (1982), Kammen and Yuille (1988), Obermayer *et al.* (1990, 1992), and Durbin and Mitchison (1990). These models focus less strongly on a biophysically realistic description of the developmental process; instead they provide mathematical abstractions of it. In the context of this review I am more concerned with the final orientation map and its relation to other features of cortical cell behavior, and therefore I will concentrate on the other group of developmental models, namely those which also include a parametrization of the size, shape, and ON–OFF substructure of the cortical receptive fields.

In the most general description, which applies to almost all such developmental models, they usually consist of a layer of cortical cells which gets afferent input from an input layer. At iteration zero some degree of intrinsic connectivity exists in the cortical layer and between afferents and cortex; however, all connection patterns and synaptic weights are rather unstructured. Orientation specificity is then generated by the modification of synaptic weights according to the correlation between the cell activities. The diameter and the general shape of the connectivity pattern are often predefined and crucially influence the formation of the cortical map.

The first model of this kind was proposed by von der Malsburg (1973) using a distance-dependent connectivity function and a Hebbian plasticity rule with constraint synaptic weights to assure competition between the units. The model required visual input for the map formation. A related model was designed by Perez *et al.* (1975).

A higher degree of biological realism is achieved in a large subgroup of models which include a polarization of the receptive fields (e.g., an ON and OFF characteristic). For example, Linsker (1986a–c) used circular symmetric correlation functions with a distance-dependent polarity change. He required that the synapses in his model can take positive or negative strengths. Thereby he introduced opposing polarities, which, however, in his model should not directly be interpreted as

ON/OFF. Nevertheless, with an additional constraint on the relation *between* the opposing synapses in a cell, he showed that within a rather small regime of the parameter space, orientation-specific cells emerge (MacKay and Miller, 1990a, b).

Other models that rely on interactions between the ON and the OFF subsystems were devised by Soodak (1987, 1991), Miyashita and Tanaka (1992), and Miller (1992, 1994). The mechanism which underlies all these models requires a competition between the ON and the OFF inputs to any given cortical cell (e.g., Fig. 4). One way to get a realistic behavior in the model of Miller (1992, 1994) is to adjust parameters such that two inputs with the same polarity (ON/ON, OFF/OFF) are positively correlated, while mixed polarities (ON/OFF) yield an anticorrelation for small distances between the input units. For an increasing distance a positive correlation is introduced between ON/OFF inputs in the model. Thus, the *difference correlation function* C_D, defined as the correlation between the like inputs minus mixed inputs, takes the shape of a Mexican hat (see Figs. 2 and 3 in Miller, 1994). He also discussed other possible shapes for C_D, but showed that a Mexican hat leads to the best results. Consequently, for close distances the development of like correlations (ON/ON, OFF/OFF) is facilitated, while for larger distances the probability for the generation of a subfield with opposite polarity increases. This particu-

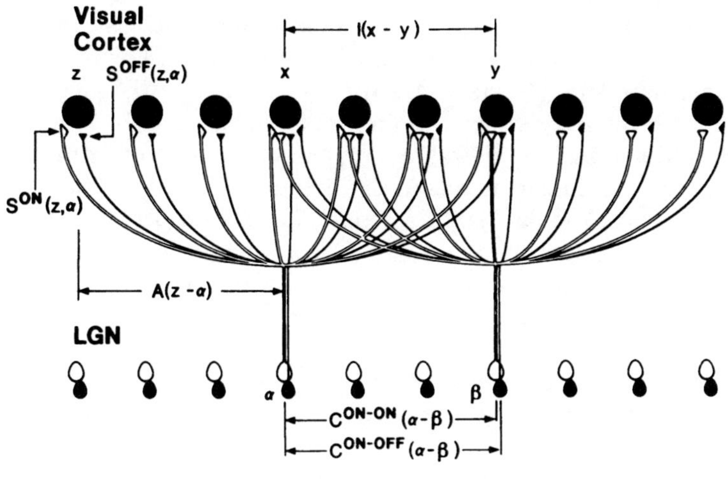

Figure 4. Structure of Miller's model for the development of the cortical orientation selectivity. Afferents from the lateral geniculate nucleus (LGN) project to the visual cortex. ON-center and OFF-center afferents (open and closed ellipses, respectively) make equivalent initial projections to the cortex. Synaptic interconnections among the cortical cells (filled circles) can be excitatory (direct connections) or inhibitory (connections via an inhibitory interneuron; interneuron not drawn). The strength of the connection between the cortical point z and geniculate point α is given by the arbor function $A(z-\alpha)$, which is zero outside of the arbor. The degree of correlation among the incoming afferents from retinotopic positions α and β is represented by the correlation function $C^{ON\text{-}ON}(\alpha-\beta)$. (The equivalent function for the OFF afferents also exists.) $C^{ON\text{-}OFF}(\alpha-\beta)$ defines the correlation between an ON-center afferent originating from α and an OFF-center afferent from β. Each synapse has a physiological strength that varies with time during development. This is illustrated by the functions $S^{ON}(z,\alpha)$ and $S^{OFF}(z,\alpha)$. Finally, there is some influence from activity at a cortical point x on the strength of synapses at a cortical point y. This is defined by the corticocortical interaction function $I(x-y)$ which may be both excitatory and inhibitory at different distances. Reprinted from Miller (1994), with kind permission from Elsevier Science.

lar correlation structure (Mexican hat) is also Miller's strongest prediction which would require experimental testing. The width of the afferent correlation is given by a so-called arbor function and additional distance-dependent correlations are introduced intracortically by a distance function *I*. Both functions limit the absolute receptive field size.

The resulting cortical receptive fields are circular and contain elongated ON and OFF subfields that are spatially segregated (Fig. 5). Orientation specificity emerges as the consequence of the elongated subfields rather than from an elongation of the total receptive field. This seems to be in accordance with experimental findings that show that the outline of the total receptive field of most simple cells is only slightly elongated, while the elongation of each subfield is much larger (Jones and Palmer, 1987a). At the same time, spatial frequency specificity occurs in Miller's model and the optimal spatial frequency is governed by the width of the subfields. Depending on the parameter settings, the model develops different preferred orientation maps, some of which bear strong resemblance to those measured experimentally in cat and monkey (Blasdel and Salama, 1986; Grinvald *et al.*, 1986; Blasdel, 1992a, b). Similar spatial scales, gradients of the preferred orientation changes, and singularities where all orientations meet (pinwheels; Bonhoeffer and Grinvald 1991, 1993) are observed in the model maps and the experimental maps. In parallel to the preferred orientation map, also a map of the orientation tuning strength is obtained from the model. In this map the orientation tuning strength drops to zero at the singularities, an observation for which, however, conflicting evidence exists from cat (Bonhoeffer and Grinvald, 1991; Bartfeld and Grinvald, 1992) and monkey data (Blasdel, 1992b).

The goal of this section was to point out a few aspects which are of relevance for the discussion of the biophysical models below. While several other models exist which utilize in part similar mechanisms (Bienenstock *et al.*, 1982; Soodak, 1987, 1991, Miyashita, and Tanaka 1992), the model of Miller (1994) can account most completely for the different phenomena observed in cortical maps and it can subserve a summary about what seems to be central to most of the correlation-based developmental models:

- Cortical maps are developed by the use of one (or more) linear differential equation which determines the changes of the synaptic strength at each iteration.
- Almost always the afferent and intracortical connectivity is determined in the models by local positive correlation, sometimes accompanied by a more distant anticorrelation (Mexican hat profile). Intracortical synapses do not change in most models.
- The map of the orientation tuning strength (if obtainable from a model) usually contains features that could be unrealistic, like zero tuning strength at the singularities.
- The development of orientation specificity requires neither an afferent nor cortical bias (e.g., orientation-biased LGN cells), nor does it require structured visual input. Initially isotropic receptive fields and a mechanism of symmetry breaking (Kammen and Yuille, 1988), which could also be introduced by unstructured (e.g., noise) stimulation (Linsker, 1986a–c), are sufficient for the map formation.

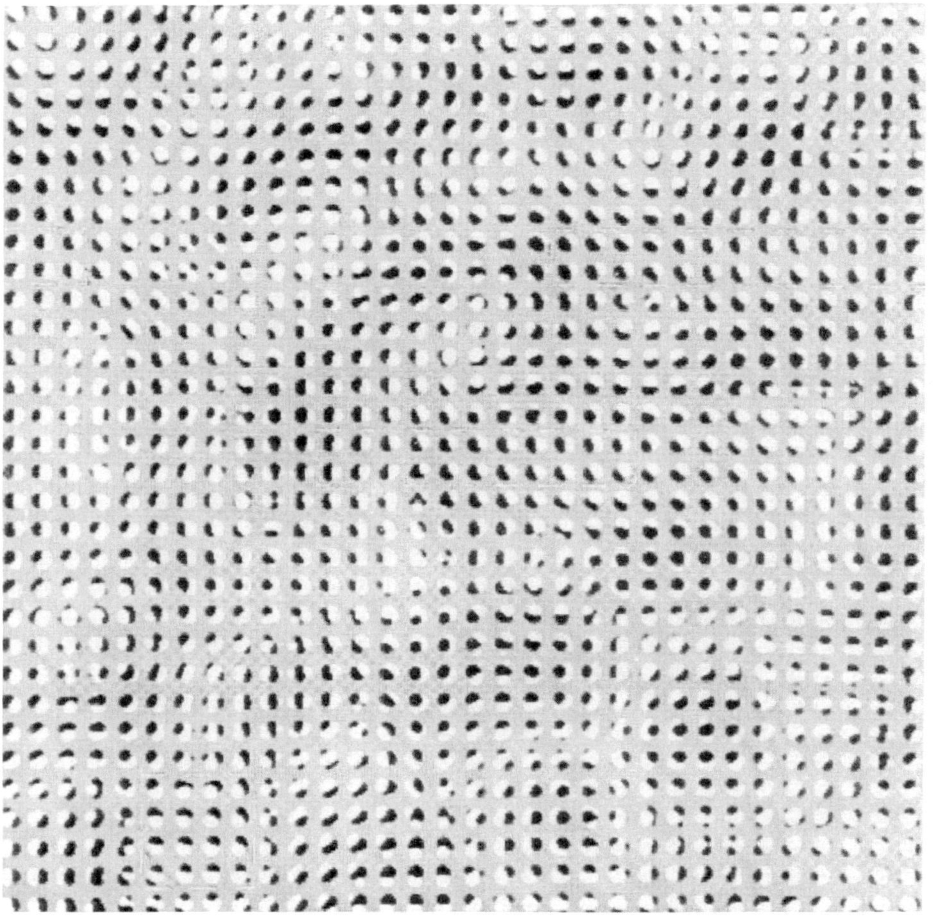

Figure 5. Final receptive field map from one of Miller's simulations. Every little blob on the gray background represents a single receptive field. The absolute receptive field size is identical for all cells and the surface area of the receptive field is disk shaped. White, ON subfields; black, OFF subfields. Even and odd symmetrical receptive fields can be observed with maximally three subfields. By blurring one's eyes an impression can be gained about zones with dominant ON or OFF response characteristic, consisting of cells with only one predominate subfield. The preferred orientation of each cell is given by the orientation of the line(s) separating adjacent ON and OFF subfield(s). The orientation tuning strength and the spatial frequency selectivity depend on the elongation and number of the subfields. Recompiled from Miller (1994).

• The map of the preferred orientation will always contain discontinuities (singularities, pinwheels).

• Realistic maps of cortical preferred orientations emerge from a multitude of mechanisms and there is often no clear decision in favor of one of the mechanisms. In some cases, the model mechanisms can be rejected from a biological viewpoint [e.g., Linsker's (1986a–c) model because of the negative synaptic strengths]. For the other models it can be suspected that the preferred orientation maps by themselves are not sufficient for a decisive exclusion of any of them and additional features (e.g., spatial frequency tuning, ON–OFF structure) need to be considered. The ease with which realistic preferred orientation maps are obtained, however, also indicates that orientation specificity develops under very weak constraints. As an additional consequence, it can be further speculated that the final connectivity used to generate orientation specificity in the mature cortex should be rather unspecific.

3.3. Structurally Defined Models

In this section I subsume all models that simply predict the final maps of the cortical orientation specificity, and try to give an intuitive and/or mathematical description of them. Obviously, this group is strongly related to the developmental models described above, the one trying to capture the developmental process, the others describing its outcome.

First suggestions about structural organization of orientation-specificity maps were made as early as 1970 by von Seelen, who suggested that the preferred orientations should be arranged on the cortical surface like spokes on a wheel. At that time such an arrangement was rather speculative, because of the lack of the data, and von Seelen had to restrict his model to a single hypercolumn, i.e., to a patch of cortex which represents all orientations (Hubel and Wiesel, 1968, 1972, 1974). Consequently, it was dismissed in favor of a simpler ice cube-like arrangement by Hubel and Wiesel (1972, 1974, 1977; see Fig. 6A). Already in 1974, however, doubts arose about whether such a structure would correctly reproduce the pattern orientation preference change in a tangential electrode penetration (Hubel and Wiesel, 1974; Albus, 1975). During such a penetration almost always the preferred orientations (PO) change gradually in clockwise or counterclockwise direction and smooth reversals of this direction are quite often observed. Less often one encounters a discontinuity where the preferred orientation jumps to a completely different value. While the ice cube model captures some of these features, the reversals of the preferred orientation changes are not reproduced and the orientation drift rates (i.e., gradient of PO along any direction) in a simulated penetration are zero for all those penetrations that travel along isoorientation slabs. In an experimental situation such flat regions are much smaller than expected from the ice cube model.

Therefore, Braitenberg and Braitenberg (1979) revived the “spokes on a wheel” idea, but unlike von Seelen (1970), they devised an elegant model of a complete cortical map (Fig. 6B) which reproduced the data from tangential electrode penetrations (Albus, 1975). As an alternative to the radial arrangement (spokes on a wheel), a concentric arrangement also was proposed. Both designs are, however, equivalent, as can be seen by rotating all POs by 90°. In the diagrammatic version of their model the orientation centers were spaced at equal distances, but already their reproductions of experimentally obtained electrode tracks showed that this assumption is too rigid and that displacements of the centers occur quite often.

Subsequently, several models similar to that of Braitenberg and Braitenberg (1979) have been proposed (Götz, 1987, 1988; Baxter and Dow, 1989; Bauer and Dow, 1991), some of which are discussed by Baxter and Dow (1989). In all these models the preferred orientations are arranged with respect to orientation centers (Figs. 6C, 6D). As soon as complete orientation preference maps became available from optical imaging (Blasdel and Salama, 1986; Grinvald *et al.*, 1986), the orientation centers were associated with those locations on the map where all preferred orientations meet (called singularities or pinwheels).

The biggest difference between these models concerns the structure of the singularities. In the "wheel-spokes model" all orientations are represented twice at a singularity. Other arrangements where each preferred orientation occurs once (e.g., Götz, 1987; see Fig. 6C) or more than twice are also possible. Singularities can be characterized by their topological index (Elsdale and Wasoff, 1976), which describes how often each preferred orientation is found in the direct vicinity of a singularity. The topological index is defined with respect to 360°, and a value of

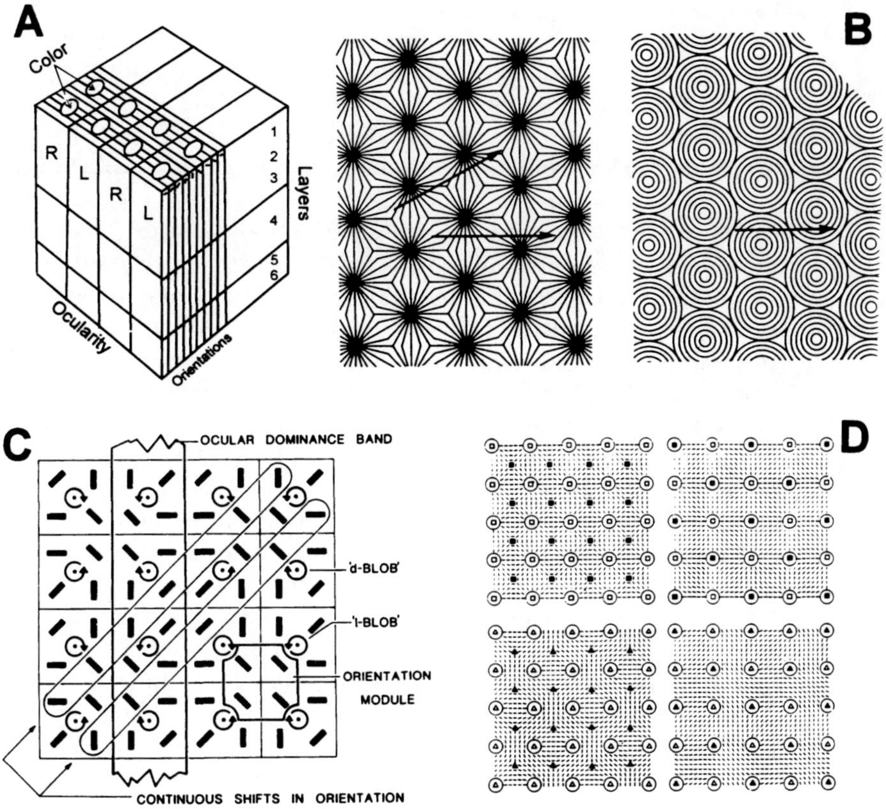

Figure 6. Different structurally defined models for the distribution of the orientation preference on the cortical surface. (A) Icecube model which incorporates orientation preference, ocular dominance, and color-sensitive blobs. (B) The model of Braitenberg and Braitenberg (1979). Preferred orientations are identical to the tangents of the lines/circles. The vorticity of both models is +1, and they can be transformed into each other by rotating all preferred orientations by 90°. Arrows indicate possible electrode tracts (Braitenberg and Braitenberg, 1979). (C) Model of Götz (1987, 1988); preferred orientations are shown as little bars. Vorticities of +1/2 ("l-blob") and −1/2("d-blob") occur. A possible relation between ocular dominance bands and orientation columns is indicated. Continuously shifting orientation preferences as observed by tangential electrode penetrations are also shown. (D) Different model classes designed by Baxter and Dow (1989); preferred orientations are indicated as tiny lines. Large circles represent cytochrome oxidase blobs, squares and triangles represent singularities. Vorticity: white squares, +1; black squares, −1; white triangles, +1/2; black triangles, −1/2.

$\pm 1/2$ represents the situation where each preferred orientation, all of which range between 0° and 180°, is only found once. A positive index indicates that the orientations rotate in the same sense as a path taken around the singularity; a negative index indicates reverse rotation. Thus, the "wheel-spokes model" has a topological index of +1. The optical imaging data (Bonhoeffer and Grinvald, 1991; Blasdel, 1992a, b), on the other hand, strongly support models with topological indices of only $\pm 1/2$ (see, e.g., Fig. 7A). In addition, optical imaging maps clearly demonstrate that the structurally defined models discussed so far strongly oversimplify the complicated map structure.

These rather simple approaches, however, were the starting point for other, more mathematically substantiated models which have been designed more recently for the description of the final map structure. In the following the arguments in favor or against a certain model will only be briefly introduced; a more detailed discussion based on some conceptional questions is given in the Appendix.

One core question which can be answered (at least in part) with the help of developmental and structurally defined models is how to define afferent and intracortical connections in order to arrive at a realistic biophysical model. This question is directly related to the problem of finding a mathematical and/or intuitive description of "cortical design principles."

In 1990 Rojer and Schwartz made the observation that the preferred orientation maps from monkey measured with optical imaging can be obtained by bandpass filtering of a white noise signal. Such a process leads to map which has an annulus-like power spectrum in the Fourier domain (Fig. 7E). In other words, a random distribution of preferred orientations and a faulty measuring system which performs a bandpass filtering would result in these maps. The limited spatial resolution of optical imaging does indeed lead to a smoothing effect (but not really a bandpass filtering) and a furious discussion arose about the relevance of these results. Three arguments support the map structures from optical imaging: (1) Several subsequently measured maps from one animal are essentially identical. (2) Similar map structures were also obtained in older studies which were based on electrode recordings (Albus, 1975; Swindale *et al.*, 1987). (3) Desoxyglucose mappings revealed maps for one (!) stimulus orientation which are also similar to those obtained from optical imaging (Hubel *et al.*, 1977; Löwel *et al.*, 1988). Thus, it is unlikely that these maps are a measuring artifact. Instead, the question arises of how this result could be interpreted in terms of cortical design. More likely Rojer and Schwartz's result points to the fact that during development the map structure arises from a selection of only one broad Fourier mode and that this process is rather isotropic, resulting in an annulus-shaped spectrum. One mode in the Fourier domain, however, represents a periodic system consisting of repeating orientation columns (modular arrangement) and the average distance between identical preferred orientations is given by the inverse of the spatial frequency of the Fourier mode. This average period is often called a hypercolumn. Since the system is isotropic, similar preferred orientations are arranged in slabs rather than in elongated stripes and the orientation columns do not have a certain direction on the cortical surface. With the help of the autocorrelation function shown in Fig. 7D one more aspect can be made visible. The curve demonstrates that the preferred orientation maps are governed by a strong local correlation (positive peak at zero) and a more distant and smaller anticorrelation (smaller negative peak).

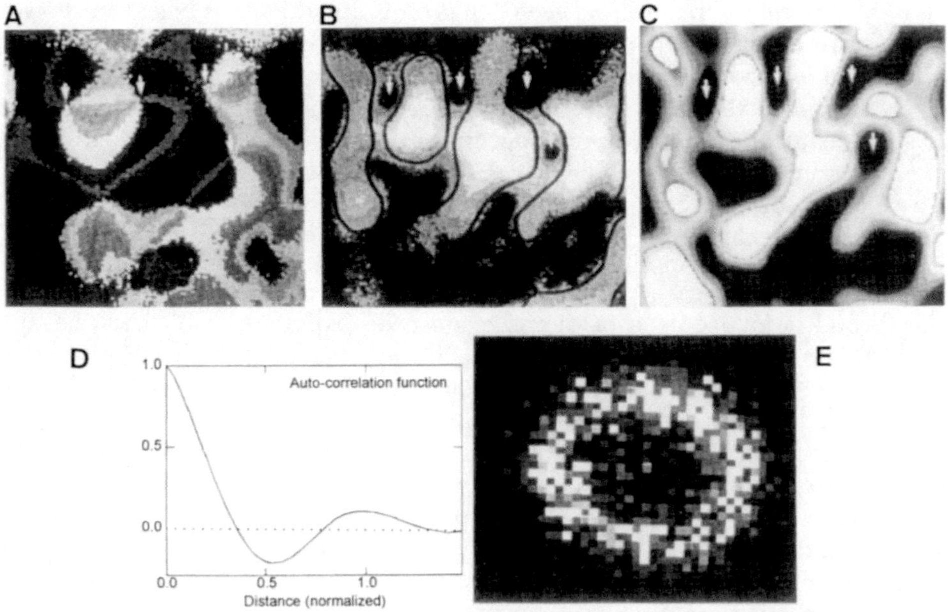

Figure 7. Maps of the preferred orientations (A) and the orientation tuning strengths (B) of a patch of cortical surface in cat (Bonhoeffer and Grinvald, 1991). (C) A computer-generted version of the orientation tuning strength map computed from the preferred orientation map by the algorithm described in the Appendix. The superimposed contours in panel B show that the maps in panels B and C are related with reasonable accuracy. Thus, the preferred orientation map can be directly transformed into the orientation tuning strength map. The gray shading in panel A indicates the different preferred orientations; regions of equal gray represent equal preferred orientation. In panels B and C light gray indicates strong orientation tuning, dark gray indicates weak tuning. Arrows point to singularities. (D) One-dimensional plot of the autocorrelation function taken from the preferred orientation map in panel A. In two-dimensions this function is circular symmetric and resembles a Mexican hat. A strong local correlation is followed by an anticorrelation (orthogonal preferred orientations) at intermediate distances and a small secondary positive peak at larger distances. (E) Annulus-shaped power spectrum of a preferred orientation map (not shown) from a monkey. Such a power spectrum can be interpreted in terms of local correlation, isotropy and modularity. Recompiled from Wörgötter and Niebur (1993) and Niebur and Wörgötter (1994).

As a simplified summary of the cortical design, one can now state that preferred orientation maps are constructed by an arrangement which is

- modular and periodic with one preferred period (single annulus!)
- locally correlated and anticorrelated at greater distances
- isotropic

This is certainly an oversimplified picture of the cortical design. For instance, it neglects species differences, because isotropy is only found in spectra obtained from monkey data, while the spectrum in cats is hemicircular. In addition, the first two features listed above are not independent of each other. A periodic representation necessarily leads to an alteration between correlated and anticorrelated regions. Nevertheless, it is helpful to keep these three instead of just two features.

The two-dimensional shape of the autocorrelation function resembles a Mexican hat. This type of function has been used in many developmental models to define the correlation pattern between cell groups. In addition it has been shown that the convolution of the preferred orientation map with its own autocorrelation function will lead to the map of the orientation tuning strength (Wörgötter and Niebur, 1993; see also Appendix). Figures 7A–7C show the measured and the computed maps. It can therefore be speculated that the Mexican hat-shaped autocorrelation function could actually represent the *average* synaptic coupling function in the mature cortex. This would indicate that in the cortex

- local excitation and more remote (lateral) inhibition should prevail

There is both evidence in favor of and against this hypothesis, which will be discussed in the sections below (see especially the Appendix). In particular, one should realize that this is a description of only the possible *average* synaptic coupling, neglecting superimposed excitatory and inhibitory influences.

The autocorrelation function, which is essentially zero at large distances, indicates that no long-range correlations exist. Recent studies of Wolf *et al.* (1993, 1994), however, provide evidence of the opposite. The correlation length in their analytical model is actually infinitely long and they show that the preferred orientation map is anchored at the singularities in the smoothest possible way minimizing an energy function (see Appendix).

Thus, one more aspect which governs the cortical design seems to be

- smoothness

There are many problems associated with these "design principles," some of which are discussed in the Appendix. Nevertheless, to be on the safe side, it seems advisable that biophysically realistic models should at least not deliberately violate these principles unless in an attempt to falsify them.

3.4. Biophysically Realistic Models

The class of biophysically realistic models can be roughly divided into three subgroups: (sub)cellular models, multicellular models, and multiareal models. The first group is concerned with (sub)cellular membrane and channel characteristics, the second group with the behavior of individual cells based on the connection pattern of the network, and the last group with higher level performance of model systems consisting of several visual areas (e.g., Tononi *et al.,* 1992).

It is generally accepted that orientation specificity is generated as the result of neuronal processing involving the LGN and the lowest visual cortical area V1. Therefore, the group of multiareal models, which sometimes utilize orientation specificity as one of their intrinsic mechanisms, shall not be considered here, because these models are concerned with interactions between higher cortical areas.

Most of the work done on (sub)cellular models is also only remotely related to cortical orientation specificity, but instead focuses on the intrinsic electrical properties of the modeled cells. Suggestions exist that cortical direction specificity could be based on dendritic shunting inhibition (Poggio, 1982; Koch and Poggio, 1985), but such a mechanism has not been tested thoroughly for orientation specificity.

However, this group also includes models in which orientation specificity is based on cell morphology. In the models of Colonnier (1964) and Tieman and Hirsch (1985) as well as Mitchison and Crick (1982; Mitchison, 1985) it is suggested that the spatial orientation and elongation of dendrites or axons might be involved in the generation of orientation specificity. In the most basic version of their model, Tieman and Hirsch (1985) claim that an oriented dendrite would sample inputs that lie along an oriented and elongated contour in the visual field, thereby generating an orientation-specific response. Alternative sampling patterns can also be devised involving intracortical inhibition. Mitchison and Crick (1982), on the other hand, suggest that excitatory links between cortical cells should predominantly contact cells which have the same preferred orientation. Considering long-range lateral connections, this would result in a system of cortical loci from which directed axon bundles arise. There is only little evidence for a pattern of oriented dendrites which would be consistent with the model of Tieman and Hirsch (1982), but axonal patches spaced roughly across a hypercolumn in distance have indeed been observed (Gilbert and Wiesel, 1979; Martin and Whitteridge, 1984; Gilbert, 1985; T'so *et al.,* 1986; Gilbert and Wiesel, 1989).

In conclusion, (sub)cellular models have only been little explored with respect to cortical orientation specificity. This is probably also due to the unspoken consensus in this field of research that orientation specificity probably arises as a network effect involving at least a few cells.

Therefore, the most appropriate level for a description of cortical orientation specificity with a biophysically realistic model is probably given by the group of multicellular models.

3.4.1. Basic Methods

Multicellular models of orientation specificity usually incorporate several thousand cells and represent a part of the primary visual pathway, i.e., retina, LGN, and visual cortex V1 including the ON and OFF subsystem (Wehmeier *et al.,* 1989; Wörgötter and Koch, 1991; Somers *et al.,* 1995). Only cortical simple cells are modeled.

Spiking behavior is simulated at the retinal level by transforming the output of the ganglion cells, which is a filtered version of the stimulus, into spike rates. To achieve this, retinal ganglion cells are simulated as "difference-of-Gaussian" filter functions G including a temporal low-pass behavior $L = \frac{1}{\tau}e^{-t/\tau}$. The image intensity function I is convolved with this set of filters:

$$r(x, y, t) = \int_0^t \int_{-\infty}^{+\infty} \int_{-\infty}^{+\infty} G(x', y')L(t')I(x - x', y - y', t - t')\, dx'\, dy'\, dt' \quad (5)$$

with

$$G(x, y) = G_{\text{center}}(x, y, t) - G_{\text{surround}}(x, y, t) \quad (6)$$

defined by

$$G_{center}(x, y) = \frac{K}{2\pi\sigma^2} e^{-(x^2+y^2)/2\sigma^2} \quad (7)$$

and similarly for $G_{surround}$, but including a temporal delay Δt. Values for the parameters (K, τ, δt, etc.) are taken from the literature (see references in Wörgötter and Koch, 1991; Somers *et al.*, 1995).

Binary action potentials are generated from this continuous response function by an approximate Poisson process given by

$$p(x, y, t) = p_0 \cdot \Delta t \cdot r(x, y, t) \tag{8}$$

where p_0 is an appropriate normalization constant.

The membrane characteristics of the postsynaptic LGN and cortical cells are modeled as improved integrate-and-fire neurons:

$$\begin{aligned} C\frac{dV_i(t)}{dt} &= \sum_{i=1}^{k} g_{\text{ex}}(t-t_i)[V_m(t)-E_{\text{ex}}] \\ &+ \sum_{i=1}^{l} g_{\text{inh}}(t-t_i)[V_m(t)-E_{\text{inh}}] + g_{\text{leak}}[V_m(t)-E_{\text{leak}}] \\ &+ g_{AHP}(t-t_{spike})[V_m(t)-E_{AHP}] \end{aligned} \tag{9}$$

where V_m is the membrane potential, C is the membrane capacitance, E is the synaptic reversal potential, g is the conductance, ex denotes excitatory influence, inh denotes inhibitory influence, leak denotes leakage, *AHP* denotes after-hyperpolarization, t_i represents the arrival times of incoming spikes, k is the number of excitatory synapses, and l is the number of inhibitory synapses.

A realistic topographic distribution of cells at all levels is assumed such that about 5° × 5° of the visual field is modeled. The models do not contain the complete number of cells which are found in such a field, but a certain degree of undersampling exists to save computational time (Fig. 8).

Usually several cortical orientation columns are modeled by defining a regular arrangement of the orientation preferences of the individual cells. As already discussed in Section 2.2 on *basic physiological findings,* evidence exists for all three possible origins of the orientation specificity of single cells: afferent bias, afferent convergence, and intracortical connectivity. Since the orientation bias of LGN cells is only strong for high-spatial-frequency stimuli, it seems less likely that this bias plays a dominant role in shaping cortical orientation specificity, and the quantitative models of Wörgötter and Koch (1991) as well as Somers *et al.* (1995) exclude this option [but see Vidyasagar (1987) for a qualitative approach].

3.4.2. Feedforward Models of the Hubel and Wiesel Type

Hubel and Wiesel (1962) proposed that cortical orientation specificity could be generated by the convergence of excitatory thalamocortical input from several LGN cells with receptive fields that cover a well-oriented region of the visual space. Nielsen (1983) designed the first quantitative model along this line (Fig. 9). Such a connection scheme can account very well for cortical orientation specificity without further requirements, and Fig. 10 shows how the aspect ratio of the converging

RETINA
ON
OFF
LGN
ON
OFF
CORTEX
CORTEX
Accum. Act.
150
250
400
550
700
850 ms

Figure 8. Run-time display of the cortex simulator. Cells receive oriented afferent input according to the model devised by Hubel and Wiesel (1962); no intracortical connections are made. Single cells are represented as little squares. A vertical bar stimulus moves from left to right across the simulated retinal patch (speed 5°/sec). Bottom panel: Intracellular potential of two cortical simple cells in a horizontal and a vertical orientation column. The hyperpolarizing potentials seen are solely caused by the after-hyperpolarization following action potential discharge (g_{AHP}), since no inhibitory intracortical inhibition was included in this run. Middle panel: Membrane potential of the five neuronal populations: ON and OFF β-type retinal ganglion cells, ON and OFF geniculate relay cells, and cortical simple cells. The color of each square represents the intracellular potential of that cell. Warm colors represent depolarization and darker blue represents hyperpolarization; the resting potential is shown in green. Spikes appear in white. Small random fluctuations of the membrane potential have been included in cortical and geniculate, but not in retinal cells. The response in the retina appears to be shifted compared to the LGN; this is due to the fact that 32×32 retinal cells project to 64×64 LGN cells. For graphical reasons the retinal cell population has been centered, so that the outer parts of the retina contain no cells. Top panels: accumulated activity in the cortex. The size of the black squares represents the total number of action potentials accumulated by that particular cell up to the time indicated. At the end of the stimulus sweep, 2.5 hypercolumns can be detected. Note that the overall activity has been artificially raised to yield clear orientation columns following a single stimulus sweep. Reprinted from Wörgötter and Koch (1991), with kind permission from Elsevier Science. UK.

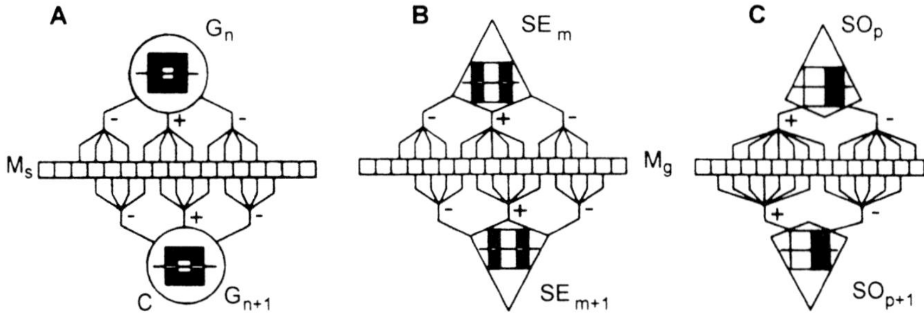

Figure 9. The model of Nielsen (1983, 1985), which essentially incorporates a Hubel and Wiesel-type feedforward structure. Cells are represented as circles (ganglion cells) and pyramids (simple cells). The figures inside the cells reflect the receptive field structure defined by the input map. One dimensional cross sections (indicated by the lines) through the receptive field input map are shown by the connecting lines and sign symbols (+, excitation; − inhibition). The structure M_s represents a row of photoreceptors, M_g a row of ganglion cells, both of which are defined as shown in panel A. Recompiled from Nielsen (1985).

LGN cells affects the cortical orientation tuning strength (Ferster, 1987; Wörgötter and Koch, 1991). the aspect ratio is defined as number of converging LGN cells along the long axis of one cortical subfield divided by the number of LGN cells on the short axis, as shown in Fig. 15A. To get the actual aspect ratio of the cortical receptive field, the spacing and overlap of individual LGN receptive fields also need

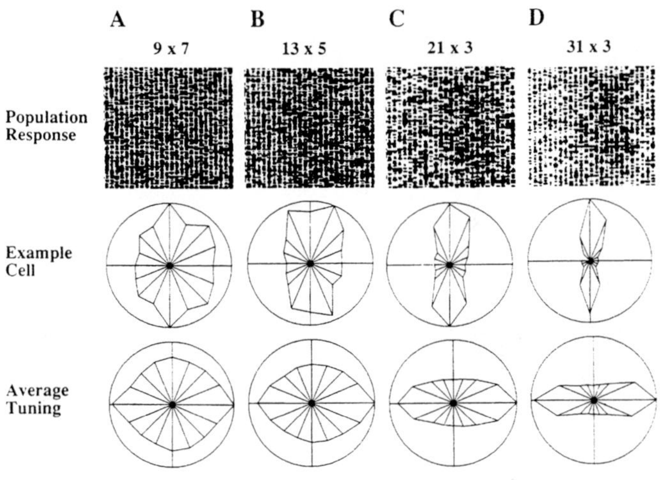

Figure 10. Orientation selectivity as a function of the aspect ratio without inhibitory corticocortical connections. Different connection patterns according to the model of Hubel and Wiesel (1962) are given by the number of geniculate cells converging onto each cortical cell (e.g., 9 rows by 7 columns of geniculate cells) resulting in receptive field width-to-length ratios of 1.16, 1.78, 3.18, and 4.40. The resulting orientation specificity is shown for an example cell (middle) and for a population average (55 cells randomly selected, bottom). The top row shows the cumulative spike count across one orientation column, accumulated during a 1000 msec-long stimulation with a moving vertical bar. Cells with an approximately horizontal preferred orientation are represented on the left side of each panel, whereas cells with vertical preferred orientation are shown on the right-hand side. Only the largest aspect ratio will lead to a reasonable orientation tuning. Reprinted from Wörgötter and Koch (1991), with kind permission from Elsevier Science.

to be taken into account. In this model, cells with similar preferred orientation share the same *X* coordinate and the orientation columns are therefore parallel straight lines. It should be noted that the strong overlap between LGN inputs applied in the model of Wörgötter and Koch (1991) will also lead to overlapping ON and OFF subfields. While this does not affect the orientation specificity in the feedforward model, such overlap is usually not found in simple cells. A solution to this would be a sparser and nonoverlapping sampling of the LGN input (e.g., Fig. 5 in Wörgötter and Koch, 1991).

The Hubel and Wiesel model is also quite potent in reproducing other features of cortical simple-cell behavior [e.g., NULL response (Movshon *et al.,* 1978), dot responses (Wörgötter and Eysel, 1989), and velocity low-pass and bandpass behavior (Movshon, 1975; Orban *et al.,* 1981)].

The conclusion at this point, however, is that the aspect ratios needed to achieve realistic orientation tuning strength with a pure feedforward model are quite high; also other unrealistic features are discussed in the following.

3.4.3. Models Including Intracortical Inhibition

A central problem that arises as the consequence of a pure feedforward model is its strongly nonlinear spatial summation behavior. Even though NULL responses could be elicited for certain cortical cells (Wörgötter and Koch, 1991), this feature, which points to a linear spatial summation behavior, ultimately arose from the linearity of the modeled retinal cells and from the adjustment of the firing thresholds. Thus, it reflects a parameter-dependent situation. The spatial summation behavior of a cell can be measured by the relative modulation of the response to sinusoidal grating stimuli. The relative modulation shows how strongly the cell response follows the stimulus modulation as compared to any stimulus-unrelated response components, e.g., maintained discharge, etc. It is therefore defined as the quotient of the amplitude of the first harmonic (i.e., the spatial frequency of the stimulus itself) of the response divided by the dc component measured from the Fourier spectrum taken from the response peristimulus–time histogram (e.g., Skottun *et al.,* 1991). It can be shown that the relative modulation is always $\pi/2$ for an ideal half-wave rectifying cortical simple cell regardless of the spatial frequency of the grating stimulus. Nonlinear summation is reflected in a significant drop of the relative modulation, which can even reach zero for very narrow gratings, and this is expected to occur as soon as cells are modeled only by feedforward afferent excitatory convergence.

Intracortical inhibition of the push–pull type, as schematically drawn in the top panels of Fig. 11 (Nelle and Wörgötter, 1993; Tolhurst and Dean, 1990), can counteract this effect. The relative modulation remains close to 1.57 regardless of the stimulus frequency for all three push–pull mechanisms shown (Fig. 11, bottom left). Cascaded inhibition, which requires very little connection specificity and is based on the inhibitory convergence of tens of cells, leads to the best performance.

This by itself implies the significance of intracortical inhibition. A push–pull mechanism can also sharpen the orientation tuning of the cells, as discussed in more detail below. The polar diagrams in Fig. 11 show the average orientation tuning of 8000 simulated cells with and without push–pull connections.

In addition, it has been shown that the orientation tuning of a cell would be contrast dependent without inhibitory contrast gain control (Wehmeier *et al.,*

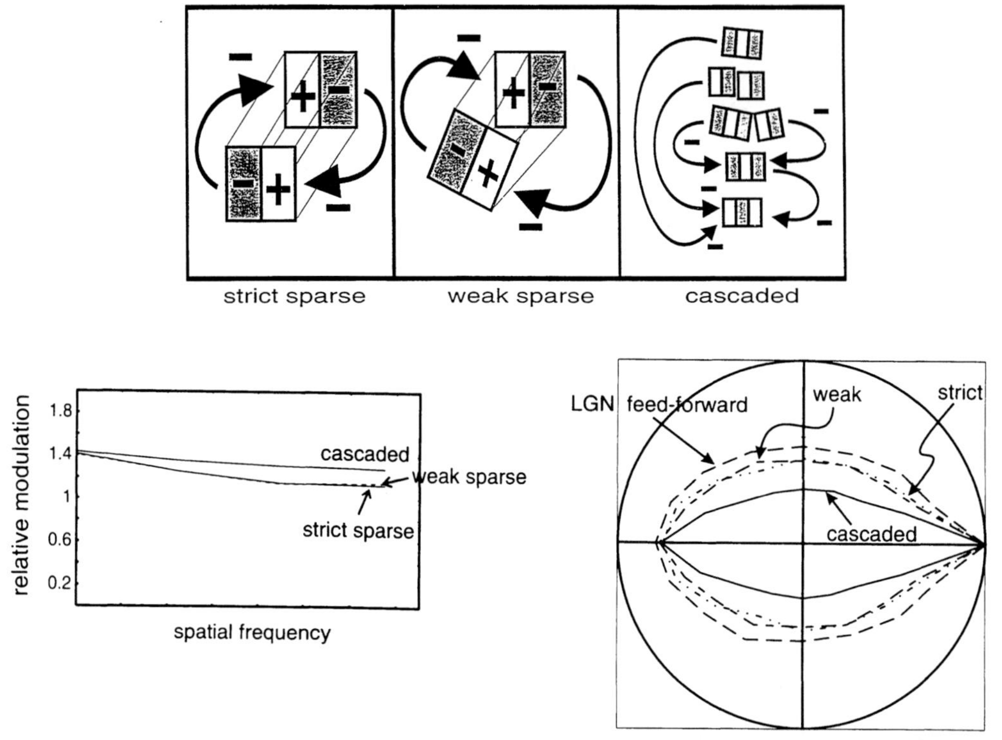

Figure 11. (Top) Different connection patterns for push–pull models. The strict version of this model requires that a cell with (almost) identical receptive field location and orientation will inhibit the target cell, whose receptive field polarity is reversed. The weak version allows for some degree of variation of this connection pattern. In the cascaded connection scheme many cells inhibit the target cell. The receptive field polarity of all source cells is inverse to that of the target cell, but the receptive fields of source and target cell match only loosely. (Bottom left) Relative modulation as a function of the spatial frequency of the stimulus measured for 8000 simulated cells. The ideal situation for a linar summation would be a flat relative modulation curve with a value of 1.57, which is best obtained with the cascaded inhibitory connection scheme. (Bottom right) Average orientation tuning for the same cell population for the different cases shown above. Cascaded inhibition produces the strongest orientation tuning.

1989), a situation which is not observed experimentally. The results of Sillito (1975, 1977, 1979) and others (Eysel *et al.*, 1990; Wörgötter and Eysel, 1991; Crook and Eysel, 1992) finally suggested that intracortical inhibition, with a connectivity pattern differing from that of the push–pull inhibition, could also be involved in shaping cortical orientation specificity.

Essentially two types of qualitative models have been proposed along this line (see also Ferster and Koch, 1987): Heggelund (1981) proposed that a lateral displacement of the receptive field of the inhibitory source cell with respect to the target cell would result in a final receptive field which is crescent shaped and thus orientation specific (Figs. 12 C–E and 12G). Note that this connection scheme does not require an initial orientation bias, but also works with totally isotropic receptive fields. In addition, note that this connection scheme bears some resemblance to the push–pull inhibition shown in Fig. 11.

Sillito (1979) as well as Morrone *et al.* (1982), on the other hand, suggested that mutual inhibition between cells with orthogonal preferred orientations should sharpen the orientation selectivity (Figs. 12A, 12B, and 12F). This mechanism has been called *cross-orientation inhibition* and requires an initial orientation bias. There is intra- and extracellular physiological evidence (Sillito, 1979; Morrone *et al.*, 1982; Volgushev *et al.*, 1993) for such a mechanism, supplemented by tracer studies (Matsubara *et al.*, 1985). Figure 12 clearly shows the similarity of the two schemes, which can be transformed into each other by changing the topographical arrangement between the cells and by altering their initial orientation tuning.

The role of inhibition in shaping the orientation tuning, however, is not unequivocally acknowledged. A series of experiments indicates that shunting (silent) inhibition probably does not significantly contribute to orientation specificity (Douglas *et al.*, 1988; Koch *et al.*, 1990; Berman *et al.*, 1991), because the measured conductance changes are rather low. It could also indicate, however that the inhibitory effects are primarily hyperpolarizing. Inhibitory postsynaptic potentials (IPSPs) have indeed been recorded, but mostly during stimulation with an orientation similar or equal to the preferred orientation of the cell; cross-orientation stimuli evoke only very weak IPSPs (Ferster, 1986, 1987, 1988; Douglas *et al.*, 1991). This would coincide with the push–pull model of intracortical inhibition, which relies on isoorientation inhibition (Allison and Bonds, 1994). It also suggests that all other inhibitory mechanisms should be rather broadly tuned (Bonds, 1989; Eysel *et al.*, 1990; Wörgötter and Eysel, 1991), which would particularly apply for any kind of "cross-orientation inhibition."

Inhibitory connection schemes like those proposed by Heggelund (1981) and Sillito (1979) and modifications of them could be tested with our biophysically realistic model (Wörgötter and Koch, 1991). Figure 13 gives a surface view of the simulated cortex and displays the inhibitory source cells (small squares) that connect onto the target cell in the center. Note that all cortical cells have an initial orientation bias imposed through a Hubel and Wiesel type of afferent convergence

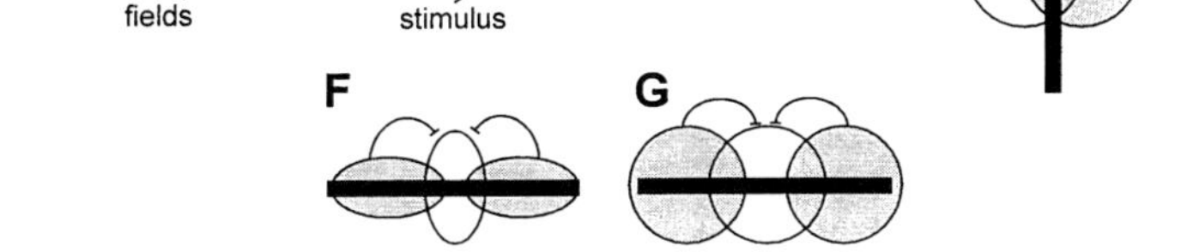

Figure 12. Diagram to schematically represent the effects of cross-orientation inhibition (A, B) and local lateral inhibition (C–E). (A, B) A cell inhibits a target cell with orthogonal preferred orientation, thereby strengthening its orientation specificity. (C–E) The inhibition between the two overlapping receptive fields leads to a cancellation of all responses from the receptive field area, which is covered by the inhibitory field. The remaining crescent-shaped portion of the receptive field is elongated and thus orientation specific. (B, C) Optimal, (A, D, E) nonoptimal stimulus situations. (F, G) Two alternative arrangements with nonoptimal stimulus. (F, G) Cross-orientation inhibition and local lateral inhibition can be transformed into each other by changing the receptive field overlap and the initial orientation tuning.

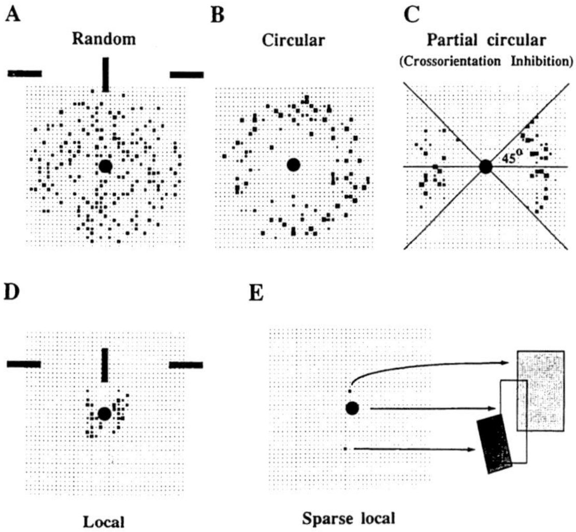

Figure 13. Different intracortical inhibitory connection schemes in a surface view. The central cell (large spot) receives inhibitory input from cells indicated by the small squares. (A, D) Sketch of the column structure as determined by the afferent LGN connections. (E) The initial receptive field structure (prior to inhibition). Reprinted from Wörgötter and Koch (1991), with kind permission from Elsevier Science.

and that cells with the same X coordinate have approximately the same preferred orientation ($\pm 10°$). Assuming an entirely linear, i.e., subtractive, action the inhibition, one can visualize the resulting receptive field shape as a three-dimensional profile (Fig. 14). Different viewing angles allow one to assess the aspect ratio of the final receptive field and give an estimate of the efficiency of the different inhibitory mechanisms. The change in the orientation tuning can be seen in Fig.

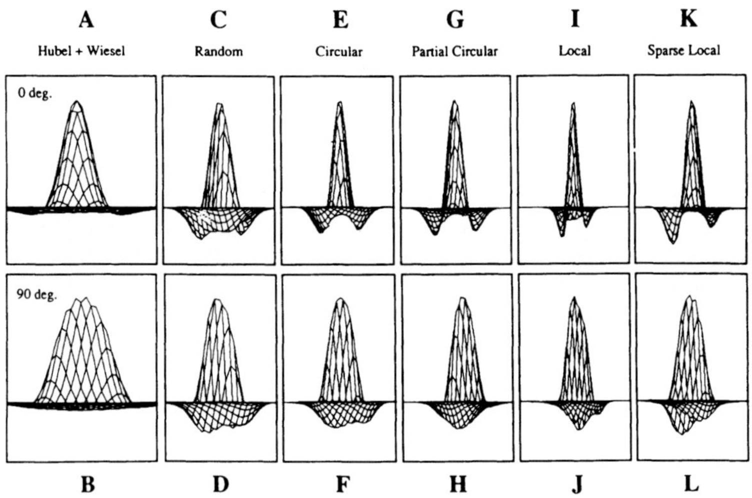

Figure 14. Effects of the different inhibitory connection schemes on the spatial shape of the receptive field of one example cell. Diagrams show receptive field activity functions computed from a superposition of basic receptive field types viewed edge on. (A, B) A superposition of DOGs determines the Hubel and Wiesel-type receptive field of the target cell. (C–K) The Hubel and Wiesel fields of the inhibitory cells are determined similarly, then those inhibitory fields are subtracted from the target cell Hubel and Wiesel field. Reprinted from Wörgötter and Koch (1991), with kind permission from Elsevier Science.

15. Assuming realistic parameter ranges, it can be observed that none of the individual mechanisms can fully account for a realistic orientation tuning strength. Almost all combinations of several mechanisms, however, will perform in the correct way if combined with appropriate ranges of synaptic weights. Intriguingly, even mechanisms with very little connection specificity, like circular inhibition combined with local inhibition, will give good results (Fig. 15N). Such a combination is very robust against parameter changes and only very little information is required to set up the connectivity pattern in the model. Thus, during cortical development such simple connection patterns would be advantageous.

Note that circular inhibition emulates a two-dimensional version of the classical lateral inhibition scheme. If the circle of converging inhibitory inputs has a diameter of about a hypercolumn, then circular inhibition acts as a very broadly tuned "cross-orientation inhibition" (Wörgötter *et al.*, 1991). Obviously, such a connection pattern must be regarded as an ideal limit case of the real lateral inhibitory connectivity in the cortex, because it is very unlikely that the source cells are located nicely on such an annulus. However, there is indeed statistical evidence for a connection pattern of large inhibitory basket cells, which roughly emulates a patchy and disrupted version of such an annular structure (Kisvárday and Eysel,

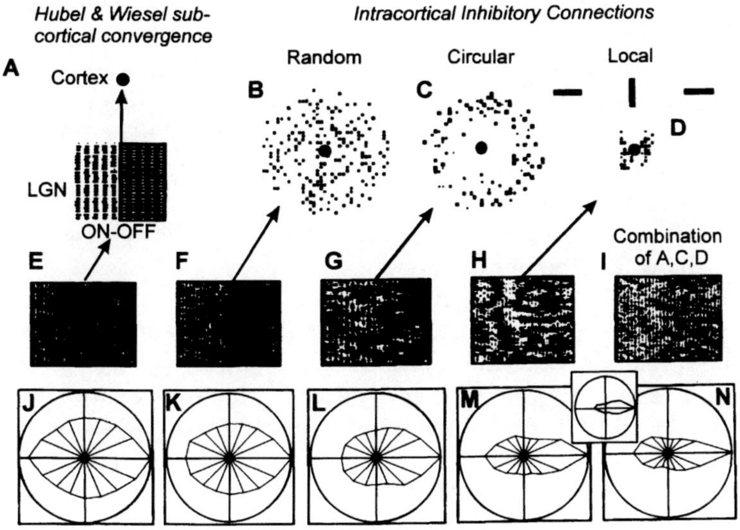

Figure 15. Orientation tuning achieved by different mechanisms in the model of Wörgötter and Koch (1991). (A–D) Connection schemes (see also Fig. 13). (A) Hubel and Wiesel-type connection pattern, (B–D) intracortical inhibitory connection pattern superimposed onto the Hubel and Wiesel connection pattern from panel A. The spacing of the orientation columns is shown in panel D. (E–I) Cumulative spike count across one orientation column, accumulated during a 1000-msec-long stimulation with a moving vertical bar. Clear orientation columns are only visible in panel I. (J–N) Average orientation tuning for a cortical cell population (other sample than in Fig. 10). Local inhibition (M) leads to a reasonable orientation tuning, but the column structure is largely destroyed (H). A combination of the mechanisms in panels A, C, and D, as shown in panels I and N, will lead to the emergence of orientation columns and to a realistic orientation tuning of the individual cells. The inset shows a population average of real cortical simple cells for comparison. Note that circular and local inhibition are combined with different synaptic weights in panels I and N; otherwise this combination would just reflect the random inhibition shown in panel B.

1993; Kisvárday *et al.*, 1993, 1994). The top panel of Fig. 16 shows the distribution of boutons of one basket cell. In the middle the average relative distributions of iso-, oblique-, and cross-oriented cells are plotted. These distributions reflect the probabilities to find cells of similar, oblique, or orthogonal orientation as a function of distance to the target cell. The bottom panel was obtained as the product of the boutonal frequency and the probability distributions. It shows that this basket cell can in principle exert local iso-orientation and medium-range oblique- and cross-orientation inhibition. It should be kept in mind, however, that this is a statistical analysis; the actual targets of the basket cell are not known. Given a random distribution of basket cells in every cortical volume element, such a distribution would amount to a weak version of circular inhibition.

In addition to its effect on the orientation tuning, circular inhibition has some more interesting features because it induces a direction selectivity bias and it is more robust than "classical" cross-orientation inhibition in the context of the underlying preferred orientation map (Wörgötter *et al.*, 1991, 1992).

While feedforward models of the Hubel and Wiesel type are able to produce a realistic orientation tuning strength, the above arguments showed that intracortical inhibition needs to be included in order to reduce the aspect ratio of the afferent convergence and to make the model more realistic with respect to the linearity of spatial summation.

At this point a link can be made to the developmental structurally defined models. Three mechanisms coexist in the biophysically realistic simulation: local excitation by afferent convergence, local inhibition, and two-dimensional lateral inhibition (circular inhibition). All mechanisms require very little connection specificity, and the total strength of the local excitation exceeds that of the local inhibition. Therefore, the resultant *average* synaptic coupling function has approximately a Mexican hat shape and directly mimics the shape of the autocorrelation function of the preferred orientation map as well as the shape of the correlation functions employed in many developmental models.

The combination of LGN feedforward excitation and intracortical inhibition with little connection specificity largely fulfills the requirements imposed by the experimental results discussed above. All these models, however, exclude the dominant source of synaptic input to a cortical cell, which arises from intracortical excitation [see Martin (1988) for a detailed comparison of the sources of synaptic inputs]. In addition, these models cannot account for a set of quite novel findings which indicate a strong involvement of intracortical excitation in the generation of orientation specificity (Nelson *et al.*, 1994).

→

Figure 16. (Top) Relative boutonal frequency of one single cortical basket cell. (Middle) Probability to encounter iso-oriented cells ($-30° <$ PO cells$\leq 30°$), oblique-oriented cells ($-60° <$ PO $\leq -30°$ or $30° <$ PO $\leq 60°$), and cross-oriented cells ($-90° <$ PO $\leq -60°$ or $60° <$ PO $\leq 90°$ and similar for the opposite quadrants) at a certain distance from any target cell with preferred orientation PO. (Bottom) Possible inhibitory impact of this basket cell at different distances. These curves are obtained as the product of the relative boutonal frequency with the probability functions and reflect therefore only a statistical measure of the inhibitory impact. Under this proviso, mostly isooriented cells will be the target of this basket cell at small distances, while at larger distances oblique- and cross-oriented cells will receive synapses.

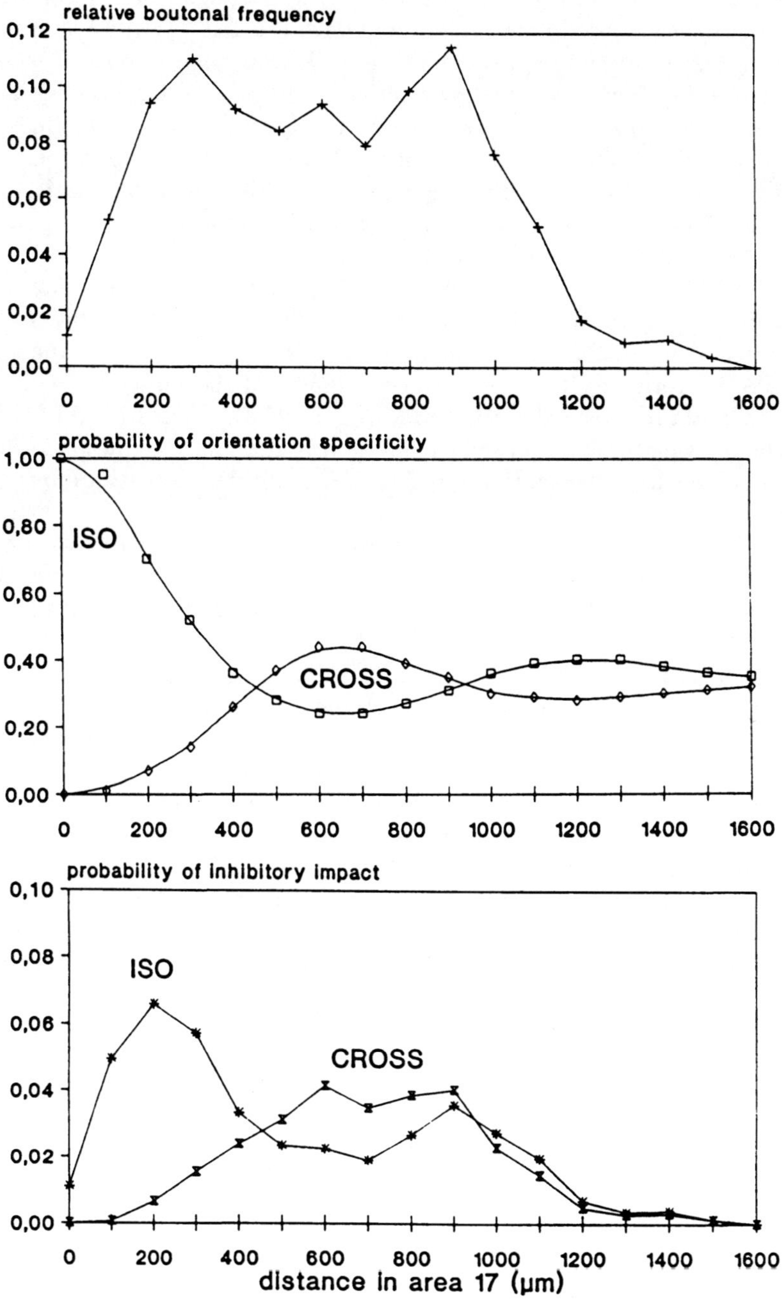
relative boutonal frequency
0,12
0,10
0,08
0,06
0,04
0,02
0,00
0
200
400
600
800
1000
1200
1400
1600
probability of orientation specificity
1,00
0,80
0,60
0,40
0,20
0,00
ISO
CROSS
0
200
400
600
800
1000
1200
1400
1600
probability of inhibitory impact
0,10
0,08
0,06
0,04
0,02
0,00
ISO
CROSS
0
200
400
600
800
1000
1200
1400
1600
distance in area 17 (μm)

3.4.4. Models Including Intracortical Excitation

Nelson *et al.* (1994) were able to disrupt intracortical inhibition by applying 4,4-diisothiocyanatostilbene-2,2-disulfonic acid (DIDS) *intracellularly* at the cell that they recorded. In order to prevent an excitatory overflow, a small hyperpolarizing current was injected and the cell was tested with a set of oriented stimuli. They showed that the orientation specificity of the cell is not altered during removal of inhibition, and they concluded that orientation tuning is predominantly shaped by intracortical excitation. Their results seem to contradict those of Sillito (1979; Sillito *et al.*, 1980), who performed an extracellular removal of inhibition *directly at* the recorded cell, which strongly altered the orientation tuning. This paradox, however, can be explained by means of a biophysically realistic model (Somers *et al.*, 1995) which permits analysis of network effects.

The model of Somers and coworkers requires all three levels of connectivity: afferent convergence from the LGN, which is mildly biased for orientation, broadly tuned intracortical inhibition, which originates from cells with maximally ±60° orientation difference to the preferred orientation of the target cell (Hata *et al.*, 1988, 1991), and intracortical excitation with a range of only ±15° (Michalski *et al.*, 1983). Synaptic weights are adjusted such that inhibition and excitation are roughly balanced in order to avoid too low impulse rates or unspecific saturation responses, respectively.

Normal orientation tuning and impulse rates achieved with this model are indistinguishable from those of visual cortical simple cells. The intracortical inhibition leads to a contrast gain control effect, so that the orientation tuning does not change with changing stimulus contrast.

The distribution of IPSPs is centered on the isoorientation, because most of the inhibitory cells are isooriented to the target cell. The distribution of cortical EPSPs is also centered on the isoorientation, but it is much narrower due to the smaller orientation range included. It is important to notice that the EPSPs from the LGN have a rather broad tuning, which cannot account for the cortical orientation specificity observed in the model cells. The cortical EPSPs, with their much narrower distribution, follow more closely the actual orientation specificity of the cells. This seems to be a paradox: The tuning of the cortical EPSPs is the "cause" and the "result" of the tuning of the cells. In models like that, however, this explanation is not circular because such systems essentially rely on the reverberation of the signal in multiple feedback loops. Somers *et al.* (1995) describe this effect as an "emergent property" of the underlying circuit. It coincides with the conclusion drawn earlier in this chapter that very little specificity is probably required for the intracortical connectivity, because the orientation tuning arises as a network effect. At the same time the emergent characteristic of cortical specificities has also been supported by the work of Douglas *et al.* (1995) focusing on recurrent excitation and Stemmler *et al.* (1995), who were able to extend similar concepts to other properties of cortical networks.

The authors perform several manipulations on the synaptic weight distributions in their model (e.g., eliminating the inhibition), and achieve predictable results. Most interestingly in this context is the comparison of the normal situation with the simulated intracellular (DIDS) or extracellular (bicuculline) blockade of inhibition (Fig. 17). As in the different experimental situations, intracellular block-

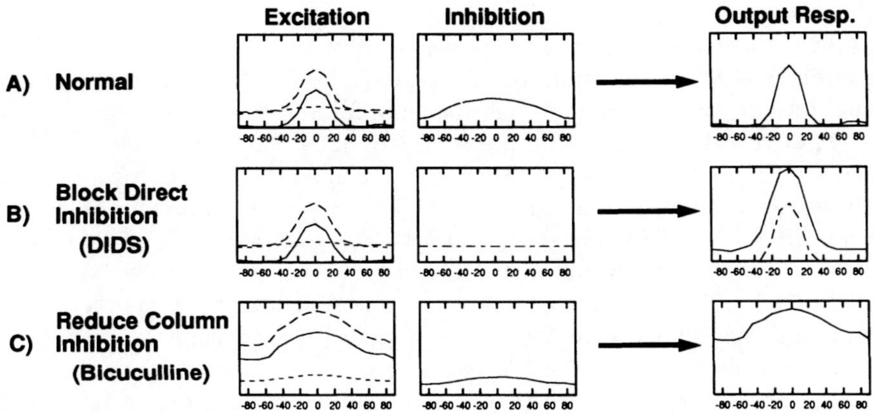

Figure 17. Summary of the emergent (recurrent excitatory) model. Sharply tuned cortical excitatory inputs provide the strongest orientation signal to cortical cells, while inhibition plays a critical, but orientation-nonspecific, role at the population level by balancing the levels of excitation. (A) Under normal conditions, sharply tuned cortical excitatory inputs (left, solid line) combine with poorly tuned LGN inputs (short-dashed line) to yield well-oriented excitatory input (long-dashed line). Although cortical inhibitory inputs (middle) are strongest at the preferred orientation, output responses are well tuned. (B) Blockade of the direct inhibitory inputs to single cells has a negligible effect on tuning of excitatory inputs, and output selectivity persists with only modest broadening (right, solid). An orientation-nonspecific, hyperpolarizing current (middle, dashed) can restore normal sharp tuning (right, dashed) without attenuating normal response amplitudes. (C) However, reduction of inhibitory inputs to all neurons in a cortical column, as may be caused by bicuculline iontophoresis, disrupts both cortical excitatory tuning (left, solid line) and output tuning (right) as stimulus responses are nonselectively amplified.

ade will not affect the orientation tuning of the cell (Nelson *et al.*, 1994), while extracellular blockade of inhibition essentially eliminates the orientation specificity (Sillito, 1979; Sillito *et al.*, 1980). The different results can be explained by the different efficacy of the blockade of inhibition. Intracellular blockade only affects the target cell, but no other cells in the network, leaving the recurrent excitatory pathways intact. Extracellular blockade was simulated in the model by a 50% reduction of the inhibitory conductances of all cells in one column, which disrupts the intracortical balance and leads to a broadening of the orientation tuning of also the excitatory cells. The rather strong reduction of the inhibitory conductances used to simulate Sillito's findings seems to be in accordance with the experimental data in the sense that much less bicuculline was necessary to disrupt the direction selectivity of a cell (Sillito, 1975, 1977) as compared to the dose needed to abolish orientation specificity (Sillito, 1979; Sillito *et al.*, 1980).

Apparently, the strong broadening of the orientation tuning results from a network effect rather than from the elimination of inhibition directly at the target cell itself. The model of Somers *et al.* (1995) therefore provides a solid and elegant explanation for the apparently paradoxical situation which arises from intra- and extracellular removal of inhibition. It also strongly emphasizes the importance of cooperativity in the network (Wörgötter and Koch, 1991). Changes at an individual cell result in a negligible effect at the cell itself, and do not alter the network behavior at all. This leads to a robust network performance, which is a highly desirable feature.

All these interesting effects, which dwell on reverberations between excitation and inhibition within the network, however, are not very intuitive. To get a better understanding, continuous variations of the critical network parameters (e.g., intracolumnar inhibitory conductance) might be helpful. It can be speculated that such curves might reveal interesting nonlinear features, like an abrupt change of the orientation specificity as soon as the network cooperativity deteriorates. Unfortunately, one of the major disadvantages of all biophysically realistic models (Wörgötter and Koch, 1991; Somers *et al.*, 1995) is their exceedingly high demand for CPU time, which often prevents such an exhaustive parameter evaluation.

The model of Somers *et al.* (1995) so far represents the last step in a sequence of biophysical models for orientation specificity with increasing realism. Only a combination of feedforward afferent excitation plus intracortical inhibition plus intracortical excitation seems to be able to account for most experimentally observed situations. Both discussed models tried to achieve cortical orientation specificity and other types of behavior as the result of a combination of "low-specificity" mechanisms. This leads to a robust network and does not require highly specialized developmental mechanisms. Therefore these simulations do not contradict developmental and structurally defined models, which mostly also rely on rather general interaction principles between their cells.

The nontrivial and emergent appearance of cortical orientation specificity in a reverberating network also justifies the trouble of designing and testing such complex biophysically realistic models.

4. Concluding Remarks

One of the major problems of modeling highly complex systems such as the visual cortex of vertebrates was outlined in the Introduction and can be rephrased as follows: Complex systems almost always justify the design of different models with different levels of abstraction even under the restriction that only one common aspect (e.g., orientation specificity) shall be emulated.

Such a diversity is rather dissatisfying with respect to the attempt toward any "unifying theory" of cortical function. In view of this, the current chapter has tried to show that the apparently divergent approaches of simulating cortical orientation specificity actually represent a situation in which a certain degree of convergence has started to develop.

In order to establish such a convergence among different models, a common ground must be defined on which all those models can be based, regardless of their strongly differing conceptual design. Given the current knowledge about intra- and extracellular cortical mechanisms, it seems reasonable to assume that cortical orientation specificity is a network property and that it is not generated intrinsically for each cell. Therefore, the common ground on which a unification of the different models could be based is the afferent and intracortical connectivity pattern, and the search for it should be guided by the largest functional subset of connection principles which is shared by all reasonably designed models.

Obviously, any attempt along this line will have to largely disregard the rich anatomical fine structure and synaptology of the visual cortex. This ideology, however, may find additional justification in the amazing robustness of cortical orientation specificity against experimental manipulations. This robustness supports a model structure in which only rather low-specificity mechanisms contribute to orientation specificity. The following summary of presumably important functional mechanisms taken from the different models discussed above therefore will be biased by this low-specificity ideology.

- *Intra-LGN mechanisms.* LGN cells display a certain degree of orientation specificity for stimuli with high spatial frequencies. Modeling studies of cortical orientation specificity have disregarded this effect and it has been shown that an LGN bias is not required to get realistic orientation tuning in either the developmental or biophysically realistic models.

- *Afferent convergence from the LGN.* The biophysical models discussed here will not produce realistic orientation tuning without an afferent bias. It should be noted, however, that this bias can be rather weak. Many developmental models are based on afferent connectivity, which, during development, will sort itself into some kind of oriented afferent input pattern.

- *Average intracortical connectivity.* The average intracortical synaptic coupling function (i.e., the envelope function) which gave the best results in developmental and biophysical models has a Mexican hat shape. This indicates that a local positive correlation (local excitation) and a more distant anticorrelation (lateral inhibition) is probably required to most easily achieve orientation specificity. The actual lateral distance from which maximal inhibitory influence is exerted is still a matter of debate. A distance of about half a hypercolumn would point to an inhibitory mechanism which is very broadly tuned, but with a weak preference for cross-orientations. A smaller distance would exclude almost all cross-orientations. The shape of the autocorrelation function measured from cortical preferred orientation maps or from realistic structurally defined models directly reflects such a possible Mexican hat-shaped synaptic coupling function.

So far, all models exclude connections which arise at distances of about a full hypercolumn. The patchy axonal distribution with such a spacing found in cat, however, supports such a connectivity pattern, which might consist of long-range isoorientation excitation. The sparseness of these connections could indicate that its contribution to cortical orientation specificity is relatively small. In analogy to this, the autocorrelation function quickly decays to zero with a small secondary positive peak at distances of about one full hypercolumn.

- *More detailed intracortical connectivity.* The average intracortical connectivity function only represents the final outcome of all excitatory and inhibitory interactions taking place at identical cortical loci. At short distances, the presumed average connectivity function indicates a predominance of excitatory influences. The approximately linear spatial summation performed by cortical simple cells, however, provides strong support for a so-called "push–pull" connectivity pattern, which includes short-range isoorientation inhibition of antagonistic receptive field polarity (see Fig. 11). It seems reasonable to assume that such an inhibitory connection scheme will be subject to statistical fluctuations of its topographic accuracy. Therefore, this connection pattern merges into the model of laterally displaced isoorientation inhibition put forward by Heggelund (1981; see Fig. 12) as soon as a displacement of about half the receptive field size occurs. Thus, push–pull connectivity and the Heggelund model might, after all, reflect a similar situation.

Regarding the actual anatomy of the visual cortex, the existence of mixed excitatory–inhibitory inputs arising from the same source distance cannot be denied and many developmental and biophysical model make use of such connection principles.

In this summary only those mechanisms were listed which are either shared or rejected (the first point discussed above) by most of the different modeling approaches. This way, a picture is painted in which the cortical connectivity seems to be based on rather generalized mechanisms which could possibly be associated with a so-called "canonical microcircuitry" (Douglas *et al.,* 1989; Douglas and Martin, 1991; Douglas *et al.,* this volume). The highly specific behavior of cortical cells (viz. orientation specificity) thereby emerges as a cooperative network property and not by individual construction at the single-cell level. While this conclusion leaves much to be desired regarding the complexity of the cortex, it can nevertheless provide a speculative working hypothesis which requires further experimental and theoretical testing.

5. Appendix

This Appendix discusses several conceptual questions which arise from the developmental and structurally defined models. The connection patterns applied in the biophysically realistic models and the mathematical description of the final map structure as described above are largely based on this discussion.

1. Are the singularities a consequence of the map structure or is the map structure a consequence of the singularities?
2. Why have (apparently) all singularities a topological index of $\pm 1/2$?
3. Are the maps of preferred orientation and orientation tuning strength related to each other?
4. Which general principles underly the design of the maps?
5. What are the relations between the structurally defined models and the developmental models?
6. What are the consequences for a connection structure in a biophysically realistic model?

5.1. Significance of the Singularities

All realistic developmental models show that no maps are generated without singularities. A map without singularities would have to be topologically equivalent to parallel stripes, which represents a highly ordered (nonrandom) structure. This is at odds with our current knowledge about the developmental mechanisms which govern the map formation. It seems that the processes which have predominant influence on the map formation are only locally correlated. This will always induce a random component in the final structure for distances exceeding the correlation length and prevent such a stripe formation. An additional reason for this, as pointed out by Durbin and Mitchison (1990), could be the dimension reduction from the visual world onto the cortical surface. Several aspects of the visual world are represented in the cortex in a maplike structure and well-established maps exist for the topography of the visual field (*X*, *Y* coordinates), the ocular dominance (OD, left eye/right eye preference), and the preferred orientation (PO). Each of these aspects can be regarded as an individual dimension of a feature space that is represented in the cortex. Thus, there are at least four (*X*, *Y*, OD, PO) dimensions that need to be represented on the two-dimensional cortical surface, because the third cortical dimension is not available due to the columnar organization. During development the system apparently tries to preserve neighborhood relationships for all these maps as well as possible; in addition, it can be assumed that the maps evolve at different rates (F. Wolf, personal communication). Durbin and Mitchison (1990) demonstrated that under these assumptions such a dimension reduction will unequivocally lead to frequent singularities.

Most of the structurally defined models explicitly or implicitly suggest that the singularities are the centers from which the map development is organized. The arguments given above, however, make it more likely that singularities are the consequence and not the reason for a certain map structure.

5.2. Topological Index

Adopting the notation of Swindale *et al.* (1987) and writing the orientation preference function $f(x, y)$ as (Wörgötter and Niebur, 1993)

$$f(x, y) = \mathrm{Re}[f(x, y)] + i\,\mathrm{Im}[f(x, y)] = \cos[2\phi(x, y)] + i\sin[2\phi(x, y)] \quad (10)$$

we obtain the preferred orientation in the form

$$\phi(x, y) = \frac{1}{2}\arctan\frac{Im[f(x, y)]}{Re[f(x, y)]} \quad (11)$$

For the numerator and denominator of Equation (11) independent traces on the two-dimensional plane exist where they are zero. Generically these traces are not parallel and a singularity occurs at locations where they intersect, which happens quite often. The map is discontinuous at these locations and the topological index of the singularity is either $+1/2$ or $-1/2$ as long as only two such traces cross. If more than two traces cross at exactly (!) the same location, a singularity with topological index unequal to $\pm 1/2$ would be observed, which is highly unlikely and a developmentally unstable situation (Swindale, 1985; Wörgötter and Niebur, 1993). During development even slight perturbations would immediately convert such a singularity into two with a topological index of $\pm 1/2$. Therefore, only the lowest order singularities are found in the visual cortex.

5.3. Relation between Preferred Orientation and Orientation Tuning Map

In the developmental models the maps of the orientation tuning strength normally show a vanishing tuning strength at the singularities. Equation (11) defines the preferred orientation as the *phase* of the function $f(x,y)$. It would thus make sense to define the orientation tuning strength by the amplitude of f. From Equation (11) it is clear, however, that this amplitude cannot be used directly, because it is unity at all map locations and undefined at a singularity.

To achieve a better description of both maps utilizing a single notational framework, the problem needs to be transferred into Fourier space. Let $F(k_x, k_y)$ be the two-dimensional Fourier transform of $f(x,y)$; then Equation (11) can be rewritten as

$$\phi(x, y) = \frac{1}{2}\arctan\frac{Im[\mathrm{IFT}(F(k_x, k_y))]}{Re[\mathrm{IFT}(F(k_x, k_y))]} \tag{12}$$

where IFT denotes the inverse Fourier transform.

Since the amplitudes of f are all unity, we instead propose to derive the map of the orientation tuning strength O as the amplitude of the convolution of the autocorrelation of f with f itself. In the Fourier domain this reads

$$O(x, y) = \sqrt{\mathrm{Re}^2[\mathrm{IFT}(L(k_x, k_y))] + \mathrm{Im}^2[\mathrm{IFT}(L(k_x, k_y))]} \tag{13}$$

where

$$L(k_x, k_y) = P(k_x, k_y)F(k_x, k_y) \tag{14}$$

This assumption essentially associates the orientation tuning strength with the *power P* of the signal, additionally including the phase information (by the use of F), necessary to anchor the tuning strength map with respect to its (unknown) origin.

Figures 7B and 7C show that the computed and the measured map of the orientation tuning strength match with reasonable accuracy. Most obviously, $O(x, y)$ vanishes at the singularities. An analysis of the quality of the match is given in Wörgötter and Niebur (1993).

This framework provides a uniform description of the preferred orientation map and the orientation tuning map in the Fourier domain and shows that the maps are related. As an additional advantage, the model is parameter-free.

As mentioned, $O(x, y)$ is equivalent to the convolution of the autocorrelation function of the preferred orientation map with the preferred orientation map itself (Wörgötter and Niebur, 1993). Figure 7D shows a one-dimensional cross section of the circular symmetrical autocorrelation function of an experimentally obtained preferred orientation map (Obermayer *et al.*, 1992), which in two-dimensions would look like a Mexican hat. The convolution of any function with such a kernel yields a filtered version of the original signal. Since the negative excursions of the autocorrelation function are rather small, in a first approximatin the kernel can be considered as a low-pass filter. Thus, the map of the orientation tuning strength is essentially obtained from a local smoothing (low-pass filtering) of the map of the preferred orientation.

This poses a severe problem: Since optical imaging always performs local smoothing due to the limited resolution of the apparatus and/or blurring by defocusing, etc., it could be possible that the map of the orientation tuning strength is a mere measuring artifact. In order to reject this, at least the necessary prerequisites should apply, one of which is that the orientation tuning strength measured experimentally (e.g., with electrode recordings) should be low at a singularity. While for the cat this does not seem to be true, data from monkey V1 support this notion (Bonhoeffer and Grinvald, 1991; Bartfeld and Grinvald, 1992; Blasdel, 1992b).

5.4. Map Design Principles

Possible map design principles have been discussed at greater length in the main text and can be summarized as follows. Local correlation, modularity, isotropy (monkey), and smoothness seem to govern the map development.

5.5. Relations to Developmental Models

Most developmental models only implement a linear algorithm for the self-organization (e.g., Obermayer *et al.*, 1990, 1992; Durbin and Mitchison, 1990; Swindale, 1982, 1992; Miller, 1994). In some of these models map formation can be interpreted as the amplification of random fluctuations in bands of initially unstable Fourier modes (Obermayer *et al.*, 1992; Durbin and Mitchison, 1990; Swindale 1982, 1992; discussed in Wolf *et al.*, 1994). By such a noise-induced symmetry-breaking process only certain bands will be selected within the range of lateral interactions and finally be stabilized by independent growth. As a result a map is generated which resembles bandpass-filtered white noise. In Miller's (1994) model there is no direct interaction between Fourier modes, but rather between a set of more generalized eigenfunctions [see Miller (1990) and Miller and MacKay (1994) for details], which, however, leads to similar map structures. The linear stage of map development does not fully reproduce the final structure of the preferred orientation map (Wolf *et al.*, 1994) and an additional nonlinear, but slower developmental stage is needed.

The maps of the orientation *tuning strength* obtained from the linear developmental models resemble those from optical imaging, but both are at odds with measurements of the orientation tuning strength using conventional electrode recordings. For example, one should expect that the orientation tuning *within* each column should be rather constant, which is not confirmed by the data from electrode recordings (Lee *et al.*, 1977; Leventhal and Hirsch, 1978). Local smoothing of the preferred orientation map leads to the orientation tuning strength map (Wörgötter and Niebur, 1993), a process which necessarily also occurs to some degree during optical imaging due to blurring effects. Such a mismatch between optical imaging and electrode recordings does not exist for the preferred orientation map (Swindale *et al.*, 1987). Therefore, it could be possible that spatial resolution of optical imaging is sufficient only to resolve the preferred orientation maps, but that the orientation tuning strength map is not correctly measured.

While this could be one possible consequence of the discussed results, a second aspect which dwells on the difference between the linear and nonlinear stages of map development could also be of relevance. It might be that the orientation tuning strength map generated from developmental models that only implement the linear stage would also reflect only the endpoint of the first stage of development in an animal. Many developmental models implement a Mexican hat-shaped afferent and/or intracortical correlation function. The results of Wörgötter and Niebur (1993) demonstrated that the convolution of the preferred orientation map with a Mexican hat-shaped kernel (the autocorelation function of the preferred orientation maps) leads to the desired smoothing and yields a good approximation of the orientation tuning strength map. It is therefore possible that the autocorrelation function directly reflects the gross afferent and/or intracortical synaptic coupling function, but only during the early, linear phase of development. Intracortical refinement of the connectivity at later stages could very well alter the orientation tuning of the cells, leading to changed preferred orientation and orientation tuning maps.

5.6. Relations to Biophysically Realistic Models

The map models suggest that cortical orientation specificity can be generated without strong requirements on the specificity of the connection structure because circular symmetric Mexican hat-shaped correlation functions are mostly sufficient to generate the maps. In addition, these models show that orientation specificity should be rather robust because wide parameter ranges lead to reasonable maps. This is also in accordance with the high robustness of orientation specificity against experimental manipulations (Wörgötter and Eysel, 1991).

It cannot be ruled out that the individual cellular fine structure adds to the performance of the system. However, at this point it seems that the link between experiment and modeling is still not strong enough to allow for an evaluation of (structurally and functionally) fine-tuned mechanisms. In view of the conclusions drawn from the map models, one should therefore be more restrictive and only implement broadband mechanisms in order to arrive at a biophysical model which does not include too many unknown degrees of freedom.

ACKNOWLEDGMENTS. The author is grateful to K. Funke and E. Nelle for valuable remarks on the manuscript. I am also indebted to D. Somers, S. Nelson, and M. Sur, who gave me access to their modeling paper prior to publication. In particular, I wish to thank D. Somers, who made Fig. 17, and commented on the section about models with intracortical excitation. Thanks are also due to K. Miller, who corrected my view of his work and helped to make this section clearer to the reader. E. Nelle made the simulations shown in Fig. 11 and helped me with the figure. Figure 16 was obtained from the data of U. Eysel and Z. Kisvárday, and the author is in particular indebted to U. Eysel, who allowed me to use the facilities of his institute. His continuous support of my work and many fruitful discussions with him are gratefully acknowledged. This work was partly supported by the Deutsche Forschungsgemeinschaft WO 388/4-2 and 5-1. This article was written in fall 1995. Literature references after this date are therefore not included.

6. References

Adelson, E. H., and Bergen, J. R., 1985, Spatiotemporal energy models for the perception of motion, *J. Opt. Soc. Am. A* **2:**284–299.

Albus, K., 1975, A quantitative study of the projection area of the central and the paracentral visual field in area 17 of the cat, *Exp. Brain Res.* **24:**181–202.

Allison, J. D., and Bonds, A. B., 1994, Inactivation of the infragranular striate cortex broadens orientation tuning of supragranular visual neurons in the cat, *Exp. Brain Res.* **101:**415–426.

Atick, J. J., and Redlich, A. N., 1990, Mathematical model of the simple cells in the visual cortex, *Biol. Cybernet.* **63:**99–109.

Bartfeld, E., and Grinvald, A., 1992, Relationships between orientation-preference pinwheels, cytochrome oxidase blobs, and ocular-dominance columns in primate striate cortex, *Proc. Natl. Acad. Sci. USA* **89:**11905–11909.

Batschelet, E., ed., 1981, *Circular Statistics in Biology*, Academic Press, New York.

Bauer, R., 1982, A high probability of an orientation shift between layers 4 and 5 in central parts of the cat striate cortex, *Exp. Brain Res.* **48:**245–255.

Bauer, R., and Dow, B. M., 1991, Local and global principles of striate cortical organization: An advanced model, *Biol. Cybernet.* **64:**477–483.

Bauer, R., Dow, B. M., Snyder, A. Z., and Vautin, R. G., 1983, Orientation shift between upper and lower layers in monkey visual cortex, *Exp. Brain Res.* **50:**133–145.

Baxter, W. T., and Dow, B. M., 1989, Horizontal organization of orientation-sensitive cells in primate visual cortex, *Biol. Cybernet.* **61:**171–182.

Benevento, L. A., Creutzfeld, O. D., and Kuhn, U., 1972, Significance of intracortical inhibition in the visual cortex, *Nature New Biol.* **238:**124–126.

Berman, N. J., Douglas, R. J., Martin, K. A. C., and Whitteridge, D., 1991, Mechanisms of inhibition in cat visual cortex, *J. Physiol.* **440:**697–722.

Bienenstock, E. L., Cooper, L. N., and Munro, P. W., 1982, Theory for the development of neuron selectivity: Orientation specificity and binocular interaction in visual cortex, *J. Neurosci.* **2:**32–48.

Bishop, P. O., Coombs, J. S., and Henry, G. H., 1973, Receptive fields of simple cells in the cat striate cortex, *J. Physiol.* **231:**31–60.

Blakemore, C., and Tobin, E. A., 1972, Lateral inhibition between orientation detectors in the cat's visual cortex, *Exp. Brain Res.* **15:**439–440.

Blasdel, G. G., 1992a, Differential imaging of ocular dominance and orientation selectivity in monkey striate cortex, *J. Neurosci.* **12:**3115–3138.

Blasdel, G. G., 1992b, Orientation selectivity, preference, and continuity in monkey striate cortex, *J. Neurosci.* **12:**3139–3161.

Blasdel, G. G., and Salama, G., 1986, Voltage-sensitive dyes reveal a modular organization in monkey striate cortex, *Nature* **321:**579–585.

Bonds, A. B., 1989, Role of inhibition in the specification of orientation selectivity of cells in the cat striate cortex, *Vis. Neurosci.* **2:**41–55.

Bonhoeffer, T., and Grinvald, A., 1991, Iso-orientation domains in cat visual cortex are arranged in pinwheel-like patterns, *Nature* **353:**429–431.

Bonhoeffer, T., and Grinvald, A., 1993, The layout of iso-orientation domains in area 18 of cat visual cortex: Optical imaging reveals a pinwheel-like organization, *J. Neurosci.* **13:**4157–4180.

Braitenberg, V., and Braitenberg, C., 1979, Geometry of orientation columns in the visual cortex, *Biol. Cybernet.* **33:**179–186.

Campbell, F. W., and Robson, J. G., 1968, Application of Fourier analysis to the visibility of gratings, *J. Physiol.* **197:**551–566.

Chapman, B., Zahs, K. R., and Stryker, M. P., 1991, Relation of cortical cell orientation selectivity to alignment of receptive fields of the geniculocortical afferents that aborize within a single orientation column in ferret visual cortex, *J. Neurosci.* **11:**1347–1358.

Colonnier, M., 1964, The tangential organization of the cortex, *J. Anat.* **98**:327–344.

Crook, J. M., and Eysel, U. T., 1992, GABA-induced inactivation of functionally characterized sites in cat visual cortex (area 18): Effects on orientation tuning, *J. Neurosci.* **12**:1816–1825.

Daugman, J. G., 1980, Two-dimensional spectral analysis of cortical receptive field profiles, *Vis. Res.* **20**:847–856.

Daugman, J. G., 1985, Uncertainty relation for resolution in space, spatial frequency, and orientation optimized by two-dimensional visual cortical filters, *J. Opt. Soc. Am. A* **2**:1160–1169.

Daugman, J., 1988, Complete discrete 2-D Gabor transforms by neural networks for image analysis and compression, *IEEE Trans. Acoustics Speech Signal Processing* **36**:1169–1179.

Daugman, J. G., 1990, Brain metaphor and brain theory, in: *Computational Neuroscience* (E. L. Schwartz, ed.), MIT Press, Cambridge, MA, pp. 9–18.

DeValois, R. L., Yund, E., and Hepler, N., 1982, The orientation and direction selectivity of cells in macaque visual cortex, *Vis. Res.* **22**:531–544.

Douglas, R. J., and Martin, K. A. C., 1991, A functional microcircuit for cat visual cortex, *J. Physiol.* **440**:735–769.

Douglas, R. J., Martin, K. A. C., and Whitteridge, D., 1988, Selective responses of visual cortical cells do not depend on shunting inhibition, *Nature* **332**:642–644.

Douglas, R. J., Martin, K. A. C., and Whitteridge, D., 1989, A canonical microcircuit for neocortex, *Neural Computation* **1**:480–488.

Douglas, R. J., Martin, K. A. C., and Whitteridge, D., 1991, An intracellular analysis of the visual responses of neurons in cat visual cortex, *J. Physiol.* **440**:659–696.

Douglas, R. J., Koch, C., Mahowald, M., Martin, K. A. C., and Suarez, H. H., 1995, Recurrent excitation in neocortical circuits, *Science* **269**:981–985.

Durbin, R., and Mitchison, G., 1990, A dimension reduction framework for understanding cortical maps, *Nature* **343**:644–647.

Elsdale, T., and Wasoff, F., 1976, Fibroblast cultures and dermatoglyphics: The topology of two planar patterns, *Rouxs Arch.* **180**:121–147.

Eysel, U. T., Crook, J. M., and Machemer, H. F., 1990, GABA-induced remote inactivation reveals cross-orientation inhibition in the cat striate cortex, *Exp. Brain Res.* **80**:626–630.

Ferster, D., 1986, Orientation selectivity of synaptic potentials in neurons of cat primary visual cortex, *J. Neurosci.* **6**:1284–1301.

Ferster, D., 1987, Origin of orientation-selective EPSPs in simple cells of cat visual cortex, *J. Neurosci.* **7**:1780–1791.

Ferster, D., 1988, Spatially opponent excitation in inhibition in simple cells of the cat visual cortex, *J. Neurosci.* **8**:1172–1180.

Ferster, D., and Koch, C., 1987, Neuronal connections underlying orientation selectivity in cat visual cortex, *TINS* **10**:487–492.

Gabor, D., 1946, Theory of communication, *J. IEE* **93**:429–457.

Gilbert, C. D., 1985, Horizontal integration in the neocortex, *TINS* **8**:160–165.

Gilbert, C. D., and Wiesel, T. N., 1979, Morphology and intracortical projections of functionally identified neurons in cat visual cortex, *Nature* **280**:120–125.

Gilbert, C. D., and Wiesel, T. N., 1989, Columnar specificity of intrinsic horizontal and cortico-cortical connections in cat visual cortex, *J. Neurosci.* **9**:2432–2442.

Götz, K. G., 1987, Do "d-Blob" hypercolumns tessellate the monkey visual cortex? *Biol. Cybernet.* **56**:107–109.

Götz, K. G., 1988, Cortical templates for the self-organization of orientation-specific d- and l-hypercolumns in monkeys and cats, *Biol. Cybern.* **58**:213–223.

Grinvald, A., Lieke, E., Frostig, R. D., Gilbert, C. D., and Wiesel, T. N., 1986, Functional architecture of cortex revealed by optical imaging of intrinsic signals, *Nature* **324**:361–364.

Hammond, P., and Pomfrett, C. J. D., 1990, Influence of spatial frequency on tuning and bias for orientation and direction in the cat's striate cortex, *Vis. Res.* **30**:359–369.

Hata, Y., Tsumoto, T., Sato, H., Hagihara, K., and Tamura, H., 1988, Inhibition contributes to orientation selectivity in visual cortex of cat, *Nature* **335**:815–817.

Hata, Y., Tsumoto, T., Sato, H., and Tamura, H., 1991, Horizontal interactions between visual cortical neurones studied by cross-correlation analysis in the cat, *J. Physiol.* **441**:593–614.

Hawken, M. J., and Parker, A. J., 1987, Spatial properties of neurons in the monkey striate cortex, *Proc. R. Soc. Lond. B* **231**:251–288.

Heeger, D. J., 1987, Model for the extraction of image flow, *J. Opt. Soc. Am. A* **4**:1455–1471.

Heeger, D. J., 1993, Modeling simple cell direction selectivity with normalized, half-squared, linear operators, *J. Neurophysiol.* **70**:1885–1898.

Heggelund, P., 1981, Receptive field organization of simple cells in cat striate cortex, *Exp. Brain Res.* **42**:89–98.

Henry, G. H., Bishop, P. O., Tupper, R. M., and Dreher, B., 1973, Orientation specificity and response variability of cells in the striate cortex, *Vis. Res.* **13**:1771–1779.

Henry, G. H., Bishop, P. O., and Drehr, B., 1974a, Orientation, axis and direction as stimulus parameters for striate cells, *Vis. Res.* **14**:767–777.

Henry, G. H., Dreher, B., and Bishop, P. O., 1974b, Orientation specificity of cells in cat striate cortex, *J. Neurophysiol.* **37**:1394–1409.

Henry, G. H., Michalski, A., Wimborne, B. M., and McCart, R. J., 1994, The nature and origin of orientation specificity in neurons of the visual pathways, *Prog. Neurobiol.* **43**:381–437.

Hubel, D. H., and Wiesel, T. N., 1959, Receptive fields of single neurons in the cat's striate cortex, *J. Physiol.* **148**:574–591.

Hubel, D. H., and Wiesel, T. N., 1962, Receptive fields, binocular interaction and functional architecture in the cat's visual cortex, *J. Physiol.* **160**:106–154.

Hubel, D. H., and Wiesel, T. N., 1963, Shape and arrangement of columns in cat's striate cortex, *J. Physiol.* **165**:559–568.

Hubel, D. H., and Wiesel, T. N., 1965, Receptive fields and functional architecture in two nonstriate visual areas (18 and 19) of the cat, *J. Neurophysiol.* **28**:229–289.

Hubel, D. H., and Wiesel, T. N., 1968, Receptive fields and functional architecture of monkey striate cortex, *J. Physiol.* **195**:215–243.

Hubel, D. H., and Wiesel, T. N., 1972, Laminar and columnar distribution of geniculocortical fibers in the macaque monkey, *J. Comp. Neurol.* **146**:421–450.

Hubel, D. H., and Wiesel, T. N., 1974, Sequence regularity and geometry of orientation columns in the monkey striate cortex, *J. Comp. Neurol.* **158**:267–294.

Hubel, D. H., and Wiesel, T. N., 1977, Functional architecture of macaque monkey visual cortex, *Proc. R. Soc. Lond. B* **198**:1–59.

Hubel, D. H., Wiesel, T. N., and Stryker, M. P., 1977, Orientation columns in macaque monkey demonstrated by the 2-desoxyglucose autoradiographic technique, *Nature* **269**:328–330.

Jones, J. P., and Palmer, L. A., 1987a, The two-dimensional spatial structure of simple receptive fields in cat striate cortex, *J. Neurophysiol.* **58**:1187–1211.

Jones, J. P., and Palmer, L. A., 1987b, An evaluation of the two-dimensional Gabor filter model of simple receptive fields in cat striate cortex, *J. Neurophysiol.* **58**:1233–1258.

Jones, J. P., Stepnoski, A., and Palmer, L. A., 1987, The two-dimensional spectral structure of simple receptive fields in cat striate cortex, *J. Neurophysiol.* **58**:1212–1232.

Kammen, D. M., and Yuille, A. L., 1988, Spontaneous symmetry-breaking energy functions and the emergence of orientation selective cortical cells, *Biol. Cybernet.* **59**:23–31.

Kato, H., Bishop, P. O., and Orban, G. A., 1978, Hypercomplex and simple/complex cell classifications in cat striate cortex, *J. Neurophysiol.* **41**:1071–1095.

Kisvárday, Z. F., and Eysel, U. T., 1993, Functional and structural topography of horizontal inhibitory connections in cat visual cortex, *Eur. J. Neurosci.* **5**:1558–1572.

Kisvárday, Z. F., Beaulieu, C., and Eysel, U. T., 1993, Network of GABAergic large basket cells in cat visual cortex (area 18). Implications for lateral disinhibition, *J. Comp. Neurol.* **327**:398–415.

Kisvárday, Z. F., Kim, D.-S., Eysel, U. T., and Bonhoeffer, T., 1994, Relationship between lateral inhibitory connections and the topography of the orientation map in cat visual cortex, *Eur. J. Neurosci.* **6**:1619–1932.

Koch, C., and Poggio, T., 1985, The synaptic veto mechanism: Does it underlie direction and orientation selectivity in the visual cortex? in: *Models of the Visual Cortex* (D. Rose and V. G. Dobson, eds.), Wiley, New York, pp. 408–419.

Koch, C., and Segev, I., eds., 1989, *Methods in Neuronal Modeling* MIT Press, Cambridge, MA.

Koch, C., Douglas, R. J., and Wehmeier, U., 1990, Visibility of synaptically induced conductance changes: Theory and simulations of anatomically characterized cortical pyramidal cells, *J. Neurosci.* **10**:1728–1744.

Kulikowski, J. J., and Bishop, P. O., 1981, Linear analysis of the responses of simple cells in the cat visual cortex, *Exp. Brain Res.* **44**:386–400.

Kulikowski, J. J., Marcelja, S., and Bishop, P. O., 1982, Theory of spatial position and spatial frequency relations in the receptive fields of simple cells in the visual cortex, *Biol. Cybernet.* **43**:187–198.
Lee, B. B., Albus, K. Heggelund, P., Hulme, M. J., and Creutzfeldt, O. D., 1977, The depth distribution of optimal stimulus orientation for neurones in cat area 17, *Exp. Brain Res.* **27**:301–314.
Leventhal, A. G., and Hirsch, H. V. B., 1978, Receptive-field properties of neurons in different laminae of visual cortex of the cat, *J. Neurophysiol.* **41**:948–962.
Leventhal, A. G., and Schall, J. D., 1983, Structural basis of orientation sensitivity of cat retinal ganglion cells, *J. Comp. Neurol.* **220**:465–475.
Levick, W. R., and Thibos, L. N., 1980, Orientation bias of cat retinal ganglion cells, *Nature* **286**:389–390.
Levick, W. R., and Thibos, L. N., 1982, Analysis of orientation bias in cat retina, *J. Physiol.* **329**:243–261.
Li, B., Wang, Y., and Diao, Y., 1994, Quantification of directional and orientational selectivities of visual neurons to moving stimuli, *Biol. Cybernet.* **70**:281–290.
Linsker, R., 1986a, From basic network principles to neural architecture: Emergence of spatial-opponent cells, *Proc. Natl. Acad. Sci. USA* **83**:7508–7512.
Linsker, R., 1986b, From basic network principles to neural architecture: Emergence of orientation-selective cells, *Proc. Natl. Acad. Sci. USA* **83**:8390–8394.
Linsker, R., 1986c, From basic network principles to neural architecture: Emergence of orientation columns, *Proc. Natl. Acad. Sci. USA* **83**:8779–8783.
Löwel, S., Bishof, H.-F., Leutenecker, B., and Singer, W., 1988, Topographic relations between ocular dominance and orientation columns in the cat striate cortex, *Exp. Brain Res.* **71**:33–46.
MacKay, D. J. C., and Miller, K. D., 1990a, Analysis of Linsker's applications of Hebbian rules to linear networks, *Network* **1**:257–298.
MacKay, D. J. C., and Miller, K. D., 1990b, Analysis of Linsker's simulations of Hebbian rules, *Neural Computation* **2**:173–287.
Maffei, L., and Fiorentini, A., 1973, The visual cortex as a spatial frequency analyzser, *Vis. Res.* **13**:1255–1267
Marcelja, S., 1980, Mathematical description of the responses of simple cortical cells, *J. Opt. Soc. Am.* **70**:1297–1300.
Marr, D., 1982, *Vision,* Freeman, San Francisco.
Martin, K. A. C., 1988, From single cells to simple circuits in the cerebral cortex, *J. Exp. Physiol.* **73**:637–702.
Martin, K. A. C., and Whitteridge, D., 1984, Form, function and intracortical projections of spiny neurones in the striate visual cortex of the cat, *J. Physiol.* **353**:463–504.
Matsubara, J., Cynader, M., Swindale, N. V., and Stryker, M. P., 1985, Intrinsic projections within visual cortex: Evidence for orientation-specific local connections, *Proc. Natl. Acad. Sci. USA* **82**:935–939.
Michalski, A., Gerstein, G. L., Czarkowska, J., and Tarnecki, R., 1983, Interactions between cat striate cortex neurons, *Exp. Brain Res.* **51**:97–107.
Miller, K. D., 1990, Derivation of linear Hebbian equations from a nonlinear Hebbian model of synaptic plasticity, *Neural Computation* **2**:321–333.
Miller, K. D., 1992, Development of orientation columns via competition between ON- and OFF-center inputs, *NeuroReport* **3**:73–76.
Miller, K. D., 1994, A model for the development of simple cell receptive fields and the ordered arrangement of orientation columns through activity-dependent competition between ON- and OFF-center inputs, *J. Neurosci.* **14**:409–441.
Miller, K. D., and MacKay, D. J. C., 1994, The role of constraints in Hebbian learning, *Neural Computation* **6**:98–124.
Mitchison, G. J., 1985, Does the striate cortex contain a system of oriented axons? in: *Models of the Visual Cortex* (D. Rose and V. G. Dobson, eds.), Wiley, New York, pp. 443–451.
Mitchison, G., and Crick, F., 1982, Long axons within the striate cortex: Their distribution, orientation, and patterns of connection, *Proc. Natl. Acad. Sci. USA* **79**:3661–3665.
Miyashita, M., and Tanaka, S., 1992, A mathematical model for the self-organization of orientation columns in visual cortex, *NeuroReport* **3**:69–72.
Morrone, M. C., Burr, D. C., and Maffei, L., 1982, Functional implications of cross-orientation inhibition of cortical visual cells. I. Neurophysiological evidence, *Proc. R. Soc. Lond. B* **216**:335–354.
Movshon, J. A., 1975, The velocity tuning of single units in cat striate cortex, *J. Physiol.* **249**:445–468.
Movshon, J. A., Thompson, I. D., and Tolhurst, D. J., 1978, Spatial summation in the receptive fields of simple cells in the cat's striate cortex, *J. Physiol.* **283**:53–77.

Murphy, P., and Sillito, A. M., 1986, Continuity of orientation columns between superficial and deep laminae of the cat primary visual cortex, *J. Physiol.* **381**:95–110.

Nelle, E., and Wörgötter, F., 1993, Cascaded intracortical inhibition: Modeling connection schemes on a large scale simulator, in: *ICANN '93: Proceedings of the International Conference on Artificial Neural Networks* (S. Gielen and B. Kappen, eds.), Springer, Berlin, pp. 157–160.

Nelson, S., Toth, L., Sheth, B., and Sur, M., 1994, Orientation selectivity of cortical neurons during intracellular blockade of inhibition, *Science* **265**:774–777.

Niebur, E., and Wörgötter, F., 1994, Design principles of columnar organization in visual cortex, *Neural Computation* **6**:602–614.

Nielsen, D. E., 1983, A functional model of the wiring of the simple cells of visual cortex, *Biol. Cybernet.* **47**:213–222.

Nielsen, D. E., 1985, Retinocortical wiring of the simple cells of the visual cortex, in: *Models of the Visual Cortex* (D. Rose and V. G. Dobson, eds.), Wiley, New York, pp. 374–379.

Obermayer, K., Ritter, H., and Schulten, K., 1990, A principle for the formation of the spatial structure of cortical feature maps, *Proc. Natl. Acad. Sci. USA* **87**:8345–8349.

Obermayer, K., Blasdel, G. G., and Schulten, K., 1992, A statistical mechanical analysis of self-organization and pattern formation during the development of visual maps, *Phys. Rev. A* **45**:7568–7589.

Orban, G. A., ed., 1984, *Studies of Brain Function*, Vol. 11, Springer-Verlag, Berlin.

Orban, G. A., Kennedy, H., and Maes, H., 1981, Response to movement of neurons in areas 17 and 18 of the cat: Velocity sensitivity, *J. Neurophysiol.* **45**:1043–1058.

Pattison, T. R., 1992, Relaxation network for Gabor image decomposition, *Biol. Cybernet* **67**:97–102.

Perez, R., Glass, L., and Shlaer, R., 1975, Development of specificity in the cat visual cortex, *J. Math. Biol.* **1**:275–288.

Poggio, T., 1982, Visual algorithms, in: *Physical and Biological Processing of Images* (O. J. Braddick and A. C. Sleigh, eds.), Springer-Verlag, Berlin, pp. 128–153.

Pollen, D. A., Lee, J. R., and Taylor, J. H., 1971, How does the striate cortex begin the reconstruction of the visual world? *Science* **173**:74–77.

Rojer, A. S., and Schwartz, E. L., 1990, Cat and monkey cortical column patterns modeled by bandpass-filtered 2D white noise, *Biol. Cybernet.* **62**:381–391.

Sherman, S. M., Watkins, D. W., and Wilson, J. R., 1976, Further differences in receptive field properties of simple and complex cells in cat striate cortex, *Vis. Res.* **16**:919–927.

Shou, T., and Leventhal, A. G., 1989, Organized arrangement of orientation-sensitive relay cells in the cat's dorsal lateral geniculate nucleus, *J. Neurosci.* **9**:4287–4302.

Sillito, A. M., 1975, The contribution of inhibitory mechanisms to the receptive field properties of neurones in the striate cortex of the cat, *J. Physiol.* **250**:305–329.

Sillito, A. M., 1977, Inhibitory processes underlying the directional specificity of simple, complex and hypercomplex cells in the cat's visual cortex, *J. Physiol.* **271**:699–720.

Sillito, A. M., 1979, Inhibitory mechanisms influencing complex cell orientation selectivity and their modification at high resting discharge levels, *J. Physiol.* **289**:33–53.

Sillito, A. M., Kemp, J. A., Milson, J. A., and Berardi, N., 1980, A re-evaluation of the mechanisms underlying simple cell orientation selectivity, *Brain Res.* **194**:517–520.

Skottun, B. C., deValois, R. L., Grosof, D. H., Movshon, J. A., Albrecht, D. G., and Bonds, A. B., 1991 Classifying simple and complex cells on the basis of response modulation, *Vis. Res.* **31**:1079–1086.

Smith, E. L., Chino, Y. M., Ridder, W. H., Kitagawa, K., and Langston, A., 1990, Orientation bias of neurons in the lateral geniculate nucleus of macaque monkeys, *Vis. Neurosci.* **5**:525–545.

Somers, D. C., Nelson, S. B. and Sur, M., 1995, An emergent model of visual cortical orientation selectivity in visual cortical simple cells, *J. Neurosci.* **15**:5448–5465.

Soodak, R. E., 1987, The retinal ganglion cell mosaic defines orientation columns in striate cortex, *Proc. Natl. Acad. Sci. USA* **84**:3936–3940.

Soodak, R. E., 1991, Reverse-Hebb plasticity leads to optimization and association in a simulated visual cortex, *Vis. Neurosci.* **6**:507–518.

Soodak, R. E., Shapley, R. M., and Kaplan, E., 1987, Linear mechanism of orientation tuning in the retina and lateral geniculate nucleus of the cat, *J. Neurophysiol.* **58**:267–275.

Stemmler, M., Usher, M., and Niebur, E., 1995, Lateral interactions in primary visual cortex: A model bridging physiology and psychophysics, *Science* **269**:1877–1880.

Swindale, N. V., 1982, A model for the formation of orientation columns, *Proc. R. Soc. Lond. B* **215**:211–230.

Swindale, N. V., 1985, Iso-orientation domains and their relationship with cytochrome oxidase patches, in: *Models of the Visual Cortex* (D. Rose and V. G. Dobson, eds.), Wiley, New York, pp. 452–461.

Swindale, N. V., 1992, A model for the coordinated development of columnar systems in primate striate cortex, *Biol. Cybernet.* **66**:217–230.

Swindale, N. V., Matsubara, J. A., and Cynader, M. S., 1987, Surface organization of orientation and direction selectivity in cat area 18, *J. Neurosci.* **7**:1414–1427.

Thompson, K. G., Leventhal, A. G., Zhou, Y., and Liu, D., 1994, Stimulus dependence of orientation and direction sensitivity of cat LGNd relay cells without cortical inputs: A comparison with area 17 cells, *Vis. Neurosci.* **11**:939–951.

Tieman, S. B., and Hirsch, H. V. B., 1982, Exposure to lines of only one orientation modifies dendritic morphology of cells in the visual cortex of the cat, *J. Comp. Neurol.* **211**:353–362.

Tieman, S. B., and Hirsch, H. V. B., 1985, Role of dendritic fields in orientation selectivity, in: *Models of the Visual Cortex* (D. Rose and V. G. Dobson, eds.), Wiley, New York, pp. 432–442.

Tolhurst, D. J., and Dean, A. F., 1987, Spatial summation by simple cells in the striate cortex of the cat, *Exp. Brain Res.* **66**:607–620.

Tolhurst, D. J., and Dean, A. F., 1990, The effects of contrast on the linearity of spatial summation of simple cells in the cat's striate cortex, *Exp. Brain Res.* **79**:582–588.

Tononi, G., Sporns, O., and Edelman, G. M., 1992, Reentry and the problem of integrating multiple cortical areas: Simulation of dynamic integration in the visual system. *Cerebral Cortex* **2**:310–335.

Ts'o, D. Y., Gilbert, C. D., and Wiesel, T. N., 1986, Relationships between horizontal interactions and functional architecture in cat striate cortex as revealed by cross-correlation analysis, *J. Neurosci.* **6**:1160–1170.

Vidyasagar, T. R., 1987, A model of striate response properties based on geniculate anisotropies, *Biol. Cybernet.* **57**:11–23.

Vidyasagar, T. R., 1992, Subcortical mechanisms in orientation sensitivity of cat visual cortical cells, *NeuroReport* **3**:185–188.

Vidyasagar, T. R., and Heide, W., 1984, Geniculate orientation biases seen with moving sine wave gratings: Implications for a model of simple cell afferent connectivity, *Exp. Brain Res.* **57**:196–200.

Vidyasagar, T. R., and Urbas, J. V., 1982, Orientation sensitivity of cat LGN neurones with and without inputs from visual cortical areas 17 and 18, *Exp. Brain Res.* **46**:157–169.

Volgushev, M., Pei, X., Vidyasagar, T. R., and Creutzfeldt, O. D., 1993, Excitation and inhibition in orientation selectivity of cat visual cortex neurons revealed by whole-cell recordings *in vivo*, *Vis. Neurosci.* **10**:1151–1155.

von der Marslburg, C., 1973, Self-organization of orientation sensitive cells in the striate cortex, *Kybernetik* **14**:85–100.

von Seelen, W., 1970, Zur Informationsverarbeitung im visuellen System der Wirbeltiere. II, *Kybernetik* **7**:86–106.

Wehmeier, U., Dong, D., Koch, C., and van Essen, D., 1989, Modeling the mammalian visual system, in: *Methods in Neuronal Modeling* (C. Koch, and I. Segev, eds.), MIT Press, Cambridge, MA, pp. 335–359.

Wolf, F., Pawelzik, K., Geisel, T., Kim, D. S., and Bonhoeffer, T., 1993, Map structure from pinwheel position, in: *ICANN '93: Proceedings of the International Conference on Artificial Neural Networks* (S. Gielen and B. Kappen, eds.), Springer, Berlin, pp. 131–135.

Wolf, F., Pawelzik, K., and Geisel, T., 1994, Emergence of long range order in maps of orientation preference, in: *ICANN '94: Proceedings of the International Conference on Artificial Neural Networks* (M. Marinaro and P. G. Morasso, eds.), Springer, Berlin, pp. 38–41.

Wörgötter, F., and Eysel, U. T., 1987, Quantitative determination of orientational and directional components in the response of visual cortical cells to moving stimuli, *Biol. Cybernet.* **57**:349–355.

Wörgötter, F., and Eysel, U. T., 1989, Axis of preferred motion is a function of bar length in visual cortical receptive fields, *Exp. Brain Res.* **76**:307–314.

Wörgötter, F., and Eysel, U. T., 1991, Topographical aspects of intracortical excitation and inhibition contributing to orientation specificity in area 17 of the cat visual cortex, *Eur. J. Neurosci.* **3**:1232–1244.

Wörgötter, F., and Koch, C., 1991, A detailed model of the primary visual pathway in the cat: Comparison of afferent excitation and intracortical inhibitory connection schemes for orientation selectivity, *J. Neurosci.* **11**:1959–1979.

Wörgötter, F., and Niebur, E., 1993, Cortical column design: A link between the maps of preferred orientation and orientation tuning strength? *Biol. Cybernet.* **70**:1–13.

Wörgötter, F., Niebur, E., and Koch, C., 1991, Isotropic connections generate functional asymmetrical behavior in visual cortical cells, *J. Neurophysiol.* **66**:444–459.

Wörgötter, F., Niebur, E., and Koch, C., 1992, Generation of direction selectivity by isotropic intracortical connections, *Neural Computation* **4**:332–340.

Zhang, J., 1990, How to unconfound the directional and orienational information in visual neuron's response, *Biol. Cybernet.* **63**:135–142.

5

The Role of Recurrent Excitation in Neocortical Circuits

RODNEY DOUGLAS, CHRISTOF KOCH, MISHA MAHOWALD, and KEVAN MARTIN

The nervous system is a computing machine which manages to do its exceedingly complicated work on a rather low level of precision: [according to the above,] only precision levels of 2 to 3 decimals are possible. This fact must be emphasised again and again, because no known computing machine can operate reliably and significantly on such a low precision level.

John von Neumann (1958)

1. Introduction

John von Neumann was a pioneer of modern electronic computing. He was involved in the design and construction of the early electronic computing machines and, as part of this process, he was led to explore the analogies between automata and the brain. Consequently, he understood very clearly some essential differences in the physical mode of design and operation of the brain and digital automata.

In the *First Draft Report of the EDVAC* (1945) von Neumann gave a formal

RODNEY DOUGLAS • Institute of Neuroinformatics, ETH/UNIZ, Zurich 8006, Switzerland; and Computation and Neural Systems Program, California Institute of Technology, Pasadena, California 91125. CHRISTOF KOCH • Computation and Neural Systems Program, California Institute of Technology, Pasadena, California 91125. MISHA MAHOWALD and KEVAN MARTIN • Institute of Neuroinformatics, ETH/UNIZ, Zurich 8006, Switzerland.

description of the design of general-purpose electronic computers that is now known as the von Neumann architecture (Fig. 1). At that time he also described a number of similarities and differences between the computer and the brain. The most obvious differences were physical: the size, power consumption, and number and speed of the components. In the early days of electronic computing this difference was particularly striking. In 1944, the original ENIAC (Electronic Numerical Integrator, Analyzer and Computer) weighed in at 30 tons and measured $100 \times 8 \times 3$ feet. By comparison, the human brain weighs about 1500 *g*, occupying a volume of about 1.5 liters. The ENIAC had less than 10^4 active components, whereas the brain contains on the order of 10^{11} neurons and 10^{14} synapses. With each of its few thousand valves consuming 5–10 *W*, the ENIAC could heat a large house. By contrast, the neuron, consuming only 10^{-9} *W* on von Neumann's estimate, remains the most efficient electronic computational element in the known universe. Forty years on, despite the development of very large scale integrated circuits (VLSI), the brain remains very compact and energy efficient compared to synthetic, silicon equivalents.

In addition to the physical differences, early digital computers were also substantially different from brains in terms of organization (Fig. 1). Von Neumann assumed that the speed of the components is a determining factor in the design of any computer. If individual computing elements were fast, then they could perform calculation very rapidly in sequence and relatively few would be required because the same element could be reused over and over again. If the computing elements were intrinsically slow, as are neurons, then the optimal architecture would be a parallel one, which enables many calculations to be performed simultaneously by a very large number of computing elements. It was this insight, together with the limitation of physical resources, and the lack of a suitable method for dealing with reconfigurable parallel processing, that led von Neumann to the architecture shown in Fig. 1.

Whatever their architecture, both digital computers and the living brain have had to solve the same basic problem that limits all computation: how to maintain precision in the face of operating (internal) noise. Von Neumann identified this inevitable deterioration of precision as the calculation progresses as the "essential interference" in the arithmetic operation of a computing machine. In serial machines even relatively simple computations require a large number of sequential operations to be performed. Von Neumann referred to the length of this sequence as the *arithmetic* or *logical depth* of the computation.

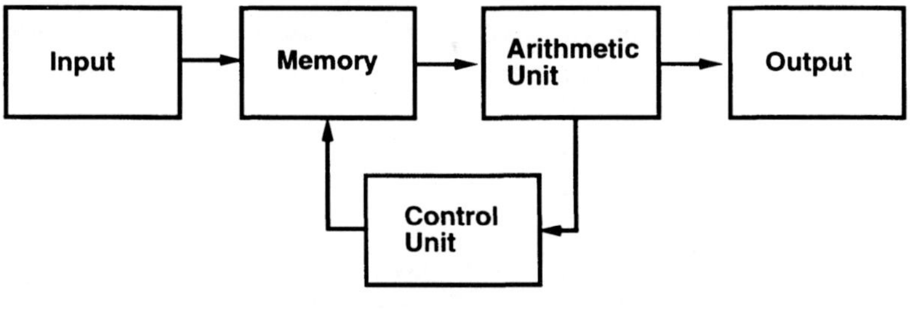

Figure 1. The organization of a general-purpose computer proposed by von Neumann. The problem to be computed is abstracted to a set of data and algorithms. This input is encoded and stored in general memory. A control unit accesses the encoded problem serially, and computes successive sections of the problem using the arithmetic unit. The result of the computation is placed in the output stage. The von Neumann architecture is radically different from that of biological nervous systems, which have strong colocalization of memory and computation, and consequently a deeply parallel computational strategy.

The deterioration in precision during a serial computation arises from two sources. The first is the accumulation of errors that arise at each stage of a computation. Because these errors occur approximately randomly, they accumulate in proportion to the square root of the number of serial operations. The second, more potent source is a multiplicative loss of precision resulting from the quantization inherent in the binary encoding of variables (roundoff error). These multiplicative errors can grow exponentially during serial computation. Von Neumann estimated that a precision of at least 10–12 decimals was necessary to counteract the corruption due to noise during the approximately 10^8 steps of a typical serial calculation.

At first sight the threat of corruption appears less for computations performed by neuronal networks, because their parallel organization permits them to compute far fewer serial steps than serial computers must. Only about 20–50 serial synaptic events are required to compute a "perceptual epoch" of 100–200 msec during which we can recognize an image or identify a sound. Nevertheless, a considerable error can accumulate over even such a short series. If we assume, conservatively, that neurons that are analog data representations have systematic imprecisions of as little as 5%, then a 50-step process results in a 10-fold multiplicative error and a 100-step calculation in a more than 100-fold error.

The obvious conclusion of von Neumann's analysis was that the nervous system uses different principles of operation than the serial processor of the digital computers that he was involved in building. Some obvious speed advantages are gained by the parallel architecture of nervous systems. However, it is not clear how these networks escape the deterioration problem.

The general constraints of neuronal hardware have rarely been used to interpret the processing strategies of the brain. Fundamentally, we do not yet know what kind of circuit architecture the neocortex has, what the constraints are on its computing elements, and what its computing strategy is. Most importantly, we also do not know whether these principles are true for all of neocortex, or whether they are locally specific. In this chapter we outline some principles based on cortical anatomy and physiology, and supported by computer modeling, whereby the cortex might avoid the problem of computational deterioration.

2. Wiring Neocortical Circuits

The simple fact that neurons connect with other neurons provides an important constraint on the design of the brain, because the size of the brain is determined not simply by the number of neurons, but also the length of 'wire' required to connect them. The amount of wiring is considerable. Each cubic millimeter of gray matter contains a total of about 4.1 km of axon (at an average diameter of 0.3 μm) and 456 m of dendrites (Braitenberg and Schüz 1991). The total length of axon is about 1000 times the length found in the equivalent volume of white matter, where the axons are usually myelinated and far thicker than in the gray matter. The implications of this wiring are crucial for the construction of cortical circuits. The total volume of the cortex depends on the number of cells and the wire required to connect them. The average connection pattern between neurons will determine how the volume of the cortex will scale with cell number. Mead and others have pointed out that the number of neurons connected to a given neuron must decrease at least as quickly as the inverse square of the distance between them, otherwise the size of the volume of the cortex would increase exponentially as more neurons as added (Mead, 1990; Mitchison, 1992). Such an increase of volume is not observed in the evolution of cortex. Instead, it seems that the neocortical neurons avoid making many long connections. Most of their connections are made locally, with their near neighbors (Fisken *et al.*, 1975; Douglas *et al.*, 1996). This local wiring strategy minimizes the amount of 'wire' added when the number of neurons in the neocortex is increased.

The thickness (about 2 *mm*), neuronal density (about 10^5 cells/mm^3), and synaptic density (about 7×10^7 synapses/mm^3) of the cortex are rather well conserved across species. Increases in cortical volume translate largely into increases in cortical area, and beneath each area of cortex the cellular composition of the cortical laminae is rather similar. Thus, the operation of making more cortex seems to emphasize local circuitry. One possibility is that evolution adds more of the same basic circuits, in which case we can hypothesize that the connections of neurons within a small area of cortex, say 1 mm^2, form the intra- and interlaminar circuits of the basic computational unit of the neocortex. It is the anatomical and physiological properties of this basic circuit that we have sought to conceptualize in our *canonical microcircuit* of neocortex (Douglas *et al.*, 1989).

3. Canonical Microcircuits

Many forms of canonical microcircuit have been proposed over the years, but Lorente de Nó was the first after Ramon y Cajal to attempt a comprehensive view of the cortical circuitry (Lorente de Nó, 1949). He proposed, on the basis of cytoarchitectonic studies and analysis of Golgi-stained material, that he could define an elementary unit, in which theoretically, the whole process of transmission of impulses from the afferent fiber to the efferent axon may be accomplished. Lorente de Nó's circuit contained excitatory neurons connected together in a recurrent chain. This elementary unit, he suggested, was an applicable to the mouse as to man.

In the subsequent revolution in concepts of cortical architecture led by Hubel and Wiesel, the concept of a module repeated throughout neocortex was retained (Hubel and Wiesel 1977), but Lorente de Nó's recurrent excitatory circuit was replaced by strictly feedforward circuits. Input from the sensory thalamus, the lateral geniculate nucleus in the case of the visual pathways, relayed signals from the sensory periphery to the middle layers of the neocortex. The recipient cortical neurons repeated this serial, feedforward processing vertically through a column of neurons. In the visual cortex these neurons shared a number of functional attributes, such as receptive field location in visual space, orientation slectivity, and eye preference. Hubel and Wiesel's inspired vision of the functional architecture of the primary visual cortex has remained for 35 years the blueprint of cortical processing. Even the subsequent extension of their view to account for the parallel processing of color, form, and motion in the primate primary visual cortex retains the essential principles of feedforward serial processing (Livingstone and Hubel, 1988).

Hubel and Wiesel extended their concept of serial processing to suggest that the processing also occurred through a hierarchy of visual areas (Hubel and Wiesel,

1965). Their experiments in the cat and monkey suggested that, in terms of their trigger features, neurons in successive tiers of the cortical areal hierarchy become more selective in their stimulus requirements. They supposed that this occurs through covergence of select patterns of excitatory and inhibitory afferents onto single cells. Subsequent research has supported their general idea: for example, the most extreme selectivity that has yet been discovered is the selectivity for faces of neurons located in the inferotemporal lobe, which is near the apex of the hierarchy for the macaque monkey's visual cortex. This, too, Tovee and colleagues (Tovee *et al.*, 1993) have suggested, occurs through a feedforward process.

The general view that the neocortical component of the visual system operates by serial processing through a hierarchy of cortical areas remains viable, although much research over the past 20 years has fulfilled von Neumann's prediction that the optimal architecture for a biological computer would be parallel. Thus, in most mammalian visual systems, each part of the visual field has access to a different part of cortex. This is visible as a map of the visual world, repeated through the visual areas. Also, at least three separate structural and functional streams can be traced from the retina through the thalamus to the primary visual cortex. Within the primary visual cortex of the monkey, further subdivisions are created. New streams of color, depth, motion, and form are created and themselves may merge and mingle as they are remapped in extraordinary ways through successive stages of the hierarchy of cortical areas (Felleman and Van Essen, 1991).

The precise nature of this parallel hierarchical architecture and its function remains unclear, however, because there are insufficient data about precisely which neurons are involved in any particular "parallel" circuit. To unravel this problem we need to have a clearer understanding of the nature of the local circuits within each visual area. But even here, the analysis has been quite restricted because of the difficulty in studying in sufficient detail the connectivity of morphologically and physiologically identified neurons. However, it seems that significant progress toward answering some of the fundamental issues raised by computational considerations could be answered by identifying the essential elements of the cortical computations. For this reason our laboratory has invested a great deal in the analysis of the detailed anatomy and physiology of the cortical microcircuits.

Our own view, based on extensive experimental and theoretical work on the cat's visual cortex, is much closer to that of Lorente de Nó. Our canonical microcircuit, illustrated in Fig. 2, attempts to capture some of the essential principles of cortical organization, viz. the columnar connectivity and the recurrent connectivity between both excitatory and inhibitory elements. The detailed experimental basis for this circuit will be described below, but first the contents of the boxes shown in Fig. 2 need to be examined. Since this can be done at a number of levels, we need first to consider the important issue of the unit of construction appropriate for this level of circuit modeling.

4. Units of Construction of the Basic Cortical Circuit

There are a number of different choices for the unit of construction. For example, one such unit could be considered to be the synapse, of which there are about 10^{15} in the human brain. The synaptic events are in some sense the elemen-

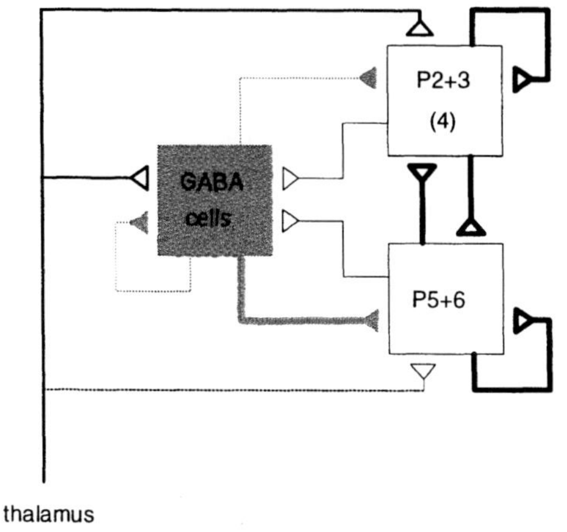

Figure 2. Simplest average cortical circuit consistent with anatomical and physiological data (Douglas and Martin, 1991). This 'canonical microcircuit' has recurrent excitatory connections between superficial pyramidal cells [P2 + 3(4)] and deep pyramidal cells (P5 + 6), and recurrent inhibitory connections with a single pool (for simplicity) of GABA cells. The line styles indicate schematically the approximate strength of the connections.

tary 'bits' of the brain. In neural modeling, each synapse performs a probabilistic multiply-add. The multiplication factor is given by the synaptic weight, while the addition occurs in the dendrite. The probabilistic element derives from the fact that at any one central synapse, a presynaptic spike only causes the release of a synaptic vesicle with a given probability *p* (which can vary between 10% and 90%). Real synapses are rather more interesting in their behavior, not only because of their dynamic properties, which are determined by the receptors and ion channels present, but also because they are modifiable. Nevertheless, dynamics and learning excepted, the possible logical operations that can be carried out in the neural circuits would have to be carried out by synaptic interactions on the dendritic tree. These synaptic interactions could involve very local operations, involving only one or a few synapses.

The simplest view is that neurons simply sum the voltages generated by the excitatory and inhibitory synapses they receive. If the sum of the excitation (pluses) and inhibition (minuses) exceeds the action potential threshold, the neuron produces an output. In such simplified model neurons, referred to as integrate-and-fire neurons, the dendritic tree acts a linear device.

The local interaction of small numbers of excitatory and inhibitory synapses can resemble logical functions of OR and AND (Shepherd and Brayton 1987), and AND-NOT (Koch *et al.*, 1982). For example, it has been shown theoretically that a 'synaptic veto' could operate by the temporally coincident activity of an inhibitory and an excitatory synapse on a single spine. Such logical operations could, in principle, account for phenomenan such as the direction or orientation selectivity in visual cortical neurons (Koch and Poggio, 1987). Other nonlinear functions such as multiplication or division can be mediated by different synaptic receptor types or voltage-dependent membrane conductances in the dendrites. For example, the *N*-methyl-*D*-aspartate (NMDA) or dendritic calcium channels can both implement a form of multiplication (Mel, 1993, 1994).

The initiation of an action potential as the result of all dendritic synaptic interactions is a more difficult matter. Neuronal outputs are sometimes modeled by a threshold rule, in which the neuron produces an action potential only if the weighted sum of the synaptic inputs exceeds a threshold. However, the operation of most neurons is considerably more complex than this. Unlike conventional logic gates, neurons have very large numbers of inputs and their thresholds are uncertain; moreover, their outputs are encoded as frequency rather than binary values.

The basic notions of arithmetic and logic have been applied, both explicitly and implicitly, to the organization and function of microcircuits in the neocortex. Thus, the formation of the main receptive field types in the primary visual cortex, the simple, complex, and hypercomplex receptive fields, has been explained on the basis of relatively straightforward linear combinations of excitatory and inhibitory synaptic inputs to single neurons. The seminal examples of such circuits are those of Hubel and Wiesel. In their hypothetical circuits, visual signals are processed in a serial fashion through individual cortical areas and through a hierarchy of visual areas. Feedforward excitation generates the basic receptive field structures. Feedforward inhibition via an inhibitory neuron generates additional features, such as the antagonism between ON and OFF subfields of simple cells, and the length inhibition of hypercomplex, or 'end-stopped' cells (Hubel and Wiesel, 1965). Any perceptible stimulus has 'trigger features' for single neurons. These trigger features in the primary visual cortex may involve one or a combination of the attributes of, e.g., direction, velocity, contrast, color, depth, and form.

A trigger implies a threshold, and so a nonlinear response. However, this is not an all-or-nothing response as in the binary coding in the digital computer. For example, the discharge rate of single neurons can code for the probability that a particular stimulus is present (Barlow 1972). In the case of primary visual cortex, it seems that many of the properties of the receptive fields can be described by approximately linear processes over a limited range (Movshon *et al.*, 1978; Jones and Palmer, 1987; Jagadeesh *et al.*, 1993). In the linear part of the range, the rate of discharge of the neuron would then be related quantitatively to the degree to which particular attributes are present in the stimulus.

The significance of the exact site of computation is this: If the activity of *single* cells reflects entirely the properties of their receptive fields, then examination the receptive fields should also reveal the essential computations. This would be convenient, because then some of the neural mechanisms predicted in the computational literature could be examined experimentally at the level of single-cell responses. This philosophical point of view has led to an extensive literature that does not require closer examination here, but will be included in our discussion of whether we need to look beyond single neurons to explain cortical computation.

For this next step we need to examine which neurons actually form the cortical circuits. For simplicity we focus on one specific layer, layer 4, which encapsulates all of the essential elements that are needed for the models that follow. In the visual cortex of the cat the link between neurons of the dorsal lateral geniculate nucleus (LGN) and the cortical recipient neurons is the site of one of the most dramatic transformations in the visual pathways. Within apparently one synapse, the monocular, nonoriented, nondirectional, center–surround receptive fields of the dLGN afferents are converted into simple receptive fields that are strongly orientation sensitive, directional, often binocular, and may be length turned as well (Hubel and Wiesel, 1962; Gilbert, 1977; Dreher, 1972). This link is made principally in layer 4, and thus layer 4 of the cat offers the most accessible model system whereby some of the basic computational principles of the neocortex can be studied. Indeed, it is from their study of simple cells in cat layer 4 that Hubel and Wiesel originally derived their influential descriptive model of cortical organization. In the sections that follow, we explore the structural and functional basis of computations performed in the input layer of the cat's primary visual cortex.

5. The Neuronal Components of Layer 4

Although cortical neurons come in a number of different types, traditional anatomical methods have shown that they divide into two basic morphological groups: those that bear spines and those that are smooth. Physiologically, too, the simplification can be made that cortical neurons are either inhibitory or excitatory in function. This functional classification has many corresponding morphological, ultrastructural, and neurochemical correlates. For example, in cortex, the inhibitory neurons are those with smooth dendrites, the excitatory neurons are those with spiny dendrites. Their axonal differences are also distinct, but more subtle: the axons of smooth neurons connect to their targets via large *en passant* boutons, whereas the axons of spiny neurons usually have mixed *en passant* and *termineaux* synaptic boutons. At an ultrastructural level, the functional and morphological separability continues to be easily visible. The excitatory neurons form synapses that have a distinctly asymmetric pre- and postsynaptic density, whereas the smooth neurons form synapses in which pre- and postsynaptic densities are symmetric on either side of the synaptic cleft. At a neurochemical level the differences are marked, since the inhibitory smooth neurons release gamma-aminobutyrate (GABA) from their terminals, whereas the exciatory spiny neurons release glutamate.

The GABAergic neurons are more easily subdivided on the basis of their axonal projections. The smooth neurons of layer 4 are the major source of the inhibitory axons. The main smooth neurons are varieties of basket neurons and neurogliaform cells (Naegele and Katz, 1990; Kisvárday *et al.*, 1985; Gabbott *et al.*, 1988). The basket cells have compact axonal arbors that are largely contained within layer 4, with smaller projection to more superficial and deeper layers. The neurogliaform neurons have tightly interwoven dendrites and axons.

The spiny neurons in layer 4 can be divided into the pyramidal neurons and the spiny stellate neurons on the basis of their dendritic morphology. The pyramidal cells are most easily distinguished by their apical dendrite, which may vary considerably in length and degree and pattern of branching. The radially symmetric arrangement of its dendrites gives the spiny stellate its name. The properties of layer 4 simple cells have been explained most simply by the interaction of the convergent pattern of feedforward excitation from the dLGN, with disynaptic feedforward inhibition from intracortical inhibitory cells (Hubel and Wiesel, 1962, 1965). At the center of this transformation is the spiny stellate neuron, which in the cat is found exclusively in layer 4. The spiny stellate is probably one of the major recipients of input from the dLGN and thus lies at the gateway of the cortical networks. The spiny stellate is not unique in being a recipient of dLGN input, for pyramidal and smooth neurons in layer 4 and other layers also receive monosynaptic input (Martin and Whitteridge, 1984). However, the spiny stellate is a particularly convenient model for examining theoretically and experimentally in a single neuron the crucial interactions that occur between the dLGN afferents and intracortical excitatory and inhibitory pathways.

The symmetric radial arrangement of spiny stellate cell dendrites provides an important simplification for structural, biophysical, and circuit studies alike. The dendrites are short with moderate diameters, so the dendritic tree is relatively compact electrically, and the membrane conductances are likely to be similar for each dendrite (Stratford *et al.*, 1996). By contrast, the active conductance found in the apical dendrites of pyramidal cells differ from that in the basal dendrites (Connors *et al.*, 1982; Stuart and Sakmann, 1994).

Because they do not have the inconvenient asymmetry of a pyramidal cell's apical dendrite, ultrastructural studies of spiny stellate cells need only consider examples of the same dendritic type. Most of their dendrites are confined to layer 4, and so the sources of their input can be more easily defined than for the pyramids, in which both basal and apical dendrites may cross laminar borders. Thus, only axons that form synapses in layer 4 could innervate the spiny stellate neurons.

5.1. Innervation of Layer 4 Neurons

The input to layer 4 is well established. Anatomical studies have shown that neurons in layer 4 neurons form synapses with the recurrent axons of layer 6 pyramids (McGuire *et al.*, 1984), with dLGN axons (LeVay, 1973), with other spiny neurons in layer 4 (J. Anderson *et al.*, 1994) (Fig. 3), and with the small GABAergic basket cells of layer 4 (Kisvárday *et al.*, 1985; J. Anderson *et al.*, 1994; Ahmed *et al.*, 1997). We were able to extend these studies by a combined physiology–anatomy experiment in which we examined the targets of single, physiologically and morphologically characterized dLGN afferents. Single dLGN afferents were identi-

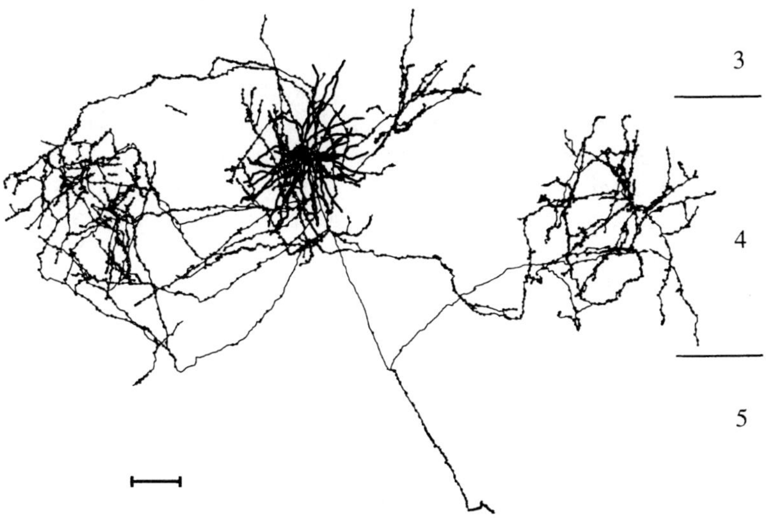

Figure 3. Spiny stellate neuron of layer 4A of cat primary visual cortex. It was a direction-selective simple cell, monocularly driven by the contralateral eye. Shown in parasagittal view. Typical clumps of axon arborizing in layer 4, and layers 3 and 5. Layers as indicated. Scale bar: 100 μm.

fied physiologically and labeled by intraxonal injections of the tracer horseradish peroxidase. After processing to reveal the dLGN axon, the same sections were then Golgi-stained to label potential postsynaptic targets of the thalamic afferents (Freund *et al.*, 1985a,b). As might have been anticipated, a great variety of neuronal types were contacted by any single dLGN axon. The significant finding was how few synapses were formed between a single dLGN relay cell and a single target cell. Typically, the number of synapses formed between any single axon and its postsynaptic neuron was very small—between one and eight synapses (Freund *et al.*, 1985a). Most previous physiological work had predicted that a single dLGN neuron would form a far higher number of synapses on its target neuron in the neocortex (Creutzfeldt *et al.*, 1974; Sillito, 1975; Heggelund, 1981; Toyama *et al.*, 1981; Vidyasagar and Heide, 1984).

The origin of the monosynaptic input from neurons other than the dLGN is much more difficult to determine physiologically. However, both extracellular and intracellular recordings from layer 4 simple cells *in vivo* reveal that all receive excitatory and inhibitory synapses from other cortical neurons. This is not surprising: in addition to the thalamic relay cells, layer 4 is the principal termination site of the axons of a subset of layer 6 pyramidal neurons (O'Leary, 1941; LeVay and Gilbert, 1976; Gilbert and Wiesel, 1979; Martin and Whitteridge, 1984), most of which also project back to the dLGN. The axons of the spiny stellates themselves also provide a major innervation of layer 4. An additional small excitatory input arises from the claustrum, which innervates all cortical layers. A recent *in vitro* study of cat visual cortex has confirmed that individual spiny stellate cells receive excitation from dLGN afferents, spiny stellate cells, and layer 6 pyramids (Stratford *et al.*, 1996).

The dLGN neurons are relatively few in number in the cat (about 0.36×10^6 cells/hemisphere) compared to the number of layer 6 pyramidal cells that project to layer 4 (4.2×10^6) and layer 4 spiny neurons (9×10^6) (Peters and Payne, 1993). It might therefore be anticipated that the dLGN afferents contribute only a fraction of the total excitatory synapses found in layer 4. Experimentally, the various degeneration and tracing techniques have produced estimates ranging from 5% (LeVay, 1986; Garey and Powell, 1971) to 22% (Einstein and Davis, 1987) to as much as 28% (LeVay and Gilbert 1976) of the asymmetric (excitatory) synapses in layer 4. This number of 28% has been used to argue that the dLGN input to cortex is functionally strong (Ferster *et al.*, 1996), but it is an unrealistic estimate, as a simple check shows. There is a total of 3.32×10^{10} asymmetric synapses in layer 4 of area 17 (P. Anderson *et al.*, 1988; Beaulieu and Colonnier, 1985). If 28% of these are contributed by the dLGN relay cells of layer A and A1, then each of the 360,000 relay cells (Madarasz *et al.*, 1978; LeVay and Ferster, 1979) must contribute over 25,000 synapses each. Counts of the synaptic boutons of individual X- or Y-type arborizations in layer 4 give average numbers of 2620 boutons for X-relay cells and 4280 for Y-type relay cells (Humphrey *et al.*, 1985). The X-relay cells are twice as numerous as the Y-type relay cells, so this means that, at best, only about 5% of the asymmetric synapses in layer 4 could possibly originate from the dLGN. Only methodological inadequacies (see, e.g., LeVay 1986) could explain the inflated estimates mentioned above.

Until recently there were no direct estimates of the proportion of dLGN synapses formed with the spiny stellate cells. The only previous estimate was from an experimental study by White in the somatosensory cortex of the mouse. White (1989) found that the thalamus provided about 13% of the synapses formed with a spiny stellate in the mouse somatosensory cortex. Ahmed *et al.* (1994) have now made estimates of the number of synapses received by different cell types in layer 4. They estimate that dLGN afferents provide only 6% of the synapses with the spiny stellates, and 13% of the synapses with the GABAergic basket cells of layer 4 (Ahmed *et al.*, 1994, 1997).

When the dominant connections between all six laminae are considered, the conclusion is inescapable that all neurons receive excitation from both feedforward and recurrent feedback connections. Thus, the soma of a given neuron will not 'know' from one instant to the next, which presynaptic neurons are the source of the inhibition and excitation that it collects. An important consideration here is the strength of the synapses involved. This has recently been reported by Stratford *et al.* (1996). They examined the fast [α-amino-3-hydroxy-5-methyl-4-isoxazole proprionic acid (AMPA)-receptor mediated] excitatory postsynaptic potential (EPSP) and found that the mean amplitude of EPSPs evoked in spiny stellate cells by stimulation of single dLGN fibers was 1.8 mV. The mean amplitude of single spiny stellate or single layer 6 pyramidal inputs to the spiny stellate cells was 1.0 and 0.3 mV, respectively. Thus, it seems that the cortical synapses of the dLGN relay cells are stronger than the other excitatory synapses that form with the spiny stellate cells. Additionally, Stratford *et al.* (1996) found that the dLGN synapses had higher release probabilities than either the spiny stellate or layer 6 pyramidal synapses. However, the dLGN synapses depresses with repetitive stimulation, whereas the layer 6 pyramidal cell synapses show strong facilitation of up to 75% of control (Ferster and Lindstrom, 1985; Stratford *et al.*, 1996). Since the major synaptic input to spiny stellate cells arises from layer 6 pyramidal cells and other spiny stellate cells (Ahmed *et al.*, 1994), it is very likely that, despite the strong dLGN synapses, the dominant excitatory current to spiny stellate neurons arises from other cortical neurons. However, almost all models of the circuits of layer 4 invoke only feedforward excitation from the dLGN afferents to provide excitation.

5.2. Theoretical Estimates of Recurrent Connectivity for Spiny Stellate Neurons

Spiny stellate neurons were recorded intracellularly *in vivo* to determine the characteristics of their thalamic inputs (i.e., X- or Y type) and receptive field types. They were identified morphologically by filling them with a marker, horseradish peroxidase, which is transported throughout the dendritic tree and intracortical axonal arbor (Fig. 3). The spiny stellates we studied projected to layer 3, as reported previously (Gilbert and Wiesel, 1979; Lorente de Nó, 1949; Lund, 1973; Martin and Whitteridge 1984). However, their collaterals also arborized within layer 4 itself and to layer 5. Occasionally spiny stellates had axons that projected into the white matter, presumably en route to other cortical areas.

From the detailed anatomical and physiological studies referred to above, we know that at least three separate excitatory pathways converge on spiny stellate neurons: from the thalamus, from layer 6 pyramids, and from layer 4 spiny neurons. Inhibitory connections arise mainly from smooth neurons within layer 4. The synaptic contribution of the three excitatory pathways onto spiny stellate cells seems to be approximately in the ratio of their average occurrence in layer 4. Thus, there is no preferential connectivity between any of these inputs and spiny stellate dendrites rather than other dendrites passing through layer 4. However, within the connections made onto spiny stellate dendrites, there is some degree of specificity in terms of the actual target sites. The axons of spiny stellates and thalamic relay cells preferentially form synapses with the spines of spiny stellates, whereas the layer 6 pyramids preferentially form synapses with the shafts of spiny stellate dendrites.

The fact that synapses onto the spiny stellate dendrite are in proportion to presynaptic availability provides the simplifying assumption that the single neurons that form the different pathways connect approximately randomly to the spiny stellate neurons. Using this assumption, we can estimate the number of spiny stellate neurons that directly excite each other (Douglas *et al.*, 1995). From this calculation we can derive an estimate of the amount of excitatory current provided by such first-order connections. This is an important preliminary calculation because it provides an estimate of the amount of positive feedback that occurs in the recurrent networks of layer 4.

Essentially, the calculation is based on the notion that the axon of a spiny stellate neurons forms an elaborate branched arborization that produces a number of 'clouds' of synaptic boutons. Some proportion of these synaptic boutons will connect to the dendrites of spiny stellate neurons. Each spiny stellate axons forms about 5,000 synaptic boutons and approximately one third of these will be in the primary cluster around the cell body of the parent neuron. As Fig. 4A shows, the primary cluster extends out about 500 μm from the cell body. Within the primary cluster, the boutons are assumed to be distributed in a spherical Gaussian distribution, which has a standard deviation of 100–120 μm. This cloud of synaptic boutons intersects with a second cloud—that of synaptic sites provided by the dendrites of neighboring spiny stellates.

The specification of the cloud of synaptic sites derives from our maps of the spiny stellate dendrites (Ahmed *et al.*, 1994). We estimate that the total number of synapses formed with the dendrites of spiny stellate neurons is 5000, of which about 1200 are formed by other spiny stellate cells. The density of spiny stellate neurons is 4×10^6 cells/mm^3 and there are 6×10^6 synapses/mm^3 asymmetric synapses in layer 4.

From the approximation of the average radial density of the synaptic boutons in the primary axonal cluster and the number of available target spiny stellate dendrites, we have calculated the number of first-order recurrent connections involving a single spiny stellate and the number of neighboring spiny stellates that participate in these recurrent connections as a function of the standard deviation σ of the primary axonal cluster (assuming that synapses have a Gaussian distribution in three dimensional space). We assume that boutons within the primary cluster do not select specific spiny stellates with which to form synapses, but connect on the basis of chance encounters. For clusters with a standard deviation of 100 μm, a given spiny stellate gives and receives synaptic connections with 117 neighbors. If the standard deviation of the primary cluster is increased to 150 μm, the average density of boutons forming the cluster decreases and the number of potential targets increase (Fig. 4B). However, the lower target density means that there is a reduced probability that a target spiny stellate will provide a connection back to the

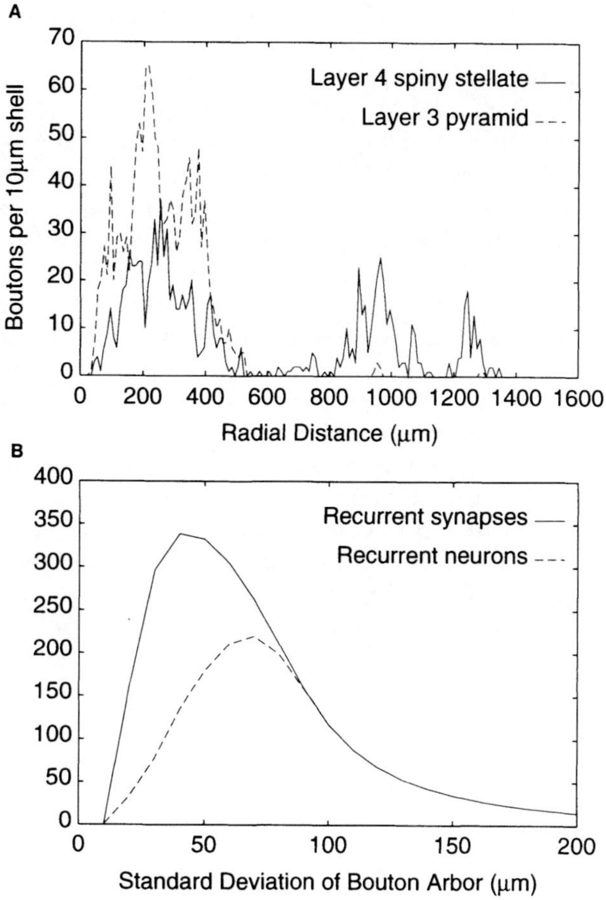

Figure 4. Recurrent synaptic connections between spiny neurons in cat visual cortex. (A) Radial density of synaptic boutons as a function of the distance from the cell body of two cells, a layer 4A spiny stellate and a layer 2/3 pyramidal neuron. In both cases the axons could not be completely reconstructed, and the total number of boutons is underestimated. Nevertheless, the data suffice to show that the primary cluster of boutons extends from the soma to a distance of about 500 μm. The primary bouton clusters of these cells can be described by a spherical Gaussian distribution with σ between 100 and 120 μm. Data were obtained by computer-assisted reconstruction of neurons in cat striate cortex that had been labeled with horseradish peroxidase during the course of physiological experiments *in vivo*. (B) Numerical estimate of the number of first-order recurrent connections made within the primary cluster of a spiny stellate neuron, and the number of neighboring spiny stellate neurons that participate in those recurrent connections as a function of the standard deviation σ of the axonal tree. For clusters with $\sigma = 100$ μm, spiny stellate cells receive 117 recurrent connections from as many neighbors. For $\sigma = 150$ μm, there are only 34 recurrent synapses. The following parameters, drawn from the literature and our own data, were used to calculate these estimates: 4×10^4 spiny stellate cells per mm^3, 6×10^8 asymmetrical (excitatory) synapses per mm,^{3} 5000 total synapses on the soma and dendrites of stellate (of which 1200 derive from other stellates). No excitatory synapses are made within 10 μm of the somata of spiny stellates. Five thousand synapses are made by the axonal arbor (of which 1200 are made onto other stellate cells). We assumed that one third of the boutons occur in the primary cluster, and that they are homogeneously distributed in three-dimensional space. From Douglas *et al.* (1995), with permission.

cell that produces the primary cluster. The net effect is that the number of such recurrent connections declines to 34. The significance of these simple calculations is that the first-order recurrent connections within the spiny stellates themselves can provide a significant source of recurrent excitation.

5.3. The Open-Loop Gain and the Role of Geometry

These first-order connections constitute the most basic positive feedback loop. It is clear that neurons within such a loop have the potential to receive more current from the recurrent connections than they are able to sink. In such a case the neurons will continually increase their firing rate until every neuron in the network reaches saturation. One way of avoiding this would be to have a rule that no neuron should give an excitatory connection to a neuron that has already given it one. This rule appears not be applied in cortex, because experimental studies in visual cortex and in the somatosensory cortex have shown that such recurrent excitatory coupling occurs relatively frequently (Mason *et al.*, 1991; Markram *et al.*, 1997). However, the fact is that cortical neurons are not routinely driven into saturation whenever they are provided with an input. Instead they respond proportionately to the magnitude of the input. This indicates that there are sufficient constraints that keep the excitation within operational bounds without the constraint of saturation. It is usually assumed that the pyramidal cells are protected from latching up in saturation by the action of the inhibitory cells, which are also linked into the recurrent loop. But we demonstrate here that recurrently connected pyramidal cells can respond stably even without inhibition.

To derive a general understanding of the operational limits of the recurrent excitatory circuits, we need to understand how input and recurrent circuits interact. In particular we need to derive the open-loop gain for the network, since if this gain is greater than one, then once excited, the neurons in the network will reexcite each other with increasing strength until they all reach a saturated firing level. If the open-loop gain is less than one, an initial amount of excitation within the network will not be sufficient to maintain or increase activity within the network, i.e., the network will be stable and will respond proportionally to its external input.

The amount of excitatory current generated by the recurrent excitatory pathways depends both on the number of synapses and the rate at which the synapses are driven. The interactions among the participants of the first-order recurrent network can be represented most simply by the analogy of an electronic circuit that is a network of recurrently connected simplified neurons (Fig. 5A). The important simplifying assumption made in this model is that all the neurons of the cooperating population have identical intrinsic properties, receive identical feedforward inputs, and form identical synapses with each other. The assumption that all the neurons form identical synapses with each other implies that all members of the cooperating population are recurrently connected to one another. With this model we can easily calculate the effect of the activity of the network on a representative neuron in the network. More detailed realistic models have many free parameters and it then becomes difficult to trace cause and effect when many different settings can be manipulated to give similar effects. In the electronic abstraction (Fig. 5B), the input conductance of the representative neuron is simplified to a single conductance, G. The output of this linear point neuron is the frequency of spike discharge F which is represented by the voltage across G. The neuron experiences a recurrent current I_{rec} generated by the other excitatory neurons in the cooperating network. The recurrent current is proportional to the output of the representative neuron. The lumped recurrent synapse acts as a current source that is controlled by the firing frequency of the neuron. Because all neurons in the cooperating population are identical (and so have the same output values), the total recurrent current I_{rec} injected via the lumped synapse is proportional to the output firing frequency F of each neuron in the network. Thus, $I_{rec} = \alpha F$. Since F is represented by a voltage, α behaves as a conductance, which we refer to as the phenomenological *network conductance.*

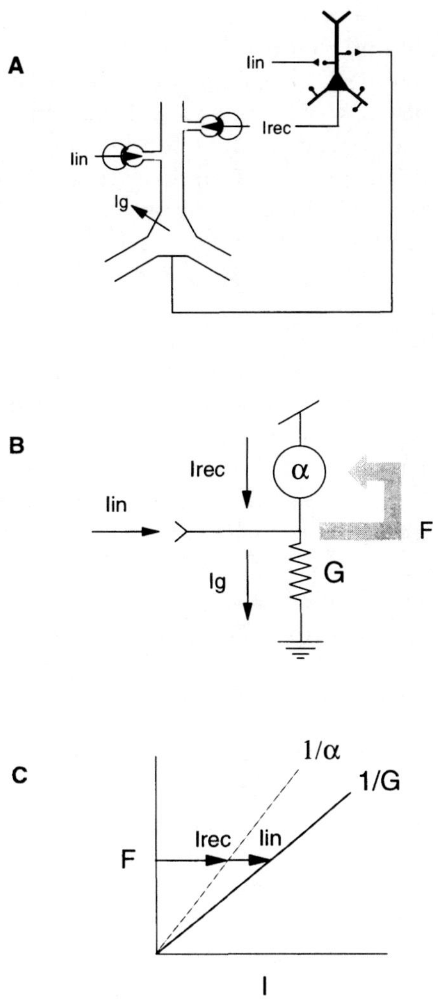

Figure 5. Feedback excitation creates a *network conductance.* (A) The output discharge of a cortical pyramidal cell drives other pyramidal neurons (shown in black), which are members of a common functional population, with similar connections. Consequently, these neurons in turn excite the source pyramidal cell. The exact mechanism of excitation is unimportant, provided that the excitatory current I_{rec} is proportional to the output of the pyramidal cell. (B) The electrical equivalent circuit of the cortical neuron in its embedding excitatory circuit. The excitatory circuit creates a *negative* network conductance, $-\alpha$, that increases the slope of the *IF* relation (C) from $1/G$ to $1/(G - \alpha)$. This mechanism gives rise to a current gain = I_g/I_{in} (see text).

Within this network model, the excitatory network conductance has the opposite sign of the neuronal conductance, i.e., it acts as a negative conductance. To see why this is so,consider the neurons without interconnections to each other. For a given current input to each of the neurons, the resultant voltages, which represent the discharge rate of the neurons in this model, will be determined solely by the conductance G of the neurons. The neuronal conductance G is increased by any process that increases the conductances of the channels in the membrane. These include the intrinsic conductances of the neuron, which include the conductances associated with the action potential, the many voltage-sensitive membrane conductances, and the calcium- and second-messenger-mediated conductances. The net effect of all these channel openings is that G increases and so more current is dissipated across the neuronal membrane. The amount of load current I_g that can be sunk by the intrinsic conductances of the neuron for a given discharge rate F determines the slope of the current–frequency curve (Fig. 5).

An observer measuring the input current and the voltage output of a neuron will find that the current–frequency relationship is described by a straight line of slope $1/G$. Consider, then, the same neuron connected into the recurrent excitatory network. For a given input current received by all of the neurons, the voltage recorded across their conductances G will be greater than for the isolated neurons because each neuron receives an additional current from the network, i.e., the total current flowing through G is $I_{ff} + I_{rec}$. For an external observer measuring the input current provided to a neuron and the voltage output, it will appear as if the conductance of the neuron has decreased. The effective conductance of the neuron G_{eff} will be $G - \alpha$ and the slope of the current–frequency relationship will now be $1/(G - \alpha)$.

As the input current to the network increases, the firing rates of the neurons in the network increase and more recurrent current is injected through the lumped excitatory synapse. In the electronic model this increase in excitation within the network is represented by the output firing frequency of the neurons, which is $I_{in}/(G - \alpha)$. Activity in the network increases the current flowing through G, which effectively increases the voltage across G. This increased voltage represents an increase in the firing rate, which in turn determines the magnitude of I_{rec}, because the output of each neuron contributes some excitatory current to each of its fellows and receives from all of them the total feedback component, i.e., I_{rec}. When the network is in steady state, conservation of current demands that the sum of the excitatory currents equals the current flowing to ground across G. In electronic engineering terms, whether the circuit is stable or not is determined by the open-loop gain, which in our formulation of the electronic model is given by the ratio of the two conductances (α/G). If a small change in the total current generates a recurrent current that exceeds the original change, the network is in uncontrollable positive feedback. The neurons will be driven relentlessly to saturation and remain there even if the initial input excitation is removed. In this case the open-loop gain α/G is greater than 1. Conversely, when the open-loop gain is less than 1, the amount of current sunk across G will be greater than the change in the total excitatory current. In this situation, the removal of the input current will result in the activity of the total network declining progressively until the recurrent current also returns to zero. The important point is that G provides the negative feedback necessary for network stability, and that a separate feedback from inhibitory neurons is not required. A population of excitatory neurons alone is *in principle* able to provide stable amplification. We are not arguing here for the complete absence of inhibition, but simply that sufficient control of stability can occur in recurrently connected excitatory networks without intervention of inhibition.

The graphical representation of the relationship of the two components that contribute to the total current the neuron receives and hence to its discharge rate is shown in Fig. 5 for a condition where the open-loop gain is less than 1. Experimentally, the current–discharge relationship for a single neuron can be determined by injecting current into the neuron. To establish the contribution of the recurrent circuit to the total current, we need to know the value of F and the value of α. The recurrent current I_{rec} plotted against F describes a line of slope $1/\alpha$. If the slope of $1/\alpha$ is greater that the slope $1/G$ (i.e., the line lies to the left of the $1/G$ line), then the network is stable and F is determined by the magnitude of I_{ff}. If the slope of $1/\alpha$ is flatter than $1/G$, then the recurrent circuit is supplying more current than can be sunk by the neuron. Consequently the firing frequency of the neuron will increase until it saturates.

We can now link this directly to the theoretical anatomy of the first-order recurrent pathways and ask how these recurrent connections might contribute to the stability or instability of the positive feedback. For this we need to determine the dependence of the open-loop gain α/G on the spread of the primary bouton cluster. The current–spike rate relation for these neurons is linear with a maximum slope of 100 spikes/sec/nA. Since G is inversely related to slope of the current–discharge curve, its magnitude corresponds to an average of 10 pA/spike/sec in steady state. Now we must calculate the current that arrives through recurrent connections. To make this estimate we assume, as in the circuit abstraction described in the previous section, that all of the neurons in the recurrent population are fully connected to each other, and all of them are receiving the same magnitude of input current. This fully reconnected network is the one for which maximum recurrent excitation is achieved with minimum feedforward input and so represents an upper bound on the magnitude of recurrent gain. The neurons provide current to each other through synapses that inject about 0.1 *pA* per presynaptic spike into the postsynaptic dendrite. The total amount of current delivered for the case of the spiny stellate in a fully reconnected population of 117 neurons (corresponding to σ of the spiny stellate axonal tree of 100 μm; see Fig. 4B) is 11.7 pA/spike/sec. This is the value of α and in this case α exceeds G by 1.7 pA/spike/sec and the network is unstable: removal of the input will have no effect on the subsequent activity because recurrent excitation will drive the network to saturation.

If the standard deviation of the Gaussian describing the distribution of synapses of the spiny stellate axonal tree is larger, for instance, σ = 150 μm (as discussed above), the number of recurrent connections is 34 (see Fig. 4B). Now the network conductance corresponds to 3.4 pA/spike/sec and G exceeds α by 6.6 pA/spike/sec. The output of the network can represent input in a proportional manner; furthermore, the network can represent input in a proportional manner; furthermore, the network will only remain active if an input current is present.

This is an overestimate of the recurrent excitation predicted by random connectivity within a cluster. However, if there is some specificity in the projection whereby only certain members of a particular class interconnect, then the density of recurrent connections between those members will approach what we have calculated here. Our estimate nevertheless demonstrates that the spatial extent of the axonal arbor is a potentially critical factor in determine the open-loop gain.

5.4. The Closed-Loop Gain

Of course, implicit in the discussion above is the possibility that the input signal can be actively and 'selectively' amplified by action of recurrent cortical excitation. In engineering terms, this is referred to as the *closed-loop gain,* which is the factor by which the circuit multiplies the external signal. In the general case, the input signal to neocortex could arise from a number of possible sources, including subcortical structures in the thalamus, and intra- as well as interhemispheric connections. In the specific case of the spiny stellate neurons in cat visual cortex, we will consider only the external input from relay neurons of the dLGN. On the basis of the microstructural investigations referred to above, we know numerically that the geniculocortical synapses constitute only 200–300 of the 5000 synapses formed with each spiny stellate cell. To simplify the present discussion we assume that all excitatory synapses are of equal functional value, regardless of their source or location on the dendritic tree.

Using the same assumption of full interconnectivity as with the estimation of the open-loop gain, we can derive the closed-loop or system gain relationship from the electronic analogue circuit shown in Fig. 5. dLGN synapses contribute a component of the total current, the feedforward current, I_{ff}. The identical neurons in the network each receive this current and in addition each receives the recurrent current, I_{rec}. The closed-loop gain of the system is given by $d(I_{ff} + I_{rec}) / dI_{ff}$. Since $I_{ff} = F(G - \alpha)$ and $I_{rec} = F\alpha$, the closed-loop gain can be expressed as

$$\frac{1}{1-(\alpha/G)} = \frac{G}{G-\alpha}$$

The greater the closed-loop gain, the smaller the input current needed to achieve a given level of activity. It is convenient that the gain can be expressed independently of the firing frequency of the neurons and is determined by the values of the conductances. The recurrent circuit will contribute more and more of the excitatory current received by any neuron as α approaches G. If α exceeds the sink conductance G in this case, the open-loop gain will exceed 1 and runaway positive feedback will follow. However, the closed-loop gain can be much greater than 1 even if the open-loop gain is less than 1.

5.5. Synaptic Inhibition

Thus far we have only considered excitatory recurrent connections in the local circuit. We have show that this system can be represented in terms of a simple electronic circuit model which indicates the bounds within which the circuit is able to amplify proportionately the input current (the open-loop gain). Somewhat counterintuitively, our analysis of the open-loop gain shows that when excitatory neurons are connected within a tight recurrent excitatory loop, they nevertheless can operate without saturation because neurons act as conductances. This view does challenge existing notions of how stabilization occurs in cortical networks, which is that powerful tonic inhibition is necessary to dampen or quench the excitatory activity of the network. However, appealing as this notion is that strong inhibition can solve the problems of a rich excitatory network in the neocortex, experimental evidence for powerful inhibition has been elusive to the point of invisibility. In short, the cumulative evidence of physiologists working on the visual cortex is that under normal operating conditions cortical inhibition is weak, but precisely targetted (Douglas *et al.*, 1988, 1991; Berman *et al.*, 1992; Ferster, 1986, 1987; Ferster and Jagadeesh, 1992).

The function of the recurrent network demonstrates why weak cortical inhibition is all that is necessary to mold cortical responses (Fig. 6). Due to recurrent

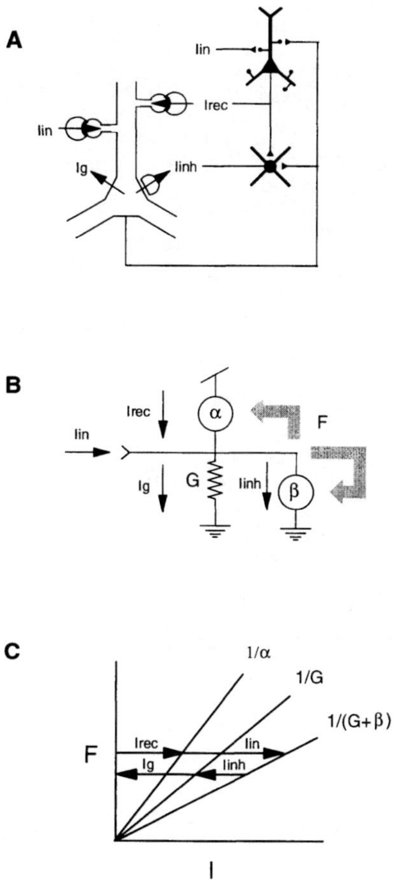

Figure 6. The output of a cortical pyramidal cell drives other pyramidal and smooth neurons (shown in black). (B) The electrical equivalent circuit of the cortical neuron in its embedding circuit. The recurrent excitatory circuit creates a network conductance α, while the recurrent inhibitory circuit creates another network conductance, β. The effective conductance of the neuron is thus $G_{eff} = G + \beta - \alpha$, and its current gain is $(G + \beta)/G_{eff}$ (see text). (C) Current–discharge relations that explain the behavior of the cortical amplifier. The $1/G$ curve expresses the current sink I_g offered by the passive membrane conductance and the averaged spike conductances. The $1/(G + \beta)$ curve indicates the increased current sink $I_g + I_{inh}$ caused by inhibition that is proportional to the output of the neurons. The $1/\alpha$ curve expresses the dependence of the feedback current measured in a particular neuron, I_{rec}, on the average output of neurons in the population. The steady-state discharge rate $F(I_{rec} + I_i n_n)$ occurs where $I_{rec} + I_i n_n + I_g + I_{inh} = 0$ is satisfied.

excitation, an input that stimulates all of the neurons in the population is amplified by a factor of $G/(G - \alpha)$, the closed-loop gain. A very small input correctly applied can thus produce a large neuronal response. However, the response can be readily controlled by a small feedforward inhibitory signal that changes the threshold of activation (Fig. 7), since it is the difference between the feedforward excitation and inhibition that is amplified by the recurrent excitatory pathways. In the case of solely feedforward excitation, a large inhibitory signal is required to overcome the large feedforward response. The threshold nonlinearity of the neurons is thus a crucial aspect of the recurrent model, because neurons that fall below threshold cannot participate in amplification loops.

In the absence of recurrent excitatory amplification, the amplitude of the subtractive inhibition has to match that of the total excitatory current. However, in contrast to the vetoing power of feedforward inhibition, feedback inhibition controls the gain of the cortical response. It is a simple matter to add an inhibitory network to the excitatory network shown above if the same principles and logic used for the excitatory loop are applied. Whereas the excitatory synapses inject current into the 'neuron' and thus act as a source of current, inhibitory synapses remove current from the neuron and thus act as an additional current sink. This

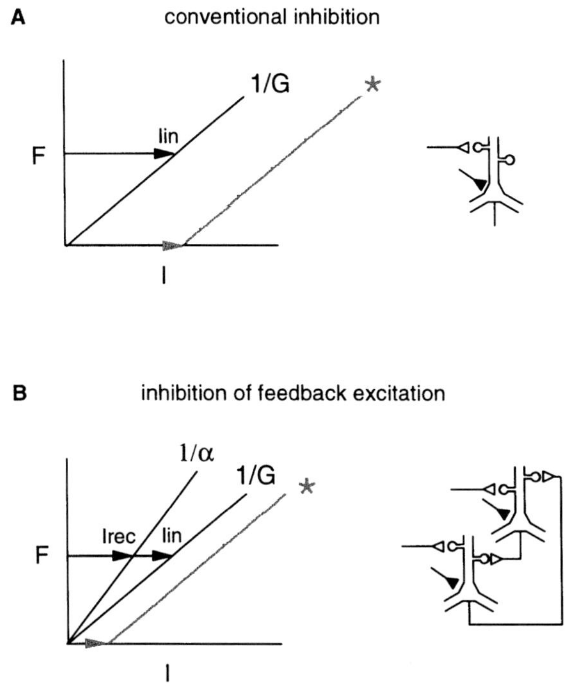

Figure 7. Neurons are more sensitive to inhibition in the presence of current gain. (A) A conventional, feedforward (inset, right), view of cortical neurons in which all of the input current I_{in} must be controlled by inhibition. In this case complete inhibition of F requires that an inhibitory current at least as large as I_{in} raise the current threshold by displacing the IF curve to the right (asterisk). (B) When the neurons are embedded in a recurrent circuit (inset, right), they cooperate to provide current amplification in each neuron. In the presence of this current gain, the same value of F is achieved with a much smaller I_{in}, because a substantial fraction of the total current is due to a recurrent current, I_{rec}. Under these conditions complete inhibition of discharge also requires that the inhibitory current be at least as large as I_{in}. But because I_{in} is smaller than in the feedforward case, so is the required inhibitory current.

can be expressed in the electronic circuit model by the addition of a recurrent circuit of inhibitory neurons, each identical to those of the recurrent excitatory neurons, but with synapses of opposite sign (Fig. 7). Since all the inhibitory neurons are identical, their synapses can be represented as a single lumped inhibitory synapse. In the electronic circuit, this lumped synapse is approximated as a current source whose amplitude is proportional to the firing rate of the point neuron, G. In place of the α of the excitatory circuit we have a second constant, β, which, like α, also acts as a network conductance, but has the opposite sign. We treat the inhibition as a linear, hyperpolarizing current, neglecting the shunting aspects of inhibitory inputs, since they have been difficult to observe.

The effective conductance of the neuron is $G - \alpha + \beta$. The magnitude of the inhibitory current is βF and the total recurrent current arriving at the soma is given by $I_{rec} = (\alpha - \beta)F$. The effective conductance G_{eff} of the neuron is now $G - \alpha + \beta$ and the output firing frequency of the neuron F is given by $I_{ff}/(G + \beta - \alpha)$. The effects of the inhibitory network on the input–output functions of the neuron can be represented graphically (Fig. 7). For an isolated neuron the current–frequency (I–F) relationship is linear with a slope $1/G$. In the presence of an inhibitory network where the inhibition applied is proportional to the output of the neuron, the total current sunk by the neurons will be the sum of the load current and the inhibitory current, $I_g + I_{inh}$. Consequently the slope of the IF curve will be $1/(G + \beta)$. Although the synapses themselves are linear, the change in slope of the IF curve during recurrent inhibition appears as a divisive process, because the strength of inhibition is proportional to the output of the neuron. Interpreted physiologically, linear feedback inhibition generates a shunting-like inhibition that changes the gain of the cortical response to a given input current.

When we apply the principles derived from the extremely abstract electronic model described above to the neuronal circuitry of the visual cortex, we can immediately appreciate that the highly selective response properties of cortical neurons compared to the relatively unselective dLGN neurons can be interpreted in terms of cortical reexcitation and its modulation by feedforward and feedback inhibition. Elsewhere, we have treated in detail (Suarez *et al.*, 1995; Maex and Orban, 1996) how these principles apply to the generation of directional selectivity in cortex (Douglas *et al.*, 1959). Here we discuss the case of orientation selectivity.

6. Computation of Orientation

Like their retinal inputs, the dLGN neurons that relay the retinal signal to the neocortex have center–surround receptive fields that are nondirectional, nonoriented, and monocular. In stark contrast, the receptive fields of spiny stellates have receptive fields called 'simple' by Hubel and Wiesel that are more elaborate than the concentric center–surround fields of the dLGN neurons. Simply receptive fields are strongly selective for the orientation of the visual stimulus. In addition they may be directional, sensitive to the size of the stimulus, and binocular. The discovery of orientation-selective cortical neurons was of major importance, for it provided the insight that the responses of single neurons could be directly related to perceptually significant events. This contrasts with the view that predominated through the 1950s that cortical computations involved mass action of neurons or electrical field-like effects. With a few exceptions, the experimental work of the time was devoted to electronencephalography or evoked potential studies, in which the responses of very large numbers of neurons were pooled. Few attempted to record from single cortical neurons and those who did concentrated on the topographic organization of the sensory maps in neocortex.

In this context, the discovery of orientation-selective simple cells was thus a startling development and signaled the beginning of the present age of prolific research on visual cortex. Thus the orientation system holds a special place in the history of cortical research. However, the mechanisms whereby the orientation-selective receptive fields of simple cells are created have been the subject of endless debate that continues to the present time (Das, 1996; Hubel, 1996; Vidyasagar *et al.*, 1996).

The first and most resilient model in terms of its longevity was proposed by the discoverers of orientation-selective simple cells. In the Hubel and Wiesel's (1962) model simple cells were created by the convergent pattern of feedforward excitatory connections from the dLGN relay neurons. The receptive fields of the dLGN neurons formed an elongated row, so that a stimulus oriented along the long axis of the row would excite the maximal possible number of dLGN cells (Fig. 8). Their activity would then be summed by the postsynaptic simple cell to form an orientation-selective ON or OFF subfield, depending on the center polarity of the LGN receptive field. The arrangement of receptive fields of the dLGN cells and the ability of the postsynaptic neuron to spatially sum the convengent EPSPs creates the machinery for the computation of orientation of a bar or grating located in the cell's receptive field. As yet, there is no direct anatomical evidence of the geometric pattern of dLGN connections to a spiny stellate neuron. But electrophysiological mapping experiments (Chapman *et al.,* 1991) and cross-correlation studies (Reid and Alonso, 1995), in which connections can be studied indirectly by correlating the occurrence of action potentials in an dLGN relay neuron with the occurrence of action potentials in a postynaptic cortical neuron, suggest that the convergent pattern is in fact that proposed by Hubel and Wiesel (1962). Nevertheless, theoretically it seems that simple cells are more selective in their orientation preferences than would be predicted on the basis of the aspect ratio of their receptive fields (Jones and Palmer, 1987; Wörgötter and Koch, 1991; Somers *et al.,* 1995). In addition, it is clear from such a feedforward system that the degree of orientation tuning will be highly dependent on the strength and, in particular, on the contrast of the LGN input and it will be very susceptible to noise. A strong stimulus presented at the nonoptimal orientation should have the same effect as a weaker stimulus presented at the optimal orientation, for example. Experimentally, orientation tuning of cortical neurons remains largely unchanged over considerable changes in the contrast of the stimulus (Sclar and Freeman 1982) and is robust in the face of potentially confounding stimuli.

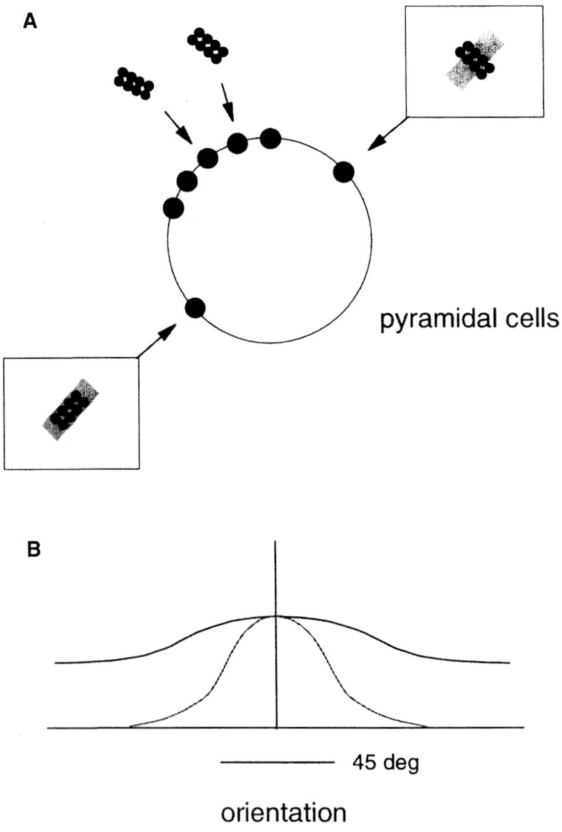

Figure 8. Simple (Hubel and Wiesel) feedforward model of orientation tuning in striate neurons. (A) Forty populations of pyramidal cells (large filled circles) each receive input from a group of LGN afferents (small filled circles), whose receptive fields are organized along a particular axis in the visual space (open rectangles) as indicated. The pyramidal cells are optimally stimulated by a bar stimulus (gray rectangle) that covers the receptive fields of, and has the same orientation as, the afferents (bottom left). If the orientation of the bar is orthogonal to that of the afferents (top right), then the cortical cells receive suboptimal stimulation and so respond less strongly. Spatially adjacent populations of neurons encode similar orientations (top left). (B) Typical orientation turning curve obtained from cortical neuron (horizontal bar) shows bandwidth of about 45 deg centered on the preferred orientation (vertical line) with no significant background activity at orthogonal orientations. This observation does not agree with the predicted orientation turning of the feedforward model (solid line), which shows a broader tuning, and substantial background activity at all orientations. The reason for this poor response is that the aspect ratio of cortical receptive fields is only about 1.6. Consequently, orthogonal bars stimulate almost as many geniculate neurons as optimally oriented bars do.

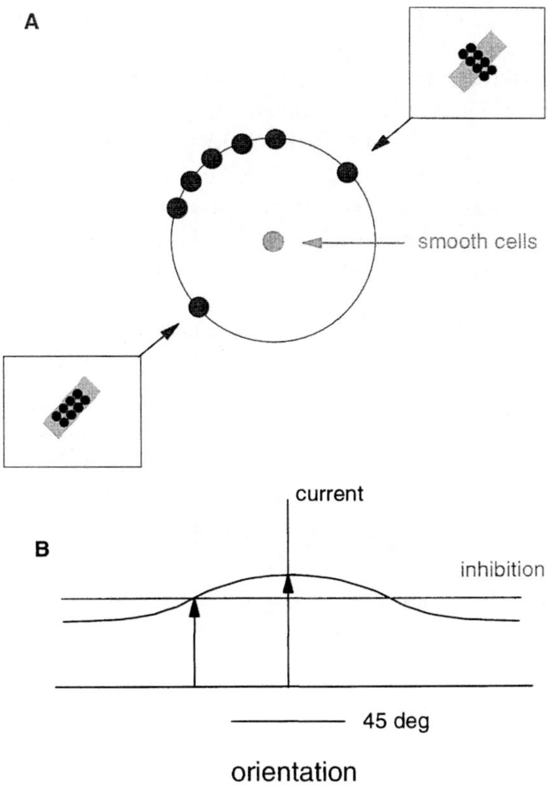

Figure 9. Inhibition improves the orientation tuning of the simple feedforward model. (A) Smooth (inhibitory) neurons receive a copy of input from all afferents to pyramidal cells, and so compute an average inhibitory current (gray straight line in panel B) that must be delivered to each pyramidal cell to limit the width of the tuning curve. This solution implies that the magnitude of the inhibitory current is much larger than the net excitatory current (large arrow versus small arrow in panel B). This prediction does not agree with our experimental findings (Douglas *et al.*, 1988) that strong inhibition is not a feature of normal selective responses in cortical cells.

The role of inhibition in determining the orientation tuning of simple cells is also a confusing issue (Fig. 9). Lengthy debates (Ferster and Koch, 1987; Das, 1996; Vidyasagar *et al.*, 1996) have ensued as to whether inhibition is involved at all; other debates have centered around exactly how inhibitory neurons might shape the receptive fields. The experimental literature has enigmas of its own. If the GABA receptors on a single neuron are blocked by intracellular manipulations, then the orientation selectivity of the treated neuron is largely unaffected (Nelson *et al.*, 1994). This result is consistent with the feedforward model. However, if a region of cortex is infused with a drug that blocks that cortical inhibition mediated by the $GABA_A$ receptors, then orientation selectivity of simple cells is reduced, even eliminated (Sillito, 1975, 1977). This is not consistent with the feedforward model. Ferster *et al.* (1996) have reported that orientation selectivity of simple cells remains unchanged when cortical synapses are inactivated by cooling.

Whatever the precise arguments for or against the particular models that have been proposed to account for orientation selectivity, it now seems that the cortical circuitry must be included in any reasonably complete account of the mechanisms (Somers *et al.*, 1995; Ben-Yishai *et al.*, 1995). Once this is accepted, we need to incorporate the additional experimental observation that the simple cells map the orientation domain in an orderly array across the two-dimensional sheet of the cortex. What the data from the cat show is that neighboring neurons have very similar orientation preferences, while more distant neurons have dissimilar orientation selectivity (Hubel and Wiesel, 1963; Albus, 1975; Bonhoeffer and Grinvald,

1991). The primary cluster of the axons of the spiny stellate neurons thus form synapses with neurons that have similar orientation selectivity. The more distant clusters of the spiny stellate axons connect to neurons that have the same orientation selectivity, i.e., isoorientation zones appear to be connected. These zones have been variously called orientation 'columns' or 'slabs' or 'pinwheels' because of their three-dimensional shape (Hubel and Wiesel, 1963; Albus, 1975; Bonhoeffer and Grinvald, 1991). In experiments where functional imaging has been applied to the cat cortex, the geometry of the orientation system is complex. From the surface of the cortex isoorientation zones appear to swirl and collide in starburst patterns, while maintaining the essential propertery that neighboring neurons have similar orientation selectivity (Bonhoeffer and Grinvald, 1993).

With these basic elements, we can now begin to link the abstractions of the electronic model of the recurrent cortical circuits with the more realistic microcircuits in order to derive a synthetic model that accounts for the generation and dynamics of orientation-selective simple cells in layer 4 of cat visual cortex. In this model, the connectivity within the spiny neurons follows the rules described above for the connections between spiny stellate neuron. The boutons in the primary cluster of the axon are distributed in a Gaussian function so that neurons that are close together connect more richly than those that are more distant. This connec-

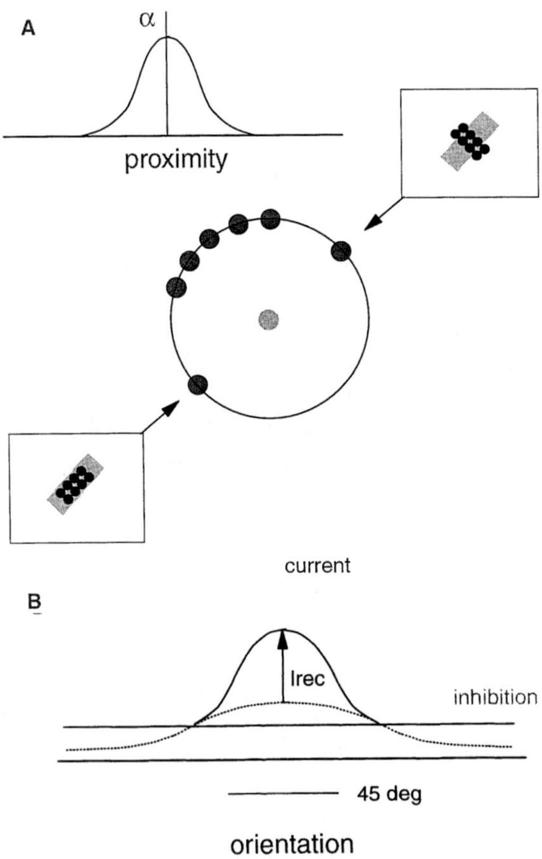

Figure 10. Recurrent (feedback) excitation improves orientation tuning. (A) the pyramidal populations are interconnected by excitatory connections. The coupling weights α_{ij} between populations are a function of spatial proximity (and consequently of similarity of preferred orientation). The recurrent connections provide current gain. This means that the afferent currents (I_{in}, broken line in panel B) are much smaller than in the feedforward case. The gain boosts the response in the preferred region by providing I_{rec}, but not at orthogonal orientations. The inhibitory neurons compute an average over the activity of all the pyramidal neurons (straight thin line in panel B), and suppress the smaller orthogonal inputs. This solution implies that the inhibitory current is smaller than the net excitatory current, and is in better agreement with our experimental findings.

tivity is represented in a circular one-dimensional array of 40 neurons that encode 180° of orientation as shown in Fig. 10. The strength of the reciprocal connections between two neurons *i* and *j* is given by α_{ij}, here assumed to be a Gaussian function of the angle between the coupled neurons *i* and *j*. The connectivity of the inhibitory neurons is different. Unlike the spiny neurons, the lateral clusters of the axons of inhibitory neurons do not connect exclusively to isoorientation zones. Instead they appear connect both to isoorientation zones and to different orientation zones. In our model, the inhibitory neurons are connected in a slightly simpler manner. They all have an identical connectivity: each receives an identical excitatory synaptic input from all the spiny neurons and each gives an identical inhibitory synaptic output to all the spiny neurons. Thus, all the neurons that form the recurrent inhibitory loop can be lumped into a single 'inhibitory neuron.'

In the orientation model, dLGN synapses again provide all the feedforward current, I_{ff}. They converge on the spiny stellates in layer 4. The pattern of the connections between the geniculocortical axons and the spiny stellates was generated by drawing a random sample from an oriented binormal spatial distribution, which represented a portion of the two-dimensional visual space. The aspect ratio of the binormal distribution was 1.6, as found experimentally by Jones and Palmer (1987). The principal axis of the binormal distribution was rotated through 180 deg and samples were taken at 40 different orientations of the distribution. These samples then provided the dLGN input to a single subfield of the simple cell. Thus each successive spiny stellate receives input from a row of dLGN relay cells whose average scatter of alignment is along a slightly different principal axis to that of its neighbor. This is illustrated in Fig. 11.

The recurrent current I_{rec} was simply expressed in the electronic model as αF, because all neurons had identical connectivity. The situation in the orientation model is more difficult, because not only do the neurons not connect to each other equally, some do not connect directly at all, and some of the neurons will be below threshold nonlinearity and will not contribute current to the network. However, the thinking in the previous, simple case provides intuition for the operation of the orientation circuit. I_{rec} is not simply αF, but the sum over the $\alpha_{ij}F_j$. The equation that the *i*th neuron in the network obeys in steady state is

$$F_i = \max\left(0, \frac{1}{G}(I_{in_i} + \sum_j \alpha_{ij} F_j - \frac{\beta}{N}\sum_{j=1}^{N} F_j)\right) \tag{1}$$

where N is the total number of neurons in the network.

Obviously, the maximum amount of current that will be delivered to the *i*th neuron will occur when all the neurons connected to *i* are themselves firing at their maximum rate. As described above, if *i* can dissipate, through its many membrane conductances, the net exitatory current it receives from the recurrent excitatory loop and the input, then the response of the neuron will be bounded without the constraint of saturation. The open-loop gain will be less than 1. However, the gain expressed in this circuit is never constant. For a given input, a given neuron will experience an amplification or gain of that input that will change as neurons connected to it pass through the threshold. Some neurons will begin to discharge,

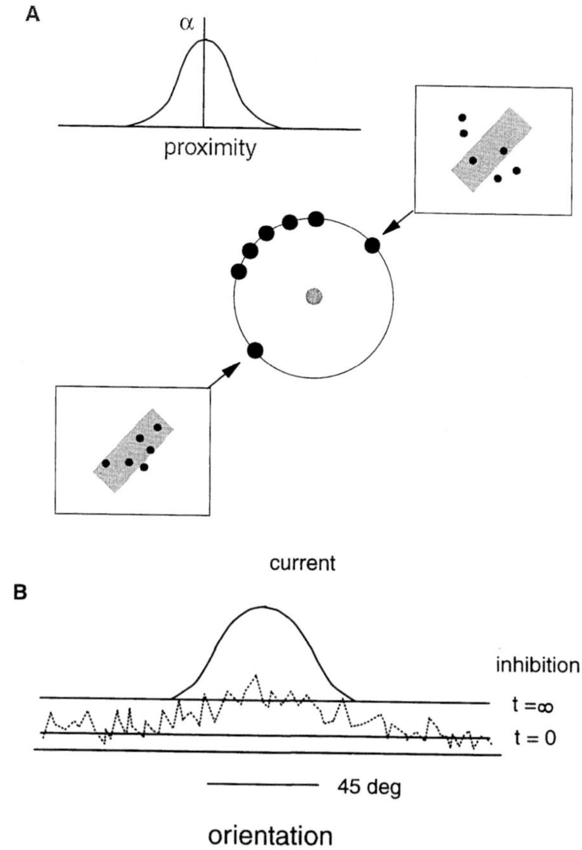

Figure 11. Orientation tuning in the presence of noisy signal and/or noisy connections. (A) Geniculocortical connections have imprecise mapping onto cortical cells (Chapman *et al.*, 1991). Other connections as for Fig. 10. Now feedforward input to cortical cells is very noisy (broken line in panel B). Initial average inhibition is insufficient to suppress noisy peaks at off orientations. Nevertheless, there is better correlation between populations near the preferred orientation, and the gain begins to increase there. The increased activity of preferred populations increases the average inhibition and suppresses more of the outliers. Elimination of the outliers improves the correlations among the survivors and enhances their gain. Finally, this process converges on a narrow-bandwidth tuning curve (solid line in panel B), indicating that this model is able to extract signal from noise. It does this by expressing the turning curve inherent in the cortical connectivity.

some will increase their existing discharge, others will decrease their discharge, and yet others will become inactive. The gain will be high when all the neurons are active, and will be low when all the neurons are below threshold. In this case, the output of the *i*th neuron is determined only by the feedforward input.

The inhibitory neurons act as a summing device and provide inhibition to all 40 spiny neurons in proportion to the total activity in the circuit. In the example shown in Fig. 11, virtually all 40 spiny neurons are excited initially, because of the feedforward dLGN cells which will respond to a stimulus of any orientation. As the total activity increases, neurons that were just above threshold will receive sufficient inhibition to be silenced and they will no longer contribute to the total excitation. However, the pattern of connectivity ensures that they will continue to receive an inhibitory input, so their threshold for activation will be raised as the total activity in the circuit increases through increasing activity in the recurrent pathway. The inhibited neurons will be those for which the presented stimulus is least optimal. The neurons that respond most strongly to the stimulus will reexcite each other more strongly and amplify the initially weak dLGN signal. As a result, the inhibition becomes more tuned to the optimal orientation, because the inhibitory neurons are driven by neurons that remain above the threshold. In addition, the inhibitory neurons are driven harder by the remaining spiny neurons and so increase the firing threshold. If we consider the *current threshold* as the amount of current required to produce a minimal discharge action potential, then it is obvious that the current threshold will vary according to the amount of recurrent inhibition at any moment.

At convergence, the orientation tuning of the inhibitory neuron is that of the net turning of the surviving active members of the excitatory group. The amount of inhibition experienced by a neuron is obviously proportional to the number of active neurons in the network. In this configuration, the role of inhibition changes over time. Initially, it is used as a threshold to extract the best estimate of the single arising from the visual input. Later, inhibition assists in stabilizing the recurrent excitation of the active population: as the excitation grows, so the inhibition grows proportionately and acts as a divisor. This proportionality is important in helping to maintain a balance between inhibition and excitation. Obviously, when the neuron produces spikes, the sum of the inhibitory and excitatory currents does not cancel out; excitation exceeds inhibition. Similarly, when the membrane potential of the neuron does reach spike threshold in the face of excitatory activity in the dLGN afferents, the cortical inhibition exceeds the excitation. However, too much inhibition would make the response to the optimal stimulus weak, while too little inhibition may broaden the tuning curve. Because of the gain provided by the recurrent excitation, small swings in the balance between excitation and inhibition produce large effects.

It may be very difficult to inhibit a neuron completely in a circuit where the excitation is provided solely by a feedforward circuit. This is because complete inhibition of the neuron's discharge requires that the inhibitory current, whether provided by a feedforward inhibition or by a feedback process, be almost as large as the excitatory current, which is entirely supplied by the one source. By contrast, in the case of the recurrent circuit there are two quite separable sources of current, one of which (I_{rec}) is dependent on and proportional to the other (I_{ff}). Inhibition of the feedforward current will therefore be extremely effective, particularly in a situation where the closed-loop gain is high, because small changes in the input current will be massively amplified by the recurrent current. Thus, the changes in the gain itself will have significant consequences for the relative effectiveness of cortical inhibition. Obviously, inhibition is most effective when it is applied simultaneously with the arrival of the excitatory feedforward current. Simultaneity in this case would require the inhibitory neurons to mediate a feedforward component of inhibition, which we have demonstrated theoretically to be effective in the example of direction selectivity (Douglas and Martin, 1991; Suarez *et al.*, 1995) and contrast adaptation (Ahmed *et al.*, 1997a). This feedforward inhibition can be mediated either through direct excitation of GABAergic inhibitory neurons by the dLGN afferents or by other cortical excitatory neurons. There is good anatomical evidence for both these pathways (Freund *et al.*, 1985a). However, the recurrent inhibitory circuit alone is clearly effective in providing the inhibition necessary.

7. Noise and Restoration

This brings us back to the point at which we began: how to design a machine that provides for robust and noise-resistant computations. This dynamic network illustrates the computational significance of recurrent connections, which we believe to be a canonical, that is, a 'most basic' feature of neocortical circuits. We have chosen to consider the computation of orientation selectivity because of its preeminent position in the history of visual physiology and because it provides an archetypical example of cortical processing. Previous models have achieved the computation of orientation either by setting up precise patterns of dLGN input, as in the model of Hubel and Wiesel (1962), or have relied on spatially specific patterns of strong inhibition [see Martin (1988) for a critical review of these models]. Both strategies do work, but they are computationally fragile and susceptible to noise. The recurrent microcircuit described here gives a more robust computation of the variable, orientation, by employing the strategy of amplifying only the portion of the incoming dLGN activity that best estimates the orientation of the visual stimulus. This 'best estimate' derives from the basis provided by the dLGN input, in the manner originally suggested by Hubel and Wiesel (1962), but avoids the problems of noise and ambiguity that are inherent in their purely feedforward system. The recurrent circuit also permits the excitation to be controlled by small levels of inhibition, provided that the inhibition is provided to all members of the excitatory circuit.

The cooperative action of the neurons provides for a degree of noise resistance. But the pattern of activity that emerges is latent in the pattern of intracortical recurrent connections: nearest neighbors reexcite each other more strongly than more distant neurons. This permits the property of signal restoration, which is implemented in a simple way in the digital computers of von Neumann. However, in the digital mode, whatever the actual voltage on the transistor is, there are only two allowable interpretations of what this voltage means: 0 or 1. In the case of the analogue computations being performed by neurons, the individual elements achieve a rate of firing that is proportionately related to the stimulus. The noise resistance of the system is achieved by 'preprogramming' the cortical circuit to amplify a particular pattern of incoming activity, and by making the influence of one neuron on another very weak. This preprogramming can be specified by genetic mechanisms or can be set up by an experience-dependent learning process. To that extent, it may not be a coincidence that the recurrent network has similarities to the content-addressable neural networks with associative memories (Hopfield, 1982). However, unlike associative memories, these cortical circuits remain sensitive to analogue information from a changing stimulus.

Our hypothesis is that all cortical areas are built according to the same fundamental blueprint. The recurrent architecture described here defines one aspect of this 'canonical' circuit and we would predict that such circuits will be found in all cortical areas. Given that 95% of the synapses in the cortex arise from cortex itself, the likelihood that receptive field properties even in primary sensory cortex are shaped by the collective behavior of large numbers of cortical neurons seems not unlikely. The principle of analogue computation using positive feedback appears to be a fundamental part of cortical computation that is unexplored in traditional computing and circuit design. Understanding the behavior of cortical circuits may lead us to the realization of von Neumann's dream.

ACKNOWLEDGMENTS. The preparation of this chapter was supported by the Office of Naval Research, Swiss National Schwerpunkt Program, and EU.

8. References

Ahmed, B., Anderson, J., Douglas, R., Martin, K., and Nelson, C., 1994, Polyneuronal innervation of spiny stellate neurons in cat visual cortex, *J. Comp. Neurol.* **341**:39–49.

Ahmed, B., Allison, J., Douglas R., and Martin, K., 1997, An intracellular study of contrast dependence of neuronal activity in cat visual cortex, *Cerebral Cortex* **7**:559–570.

Ahmed, B., Anderson, J., Martin, K., and Nelson, J., 1997b, A map of the synapses onto layer 4 basket cells of the primary visual cortex of the cat, *J. Comp. Neurol.* **379**;1–13.

Albus, K., 1975, A quantitative study of the projection area of the central and the paracentral visual field in area 17 of the cat. ii. The spatial organization of the orientation domain, *Exp. Brain Res.* **24**:18–202.

Anderson, J., Douglas, R., Martin, K., Nelson, C., and Whitteridge, D., 1994, Synaptic output of physiologically identified spiny neurons in cat visual cortex, *J. Comp. Neurol.* **341**:16–24.

Anderson, P., Olavarria, J., and Van Sluyters R. C., 1988, The overall pattern of ocular dominance bands in cat visual cortex, *J. Neurosci* **8**:2183–2200.

Barlow, H. B., 1972, Single units and sensation: A neuron doctrine for perceptual psychology?, *Perception* **1**:371–394.

Beaulieu, C., and Colonnier, M., 1985, A laminar analysis of the number of round-asymmetrical and flat-symmetrical synapses on spines, dendritic trunks, and cell bodies in area 17 of the cat, *J. Comp. Neurol.* **231**:180–189.

Ben-Yishai, R., Lev Bar-Or, R., and Sompolinsky, H., 1995, Theory of orientation tuning in visual cortex, *Proc. Natl. Acad. Sci. USA* **92**:3844–3848.

Berman, N., Douglas, R., and Martin, K., 1992, GABA-mediated inhibition in the neural networks of visual cortex, *Prog. Brain Res.* **90**:443–476.

Bonhoeffer, T., and Grinvald, A., 1991, Iso-orientation domains in cat visual cortex are arranged in pinwheel-like patterns, *Nature* **353**:429–431.

Bonhoeffer, T., and Grinvald, A., 1993, The layout of iso-orientation domains in area 18 of cat visual cortex: Optical imaging reveals a pinwheel-like organization, *J. Neurosci* **13**:4157–4180.

Braitenberg, V., and Schüz, A., 1991, *Anatomy of the Cortex: Statistics and Geometry*, Springer-Verlag, New York.

Chapman, B., Zahs, K., and Stryker, M., 1991, Relation of cortical cell orientation selectivity to alignment of receptive fields of geniculocortical afferents that arborise within a single orientation column in ferret visual cortex, *J. Neurosci* **11**:1347–1358.

Connors, B., Gutnick, M., and Prince, D., 1982, Electrophysiological properties of neocortical neurons *in vitro*, *J. Neurophysiol.* **48**:1302–1320.

Creutzfeldt, O., Kuhnt, U., and Benevento, L., 1974, An intracellular analysis of visual cortical neurones to moving stimuli: Responses in a co-operative neuronal network, *Exp. Brain Res.* **21**:251–274.

Das, A., 1996, Orientation in visual cortex: A simple mechanism emerges, *Neuron* **16**:477–480.

Douglas, R., and Martin, K., 1991, A functional microcircuit for cat visual cortex, *J. Physiol.* (Lond.) **440**:735–769.

Douglas, R., Martin, K., and Whitteridge, D., 1988, Selective responses of visual cortical cells do not depend on shunting inhibition, *Nature* **332**:642–644.

Douglas, R., Martin, K., and Witteridge, D., 1989, A canonical microcircuit for neocortex, *Neural Computation* **1**:480–488.

Douglas, R., Martin, K., and Whitteridge, D., 1991, An intracellular analysis of the visual responses of neurones in cat visual cortex, *J. Physiol.* (Lond.) **440**:659–696.

Douglas, R., Koch, C., Mahowald, M., Martin, K., and Suarez, H., 1995, Recurrent excitation in neocortical circuits, *Science* **269**:981–985.

Douglas, R., Mahowald, M., Martin, K., and Stratford, K., 1996, The role of synapses in cortical computation, *J. Neurocytol.* **25**:893–911.

Dreher, B., 1972, Hypercomplex cells in the cat's striate cortex, *Invest. Opthalmol. Vis. Sci.* **11**:355–356.

Einstein, G., and Davis, T. P. S., 1987, Ultrastructure of synapses from the a-laminae of the lateral genicualte nucleus in layer iv of the cat striate cortex, *J. Comp. Neurol.* **260**:63–75.

Felleman, D., and Van Essen, D., 1991, Distributed hierarchical processing in the primate cerebral cortex, *Cerebral Cortex* **1**:123–200.

Ferster, D., 1986, Orientation selectivity of synaptic potentials in neurons of cat primary visual cortex, *J. Neurosci.* **6**:1284–1301.

Ferster, D., 1987, Origin of orientation-selective EPSPs in simple cells of cat visual cortex, *J. Neurosci.* **7**:1780–1791.

Ferster, D., and Jagadeesh, B., 1992, EPSP–IPSP interactions in cat visual cortex studied with *in vivo* whole-cell patch recording, *J. Neurosci.* **12**:1262–1274.

Ferster, D., and Koch, C., 1987, Neuronal connections underlying orientation selectivity in cat visual cortex, *TINS* **10**(12):487–492.

Ferster, D., and Lindstrom, S., 1985, Synaptic excitation of neurones in area 17 of the cat by intracortical axon collaterals of cortico-geniculate cells, *J. Physiol.* (Lond.) **367**:233–252.

Ferster, D., Chung, S., and Wheat, H., 1996, Orientation selectivity of thalamic input to simple cells of cat visual cortex, *Nature* **380**:249–252.

Fisken, R., Garey, L., and Powell, T., 1975, The intrinsic and commissural connections of area 17 of the visual cortex, *Proc. R. Soc. Lond. B* **272**:487–536.

Freund, T., Martin, K., Somogyi, P., and Whitteridge, D., 1985a, Innervation of cat visual areas 17 and 18 by physiologically identified X- and Y-type thalamic afferents. II. Identification of postsynaptic targets by GABA immunocytochemistry and Golgi impregnation, *J. Comp. Neurol.* **242**:275–291.

Freund, T., Martin, K., and Whitteridge, D., 1985b, Innervation of cat visual areas 17 and 18 by physiologically identified X- and Y-type thalamic afferents. I. Arborization patterns and quantitative distribution of postsynaptic elements, *J. Comp. Neurol.* **242**:263–274.

Gabbott, P., Martin, K., and Whitteridge, D., 1988, Evidence for the connections between a clutch cell and a corticotectal neuron in area 17 of the cat visual cortex, *Proc. R. Soc. Lond. B* **233**:385–391.

Garey, L., and Powell, T., 1971, An experimental study of the termination of the lateral geniculo-cortical pathway in the cat and monkey, *Proc. R. Soc. Lond. B* **179**:21–40.

Gilbert, C., 1977, Laminar differences in receptive field properties of cells in cat primary visual cortex, *J. Physiol.* (Lond.) **268**:391–421.

Gilbert, C., and Wiesel, T., 1979, Morphology and intracortical projections of functionally characterised neurons in the cat visual cortex, *Nature* **280**:120–125.

Heggelund, P., 1981, Receptive field organization of complex cells in cat striate cortex, *Exp. Brain Res.* **42**:99–107.

Hopfield, J., 1982, Neural networks and physical systems with emergent collective computational abilities, *Proc. Nat. Acad. Sci. USA* **79**:2554–2558.

Hubel, D., 1996, A big step along the visual pathway, *Nature* **380**:197–198.

Hubel, D., and Wiesel, T., 1962, Receptive fields, binocular interaction and functional architecture in the cat's visual cortex, *J. Physiol.* **160**:106–154.

Hubel, D. H., and Wiesel, T. N., 1963, Shape and arrangement of columns in cat's striate cortex, *J. Physiol.* (Lond.) **165**:559–568.

Hubel, D. H., and Wiesel, T. N., 1965, Receptive fields and functional architecture in two non-striate visual areas (18, 19) of the cat, *J. Neurophysiol.* **28**:229–289.

Hubel, D., and Wiesel, T., 1977, The functional architecture of the macaque visual cortex. The Ferrier Lecture, *Proc. R. Soc. Lond. B* **198**:1–59.

Humphrey, A., Sur, M., Ulrich, D., and Sherman, S., 1985, Projection pattern of individual x- and y-cell axons from the lateral geniculate nucleus to cortical area 17 in the cat, *J. Comp. Neurol.* **233**:159–189.

Jagadeesh, B., Wheat, H., and Ferster, D., 1993, Linearity of summation of synaptic potentials underlying direction selectivity in simple cells of the cat visual cortex, *Science* **262**:1901–1904.

Jones, J., and Palmer, L., 1987, The two-dimensional spatial structure of simple receptive fields in cat striate cortex, *J. Neurophysiol.* **58**:1187–1211.

Kisvárday, Z., Martin, K., Whitteridge, D., and Somogyi, P., 1985, Synaptic connections of intracellularly filled clutch neurons, a type of small basket neuron in the visual cortex of the cat, *J. Comp. Neurol.* **241**:111–137.

Koch, C., and Poggio, T., 1987, Biophysics of computation: Neurons, synapses, and membranes, in: *Synaptic Function* (G. M. Edelman, W. E. Gall, and W. M. Cowan, eds.), Wiley, New York, pp. 637–697.

Koch, C., Poggio, T., and Torre, V., 1982, Non-linear interaction in a dendritic tree: Localization, timing, and role of information processing, *Proc. Natl. Acad. Sci. USA* **80**:2799–2802.

LeVay, S., 1973, Synaptic patterns in the visual cortex of the cat and monkey. Electron microscopy of Golgi preparations, *J. Comp. Neurol.* **150**:53–86.

LeVay, S., 1986, Synaptic organization of claustral and geniculate afferents to the visual cortex of the cat, *J. Neurosci.* **6**:3564–3575.

LeVay, S., and Ferster, D., 1979, Proportions of interneurons in the cat's lateral geniculate nucleus, *Brain Res.* **164**:304–308.

LeVay, S., and Gilbert, C., 1976, Laminar patterns of geniculocortical projection in the cat, *Brain Res.* **113**:1–19.

Livingstone, M., and Hubel, D., 1988, Segregation of form, color, movement and depth: Anatomy, physiology and perception, *Science* **240:**740–749.

Lorente de Nó, R., 1949, Cerebral cortex: Architecture, intracortical connections, motor projections, in *Physiology of the Nervous System* (J. Fulton, ed.), Oxford University Press, Oxford, pp. 288–315.

Lund, J., 1973, Organization of neurons in the visual cortex, area 17, or the monkey (*Macaca mulatta*), *J. Comp. Neurol.* **147:**455–496.

Madarasz, M., Gerle, J., Hajdu, F., Somogyi, G., and Tombol, T., 1978, Quatitative histological studies on the lateral geniculate nucleus in the cat. II. Cell numbers and densities in the several layers, *J. Hirnforsch.* **19:**159–164.

Maex, R., and Orban, G., 1996, A model circuit of spiking neurons generating directional selectivity in simple cells, *J. Neurophysiol.* **75:**1515–1545.

Markram, H., Lübke, J., Frotscher, M., Roth, A., and Sakmann, B., 1997, Physiology and anatomy of synaptic connections between thick tufted paramidal neurons in the developing rat neocortex, *J. Physiol.* (Lond.) **500:**409–440.

Martin, K., 1988, The Wellcome Prize Lecture: From single cells to simple circuits in the cerebral cortex, *Q. J. Exp. Physiol.* **73:**637–702.

Martin, K., and Whitteridge, D., 1984, Form, function and intracortical projection of spiny neurones in the striate visual cortex of the cat, *J. Physiol.* (Lond.) **353:**463–504.

Mason, A., Nicoll, A., and Stratford, K., 1991, Synaptic transmission between individual pyramidal neurons of the rat visual cortex *in vitro*, *J. Neurosci.* **11:**72–84.

McGuire, B., Hornung, J. P., Gilbert, C., and Wiesel, T., 1984, Patterns of synaptic input to layer 4 of cat striate cortex, *J. Neurosci.* **4:**3021–3033.

Mead, C., 1990, Neuromorphic electronic systems, *Proc. IEEE* **78:**1629–1636.

Mel, B., 1993, Synaptic integration in excitable dendritic trees, *J. Neurophysiol.* **70:**1086–1101.

Mel, B., 1994, Information processing in dendritic trees, *Neural Computation* **6:**1031–1085.

Mitchison, G., 1992, Axonal trees and cortical architecture, *TINS* **15:**122–126.

Movshon, J., Thompson, I., and Tolhurst, D., 1978, Spatial summation in the receptive fields of simple cells in the cat's striate cortex, *J. Physiol.* (Lond.) **283:**53–77.

Naegele, J., and Katz, L., 1990, Cell surface molecules containing *N*-acetylgalactosamine are associated with basket cells and neurogliaform cells in cat visual cortex, *J. Neurosci.* **10:**540–557.

Nelson, S., Toth, L., Sheth, B., and Sur, M., 1994, Orientation selectivity of cortical neurons during intracellular blockade of inhibition, *Science* **265:**774–777.

O'Leary, J., 1941, Structure of the area striata of the cat, *J. Comp. Neurol.* **75:**131–164.

Peters, A., Payne, B., 1993, Numerical relationships between geniculocortical afferents and pyramidal cell modules in cat primary visual cortex, *Cerebral Cortex* **3:**69–78.

Reid, C., and Alonso, J. M., 1995, Specificity of monosynaptic connections form thalamus to visual cortex, *Nature* **378:**281–284.

Sclar, G., and Freeman, R., 1982, Orientation selectivity in the cat's striate cortex is invariant with stimulus contrast, *Exp. Brain Res.* **46:**457–461.

Shepherd, G., and Brayton, R., 1987, Logic operations are properties of computer-simulated interactions between excitable dendritic spines, *Neuroscience* **21:**151–166.

Sillito, A., 1975, The effectiveness of bicuculline as an antagonist of GABA and visually evoked inhibition in the cat's striate cortex, *J. Physiol.* (Lond.) **250:**287–304.

Sillito, A., 1977, Inhibitory processes underlying direction specificity of simple, complex, and hypercomplex cells in cat's striate cortex, *J. Physiol.* (Lond.) **271:**699–720.

Somers, D., Nelson, S., and Sur, M., 1995, An emergent model of orientation selectivity in cat visual cortical simple cells, *J. Neurosci.* **15:**5448–5465.

Stratford, K., Tarczy-Hornoch, K., Martin, K., Bannister, N., and Jack, J., 1996, Excitatory synaptic inputs to spiny stellate cells in cat visual cortex, *Nature* **382:**258–261.

Stuart, G., and Sakmann, B., 1994, Active propogation of somatic action potentials into neocortical pyramidal cell dendrites, *Nature* **367:**69–72.

Suarez, H., Koch, C., and Douglas, R., 1995, Modeling direction selectivity of simple cells in striate visual cortex using the canonical microcircuit, *J. Neurosci* **15:**6700–6719.

Tovee, M., Rolls, E., and Treves, A. R. B., 1993, Information encoding and the responses of single neurons in the primate temporal visual cortex, *J. Neurophysiol.* **70:**640–654.

Toyama, K., Kimura, M., and Tanaka, K., 1981, Cross-correlation analysis of interneuronal connectivity in cat visual cortex, *J. Neurophysiol.* **46:**191–201.

Vidyasagar, T., and Heide, W., 1984, Geniculate orientation biases seen with moving sine wave gratings: Implications for a model of simple cell afferent connectivity, *Exp. Brain Res.* **57**:196–200.

Vidyasagar, T., Pei, X., and Volgushev, M., 1996, Multiple mechanisms underlying the orientation selectivity of visual cortical neurones, *TINS* **19**:272–2777.

von Neumann, J., 1958, *The Computer and the Brain,* Yale University Press, New Haven, CT.

White, E., 1989, *Cortical Circuits* Birkhäuser, Boston, MA.

Wörgötter, F., and Koch, C., 1991, A detailed model of the primary visual pathway in the cat: Comparison of afferent excitatory and intracortical inhibitory connection schemes for orientation selectivity, *J. Neurosci.* **11**:1959–1979.

6

Neural Mechanisms Underlying the Analysis of Moving Visual Stimuli

PHILIP S. ULINSKI

1. Introduction

Animals analyze objects moving in the visual world in order to follow the movements of predators and prey (Ewert, 1991), control the orientation of their bodies as they move through their environments (Lee, 1980; Roy and Wurtz, 1990), calculate the range of distant objects in the world (Srinivasan, 1992), help determine the three-dimensional shapes of objects (Wallach and O'Connell, 1953; Johansson, 1971), and decipher the relations of objects to their backgrounds (Braddick, 1993). Because of the biological importance of visual motion analysis (Nakayama, 1985), the mechanisms used to detect and analyze moving stimuli have been studied in a wide range of species including arthropods (Borst and Egelhaaf, 1989; Franceschini *et al.*, 1989; Egelhaaf and Borst, 1987, 1993a, b, 1994), rabbits (Oyster, 1968), turtles (Ulinski *et al.*, 1991; Granda and Sisson, 1992), cats (Orban, 1984), and primates (Orban, 1994). One goal for such work is to understand how animals are able to implement the computations that underlie motion analysis using the "wetware" available in their nervous systems. In the case of vertebrates, this involves under-

PHILIP S. ULINSKI • Department of Organismal Biology and Anatomy, University of Chicago, Chicago, Illinois 60637.

standing the role of neural circuitry embedded both in the retina and in components of the central nervous system. Consistent with the theme of the volume, this chapter focuses on understanding how the biophysical, synaptic, and connectional properties of neurons in visual cortex are used to effect the computations that allow animals to extract a rich picture of the external world from patterns of light moving past their eyes.

Modern approaches to this question derive largely from three sets of studies carried out in the early 1960s. Werner Reichardt and colleagues used visual behavior in flies to study motion analysis and selectivity (Reichardt, 1961; Poggio and Reichardt, 1976; Reichardt and Poggio, 1976). Neurons are *motion selective* if they respond preferentially to stimuli moving within a particular range of directions and speeds. Reichardt's model for motion selectivity involves sampling visual information from two spatial channels (Fig. 1A). The output from one channel is delayed before it is combined with the output from the second channel. The delay can be an actual temporal delay or a low-pass filtering of the input to the first channel. The interaction is described mathematically by multiplying the output from one spatial channel by the time-delayed output from the second channel and then integrating the product over time. This model is often referred to as a *Reichardt detector* and described as being *bilocal* (Fredericksen *et al.*, 1993). The two spatial channels are separated by a distance referred to as the *sampling base* and involve a particular

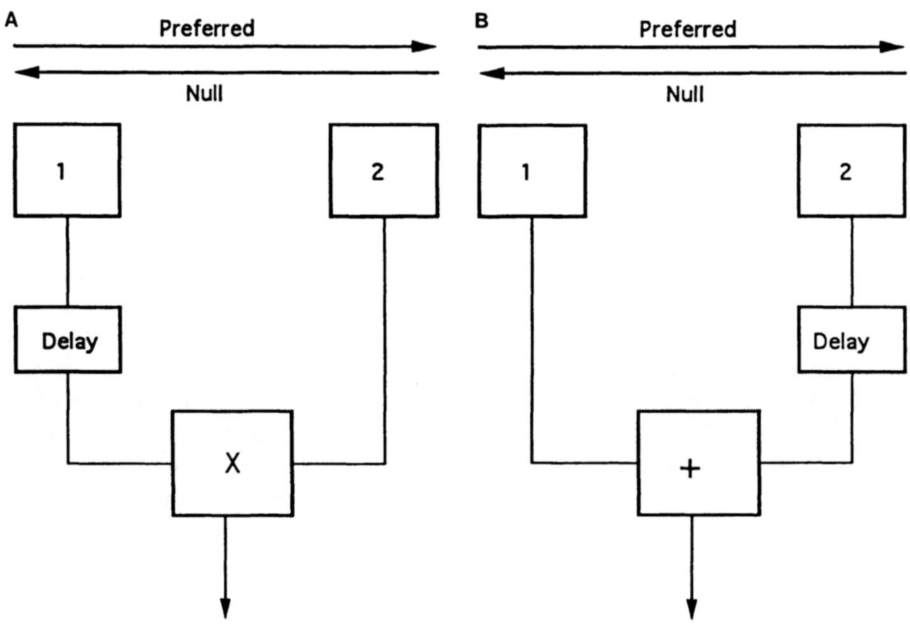

Figure 1. Motion detectors. Models of visual motion detection and analysis typically involve combined inputs from two spatial channels (boxes 1 and 2). (A) In the model proposed by Reichardt, output from the first spatial channel is delayed or low-pass filtered and then combined in a nonlinear or multiplicative step with the input from the second spatial channel. The output from the detector is larger for a stimulus moving in the preferred direction (left to right) than from a stimulus moving in the nonpreferred or null direction. (B) In a class of models proposed by Barlow and Levick, output from the spatial channel is delayed and then combined in a linear or additive step with the input from the first channel. Barlow–Levick models can involve either excitatory or inhibitory interactions.

temporal delay. The detector thus samples a local region of space and a local region of time. Reichardt's approach was to model motion-selective neurons as a mathematical filter that makes no specific statements about the cellular or synaptic mechanisms involved.

Barlow and Levick (1965), by contrast, used the retinal ganglion cells of rabbits to investigate the synaptic mechanisms that allow neurons to respond selectively to moving stimuli. They recorded the firing patterns of ganglion cells to pairs of stimuli presented with specific spatial and temporal separations. One stimulus in the pair was a conditioning stimulus, while the other was a test stimulus presented on either side of the conditioning stimulus. Their experiments led to a model for direction selectivity that relies upon a mechanism for temporal delay between the input from spatially separated groups of photoreceptors followed by a summation of excitatory and inhibitory synaptic inputs by the ganglion cell (Fig. 1B). Postsynaptic inhibition allows a ganglion cell to respond to a stimulus moving in one direction (the *preferred direction*), but to remain silent when a stimulus moves past the same cell from the opposite (*nonpreferred* or *null*) direction.

Hubel and Wiesel pioneered the analysis of visual motion analysis at the cortical level. They demonstrated that neurons in the primary visual cortex of cats (Hubel and Wiesel, 1962) and monkeys (Hubel and Wiesel, 1968) respond preferentially to bar-shaped stimuli moving in specific orientations and directions. They proposed a model for the generation of the orientation-specific responses of cortical neurons based upon the summation of excitatory inputs from rows of geniculate afferents; and the generation of direction-selective responses by asymmetric patterns of inhibitory connections within the cortex.

A complete understanding of how neuronal interactions are used by visual systems to analyze moving objects has continued beyond these three early studies and this remains an active area of inquiry. It is achieving greater significance as we learn more about the sequence of steps involved in visual motion analysis and has been subjected to analysis by a variety of mathematical techniques based on systems-engineering approaches. At the same time, rapid increases in our understanding of the voltage-gated conductances, neurotransmitters, neuromodulators, and second messenger systems implicated in the integrative physiology of neurons provides the opportunity to obtain a detailed understanding of how neural circuits actually implement the computations involved in visual motion analysis.

These parallel developments pose a technical challenge (Douglas *et al.*, 1991). The paradigms generated by Reichardt, by Barlow and Levick, and by Hubel and Wiesel involve attempts to deduce the nature of the neural mechanisms involved in visual motion analysis from the firing patterns of cortical neurons using extracellular recording methods. However, it is now clear that a complete understanding of the integrative biology of visual motion analysis will require the kinds of information available from intracellular recording methods carried out in slice preparations of the cortex. The difficulty is that the preparation of the slice necessitates the interruption of the fiber systems that carry visual information to the cortex and mediate the interplay of activity between cortical areas, making it impossible to relate the responses of individual neurons to the underlying neural mechanisms. Two preparations are currently being used to circumvent this difficulty and study the cortical circuitry involved in motion selectivity. One involves the use of intracellular recording methods in *in vivo* cat preparations (e.g., Ferster, 1986, 1987), most recently with whole-cell patch clamp methods (Ferster and Jagadeesh, 1992; Ferster *et al.*, 1996). This approach allows a direct analysis of the synaptic responses of cortical neurons to natural visual stimuli. The other preparation is an *in vitro* preparation of the visual system of turtles that allows the morphology, biophysics, and synaptic physiology of neurons to be related directly to their responses to visual stimuli (e.g., Ulinski *et al.*, 1991; Mancilla *et al.*, 1998).

This chapter examines the impact that work on cortical circuitry is having on our general concepts of visual motion analysis by focusing on four issues, three of which were identified by Reichardt, Barlow and Levick, and Hubel and Wiesel. The first is to understand the basic structure of *spatiotemporal receptive fields.* The Barlow–Levick and Reichardt models require an interaction between channels that are separated both temporally and spatially. The essence of analyzing movement is that one snapshot of the visual scene must be stored before it can be compared to a second, later snapshot of the scene. Attempts at understanding how the nervous system stores information have been dominated recently by interest in the long-term storage involved in learning and memory, and relatively little is known about the mechanisms by which the nervous system stores information in order to analyze time-varying events in the external world (Sejnowski, 1986; Buonomano and Merzenich, 1995; Llinás, 1993; Merzenich *et al.*, 1993). Temporal storage for the tens of milliseconds needed to analyze stimuli moving over small spatial distances can be achieved by using low-pass filters or by using the delays inherent in axonal conduction times (Carr, 1993). How can neural systems store information for the hundreds of milliseconds required for many naturally occurring movements, and how is temporal information combined with spatial information to fashion spatiotemporal receptive fields capable of extracting information from moving stimuli?

A second issue concerns a particular aspect of the interaction between spatial channels. Reichardt's concept of motion analysis requires a multiplication of the output from one spatial channel by that of a second spatial channel, and has focused interest on determining the relative contributions of *linear* and *nonlinear* mechanisms to motion analysis. A mechanism is linear if its response to two stimuli presented simultaneously equals a weighted sum of the same two inputs presented separately. Otherwise, the mechanism is considered to be nonlinear. Linear mechanisms are attractive subjects for investigation because they can be approached via an array of mathematical and engineering methods. Can visual motion analysis be effected using only linear interactions between neurons, and what neural mechanisms would underlie multiplicative (or divisive) interactions between neurons?

Finally, the early studies raise the issue of discerning the relative roles of *excitatory* and *inhibitory* interactions in motion analysis. The Barlow–Levick and Hubel–Wiesel models suggest that inhibition is important in suppressing the responses of neurons to stimuli moving in nonpreferred directions, while the Reichardt model involves only excitatory interactions. The three early studies were conducted when the synaptic mechanisms involved in inhibition were just being elucidated (Eccles, 1969), but it has since become clear that most of the inhibitory interactions in cortical structures are mediated via two subtypes of receptors for the inhibitory neurotransmitter γ-aminobutyric acid (GABA). There is also a considerable diversity in inhibitory neurons and their interactions with excitatory cells (e.g., Buhl *et al.*, 1994). Is inhibition necessary for visual motion analysis, and, if so, what are the relative contributions of different inhibitory receptors and interneurons?

A fourth issue of current interest—not explicitly articulated in the early work—is that visual motion analysis occurs in a series of steps. The first step involves neurons referred to as *local motion detectors* that analyze stimuli moving with specific directions and speeds. The second step involves the combination of inputs from arrays of local motion detectors by other neurons referred to as *global motion detectors*. They are responsible for generating information about the movement of real objects across larger domains of visual space. What are the basic properties of local and global motion detectors and to what extent are similar neural mechanisms involved in local and global motion analysis?

The chapter continues in Section 2 with a primer of basic concepts involved in studying visual motion analysis. Section 3 reviews work dealing with the neural mechanisms underlying visual motion analysis in local and global motion detectors in cats and monkeys. Section 4 reviews work dealing with the neural mechanisms underlying motion-selective neurons in the visual cortex of turtles. Section 5 draws together the information currently available on the neural mechanisms underlying cortical motion analysis in cats, monkeys, and turtles and considers its relationship to the issues raised above.

2. A Primer of Basic Concepts

Complete discussions of the fundamentals of visual motion analysis can be found in earlier books and review articles (Marr, 1982; Koch *et al.*, 1986; Ullman, 1984; Nakayama, 1985; Hildreth and Koch, 1987; Grzywacz and Koch, 1987; Grzywacz *et al.*, 1990; Sekuler *et al.*, 1990; Sereno, 1993; Regan, 1991a; Bock and Goode, 1994; Kosslyn and Andersen, 1995; Wandel, 1995). Models of visual motion analysis are discussed in several recent articles, reviews, and books (Borst and Egelhaaf, 1989; Grzywacz *et al.*, 1990; Jin and Srinivasan, 1990; Mizunami, 1990; Zanker, 1990; Landy and Movshon, 1991; Pinter and Nabet, 1992). This section reviews basic ideas related to visual motion analysis, beginning with a discussion of the kinds of visual stimuli commonly employed in investigations of visual motion analysis.

2.1. Apparent Motion, Sinusoids, and Random-Dot Kinetograms

Objects in the visual world move continuously along trajectories that cover tens or hundreds of degrees in visual space for durations of hundreds of milliseconds. As they move, they produce a time-varying image on the photoreceptor arrays of the two eyes. This image can be represented by a light intensity function $I(x,y,t)$ that specifies the intensity or luminance of light at a position (x,y) in the visual field at time t (Fig. 2). Light intensity is usually expressed in terms of numbers of photons or energy units upon a given area per unit time (Rodieck, 1973; Davson, 1980; Wandel, 1995). The intensity function is continuous, but the information about the moving object that is available to the animal is sampled from discrete loci in visual space whose sizes and positions are determined by the photoreceptors. The photoreceptors thus define a matrix of *spatial channels* and the task in studying visual

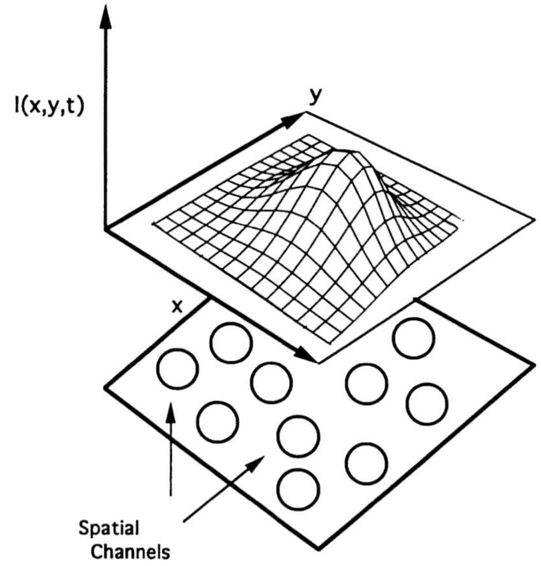

Figure 2. Motion analysis. The upper plot shows the input to the visual system, which is a continuous light intensity function $I(x,y,t)$ that varies as a function of $\{x,y\}$ position in visual space and with time t. The retina contains a discrete matrix of sampling elements defined by the photoreceptor array that samples loci in the intensity function. These elements form spatial channels that are the initial inputs to motion detectors (see Fig. 12).

motion analysis is to understand how animals can extract information about the properties of the moving object from variations in the light intensity function sampled by the different spatial channels at different time intervals. Characterizing the responses of the animal or of neurons in the visual system to naturalistic stimuli such as real moving objects has limited analytic utility because it is difficult to be certain which aspects of the stimulus (e.g., size, shape, direction of movement, speed of movement, etc.) are modulating the response. It is preferable to abstract from the stimulus particular aspects of movement that may be used by the visual system to effect specific computations. Three major types of stimuli are used to analyze the properties of motion analysis systems.

Apparent motion paradigms (Anstis, 1980; Braddick, 1980) utilize stimuli presented at disjunct points in visual space with relatively short temporal delays. The stimuli have classically been small spots or bars. They appear to the observer as continuously moving if they are presented with a frequency above the flicker fusion frequency of the animal and if their spatial displacement is less than some maximal value, D_{max}. Sequences of several stimuli evoke stronger perceptions of motion than does a single pair of stimuli. Apparent motion takes place on two different spatial and temporal scales (Braddick, 1974, 1980). *Short-range processes* are responsible for analyzing inputs from spatial channels separated by up to 15 min of visual arc, while *long-range processes* are responsible for analyzing inputs from spatial channels separated for many degrees of visual arc. Short-range processes can utilize stimuli delivered to one eye, while long-range processes can utilize binocular stimuli.

Sinusoidal contrast gratings are a popular stimulus for analyzing the linearity or nonlinearity of the neural mechanisms involved in visual motion analysis. These stimuli appear as a field of dark bars on a lighter background, or vice versa, presented on a video monitor. The bars do not have sharp edges and are described instead by a light intensity function

$$L(x) = L_0[\sin(2\pi kx)] \qquad (1)$$

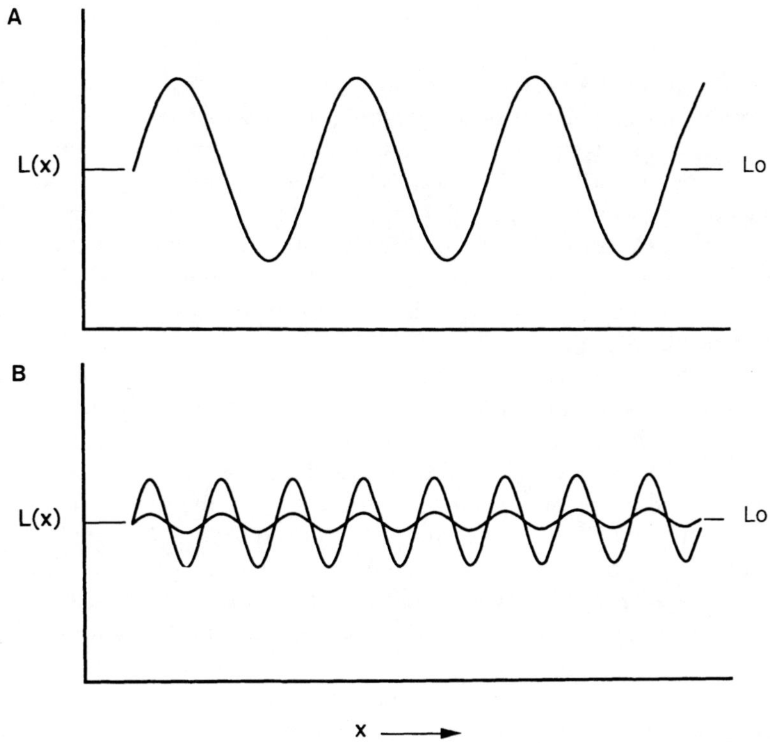

Figure 3. Sinusoidal stimuli. (A) A sinusoidal stimulus consists of a light intensity or luminance function $L(x)$ that varies sinusoidally along the x axis around a mean level of L_0. The stimulus can be fixed or drifting, in which case the sinusoid would be to the right or left with a fixed speed. (B) Contrast reversal stimuli are formed by summing two sinusoids, like the one shown in panel A, that are moving in opposite directions or by multiplying the amplitude of a fixed sinusoid by a sinusoidal function of time. This plot shows a contrast reversal stimulus at different time values.

where $L(x)$ is the luminance at point x along one axis of visual space, L_0 is the mean luminance of the background, and k is the spatial frequency of the sinusoid (Fig. 3A). The *contrast* of the sinusoid is given by

$$C = (L_{max} - L_{min})/(L_{max} + L_{min}) \qquad (2)$$

where L_{max} is the maximum luminance of the sinusoid and L_{min} is the minimum luminance. The spatial and temporal frequency of the sinusoid can be varied and the contrast pattern can be moved (or *drifted*) past the observer with some speed. A slightly more complex stimulus is a *contrast reversal grating,* which is made up of a pair of sinusoids moving in opposite directions past the observer (Fig. 3B). Here

$$L = L_0[1 + C\cos[2\pi(kx \pm ft)] \qquad (3)$$

where k is the spatial frequency of the sinusoid and f is its temporal frequency. The plus and minus signs specify the direction in which the sinusoid is drifting. Contrast reversal gratings are mathematically equivalent to a single grating with temporally modulated contrast:

$$L = L_0[1 + C\cos(2\pi kx + \phi)\cos(2\pi ft)] \tag{4}$$

where ϕ is the spatial phase of the sinusoid.

Yet another variety of sinusoidal stimulus consists of two contrast gratings oriented at some angle to each other. These stimuli appear to the observer as *plaids* on an oscilloscope or computer monitor. Since the movements of the two component gratings can be varied independently, plaids provide a useful means of studying the integration of information from two different local motion channels. The general value of sinusoidal stimuli derives from the fact that a large body of literature in systems engineering deals with the use of sinusoidal stimuli for the analysis of linear and nonlinear systems.

Much recent work uses fields of random, moving dots presented upon a video monitor. This type of stimulus was introduced by Bela Julesz for studies of binocular disparity (Julesz, 1971). *Random-dot kinetograms* used in studies of motion analysis consist of a field of dots moving in random directions. One or more dots moving in a nonrandom sequence of pixels can be introduced into the field. The percentage of dots moving in the same direction that are required for an observer to perceive a cohesive motion can be determined. Random-dot kinetograms can be used for psychophysical experiments in humans or animals and are particularly useful when they are used in monkeys for both psychophysical experiments and as stimuli for extracellular recording experiments. This combination of approaches makes it possible to relate the responses of individual cortical neurons to sensory stimuli and to the perceptual experiences of the animal. An important result from the use of random-dot kinetograms is that stimuli in which there is no overall movement of an intensity function can sometimes elicit the perception of movement. These kinds of movements are referred to as *non-Fourier* (or *second order*) movement (e.g., Zanker, 1995), in contrast to the situation in which the perception of movement depends upon the spatial and temporal frequency of an intensity function (so-called *Fourier* or *first-order* movement). Non-Fourier movement is beyond the scope of this chapter, but is considered elsewhere in this volume by Wilson and is discussed in many articles (e.g., Victor and Conte, 1992; Mather and West, 1993; Fleet and Langley, 1994; Gorea, 1995; Petersik, 1994; Sutter *et al.*, 1995).

2.2. Classical and Nonclassical Receptive Fields

Both apparent motion and continuously moving stimuli can be used to characterize the receptive fields of neurons involved in motion analysis. The classical definition of the *receptive field* for a neuron in the visual system is the area in visual space which, when stimulated, alters the response pattern of the cell (for a historical review see Enroth-Cugell, 1993). The utility of the concept is that it links (Teller, 1990) events in the visual world to neuronal events such as trains of action potentials, excitatory postsynaptic potentials (EPSPs), or inhibitory postsynaptic potentials (IPSPs). Neurons in the retina and in the central nervous system vary in their selectivity to stimuli moving through their receptive fields. Some respond well to stationary visual stimuli and do not vary their response patterns as a function of specific characteristics of a moving stimulus such as its direction or speed. Other neurons respond preferentially to moving stimuli and vary their responses as the stimulus moves with a particular direction or speed. These neurons can be thought of as *motion detectors.*

The first systematic study of the receptive field properties of motion detectors in vertebrates was the work of Lettvin *et al.* (1959), who recorded from the axons of frog retinal ganglion cells while moving objects through the frog's visual space. They found that some ganglion cells respond selectively to stimuli, with specific sizes and shapes moving within spatially restricted receptive fields. Ewert and colleagues carefully quantified the sizes, shapes, directions, and speeds of moving stimuli required to activate motion-selective ganglion cells and to elicit orienting movements by a frog or toad, thereby establishing a relationship between the behavior of the animal and the properties of neurons in its visual system (Ewert, 1991). Lettvin and colleagues also attempted to relate the receptive field properties of ganglion cells to their anatomical characteristics as seen in Golgi preparations. They raised the possibility that the physiological properties of ganglion cells are determined in part by the sizes and shapes of their dendritic fields (Maturana *et al.*, 1960). This work was extended by other workers to include characterizations of retinal ganglion cells in a range of vertebrates including pigeons, rabbits, and ground squirrels (for review see Rodieck, 1973). Retinal ganglion cells in some animals have distinct motion-selective properties, while those in other animals do not respond preferentially to moving stimuli. Whether or not a particular animal has motion-selective ganglion cells is more a function of its retinal circuitry and visual lifestyle than its phylogenetic relationships. In particular, some mammals have motion-selective ganglion cells, while others do not.

Attempts to understand the detailed structure of receptive fields began with the work of Stephen Kuffler (1953), who demonstrated that ganglion cells in cats have receptive fields with antagonistic subfields. He recorded from the axons of retinal ganglion cells while stimulating them with relatively small spots of light and found that the receptive fields could be subdivided into a circular or oval center and an annular surround. Stimulation of the center of some cells increased the firing rate of the cell, while stimulation of the surround decreased the firing rate. These cells were characterized as having an ON-center and an OFF-surround. Other cells had receptive fields with the opposite arrangement and were characterized as OFF-center and ON-surround. The center and surround interact with each other so that simultaneous stimulation of the center and surround produces relatively little response.

This account gives the impression that inhibitory and excitatory influences are restricted to specific regions of the receptive field. However, careful analysis of the receptive fields of cat ganglion cells shows that both excitatory and inhibitory influences are present throughout the receptive field (e.g., Rodieck and Stone, 1965; Troy, 1993). The excitatory and inhibitory components of the receptive field can be accurately described as Gaussian functions with the amplitudes and widths of the functions determining whether the receptive fields have ON-centers or OFF-centers. The responses of cat ganglion cells to stationary spots of light can thus be described as a difference of two Gaussian functions. Explaining the responses of the cells to moving stimuli, however, requires a temporal factor that Rodieck and Stone modeled as an exponential function. The response of the cell to a stimulus at some point in the receptive field could then be obtained by multiplying together the separate spatial and temporal response functions. The receptive fields were linear in that the response of the cell to a stimulus moving through the receptive field could be obtained by adding together the responses of the cell to stimuli presented first at one spot in the field and then at another spot.

This line of thought was extended to a general consideration of the linearity or nonlinearity of the receptive fields by Enroth-Cugell and Robson (1966). They used psychophysical methods to characterize the responses of individuals to sinusoidally varying contrast patterns, and used neurophysiological methods to characterize the responses of ganglion cells in cats to the same stimuli. Ganglion cells could be placed into several groups that have distinctive responses to sinusoidal stimuli. X-cells have small receptive fields and demonstrate linear responses to sinusoids, while Y-cells have larger receptive fields and show nonlinear responses. At about the same time, Boycott and Wässle (1974) used Golgi methods to study the morphology of cat retinal ganglion cells. They placed ganglion cells into four groups based on differences in the sizes and shapes of their dendritic fields. Two of the groups, the alpha and beta cells, could be related to the X- and Y-cells described with physiological methods. This relationship was subsequently confirmed using intracellular filling methods to visualize both the physiological responses and the morphology of the same ganglion cell. The general conclusion is that the X- and Y-cells represent about half of the ganglion cells in cat retina and show relatively few motion-selective properties. The remaining half of the cells—often lumped together as W-cells—contain many neurons with some motion-selective abilities (for a review see Stone, 1983).

Hubel and Wiesel (1962) were the first workers to detail the receptive field properties of neurons in a visual cortex. They found that neurons in the primary visual cortex of cats and monkeys often respond to oriented bars of light moving in particular directions or with particular speeds. Cortical neurons could be placed into two basic groups called simple cells and complex cells (e.g., Henry, 1985). Simple cells have receptive fields that can be divided into subfields, some that increase the firing rate of the cell when stimulated and other fields that decrease the firing rate of the cell when stimulated. As in retinal ganglion cells, there is an antagonism between the subfields so stimuli presented simultaneously in the excitatory and inhibitory subfields of the receptive field can cancel each other out. However, the receptive field structure of simple cells is more complicated than a difference-of-Gaussians configuration in that there can be multiple subfields (Fig. 4A). One approach to modeling them is to employ a Gabor function, which is the product of a difference of Gaussians with either a sinusoidal or a cosinusoidal function (e.g., Henry, 1985)

$$P(x) = \mathrm{Exp}[-(x - x_m)^2/4\,\sigma^2]\,\cos[2\pi\, f_0(x - x_m)] \qquad (5)$$

where $x - x_m$ is the distance from point x to the center of the receptive field x_m, f_0 is the optimal spatial frequency, and σ is the width of the Gaussian function. Gabor functions produce the multiple subfields seen in simple cells due to the periodic nature of the trigonometric functions (Fig. 4B).

The linearity of simple and complex cells was examined by studying their responses to sinusoidal contrast gratings in the manner used by Enroth-Cugell and Robson to characterize the linearity of X-type ganglion cells. Simple cells typically show a linear summation of spatial stimuli. Complex cells, on the other hand,

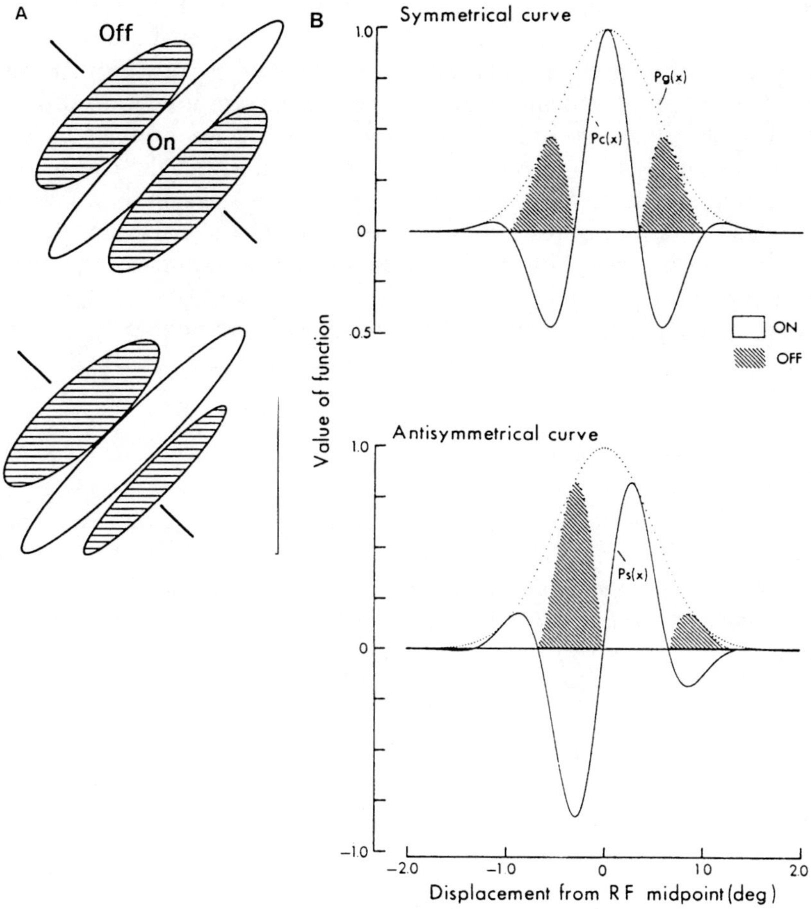

Figure 4. Gabor functions. (A) Spatial receptive fields of cortical cells are illustrated by excitatory or ON regions (open ellipses) and inhibitory or OFF regions (shaded ellipses). The receptive field shown at the top of the panel has a central ON subfield and symmetric inhibitory subfields. The field shown at the bottom of the panel has a central excitatory field and asymmetric inhibitory fields. Lines show transects through the receptive fields, which are plotted in panel B. (B) Transects through the two receptive fields are represented by symmetric (top) and asymmetric (bottom) Gabor functions. ON subfields are clear and OFF subfields are shaded. Panel B is from Henry (1985).

cannot be divided into subfields in this way and their responses cannot be characterized by linear interactions between stimuli presented at different positions in the receptive field. Instead, pairs of bars must be used to fully characterize their responses (Movshon *et al.*, 1978b). Attempts at modeling the spatial structure of simple and complex cells are considered in detail in the chapter by Wörgötter elsewhere in this volume.

The classical definition of a receptive field implies that neurons do not respond to stimuli presented outside of the spatial extent of their receptive fields. However, it is now recognized that the responses of some neurons can be modulated by stimuli presented outside of their classical receptive fields (Allman *et al.*, 1985b). Neurons in the optic tectum of pigeons, for example, have relatively large, but spatially restricted receptive fields, and respond selectively to stimuli moving with specific directions. The responses of tectal neurons can be augmented by stimuli presented outside of this classical receptive field (Frost, 1982; Frost and Nakayama, 1983). Stimuli moving parallel to the preferred direction suppress the responses of tectal neurons, while those moving in the nonpreferred direction augment the response of a neuron to stimuli presented within the classical receptive field. Similarly, neurons in the primate cortical area MT have large classical receptive fields that are responsive to dot patterns moving in particular directions. Large fields of dots presented outside of the *classical receptive field* can modulate the response of the cell to stimuli presented within the classical field (Allman *et al.*, 1985a). They can suppress the classical response, with the degree of suppression depending upon the direction and speed of the background pattern.

2.3. Motion Analysis Pathways

A 19th century view of vertebrate brain organization was that the brains of vertebrates evolved in a linear sequence in which the sensory thalamus and sensory cortical areas were formed relatively late in the evolutionary process (Hodos and Campbell, 1969; Ulinski, 1997). This view holds that the visual nuclei of the thalamus and visual areas within the cerebral cortex are mammalian characteristics not present in other vertebrates such as reptiles and birds. It implies there are functional differences between the visual systems of mammals and nonmammalian vertebrates. It was thought, for example, that visual analysis in nonmammals is carried out by projections from the retina to the optic tectum or roof of the midbrain. In mammals, visual information was relayed from the retina to the visual thalamus (the dorsal lateral geniculate nucleus) and then to the visual cortex. The roof of the midbrain (the superior colliculus in mammals) was responsible for general visual behaviors such as orienting to visual stimuli, while the visual cortex was held responsible for more sophisticated visual tasks such as pattern vision.

This view is incorrect. Work with modern axonal tracing techniques during the 1970s established that the brains of all of the vertebrates demonstrate a common organizational plan (Ulinski, 1997) that includes a basic design to the visual system (Ulinski, 1980). The retinas of all of the vertebrates have photoreceptors and bipolar, horizontal, amacrine, and ganglion cells. The ganglion cells project to a similar array of central visual nuclei in all of the vertebrates, including projections to the optic tectum or superior colliculus and to a thalamic nuclear complex on the lateral surface of the diencephalon (Fig. 5). This latter nucleus projects to a visual area in the telencephalon, establishing a *thalamofugal pathway*. The optic tectum/ superior colliculus projects to a visual nucleus situated in the central thalamus, which projects in turn to a second telencephalic visual area, establishing a *tectofugal pathway*. The cerebral hemispheres of vertebrates contain at least two visual areas (and often many more), although they are not always arranged in a cortical configuration consisting of alternating layers of cell bodies. Although these elements of the visual system are universally present in all of the vertebrates, there are significant variations in the functional organization of the various components of the system that reflect the different visual lifestyles of individual species of vertebrates.

The difference between the 19th century and contemporary views of visual

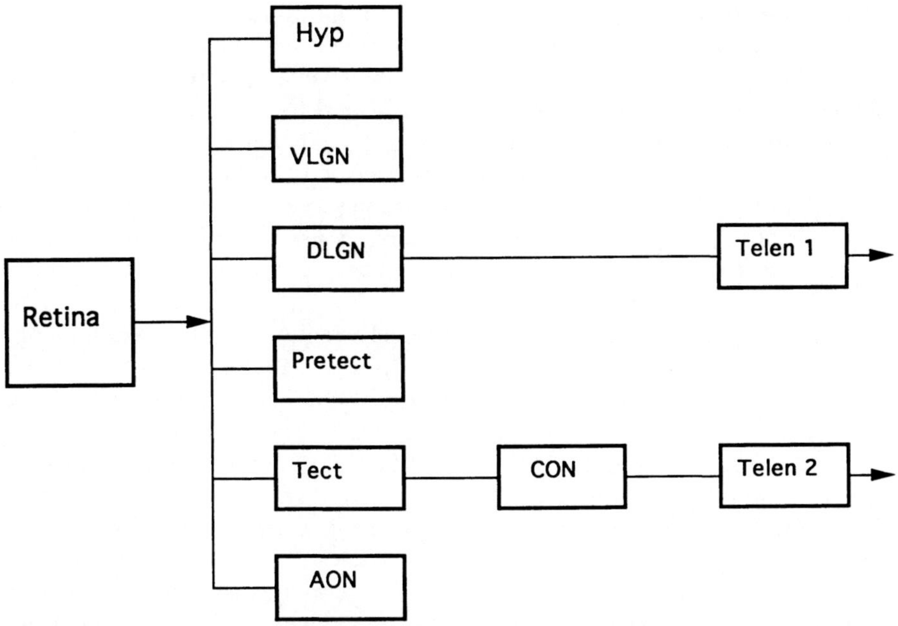

Figure 5. Visual pathways. This is a simplified diagram of the visual pathways involved in visual motion analysis in vertebrates. Ganglion cells in the retina project to several central targets: the hypothalamus (Hyp), ventral lateral geniculate nucleus (VLGN), dorsal lateral geniculate nucleus (DLGN), pretectum (Pretect), optic tectum (Tect), which is called the superior colliculus in mammals, and the accessory optic nuclei (AON). The accessory optic nuclei are involved in stabilization of the eyes and retina, but are probably not heavily involved in the perception of visual targets. These functions are subserved by two parallel, ascending pathways. The thalamofugal pathway involves projections from the dorsal lateral geniculate nucleus to one telencephalic visual area (Telen 1). The tectofugal pathway involves projections from the tectum to a visual nucleus situated centrally in the thalamus (called the central optic nucleus or CON, here), which projects in turn to a second telencephalic visual area (Telen 2). The two telencephalic visual areas are typically interconnected by telencephalic connections. All vertebrates have two telencephalic visual areas, but many species have as many as 20 additional visual areas in the telencephalon.

system organization deserves considerable emphasis because the seminal studies of cortical organization by Hubel and Wiesel were done before the modern view of visual system organization was elaborated. Hubel and Wiesel were thus operating on the premise that the primary visual cortex is the sole route by which visual information reaches the cortex. The textbook concepts of cortical organization reflect in large measure the organization of primary visual cortex, but it is now clear that this cortex is responsible for only the initial steps in processing visual information. The extrastriate areas of visual cortex that seem responsible for the more sophisticated aspects of visual motion analysis can have organizational features quite different from primary visual cortex and receive inputs via the superior colliculus as well as from the primary visual cortex.

Insights into the functional organization of the central visual pathways were gained by applying the methods of receptive field analysis developed by Lettvin *et al.* and by Hubel and Wiesel to visual system neurons in a range of species and in a wide range of animals. The general pattern seems to be that specific pathways are dedicated to the analysis of particular aspects of the stimulus, such as those involved in motion or form or color. There was a tendency to think initially in terms of parallel visual pathways, each devoted to analysis of a specific aspect of visual stimuli. Thus, one visual pathway might be dedicated to form, another to motion, and still another to color vision. It is now recognized that sequences of neurons involved in visual motion analysis are also involved in other aspects of vision (Merigan and Maunsell, 1993), but the concept that sequences of neurons play a dedicated role in analyzing moving stimuli remains a useful fiction. *Motion analysis pathways* in vertebrates have been best characterized in cats, monkeys, and turtles.

The elaboration of motion-selective receptive fields occurs along two parallel routes in cats (Fig. 6). Many retinal ganglion cells included in the W-cell category have motion-selective properties. The W-cells are distributed along the nasotemporal axis of the retina in an area of increased ganglion cell density called the *visual streak.* They project to several structures in the central visual system via branching axonal collaterals. One of these is the dorsal lateral geniculate complex, which is divided in many mammals into a series of distinct cytoarchitectonic layers that receive inputs selectively from different populations of ganglion cells (for review see Jones, 1985). Carnivores—including cats—have a pattern of geniculate lamination in which the major layers are named the A layers, the C layers, and the medial interlaminar nucleus. The W-cells project to the C layers and to the medial interlaminar nucleus. Neurons in these structures project to regions of the visual cortex located on the lateral surface of the hemispheres and known as the lateral suprasylvian areas. These cortical areas contain topographically organized representations of visual space in which the visual streak occupies a disproportionately large area (for review see Rosenquist, 1985). In addition to the dorsal lateral geniculate nucleus, W-cells have axon collaterals that project to the superior colliculus on the roof of the midbrain. Neurons in the superior colliculus conduct visual information

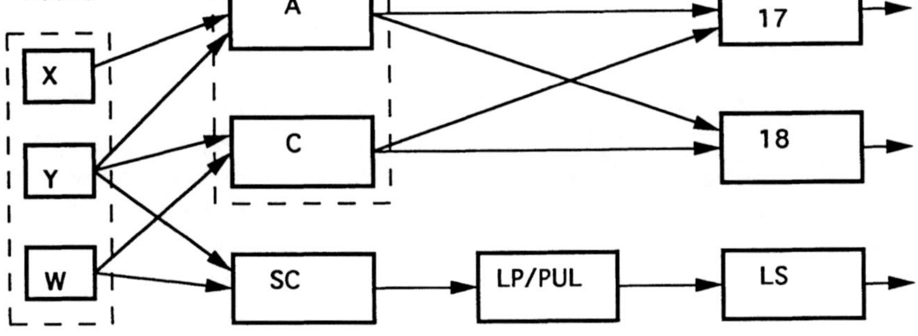

Figure 6. Motion analysis pathways in cats. Cat retina contains three principal ganglion cell classes, the W-, X-, and Y-cells. The X-cells project to the A layers of the dorsal lateral geniculate complex (DLGN). The Y-cells project to the C layers of the DLGN and to the superior colliculus. Both sets of geniculate layers project to the primary visual cortex, area 17 or striate cortex, and to the secondary visual cortex, area 18. W-cells project to the C layers of the geniculate and to the superior colliculus (SC). The superior colliculus projects to the lateral posterior (LP) and pulvinar (PUL) nuclei of the central thalamus, which project to the several lateral suprasylvian areas of the cortex (LS). Projections between area 17 and the lateral suprasylvian areas exist, but are not shown. This is a highly simplified view of cat visual pathways. References in the text should be consulted for more detailed treatments.

to a second group of thalamic nuclei called the lateral posterior nuclei and pulvinar. They project to a variety of cortical areas that include the lateral suprasylvian areas. Thus, the cerebral cortex receives inputs from the W-cells of the retina via two routes; one involves a thalamofugal pathway including the dorsal lateral geniculate nucleus and the other involves a tectofugal pathway including the superior colliculus. The relative contributions of the two pathways to the receptive field properties of neurons in the lateral suprasylvian areas remain a matter of continued investigation.

The X-cells in the retina show little in the way of motion-selective properties and are located in a second region of high ganglion cell density, the area centralis (Rodieck *et al.*, 1993). They project to the A layers of the lateral geniculate, and neurons in these layers project to the primary visual area of the cortex (for reviews see Stone, 1983; Rosenquist, 1985). This area is also called the striate cortex, the primary visual cortex, visual area V1, and area 17. Like the lateral suprasylvian areas, the primary visual area contains a topographic representation of visual space, but it shows an overrepresentation of the area centralis instead of the visual streak. The Y-cells have axons that bifurcate and project to several targets in the geniculate, including the A-laminae and medial interlaminar nucleus, as well as to the superior colliculus. The primary visual cortex and lateral suprasylvian areas are connected by intracortical connections so that the W-cell and X/Y-cell pathways interact with each other at the cortical level. The primary visual cortex contains many neurons that respond to stimuli such as oriented bars moving in specific directions, and many neurons are tuned to stimuli moving within specific speed ranges (Orban, 1984, 1986).

In contrast to the situation in cats, there is relatively little evidence for neurons with motion-selective properties in the retina or dorsal lateral geniculate of primates (Fig. 7) and most motion analysis seems to occur at the cortical level (Orban, 1994; Allman *et al.*, 1994). Retinal ganglion cells in primates can be placed into several groups on the basis of their morphology and physiological characteristics (Rodieck, 1988; Lennie, 1993). The midget cells have very small dendritic and receptive fields and project selectively to layers of neurons (the parvocellular layers) in the dorsal lateral geniculate nucleus that contain neurons with relatively small somata. The parasol cells have larger dendritic and receptive fields and project to layers of geniculate neurons (the magnocellular layers) that contain neurons with large somata. Neurons in the primary cortical area, V1, receive visual information from both the parvocellular and magnocellular layers of the dorsal lateral geniculate nucleus via very selective projections (Casagrande and Kaas, 1994). As in the primary visual cortex of cats, neurons in primate V1 have small receptive fields and are often tuned to stimuli oriented in specific directions, moving in specific directions (typically orthogonal to the preferred orientation), and moving within a specific range of speeds (Orban, 1994).

Neurons in V1 project via a complex sequence of intracortical projections to other cortical areas such as the medial temporal (MT) area, medial superior temporal (MST) area, and areas in the posterior parietal lobe located deeper in the so-called *motion pathway* (Zeki, 1976; DeYoe and Van Essen, 1988; Wurtz *et al.*, 1990). Neurons in these areas have larger receptive fields—on the order of 30 deg of visual arc in MT—and larger still in the MST and the posterior parietal complex. They respond to more complex aspects of motion such as rotations or expanding or

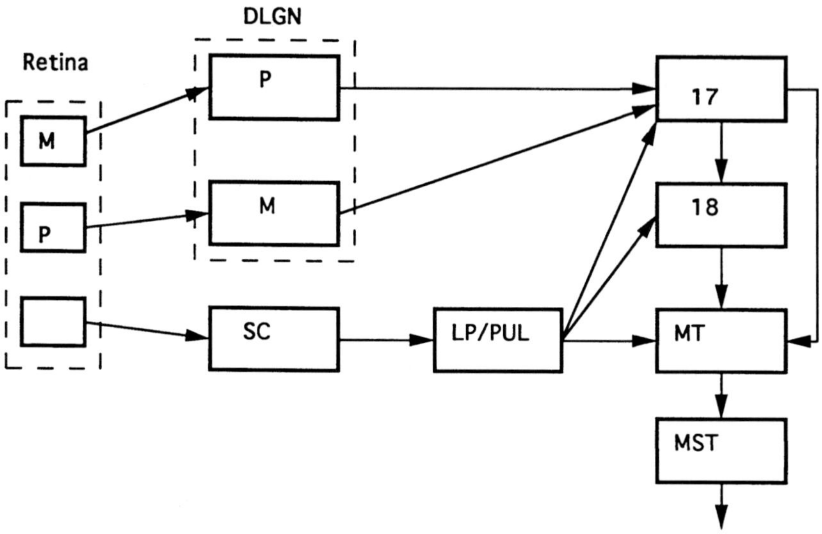

Figure 7. Motion analysis pathways in primates. Midget (M) and parasol (P) cells in the retina project to the parvocellular and magnocellular layers of the dorsal lateral geniculate (DLGN), respectively. Both of these layers project to primary visual cortex, which is also known as area 17 or striate cortex. The magnocellular geniculate layers project preferentially to motion analysis pathways. Several groups of ganglion cells project to the superior colliculus, which projects to the lateral posterior (LP) and pulvinar (PUL) nuclei in the central thalamus. The LP/PUL nuclei project to both striate and extrastriate cortical areas. Area 17 projects to the medial temporal area (MT) directly and indirectly. Area MT projects, in turn, to the medial superior temporal area (MST). This is a highly simplified diagram of visual pathways in primates. References in the text should be consulted for more complete diagrams of these pathways.

contracting objects (Saito *et al.,* 1986; Roy and Wurtz, 1990; Graziano *et al.,* 1994). Lesions of these cortical areas produce deficits of selective aspects of motion analysis and certain aspects of the receptive field properties of their neurons can be associated with more cognitive aspects of motion analysis such as attention (Wurtz *et al.,* 1990).

There are also retinal projections to the superior colliculus in primates, but the specific populations of ganglion cells participating in these projections are not completely characterized. The superior colliculus projects to the pulvinar, which is the largest nucleus in the thalamus of primates (for review see Jones, 1985). The pulvinar projects to the extrastriate areas of the cortex so that—as in cats—visual information reaches these areas both through the dorsal lateral geniculate complex and primary visual cortex and through the superior colliculus and pulvinar. The relative contributions of the two sources of input to the receptive field properties of neurons in the extrastriate cortex is not well understood and may vary among species of primates (for review see Kaas and Krubitzer, 1992). Given the size of the pulvinar and the extensive projections from the pulvinar to the extrastriate areas, it would be reasonable to guess that vision in primates is dominated by the tectofugal system.

Freshwater turtles are the only nonmammalian vertebrates for which information on the organization of cortical motion analysis systems is available (Fig. 8). Like cats, rabbits, and ground squirrels, many of the retinal ganglion cells of turtles respond selectively to stimuli moving with specific directions and speeds (for review see Granda and Sisson, 1992). They occur with the greatest density in a visual streak and the majority have axons that bifurcate in the optic tract to terminate in both the optic tectum (the nonmammalian homologue of the superior colliculus) and in the dorsal lateral geniculate nucleus (Sjöström and Ulinski, 1985). Neurons in the optic tectum project to nucleus rotundus in the thalamus, which projects to a visual area in the dorsal ventricular ridge (Ulinski, 1983) of the telencephalon. This is a tectofugal pathway that is comparable to the projections through the pulvinar to the extrastriate cortical areas of mammals. Neurons in the dorsal lateral geniculate complex project to a visual area in the cerebral cortex (Heller and Ulinski, 1987; Mulligan and Ulinski, 1990). This is a thalamofugal pathway comparable to the geniculocortical pathway of mammals. Neurons in both the dorsal ventricular ridge (Dünser *et al.*, 1981) and in visual cortex (Mazurskaya, 1974) have large receptive fields and respond well to stimuli moving with specific speeds across visual space.

A contemporary alternative to the 19th century view of visual pathway organization can be formulated by recognizing three motion analysis tasks. The first involves stabilizing the visual image on the retina and is universally subserved by the central visual pathways in the accessory optic system (for review see Wallman and Fite, 1985). The second involves the control of basic visual orienting tasks such as stabilizing the position of the body in visual space as the animal moves through the visual world, avoiding collision with objects in the external world, and orienting to and following objects moving in the visual space. This task does not depend upon precise spatial acuity and can be accomplished by input from retinal ganglion cells with motion-selective properties. These ganglion cells project in parallel through the tectofugal and thalamofugal pathways to two or more visual areas in the telencephalon. The third task involves spatially acute analyses of moving objects. It is particularly important for animals that use binocular stereopsis as a mechanism of determining the distance of objects in the external world because stereopsis relies

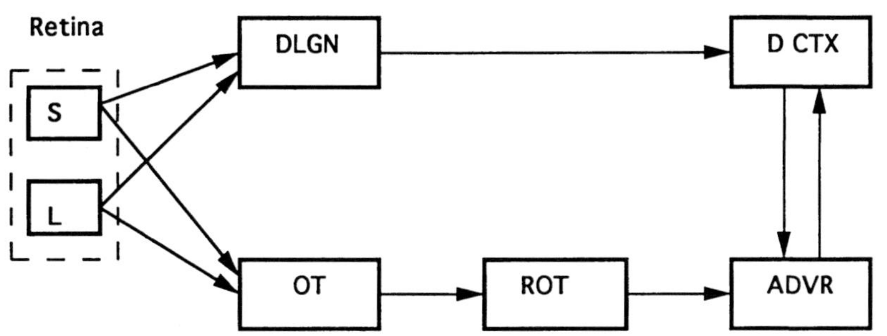

Figure 8. Motion analysis pathways in turtles. Retinal ganglion cells in turtles can be grouped for heuristic purposes into small (S), large (L), and very large classes, each of which contains motion-selective cells. The very large ganglion cells project selectively to the accessory optic system. The small and large ganglion cells project to both the dorsal lateral geniculate complex (DLGN) and the optic tectum (OT). The dorsal lateral geniculate projects to the dorsal cortex (DCTX), a visual area in the cerebral cortex. The optic tectum projects to nucleus rotundus (ROT) in the central thalamus, which projects to a visual area in the anterior dorsal ventricular ridge (ADVR) of the cerebral hemispheres. The two telencephalic visual areas have reciprocal connections with each other. Descending projections from the dorsal cortex and ADVR to the lower stations in the motion pathways exist, but are not shown.

upon comparing two images of the visual world as obtained by the two eyes and requires two very precise retinotopic maps of visual space. The task relies upon ganglion cells that have spatially restricted receptive fields and consequently can subserve acute spatial vision. A system with these characteristics is a necessity for animals with visual lifestyles that depend upon stereoscopic visual cues, and involves projections via the thalamofugal pathway to one or more visual areas in the telencephalon. Motion analysis is thus carried out via multiple, parallel pathways to the telencephalon in all of the vertebrates for which there is sufficient information to make a statement.

Details of the functional organization of the pathways differs among animals and is related to the visual lifestyle of a particular species. Turtles, for example, have visual lifestyles that do not depend upon spatially acute vision. Their visual system is dominated by ganglion cells with motion-selective properties organized in a visual streak. Old World primates (e.g., macaques and humans) have visual lifestyles that emphasize very precise visual control of eye and limb movements. Their visual system is dominated by an area centralis containing neurons with very small, classical receptive fields that permit fine, spatial acuity and support binocular stereopsis. Cats have visual lifestyles that take advantage of both approaches and their visual system shows aspects of both the turtle and primate patterns. They have both a visual streak with motion-sensitive W-cells that project in parallel to the midbrain and to the dorsal lateral geniculate nucleus, as well as an area centralis with X-cells that project to the primary visual cortex. It is not surprising to learn, then, that owls have a telencephalic visual area with a precise retinotopic map of visual space and neurons responsive to binocular cues and oriented stimuli in the same way as neurons in the primary visual cortex of cats and primates (Pettigrew, 1979). Like cats and primates, owls have a visual space that includes a large region of frontal, binocular overlap and a visual lifestyle in which they use their limbs for visually directed reaching.

2.4. Local and Global Motion Detectors

These brief accounts of motion analysis pathways do not do justice to the amount of information that has been gained on the organization of motion analysis pathways (particularly in primates), but they do support the generalization that motion analysis involves at least two distinct steps that are accomplished by different means in different animals. Regardless of interspecific differences in the detailed organization of motion pathways and visual lifestyles, the neurons involved in these two distinct steps have common functional properties and can be called local and global motion detectors.

2.4.1. Definition of Local and Global Motion Detectors

The first step in analyzing moving visual stimuli is universally an analysis of stimuli moving through relatively small regions of visual space by *elementary motion* or *local motion detectors* (Fig. 9). The local motion detectors receive convergent input (often indirectly) from several or many photoreceptors and therefore have the task of comparing inputs sampled by the photoreceptors from various loci in visual

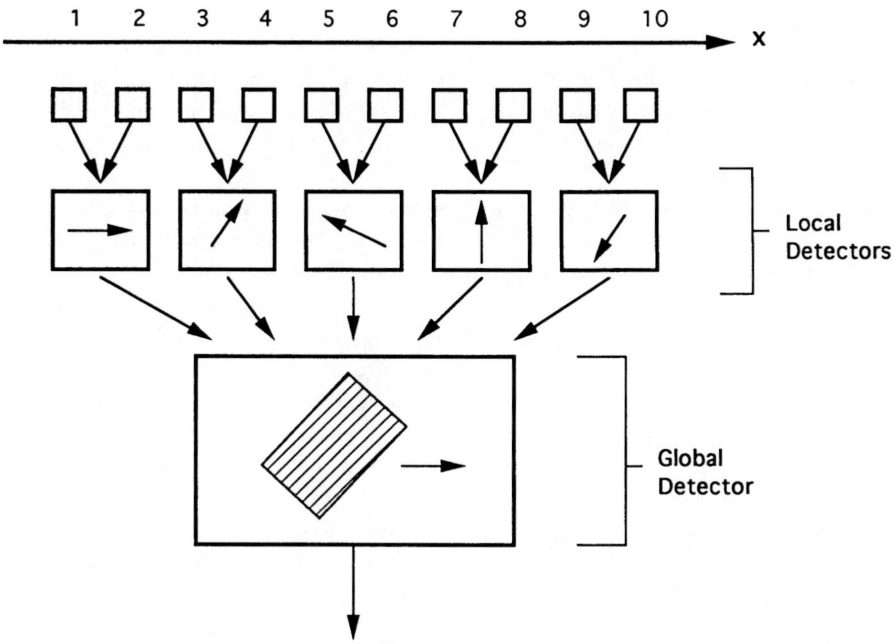

Figure 9. Local and global motion detectors. Local motion detectors sample local regions of space situated at different loci along the *x* axis. Each detector in an array of local detectors receives inputs from a pair of spatial channels separated by a spatial separation called the sampling base. Each detector responds preferentially to a stimulus moving with a preferred speed or direction (indicated by the vectors in the local detectors). The global motion detector samples a larger region of space. It receives inputs from an array of local detectors and responds preferentially to a stimulus moving with a preferred speed and direction. Since the stimulus is large, the global motion detector requires inputs from several local detectors to resolve ambiguities arising from the local motion signals created by its edges, each of which is moving in a different apparent direction.

space at slightly different times. They use this information to carry out the freshman-physics exercise of calculating velocity vectors that have direction and magnitude (the speed of the moving object). Local motion detectors thus respond to stimuli moving within a particular range of directions and speeds through small regions of visual space. Depending upon the species being considered, "small" means between a few degrees of visual arc in some animals and a fraction of a degree of visual arc in others. Local motion detectors occur in the retina and primary visual cortex of cats, in the primary visual cortex of primates, and in the retina of turtles.

The second step involves analysis of stimuli moving across larger regions of visual space. Again depending upon the species, "large" means between tens of degrees of visual arc and the entirety of binocular visual space. The *global motion detectors* responsible for the second step receive convergent input from arrays of local motion detectors and therefore serve to combine or integrate information from the local motion detectors. They are found in the temporal and parietal visual areas of primates, the lateral suprasylvian areas of cats, and the visual cortex of turtles. Since local motion detectors can calculate the direction and speed of a moving object, it may not be immediately apparent why global motion detectors are also needed. However, local motion detectors alone cannot provide unambiguous information about an object moving past the observer because each local detector is privy to only a restricted view of the visual world and, in most cases, will actually be stimulated by only a part of a moving object. This leads to several perceptual difficulties that can be resolved by global motion detectors situated deeper in motion analysis pathways.

The first is called the *correspondence problem.* Imagine some relatively large object (a rectangle) is moving past an observer. Any one local detector sees only a part of the rectangle at any instant in time. Thus, a local detector may be activated once as the leading edge of the rectangle passes through its receptive field and a second time as the trailing edge passes through its receptive field. A second detector some distance away will be activated by the leading edge of the rectangle at a later time. The situation is ambiguous because there is no way to tell whether the same leading edge is passing over both local detectors versus a situation in which two different edges pass independently over the two detectors, or whether the leading and trailing edges are parts of the same or different objects. A second ambiguity is the *aperture problem.* It arises when a large edge moves across a small receptive field (or aperture) so that the direction of movement cannot be unambiguously determined. A third problem involves the *segmentation* of the visual world into regions containing objects moving with different speeds and directions (Braddick, 1993). Looking out the window of a moving train, for example, can produce a situation in which an edge (an object inside the train) is moving in one direction while objects (trees, telephone poles, etc.) outside the window are moving in the opposite direction. A cluster of local motion detectors therefore receives simultaneous inputs from one object moving parallel to the motion of the train and a second object moving antiparallel to the motion of the train. Finally, there is the situation that occurs when objects move past each other. The visual system must interpret local variations in contrast (and in some cases binocular disparity) to determine which part of the visual signal is associated with which object. This results in the perception of *motion transparency* in which one object appears to move behind another. Resolution of all of these difficulties is thought to involve global motion detectors.

The remaining parts of this section consider the functional properties expected of local motion detectors and summarizes classes of models that have been used to think about these properties.

2.4.2. Properties and Models of Local Motion Detectors

The essential computations involved in visual motion analysis require spatiotemporal interactions. It is convenient, however, to begin by setting aside these interactions and to focus initially on the issue of how neurons in visual structures can store temporal information. This can be done by adopting the approach common in systems engineering and considering visual structures as "black boxes" or filters defined by their input–output relations (Regan, 1991b). The approach ignores the detailed interactions known to take place between neurons in cortical areas and emphasizes instead the overall dynamic properties of the neurons. The temporal storage properties of structures viewed in this way can be analyzed by two standard engineering approaches.

One is to think about local motion detection in the context of an apparent-motion paradigm and characterize the responses of elements of the visual system to short pulses of light presented to one spatial channel. Understanding the response of a spatial channel to a pulse of light can be particularly useful in that linear systems are completely described by their responses to short-stimulus pulses. The response of any system to an infinitely short impulse [represented mathematically by a Dirac delta function, $\delta(t)$] presented at time t is called the *impulse response function*, $h(t)$, of the system (Fig. 10). If the system is linear, its response to any arbitrary input [described by an input function $x(t)$] can be obtained by viewing the input function as made up of a series of inputs (or impulses) presented at discrete time intervals, and adding together the responses of the system to this sequence of inputs. Mathematically, this calculation is equivalent to evaluating the *convolution integral*

$$y(t) = \int_0^t h(u)x(t-u)\,du \tag{6}$$

where $y(t)$ is the time-varying output of the system.

The impulse response function may be difficult to measure directly, and it is often more convenient to characterize the temporal storage or memory properties of a visual system structure by measuring a *temporal integration function*. A system has a *memory* if its behavior at one instant of time depends upon the history of the system. A system that gives the same response to each stimulus in a sequence of stimuli regardless of the history of preceding stimuli is a *zero-memory system*. The temporal integration function of a system characterizes its memory and can be measured by conducting a *two-pulse experiment* (Fig. 11) in which two stimuli (S1 and S2) are presented in the sequence S1 followed by S2 with a range of temporal separations (Marmarelis and Marmarelis, 1978). The temporal integration function, $T(\Delta t)$ is defined by plotting the amplitude of the response of the system, $A(\Delta t)$, when S2 is presented after S1 with a temporal separation of $\Delta t = t_2 - t_1$ divided by the amplitude of the response to S2 presented alone, $A(0)$. Thus, $T(t) = A(\Delta t)/A(0)$. A zero-memory system has a temporal integration function of unity: $T(\Delta t) = 1$. A system with a memory typically has a temporal integration function with one or more peaks greater than 1.

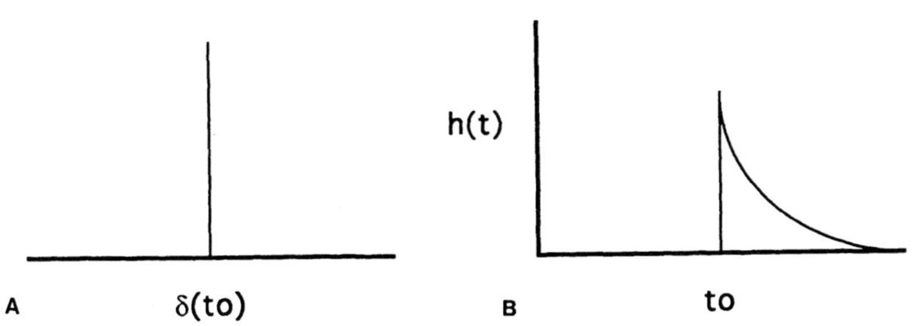

Figure 10. Impulse response function. (A) A Dirac delta function $\delta(t)$ is a function that is zero everywhere except at t. The plot shows the delta function $\delta(t_0)$. (B) An impulse response function $h(t)$ is the response of a system to a delta function input. In the case shown, the impulse response function is a declining exponential function.

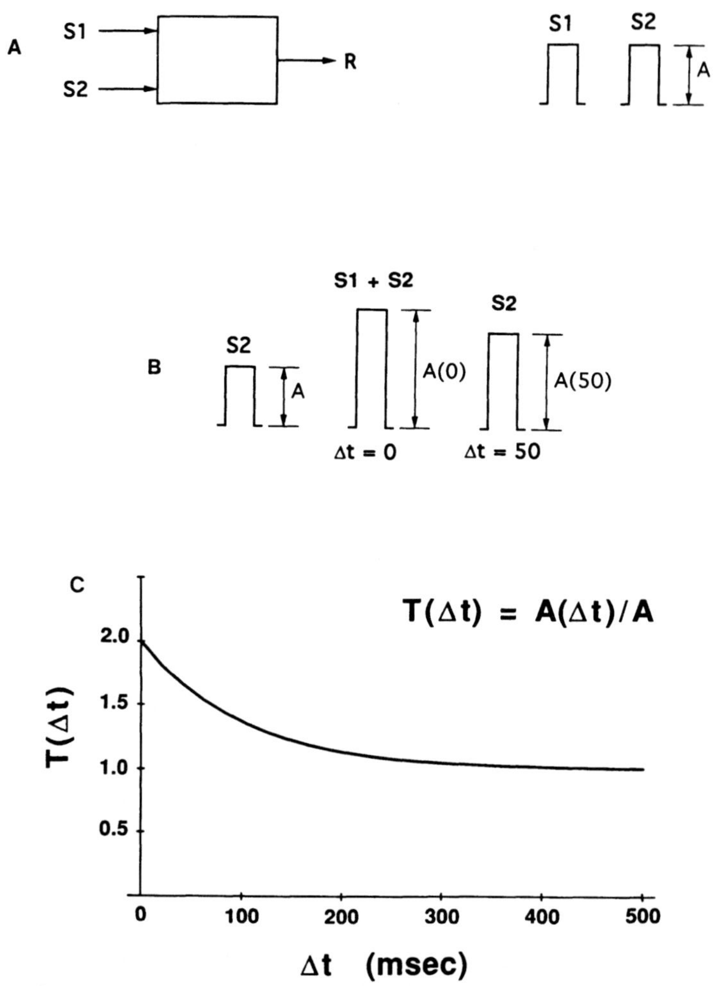

Figure 11. Two-pulse experiments. (A) A system is depicted as a black box with two input channels (S1 and S2) and one response. In this example, activation of either input channel in isolation produces the same response, with amplitude A. (B) Simultaneous activation of both channels produces a response with amplitude $A(0)$ which is greater than the amplitude of the response obtained by activation of channel S2 in isolation. When S1 and S2 are activated sequentially (e.g., with a 50-msec temporal separation), the response to activation of S2, $A(50)$, is greater than A, but less than $A(0)$. (C) Temporal integration function obtained by plotting $A(\Delta t)$ as a function of Δt. In this case, the temporal integration is a declining exponential with a time constant of 100 msec. The system is linear, $A(0) = 2A$. A two-pulse experiment can be used to describe both linear and nonlinear systems. If the system is linear, the temporal integration function equals the impulse response function and completely describes the system.

To take a simple example, consider an electrical circuit consisting of a resistor R and a capacitor C arranged in parallel. This system has a temporal integration function that is a declining exponential: $\exp[-\Delta t/RC]$. The response of the system to the second of a pair of electrical pulses varies as a function of the time that has elapsed since the first pulse, and the temporal integration function characterizes this memory. It is a linear system and the response of the system to any arbitrary input can be obtained by convolving the input and the temporal integration function, which is equivalent to the impulse response function for linear systems. A first step in characterizing local motion detectors, then, will be to measure their temporal integration functions.

The next step is to consider the response of the motion detector to two spots of light and allow both the temporal separation Δt and the spatial separation Δx between channels to vary. This is, of course, exactly what both Barlow–Levick and Reichardt did, and the process leads to the concept of *spatiotemporal receptive fields* (also called *space–time plots*). These provide a convenient way to think about the response properties of motion-selective neurons (Adelson and Bergen, 1985). Spatial separation can depend upon both the x and y coordinates of visual space, but matters can be simplified by considering an object moving along the horizontal meridian of visual space at a constant speed v, and representing spatial separation solely as differences in position along the x axis. The trajectory of an object demonstrating one-dimensional (1D) motion is then a line in a plot of object position as a function of time (Fig. 12A). The object is located at point x_1 at time t_1, at point x_2 at time t_2, etc. Its trajectory can also be represented by a line in a space–time plot (Fig. 12B). If the spatial separation Δx is given by $\Delta x = x_2 - x_1$ and the temporal separation Δt is given by $\Delta t = t_2 - t_1$, then the speed of the object is given by the slope of the line, $s = \Delta x / \Delta t$, in the space–time or $\Delta x\,\Delta t$ plane. An object moving in the opposite direction with the same speed is represented by a line perpendicular to the first vector in the space–time plot.

Suppose now that we have a local motion detector with a sampling base Δx (Fig. 13A). The detector receives its input from two spatial channels that sample visual inputs at points x_1 and x_2. A moving object would then activate first one channel and then the other with some temporal delay. By definition, the detector should respond selectively to an object moving with a particular speed and in a particular direction. It should respond preferentially, for example, to an object moving from left to right as compared to an object moving from right to left. To characterize the local motion detector, we can modify the two-pulse experiment performed above by presenting the detector with two pulses, one to spatial channel 1 and the second to spatial channel 2. As before, the temporal separation of the two inputs would be varied. A temporal separation of $\Delta t = 0$ means that the two input stimuli are presented simultaneously, while nonzero values of temporal separation mean that one stimulus precedes the other in time. An expectation for a motion detector is that it should not respond well to simultaneous presentation of stimuli to the two input channels. Thus, a temporal integration function that is a declining exponential would be excluded because it would produce a maximal response for stimuli presented simultaneously. Rather, we would expect that the maximal response would occur for stimuli presented with some optimal delay. This delay would determine the preferred speed of the detector because detectors tuned to stimuli moving with low speeds would have longer optimal delays, while those tuned to high speeds would have shorter delays. Thus, we would expect a temporal integration function that increases from a relatively low value at $\Delta t = 0$ to some maximal value at the optimal temporal separation, and then declines (Fig. 13B). The optimal delay depends on the sampling base: detectors with larger sampling bases, but tuned to stimuli moving with the same speed, must have longer optimal delays.

This idea can be used to follow the trajectory of a moving object by considering

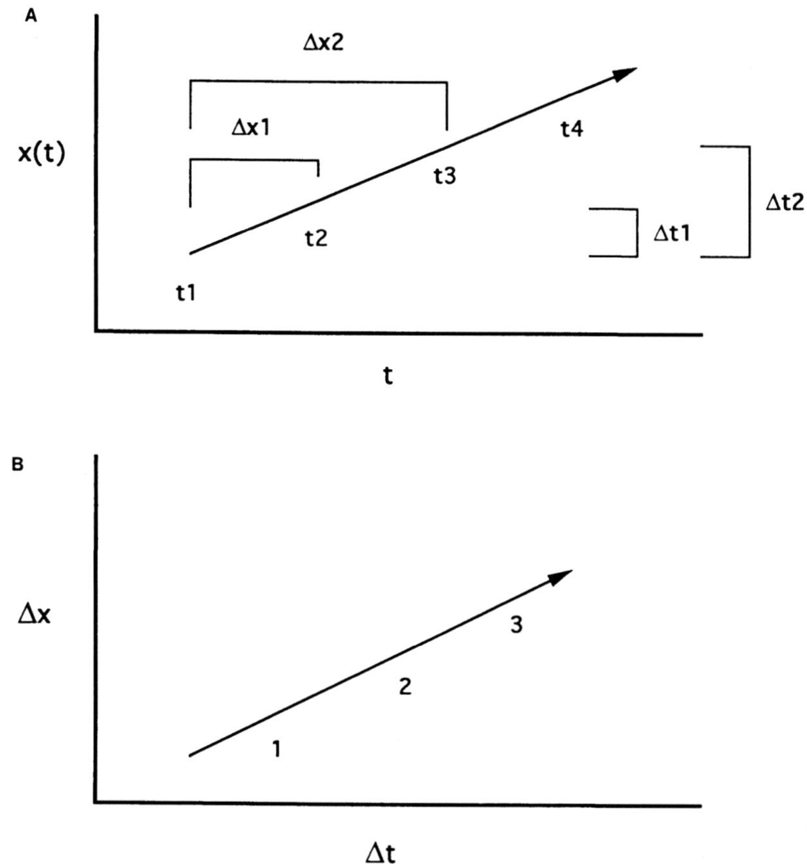

Figure 12. Space–time plots. (A) Trajectory of a stimulus moving with a constant speed. The position of the stimulus along the *x* axis, $x(t)$, is plotted as a function of time elapsed from the start of the movement. Location of the stimulus at four instants ($t1$, $t2$, $t3$, and $t4$) is shown. These points can be used to calculate spatial ($\Delta x1$, $\Delta x2$, etc.) and temporal ($\Delta t1$, $\Delta t2$, etc.) separations. (B) Space–time plot for the same trajectory. Spatial separations are now plotted as a function of temporal separations. A stimulus moving with fixed speed produces a straight line in the Δx, Δt plane.

an orderly array of local motion detectors with different spatial and temporal scales. If these are arranged so that the first spatial channel in all of the individual detectors samples the visual input at time $t = 0$, then the trajectory of the object can be specified by the outputs from the second spatial channel in each of the detectors. The behavior of a local motion system can be characterized by presenting it with pairs of inputs for which both the temporal and spatial separations are varied in a systematic way, because the outputs of the individual detectors will be the same for an object moving continuously past them or for a pair of stimuli presented with the correct spatial and temporal separations. The expectation is that the response of the detector will show an orderly relationship between spatial and temporal separation that is related to the preferred speed of the detector.

The result of such an analysis is that a motion-selective neuron can be represented by a *spatiotemporal receptive field* or *response function* defined on the space-time plane (Fig. 14A). The function $R(\Delta x,\Delta t)$ can be visualized either as a three-dimen-

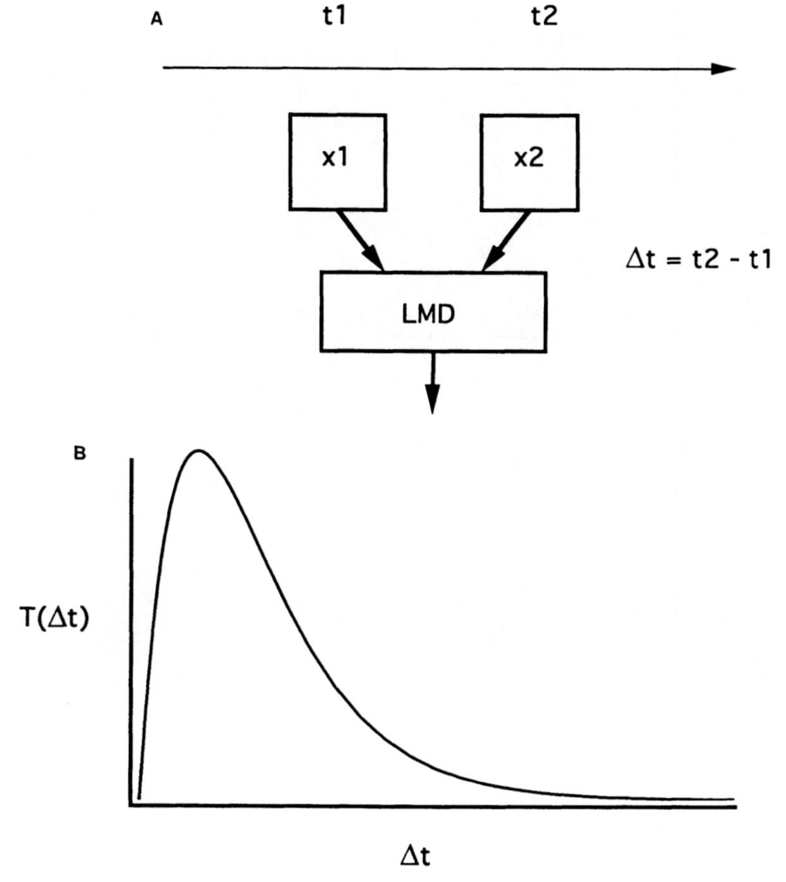

Figure 13. Temporal integration function for a local motion detector. This figure illustrates the shape expected for a temporal integration function for a local motion detector. (A) The detector receives inputs from spatial channels $x1$ and $x2$. A stimulus moving from left to right with a speed v activates channel $x1$ at time $t1$ and channel $x2$ at time $t2$ ($v = \Delta x/\Delta t = x2 - x1 / t2 - t1$). (B) The detector has a temporal integration function $T(\Delta t)$ which is zero when $\Delta t = 0$, so that the detector does not respond to simultaneous presentation of two stimuli to the two spatial channels. It has a peak near temporal separations corresponding to the preferred speed. The width of the peak is related to the width of the speed tuning curve.

sional or a contour plot. It is some measure of the neuron's activity such as its firing rate in an extracellular recording experiment or the amplitude of a postsynaptic potential in an intracellular experiment. A motion-selective cell will have a spatiotemporal receptive field that is *oriented* in the Δx, Δt plane. That is, it has a crest of maximal values oriented at an angle oblique to the Δt axis. The slope of this crest is the preferred speed of the cell and the cell gives its strongest response when the object is moving along a line $\Delta x(t) = t_0 + v\,\Delta t$ in the Δx, Δt plane. An object moving with a trajectory colinear with the crest of $R(\Delta x,\Delta t)$ will have a speed and direction preferred by the cell and produce a strong response. An object moving with the same speed in the opposite direction [along a trajectory $\Delta x(t) = t_0 - v\,\Delta t$] will have a trajectory at right angles to the preferred trajectory and will produce a weaker response. Cells that lack motion-selective properties should have spatiotemporal

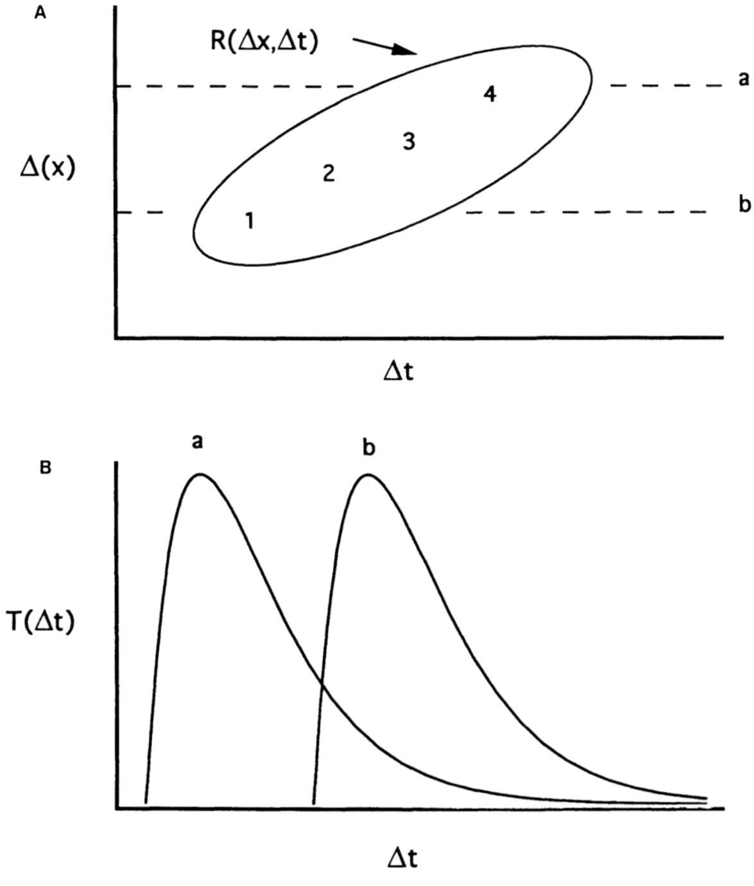

Figure 14. Spatiotemporal response function. (A) A spatiotemporal response function $R(\Delta x, \Delta t)$ is shown as a contour plot in the Δx, Δt plane. This function is oriented so that it will respond maximally to a stimulus moving with the trajectory shown in Fig. 12. Pairs of inputs with spatial and temporal separations (1, 2, 3, and 4) falling within the contour will produce strong responses in the cell. Stimulus pairs with separations falling outside of the contour will produce weak responses. For example, a stationary stimulus would appear as a vertical line in the Δx, Δt plane and produce only a weak response from the function. This response function has only excitatory regions, but many functions have both excitatory and inhibitory regions. Temporal integration functions like those shown in Fig. 13 can be obtained by taking transects through the response function parallel to the Δt axis for fixed spatial separations (e.g., *a* and *b*). If the response function is optimally responsive to a stationary stimulus, it will appear as a vertical profile in the Δt, Δt plane and both transects will be the same. If the function is responsive to moving stimuli, it will be oriented in the Δx, Δt plane and the two transects will produce temporal integration functions with different peak values for the two spatial separations. (B) Temporal integration functions obtained from the transects shown in panel A. Notice that the peak of the function shifts along the Δt axis because of the tilt of the response function.

receptive fields that are not oriented obliquely in the Δx, Δt plane, and are referred to as *unoriented.* The extent of direction and speed tuning is represented by the width of the response function. A very narrow response function indicates that the cell responds only to a narrow range of speeds and directions, while a broad response function indicates that the cell responds to a larger range of speeds and directions.

A mathematical characteristic of an unoriented response function is that it can be factored into two functions: $R(\Delta x,\Delta t) = T(\Delta t)\,S(\Delta x)$. The response function is said to be *separable in space and time* in this case and *inseparable* if it cannot be factored in this way. The function $T(\Delta t)$ is the temporal integration function defined above and the function $S(\Delta x)$ is a *spatial integration* or *line weighting function* that describes the response of the system to two stimuli presented with a fixed temporal separation, but varying spatial separations. Temporal and spatial integration functions can also be defined for many motion-selective neurons, but typically cannot be obtained by factoring the response function. Rather, $T(\Delta t)$ is obtained by taking a transect through $R(\Delta x,\Delta t)$ at some particular value of spatial separation (Fig. 14B), and $S(\Delta x)$ is obtained by taking a transect through $R(\Delta x,\Delta t)$ at some fixed value of temporal separation (Fig. 15). Since $R(\Delta x,\Delta t)$ is oriented obliquely in the $\Delta x,\Delta t$ plane for motion-selective neurons, temporal integration functions measured for two different values of spatial separation will not be the same in motion-selective neurons. Rather, peaks in their temporal integration functions will shift systematically as a function of spatial separation, with the magnitude of the shift determining the preferred speed of the cell (Fig. 15B). This is a mathematical expression of the concept that spatiotemporal interactions are essential to motion analysis.

The receptive fields of X-cells from cat retina and simple cells from primary visual cortex can be taken as examples. We have already seen that Rodieck and Stone (1965) successfully modeled the receptive fields of cat retinal ganglion cells as a product of a spatial integration function (described as a difference of Gaussians) and a temporal integration function (described as an exponential function). The spatiotemporal response function is thus separable and appears as a vertical plot in the space–time plane because the shape of the spatial component of the receptive field does not depend on the temporal separation of the stimuli. The same holds for neurons in the dorsal lateral geniculate complex. Simple and complex cells in visual cortex, by contrast, prefer stimuli such as oriented bars and have spatial integration functions described by Gabor functions. They often appear as oriented profiles in the space–time plane. Whether or not the profile is oriented depends on whether or not the cell is speed-selective. Simple cells that are speed-selective would have response functions that are obliquely oriented in the space–time plane, with the slope of the function determining its preferred speed.

The spatiotemporal response function for a neuron can be measured in an apparent motion paradigm as outlined above by presenting stimuli with varying spatial and temporal separations and plotting the response of the neuron as a function of each pair of stimuli. However, this procedure can be tedious, and most experimental descriptions of spatiotemporal response functions have relied upon more sophisticated methods derived from nonlinear systems theory. The mathematician Volterra provided a way to generalize the convolution integral of linear systems theory by expressing the input–output relationship of a general system as a series of integrals now known as a *Volterra series* expansion of the system,

$$y(t) = \int_0^\infty k_1(u_1)x(t-u_1)\,du_1 + \int_0^\infty\int_0^\infty k_2(u_1,u_2)x(t-u_1)x(t-u_2)\,du_1\,du_2 + \ldots \qquad (7)$$

Each integral in the series contains the input function and a response function that is referred to as the *kernel* of the integral (e.g., Victor, 1992). The first kernel (the

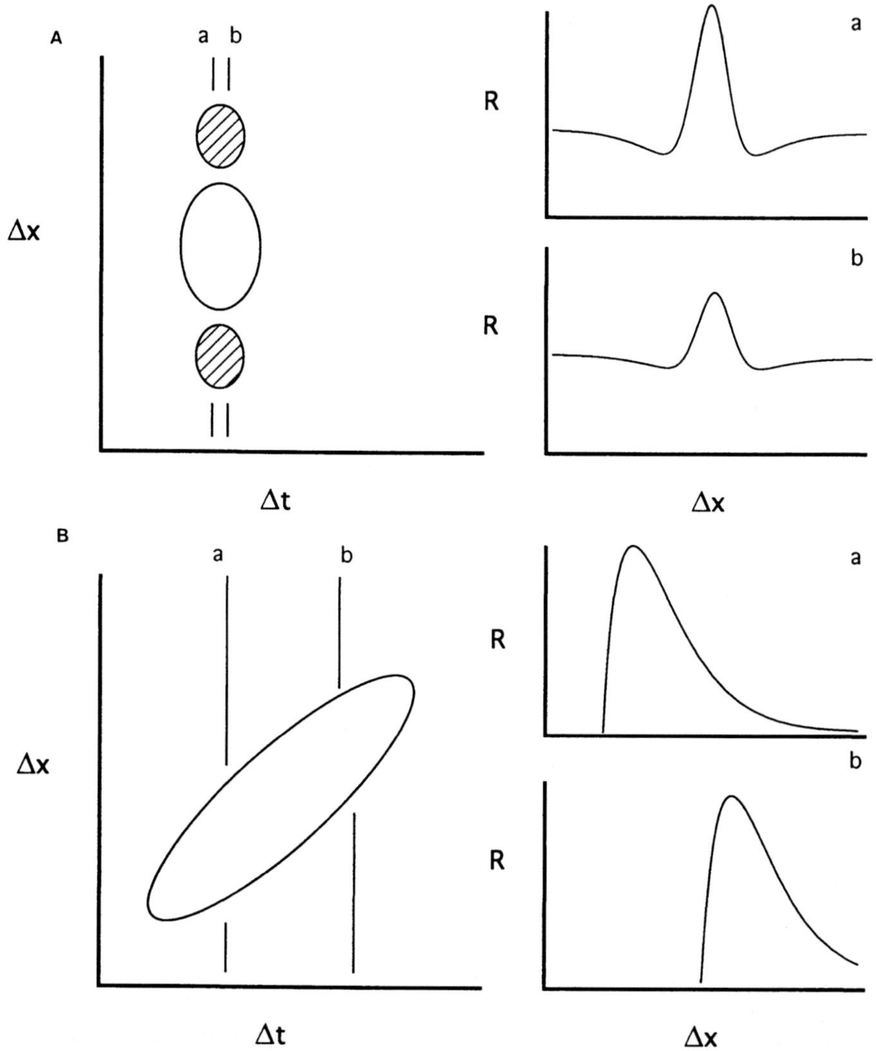

Figure 15. Separable and inseparable spatiotemporal response functions. This figure shows how the spatial integration function varies as a function of time for separable and inseparable spatiotemporal response functions. (A) The plot to the left is a space–time plot for a separable or unoriented spatiotemporal response function. It has an excitatory region (open oval) flanked on both sides by inhibitory surrounds (hatched ovals). (a, b) Two transects through the function taken at different values of temporal separation Δt. The two plots on the right show the spatial integration functions obtained from these two transects. Notice that the two functions have the same shape, but differ in amplitude. This response function can be obtained by multiplying a difference-of-Gaussians spatial integration function by a temporal integration function that is a declining exponential function of time, as for cat retinal ganglion cells. (B) The plot to the left is a space–time plot for an inseparable or oriented spatiotemporal response function. For visual simplicity, only the excitatory center of the function is shown. (a, b) Two transects through the function taken at different values of temporal separation. The two plots on the right show the spatial integration functions obtained from these two transects. The two functions have the same shape and amplitude, but are shifted along the Δx axis due to the slope of the response function.

first-order kernel) is the impulse response function and describes the response of the system to a single impulse. All of the other kernels are zero for a linear system and the Volterra series reduces to the convolution integral. In nonlinear systems, however, the second-order kernel describes the response of the system to pairs of impulses, the third-order kernel describes the response of the system to triplets of impulses, etc. The Volterra series provides a mathematical description of a general, nonlinear system so that the spatiotemporal response function can be expressed as a sum of first- and higher order kernels.

Measuring the second- and higher order kernels in an experimental situation can be difficult, and an important advance was made by Norbert Wiener, who developed a method of measuring the kernels of nonlinear systems using a sequence of random inputs such as white noise. For the visual system, the "snow" on a malfunctioning television set is a white noise stimulus. A number of more sophisticated white noise stimuli have been developed in recent years. They can involve presenting stimuli such as bars of light or pixels on a video monitor, randomly varying spatial and temporal separations, and recording the responses of the neuron in a *forward correlation method* in which stimuli are correlated to the subsequent responses of the neuron. Alternatively, a *reverse correlation method* in which individual action potentials are recorded and related to the stimuli that preceded them can often be used (Jones and Palmer, 1987). A complete discussion of these methods is beyond the scope of this chapter, but the book by Marmarelis and Marmarelis (1978) provides a basic introduction to the subject and the multiauthored book edited by Pinter and Nabet (1992) is a recent review of applications of the methods of nonlinear systems analysis to the visual system.

A general finding is that only the first few kernels are required to describe the response properties of neurons in the visual system (e.g., Emerson *et al.*, 1992a). The kernels will be functions of the x and y spatial coordinates and time and are therefore impossible to visualize directly. However, they can be related to spatiotemporal response functions by plotting a transect through the kernel in the Δx, Δt plane for fixed values of Δy. The kernels are then represented as contour plots in the Δx, Δt plane that resemble spatiotemporal response functions (Fig. 16). The first-order kernel describes the response of the cell to single stimuli such as bars of light presented at different loci within the receptive field. The second-order kernel describes the response of the cell to pairs of stimuli such as two bars presented with varying spatial and temporal separations. The first- and second-order kernels can be viewed (approximately) as separating the spatiotemporal response function into linear and nonlinear components. The linear and nonlinear components of the spatial and temporal integration functions can be obtained by taking transects through the kernels for specified temporal or spatial separations, respectively. Estimation of the first- and second-order kernels for neurons in the geniculocortical pathway thus determines both whether or not the neurons have inseparable spatiotemporal response functions (and are, presumably, motion-selective) and whether or not the neurons are behaving as linear systems (in which case their second- and higher order kernels would be zero). This approach has led to several recent models of the neural mechanisms underlying visual motion analysis.

One interesting class of models called *motion energy models* represent the generation of motion-selective receptive fields of neurons in the primary visual cortex of cats as a sequence of black boxes or spatiotemporal filters (Adelson and Bergen,

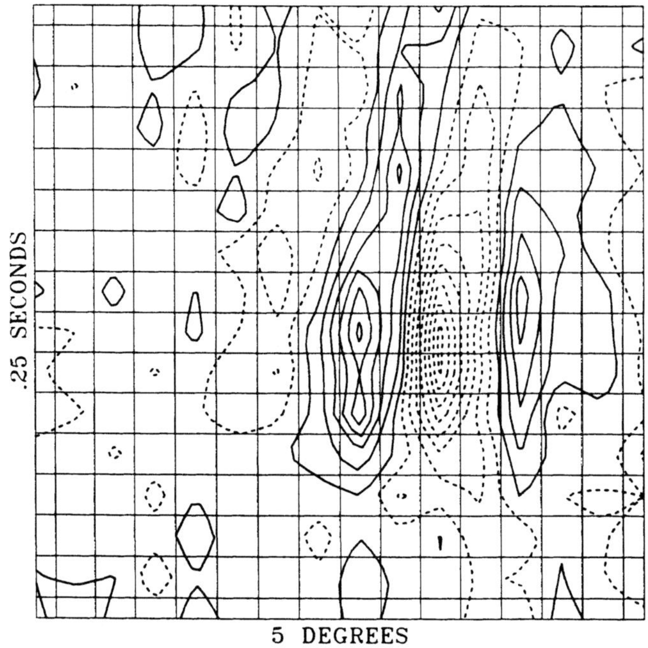

Figure 16. First-order kernel for a simple cell from cat area 17. Space–time plot of the first-order kernel for a simple cell. The time and space origins are located at the lower left corner of the graph. Note that this plot is rotated as compared to the schematic plots shown in earlier figures. Each contour stands for an increment in the magnitude of the response by 0.06 impulse/sec. The dashed contours are excitatory responses to decrements, while the solid contours are excitatory responses to increments. The tilt in the plot, especially at long times, is an indication of the space–time inseparability of the first-order kernel. From Shapley *et al.* (1991), with kind permission of the publisher.

1985; van Santen and Sperling, 1985; Watson and Ahumada, 1985). Each filter represents neurons in one stage of the geniculocortical pathway and is characterized by its input–output relations. The appropriate description of the input–output relation for a filter is its response to a single visual stimulus or to pairs of visual stimuli, so it is described by a spatiotemporal response function or by first- and second-order kernels. Motion energy models borrow from Reichardt's work the idea that a time delay and nonlinear interaction between spatial channels are essential for the generation of motion-selective receptive field properties, but they have a structure that more closely resembles the anatomical organization of the geniculocortical pathway in cats and primates (Fig. 17). Motion energy models typically consist of a pair of oriented response functions that are 90° out of phase with each other. The two filters are said to be in *spatiotemporal quadrature* to each other, or to form a *quadrature pair.* The structure of the quadrature pair is responsible for the time delay because the two spatial channels are arranged so that a sinusoidal input to the first channel appears as a delayed sinusoid (or cosine function) in the second channel. This would occur if the sinusoid is drifting past the two channels with a speed appropriate to their spatial separation. The output from each oriented filter is squared and the two outputs are then summed to produce a final output known as motion energy. Slightly more complex versions of this motion energy model have been shown to reproduce faithfully the responses of cat cortical neurons (Emerson *et al.*, 1992a).

Although motion energy models do account for many of the phenomena associated with local motion detectors in cat primary cortex, they do not account for a variety of nonlinear effects that can be demonstrated in cortical neurons. For example, the responses of cortical neurons typically vary as a function of the stimulus contrast (Heeger, 1991, 1992; Carandini and Hegger, 1994). The neuronal response shows an expansive nonlinearity (Fig. 18) at low stimulus contrasts and a compressive nonlinearity at high contrast values, so that the contrast response functions for cortical neurons are sigmoidal in shape (Fig. 18A). A related phenomenon is that of *contrast normalization.* This means that the contrast response function is constant over a wide range of background luminance levels, suggesting that the visual system adapts to the mean level of background luminance in a way that allows the animal to make contrast discriminations accurately across several logarithmic units of ambient luminance. A mechanism responsible for contrast normalization and other nonlinear phenomena has recently been incorporated into motion energy models. These *normalization models* involve a feedback to the neuron under consideration from other neurons in the cortex (Fig. 19B). Activity from a population of cortical neurons is pooled and used as a source of negative (or inhibitory) feedback to the neuron. The feedback is responsible for the compressive nonlinearity at high contrast levels and satisfactorily accounts for contrast normalization. Normalization models are considered in more detail in the chapter by Carandini *et al.* in this volume.

2.4.3. Properties and Models of Global Motion Detectors

Global motion detectors combine inputs from arrays of local motion detectors. They can be represented formally as a black box or filter that receives inputs from

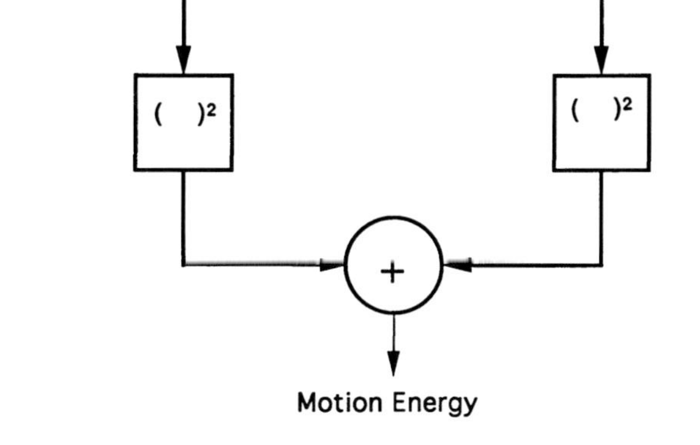

Figure 17. Motion energy model. Input to the model consists of a pair of linear filters which are 90 deg out of phase with each other. These filters are called a quadrature pair. Output from each filter in the pair is squared and then summed to produce an output that is called motion energy.

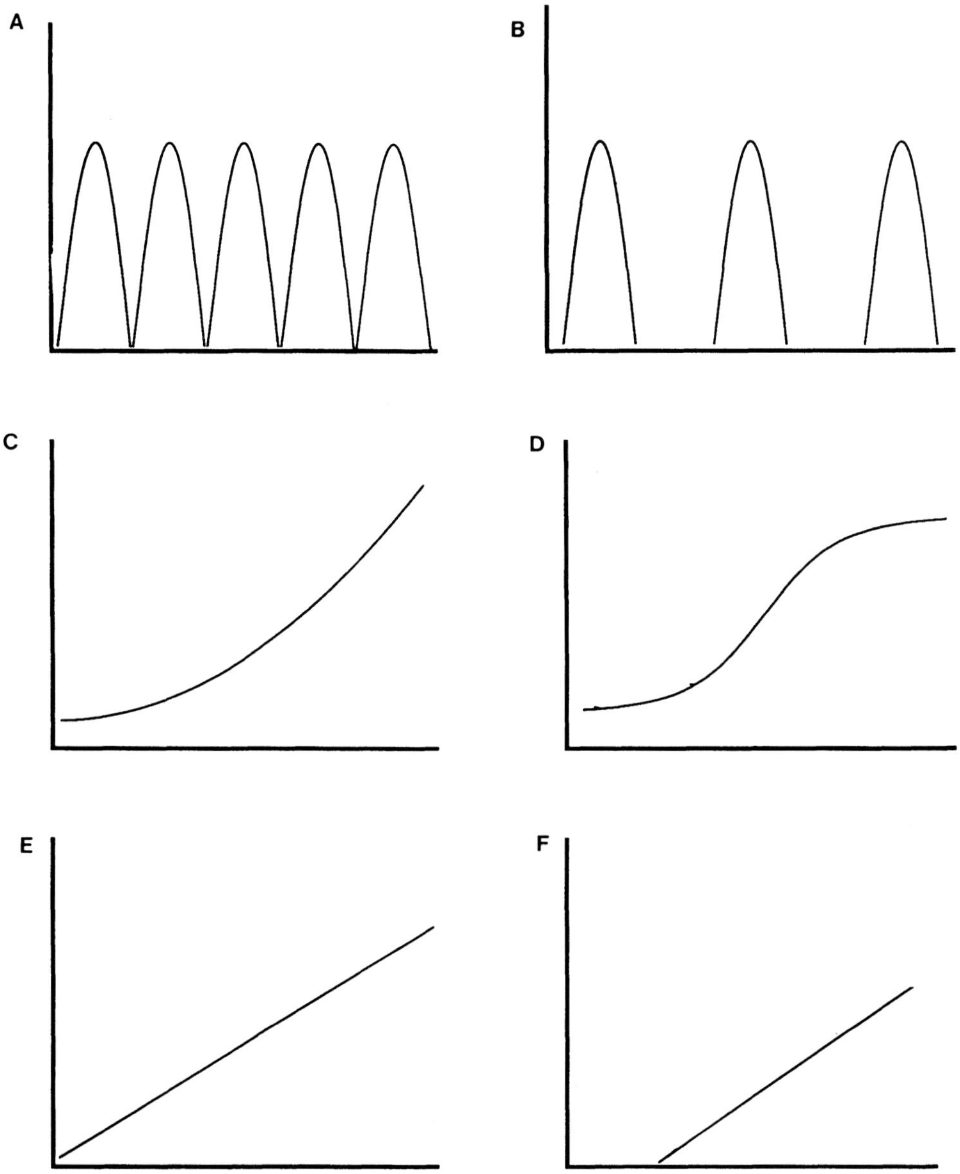

Figure 18. Classification of nonlinearities. (A) Full-wave rectification. A sinusoidal input produces an output in which both the negative and positive lobes of the sinusoid result in positive output. (B) Half-wave rectification. Only the positive lobes of the sinusoid result in nonzero outputs. (C) Expansive nonlinearity. Output is an increasing function of input. Specific examples of expansive nonlinearities are squaring, in which the output of the filter is equal to the input squared, or half-squaring, which results if only the positive lobes of an input sinusoid result in nonzero outputs. (D) Saturating nonlinearity. The magnitude of the output reaches a constant value as the input increases. A specific example of a saturating nonlinearity is a sigmoidal function. (E) Rectification. The output is zero for negative values of the input and increases linearly for positive values of the input. (F) Overrectification. Output is zero for input values less than a threshold value and then increases linearly for higher input values.

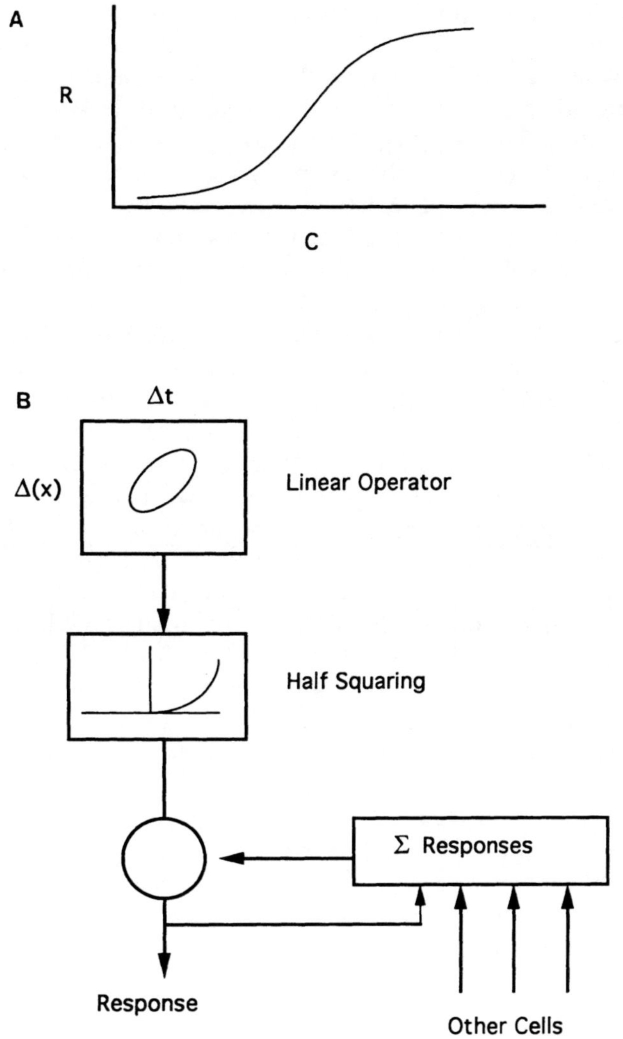

Figure 19. Normalization model. (A) Cartoon of a nonlinear contrast response function. Neural response *R* is plotted as a function of stimulus contrast *C*, producing a sigmoidal function or saturating nonlinearity. (B) Structure of normalization model. Output from an oriented, linear filter or operator is passed through a half-squaring filter. The resulting output is combined with a weighted sum of responses from other cortical cells to produce a final output.

two or more local motion detectors (Fig. 9). Each local motion detector samples stimuli occurring in a specific region of visual space (x_1 and x_2). The global motion detector will thus have a spatial receptive field that is larger than that of each local detector because it is receiving a convergent input from more than one local detector. The local detectors will not respond to every stimulus moving within their receptive fields because each is tuned preferentially to stimuli moving in specific directions with specific speeds. The two local detectors could have very different tuning curves and have different preferred directions and speeds. The functional properties of the global detectors will thus depend upon a combination of the tuning curves for the local detectors.

An important difference between local and global motion detectors is that global detectors are not bilocal. Although they can have restricted receptive fields and respond to stimuli within some (relatively large) locale of visual space, they may respond to stimuli occurring simultaneously over the two local detectors for extended periods of times. This happens because real objects moving in visual space will usually be larger than the receptive fields of local detectors and have trajectories that extend for an order of magnitude longer than those sampled by local motion detectors. A global motion detector, for example, might receive inputs from two local detectors that are providing information about a rectangular object that is passing over both detectors. Each local detector could be simultaneously processing information about the speed and direction of two different edges of the rectangle, which will be moving with different apparent directions. A task for the global motion detector is to extract information about the speed and direction of the object as a whole as opposed to the speed and direction of its components.

To think about this, it is useful to introduce the concept of a *velocity flow field* (Fig. 20). Velocity is a vector quantity that has both a direction and a magnitude. The visual world can consequently be represented as a vector field in which each point in visual space is plotted as a vector with the appropriate direction and magnitude at each $\{x, y\}$ point in visual space. The vectors, of course, change their directions and magnitudes as a function of time. Imagine that you are walking down a crowded city street toward a busy highway on which traffic is moving from left to right. As you walk, stationary objects on the street will have velocity vectors arranged radially around the point toward which you are walking (Fig. 21A). Distant objects will seem to move more slowly and will produce velocity vectors with small magnitudes, while near objects will seem to move faster and produce velocity vectors with larger magnitudes. People on the street will produce vectors that depend upon their direction and speed. Automobiles on the highway will produce vectors oriented from left to right and magnitudes equal to the speed of the automobiles (Fig. 21B). Interpreting the velocity field depends upon the ability of global motion detectors to segment the velocity field into local regions in which the velocity vectors have some orderly relationship to each other. Segmenting out the automobiles moving on the highway is the easiest part of the task because the automobiles produce a very coherent component of the velocity flow field.

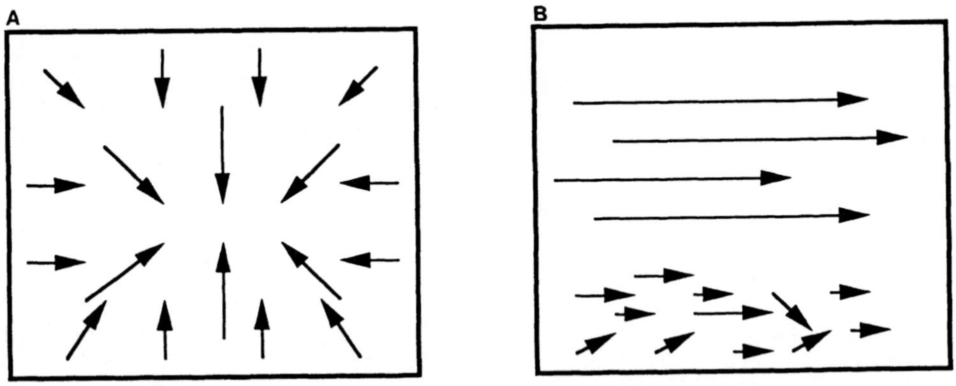

Figure 20. Velocity flow fields. Cartoons of two velocity flow fields discussed in the text. (A) Velocity flow field produced by the observer moving toward a fixed object. (B) Velocity flow field seen by an observer standing on a street with pedestrians and traffic both moving to the right. Length and orientation of the velocity vectors represent the apparent speed and directions, respectively, of objects in the environment.

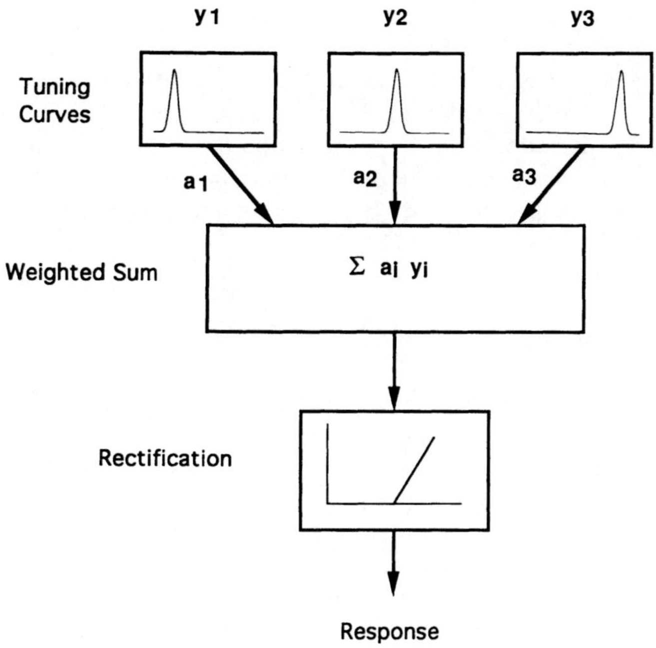

Figure 21. Line element model. The model involves computing the weighted sum of inputs from three filters (*y*1, *y*2, and *y*3) represented by tuning curves with different preferred speeds or directions. Output from the summation is passed through a filter that overrectifies the weighted sum to produce a response.

One approach to modeling a global motion detector is to construct a filter that adds together the inputs from the two local detectors. The global detector would respond to activation of either or both of two velocity channels. This concept is the basis of *line element models* (Fig. 21). The principle is to recognize that the local motion detectors will be tuned to some range of preferred directions and speeds as opposed to responding to one specific direction and speed. This can be modeled by including a tuning function that characterizes each local motion detector. Each local motion detector can be described by a Gaussian function of speed or direction and the global motion detector described by a weighted sum of the various Gaussian functions followed by a rectification.

A second approach that includes potential interactions between velocity channels is provided by *cooperative models.* Like line element models, cooperative models can be constructed around a weighted sum of tuning functions, but they are more complex in that there can be inhibitory interactions between some or all of the velocity channels. Activation of one velocity channel can then lead to activation of additional channels (via excitatory interactions), but also to a suppression of activity in other channels (via inhibitory interactions). A common feature of many cooperative models is that they have *winner-take-all* features in which activity converges upon a particular velocity channel. This happens due to excitatory interactions between the channels tuned to the same direction and inhibitory interactions between channels tuned to different directions.

A third approach is to build into the model a function that is more complex than linear excitatory or inhibitory interactions between velocity channels. A common strategy in artificial intelligence approaches is to define a function on the velocity flow field that optimizes some parameter of the trajectory of a moving object. For example, Yuille and Grzywacz (1988) provide a theory for the perception of coherent motion in which the global motion detector would minimize a function defined on the velocity flow field and the true image velocity. Such theories and the general field of computational vision are a subject in themselves and will not be pursued here because they are not generally phrased in terms of neural mechanisms.

2.5. Summary

The visual system of vertebrates has an organization that is constant in its fundamental features across species. All of the vertebrates have central visual pathways in which visual information reaches the telencephalon in parallel tectofugal and thalamofugal pathways. The organization of the telencephalic visual areas varies, but turtles and mammals have visual cortices that play a role in visual motion analysis. Functional features of neurons in the motion analysis pathways are defined in terms of their receptive field properties, and two distinct types of motion-selective neurons are required. Local motion detectors provide information about the speed and direction of stimuli moving within restricted regions of visual space; global motion detectors integrate the inputs provided by arrays of local motion detectors to elaborate the perceptions of objects moving through the visual environment. These elements of visual system organization are present in all vertebrates, but each species shows a variation upon the basic theme that is consistent with its visual lifestyle.

This thumbnail sketch of visual motion pathways is based upon perspectives drawn from the anatomy of the visual pathways and descriptions of the response properties of neurons as judged from alterations in their firing rates when the animal is presented with specific classes of visual stimuli. It does not, however, provide much information on the neural mechanisms that underlie the design of local and global motion detectors. The problem of understanding the neural mechanisms underlying visual motion analysis is addressed in the following two sections, which consider the neural architecture of motion detectors in mammals and in turtles.

3. Neural Mechanisms: Mammals

Virtually all available information on visual motion analysis in mammals is derived from work on cats or monkeys. Studies on cats have focused to a large extent on the receptive field properties of neurons in the primary visual cortex. Studies on monkeys have tended, particularly in recent years, to emphasize the properties of neurons in the extrastriate areas of the motion analysis pathways. Neurons in the primary cortex of cats and monkeys serve as local motion detectors, while those in the extrastriate areas include those involved in global motion detection.

Motion energy and normalization models of local motion detection capture basic aspects of receptive field organization in mammals, so most work over the past decade has been concerned with examining neural mechanisms related to these models.

3.1.1. Spatiotemporal Receptive Fields

Motion energy models predict that direction-selective neurons have spatiotemporal receptive fields that are oriented or inseparable in the space-time plane, with the slope of the response function reflecting the preferred speed of the neuron. Do real neurons have this property? One point of view is that motion-selective neurons have circular or vertically elongate spatiotemporal response fields that are not obliquely oriented in the space-time plane and respond best to stimuli presented with an optimal separation or D_{opt}. Baker and Cynader (1986) measured D_{opt} for simple and complex cells in cat primary visual cortex by presenting pairs of bars within their receptive fields. The second bar was presented 100 msec after the first bar first appeared in the receptive field; it is said to have a stimulus onset asynchrony (SOA) of 100 msec. Cells responded to bars presented with a range of spatial separations, but gave their best responses in an apparent motion paradigm when the bars were presented with a spatial separation equal to D_{opt}. The value of D_{opt} was directly proportional to the receptive field size for both simple and complex cells. Experiments in which spatial separation and SOA were systematically varied showed that cells typically have optimal responses to SOAs of 60–90 msec and spatial separations of 0.3–0.6 deg (Baker and Cynader, 1988). Spatiotemporal response functions were generally circular in shape or showed only a slight degree of orientation in the space–time plane. The preferred speed of an individual cell (measured in response to a bar moving continuously in the preferred direction) was directly proportional to its D_{opt} and receptive field size (Baker, 1988). Similarly, preferred speed was related to the preferred spatial and temporal frequencies as measured with drifting sinusoidal gratings (Baker, 1990). Neurons preferring higher speeds of bar motion also preferred lower spatial and higher temporal frequencies. This line of thought suggests that speed preference is coded by populations of cortical neurons: neurons with larger receptive fields code for larger velocities than do neurons with small receptive fields.

A contrary point of view is that both simple and complex cells have response functions that are oriented in the space–time plane as predicted by motion energy models. The most detailed studies suggest that the response function has two segments, one at relatively short temporal separations and one at longer temporal separations, resulting in a bend or "kink" in the response function. McLean and Palmer (1989) examined the relationship between the shape of the response function and motion selectivity in simple cells in primary visual cortex of cats. They measured response functions using a reverse-correlation method and stimuli consisting of fields of stationary black dots in which the positions of the dots were randomized between trials. They also studied the speed selectivity of the same cells using moving bars. The majority of cells had oriented spatiotemporal response functions. Cells with unoriented response functions showed little speed selectivity, while the shape of response functions for cells with oriented space–time plots accurately predict the preferred speed of the cell. Comparable results have been obtained for simple cells using stimuli consisting of randomized bars (Emerson and Citron, 1992; McLean *et al.*, 1994; McLean and Palmer, 1994) and squares (DeAngelis *et al.*, 1993a, b, 1996) and for complex cells using randomized bars (Emerson *et al.*, 1992a).

The source of the discrepancy between the two sets of studies is not entirely clear. There is a difference in emphasis in that Baker and colleagues call particular attention to D_{opt}, which is the maximal value in the response function. This is analogous to describing the response functions of neurons in the auditory system by their best or characteristic frequencies (which are the peaks in their tuning curves) and does not imply that the cell responds only to the optimal spatial separation. However, there is a fundamental difference between the two sets of studies with regard to motion energy models. These models predict that D_{opt} should shift as a function of temporal separation and that the extent of this shift is related to the preferred speed of the unit. Baker and Cynader (1988) examined this point and found no significant correlation, while other workers found the predicted relationships. All of the work was done on cats, using standard criteria for simple and complex cells, but used different stimulus patterns. At this time, evidence from three laboratories supports the concept that motion-selective neurons have oriented spatiotemporal response functions for Fourier or first-order motion.

Spatial and temporal integration functions can be obtained by taking transects through the response functions at fixed temporal or spatial intervals, respectively. The spatial integration functions obtained in this way resemble those obtained by mapping receptive fields with flashing spots or bars and generally resemble Gabor functions (e.g., Baker and Cynader, 1986; Baker, 1988). The temporal integration functions (e.g., McLean *et al.*, 1994) are typically biphasic with one facilitory peak at about 75 msec and a second inhibitory peak range between 100 and 400 msec. Functions with comparable shapes have been employed in constructing motion energy and related models. Adelson and Bergen (1985), for example, used the function

$$f_n(t) = (kt)^n \operatorname{Exp}(-kt)\ [1/n! - (kt)^2/(n+2)!] \tag{8}$$

with values of $n = 3$ and $n = 5$ for the temporal integration function in their motion energy model. Watson and Ahumada (1985) used the function

$$f_i(t) = [u(t)/t_u(n_i - 1)!][(t/t^{ni})(t/(t_i)^{ni} - 1]\ \operatorname{Exp}(-t/t_i) \tag{9}$$

in their model. Here, $u(t)$ is a unit step function. These functions are derived from previous psychophysical (Robson, 1966) and neurophysiological experiments that attempt to measure the sensitivity of ommatidia in the horseshoe crab, *Limulus* (Fuortes and Hodgkin, 1964). They are appropriate in shape for the temporal integration functions for motion-selective neurons or filters. They are zero or have small values for $\Delta t = 0$, so that the cell does not respond strongly to stimuli presented simultaneously to the two spatial channels. They increase to a peak, so that the cell gives a bilocal response. If this peak were constant, it would correspond to a local motion detector tuned to a D_{opt}. If the peak shifts as a function of spatial separation, it would correspond to a local motion detector that is oriented in the space–time plane. Studies that analyze the response functions of simple or complex cells give descriptions of the temporal integration functions, but do not say anything about the neural mechanisms that underlie them.

3.1.2. Linear and Nonlinear Mechanisms

Like Reichardt detectors, motion energy models suggest that motion selectivity requires a nonlinear step. Is a nonlinear mechanism actually required for the generation of motion-selective receptive fields and, if so, what is the source of the nonlinearity? Phrased in terms of nonlinear systems theory, the prediction is that motion-selective cells have second-order kernels that are oriented in the space–time plane. This issue has been examined for simple and complex cells with white noise methods and with sinusoidal stimuli. White noise methods show, as expected, that complex cells have both first- and second-order kernels, but that both kernels can have oriented regions when plotted as response functions in the space–time plane (Emerson *et al.,* 1992a). As indicated above, simple cells show first-order kernels that may or may not have oriented regions in the space–time plane, but those cells with oriented response functions are motion-selective and have preferred speeds that can be predicted from the slope of the response function (Emerson and Citron, 1992; DeAngelis *et al.,* 1993a; McLean *et al.,* 1994; McLean and Palmer, 1994). These results suggest that motion selectivity derives from both linear and nonlinear components.

Quantitative analysis with a variety of sinusoidal stimuli supports this conclusion. The response of a linear system to a sinusoidal input is viewed classically as a sinusoid that is shifted in phase an invariant amount relative to the input. However, Movshon *et al.* (1978a) showed that the output from a linear system can be a sinuosoid with a phase that is a continuously varying function of the spatial position of the input if the temporal properties of the system vary as a function of position. Reid *et al.* (1987, 1991) examined this idea for simple cells in cat visual cortex, using extracellular recording techniques. They found that cells lacking direction selectivity have responses to sinusoidal inputs with invariant phase, while directionally selective cells show responses with continuously varying phase. This is consistent with a purely linear mechanism underlying directional selectivity. However, the same type of experiment also provides evidence for a nonlinearity when the responses to stimuli moving in the nonpreferred direction are considered (Reid *et al.,* 1987, 1991; Shapley *et al.,* 1991). The relationship between the amplitude and phase of the response of a system to a sinuosoidal input can be represented by plotting response amplitude as a function of phase in a polar plot (Movshon *et al.,* 1978a). The result is typically an ellipse or circle for direction-selective cells and a straight line (a degenerate ellipse) for cells that are not direction-selective. It is possible in this format to compare the actual responses of simple cells to contrast reversal gratings with the response predicted assuming a purely linear mechanism for direction selectivity. The results are ellipses that have pinched profiles ("wasp-waisted" ellipses) for stimuli moving in the nonpreferred direction. This indicates that the model underestimates the response of the cell to stimuli moving in the nonpreferred direction, and raises the possibility of a suppressive or inhibitory nonlinearity operating in the nonpreferred direction.

Experiments conducted with intracellular recording methods provide evidence that the subthreshold mechanisms underlying motion selectivity are linear in the spatial domain, but nonlinear in the temporal domain. Postsynaptic potentials evoked in simple cells using drifting or contrast reversal gratings (Ferster *et al.*, 1994; Jagadeesh *et al.*, 1993) show a general resemblance to the results obtained with extracellular data in that the potentials are sinusoidally modulated and show a continuously varying phase (Fig. 22). The extent of the phase shift is related to the degree of direction selectivity, indicating that the phase shift is an important determinant in direction selectivity. The postsynaptic potentials produced by stimuli moving in both the preferred and nonpreferred directions can be predicted by summing together the phase-shifted responses to stationary stimuli. Ferster and colleagues suggest that the difference between the intracellular and extracellular data is due to the threshold nonlinearity resulting from the spike threshold. Consistent with this, the directional indices calculated from the firing rates of cells underestimate the responses of the cells in the nonpreferred direction (Ferster, 1988; Nelson *et al.*, 1994). The suggestion was that nonlinear mechanisms related to spike

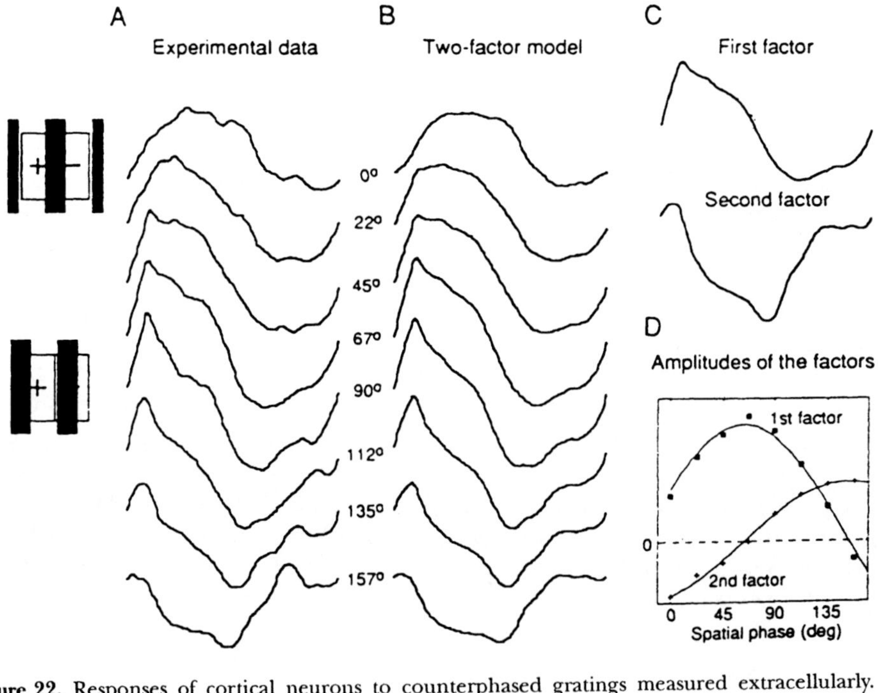

Figure 22. Responses of cortical neurons to counterphased gratings measured extracellularly. This figure illustrates an analysis of the responses of neurons in cat area 17 to counterphased gratings. (A) Insets show the receptive field of a directionally selective simple cell with grating stimuli superimposed. Each trace represents the intracellular response of the cell to a counterphased flicker (2 Hz) of a stationary sinusoidal grating with 60% contrast in the spatial phase shown to the left. The units of the response are arbitrary, but the same across all recordings. (B) The data fit by a two-factor model that accounts for 98.8% of the data variance. (C) The two main factors of the model. These factors are components in an orthogonal basis in the two-dimensional space of the response profiles. (D) Amplitudes of the two main factors versus spatial phase. The good fit to sinusoidal variation indicates the linearity of the subunits to local spatial contrast. Data from Jagadeesh *et al.* (1993). Figure from Kontsevich (1995), with kind permission of the publisher.

generation accentuate, but are not necessary for, direction selectivity. Kontsevich (1995) extended the analysis of these data by expressing the postsynaptic potentials in terms of a set of orthogonal basis functions. The synaptic responses could be expressed as a weighted sum of two functions or factors, confirming the suggestion that thalamic inputs are summed linearly in the spatial domain. However, the functions were nonlinear in the temporal domain.

Most workers have accepted the idea that the generation of action potentials is a source of nonlinearity, but have tended to view it from somewhat different perspectives. One idea is that the nonlinearity is a half-wave rectification. This concept derives from the fact that simple cells tend to have very low spontaneous firing rates. Their responses to sinusoidal inputs will therefore not be complete sinusoids, but will reflect only increases in firing rate due to contrasts that lie above the mean luminance of the background. The cell cannot respond to contrasts below the mean luminance by decreasing its firing rate because of its low spontaneous rate. Another point of view is that the nonlinearity is a threshold effect with the threshold either at zero (i.e., rectification) or at some membrane potential that corresponds to the spike threshold (i.e., overrectification). Both rectification and overrectification reflect the idea that the spike rate is zero below the threshold and then increases linearly with increasing depolarization. Such a linear relation is seen in physiological data in that the interspike interval between the first and second action potentials generated by injecting depolarizing currents into a neuron is often a linear function of the amount of current injected (e.g., Connors and Gutnick, 1990). Finally, some authors have suggested that the nonlinearity is either a squaring operation (as in the original motion energy models) or an expansive nonlinearity (common in normalization models). There have been attempts to determine the nature of the nonlinearity by comparing response functions of real cells to alternative models (Emerson and Citron, 1992) and by using techniques for identifying nonlinearities from systems engineering (Emerson *et al.*, 1992b). These studies support a squaring nonlinearity as important in the generation of direction selectivity. However, the approach of describing simple and complex cells in terms of spatial and temporal filters is subject to the general limitation of systems theory: although systems theory can describe a biological system as a cascade of filters, it cannot uniquely determine the nature of the underlying processes [see Regan (1991a) for a discussion of this point]. The intracellular recording studies conducted by Ferster and his colleagues are then of particular value because they allow a direct comparison between the subthreshold events and the firing patterns of the neurons. It appears from this approach that some nonlinearity is introduced both in the temporal integration of postsynaptic potentials and in spike generation.

In addition to the problem of determining the neural mechanisms responsible for nonlinearities associated with motion selectivity, a second issue raised by the experiments with contrast reversal gratings is to understand the mechanisms responsible for the continuously varying phase relationships reported in both extracellular and intracellular experiments. One possibility is that the necessary temporal delay is introduced at the level of the dorsal lateral geniculate complex. The X and Y retinal ganglion cells of cats project to the dorsal or A layers of the dorsal lateral geniculate complex of the thalamus, and these neurons project in turn to the primary visual cortex. Segregation of information from the X- and Y-cells is maintained at the level of the geniculate, so that there are both X-like geniculate neurons and Y-like geniculate neurons (for a review see Stone, 1983). A further distinction was made by Mastronarde (1987a, b), who suggested that X-like geniculate neurons can be placed into two groups based on the temporal characteristics of their responses to visual stimuli. One group of X-like cells respond to brief presentations of spots or bars of light by a decrease in firing rate during the period of retinal input that is followed by an increase in firing rate when retinal input ceases. These cells thus show a response to a visual stimulus that lags the onset of the response, and Mastronarde termed these *lagged cells* (Fig. 23). Other cells show only a transient increase in firing at the onset of the stimulus and are called *nonlagged cells*. The distinction between lagged X-like cells (X_L) and nonlagged X-like cells (X_N) has been confirmed (Humphrey and Weller, 1988a), shown to represent morphologically distinct populations of geniculate neurons (Humphrey and Weller, 1988b), and extended to include Y-like geniculate neurons (Mastronarde *et al.*, 1991). Lagged cells have thinner axons and slower conduction velocities than do nonlagged cells (Humphrey and Weller, 1988a, b).

The synaptic mechanisms underlying the initial decrease and subsequent increase in firing seen in lagged cells have not been studied in detail. The decrease in firing is believed to result from a feedforward inhibition of the lagged cells by inhibitory interneurons within the geniculate (Mastronarde, 1987a, b; Humphrey and Weller, 1988a, b). The inhibition is mediated by the $GABA_A$ subtype of GABA receptor (Heggelund and Hartveit, 1990). Retinal ganglion cell terminals are known to access both the 2-amino-3-hydroxy-5-methyl-4-isoxazole propionic acid (AMPA) and NMDA subtypes of glutamate receptors on geniculate neurons, but lagged cells show a greater dependence upon excitation mediated via the NMDA subtype of glutamate receptor than do nonlagged cells (Heggelund and Hartveit, 1990; Kwon *et al.*, 1992). McCormick (1990) reports that geniculate neurons have two transient potassium conductances that inactivate with different time constants. Activation of a cell that is hyperpolarized to about −65 mV results in a lagged response. Activation of the cell at more-depolarized membrane potentials produces a nonlagged response due to inactivation of the potassium conductance.

Regardless of the biophysical and synaptic mechanisms underlying the properties of the two types of geniculate neurons, lagged and nonlagged cells can contribute to the phase relationships seen in motion-selective cortical neurons because a quadrature pair could consist of an X_L and an X_N geniculate neuron with spatial receptive fields that are displaced from each other and converge upon the same cortical neuron. A moving stimulus would produce a rapid increase in firing from the X_N neuron and a delayed response from the X_L neuron. Linear summation of inputs from the X_N and X_L cells by a cortical simple cell could result in a spatiotemporal response function with a first-order or linear kernel that is oriented in the space–time plane, but the temporal delay would be a property of the intrinsic circuitry of the geniculate neurons. Consistent with this, Saul and Humphrey (1990) showed that X_N-cells respond to a sinusoidal stimulus with a maximal response at the maximal luminance of the grating, while the X_L-cells show a maximal response that is delayed by a quarter of a cycle. They also showed that direction-selective responses in cortical neurons can be modeled as a sum of X_L- and X_N-cells with the proper spatial separation. The X_L- and X_N-cells have different frequency response profiles, and Saul and Humphrey (1992a, b) subsequently showed that simple cells in areas 17 and 18 have frequency response profiles consistent with the

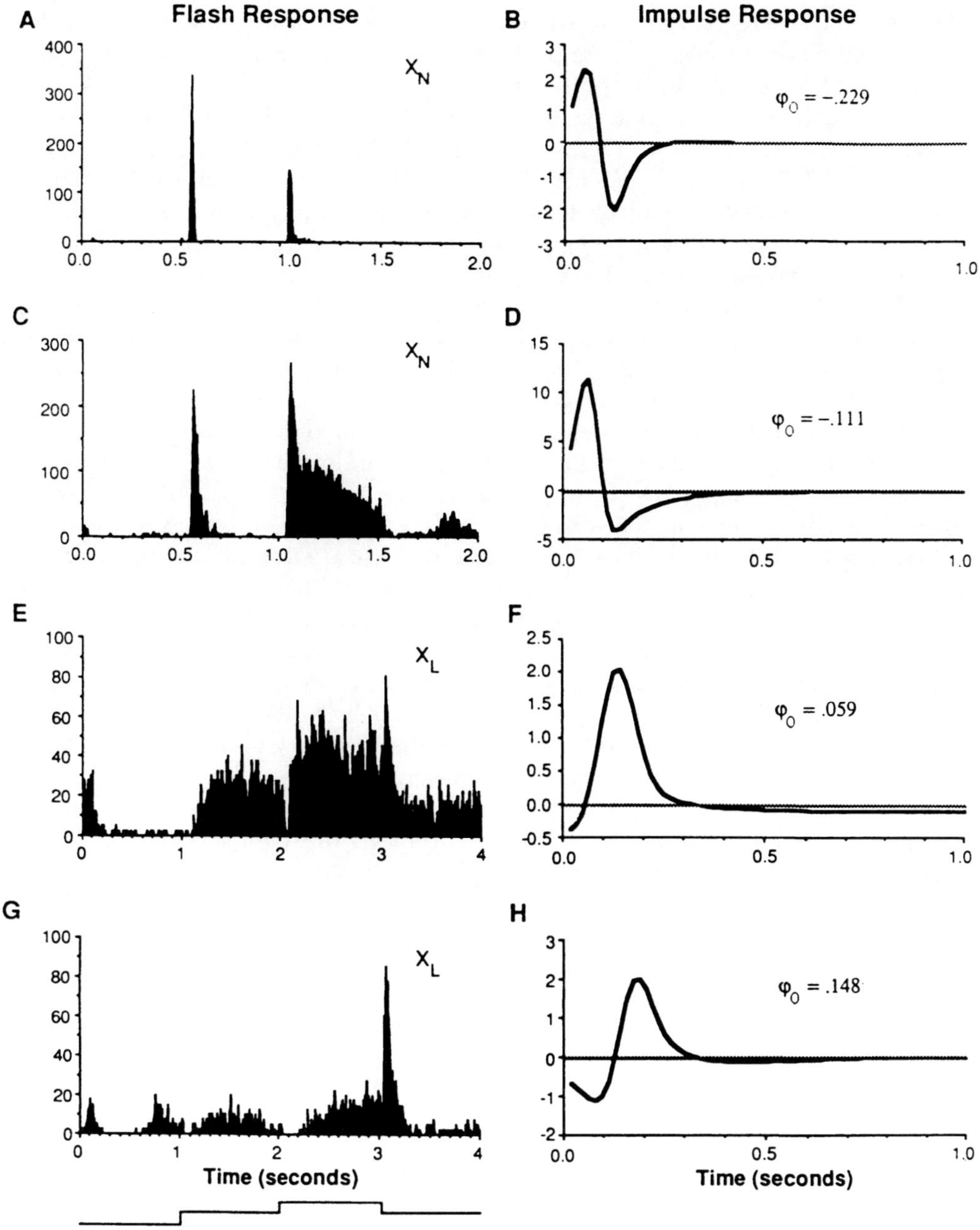

Figure 23. Responses of lagged and nonlagged geniculate cells. (A, C, E, G) Responses of four X-like cells from cat dorsal lateral geniculate nucleus. Two are nonlagged cells (X_N) and two are lagged cells (X_L). The time course of a four-part flashing spot stimulus is shown in the bottom trace. Notice that the two lagged cells show significant delayed or lagged responses. Vertical scales are firing rates in spikes per second. (B, D, F, H) Impulse response functions calculated for the four cells. Vertical scales are in arbitrary units. From Saul and Humphrey (1990), with kind permission of the publisher.

hypothesis that they receive convergent inputs from X_L- and X_N-cells. Further, detailed exploration of the receptive fields of simple cells provided evidence for lagged and nonlagged inputs to subunits within the receptive field based on the differences in frequency response profiles demonstrated by X_L- and X_N-cells (Saul and Humphrey, 1992a, b; Humphrey and Saul, 1993).

Although the variations in phase with stimulus position that are seen in cortical cells may well reflect a convergent input from X_L- and X_N-cells, this does not rule out the possibility of a concomitant contribution from intracortical circuitry. There do not seem to have been direct experimental investigations of this point, but two recent models emphasize a potential role for horizontal connections within the cortex in generating direction-selective properties in cortical neurons. Wörgötter and Holt (1991) propose a model in which inputs from a row of geniculate neurons (with receptive fields modeled as a difference of Gaussians multiplied by an exponential temporal integration function) distribute activity to an array of cortical neurons. These neurons are not speed-selective, but converge upon a second-order cortical neuron that has an inseparable response function and is speed-selective. They argue that the spatial distribution of activity within the cortex is a key element to establishing directional selectivity. Maex and Orban (1992) propose a model in which an ON-center and OFF-center pair of geniculate neurons project to both excitatory and inhibitory cortical cells. The inhibitory cells inhibit each other and both excitatory cells; each excitatory cell excites itself. The model explores mechanisms by which the velocity response curves of low-pass cortical cells are established. It suggests that feedback excitation within the cortex is important in generating the low-pass behavior of cortical cells, particularly because the time course of intracortical systems includes an EPSP (presumably mediated by NMDA receptors) that has a relatively long time course and extends the time window over which geniculate inputs can coincide with intracortical excitation.

This leaves open the possibility that the orientation of the spatiotemporal response functions of motion-selective neurons in the cortex result from both geniculate and intracortical processes. In fact, some elements of intracortical processing are obligatory because the cortical neurons must have temporal integration functions with sufficient memory to store the input from the nonlagged neuron so that it can be combined with the input from the lagged neuron. The cortical circuitry needs its own way of establishing a temporal integration function with a longer memory. The neural mechanisms that underlie such storage processes have not been explored for neurons in cat cortex.

3.1.3. Excitatory and Inhibitory Mechanisms

Models for motion selectivity differ in whether or not inhibition is considered essential for the generation of motion selectivity. Reichardt detectors and motion energy models invoke multiplicative nonlinearities that do not necessarily require inhibitory interactions. Barlow–Levick models, by contrast, can include an inhibitory mechanism acting in the nonpreferred direction. Several groups have extended the Barlow–Levick paradigm to simple and complex cells in cat striate cortex (Goodwin *et al.*, 1975; Emerson and Gerstein, 1977; Ganz and Felder, 1984). The experiments consisted in presenting a priming stimulus followed by test stimuli presented at specific loci in the preferred or nonpreferred directions. Results from all of the experiments were consistent with an inhibitory mechanism playing a significant role in the null direction. Evidence for a facilitatory mechanism acting in the preferred direction was less consistent. A prediction from a Barlow–Levick model is that application of a pharmacological agent that blocks inhibitory synapses should reduce or abolish orientation or direction selectivity. Experiments of this type have been carried out by recording the receptive field properties of neurons in cat primary visual cortex, using extracellular electrodes, and then iontophoretically applying a $GABA_A$-receptor antagonist such as bicuculline (Sillito, 1975, 1977, 1992; Tsumoto *et al.*, 1979). Bicuculline reduces both the orientation and direction selectivity of cortical neurons, suggesting that $GABA_A$-mediated inhibition is essential to the elaboration of receptive field properties. The effects of applying $GABA_B$-receptor antagonists on the stimulus specificity of cortical neurons has not been extensively investigated, but two studies suggest that blockade of $GABA_B$ receptors has little (Baumfalk and Albus, 1988; Pflegler and Bonds, 1995) or only a minimal effect on orientation selectivity.

The implication is that stimuli moving in the nonpreferred direction should elicit IPSPs, while stimuli moving in the preferred direction should elicit EPSPs. However, intracellular recording studies have consistently shown that both EPSPs and IPSPs are elicited by stimuli moving in either direction. Creutzfeldt *et al.* (1974) found that stimuli moving in both the preferred and nonpreferred directions elicited a combination of EPSPs and IPSPs in cat simple and complex cells. They discovered that the balance of excitation and inhibition was altered by the direction of the stimulus, so that stimuli moving in the nonpreferred direction produced a greater number of IPSPs. Innocenti and Fiore (1974) found that both EPSPs and IPSPs were present in the responses of neurons in area 17 to a range of stimulus speeds, although IPSPs were present in responses to higher speeds than were EPSPs, consistent with the tendency of neurons in area 17 to respond to relatively slow speeds. Ferster (1986, 1988) found that oriented bars moving in either the preferred or nonpreferred directions produce both EPSPs and IPSPs in cat simple cells. His results differ from those of Creutzfeldt *et al.* (1974) in that both IPSPs and EPSPs were more frequent for stimuli moving in the preferred direction. All of these studies suggest that both stationary and moving stimuli result in a balance of excitation and inhibition. Cells are activated in a push–pull manner with excitation and inhibition being recruited in parallel. Ferster (1986) points out that this is consistent with the known anatomy of projections within cat primary visual cortex because geniculate afferents synapse upon both excitatory and inhibitory interneurons in parallel.

In an attempt to reconcile the extracellular and intracellular experiments, Nelson *et al.* (1994) used intracellular dialysis of cortical neurons with agents that block the chloride channels involved in the activation of $GABA_A$ receptors and the potassium channels involved in $GABA_B$-mediated inhibition. They found that orientation and direction selectivity show minimal change when inhibition is blocked in an *individual* cell and suggest that iontophoretic application of bicuculline alters the overall balance of inhibitory activity in *populations* of cortical neurons. Somers *et al.* (1995a, b) examined this idea in some detail by constructing a network model of visual cortex in which individual neurons are represented by compartmental models of neurons. They were able to simulate the results obtained by both iontophoretic application of bicuculline to populations of neurons within the network and the results obtained by blocking inhibition to a single neuron. Consistent with both sets of experimental studies, they found that the selectivity of neurons to stimuli with preferred orientations was decreased by reducing the inhibition acting on populations of neurons, but not by abolishing inhibition to individual neurons.

This idea is consistent with data obtained by Bonds, who used extracellular recording methods to characterize the responses of cat cortical cells to a variety of stimuli (Bonds, 1989). He found that cortical inhibition tends to be global or nonspecific in nature, in contrast to being evoked by stimuli moving specifically in a null direction. For example, experiments in which contrast gratings were presented with differing orientations demonstrated that a mask grating can reduce the response of a test grating regardless of its orientation (Bonds, 1989; Bauman and Bonds, 1991). These findings are interpreted as evidence of a general or *global inhibition,* consistent with the finding that application of bicuculline results in a general hyperexcitability of cortical neurons as well as decreased orientation selectivity (Sillito, 1975, 1977). The temporal dynamics is relatively complicated in that presentation of stimuli with increasing and then decreasing contrasts produces a hysteresis effect suggestive of inhibitory effects that persist for up to 54 sec. These effects most likely play a role in contrast gain control of the kind addressed by normalization models.

The normalization models suggest that the compressive nonlinearity underlying normalization takes place at the level of individual neurons and involves a divisive or *shunting inhibition.* This type of inhibition has been documented in some systems and occurs when the resting membrane potential of the neuron is near the chloride reversal potential of about −70 mV. Activation of $GABA_A$ receptor/channel complexes does not produce an inhibitory conductance because the chloride ions are subjected to little or no driving potential. However, the open channels can shunt current from excitatory synapses if the inhibitory and excitatory synapses are situated near each other and are active at roughly the same time. The interaction between excitatory and inhibitory synapses is nonlinear in this case and the inhibitory synapses have the effect of dividing the amplitude of the compound EPSPs (Blomfeld, 1974).

Shunting inhibition is an attractive concept from a theoretical point of view and has inspired considerable discussion. Although known synaptic mechanisms were adequate to account for the interactions between IPSPs and EPSPs required for the Barlow–Levick model, it was more difficult to think of plausible synaptic mechanisms that could account for the nonlinear interactions required for the Reichardt model. Koch and Poggio addressed this issue in several theoretical papers (e.g., Koch and Poggio, 1983a, b) that argue shunting inhibition could serve as a substrate for the kinds of nonlinear interactions required. However, attempts to verify the role of shunting inhibition in motion selectivity face several technical difficulties. The classical form of postsynaptic inhibition is detectable with intracellular recording electrodes as a hyperpolarization of the cell's membrane or IPSP. Shunting inhibition, by contrast, is "silent" in that it produces no hyperpolarization because the cell membrane is at the equilibrium potential for chloride ions, and the opening of $GABA_A$ receptors by GABA consequently produces no flux of chloride ions due to the absence of a driving potential for chloride ions. Shunting inhibition, however, can be measured as an increase in the conductance of the cell due to the opening of chloride channels. This conductance change might be detectable by a recording electrode situated in the soma of the cell if the inhibitory synapse were located relatively close to the soma. It would be more difficult to detect if the synapse were more remote.

Several laboratories have made concerted efforts to find evidence for shunting inhibition in cortical neurons. Martin, Douglas, and colleagues used small intracellular current pulses to search for shunting inhibition using an *in vivo* cat preparation (Berman *et al.*, 1991; Douglas *et al.*, 1991). They found that stimuli moving in the nonpreferred direction did not produce detectable conductance changes, and thus found no evidence to support a role for shunting inhibition in direction selectivity. They considered the possibility that the shunting could involve excitatory and inhibitory synapses upon dendritic spines. The necks of the spines might serve to electrically isolate the spine heads from the shafts of the dendrites making conductance changes difficult to detect at the soma. However, serial reconstruction of the synapses upon dendritic spines failed to reveal the necessary spatial relationships between presumed excitatory and inhibitory synapses (Dehay *et al.*, 1991). Koch *et al.* (1990) carried out theoretical calculations of the visibility of shunting on distal dendrites from the soma and concluded that it should be possible to detect physiologically significant shunting with intracellular electrodes in the soma. Ferster and Jagadeesh (1992) examined the possibility that shunting inhibition plays a role in orientation selectivity by using electrical stimulation of geniculate fibers. Test EPSPs were elicited in cortical neurons by electrical stimulation of geniculate afferents at various times during the presentation of visual stimuli. The test EPSPs consequently appeared against a background of visually evoked EPSPs and IPSPs. The amplitudes of the test EPSPs were found to vary linearly as a function of the membrane voltage when the visual stimulus was moving in both the preferred and nonpreferred directions. Shunting inhibition should produce nonlinear reduction of the amplitude of the test EPSPs by stimuli moving in the nonpreferred direction, and Ferster and Jagadeesh were unable to support a shunting inhibition model. There is no doubt that the kinds of nonlinearities that are explained by normalization models occur and that some form of feedback nonlinearity is required. At this point, though, there is no indication that shunting inhibition is a good candidate for the neural mechanism underlying this nonlinearity.

3.2. Global Motion Detectors

Work on the neural mechanisms underlying global motion detection is limited by the absence of studies on the physiological properties of global motion detectors using intracellular recording methods, but some ideas can be developed from work with extracellular recording methods and psychophysical experiments.

3.2.1. Spatiotemporal Receptive Fields

Integration of a stimulus moving in one direction by neurons in the extrastriate area MT of macaques was measured by Mikami *et al.* (1986a) using a stroboscopic presentation of bars. This is an extension of the logic of the Barlow–Levick experiment, which used the presentation of a bar in one region of the receptive field of rabbit ganglion cells to suppress or enhance the response of the cell to stimuli presented at a different point of space and time. Mikami *et al.* plotted the responses of MT neurons as measured by peak firing rate as a function of the spatial and temporal separation of a sequence of bars. The space–time plots showed that MT neurons were not responsive to stimuli presented with temporal separations greater than 320 msec, but showed a prominent facilitation for temporal separations of 80 msec or less. The maximal spatial separations for which a displacement appeared as a movement were larger (0.2–6.0 deg) for neurons in MT than (0.2–1.3 deg) in V1 (Mikami *et al.*, 1986b). By using the same stimuli in physiological recordings from V1 and MT in macaques and in psychophysical studies with humans, Newsome *et al.* (1986) determined that neurons in both V1 and MT could be involved in analysis of stimuli moving over a range of speeds from slow to fast, but that MT may be more heavily involved in analyzing stimuli moving with faster speeds.

These findings could be viewed as demonstrating that neurons in MT simply have larger receptive fields, and sample motion at a coarser spatial scale than do neurons in V1. However, experiments with random-dot kinetograms suggest that neurons involved in global motion detection combine inputs from many spatial channels with smaller spatial scales. The kinetograms consist of a field of dots moving in various directions. Within certain limits, the field appears to move as a whole in a direction equal to the mean direction of the individual dots (Williams and Sekuler, 1984). Subjecting the kinetograms to a high-pass filter does not impair the ability of subjects in psychophysical experiments to perceive a coherent motion. This suggests that global motion detection does not depend upon sampling motion at coarse spatial scales (A. T. Smith *et al.*, 1994). Watamaniuk *et al.* (1989) were able to account for the ability of observers to see coherent movement in the random-dot kinetograms using a line element model in which the responses of individual tuned velocity channels were summed. The same models could also account for experiments in which a single dot moved along a continuous trajectory within a field of dots with random trajectories (Watamaniuk *et al.*, 1995). Ability to detect the trajectory of the dot improves with increases in the duration of the trajectory up to durations of 250–400 msec. The trajectory of the dot was reliably detected when it changed its direction a small amount—about 30 deg or less—per frame. These findings suggest that trajectory motion is detected in a background of motion noise because the motion detectors are enhanced when local motion detectors with similar directional tuning are stimulated in a sequence along their preferred directions.

Snowden and Braddick (1989, 1991) used similar experiments to suggest that dots moving continuously in the same direction would recruit the activity of motion detectors with increasingly larger sampling bases and lead to an improvement of the signal-to-noise ratio of the motion detection process. A clever attempt to learn more about the signal-to-noise ratio in motion detectors involves using stimuli consisting of a rectangular frame of coherently moving dots in a background of randomly moving dots (Frederiksen *et al.*, 1994a, b). The strategy is to manipulate the number of individual motion detectors activated by the stimulus by keeping the area of the coherent frame constant, but varying its height and width. The data were consistent with a model in which each individual detector is subject to a certain amount of noise and the signal-to-noise ratio is improved by averaging the output of a larger number of individual detectors.

Work in which cats are raised in an environment that alters neurons in the lateral suprasylvian areas supports a role for global motion detectors in improving the signal-to-noise ratio in motion analysis tasks (Pasternak *et al.*, 1985). Kittens were allowed to see only stroboscopic presentations of moving stimuli while they were growing up. The regime reduces the percentage of direction-selective neurons in the lateral syprasylvian areas from a normal value of about 64% to 5% (Pasternak *et al.*, 1990). Experimentally modified cats were able to discriminate the direction of a field of coherently moving dots in the presence of low levels of dots moving in random directions, but performed well below normal cats as the noise level increased. Experimental animals also had difficulty discriminating small differences in direction. Surprisingly, however, experimental animals were better than normal animals in discriminating the directions of movements taking place over larger spatial scales (Rudolph *et al.*, 1994). The latter findings were consistent with modified line element models.

Ability to detect moving dots in random-dot kinetograms is linked directly to neurons in MT by combined psychophysical and neurophysiological experiments on monkeys. Lesions of MT impair the ability of a monkey to detect coherently moving dots in kinetograms (Newsome and Pare, 1988). Extracellular recordings from MT neurons demonstrate that the direction selectivity of the neuron varies as a function of the percentage of coherently moving dots in the kinetogram in a way that closely parallels the ability of the same monkey to determine the direction of the dots in psychophysical experiments (Newsome *et al.*, 1989; Britten *et al.*, 1993). These experiments were quantified by constructing psychometric functions in which the percentage of correct choices is plotted as a function of the percentage of dots in the kinetogram moving in a correlated fashion. A relationship between neuronal activity and the psychometric function could be established by using electrical stimulation of neurons in MT (Salzman *et al.*, 1990) and MST (Celebrini and Newsome, 1995) to shift the psychometric function that is calculated based on psychophysical experiments in the same monkey. These experiments not only support roles for MT and MST neurons as global motion detectors, but suggest that individual neurons have motion analysis capabilities that match those of the animal in terms of sensitivity.

All of this work leads to the conclusion that neurons in areas such as MT, MST, and the lateral suprasylvian areas are functioning as global motion detectors that combine inputs from individual local motion detectors serving as velocity channels. The integration of information across velocity channels allows the detector to calculate global features of stimulus velocity and to improve the signal-to-noise ratio of the system. An interesting approach to studying the interactions *between* velocity channels in global motion detectors is to examine the responses of neurons to plaid stimuli (Movshon *et al.*, 1986). A plaid is composed of two sinusoidal or rectangular contrast gratings moving with different orientations. The perception evoked by the plaids depends upon the stimulus conditions, so that the two gratings are seen under some circumstances to move coherently as a dark plaid upon a lighter background, while they appear as two separate gratings sliding past each other (*transparent motion*) under other circumstances. Movshon *et al.* (1986) used psychophysical methods to define the conditions under which coherent motion was observed by humans, and used extracellular recording methods to characterize the responses of neurons in areas VI and MT in macaques for stimuli that appeared coherent or transparent to humans. The direction in which the plaid was moving was different from the directions of the two component sinusoids. In the physiological experiments, they found that neurons in V1 behaved as local motion detectors and responded to the motion of the component sinusoids if the sinusoids moved in the preferred direction of the neuron. These neurons displayed what was called *component motion selectivity.* Many neurons in MT, by contrast, did not respond well to individual components of the plaid, but did respond robustly when the two gratings were moving under conditions that evoked the perception of coherent movement in human observers. They displayed *pattern motion selectivity.* This suggests that MT neurons combine information from two different velocity channels to provide information about a stimulus perceived as a coherent object moving with a direction different than either component.

The psychophysical experiments showed that the two gratings were more likely to be perceived as a coherently moving plaid if they were similar in features such as contrast, spatial frequency, and direction. The problem of predicting the direction of pattern motion for a coherent plaid from the directions of the two components is related to the aperture problem because the local motion detectors (e.g., in V1) will have only restricted views of the movements of individual edges in the plaid and cannot unambiguously determine the actual movement of the plaid. Movshon *et al.* (1986) suggest that the global motion detectors could resolve the ambiguity by carrying out a vector addition in velocity space. The two component gratings define two *constraint lines* in velocity space and the motion of the coherent plaid is determined by the *intersection of the constraint lines.* Since the velocity of the plaid is a vector sum of the two component vectors, it follows that the speed of the plaid will be greater than the speed of either component. Welch (1989) studied the ability of human observers to judge the speed of plaids and found that they were able to detect the speed of the plaid rather than the speeds of the two components.

This work on component versus pattern motion is significant here in that it provides strong psychophysical and physiological evidence for the concept of local and global motion detectors. It argues that visual motion analysis involves at least two sequential stages and that the global motion detectors combine inputs from several local motion detectors. The work also makes a specific proposal as to the rules underlying the intersection-of-constraints construction in velocity space. A question, then, is whether the intersection-of-constraints concept holds generally. The rules that govern whether pairs of sinuosoidal or rectangular gratings elicit coherent versus transparent motion have been the subject of intensive investigation in the psychophysical community during the past decade, and it is now clear that there are a number of interacting variables. It makes a difference, for example, whether or not the velocity vectors of the two components lie on the same or opposite sides of the pattern vector (Ferrera and Wilson, 1990). The luminance of the squares occurring at the intersections of the two gratings matters (Stoner *et al.*, 1990; Stoner and Albright, 1992), as do the relative speeds of the components and the pattern.

A second question is how the combination of the two velocity signals might be implemented in terms of neural interactions by the global motion detectors. Given that the transition from coherent to transparent motion depends upon more than just the speeds of the two components, current models of pattern motion must incorporate elements beyond simple vector summation. Wilson and Kim (1992,

1994) proposed models that predict the response of a pattern-sensitive unit to component stimuli. Input to each pattern unit depends upon excitatory inputs from the component units, each of which is expressed relative to the preferred direction of the pattern unit. The firing of each pattern unit is also influenced by inhibitory connections between neighboring pattern units with other preferred directions. The 1994 model includes both Fourier and non-Fourier inputs, but the key to both models is that the response of pattern units is determined by a relative balance of excitatory inputs from component units and inhibitory inputs from other pattern units. This balance determines the response of pattern units to two contrast gratings moving with different speeds and directions, contrasts, etc., and results in a winner-take-all strategy.

Qian and Andersen (1994) and Qian *et al.* (1994a) recently examined the way in which the balance of activity in opposing motion channels influences the perception of motion transparency, using a variety of stimuli such as counterphased gratings and random-dot kinetograms. Counterphased gratings consisted of two sinusoidal gratings moving past each other in opposite directions and failed to show motion transparency over a wide range of spatial and temporal frequencies. By contrast, two kinetograms consisting of fields of random dots or thin, vertical lines distributed randomly in space and moving in opposite directions typically produced a perception of motion transparency. Streams of dots moving in opposite directions in different regions of visual space were perceived as independent velocity flow fields. These observations suggested the existence of a suppression stage in the motion pathway at which motion signals in different directions in each small region of space locally inhibit each other. Motion transparency corresponds to the detection of locally unbalanced motion signals. This concept was explored by recording the responses of neurons in V1 and MT of macaques to transparent stimuli (Snowden *et al.,* 1991; Qian *et al.,* 1994b). Neurons in V1 responded to fields of dots moving in their preferred directions, whereas neurons in MT showed suppressed responses under stimulus conditions that led to perceptions of motion transparency in human observers. MT neurons did not respond well to flickering noise, which would produce a locally balanced signal. Qian *et al.* (1994b) analyzed these results using a model for MT cells that consisted of inputs from two V1 units. Each V1 unit was modeled by a quadrature pair of Gabor filters with opposing orientations in the space–time plane. The output from each of the filters in the pair was squared and they were added together to produce a motion energy mechanism. Finally, the two motion energy mechanisms suppress each other via either a subtractive or a divisive inhibition. The subtractive inhibition adequately accounted for most psychophysical and physiological results, except that the divisive inhibition was needed to account for contrast normalization.

In addition to a role in motion transparency, there is some indication that MT neurons are involved in the initial stages of segmenting velocity flow fields because monkeys with MT lesions perform poorly on a shape discrimination task based on the detection of kinetic boundaries (Marcar and Cowey, 1992). Since the classical receptive fields of MT neurons contain centers as well as antagonistic surrounds (e.g., Raiguel *et al.,* 1995) that respond to stimuli moving in the antipreferred direction of the center, these could be situations in which objects are moving in opposite directions. Marcar *et al.* (1995) thus studied the responses of MT neurons to kinetic boundaries defined by motion vectors moving in opposite directions. MT neurons were insensitive to the position and orientation of kinetic boundaries, suggesting that MT is involved only in the initial stages of segmenting velocity flow fields by providing neurons deeper in the motion analysis pathway with information on stimulus direction.

The observations on the interaction of velocity signals within the classical receptive fields of MT neurons are an interesting contrast to the observations of interactions between the classical and nonclassical fields of neurons in MT (Allman *et al.*, 1985a, b) and superior temporal sulcus (STS) (Tanaka *et al.*, 1986). Opposing velocity signals occurring within a local region of visual space inside of the classical receptive field tend to suppress each other and lead to a perception of motion transparency. However, a velocity signal moving in the antipreferred direction of the nonclassical surround can enhance the response of an MT neuron to a stimulus moving in the preferred direction within the classical receptive field. Thus, MT neurons may have complex biological functions and play a role in both motion transparency and the segmentation of the velocity flow field, depending on how velocity signals are distributed in space. The receptive field properties seem to result from a complicated spatial arrangement of excitatory and inhibitory connections within the classical and nonclassical fields. Individual neurons participating in different velocity channels seem to have strong inhibitory interactions within the confines of the classical field, but to have excitatory interactions across the border of the classical receptive field.

An initial clue as to the anatomical substrates that underlie the formation of such receptive fields comes from work in owl monkeys (Born and Tootell, 1992). Exposing monkeys to a large-field random-dot stimulus moving coherently at systematically varied directions, speeds, and monocular disparities failed to produce uniform uptake of radioactively labeled 2-deoxyglucose within MT. Instead columnar patterns of bands and interbands were consistently obtained. Neurons in lightly labeled interbands had receptive fields with antagonistic surrounds. Their responses to a centrally placed moving stimulus were suppressed by motion in the surround. Neurons in more densely labeled bands, by contrast, had surrounds that reinforced the center response so that they integrated motion cues over large areas of the visual field. This suggests that the geometric pattern of intracortical connections is an important determinant in receptive field structure.

Less is known about global motion detection in cats than in primates. However, Rauschecker *et al.* (1987) and Brenner and Rauschecker (1990) reported that neurons in the anterior lateral suprasylvian area tend to prefer stimuli moving along a trajectory that originates at the center of the cat's center of gaze and passes outward through the receptive field center. These directions coincide with the optic flow pattern created when the cat moves toward the fixation point. Sherk *et al.* (1995) found that neurons in the posterior lateral suprasylvian area respond preferentially to stimuli moving orthogonal to these "optic flow" directions.

3.2.2. Excitatory and Inhibitory Mechanisms

In addition to measuring the temporal integration functions for MT neurons, Mikami *et al.* (1986a) used the stroboscopic presentation of bars to investigate the relative contributions of inhibition and excitation to direction and speed tuning in MT neurons. They found that all of the MT neurons showed a strong inhibition in the nonpreferred direction that extends for spatial separations up 3.2 deg and temporal intervals of up to 160 msec. Some of the neurons also showed a strong facilitatory effect for stimuli moving in the preferred direction. Suppressive effects acting in both the preferred and nonpreferred directions could be demonstrated by stimuli moving at both the low and high ends of the velocity tuning curve, suggesting that inhibitory mechanisms contribute to both the low- and high-speed cutoff properties of the units.

Similar results on a still larger spatial and temporal scale were demonstrated for neurons in the posterior parietal areas of monkeys. These neurons have large, often bilateral visual receptive fields and respond strongly to stimuli moving along radii arranged around the central zone of the receptive field (Motter *et al.*, 1987; Steinmetz *et al.*, 1987). Stimuli moving along the radii show preferred and non-preferred directions. Such neurons could play a role in analyzing the velocity flow field created when the animal moves relative to its surroundings. Use of a Barlow–Levick condition-test paradigm demonstrated a strong feedforward inhibition that preceded the stimulus. The inhibition extended for 10–20 deg in front of the leading edge of the moving stimulus and lasted for more than 600 msec (Motter *et al.*, 1987). The receptive fields of these neurons could be described as a difference of Gaussians in which the inhibitory Gaussian is situated in an asymmetric position relative to the excitatory Gaussian.

4. Neural Mechanisms: Turtles

This section discusses the neural mechanisms underlying motion selectivity in turtle visual cortex. Since the organization of the visual system in turtles will be unfamiliar to many readers, it begins with an overview of the geniculocortical pathway in turtles.

4.1. Geniculocortical Pathway

The geniculocortical pathway involves projections of retinal ganglion cells to the dorsal lateral geniculate complex of the thalamus and subsequent projections to the visual cortex. Neurons in visual cortex function as global motion detectors: they have large, binocular receptive fields and respond to small, moving stimuli. Elaboration of these receptive fields can be understood by tracing the flow of information from the retina to the cortex.

4.1.1. Retina and Local Motion Detection

The literature on retinal neurons in turtles is reviewed by Granda and Dvorak (1977), Peterson (1992), and Granda and Sisson (1992). Most work is done with freshwater turtles of the family Emydidae, which includes *Pseudemys scripta* and *Chrysemys picta* in North American and *Emys orbicularis* in Europe.

The eyes in *Pseudemys* and *Chrysemys* measure approximately 8.1 mm in horizontal diameter and 7.7 mm in vertical diameter (Peterson and Ulinski, 1979); the retina has horizontal and vertical diameters of about 13 mm (Peterson and Ulinski, 1979). The acceptance angle of the eye is approximately 165–192 deg in *Pseudemys,* depending upon the state of accommodation (Northmore and Granda, 1991b). One degree of visual space extends across 83–98 μm when projected onto the retinal surface (Peterson and Ulinski, 1979; Northmore and Granda, 1991a). Estimates based on visual evoked potentials recorded from the surface of the optic tectum suggest that *Pseudemys* is capable of spatial resolution of 6 cycles/deg and 10 cycles/deg with contrast thresholds of about 1% (Northmore and Granda, 1991a). The eyes are oriented to form a frontal binocular field and an extensive binocular visual field that extends above and behind the animal. An object positioned above the midline of the head will thus be seen by the inferior edges of both retinas.

The retina contains rods and several types of cones, which occur as single cones or as double cones (for review see Peterson, 1992; Schneeweiss and Green, 1995). Photoreceptors have oil droplets that aid in light collection and shift their spectral sensitivities. In *Pseudemys,* red oil droplets are associated with red-sensitive (620 nm) single cones, orange oil droplets with the red-sensitive chief members of double cones, and yellow droplets with green-sensitive (518–540 nm) single cones. Large, pale-green fluorescent droplets are associated with single blue-sensitive cones, and small, nonfluorescent, clear droplets are associated with ultraviolet-sensitive single cones. The proportion of photoreceptor types varies between species and is related to their visual lifestyles. *Pseudemys* has a cone-dominated retina in which red-sensitive and green-sensitive cones or double cones predominate. The total numbers and proportions of photoreceptors vary as a function of position on the retina (Brown, 1969; Peterson, 1992). Red cones are distributed with a high density in a *visual streak* that extends along the nasotemporal axis of the retina and then bends slightly into the inferior temporal retina (Brown, 1969). The visual streak is situated in the retina so that it is maintained at an acute angle to the horizontal meridian. The visual streaks in the two eyes are maintained with this orientation by vestibular reflexes, but the turtle can orient them toward specific objects in the environment (Brown, 1969).

The retina contains approximately 380,000 ganglion cells, which occur with the greatest density, about 18,000 cells/mm^2, in the visual streak (Fig. 24; Peterson and Ulinski, 1979). The density of ganglion cells drops precipitously above the streak and more gradually below it to about 1000 cells/mm^2 at the periphery of the retina. There is a gentle gradient in ganglion cell density along the nasotemporal extent of the streak. An oval area of peak ganglion cell density, with a density 20,000 cells/mm^2, is embedded within the streak and positioned on the retina so as to face laterally when the eyes are in their natural positions. Comparing values for the density of photoreceptors and ganglion cells at specific loci on the retinal surface suggests a convergence ratio of 4–6 receptors/ganglion cell within the streak and 2–3 receptors/ganglion cell in the peripheral retina. The function of the visual streak is not known, but these convergence ratios suggest it is not a retinal specialization designed to enhance visual acuity.

The sizes of both the somata and dendritic arbors of ganglion cells vary as a function of position in the retina. Ganglion cells situated in the streak have soma diameters of 6–10 μm (Peterson and Ulinski, 1982). Soma size increases systematically above and below the streak to 9–10 μm in the dorsal and ventral peripheries of the retina. Soma size also increases along the nasal to temporal axis of the retina, with temporal ganglion cells being larger than those in the nasal retina. A small population of ganglion cells with soma sizes in excess of 15 μm occurs in the temporal retina. There is an inverse relationship between soma size and density along the dorsal–ventral axis of the retina, so the streak has a high density of small ganglion cells. The few data available (Kolb, 1982; Ammermuller *et al.*, 1995) suggest that dendritic field size increases as a function of distance from the streak. Cells located close to the streak tend to have dendritic fields shaped like ovals with their long axes oriented parallel to the streak.

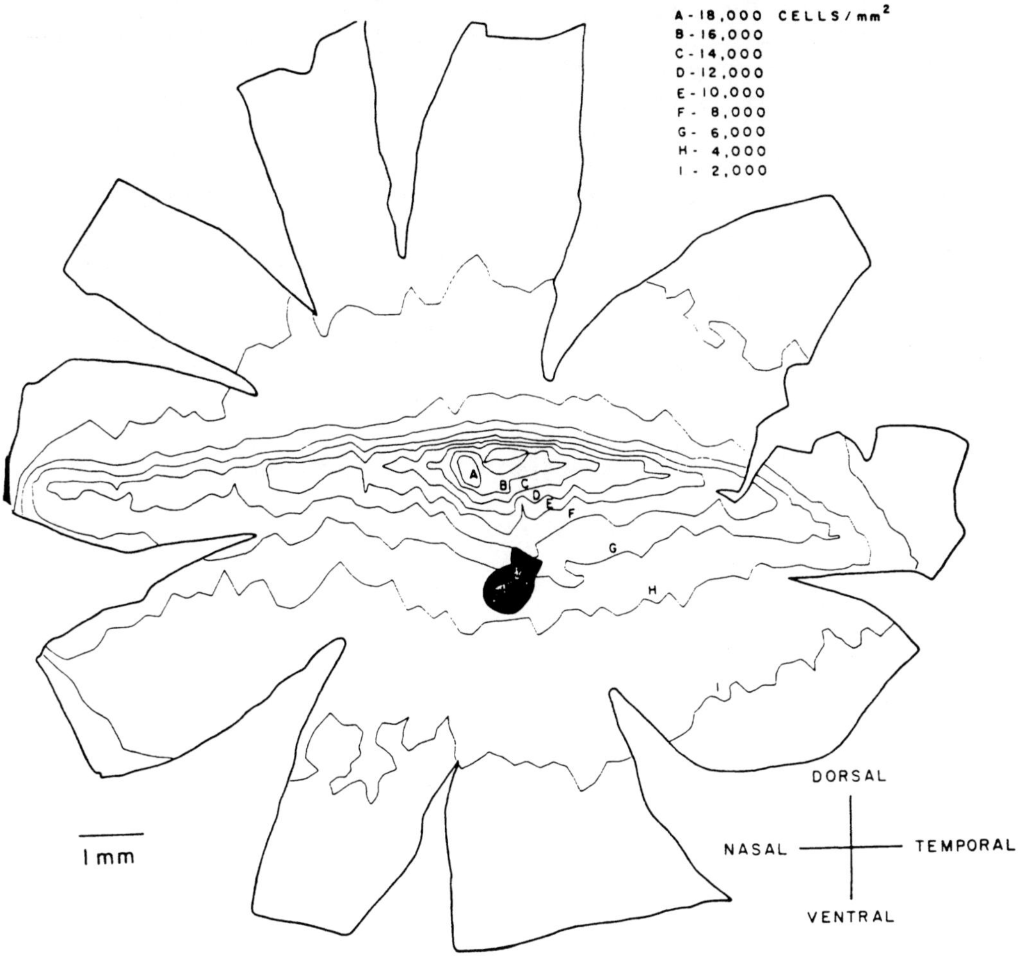

Figure 24. Retinal ganglion cell isodensity map. This figure shows the spatial distribution of ganglion cells on a flattened, wholemount preparation of *Pseudemys* retina. The density of ganglion cells was sampled in 100 μm² boxes every 200 μm over the retinal surface. Isodensity lines depict points at which neuron density decreased from highest (A) to progressively lower densities. For example, in the area between isocount lines I and H, there are more than 2000 and less than 4000 presumed neurons per square millimeter. The isodensity lines form a series of concentric ellipses. At higher densities, these ellipses become increasingly eccentric in shape to form a visual streak. The isodensity lines converge upon a peak density area which is nearly circular and is located at approximately the center of the retina, less than 2 mm above the optic disk (black). The retina is depicted as seen in the microscope, i.e., reversed left to right. From Peterson and Ulinski (1979).

There is no general agreement as to the number of morphologically distinct types of ganglion cells in turtles. Kolb and her colleagues (Kolb, 1982; Kolb *et al.*, 1988; Ammermuller and Kolb, 1995; Ammermuller *et al.*, 1995) have identified 24 types based on differences in soma size and dendritic branching patterns. Marchiafava and colleagues (Marchiafava and Weiler, 1980; Marchiafava and Wagner, 1981; Marchiafava *et al.*, 1983) introduced a dichotomous classification of ganglion cells in *Pseudemys* retina based upon studies in which ganglion cells were characterized physiologically in an eye cup preparation and then filled with Lucifer yellow for morphological study. Cells in the two categories vary in soma size, dendritic branching pattern, axon size, axon conduction velocity, and receptive field properties (Fig. 25). Type A ganglion cells have large somata about 15 μm in diameter. Their dendrites form wide arbors, averaging 400 μm in diameter, that stratify within a particular depth of the inner plexiform layer. Their axons have conduction velocities of 2 ± 1 m/sec. They have color-opponent characteristics and respond to stimuli with one particular spectral characteristic (e.g., red light) with an ON response and to stimuli with another spectral characteristic (e.g., green light) with an OFF response. Their responses are transient and generate up to 150 spikes/sec. Type B ganglion cells have small somata about 10 μm in diameter.

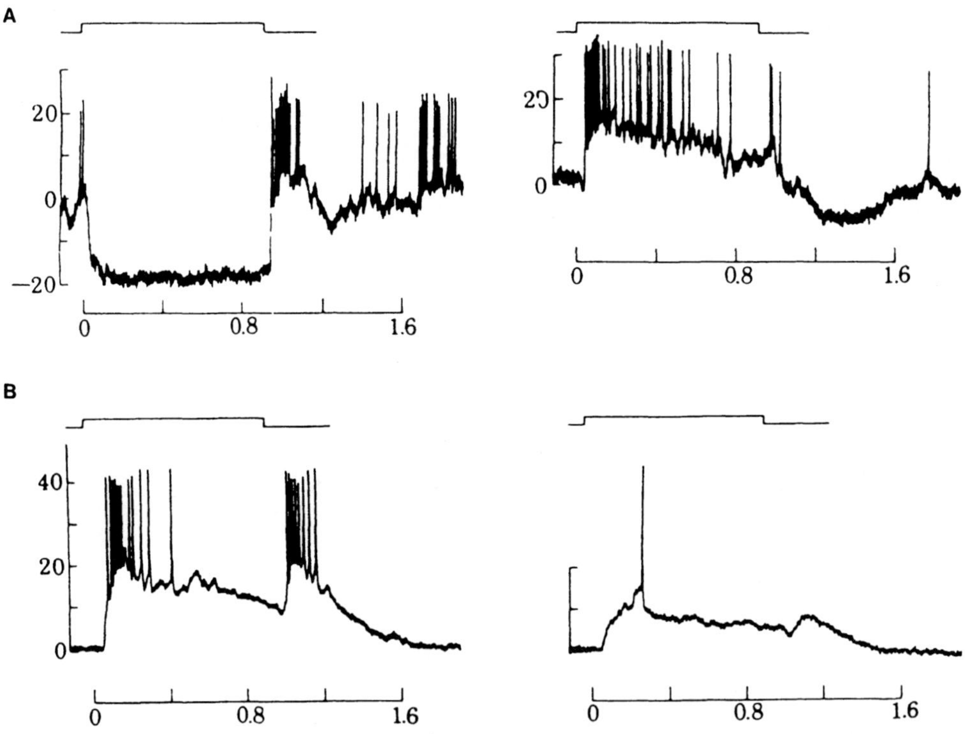

Figure 25. Type A and B ganglion cells. Responses of turtle retinal ganglion cells to illumination by spots and annuli. (A) Responses of a type A ganglion cell to a 200-μm-radius light spot (left trace) and to an annulus with internal and external radii of 180 and 800 μm, respectively (right trace). (B) Responses of a type B ganglion cell to a spot (left trace) and to an annulus. Vertical scale is response amplitude in millivolts. Horizontal scale is time in seconds. From Marchiafava and Weiler (1980), with kind permission of the publisher.

Their dendrites form narrow dendritic arbors that average 250 μm in diameter and extend throughout the depth of the inner plexiform layer. Their axons have conduction velocities of 0.4 ± 0.3 m/sec. They have sustained responses that seldom generate more than 60 spikes/sec. In addition to the type A and type B ganglion cells, there is clearly a small percentage of ganglion cells with very large somata, in excess of 15 μm in diameter, and extensive dendritic fields (Reiner, 1981).

Estimates of the number of axons in the optic nerve range from 390,000 to 500,000 (Bowling, 1980; Davydova *et al.*, 1975; Geri *et al.*, 1982), some of which are centrifugal fibers from the brain to the retina (Marchiafava and Torre, 1977). Axons in the optic nerves in *Pseudemys* have diameters of 0.4–3.0 μm; about 80% are myelinated (Woodbury and Ulinski, 1986). Conduction velocities of axons in the optic nerves fall into three underlying conduction velocity groups (Woodbury and Ulinski, 1986). The fastest group has a mean conduction velocity of 13 m/sec; the intermediate group has a mean velocity of 3 m/sec; the slowest group has a mean velocity of 1 m/sec. Reconstruction of the evoked responses from the size distribution of axons in the nerve suggests that the fast group results from a relatively small number of very large axons, while the slow group results from a large number of very small axons.

Receptive fields of retinal ganglion cells have been studied by projecting visual stimuli upon a tangent screen and recording from axons in the optic nerve using *in vivo* preparations (Lipetz and Hill, 1970; Bowling, 1980; Fulbrook, 1982; Granda and Fulbrook, 1989) or *in vitro* eye cup preparations in which visual stimuli are projected directly upon the retinal surface (Marchiafava and Weiler, 1980; Jensen and DeVoe, 1983; Ammermuller *et al.*, 1995). Ganglion cells have receptive fields with circular or oval excitatory centers ranging from 3 to 22 deg of visual arc in diameter. Detailed studies of the relationship between receptive field size and retinal eccentricity have not been carried out, but ganglion cells within the streak have relatively small receptive fields, while those in the periphery have larger fields. As expected from the presence of cones with a range of spectral sensitivities, ganglion cells respond to chromatic stimuli and most have strong responses to long-wavelength or red stimuli. The responses of ganglion cells to stimuli consisting of sinusoidally varying contrasts always contain nonlinear contributions (Armington *et al.*, 1991).

The firing patterns of ganglion cells vary as a function of stimulus luminance. As in other vertebrates (Rodieck, 1973), turtle ganglion cells show an on-going or maintained discharge of less than 1 spike/sec when the animal or preparation is maintained in the dark (Granda and Fulbrook, 1989). This discharge is believed to result from the spontaneous activation of ion channels in the outer segments of the photoreceptors (Lamb and Simon, 1977). Also as in other species (e.g., Penn *et al.*, 1994), the discharges occur in organized waves of ganglion cells during early developmental stages, which are replaced by more random discharges in adult turtles (Sernagor and Grzywacz, 1995). Burgi and Grzywacz (1994) suggest that the organized waves in young animals result from accumulation of extracellular potassium following a burst of activity in a local population of ganglion cells. A function for the waves is not certain, but they may play a role in the refinement of retinotopic order during retinal development. The ongoing discharge in adult animals would limit the sensitivity of the retina at low levels of illumination, but Lamb and Simon (1976) found that the level of noise present in photoreceptors and ganglion cells in the dark is reduced by low levels of ambient illumination. This phenomenon is attributed to a coupling of horizontal cells via electrical synapses that improves the signal/noise ratio of the photoreceptor–bipolar–ganglion cell pathway at the expense of some spatial resolution.

Activity in photoreceptors (Burkhardt, 1995) and ganglion cells (Granda and Fulbrook, 1989) is a sigmoidal function of stimulus luminance. Information transfer in the photoreceptor–bipolar cell–ganglion cell chain was detailed in elegant studies by Baylor and Fettiplace (1976, 1977a, b), who recorded intracellularly from a ganglion cell while using a very focused slit of light or electrical currents to stimulate an overlying photoreceptor. Photoreceptors had an integration time of more than 200 msec, meaning that the response of the photoreceptor light flash develops over a time interval of hundreds of millisecond. Ganglion cells thus can have latencies of 250 msec or more to light flashes (Baylor and Fettiplace, 1977b; Bowling, 1980; Granda and Fulbrook, 1989).

Some ganglion cells respond to moving stimuli without preference for a particular direction or speed, but a significant number are direction- or speed-selective and serve as local motion detectors (Lipetz and Hill, 1970; Marchiafava and Weiler, 1980; Bowling, 1980; Fulbrook, 1982; Jensen and DeVoe, 1983; Granda and Fulbrook, 1989; Ammermuller *et al.*, 1995). Estimates of the number of motion-selective ganglion cells range from 20% to 40%. Sernagor and Gryzwacz (1995) characterized the development of direction selectivity in embryonic and adult individuals. Direction tuning plots of early embryonic stages show several lobes, but later embryonic stages show a reduction in direction selectivity that reappears in late embryonic and adult stages. Ganglion cells in adult turtles have direction tuning curves that are about 120 deg wide (Fig. 26). Individual ganglion cells have preferred directions distributed over much of visual space, but there is some indication that the overall population of ganglion cells shows two preferred directions oriented parallel to and orthogonal to the visual streak. Some direction-selective ganglion cells have oval or circular receptive fields; others have elongate or bar-shaped fields. Cells with elongate fields can be orientation-selective and respond preferentially to stimuli moving perpendicular to the long axis of the receptive field. Direction selectivity can occur in type A cells, type B cells, and those ganglion cells with large somata and extensive dendritic fields. Jensen and DeVoe (1983) and Ammermuller *et al.* (1995) showed that motion-selective ganglion cells include those with stratified as well as narrow dendritic fields. Ariel and Adolph (1985) found that direction selectivity was abolished in ganglion cells by GABA antagonists, but R. D. Smith *et al.* (1996) recently reported that direction selectivity in the majority of ganglion cells is not completely abolished by the $GABA_A$-receptor antagonist picrotoxin.

Ganglion cells often show some preference for stimuli moving with particular speeds. Fulbrook (1982) found cells with speed tuning curves with peak sensitivities for stimuli moving from 9 to 200 deg/sec. Ammermuller and Kolb (1995) did not construct formal tuning curves, but do show examples of ganglion cells responding to stimuli moving with a comparable range of speeds. Orban (1986) suggested that speed-selective cells in cat striate cortex can be classified into low-pass, bandpass, and high-pass units that prefer low stimulus speeds, stimulus speeds distributed around a specific best speed, and high stimulus speeds, respectively. Turtle ganglion cells can be placed into the same general categories (Fulbrook, 1982), but the sample of ganglion cells studied to date is not sufficient to determine whether the

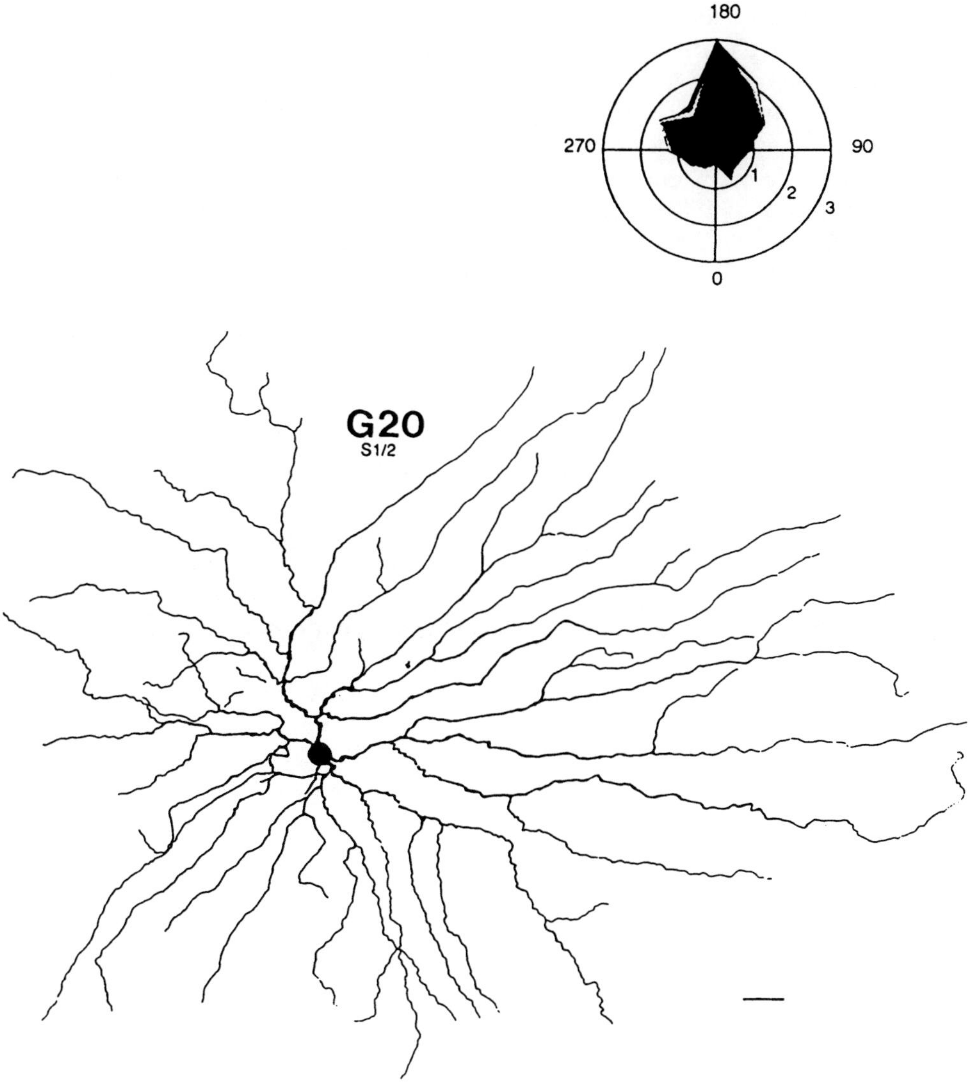

Figure 26. Direction selective-ganglion cell. Camera lucida drawing of an HRP-labeled turtle retinal ganglion cell. Inset is a tuning curve for the cell constructed by recording the responses of the cell to a moving bar with dimensions of 2.4 × 0.2 mm and a speed on the retinal surface of 1.7 mm/sec. Movement perpendicular to the visual streak in the ventral direction was defined as 0 deg. Movement in the nasal direction was 270 deg. Scale bar equals 25 μm. Modified from Ammermuller *et al.* (1995).

ganglion cells fall into discrete groups as opposed to showing preferred speeds distributed with some continuous function across the range of speeds. Some cells appear to be tuned to stimuli moving with both specific directions and speeds, while others are tuned to the direction or speed of the stimulus, but not both.

The history of the W/X/Y classification of ganglion cells in cats demonstrates the heuristic value of having a characterization of ganglion cells that can be followed to cortical levels. Less information is available on the ganglion cells of turtles, but it seems useful at this point to recognize small, large, and very large ganglion cells. Small ganglion cells have small somata, narrow dendritic fields that extend through the inner plexiform layer, and relatively small axons with slow conduction velocities. Large cells have relatively large somata, wide dendritic fields that are stratified in the inner plexiform layer, and medium-size axons with faster conduction velocities. Very large cells have very large somata that are often displaced to the inner plexiform layer, extensive dendritic trees, and very fast conduction velocities. All three groups of ganglion cells appear to contain types of cells that are motion-selective. The next section follows the flow of information from these three groups of ganglion cells into the central visual pathways. However, it should be stressed that this classification is nothing more than a provisional scheme intended to generate testable hypotheses about the organization of visual pathways in turtles.

4.1.2. Retinogeniculate Projection

Retinal ganglion cells in turtles project bilaterally to a series of central visual structures in the midbrain, pretectum, and diencephalon (Kosareva, 1967; Knapp and Kang, 1968a, b; W. C. Hall and Ebner, 1970a; Belekhova and Kosareva, 1971; Bass and Northcutt, 1981a, b; Ulinski and Nautiyal, 1987; Hergueta *et al.*, 1992, 1994; for review see Repérant *et al.*, 1992). One of these is the *dorsal lateral geniculate complex,* which is situated in the dorsal thalamus internal to the optic tract (Rainey and Ulinski, 1986).

The dorsal lateral geniculate complex receives inputs from the small and large ganglion cell groups, but not from the very large ganglion cells (Woodbury and Ulinski, 1986). Studies based on the caliber of degenerating axons in Fink–Heimer preparations following retinal ablation show very large caliber axonal fragments distributed around the periphery of the optic nerves, consistent with electron microscopic observations of axon diameters within the nerves. These large-caliber fragments can be traced to the basal optic nucleus and pretectal nuclei of the brainstem. Small- and medium-sized axons are situated in the central parts of the nerves and can be traced into the principal component of the optic tract (the marginal optic tract) and subsequently to the tectum and dorsal lateral geniculate complex. Axons in the optic tract that have been filled with horseradish peroxidase typically bifurcate into a relatively thick branch that continues caudodorsally into the optic tectum and a thinner branch that continues dorsally into the optic tract as it overlies the dorsal lateral geniculate complex (Sjöström and Ulinski, 1985). This suggests that information from the small and medium ganglion cells is transmitted in parallel to the optic tectum and the dorsal lateral geniculate complex, consistent with experiments in which injections of marker substances in the optic tectum label both small- and medium-sized somata in the ganglion cell layer of the retina (Peterson, 1978; Guiloff and Kolb, 1992).

The entire surface of the retina is mapped topographically upon the contralateral dorsal lateral geniculate complex (Ulinski and Nautiyal, 1987). The dorsal lateral geniculate complex in *Pseudemys* is a rostrocaudally elongate structure that measures approximately 2.0 mm in length and 0.5 mm in dorsoventral extent (Rainey and Ulinski, 1986). The nasotemporal axis of the retina projects along the rostrocaudal axis of the geniculate; the polarity of the projection of the dorsoventral axis of the retina upon the geniculate is not certain (Fig. 27). When the effect of the lens is taken into account, the topography of the retinogeniculate

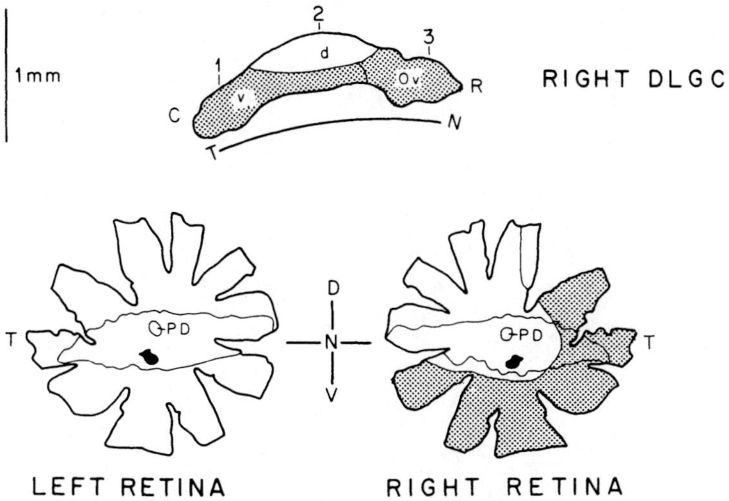

Figure 27. Retinogeniculate projection. This figure summarizes the principal features of the retinogeniculate projection in turtles. The lower part of the figure shows the right and left retinas displayed as retinal wholemounts. The dorsoventral (D–V) and nasotemporal (N–T) axes of each retina are indicated. The orientation of the visual streak is shown by the horizontally elongated region in each retina. The thin line represents the 6000-cells/mm^2 isodensity line. The closed contour embedded in the visual streak is the peak density area (PD). The optic disc is shown in black. The stippled area represents the region of the right retina that projects to the right dorsal lateral geniculate complex (DLGC). The top part of the figure shows a lateral view of the right DLGC. The boundaries of the subnucleus ovalis (Ov) and the dorsal (d) and ventral (v) subnuclei are indicated. The rostral (R) and caudal (C) poles of the complex are also indicated. The curved line designated T–N shows the relation of the nasotemporal axis of each retina to the geniculate complex. There are projections from the nasal poles of both right and left retinas to the rostral pole of the geniculate complex and from the temporal poles of both retinas to the caudal pole of the geniculate complex. The stipple on the complex shows the parts of the geniculate complex that receive binocular input. Notice that the retinas and the geniculate complex reconstructions are drawn to different scales. Modified from Ulinski and Nautiyal (1987).

projection results in a retinotopic map of the horizontal meridian of visual space upon the geniculate with the frontal, binocular visual field projecting to the caudal pole of the geniculate and the peripheral, monocular segment of the visual field projecting to the rostral pole of the geniculate. The ipsilateral retinogeniculate projection originates from ganglion cells situated in the inferior and temporal edges of the retina. Thus, *eccentricity* along the horizontal meridian of binocular visual space is mapped along the rostrocaudal axis of the geniculate. Individual vertical lines in visual space (*azimuth lines*) are mapped as dorsoventral lines on the geniculate.

The geniculate complex has two distinct layers, each of which contains a morphologically distinct population of neurons that effects different spatial relationships with the arbors of retinal ganglion cells and consequently must perform different integrations of spatial information (Rainey and Ulinski, 1986; Kenigfest *et al.*, 1995). The medial face of the geniculate complex is formed by a plate of cells with densely packed somata. The dendrites of the most medially placed *cell plate neurons* are contained within the cell plate and have dendrites that run concentric

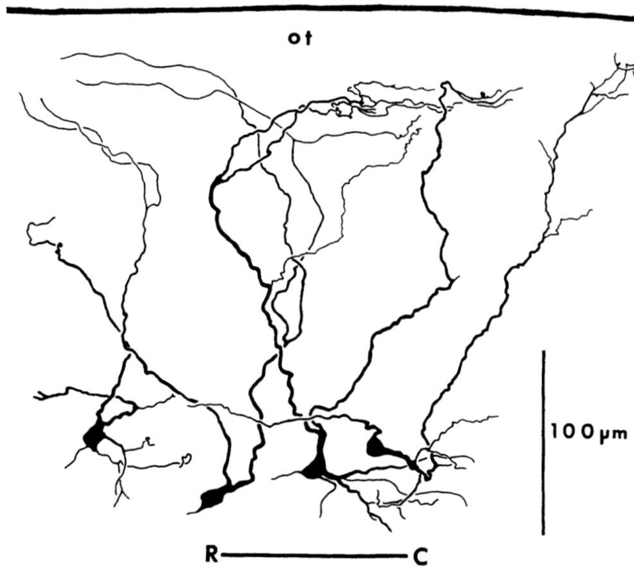

Figure 28. Cell plate neurons from the dorsal lateral geniculate complex of a turtle. This figure shows four examples of neurons in the cell plate of the dorsal lateral geniculate complex that were filled with horseradish peroxidase. Each neuron has dendrites that extend laterally into the neuropile and reach the inner face of the optic tract (ot). From Rainey and Ulinski (1986).

with the surface of the geniculate (Fig. 28). Cells with somata located more laterally have dendrites that extend in bowl-shaped arbors into the lateral neuropile of the geniculate. The dendrites branch and ultimately end in arbors of dendrites consisting of several fingerlike dendritic segments. The neuropile layer of the geniculate is divided into an inner neuropile that contains the relatively straight and thick proximal dendrites of the cell plate cells and an outer neuropile containing their terminal arbors. The neuropile also contains the somata of *neuropile cells* whose dendrites extend concentrically with the lateral surface of the geniculate within the neuropile (Fig. 29). There are systematic variations in the sizes of the somata and dendritic fields of both the cell plate and neuropile cells within the geniculate complex (Rainey and Ulinski, 1986). Soma size increases steadily along the rostrocaudal axis of the geniculate, with cells at the caudal pole of the complex having significantly larger somata than those at the rostral pole. However, the dendritic arbors of geniculate neurons at the rostral and caudal poles of the complex have larger diameters than do those neurons situated in the intermediate region of the geniculate. Dendritic field size thus first decreases and then increases along the caudorostral axis of the geniculate.

The receptive field sizes of geniculate neurons will be determined by the dimensions of the dendritic fields of the geniculate neurons and the relative sizes of the terminal arbors of retinal ganglion cells. The arbors fall into three morphological groups (Sjöström and Ulinski, 1985). Axons in the first group are relatively unbranched and course dorsally into the geniculate from the optic tract. Axons in the other two groups branch into distinct arbors within the geniculate (Fig. 30). Those with relatively large diameters form arbors with a mean volume of about 12.8×10^3 μm^3 and about 40 varicosities which average 1.69 μm in length. Axons

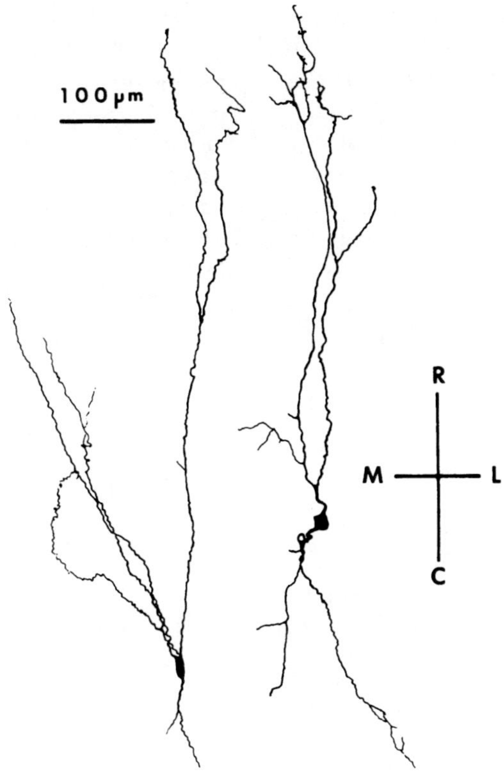

Figure 29. Neuropile cells. This figure shows two cells from the neuropile layer of turtle dorsal lateral geniculate nucleus. The dendrites of the cells are oriented along the rostral (R) to caudal (C) axis of the geniculate complex. From Rainey and Ulinski (1986).

with smaller diameters form arbors that are spatially more restricted, with a mean volume of about 6.4×10^3 μm^3 and about 53 smaller varicosities measuring about 0.92 μm in length. The large-diameter axons arborize preferentially within the inner neuropile; the small-diameter axons arborize preferentially within the outer neuropile. Electron microscopic examination of the geniculate complex following retinal lesions demonstrates that ganglion cell axons terminate upon both the dendritic shafts and terminal arbors of geniculate neurons (Ulinski and Nautiyal, 1987). The relative dimensions of the retinogeniculate arbors and the dendritic fields of the geniculate neurons suggest that each geniculate neuron receives a convergent input from several ganglion cells, but the information needed to calculate the convergence ratio is not available.

Although there are significant gaps in our knowledge of the geniculocortical pathway in turtles, the anatomical data permit some inferences about the convergence of information that takes place in the projection of the retinal ganglion cells onto geniculate neurons. One is that there is a convergence of most of the ganglion cell types upon neurons in the geniculate complex. This is in contrast to the separation of information from different ganglion cell types upon neurons in the geniculate complex seen in cats and primates. The medial cell plate cells appear to receive synaptic contacts from both large- and small-caliber axons on different parts of their dendritic trees, suggesting that medial cell plate neurons receive inputs from both sustained type A cells and phasic type B cells (Ulinski, 1986a). The information processing that follows may be different for the two types of ganglion cells because they effect differing synaptic relationships with the geniculate neurons. The large retinal terminals in the medial neuropile make simple, axodendritic contacts upon the radially directed dendrites of the lateral cell plate cells (Ulinski and Nautiyal, 1987). These terminals contain large, clear synaptic vesicles that are consistent with an excitatory input from the retinal ganglion cells to the geniculate neurons. The small retinal terminals in the lateral neuropile participate in more complex synaptic arrangements within glomeruli consisting of retinal terminals and the fingerlike processes of the lateral cell plate cells (Fig. 31). The retinal terminals contain large, clear, round synaptic vesicles that are presynaptic to the fingerlike processes. The fingerlike processes, however, contain small, clear, pleomorphic vesicles that suggest inhibitory interactions between the geniculate

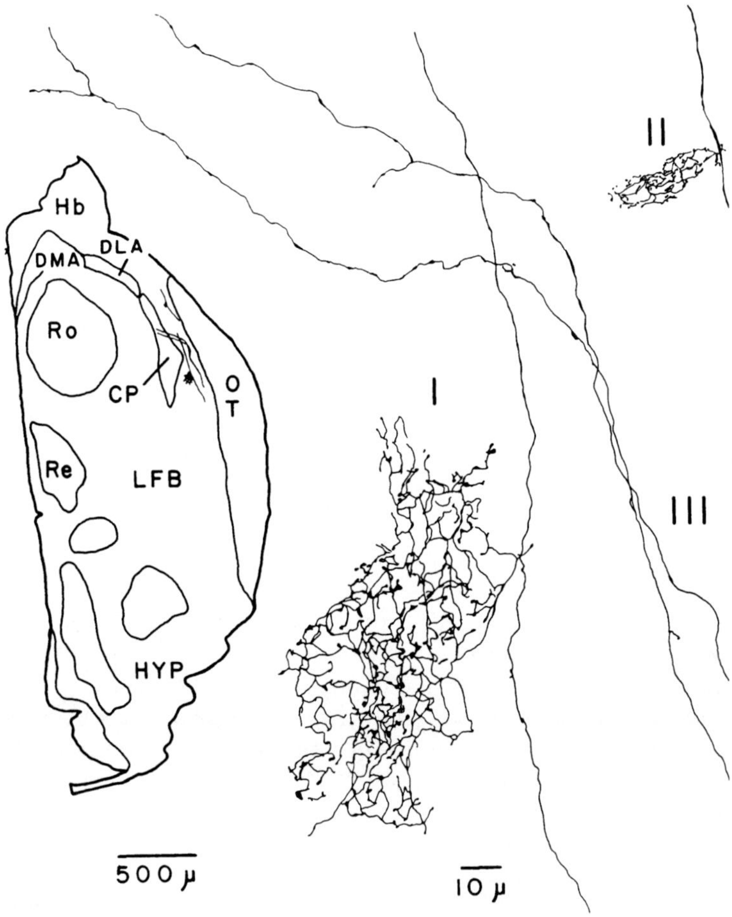

Figure 30. Retinogeniculate arbors. This figure shows the arbors of retinal ganglion cells in the dorsal lateral geniculate complex. Arbors were filled with horseradish peroxidase (HRP) following HRP injections in the optic tract (OT). The inset shows transverse sections through the diencephalon. Three types of arbors are seen. Type I arbors are relatively large. Type II arbors are relatively small. Type III arbors course from ventral to dorsal in the geniculate complex. CP, Cell plate of the geniculate complex; HYP, hypothalamus; Hb, habenula; DMA, dorsomedial anterior nucleus; DLA, dorsolateral anterior nucleus; LFB, lateral forebrain bundle; RE, nucleus reuniens; RO, nucleus rotundus. From Sjöström and Ulinski (1985).

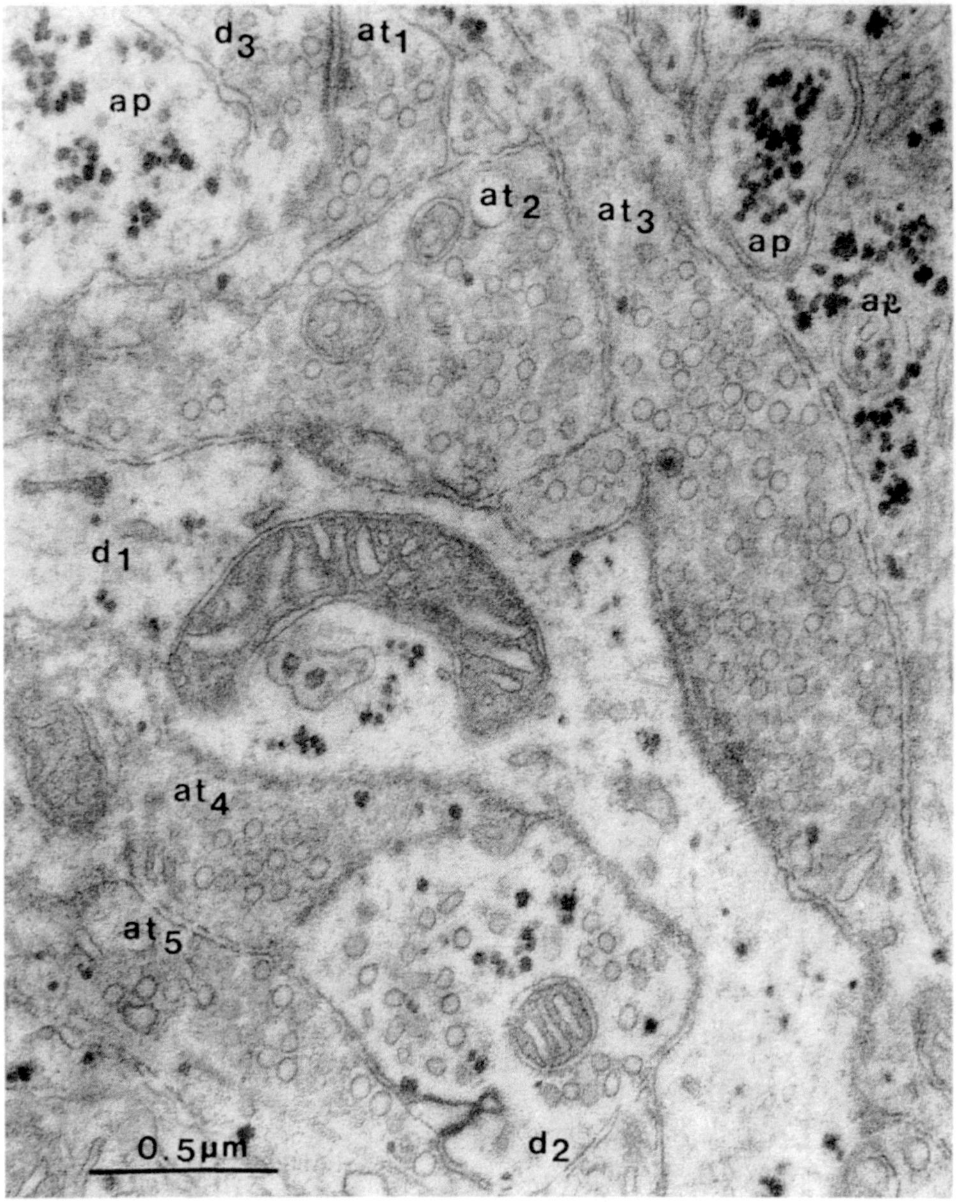

Figure 31. Ultrastructure of the dorsal lateral geniculate complex. Electron micrograph of a synaptic nest in the dorsal lateral geniculate complex. The nest contain centers around two dendrites (d_1 and d_2). The dendrites marked d_2 contain clear pleomorphic vesicles. The dendrites are surrounded by several axon terminals (at_1–at_5), but only at_3 shows any sign of a junctional complex in the plane of section. Notice that at_1 at the top left is presynaptic to a vesicle-containing profile. Astrocytic processes (ap) partially encapsulate the synaptic nest. From Ulinski (1986a).

neurons. These terminals show GABA-like immunoreactivity (Kenigfest *et al.*, 1995). Activity in the retinal ganglion cells should thus produce a complex sequence of excitatory and inhibitory interactions. The correlations that have been established between ganglion cell morphology, axon size, and the morphology of retinal terminals in the geniculate lead to the inference that the simple contacts on the dendrites of the lateral cell plate cells originate in the large group of ganglion cells, while the complex contacts on the fingerlike processes originate in the small group of ganglion cells (Fig. 32).

The relative sizes of the dendritic fields of the geniculate neurons and of the axonal arbors of the ganglion cells suggest that geniculate neurons have receptive field sizes that vary significantly. The cell plate cells in the intermediate segment of the geniculate are situated within the representation of the central retina and should have relatively small receptive fields. The neuropile cells, by contrast, have dendrites that extend considerable distances along the geniculate surface, presumably receiving inputs from more spatially separated ganglion cells, and would be expected to have rather large receptive fields. The single study of receptive field sizes in the dorsal lateral geniculate complex with physiological methods is consistent with these expectations (Boiko, 1980). Some geniculate neurons have large receptive fields, sometimes encompassing all of visual space, consistent with the anatomy of neuropile cells. Other neurons have small (less than 11 deg) and medium-sized (11–30 deg) receptive fields, consistent with the anatomy of cell plate cells. Some neurons in both groups showed clear direction selectivity, consistent with the occurrence of direction selectivity in both small and large ganglion cells and the apparent projection of both groups of ganglion cells to the geniculate. An untested hypothesis is that the neuropile cells should show more-phasic responses, while the cell plate cells should show more-sustained responses.

Several studies have retrogradely labeled geniculate neurons following injections of horseradish peroxidase (HRP) in the visual cortex (J. A. Hall *et al.*, 1977; Belekhova *et al.*, 1979; Desan, 1984; Kenigfest *et al.*, 1995). All of these studies labeled cell plate neurons, but the issue of whether or not neuropile cells also project to the visual cortex has not been resolved. Mulligan and Ulinski (1990) were able to retrogradely label neuropile cells following restricted injections of HRP in the visual cortex, but Kenigfest *et al.* (1995) did not find retrogradely labeled neuropile cells. Since the neuropile cells show GABA-like immunoreactivity, Kenigfest *et al.* suggest that they are interneurons. It remains possible that the neuropile cells both send axons to the cortex and effect GABAergic, dendrodendritic synapses within the geniculate. This would imply that the neuropile cells make glutaminergic contacts via their axonal synapses and GABAergic synapses via their dendrites. A related issue is whether or not there are additional GABAergic inputs from the ventral thalamus to the geniculate in turtles analogous to the projection of the reticular nucleus to the geniculate in mammals. In support of this, Ulinski (1987) found that neurons in the ventral thalamus of *Pseudemys* show high-affinity uptake of tritiated GABA and have axons that can be traced toward the geniculate.

4.1.3. Geniculocortical and Corticogeniculate Projections

The axons of geniculate neurons course ventrally from the geniculate complex and enter the lateral forebrain bundle (W. C. Hall and Ebner, 1970b; Heller and

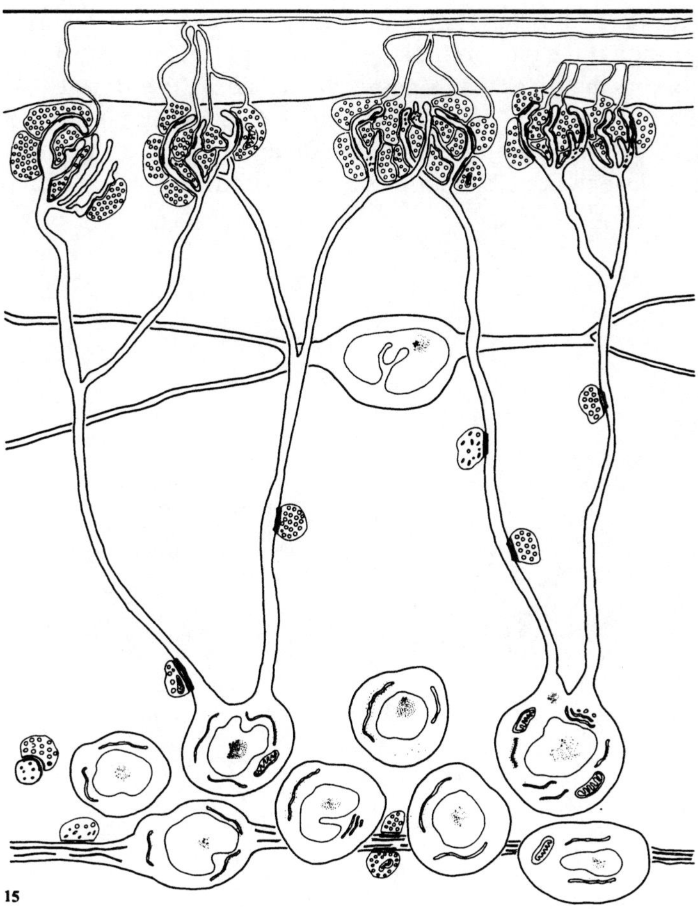

Figure 32. Synaptic organization of the dorsal lateral geniculate complex. This diagram summarizes the organization of the geniculate complex. It represents a section through the geniculate complex. The cell plate (CP) consists of densely packed somata of cell plate neurons which receive few synaptic contacts (none are shown here). The dendrites of cell plate cells run some distance in the cell plate and are postsynaptic to axon terminals containing only clear, round vesicles, clear, pleomorphic vesicles, or clear, round and dense-cored vesicles. Some cell plate cells have dendrites that run obliquely through the inner neuropile (INp). They appear in electron micrographs as rectilinear profiles oriented radially to the optic tract. They are postsynaptic to the same three types of axon terminals. Neuropile cells have both somata and dendrites situated in the neuropile. As the dendrites of the cell plate cells approach the outer one half to one third of the neuropile (ONp) they begin to branch extensively into distal dendrites which form varicose, fingerlike processes. These are involved in dendrodendritic synapses. They contain clear pleomorphic vesicles. The dendrites are postsynaptic to axon terminals which contain clear, round vesicles and are involved in asymmetric junctional complexes. The distal dendrites of cell plate neurons and the terminal arbors of the retinal ganglion cells together form synaptic nests in the outer neuropile (ONp). From Ulinski (1986a).

Ulinski, 1987). This is the reptilian equivalent of the internal capsule of mammals and serves to carry axons in both directions between the thalamus and cortex (Fig. 33). The lateral forebrain bundle is divided into dorsal and ventral fascicles or peduncles that course between the brainstem and dorsal area of cortex. Geniculate axons enter the dorsal peduncle (Heller and Ulinski, 1987). They are myelinated, while they run through the basal forebrain and can be placed into two groups based on variations in axon diameter. Thin-caliber fibers lie in the dorsal part of the dorsal peduncle and have diameters of less than 1 μm. Medium-caliber fibers lie more ventrally in the dorsal peduncle and have diameters of 1–2 μm. Both groups of geniculate axons lose their myelination as they reach the lateral margin of dorsal

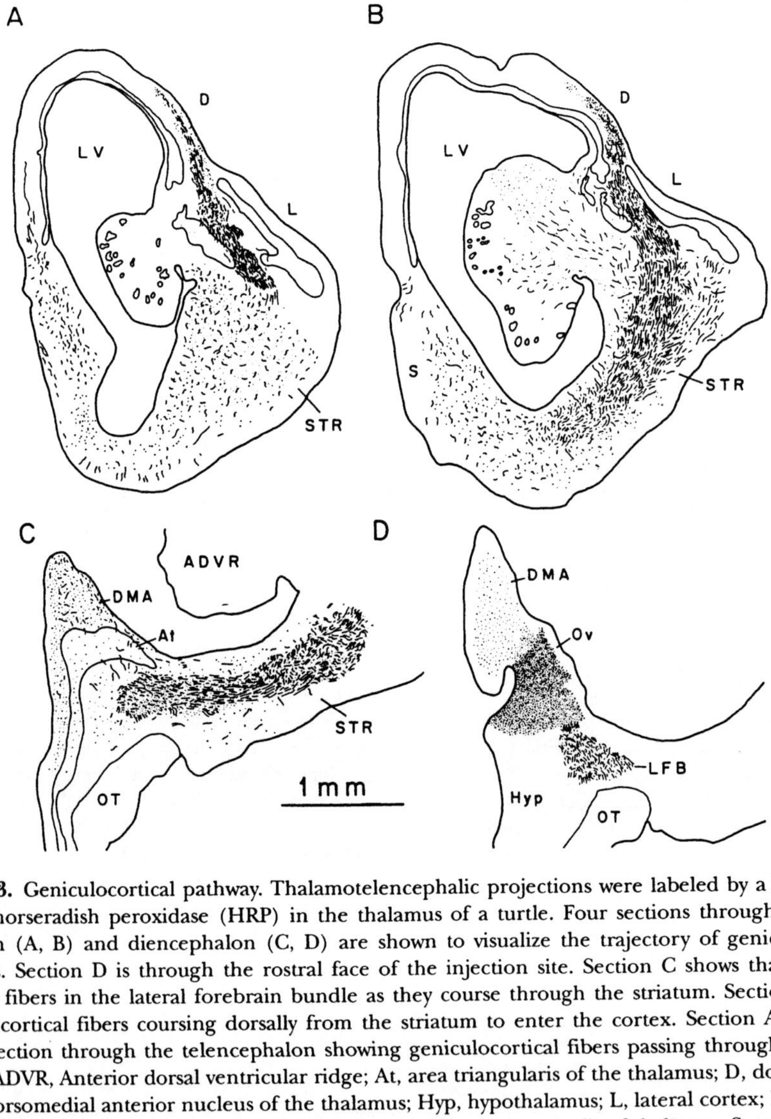

Figure 33. Geniculocortical pathway. Thalamotelencephalic projections were labeled by a large injection of horseradish peroxidase (HRP) in the thalamus of a turtle. Four sections through the telencephalon (A, B) and diencephalon (C, D) are shown to visualize the trajectory of geniculocortical afferents. Section D is through the rostral face of the injection site. Section C shows thalamotelencephalic fibers in the lateral forebrain bundle as they course through the striatum. Section B shows geniculocortical fibers coursing dorsally from the striatum to enter the cortex. Section A is a more rostral section through the telencephalon showing geniculocortical fibers passing through the visual cortex. ADVR, Anterior dorsal ventricular ridge; At, area triangularis of the thalamus; D, dorsal cortex; DMA, dorsomedial anterior nucleus of the thalamus; Hyp, hypothalamus; L, lateral cortex; LFB, lateral forebrain bundle; LV, lateral ventricle; OT, optic tract; Ov, nucleus ovalis of thalamus; S, septum, STR, striatum. From Heller and Ulinski (1987).

cortex and run as thin (less than 1 μm), unmyelinated axons from lateral to medial across the visual cortex.

The cortex as a whole (Fig. 34) has three layers, with the outer layer 1 containing scattered neurons, the intermediate layer 2 containing many, densely packed neurons, and the inner layer 3 containing a moderate number of scattered neurons (Desan, 1984, 1985; Davydova and Goncharova, 1979; Ulinski, 1988, 1990; Blanton and Kriegstein, 1991a). Layer 2 contains the majority of neurons that project between areas within the cortex or to subcortical structures. These are pyramidal cells that have apical dendrites extending into layer 1 and basal dendrites extending into layer 3. Layers 1 and 3 contain several populations of local circuit neurons, many of which utilize GABA. These are cells with a stellate morphology or horizontal cells that have dendrites extending concentrically with the ventricle in layer 3. Four cytoarchitectonic areas are identified in the cortex based on variations in the configuration of the three layers. They are named the medial, dorsomedial, dorsal, and lateral areas according to their positions on the hemisphere. Orrego (1961) demonstrated that electrical stimulation of the optic nerves or presentation of flashes of light in *in vivo* preparations produced surface negative potentials in the dorsolateral cortex, and it is now clear that this visual area corresponds to the *dorsal area* (Heller and Ulinski, 1987). Desan (1984, 1985) used a different nomenclature for the cytoarchitecture, recognizing two dorsal areas, D1 and D2. Desan's area D2 is the more laterally placed of the two and corresponds to the dorsal area of Heller and Ulinski; D1 is part of the dorsomedial cortex. Dorsal cortex is composed of two cytoarchitecturally distinct parts. Its lateral part was formerly called the pallial thickening and is characterized by the presence of large clusters of neurons; its medial part has smaller cell clusters (Johnston, 1915; Heller and Ulinski, 1987).

The geniculocortical axons have very precise relationships within dorsal cortex that can be demonstrated by filling fascicles of geniculocortical axons with HRP in *in vitro* preparations (Heller and Ulinski, 1987; Mulligan and Ulinski, 1990). Most axons extend across the full mediolateral extent of the dorsal cortex bearing varicosities *en passant* (Fig. 35). These varicosities undoubtedly represent the presynaptic elements of the geniculate axons. They contain clear, round synaptic vesicles (Ebner and Colonnier, 1975, 1978; L. M. Smith *et al.*, 1980) that are presynaptic to the spines and shafts of pyramidal neurons and the smooth shafts of interneurons. The geniculate fibers generate EPSPs (Fig. 36) that utilize one of the excitatory amino acids and access the AMPA or non-NMDA subtype of glutamate receptors on their postsynaptic targets (Larson-Prior *et al.*, 1991; Blanton and Kriegstein, 1991b, c, 1992).

Fascicles of geniculate axons originating at one rostrocaudal level of the geniculate complex enter the dorsal cortex together and extend across the mediolateral extent of the dorsal area (Mulligan and Ulinski, 1990). Thus, a band of cortical tissues positioned at one rostrocaudal level of dorsal area receives input from a fascicle of geniculocortical axons. Since this fascicle originates from neurons within the representation of an individual azimuth within the geniculate, it defines an *isoazimuth lamella* or band of cortical neurons receiving input from a single azimuth line in visual space. This concept is consistent with retrograde tracing experiments (Mulligan and Ulinski, 1990) in which restricted HRP injections were placed into different regions of dorsal area (Fig. 37). The experiments demonstrate that the rostrocaudal axis of the geniculate complex is mapped along the caudorostral axis of dorsal area. HRP injections in the caudal dorsal area retrogradely label neurons in the rostral pole of the geniculate complex and vice versa. Visual space is thus mapped along the rostrocaudal axis of the cortex as a series of overlapping isoazimuth lamellae of differing horizontal eccentricity.

Since each fascicle of geniculocortical axons courses from lateral to medial

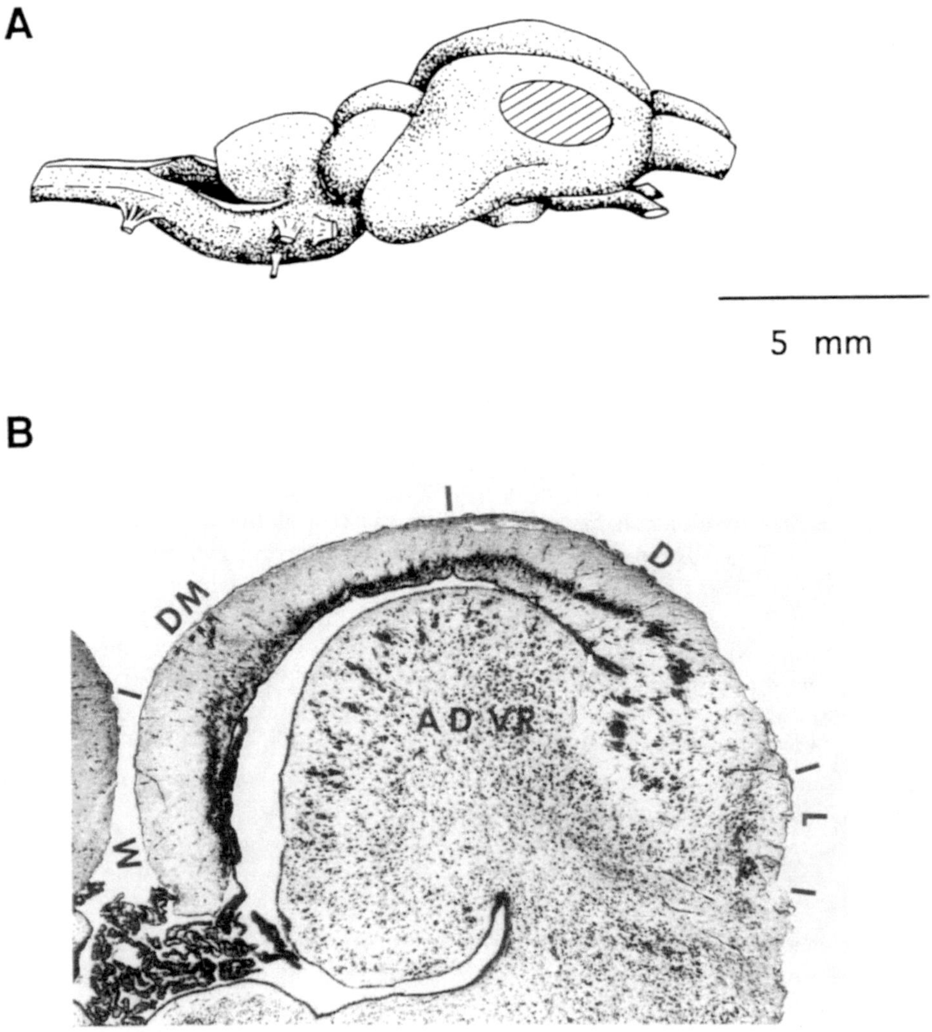

Figure 34. Cytoarchitecture of turtle cortex. (A) Lateral view of turtle brain with the visual cortex indicated by the shaded ellipse. (B) Transverse section through the right cerebral hemisphere illustrating the cytoarchitecture of the cortex. The cortex is divided into medial (M), dorsomedial (DM), dorsal (D), and lateral (L) areas. The medial and dorsomedial areas are limbic in nature; the lateral area is olfactory cortex. The dorsal area is the visual cortex. It is divided into lateral and medial parts on the basis of the cell clusters in the second, or intermediate, layer. The lateral part of dorsal cortex has large cell clusters; the medial part has small cell clusters. The dorsal ventricular ridge (ADVR) protrudes into the lateral ventricle.

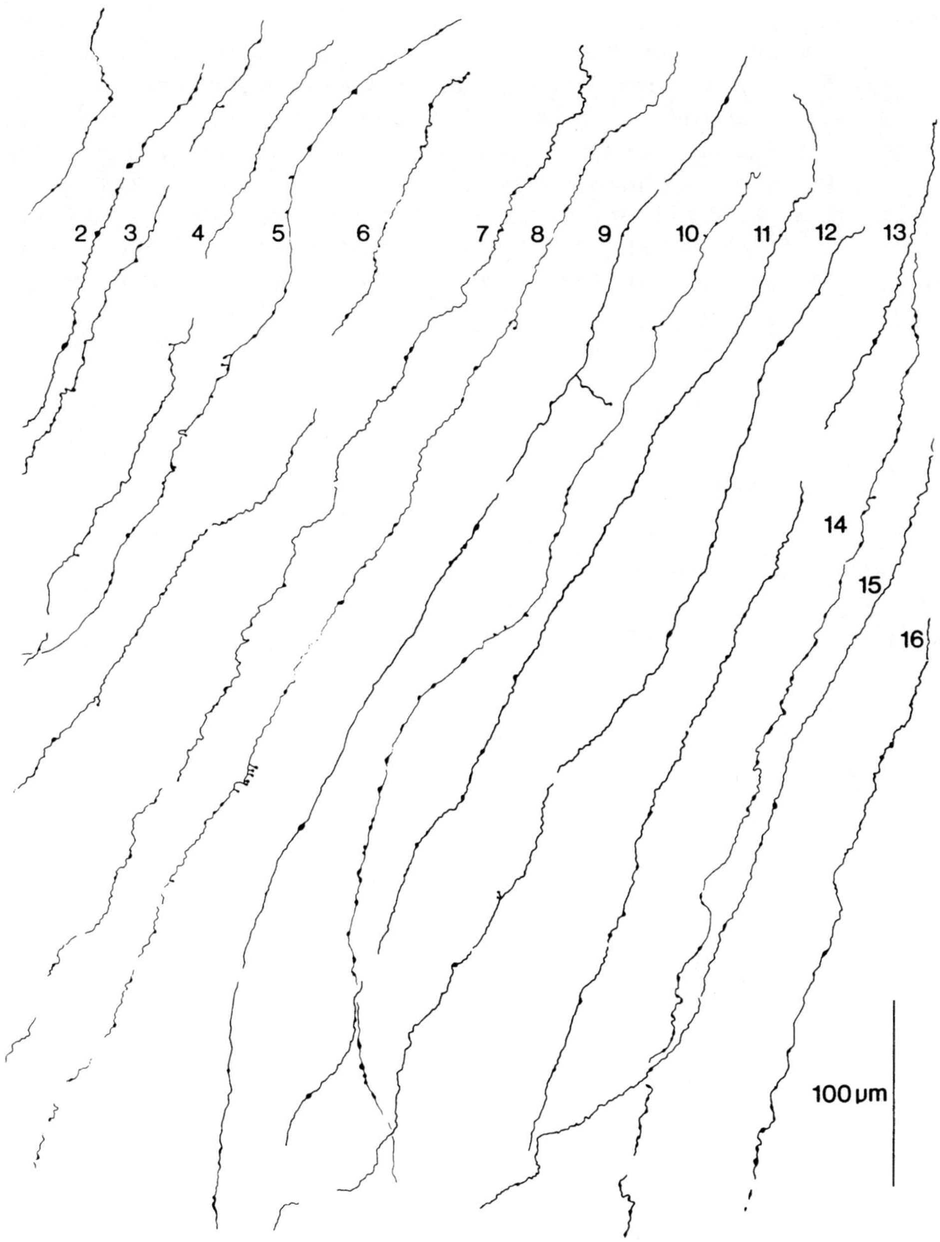

Figure 35. Morphology of geniculocortical afferents. These are drawings of individual geniculocortical axons anterogradely filled with HRP injections in the lateral forebrain bundle. From Mulligan and Ulinski (1990).

across the two cytoarchitectonic parts of the dorsal area, visual information should propagate across the dorsal area with a finite velocity. The axons are thin (less than 1 mm; Mulligan and Ulinski, 1990) within the cortex and are expected to have slow conduction velocities. This intracortical conduction velocity has been estimated at about 0.01 mm/msec by comparing the latencies of visually evoked postsynaptic potentials in pairs of cortical neurons situated at different points on the cortex (Colombe and Ulinski, 1996). The latency of a visually evoked response in a cortical neuron is determined in part by conduction delays within the cortex, but also depends upon conduction delays in all of the stages within the geniculocortical pathway. Because of the gradient in cell sizes seen in both the retinal ganglion cells and geniculate neurons, some variation in conduction time as a function of the positions of visual stimuli in space is expected. Intracellular recordings show that the latency of visually evoked PSPs in pyramidal cells varies along the rostrocaudal axis of the cortex. The visual cortex consequently has a family of *isolatency contours* that run across the isoazimuth lamellae.

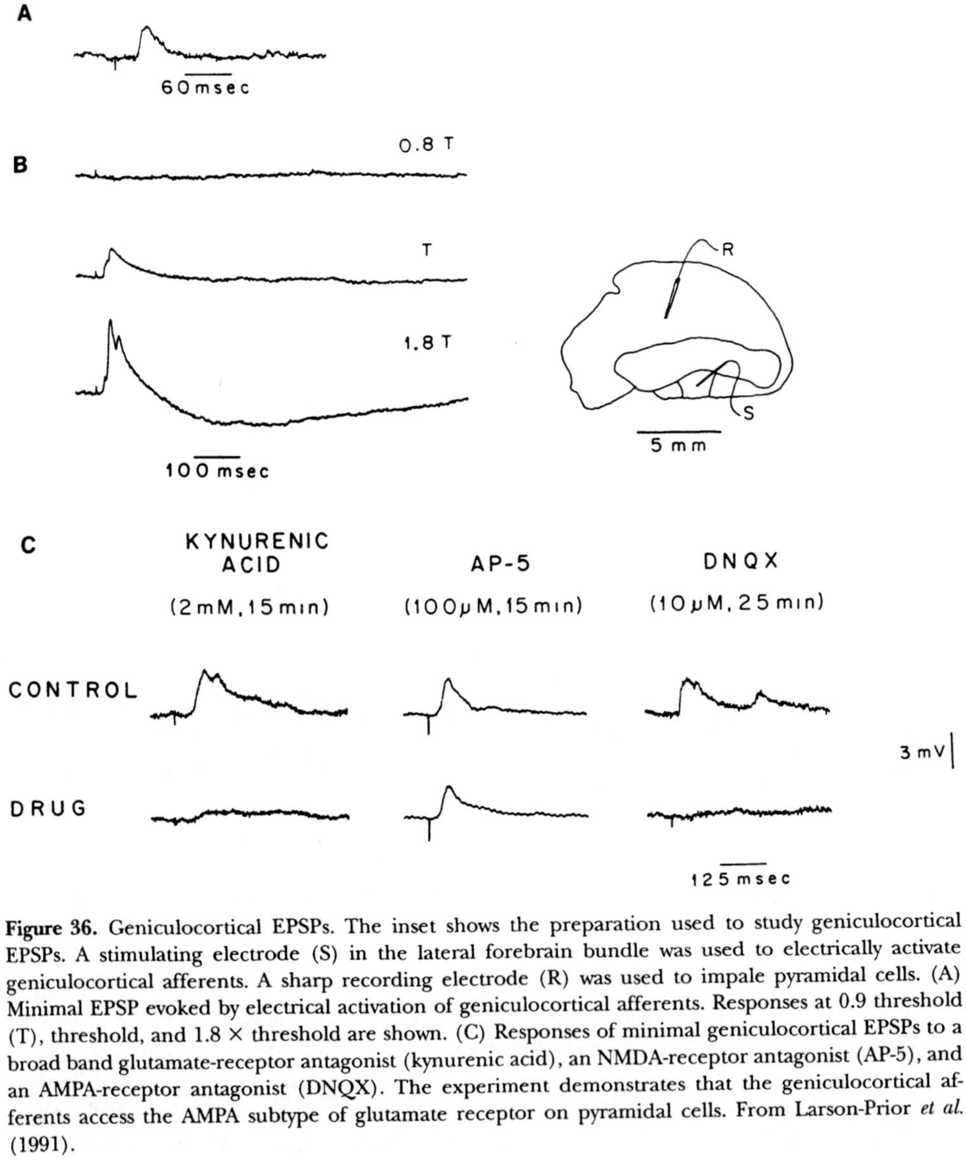

Figure 36. Geniculocortical EPSPs. The inset shows the preparation used to study geniculocortical EPSPs. A stimulating electrode (S) in the lateral forebrain bundle was used to electrically activate geniculocortical afferents. A sharp recording electrode (R) was used to impale pyramidal cells. (A) Minimal EPSP evoked by electrical activation of geniculocortical afferents. Responses at 0.9 threshold (T), threshold, and 1.8 × threshold are shown. (C) Responses of minimal geniculocortical EPSPs to a broad band glutamate-receptor antagonist (kynurenic acid), an NMDA-receptor antagonist (AP-5), and an AMPA-receptor antagonist (DNQX). The experiment demonstrates that the geniculocortical afferents access the AMPA subtype of glutamate receptor on pyramidal cells. From Larson-Prior *et al.* (1991).

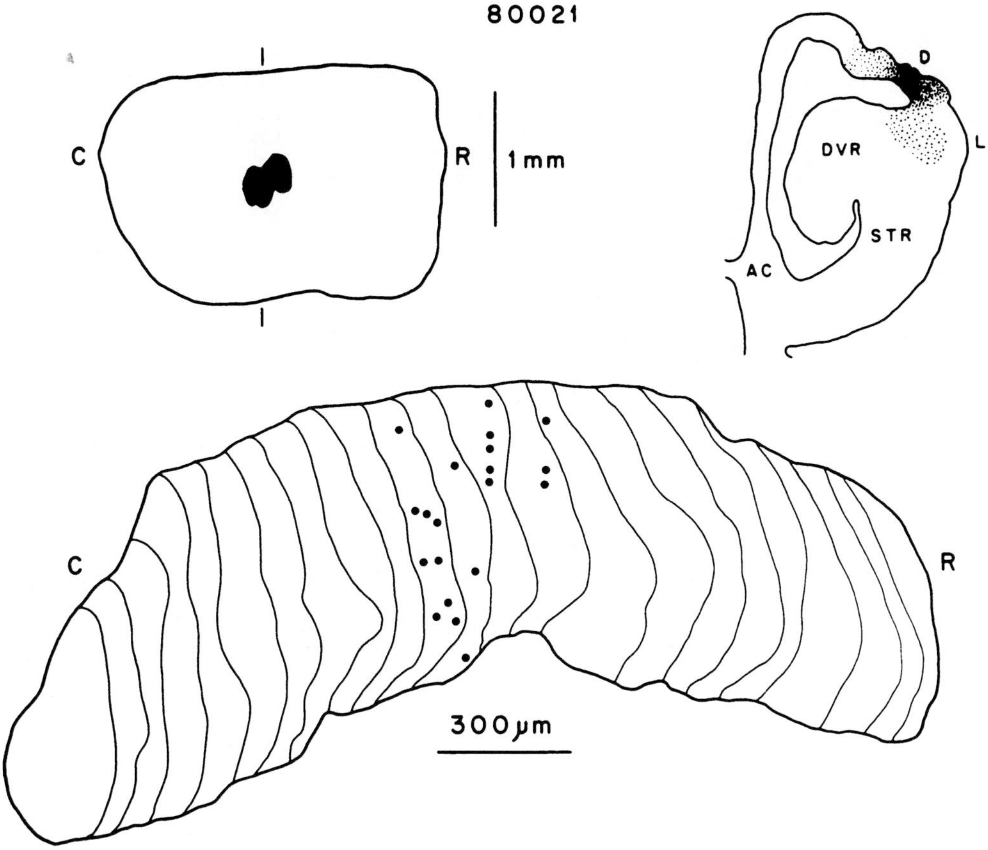

Figure 37. Retrograde transport experiment. This is the result of an experiment in which the retrograde transport of HRP was used to investigate the organization of the geniculocortical pathway. The drawing on the top left is a surface view of turtle visual cortex with the black profile indicating the size and position of an HRP injection. The caudal (C) and rostral (R) edges of the cortex are indicated, and the vertical lines indicate the plane of the section shown at the top right. The focus and surround of the injection site are shown by the black and stipple, respectively. The main part of the figure is a reconstruction of the dorsal lateral geniculate complex with the caudal and rostral poles indicated. Dots represent the positions of geniculate neurons retrogradely labeled by the cortical injection. Injections in the center of the cortex labeled neurons in the center of the geniculate complex. Injections in the caudal cortex labeled neurons in the rostral geniculate. Injections in the rostral cortex labeled neurons in the caudal geniculate. The experiments demonstrate that the rostral–caudal axis of the geniculate projects along the caudal–rostral axis of the cortex. From Mulligan and Ulinski (1990).

Neurons in the visual cortex give rise to a reciprocal projection to the dorsal lateral geniculate complex that is organized in terms of the isoazimuth lamellae (Gaidenko, 1978; J. A. Hall *et al.,* 1977; Ulinski, 1986b). The projection originates from neurons in the second and third layers of the cortex. The second layer contains many pyramidal cells that show glutamate-like immunoreactivity (Reiner, 1991); the third layer contains many neurons that show GABA-like immunoreactivity (Reiner, 1991; Nicolaus and Ulinski, 1991). The transmitters of neurons involved in the corticogeniculate projection are not known, but it is possible that the projection includes both excitatory and inhibitory components. Axons of corticogeniculate neurons leave the lateral edge of the cortex and descend through the striatum (Ulinski, 1986b). They course into the thalamus via the ventral peduncle of the lateral forebrain bundle and turn dorsally into the dorsal lateral geniculate complex. Individual axons course from ventral to dorsal through the nucleus, intersecting the dendrites of cell plate neurons within the inner neuropile zone. The corticogeniculate projection is arranged so that cells in the rostral cortex project to neurons in the caudal geniculate complex, etc.

4.1.4. Intracortical Projections

As mentioned above, the geniculocortical projection carries information from different isoazimuth bands within the geniculate to isoazimuth lamellae within the dorsal area of cortex. Intracortical connections then provide a convergence of information between isoazimuth lamellae within the dorsal area itself. They are effected by the axon collaterals of pyramidal cells with somata in layer 2 (Connors and Kriegstein, 1986). Individual pyramidal cells have axons that descend from their somata into layer 3, where they bifurcate into two branches, one directed medially toward the limbic cortex and the other laterally toward the striatum, lateral geniculate, and optic tectum (Ulinski, 1986b). The parent axon additionally bears a wreath of collaterals that ascend into layers 2 and 3 (Connors and Kriegstein, 1986). Each collateral bears varicosities and is positioned so that it can potentially synapse upon the dendrites of other pyramidal cells, as well as on inhibitory neurons. The spatial extent of these collateral projections can be visualized in HRP tracing experiments carried out *in vitro* with wholemount preparations of visual cortex (Cosans and Ulinski, 1990). Restricted injections of HRP in cortical wholemounts produced a halo of orthogradely labeled axons that extended in all directions from the injection site. They showed a distinct tendency to run mediolaterally along isoazimuth lamellae, but also extended for roughly one third the length of the cortex, orthogonal to the lamellae, bearing varicosities at irregular intervals (Fig. 38). Thus, any individual pyramidal cell in dorsal cortex can receive information from pyramidal cells located at any horizontal eccentricity via a minimum of three intracortical relays.

The physiology of the intracortical projections has been studied by recording intracellularly from pyramidal cells while activating intracortical afferents with a stimulating electrode at a distant point in the cortex (Larson-Prior *et al.*, 1991). Stimulation produced a complex EPSP in pyramidal cells that has three definable components (Fig. 39). A fast component was not abolished by bath application of kynurenic acid, a broadband antagonist to excitatory amino acids. The neurotransmitter involved in generating this component is not known. A second component with intermediate latency was abolished by bath application of quinoxolinediones such as cyanonitroquinoxolenedione (CNQX) and dinitroquinoxolenedione (DNQX) that are specific antagonists to the quisqualate/kainate or AMPA subtypes of glutamate receptor. The slow component was abolished by bath application of 5-phosphonovaleric acid (APV), a specific antagonist of the *N*-methyl-D-aspartate (NMDA) subtype of glutamate receptor.

4.1.5. Spatial and Temporal Organization of Visual Cortex

Anatomical characterization of the geniculocortical pathway and the pattern of connections within visual cortex points to an extensive convergence of informa-

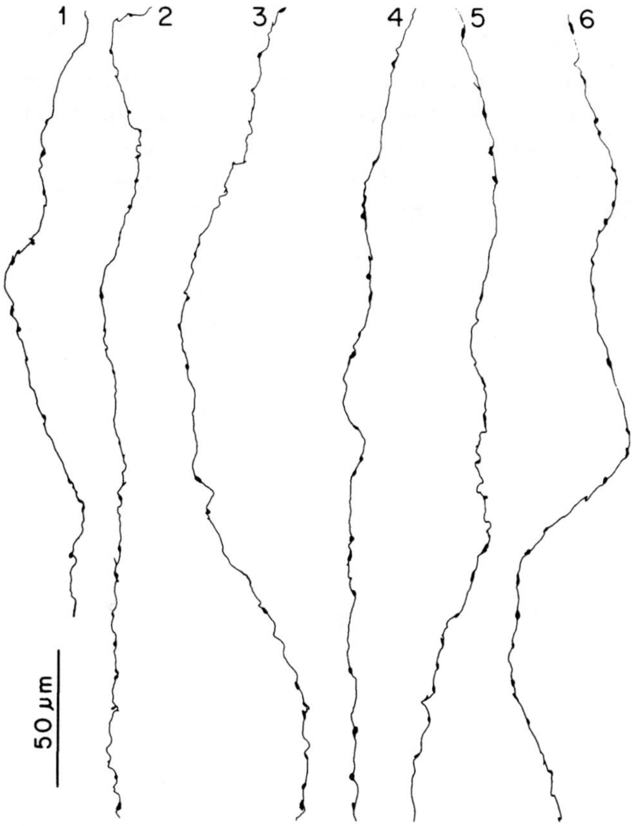

Figure 38. Morphology of pyramidal cell axons. These are drawings of individual intracortical axon collaterals to show details of their morphology. They were visualized following focal intracortical injections of HRP. From Cosans and Ulinski (1990).

tion from points throughout visual space upon individual neurons within the visual cortex (Fig. 40). This is supported by studies of the spatial organization of visual cortex using physiological methods in alert, paralyzed turtles (Mazurskaya, 1972). The animal's head was positioned in the center of a hemisphere and lights measuring 0.5 deg in diameter were flashed for 100–500 msec. Evoked potentials were recorded from the visual cortex using glass microelectrodes. Stimuli presented at any point in binocular visual space produced evoked responses at anterior, posterior, medial, lateral, and central loci in the visual cortex. Presentation of stimuli at 10-deg intervals throughout visual space suggested that a region of binocular representation extends for 50–55 deg along the horizontal meridian in front of the animal and expands to 85–90 deg behind the animal. Although neurons at each cortical locus responded to stimuli throughout visual space, the latencies and the amplitudes of the responses varied as a function of stimulus position. Responses with the shortest latency and greatest amplitude were obtained at the posterior and middle regions of the visual cortex when points in the nasal quadrant of visual space were stimulated. Responses with the shortest latency and greatest amplitude were obtained at the anterior recording site when stimuli were presented to the temporal visual field. Assuming that the shortest latency responses result from direct geniculocortical projections, these results are consistent with the geometry of the geniculocortical path.

Prechtl and Bullock (1995) used evoked potential methods to record extracellularly from multiple sites on the dorsal cortex of intact turtles. The recordings showed that visual stimuli activate neurons throughout dorsal cortex. Evoked responses oscillated or "fluttered" with a characteristic frequency of about 20 Hz (Prechtl, 1994). Waves of synchronized activity propagated along the rostrocaudal

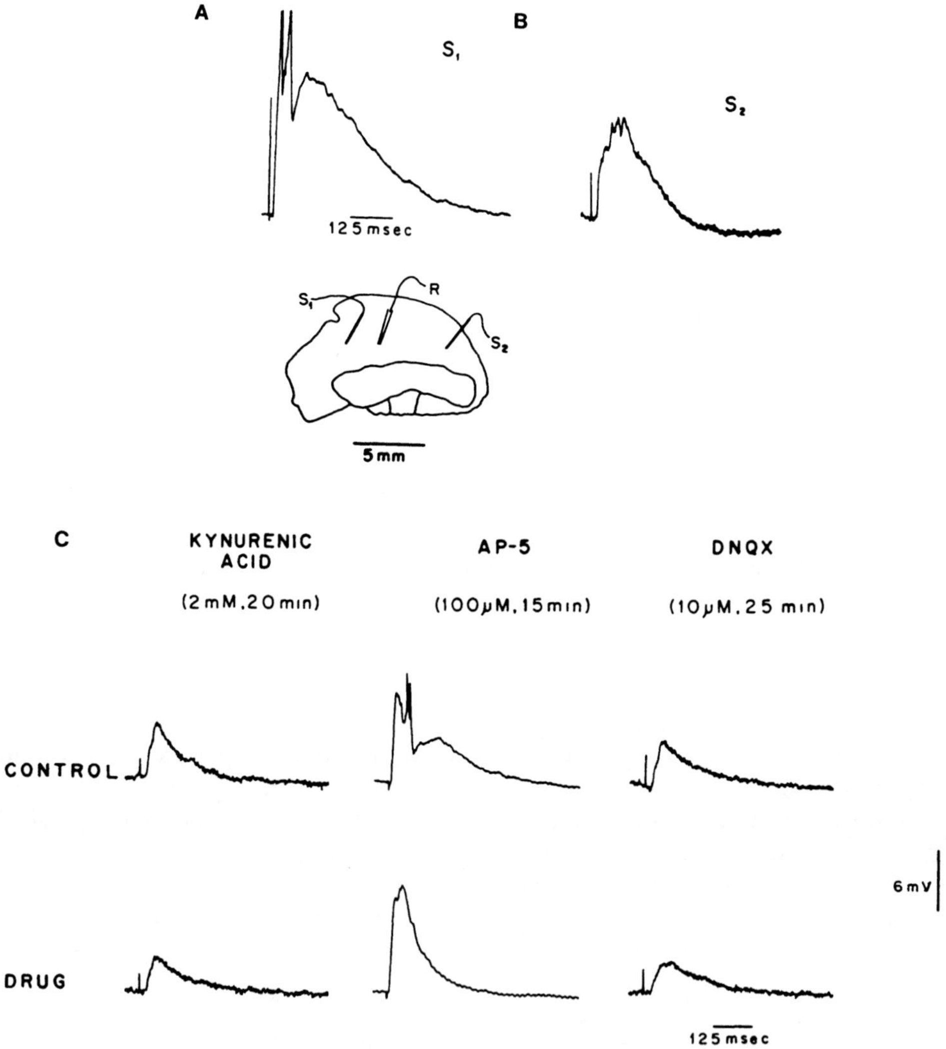

Figure 39. Intracortical EPSPs from turtle visual cortex. The inset shows the preparation used to evoke EPSPs mediated by intracortical afferents. Stimulating electrodes (S_1 and S_2) are placed in the cortex rostral and caudal to a recording electrode (R). Activation of the electrodes produces compound PSPs (A, B). The PSP produced by activation of S_2 is large enough to generate two action potentials in the pyramidal cell. Shown are responses of the pyramidal cells following bath application of the broad band glutamate-receptor antagonist kynurenic acid, the NMDA-receptor antagonist AP-5, and the AMPA-receptor antagonist DNQX. From Larson-Prior *et al.* (1991).

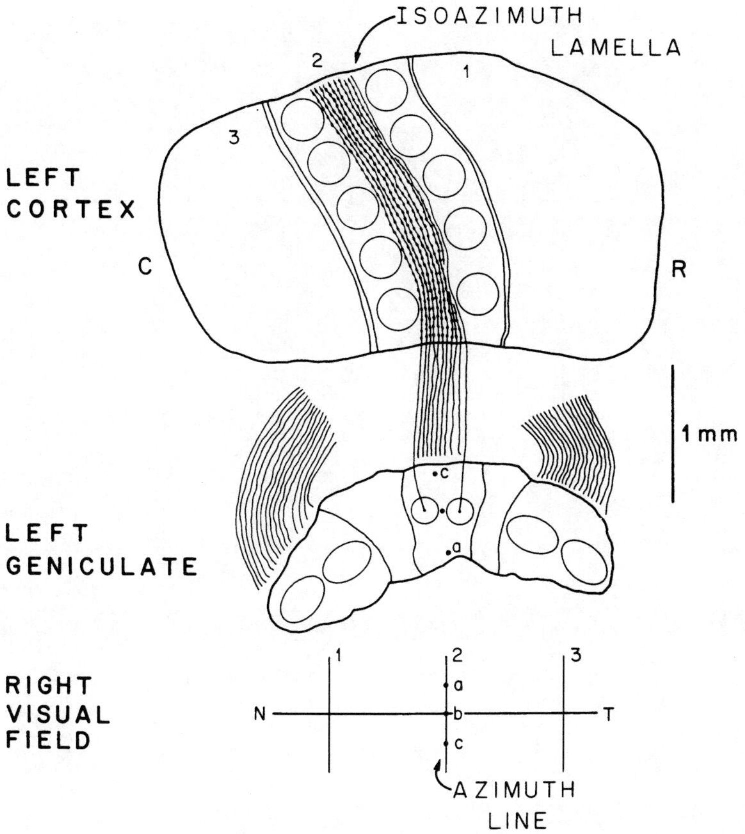

Figure 40. Organization of geniculocortical pathway. This figure summarizes the overall spatial organization of the geniculocortical projections. The lower part shows the orientation of the right visual field. The nasal–temporal axis (N–T) is indicated and three azimuth lines (1, 2, 3) are shown. Three points of different elevation (a, b, c) are indicated on azimuth line 2. The center of the figure shows a lateral view of the left dorsal lateral geniculate complex. Caudal is to the left and rostral is to the right. From Mulligan and Ulinski (1990).

axis of the cortex with a phase lag of 0.05–0.3 mm/msec. The waves occurred following the introduction of salient stimuli, such as moving objects, or when the animal shifted its gaze by making a visual saccade (Prechtl and Bullock, 1995).

These electrophysiological studies are consistent with work using voltage-sensitive dyes in *in vitro* preparations of the eye and brain (Senseman, 1995a, b, 1996). The isolated cortex was bathed with a dye that permits visualization of changes in the membrane potential of populations of neurons. The relationship between the visual signal and changes in membrane potential was confirmed by simultaneously recording optical changes and recording with intracellular methods from a single pyramidal cell. The optical recordings demonstrated that depolarizing waves propagate along the rostrocaudal axis of the cortex at a speed between 0.02 and 0.04 mm/msec (Fig. 41). All neurons in the cortex are thus activated by a visual stimulus in a spatiotemporal pattern that results in different neurons responding with different latencies and amplitudes.

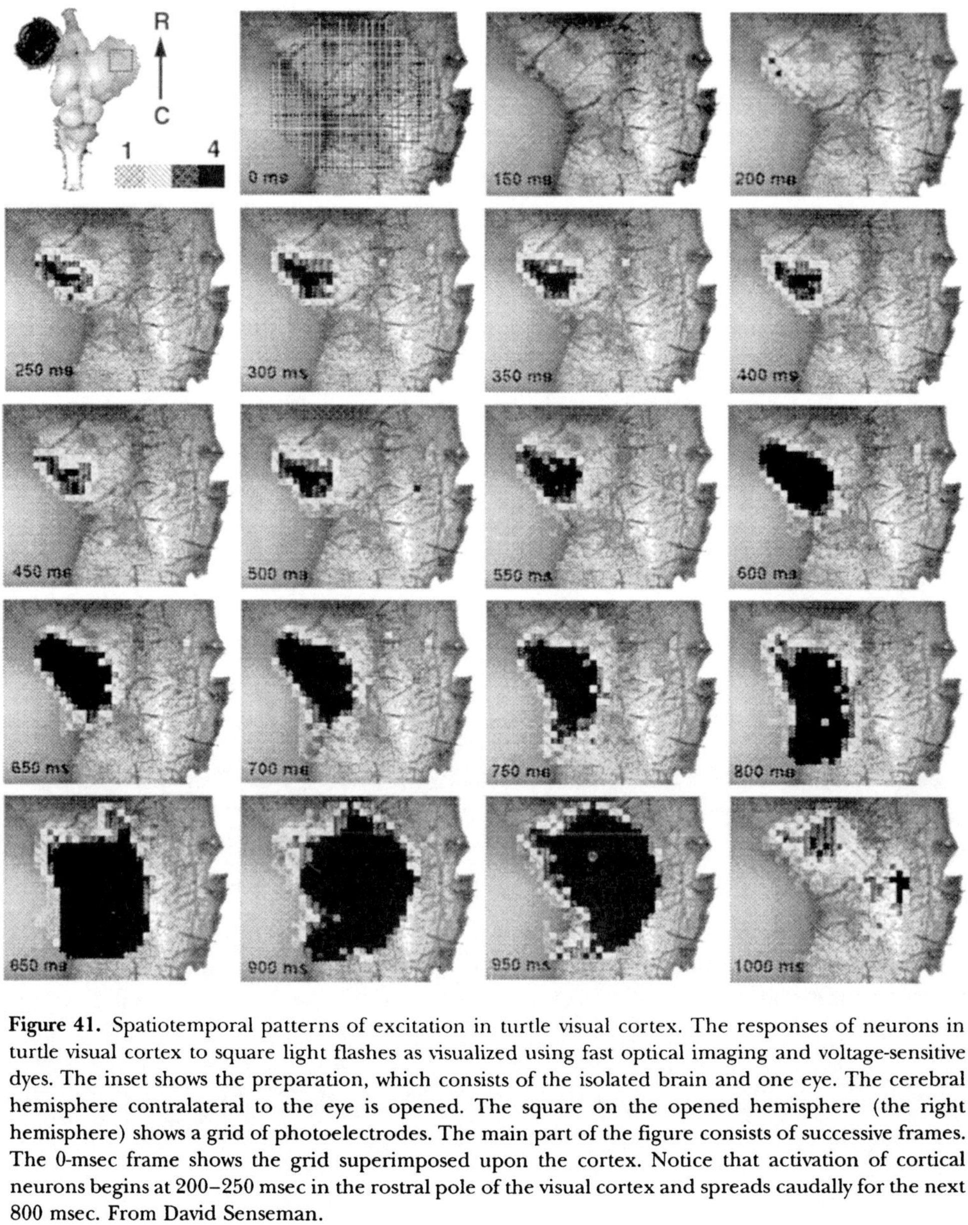

Figure 41. Spatiotemporal patterns of excitation in turtle visual cortex. The responses of neurons in turtle visual cortex to square light flashes as visualized using fast optical imaging and voltage-sensitive dyes. The inset shows the preparation, which consists of the isolated brain and one eye. The cerebral hemisphere contralateral to the eye is opened. The square on the opened hemisphere (the right hemisphere) shows a grid of photoelectrodes. The main part of the figure consists of successive frames. The 0-msec frame shows the grid superimposed upon the cortex. Notice that activation of cortical neurons begins at 200–250 msec in the rostral pole of the visual cortex and spreads caudally for the next 800 msec. From David Senseman.

4.1.6. Cortex and Global Motion Detection

Work with both single-unit recording and behavioral methods indicates that neurons in dorsal cortex meet the definition of global motion detectors. In the electrophysiological work (Mazurksaya, 1972, 1974), stimuli consisting of black stripes, circles, or squares of different sizes were moved in different directions with different speeds. Units varied in receptive field size and depth in the cortex. *Superficial units* had relatively small receptive fields (2–40 deg) and were probably geniculate afferents. They were direction-selective and showed a preference for radial motion. A stimulus moving toward the center of visual space anywhere in the receptive field evoked neural activity in some units, while centrifugal movements did not. In other units, only stimuli moving centripetally were effective. *Intermediate units* were located between 200 and 300 μm from the surface of the cortex and are probably layer 1 interneurons. They responded to moving stimuli at all points in visual space. They also responded to changes in diffuse illumination, but did not respond well to local changes in light intensity. *Deep units* were located from 800 to 1000 μm from the pia and were most likely layer 2 pyramidal cells. In contrast to intermediate units, deep units did not respond well to diffuse changes in illumination. The most effective stimulus for these units was movement of an edge. Comparable responses were obtained when a black edge was moved across a white background as when a white edge was moved against a black background. Responses could be obtained by moving the stimulus anywhere in the visual field. Best responses could be obtained when the stimulus moved with a speed of 5–8 deg/sec. The direction of the movement was "unimportant." Units showed a marked habituation to repeated (five to seven) presentations of the same stimulus. They did not recover when the direction of movement was changed or when the target was moved to another locus in the visual field. Units were sensitive to small displacements (less than 0.5 deg) of the stimulus.

Cortical neurons show speed tuning. Gusel'nikov *et al.* (1972) used stimuli of varying sizes moving across the visual field of the animal with a variety of speeds. Neurons had background activity in the light and the dark. Units in the medial part of dorsal cortex had background discharge rates of 0.2–1.0 spikes/sec, while those in the lateral part (or pallial thickening) had slightly higher rates of 0.3–1.9 spikes/sec. Units responded to a black circle with a diameter of 200 deg over a range of speeds between 6 and 40 deg/sec, with a mean preferred speed of 14.5 deg/sec.

Finally, cortical neurons show interactions between moving stimuli presented at different loci in visual space. This was investigated in experiments in which stimuli were moved from nasal to temporal along the horizontal meridian of visual space with a speed of 40 deg/sec (Gusel'nikov and Pivovarov, 1978). A background stimulus consisting of a strip measuring 6 deg × 60 deg oriented orthogonal to the horizontal meridian was moved across the visual field at 20-sec intervals. After the tenth presentation of this background stimulus, test stimuli consisting of a 60 deg × 60 deg strip oriented 135 deg to the streak, or a strip measuring 60 deg × 20 deg oriented 90 deg to the strip, were introduced. Both stimuli evoked potentials, and PSPs showed a reduction in amplitude during successive presentations of the test stimulus. Presentation of test stimuli led to an increase in the response of the unit, but habituation to presentation of test stimuli occurred after repeated presentations.

In behavioral studies, Killackey *et al.* (1972) found that lesions of dorsal cortex impaired the ability of turtles to habituate to a looming stimulus or objects moving toward the animal. Turtles had their heads restrained while they faced a ping-pong ball that fell toward, but did not contact, their noses. They initially reacted to the ball by withdrawing their heads, but gradually stopped responding upon repeated presentations. However, lesions of dorsal cortex impaired the animals' ability to habituate, and lesioned animals continued to withdraw their heads. Powers and

Wojcik (1992) repeated these experiments using a looming stimulus that consisted of a circle projected upon a screen. The circle expanded from 7.5 cm to 22.5 cm in diameter within 1 sec. Again, lesions of dorsal cortex impaired the ability of the turtles to habituate to the stimulus. It appears that dorsal cortex cannot be the sole location of "looming detectors" within the brain, or else the lesioned animals would not be able to detect and respond to looming stimuli. However, the experiments suggest a role for dorsal cortex in processing sequences of complex visual stimuli over time.

To summarize, the information available from anatomical, electrophysiological, and behavioral studies indicates that turtle dorsal cortex contains neurons with large receptive fields that are involved in analyzing movements of visual stimuli throughout visual space over periods of time lasting hundreds of milliseconds or seconds. The remainder of this section discusses the neural mechanisms underlying the generation of these receptive fields.

4.2. Spatiotemporal Receptive Fields

Temporal integration functions for neurons in turtle visual cortex have been measured using extracellular recording methods with *in vivo* preparations (Mazurskaya, 1974). Stimuli were small spots of light presented sequentially at two disjunct points in visual space with varying temporal separations. Temporal separations of $\Delta t = 0$ represent simultaneous presentation of the two stimuli. The response of the cell was measured as the firing rate of the cell in action potentials per second averaged over ten stimulus presentations. The functions showed two distinct peaks, one occurring with temporal separations of about 50 msec and a second, more variable peak occurring with temporal separations of 300–800 msec (Fig. 42). An important difference between neurons located in the superficial layer of the cortex (layer 1) and those located in the deep layer (layer 2) was that layer 1 neurons were facilitated by simultaneous presentation of the stimuli, while those in layer 2 were inhibited by simultaneous presentation. Since layer 1 neurons likely include inhibitory interneurons, while layer 2 neurons are probably pyramidal cells, these differences are important in considering the balance of excitation and inhibition generated by a moving stimulus and will be considered further in Section 4.4. The response properties of the pyramidal cells suggest they have spatiotemporally inseparable receptive fields because the peaks in their temporal integration functions shifted as a function of the spatial separation of the stimulus pair. A sequential pair of stimuli separated by 10 deg of visual arc produced a temporal integration function with peaks at 70 and 500 msec, while a stimulus pair separated by 20 deg of visual arc produced a temporal integration function with peaks at 200 and 800 msec. The spatiotemporal receptive field of the neuron is thus oriented in the space–time plane and has two segments connected by a "kink." The first segment would correspond to a preferred speed of about 77 deg/sec and the second to a speed of about 33 deg/sec.

These *in vivo* experiments determine the shapes of the temporal integration functions for cortical neurons, but have two limitations. First, measurement is made in intact animals, so it is difficult to know whether or not the temporal integration observed is a property of the cortical circuitry, occurs at a subcortical level, or reflects interactions between a series of structures. A second difficulty is that the functions were measured with extracellular recording methods, making it difficult to say anything about the cellular mechanisms that underlie temporal integration because the firing pattern of the cell is a complex matter that can reflect the biophysical properties of the neuron as well as its synaptic interactions.

These difficulties can be circumvented by taking advantage of an *in vitro* preparation (Fig. 43) of turtle visual cortex (Kriegstein, 1987; Ulinski *et al.*, 1991). Turtles have a diving physiology that endows their brain and eyes with a unique and remarkable tolerance to anoxia (Hounsgaard and Nicholson, 1990). The entire visual system can be maintained intact in an *in vitro* preparation and used for studies of the physiology of neurons in both the retina (e.g., Rosenberg and Ariel, 1990, 1991) and in central visual structures (Kriegstein, 1987). Physiological re-

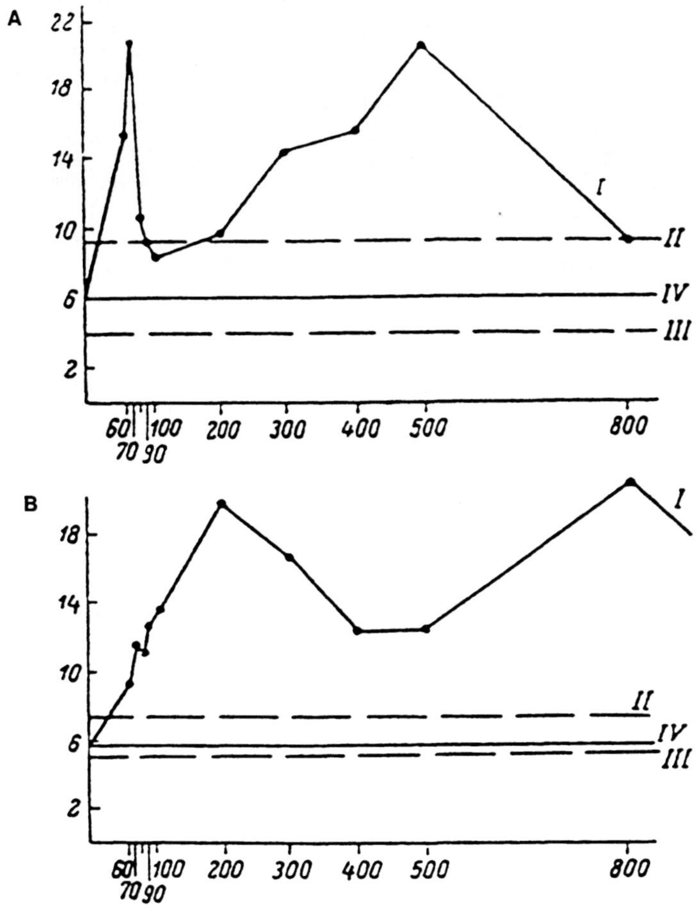

Figure 42. Temporal integration curves for turtle cortical neurons. Temporal integration curves for presumed pyramidal cells in turtle visual cortex obtained with an *in vivo* preparation and extracellular recording methods. The firing rate of the cell in spikes per second averaged over ten presentations is plotted as a function of the temporal separation of two spots of light presented at spatial separations of (A) 10 deg and (B) 20 deg. Note that both curves show fast and slow facilitatory peaks and that the peaks shift as a function of spatial separation. From Mazurskaya (1974).

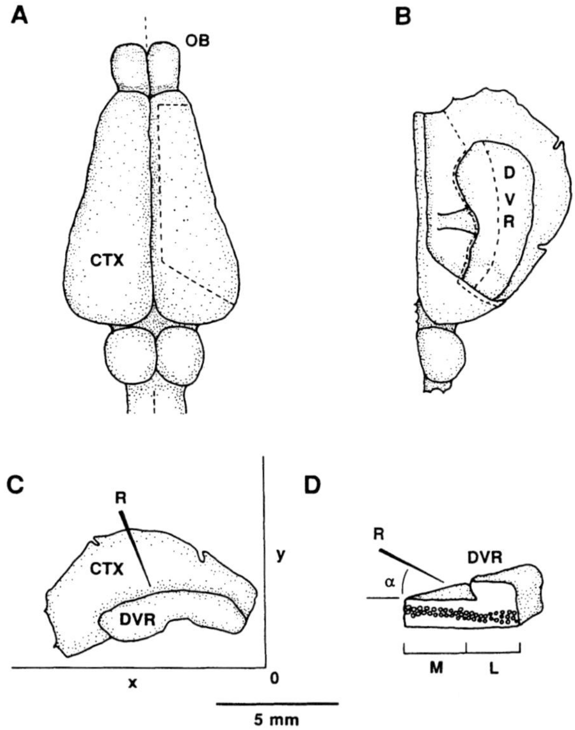

Figure 43. *In vitro* preparation of turtle visual cortex. (A) Dorsal view of intact turtle brain with the cortex (CTX) and olfactory bulb (OB) labeled. The dashed lines show the cuts used to bisect the brain on the midline and to open the cerebral hemispheres. (B) Preparation with the cortex unfolded to expose the dorsal ventricular ridge (DVR). Dashed lines show the cuts made to isolate the cortex. (C) Cortex placed upon an *x,y* grid used to record the position of the recording electrode (R). Side view of a block of cortical tissue to illustrate the orientation of the recording electrode to the cortical layers in the lateral (L) and medial (M) parts of the cortex. The angle of the electrode to the surface of the preparation is used to calculate the depth of an impaled cell from the surface of the preparation.

sponses can be obtained using extracellular methods (Rosenberg and Ariel, 1991) or intracellular methods with sharp (Larson-Prior *et al.*, 1991) or patch-clamp electrodes (Blanton *et al.*, 1989). Recordings can be made from neurons in the retina or central visual pathways while visual stimuli are presented to the retina, making it possible to characterize the receptive field properties of central neurons while carrying out the kinds of biophysical and pharmacological studies that are done in slice preparations of mammalian neocortex.

Since the geniculate axons maintain their topography as they approach the cortex, fascicles situated at different loci in the lateral forebrain bundle carry information about different azimuth lines of visual space. Electrical stimulation of geniculate afferents at different loci in the lateral forebrain bundle of an *in vitro* preparation of isolated cortex would then activate geniculate fascicles that carry information about disjunct regions of visual space. A two-pulse experiment can be carried out by delivering two stimuli S1 and S2 with a range of temporal separations. Activation of either electrode elicits a complex postsynaptic potential (PSP) in a cortical neuron that has been impaled with a sharp microelectrode (Fig. 44). Sequential activation of the two electrodes with interstimulus delays Δt ranging from 25 to 1000 msec allows the calculation of a temporal integration function. These functions typically show two facilitatory peaks, one at about 50 msec and a second at about 800 msec. A disfacilitatory trough occurs between the two peaks at about 125 msec. The temporal integration functions obtained for PSPs in cortical neurons can be represented as a sum of three functions of the form $a_i(\Delta t)$ $\mathrm{Exp}(-b_i\,\Delta t)$:

$$T(\Delta t) = 1 + \sum_{i=1}^{3} a_i(\Delta t)\,\mathrm{Exp}(-b_i \Delta t) \qquad (10)$$

These are alpha functions of the kind that are often used to model synaptic currents. The times-to-peak of the peaks and troughs are given by the parameters $(b_i)^{-1}$. The weighting coefficients a_1 and a_3 are positive for the facilitatory peaks; the coefficient a_2 is negative and specifies the disfacilitatory trough.

The neural mechanisms underlying these temporal integration functions are not fully understood, but several candidate mechanisms can be mentioned. First, some temporal summation of PSPs is involved because PSPs elicited by activating electrodes S1 and S2 overlap in time for a wide range of temporal separations. The summation is not linear; PSPs produced by simultaneous activation of the electrodes are on average 66% of the algebraic sum of the two individual PSPs. This may reflect a compressive nonlinearity due to the reduction of the synaptic driving potentials that occurs when two or more EPSPs are generated in a cell within several milliseconds of each other. The membrane depolarization resulting from the first EPSP reduces the driving potential of the second EPSP, causing a nonlinear interaction of the two EPSPs. A second class of candidate mechanisms involves presynaptic facilitation. They assume that the two spatial channels are converging upon a common pool of neurons situated presynaptic to the neuron being studied. Thus, activation of the S2 channel activates elements presynaptic to the pyramidal cell under consideration which have elevated levels of intraterminal calcium due to their prior activation by the S1 channel. Two-pulse experiments conducted at the synaptic level show that presynaptic facilitation has two facilitatory phases with times-to-peak that correspond roughly to the peaks seen in temporal integration functions (Katz and Miledi, 1968; Mallart and Martin, 1967; Magelby and Zengel, 1982). Finally, the two facilitatory phases of the temporal integration curves may reflect the kinetics of glutamate receptors (Trussell *et al.*, 1988; Trussell and Fischbach, 1989; D. O. Smith *et al.*, 1991; Lester and Jahr, 1992). Both geniculocortical and intracortical axons are glutaminergic and access glutamate receptors on pyramidal cells. Consistent with this, temporal integration functions measured in the presence of APV show an alteration in the amplitudes of the two facilitatory phases.

Regardless of the mechanisms underlying the temporal integration functions, experiments with the isolated cortex preparation demonstrate two important points. The first is that the shape of the temporal integration functions is determined within the cortex itself, because the preparation isolates the circuitry of visual cortex from interactions with subcortical systems or other cortical areas. The second is that the spike-generating mechanism of cortical neurons plays no essential role in determining the shape of the temporal integration functions, because temporal integration functions for subthreshold PSPs have the same shape as those obtained with extracellular recording methods. The two kinds of curves can be compared directly by recording intracellularly from a pyramidal cell and measuring two temporal integration functions (Ulinski *et al.*, 1993). The first is obtained while the cell is slightly hyperpolarized and therefore does not generate action potentials. The second curve is obtained at the normal resting membrane potential of the cell and shows the action potentials elicited by activation of the S1 and S2 channels.

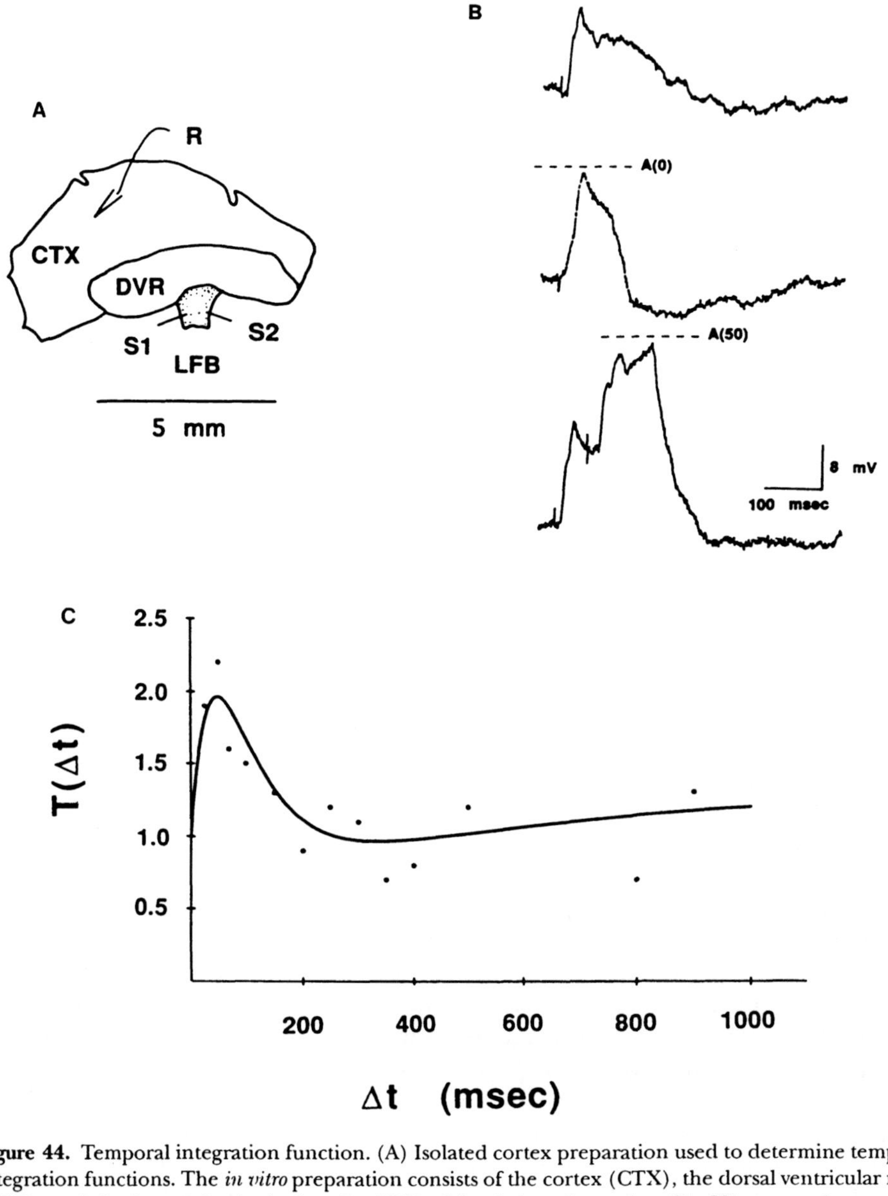

Figure 44. Temporal integration function. (A) Isolated cortex preparation used to determine temporal integration functions. The *in vitro* preparation consists of the cortex (CTX), the dorsal ventricular ridge (DVR), and the lateral forebrain bundle (LFB). Stimulating electrodes (S1, S2) are used to activate geniculate afferents in fascicles carrying information from two different regions of visual space. A sharp electrode is used to record intracellular responses from a pyramidal cell in visual cortex. (B) Response of the pyramidal cell to activation of electrode S2 (top trace); response to simultaneous activation of the S1 and S2 electrodes (middle trace); and response to activation of electrode S1 followed by S2 with a temporal separation Δt of 50 msec. (C) Temporal integration function $T(\Delta t)$ determined for a range of temporal separations. Notice a distinct, fast facilitatory peak at 50 msec and a less distinct, slow facilitatory peak from 400 to 1000 msec.

The temporal integration functions lead to some ideas of how spatiotemporal receptive fields are generated (P. S. Ulinski, in preparation). The simplest case would be if the response functions $R(\Delta x,\Delta t)$ for turtle cortical neurons were separable functions of Δx and Δt. The response functions could then be obtained by multiplying together the spatial and temporal integration functions: $R(\Delta x,\Delta t) = T(\Delta t)\,S(\Delta x)$. Work with the isolated cortical preparation gives no indication of whether or not the response functions are separable, because there is no way to determine adequately the spatial separation of the two spatial channels being activated. However, the temporal integration functions obtained with extracellular recording methods (Mazurskaya, 1974) show that the response functions are inseparable because the functions obtained by presenting visual stimuli at two disjunct points in visual space vary as a function of spatial separation. This experimental finding can be represented formally by making the spatial integration function depend upon both spatial and temporal separation: $R(\Delta x, \Delta t) = T(\Delta t)\,S(\Delta x, \Delta t)$.

To understand why this is reasonable, consider again the case of an object moving with some fixed speed v along the horizontal meridian of visual space. The time it takes for information to flow from the retina to the cortex depends upon the position of the object in visual space and there appears to be a gradient in the conduction velocities in the geniculocortical system as a function of position along the horizontal meridian. However, matters can be simplified by considering initially cortical neurons lying along an isolatency contour within the cortex. Activation of a sequence of spatial channels by the moving object will elicit activity in a sequence of cortical neurons arranged along the contour. An object appearing at point x_1 in the visual system at time t_1 will activate a band of cortical neurons lying within the isoazimuth lamella that is the line image of point x_1 at time $t_1 + c$, where c is the conduction time or latency from the retina to neurons along the isolatency contour. Some neurons along the contour will reach threshold and fire action potentials, evoking a volley of action potentials that travels away from the neuron toward the caudal pole of visual cortex. The propagation velocity p with which information is traveling along the cortex is available from voltage-sensitive dye experiments (Senseman, 1996). Such experiments suggest that values of $p = 0.03$ mm/msec are reasonable. It is useful to express the propagation velocity in terms of degrees of visual arc/msec, which can be done using a magnification factor that specifies the number of degrees of visual space represented in 1 mm of cortex. Adequate information on the magnification factor is not available, but assuming that there is an overrepresentation of the visual streak in the geniculate and cortex leads to a guess of about 2 deg/mm, and the propagation velocity would be about 60 deg/sec.

The probability that neurons situated in isoazimuth lamella x_1 will activate neurons in other isoazimuth lamellae is determined by the spatial integration function, $S(\Delta x,\Delta t)$. Precise determination of this function is difficult, but information on the spatial distribution of intracortical axons (Cosans and Ulinski, 1990) suggests that the probability will decrease as a function of distance from isoazimuth x_1. Trial simulations show that the form of the spatial integration function does not strongly influence the response function, and a reasonable first approximation is to consider the spatial integration function as a Gaussian function of the form

$$S(\Delta x, \Delta t) = \mathrm{Exp}[-(\Delta x - p\,\Delta t)^2/s^2] \qquad (11)$$

where s is a spatial scale factor that determines the range over which neurons at one cortical locus can affect neurons at other cortical loci. The data reviewed above suggest a value of roughly 40 deg for s. Visual stimuli at x_1 in visual space will therefore activate cortical neurons in isoazimuth lamella x_1 at time $t_1 + c$ via the geniculocortical afferents, but will also activate neurons in lamella x_2 at time $t_1 + c + \Delta x/p$ via the intracortical afferents. The intracortical inputs to neurons at x_2 can be subthreshold, but the temporal integration functions for cortical neurons demonstrate that the effects of synaptic inputs to neurons at x_2 will persist for hundreds of msec and show two facilitatory phases, one with Δt values of about 50 msec and a second with values of about 800 msec. A subthreshold geniculocortical input to neurons at x_2 could thus fire the neurons if the arrival of the second geniculate volley at time $t_2 + c$ coincides with one of the two facilitatory phases. The neurons at x_2 show an increased probability of firing if they are sequentially activated by a second fascicle of geniculocortical afferents within one of these facilitatory periods. They will be tuned to stimuli moving along the horizontal meridian of visual space with a speed determined by the conduction velocity of the intracortical volley.

Spatiotemporal receptive fields for these neurons can be modeled by multiplying together the temporal integration and spatial integration functions given in Equations (10) and (11). Figure 45 is such a function plotted as both three-dimensional and contour plots. The plots show several major features. First, the response functions have two distinct maximal regions, reflecting the two excitatory phases of the temporal integration function. These resemble the "kinks" that are frequently reported for spatiotemporal receptive fields in cat striate cortex. Second, both of these regions are oriented in the Δx, Δt plane. The second region has a distinctly oblique orientation corresponding to a preferred speed of 70 deg/sec, but the first region has a nearly vertical orientation, so that the cell will show relatively little

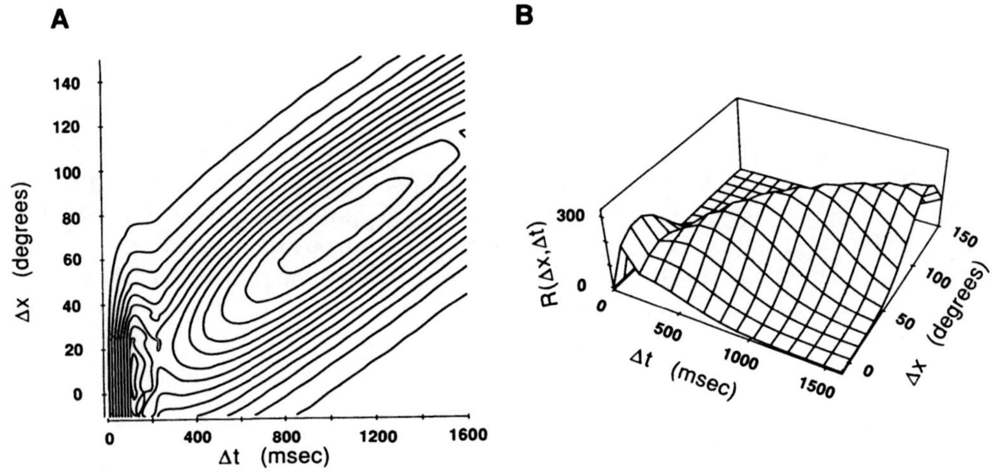

Figure 45. Spatiotemporal response function. A simulated spatiotemporal response function for a pyramidal cell in turtle visual cortex produced by the model described in the text. (A) The function in a contour plot; (B) the function $R(\Delta x, \Delta t)$ plotted as a function of spatial (Δx) and temporal (Δt) separation. Notice that the function has two segments, one between temporal separations of 0 and 100 msec with a relatively vertical orientation, and one between 100 and 1000 msec with a more oblique orientation.

speed tuning for stimuli moving over small spatial and temporal separations. This could reflect either a purely speed-selective process, or an orientation-selective unit with little speed selectivity. Oriented spatiotemporal receptive fields may then originate from an interaction between the geometric organization of the geniculocortical and intracortical systems and the properties of glutaminergic synapses.

4.3. Linear and Nonlinear Mechanisms

The two-pulse experiment described in the preceding section demonstrates significant nonlinearities in the interaction between two spatial channels. Establishing the details of the interactions leading to these nonlinearities is difficult with this experimental paradigm because the PSPs produced by electrical activation of geniculocortical fascicles results in contributions from both the geniculocortical and intracortical systems. It is also difficult to relate these results to the processes involved in analysis of real moving stimuli because electrical activation produces an unnatural synchronization of geniculocortical afferents. These difficulties can be circumvented by analyzing the responses of pyramidal cells to natural visual stimuli so that the pyramidal cells are subjected to inputs with naturalistic temporal dynamics. The ultimate goal is to understand the responses of pyramidal cells to moving stimuli, or to sequential stimulation of two spatial channels. However, an important first step is to understand in some detail the sequence of events that occur when a pyramidal cell responds to activation of a single spatial channel. This can be accomplished by studying the responses of cortical neurons to either focused or diffuse light flashes.

4.3.1. Responses of Pyramidal Cells to Light Flashes

Responses of pyramidal cells to square light flashes have been obtained for both spots of light and to whole-retina illumination with *in vitro* preparations that include the eye and whole brain or in preparations that are reduced to include just the geniculocortical pathway (Mancilla *et al.*, 1998). Flashes varied in amplitude from 0 to 10^4 photons/μm^2/sec and were 1 sec in duration. Pyramidal cell responses to light flashes were qualitatively similar regardless of whether spots of light or whole-retinal flashes are used.

Pyramidal cells show maintained synaptic potentials with the preparation maintained in the dark (Fowler and Ulinski, 1993; Fig. 46). Comparable potentials do not occur in isolated cortex preparations or when the optic nerve has been cut in the geniculocortical pathway preparation. This indicates that the maintained PSPs do not result from some process intrinsic to the cortex itself, such as spontaneous release of neurotransmitter from axons in the cortex. It is likely, instead, that they result from the maintained discharge of retinal ganglion cells in the dark (see Section 4.1.1). The PSPs include both hyperpolarizing and depolarizing events, but hyperpolarizing events are the most frequent. The amplitudes of the events vary as a function of the membrane potential of the neuron. The hyperpolarizing events reverse when the membrane potential approaches −70 mV, which suggests that they are IPSPs mediated by the $GABA_A$ subtype of GABA receptors. The depolarizing events reverse when the membrane potential approaches 0 mV, suggesting that they are EPSPs mediated by an excitatory receptor system such as one of the glutamate receptor subtypes. Since pyramidal cells seldom generate action potentials in the dark, these EPSPs result from activation of either geniculocortical afferents or from activation of excitatory cortical interneurons. The maintained PSPs vary in shape. Some have relatively simple waveforms consistent with their being unitary synaptic events or PSPs produced by activation of a single-axon bouton. Other PSPs have more complex shapes that include multiple inflection points or notches, indicating that they are compound events composed of several unitary events occurring within some particular time window (Mancilla and Ulinski, 1996a). Pyramidal cells are thus subjected to an ongoing barrage of EPSPs and IPSPs in the dark with the balance of excitation and inhibition leaning toward a tonic inhibition.

Responses of pyramidal cells to light flashes occur with a latency of about 150 msec (Mancilla *et al.,* 1998b; Fig. 47). Of this, approximately 100 msec is accounted for by the photoreceptor–bipolar cell–ganglion cell chain. The responses vary in a stereotyped manner with light intensity. Flashes with low intensity levels produce smaller numbers of PSPs with smaller amplitudes. The balance of excitation and inhibition converging on the pyramidal cell shifts toward a net excitation as the light intensity increases until relatively large-amplitude EPSPs predominate, and EPSPs that are large enough to bring the cell to its firing threshold occur as the light intensity is increased. The cell usually generates a single action potential. Increasing the light intensity even further leads to a suppression of action po-

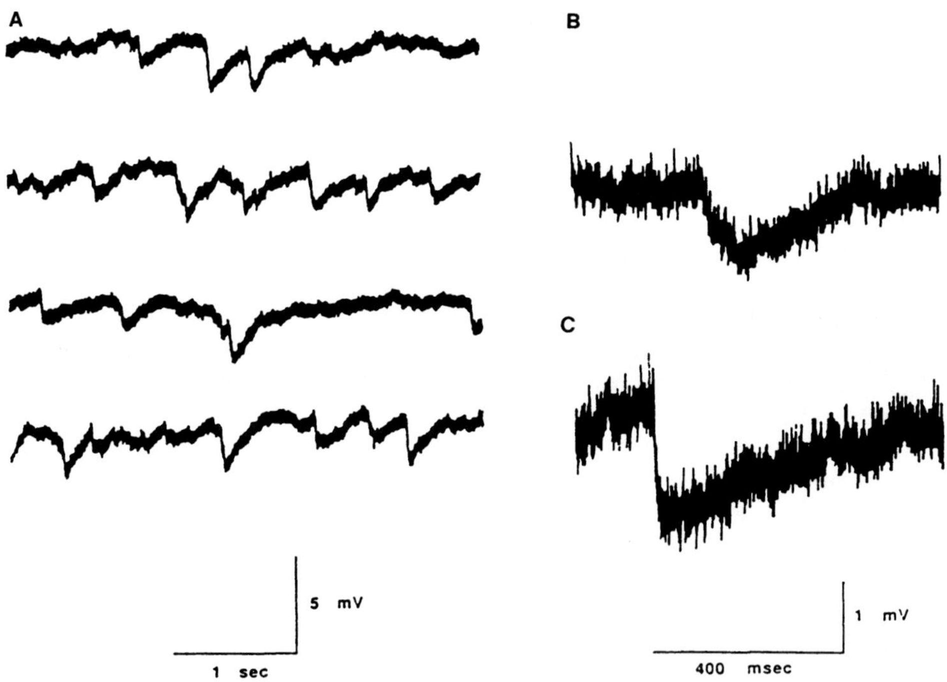

Figure 46. Postsynaptic potentials recorded in the dark. (A) Examples of trains of PSPs recorded with the *in vitro* geniculocortical preparation in the dark. (B, C) examples of IPSPs shown at higher gain to illustrate variation in the shape parameters of IPSPs.

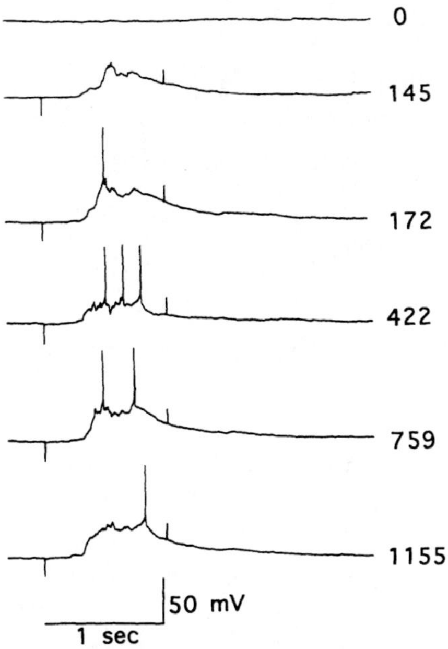

Figure 47. Light responses of cortical neurons. Responses of a pyramidal cell in turtle visual cortex to 1-sec square pulses of 640-μm light. Numbers at the right of the traces are the light intensities of flashes expressed in photons/μm²/sec. The response to 145 photons/μm²/sec produces is subthreshold. Responses to higher light intensities are suprathreshold. Notice that the number of action potentials is reduced in the lower two traces. These responses are the basis of the intensity–response function shown in the next figure.

tentials that is produced by a hyperpolarizing trough on the crest of the underlying EPSP. The timing of this trough is such that the cell would likely generate an action potential if the trough were not present. Increasing the light intensity still further increases the depth of the trough and produces a late hyperpolarization that increases in magnitude with very high intensity levels. It is not uncommon for the cell to show a depolarization following the hyperpolarization that can be strong enough to trigger an action potential. The responses can be quantified by constructing an intensity–response curve in which the peak amplitude of the PSP or the firing rate of the cells is plotted as a function of light intensity (Fig. 48). Such curves can be described by sigmoidal functions.

The full sequence of events leading to cortical activation has not yet been studied, but some information is available on the events leading up to the generation of the first action potential by a pyramidal cell. These depend upon the electrotonic structure of the pyramidal cells, properties of the geniculocortical synapses, and the voltage-gated conductances present on the neuron.

4.3.2. Electrotonic Properties of Pyramidal Cells

Pyramidal cells in the lateral and medial parts of dorsal cortex show distinct differences in their morphologies that reflect different relationships to geniculocortical inputs (Fig. 49). The basal arbors of the lateral pyramidal cells are approximately as large as their apical arbors, so these cells have roughly symmetric dendritic trees (Connors and Kriegstein, 1986; Desan, 1984, 1985). However, the basal arbors of the medial pyramidal cells are smaller than their apical arbors and become almost vestigial at the medial border of dorsal cortex. An individual geniculate afferent synapses upon the soma and proximal dendritic shafts of a lateral pyramidal cell and then proceeds medial to synapse upon the distal dendrites of a medial pyramidal cell. Assuming that the physiology of synaptic boutons along the

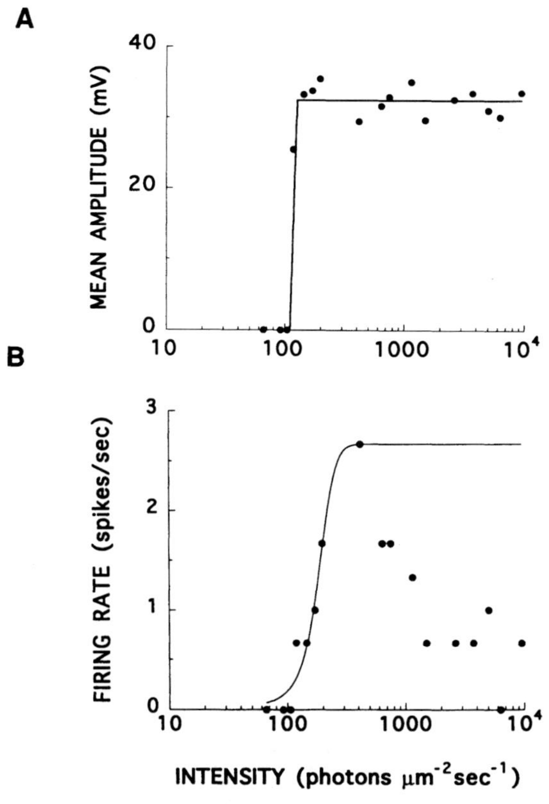

Figure 48. Intensity–response functions of pyramidal cells. (A) Maximal amplitude of the compound EPSPs produced in a pyramidal cell of turtle visual cortex by 1-sec square pulses of 640-μm light presented to the retina. Light intensity is measured in photons/μm²/sec. The intensity–response function is well fit by a sigmoidal function. (B) The firing rate of the cell in spikes/sec plotted as a function of light intensity. The slope of the rising phase of the sigmoid used to fit the firing rate to light intensity is smaller than that for the subthreshold events shown in panel A. Points at higher light intensities show a marked supersaturation.

length of the axon is the same, the input to the two morphologically quite distinct pyramidal cells would be the same. However, classical views of the properties of dendrites predict that EPSPs recorded at the somata of the two cells would differ significantly due to electrotonic filtering of the excitatory postsynaptic currents (EPSCs) by their dendrites.

Linear cable theory predicts that a steady current injected into a dendrite at a distance x from a recording electrode produces a voltage that decreases exponentially as a function of the distance and the space constant λ of the cable (Rall, 1977; Spruston *et al.*, 1994). The morphological dimensions of the dendritic tree can thus be replaced by an electrotonic coordinate system in which the distance of the synapse from the recording electrode is measured by its electrotonic distance, $X = x/\lambda$. Real dendritic trees have a finite anatomical length l and can be described under certain, well-understood circumstances as equivalent cables of electrotonic length L, where $L = l/\lambda$. Calculations based on neurons modeled as equivalent cables predict that the waveform of an EPSP will vary as a function of the anatomical and electrotonic distance of its synapse from the recording electrode. The amplitude of the EPSP will decrease as the distance of the synapse from the electrode increases, and the time-to-peak and width of the EPSP at half-maximal amplitude will increase. The expectation from linear cable theory is that the same geniculate fiber would elicit EPSPs with significantly different shape parameters in lateral and medial pyramidal cells.

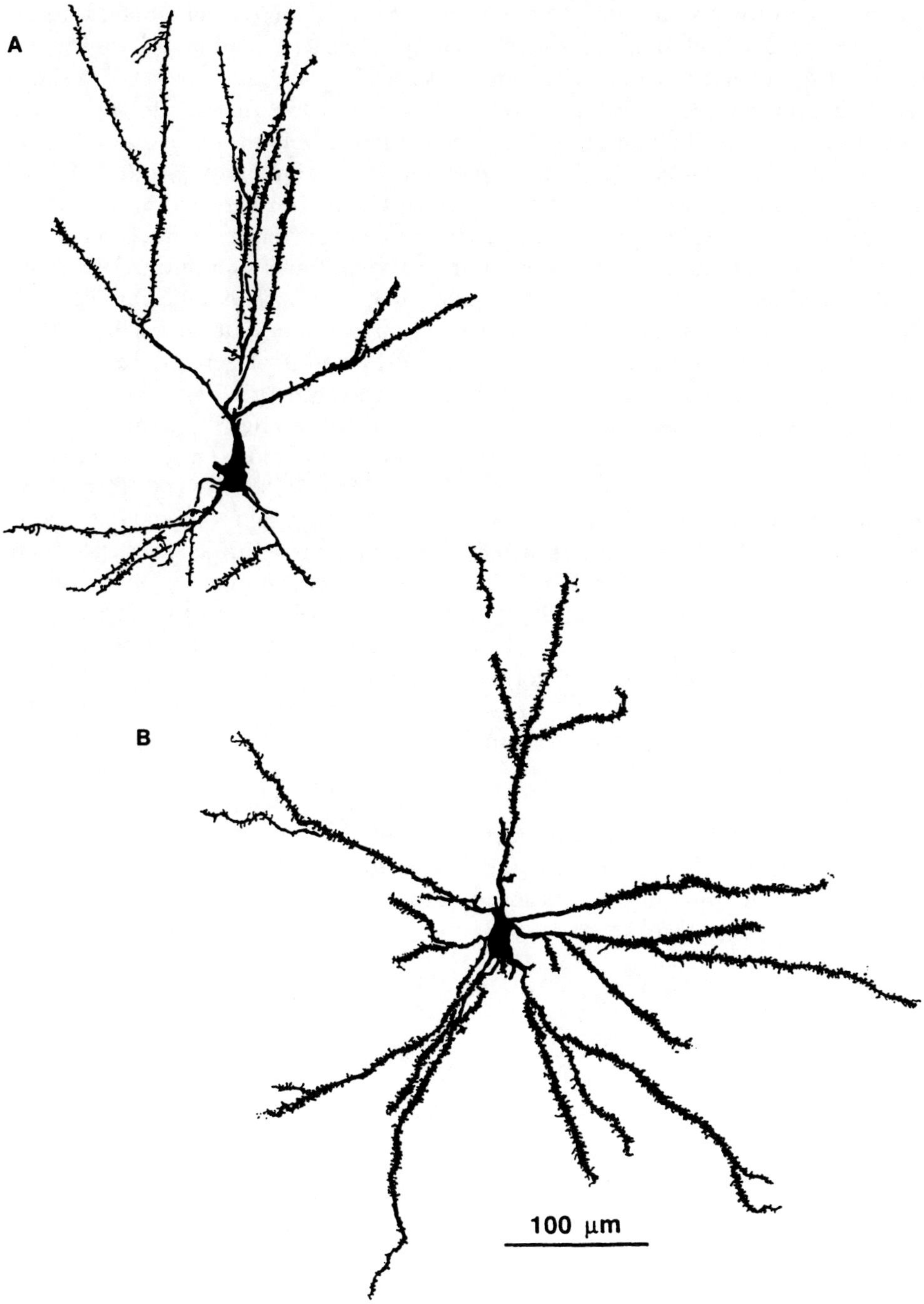

Figure 49. Lateral and medial pyramidal cells from the medial (A) and lateral (B) parts of turtle visual cortex. Drawn from Golgi preparations.

This can be tested by recording the EPSPs evoked by electrical activation of the geniculate fibers in the dorsal peduncle of the lateral forebrain bundle using the *in vitro* isolated cortex preparation. The response of the pyramidal cell depends upon the intensity of stimulation. Very low levels of electrical stimulation within the fascicles of geniculate fibers produce no response in pyramidal cells. A threshold level of stimulation often produces a relatively simple EPSP that is referred to as a *threshold* or *minimal EPSP* because it is the smallest amplitude event that can be produced by electrical stimulation. Higher levels of stimulation produce either complex EPSPs or IPSPs or cause the pyramidal cell to generate action potentials. Minimal EPSPs in lateral and medial pyramidal cells have times-to-peak and half-widths that are significantly different, but the amplitudes of minimal EPSPs in lateral pyramidal cells are only slightly smaller than those of medial pyramidal cells and do not differ statistically (Fig. 50). This observation deviates from the predictions of linear cable theory, given the different locations of geniculocortical synapses upon lateral and medial pyramidal cells, and requires explanation.

One possibility is that the two morphologically different groups of pyramidal cells have electrotonic properties that compensate for the more distal placement of geniculocortical afferents on the medial cells (Fowler *et al.*, 1992). The electrotonic lengths of neurons can be measured by standard biophysical methods that involve measuring the voltage response of the cell to somatic current injections and calculating a measure of electrotonic length known as L_{peel} (Holmes *et al.*, 1992). Lateral pyramidal cells have an L_{peel} of 2.6 ± 0.4, while medial pyramidal cells have a significantly smaller L_{peel} of 1.5 ± 0.1. An alternative approach to understanding the electrotonic structure of a neuron is to construct a compartmental model that represents the neuron as a series of isopotential compartments. Compartmental models of representative lateral and medial pyramidal cells have overall electrotonic lengths of 0.7 and 0.5, respectively (Fig. 51). The models are thus consistent with the estimates of L_{peel} obtained from real neurons in that medial pyramidal cells are electrotonically more compact than are lateral pyramidal cells.

It appears, then, that the amplitudes of minimal EPSPs in medial pyramidal cells have roughly the same amplitudes as those in lateral pyramidal cells because the more compact electrotonic structure of the medial pyramidal cells partially

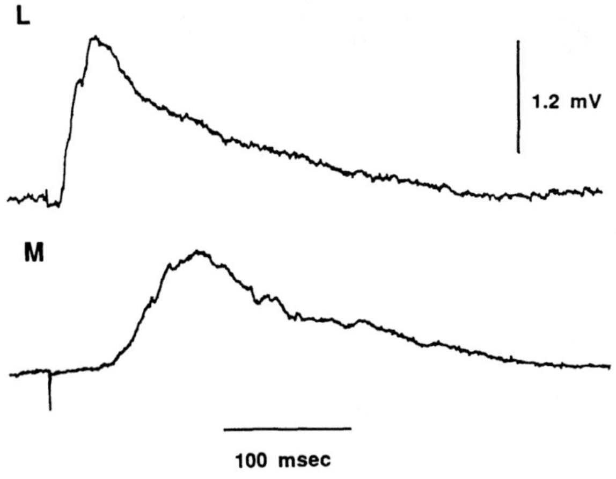

Figure 50. Minimal EPSPs. Compound EPSPs from lateral (L) and medial (M) pyramidal cells in turtle visual cortex produced by minimal activation of geniculocortical axons in the lateral forebrain bundle. Notice that the two EPSPs have approximately the same amplitude, but differ significantly in their latencies, times-to-peak, and widths at half-maximal amplitudes.

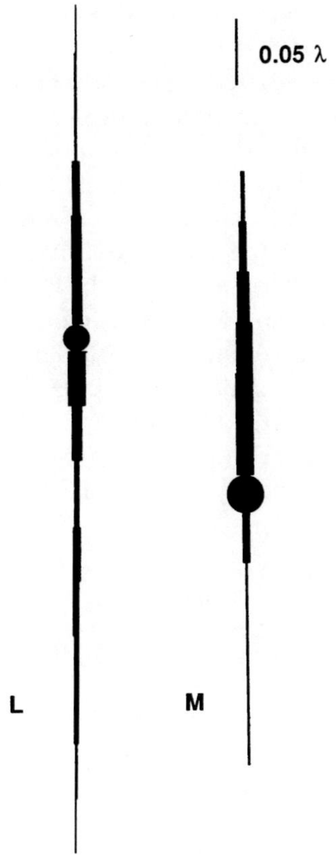

Figure 51. Reduced compartmental models. Reduced compartmental models of lateral (L) and medial (M) pyramidal cells are shown in a format in which the length of each compartment represents its electrotonic length and the width represents its surface area.

compensates for the more distal placement of the geniculocortical synapses. The differences in waveforms observed between minimal EPSPs in the two types of pyramidal cells occur because the minimal EPSPs are compound events resulting from the summation of several unitary EPSPs produced by *en passant* synapses of fascicles of geniculocortical afferents upon the somata and dendrites of the pyramidal cells. Since geniculocortical fascicles intersect the somata and proximal dendrites of lateral pyramidal cells, but the distal dendrites of medial pyramidal cells, minimal EPSPs in lateral cells will occur over shorter time windows than those in medial cells (Colombe and Ulinski, 1996). The consequence of the distal placement of geniculate synapses upon medial cells is that these cells integrate inputs over longer time windows than do lateral cells. The functional implication of this difference is not yet clear.

4.3.3. Interactions between EPSPs

Compartmental models can be used to understand how two excitatory synapses situated at different points on the neuron and activated at different times interact with each other. This requires information on the synaptic currents produced by individual synapses, but the minimal EPSPs elicited by electrical activation of geniculate afferents cannot be used for this purpose because they are compound events. What is needed is a characterization of *unitary EPSPs* resulting from the activation of individual geniculocortical boutons. Analysis of spontaneous events recorded with tetrodotoxin (TTX) in the bath medium is not possible in the case of the isolated cortex preparation of turtle visual cortex because spontaneous PSPs are so rare in adult animals as to preclude analysis. However, some of the ongoing PSPs seen with the geniculocortical pathway preparation in the dark are potentially unitary events resulting from activation of transmitter by a single axon bouton. The synaptic conductances underlying unitary events can therefore be approximated by using the compartmental models to vary the kinetic parameters of the synaptic currents to match the waveforms of presumed unitary EPSPs and IPSPs in real neurons. Models of lateral and medial pyramidal cells can then be used to investigate interactions between geniculate synapses by simulating the activation of two or more geniculate inputs at various time intervals or at various positions on the dendritic tree (Fig. 52). One general point that emerges from such simulations is that activation of two synapses at time intervals of less than 4 msec produces an apparently unitary EPSP that is a linear combination of the two individual EPSPs.

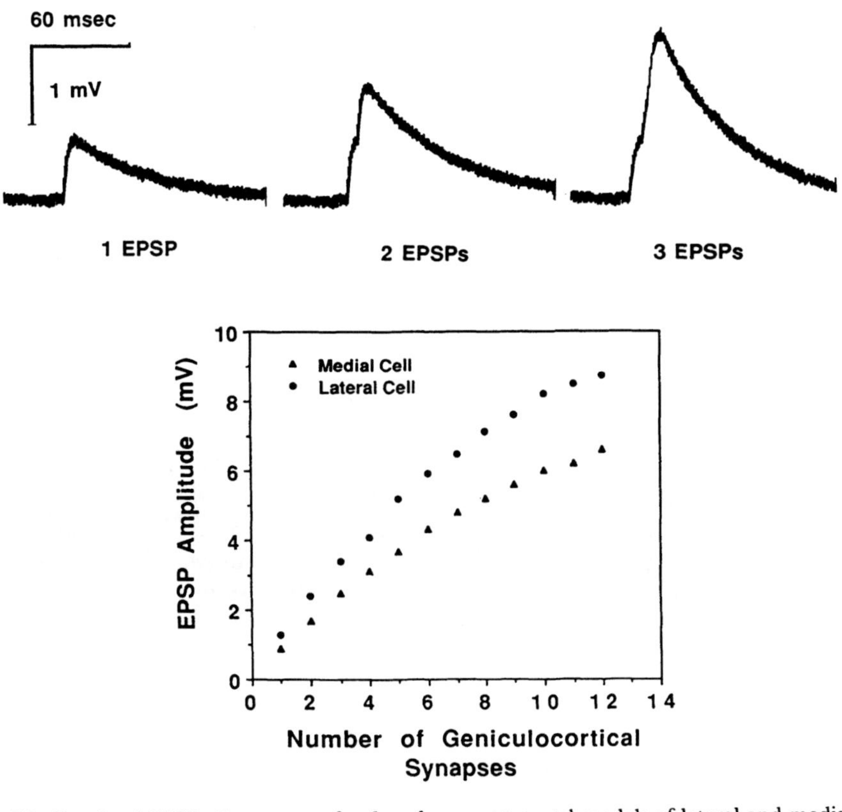

Figure 52. Simulated EPSPs. Responses of reduced compartmental models of lateral and medial pyramidal cells to simulated geniculocortical synapses. The traces at the top show simulated responses to one geniculocortical synapse (left trace), to two synapses separated by a 5-msec interval (middle trace), and one synapse followed by two synapses separated by a 5-msec interval. The graph shows the amplitude of the second set of EPSPs plotted as a function of the number of synapses in the set for medial and lateral pyramidal cells. Notice that a compressive or saturating nonlinearity occurs in the amplitudes of the compound EPSPs in both cells.

Activation at time intervals greater than 4 msec produces a compound EPSP with a distinguishable inflection point on its rising phase. The interaction between asynchronous unitary events is nonlinear because the first EPSP depolarizes the dendritic membrane. The second EPSP is consequently generated on a membrane that is closer to the reversal potential of the EPSP and the driving potential for the second EPSP is less than that for the first. Activation of trains of EPSPs at intervals greater than 4 msec in both lateral and medial pyramidal cells thus leads to a compressive nonlinearity or saturation effect and a maximum depolarization possible at about 12 asynchronously activated synapses.

4.3.4. Action Potential Generation

The depolarization produced by geniculocortical inputs activates the voltage-gated conductances responsible for action potential generation. The properties of these conductances are typically characterized by recording the responses of cells to long, depolarizing current pulses. Neurons in turtle visual cortex show two distinct firing patterns typical of cortical neurons. Neurons classified as pyramidal cells on the basis of either their location in layer 2 or on their morphology when filled with intracellular markers show a pattern designated as *regular spiking* (Connors *et al.*, 1982; Connors and Gutnick, 1990). A current pulse that depolarizes the neuron above its firing threshold produces several action potentials (Fig. 53). Current pulses with larger amplitudes decrease the interval between the first two action potentials, so plots of the firing rate of the cell as measured by the first interspike interval versus the injected current are linear. The interspike interval, however, varies during the course of a long current injection and the neuron shows a significant adaptation of its firing rate. Firing rate–current curves (*f–I* curves) for successive pairs of spikes in the train are lines with decreasing slopes. Neurons in layers 1 and 3 typically display a second firing pattern known as *fast spiking* because the individual action potentials are relatively narrow or fast. These neurons show little tendency to adapt and can maintain essentially the same firing rate over the duration of a 1-sec current pulse. A third or bursting firing pattern is sometimes encountered in turtle visual cortex.

The voltage-gated conductances responsible for these differences in firing pattern have been only partially characterized for turtle cortical neurons. Action potentials are mediated principally by a sodium conductance that is blocked by bath application of TTX (Connors and Kriegstein, 1986). The kinetics of the conductance has not been studied with voltage-clamp methods, but it undoubtedly corresponds to the fast sodium conductance found on pyramidal cells in other cortices. Application of TTX reveals an underlying component that has a lower amplitude and is slower than the sodium current. It is blocked by cadmium, suggesting it is mediated by calcium channels. The width and spacing of action potentials in neurons is typically determined by the presence of one or more types of potassium channel (e.g., Johnston and Wu, 1995). There have been no studies of the potassium channels present on turtle cortical neurons, but the differences in the widths of action potentials in neurons showing the regular-spiking and fast-spiking patterns suggest the presence of a traditional delayed-rectifier conductance, I_{DR}, on both types of neurons as well as a more transient conductance such as the A current, I_A, on fast-spiking neurons. The firing patterns seen in turtle cortical

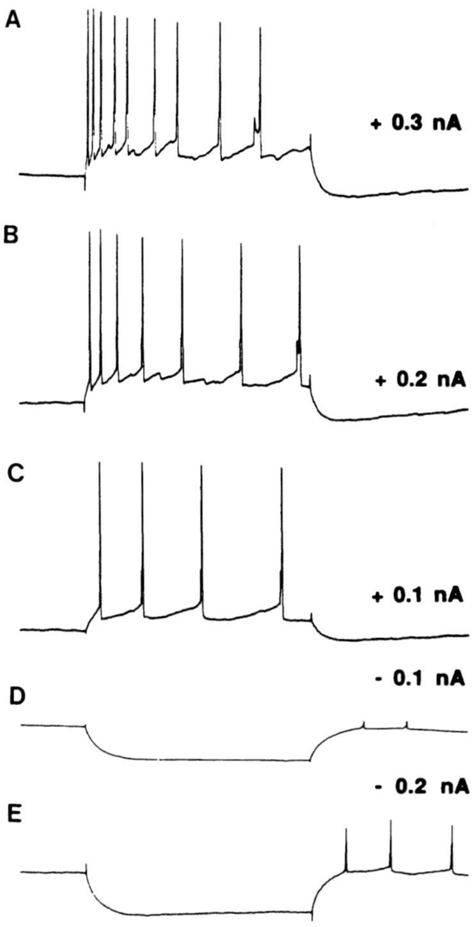

Figure 53. Pyramidal cell firing. Responses of a pyramidal cell in turtle visual cortex to intrasomatic injection of square current pulses. (A) A −0.1 nA current pulse produces an apparently passive response. The response to the −0.2 nA current pulse shows some sag or tendency to depolarization during the current pulse and three rebound spikes. The responses to +0.1, +0.2, and +0.3 nA current pulses produces trains of action potentials that show a distinct adaptation in firing rate during the course of the spike train. Each train shows a small action potential or "spikelet." The spikelet preceding the last large action potential in the 0.3-nA response is the most obvious.

neurons thus suggest they have the same menu of voltage-gated conductances present on neocortical neurons that are present in mammals, but it will be important to characterize properly the conductances using voltage-clamp methods.

The relationship between the amplitudes of currents generated by square current pulses and the firing rates of cortical neurons can be reliably described by linear *f–I* curves, but the synaptic currents that generate action potentials in physiological circumstances have more complex waveforms and occur in specific temporal patterns. Attempts at deducing the rules that govern the firing rates of neurons under these circumstances using ramps of injected current have not been successful, but simulations with compartmental models incorporating fast sodium conductances adequately account for these events (Peiper and Ulinski, 1996a). Activation of the spike-generating mechanism is expressed as the magnitude of the conductance of the fast sodium channels, $g_{Na}(V,t)$. The relationship between EPSP properties and g_{Na} can be investigated through simulations with compartmental models because the value of g_{Na} at each point in time is available in the calculations that underlie the simulation. As expected, a larger number of synapses is needed to reach firing threshold if synapses are placed at greater electrotonic distances from the soma (Fig. 54). The magnitude of g_{Na} produced by asynchronously activated

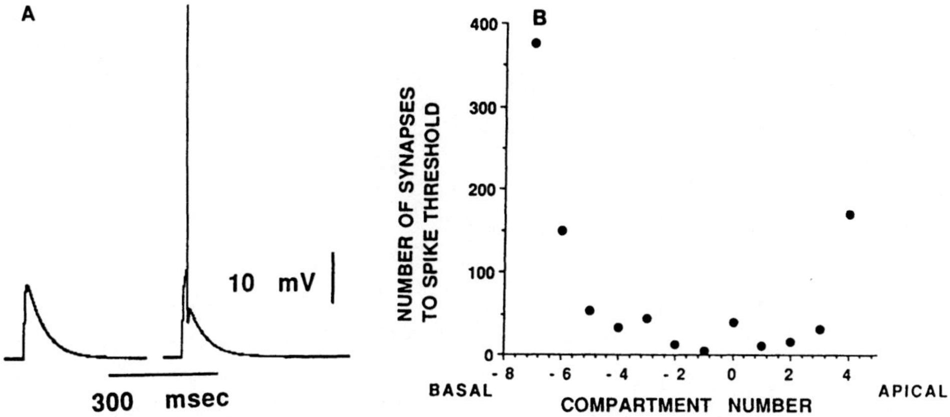

Figure 54. Number of synapses required to fire a pyramidal cell. A reduced compartmental model of a lateral pyramidal cell from turtle visual cortex was used to study the number of synchronously active geniculocortical synapses needed to generate an action potential. (A) Response of the model to synchronous activation of one less than the number of synapses required to bring the cell to spike threshold (left trace) and of the number of synapses required to fire the cell (right trace). (B) Number of synapses required to generate an action potential as a function of position on the neuron. Relatively small numbers of synapses are required to generate action potentials when synapses are located on those regions of the somadendritic surface of the neuron that actually receive geniculate input. From Peiper and Ulinski (1996b).

synapses can be studied by activating synapses in two groups separated by varying intervals. This paradigm indicates the existence of a *spike threshold window* during which sequences of EPSPs can interact to generate an action potential (Fig. 55).

4.3.5. Summary

Analyses of the responses of pyramidal cells to electrical and natural activation of geniculocortical afferents document the existence of a series of nonlinear steps that link synaptic activation of the cells to action potential generation. These include a saturating nonlinearity in the interaction between unitary EPSPs and interactions between EPSPs and voltage-gated conductances on the soma and dendrites of pyramidal cells. As reviewed in the previous section, the synaptic events underlying interactions between spatial channels are also nonlinear. Almost paradoxically, the relationship between synaptic current and instantaneous firing rate is approximately linear, so that the process of spike generation results in a roughly linear transformation of subthreshold events.

4.4. Excitation and Inhibition

The model for spatiotemporal receptive fields developed in Section 4.2 suggests that the basic shape of the spatiotemporal response function of pyramidal cells is determined by excitatory processes. However, the responses of pyramidal cells to light flashes suggest that both excitatory and inhibitory components occur when cortical neurons respond to natural visual stimuli. It is possible, then, that

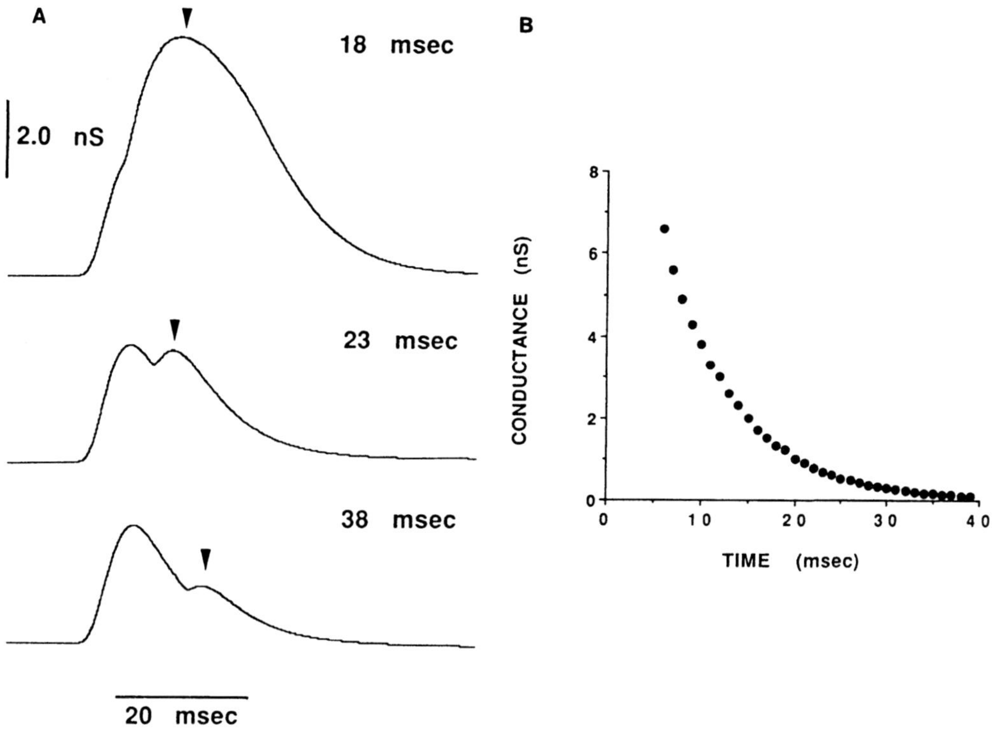

Figure 55. Spike threshold window. A reduced compartmental model of a lateral pyramidal cell from turtle visual cortex was used to study the time window over which successive activation of geniculocortical synapses could generate a somatic action potential. (A) Sodium conductance resulting from sequential activation of two sets of geniculocortical EPSPs. The first EPSP was a compound EPSP consisting of one less than the number of synapses required for spike activation. The second synapse was a single EPSP. The traces show the simulated sodium conductances resulting when the first and second EPSPs were separated by 18, 23, and 38 msec. The arrows indicate the maximal amplitudes of the conductance transient resulting from the second EPSP. (B) Maximal amplitudes of second sodium transient as a function of temporal separation of EPSPs. EPSP pairs occurring within 6 msec of each other produce an action potential with a maximal conductance in the mS range (which is off the scale of the plot). Larger temporal separations produce conductance transients in the nS range. Thus, this cell has a spike threshold window of 6 msec for synapses at this particular dendritic location. From Peiper and Ulinski (1996b).

inhibitory processes modulate the shapes of spatiotemporal receptive fields during naturally occurring visual stimuli. This can be explored by varying the relative balance of inhibition and excitation in the model for spatiotemporal receptive fields. Increasing the extent of inhibition in the model decreases the amplitude of the response function, indicating that inhibitory processes decrease the general excitability of pyramidal cells. Increasing the balance of inhibition also shapes the direction selectivity and sharpens the speed tuning curve of the cell.

The functional organization of the inhibitory circuits responsible for these effects is complicated in two ways. The first is that there are both feedforward and feedback inhibitory pathways that likely function with different time courses and may well modulate different features of the pyramidal cell responses. Feedforward pathways involve interneurons in layer 1 that receive direct geniculate input and have axons that synapse upon pyramidal cells (Fig. 56). Electron microscopic studies following thalamic lesions (L. M. Smith *et al.*, 1980) demonstrate that thalamic fibers (presumably including geniculate afferents) terminate on the dendrites and somata of layer 1 neurons with smooth dendrites. Since most of the spine-free neurons in layer 1 show GABA-like immunoreactivity (Nicolaus and Ulinski, 1991; Reiner, 1991) and stain positively for markers of GABAergic systems (Blanton *et al.*, 1987), they are candidates for elements in feedforward inhibitory pathways. Feedback pathways involve interneurons in layer 1 as well as layer 3 neurons that also show GABA-like immunoreactivity. In addition to geniculate afferents, the layer 1 neurons can receive excitatory inputs from pyramidal cell collaterals arborizing in layer 1. They will therefore receive a stronger excitatory drive as pyramidal cells become active. Layer 3 neurons do not receive geniculate inputs and would receive excitatory inputs only as the pyramidal cells become active. In contrast to the pyramidal cells, it is known that the layer 3 horizontal cells show a fast-spiking pattern and little tendency to adapt (Connors and Kriegstein, 1986).

A second source of complexity in the organization of cortical inhibitory pathways is that both $GABA_A$ and $GABA_B$ receptor subtypes are known to be present in turtle cortex (Larson-Prior and Slater, 1988). IPSPs evoked by electrical stimulation frequently display two distinct components. A fast component is blocked by the $GABA_A$-receptor antagonist bicucculine and the slow component is blocked by the $GABA_B$-receptor antagonist phaclofen (Kriegstein and Connors, 1986; Blanton and Kriegstein, 1992). The pattern of IPSPs seen in the *in vitro* geniculocortical pathway preparation differs significantly from IPSPs following electrical activation. Trains of

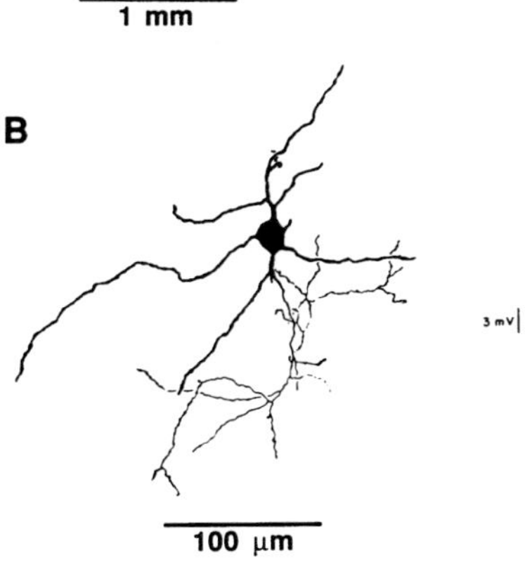

Figure 56. Smooth stellate cell from turtle visual cortex. (A) Low-power and (B) high-power drawings of a stellate cell with smooth dendrites situated in layer 1 of the lateral part of turtle visual cortex. Neurons of this type are presumed to be GABAergic inhibitory interneurons.

on-going IPSPs recorded with the geniculocortical pathway preparation in the dark contain IPSPs that have amplitudes and times-to-peak consistent with those of $GABA_A$-mediated IPSPs.

The absence of apparent, spontaneous $GABA_B$-mediated IPSPs is consistent with observations made in the hippocampus, neocortex, and pyriform of mammals and suggests that some poorly known factors are involved in the activation of $GABA_B$ receptors by GABA (e.g., Mody *et al.*, 1994). One possibility is that different populations of interneurons access $GABA_A$ and $GABA_B$ receptors. This implies that the tonic levels of geniculate drive present in the geniculocortical preparation selectively activate interneurons that access $GABA_A$ receptors, while synchronous electrical activation of either the geniculocortical or intracortical systems activates other populations of interneurons that access $GABA_B$ receptors. Analysis of the shape parameters of the ongoing IPSPs in the geniculocortical preparation provides some information relevant to this idea. The majority of IPSPs have times-to-peak that are distributed toward the small end of the total distribution of times-to-peak, suggesting $GABA_A$ receptors are preferentially located at electrotonic distances relatively near the soma (Fowler *et al.*, 1994). Since axon terminals with clear, pleomorphic vesicles and symmetric active zones are found along the entire lengths of the dendrites (Ebner and Colonnier, 1975, 1978), it follows that $GABA_B$ receptors may be located preferentially toward the distal ends of the dendritic trees. This does not exclude the possibility that $GABA_A$ and $GABA_B$ receptors are colocalized at the sites of many synaptic contacts, but that $GABA_B$ receptors require a greater concentration of GABA in the synaptic cleft for activation.

The contributions of feedforward and feedback circuits and the relative roles of $GABA_A$ and $GABA_B$ receptors can be studied using the responses of pyramidal cells to light flashes while holding the cells at a variety of membrane potentials. These experiments yield information about feedforward and feedback mechanisms because pyramidal cells seldom fire spontaneously at low light intensities. Responses recorded in pyramidal cells at subthreshold light intensities are most likely due to feedforward excitatory and inhibitory pathways. Responses that occur after action potential generation will include contributions from feedback pathways. Information about the two subtypes of GABA receptors is obtained because the $GABA_A$ and $GABA_B$ receptors have different voltage dependences. The amplitudes of IPSPs mediated by $GABA_A$ and $GABA_B$ receptors both decrease as the membrane potential approaches the reversal potential of the receptor. However, $GABA_A$-mediated IPSPs reverse near the chloride reversal potential of about −70 mV, while $GABA_B$-mediated IPSPs reverse near the potassium reversal potential of about −90 mV. Figure 57 shows the results of an experiment in which a light flash was used to evoke a PSP in a pyramidal cell. This intensity was sufficient to suppress the action potential generated at lower light intensities. The response is shown for two holding potentials and illustrates the expected variation in amplitude as a function of membrane potential. The reversal potential of about −45 mV lies between the reversal potential of $GABA_A$-mediated IPSPs and EPSPs mediated by glutaminergic synapses, suggesting that the PSP evoked by the light flash results from both EPSPs and $GABA_A$-mediated IPSPs. Similar analyses can be carried out for a range of time intervals, suggesting the light-evoked PSP results from several synaptic mechanisms with differing time courses. Initially, the light-evoked PSP appears to contain both EPSPs and $GABA_A$-mediated IPSPs, with the balance shifting toward excitation as

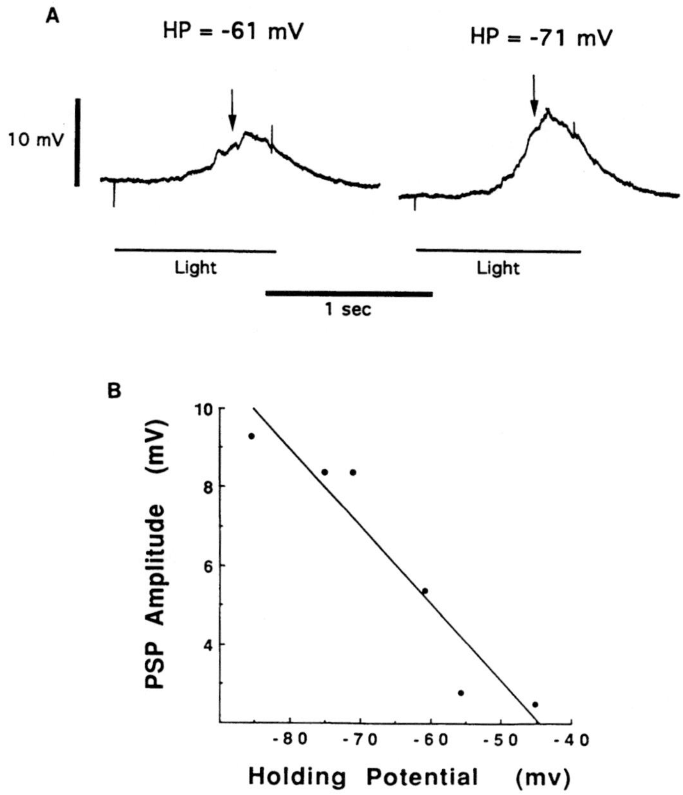

Figure 57. Voltage dependence of light-evoked PSPs in turtle pyramidal cells. (A) PSPs evoked in cortical pyramidal cells by 1-sec light flashes presented to the retina at two different membrane holding potentials (−61 and −71 mV). Notice that the amplitude of the PSP is larger when the membrane is held at −71 mV than at −61 mV. Arrows indicate the time at which the amplitude of the PSPs was measured to construct the plot shown in panel B. (B) PSP amplitude as a function of membrane holding potential. The amplitude of the light-evoked PSPs was measured 400 msec after PSP onset at a series of holding potentials. The regression line has a reversal potential of about −45 mV, consistent with a mixture of glutaminergic and $GABA_A$ergic receptors contributing to the PSP.

the cell nears its firing threshold. This trend reverses after the time period during which pyramidal cells are likely firing and leads to very negative reversal potentials during the period of peak hyperpolarization. Reversal potentials negative to −90 mV are consistent with IPSPs mediated by $GABA_B$ receptors localized on distal dendrites. Because of their distal placement, hyperpolarizing currents larger than those needed to hold the soma membrane at −90 mV would be needed to spread enough current to hold the dendritic membrane at −90 mV. This experiment, then, suggests that the initial responses of pyramidal cells to light responses are mediated by a balance of excitation and $GABA_A$-mediated inhibition via feedforward circuits, while the later phases of the response include $GABA_B$-mediated inhibition that involves feedback circuits.

The experiment is also consistent with work on cat visual cortex that fails to support a role for shunting inhibition in the generation of visual receptive fields. The linear dependence of the PSP amplitude upon membrane holding potential seen in the experiment is consistent with a linear addition of EPSPs and IPSPs.

Shunting inhibition would appear in this paradigm as a nonlinearity near the chloride reversal potential, but there is no sign of such a nonlinearity. Evidence for shunting inhibition is often sought by looking for conductance changes near the chloride reversal potential. However, turtle cortical neurons have two inwardly rectifying conductances active in the voltage range around −70 mV (Nicolaus and Ulinski, 1994). One appears to be a potassium conductance that hyperpolarizes the membrane; the other appears to be a mixed cation conductance that depolarizes the membrane. These are quite significant and would tend to obscure the conductance underlying a shunting inhibition if it were present. Suppression of the action potentials at high light intensities is effectively a form of contrast normalization. As mentioned above, shunting inhibition is commonly suggested as a mechanism responsible for the nonlinearity that generates normalization processes. A final conclusion from studies of the voltage dependence of the light-evoked response is that normalization depends upon the recruitment of inhibitory circuits by pyramidal cell collaterals and, in particular, activation of $GABA_B$-mediated inhibition. This has no effect on models that utilize feedback circuits to generate normalization processes because the models only require the presence of some source of nonlinearity, not necessarily a nonlinearity mediated by shunting inhibition.

5. Conclusions and Future Directions

The roughly three decades of work that followed the early studies of visual motion selectivity by Barlow and Levick, by Reichardt, and by Hubel and Wiesel have yielded significant progress toward a general understanding of the neural mechanisms that underlie motion analysis in cortical structures. This final section briefly summarizes some of the principal points that have been established and highlights issues that will require further attention in the years to come.

5.1. Spatiotemporal Receptive Fields

The idea of representing the motion-selective properties of cortical neurons by spatiotemporal receptive fields has had considerable heuristic value. The prediction that motion-selective neurons have spatiotemporal receptive fields that are obliquely oriented in the space–time plane appears to hold for at least those cortical neurons that respond to Fourier or first-order stimuli in many cases. Further, the slope of the receptive field correlates well with the preferred speed of the neuron. Work in cat striate cortex has been carried out exclusively with extracellular methods and provides no direct information on the mechanisms that underlie the generation of oriented spatiotemporal response functions. One suggestion is that oriented response functions are generated by a summation of lagged and nonlagged geniculate neurons, but this mechanism of response function generation would still require cortical mechanisms that are matched to the temporal characteristics of the inputs of the two types of geniculate neurons. Work on turtle visual cortex with extracellular methods suggests that neurons in that cortex also have oriented response functions because their temporal integration functions have peaks that shift as a function of the spatial separation of spots of light presented in two-pulse experiments. Studies using the isolated cortex preparation demonstrate explicitly that temporal integration curves with comparable shapes can be recorded using intracellular recording methods. The mechanisms that underlie temporal integration functions are thus intrinsic to the cortex and depend in part upon the kinetic properties of glutaminergic synapses. These temporal integration functions can be used together with reasonable estimates of the form of spatial integration functions to explain oriented spatiotemporal response functions. Since the properties of AMPA and NMDA receptors are the same in mammals and in turtles, the same basic mechanisms are likely responsible for temporal integration in both turtle and mammalian visual cortex. This hypothesis can be tested by using intracellular recording methods in cats to construct subthreshold temporal integration curves.

5.2. Linearity and Nonlinearity

The issue of whether or not motion-selective properties of cortical cells depend upon nonlinear mechanisms as suggested by Reichardt appears to have been resolved through the analysis of simple and complex cells in cat visual cortex using nonlinear systems analysis. Both the first-order (or linear) and the second-order (or nonlinear) kernels show some orientation in the space–time plane, indicating that motion selectivity is the result of both linear and nonlinear mechanisms. The cellular mechanisms responsible for the nonlinear contributions to motion selectivity have not been explicitly analyzed in mammals. A common suggestion has been that the nonlinearity is a consequence of the spike-generating mechanism. This process is often modeled as a rectification or quadratic nonlinearity. However, the temporal integration functions measured in turtle visual cortex do not support this idea because the temporal integration functions obtained with intracellular and extracellular methods have the same general shapes. Similarly, postsynaptic potentials recorded in neurons in cat striate cortex show some direction selectivity, although directional tuning curves based on spike rate are often more sharply tuned. The processes responsible for motion selectivity thus precede spike generation in the cascade of events leading to the firing of cortical pyramidal cells, but spike generation can enhance the degree of motion selectivity. Work in turtles shows that some steps in the cascade involve linear processes, while others are nonlinear. The complexity of the interactions diminishes the value of trying to categorize the overall process as either linear or nonlinear. A blindspot in discussions of receptive field generation has been the absence of a discussion of voltage-gated conductances on cortical neurons. It is well established that different kinds of cortical neurons have different firing patterns that are determined by the types of voltage-gated conductances present on their somata and dendrites. It will be important to include them in future discussions of the neural mechanisms underlying motion selectivity.

5.3. Excitation and Inhibition

Just as the debate on linear versus nonlinear mechanisms in the generation of motion selectivity has reached the conclusion that both types of processes are important, it now appears established that both excitation and inhibition are involved in motion selectivity. Studies with intracellular recording methods in cats and turtles consistently demonstrate that pyramidal cells are subjected to a balance of excitatory and inhibitory processes. The firing patterns of individual neurons seem to be controlled by constant trains of both inhibitory and excitatory postsynaptic potentials. This is consistent with the known anatomy of cortical circuits in which thalamic fibers synapse upon both excitatory and inhibitory neurons so that activation of thalamic fibers by natural stimuli produces sequences of both IPSPs and EPSPs.

An important advance during the past 30 years has been a more complete understanding of the diversity of receptor mechanisms in the cortex. Both $GABA_A$ and $GABA_B$ receptors are universally present in visual cortex, and electrical activation of thalamic fibers typically results in biphasic IPSPs mediated by both receptor types. Similarly, glutaminergic synapses access several subtypes of receptors. AMPA and NMDA receptors are found in all visual cortices and are both activated by electrical stimulation of thalamic fibers. The use of electrical activation to study synaptic mechanisms in slice preparations seriously limits our understanding of their functional roles because this form of stimulation imposes an entirely unrealistic temporal regime on the cortex. A working hypothesis is that $GABA_A$ and AMPA receptors are involved in the responses of cortical circuits to more-phasic changes in the visual scene, while $GABA_B$ and NMDA receptors are involved in more-sustained responses. In particular, NMDA receptors are implicated in the generation of motion-selective properties of cortical neurons through an involvement in the generation of oriented spatiotemporal response functions.

5.4. Local and Global Motion Detectors

The early work focused on developing concepts of how the initial stages of visual motion analysis could be implemented by neural circuitry, but it was subsequently established that motion analysis occurs in a series of steps in the motion analysis pathways. The motion detectors studied initially are now recognized as local motion detectors that provide information on the direction and speed of objects moving within relatively restricted regions of visual space. Analysis of inputs from arrays of local motion detectors by global motion detectors is achieved at subsequent stages in the motion pathways. The global motion detectors serve to resolve some of the ambiguities inherent in local motion detection and to provide information about the movement of biologically significant objects in visual space.

These two stages of analysis are obligatory and occur in all of the vertebrates that have been studied. However, the details of where local and global motion detectors are situated in the pathways and of their receptive field properties vary significantly among vertebrates in ways related to their visual lifestyles. A major variable is whether local motion analysis is achieved at the retinal level or postponed to the cortical level, and different animals use different computational strategies. The advantages and disadvantages of the various schemes are not yet well understood. It seems likely, though, that the dichotomy between local and global motion analysis is too simple and that there must be subsequent stages that have not been fully characterized. All animals, for example, respond to velocity flow fields, and neurons deep in the visual motion pathways respond to stimuli that are contracting or expanding or to rotations or translations of the visual field. These neurons are important functionally because of an involvement in the visual guidance of locomotion or orienting behaviors. It is not certain how many steps must or can be involved in generating neurons with these types of receptive fields.

An important goal is to understand the extent to which comparable neural mechanisms are involved in generating the various types of receptive fields involved in motion analysis. More information will be required, but at the moment it appears that the same synaptic mechanisms are involved in the generation of local and global motion detectors. The temporal integration functions of the two types of motion detectors are nearly identical and the work on turtle cortex suggests that the properties of these curves are determined by the synaptic properties of cortical circuitry. NMDA receptors may, for example, be used to store information for relatively long periods of time in both local and global motion detectors. This implies that the differences between local and global motion detectors derive from differences in the spatial patterns of thalamic and intracortical connections.

ACKNOWLEDGMENTS. Matt Calef, Jeff Colombe, Michael Fowler, Linda Larson-Prior, and Jaime Mancilla have all contributed to recent work in the author's laboratory. David Senseman made useful comments on the author's work. The Public Health Service, the Brain Research Foundation of Chicago, and the National Science Foundation provided financial support.

6. References

Adelson, E. H., and Bergen, J. R., 1985, Spatiotemporal energy models for the perception of motion, *J. Opt. Soc. Am. A* **2**:284–299.

Allman, J., Jeo, R., and Sereno, M., 1994, The functional organization of visual cortex in owl monkeys, in: *Aotus: The Owl Monkey,* Academic Press, San Diego, CA, pp. 287–320.

Allman, J., Miezin, F., and McGuiness, E., 1985a, Direction- and velocity-specific responses from beyond the classical receptive field in the middle temporal visual area (MT), *Perception* **14**:105–126.

Allman, J., Miezin, F., and McGuiness, E., 1985b, Stimulus specific responses from beyond the classical receptive field: Neurophysiological mechanisms for local–global comparison in visual neurons, *Annu. Rev. Neurosci.* **8**:407–430.

Ammermuller, J., and Kolb, H., 1995, The organization of the turtle inner retina. I. ON- and OFF-center pathways, *J. Comp. Neurol.* **358**:1–34.

Ammermuller, J., Muller, J. F., and Kolb, H., 1995, The organization of the turtle inner retina. II. Analysis of color-coded and directionally selective cells, *J. Comp. Neurol.* **358**:35–62.

Anstis, S. M., 1980, The perception of apparent movement, *Phil. Trans. R. Soc. Lond. B* **290**:153–168.

Ariel, M., and Adolph, A. R., 1985, Neurotransmitter inputs to directionally sensitive turtle retinal ganglion cells, *J. Neurophysiol.* **54**:1123–1143.

Armington, J. C., Adolph, A. R., and Wu, S., 1991, Spatial properties of ganglion cell activity in the turtle, *Vis. Neurosci.* **6**:439–450.

Baker, C. L., Jr., 1988, Spatial and temporal determinants of directionally selective velocity preference in cat striate cortex neurons, *J. Neurophysiol.* **59**:1557–1574.

Baker, C. L., Jr., 1990, Spatial- and temporal-frequency selectivity as a basis for velocity preference in cat striate cortex neurons, *Vis. Neurosci.* **4**:101–113.

Baker, C. L., Jr., and Cynader, M. S., 1986, Spatial receptive-field properties of direction-selective neurons in cat striate cortex, *J. Neurophysiol.* **55**:1136–1152.

Baker, C. L., Jr., and Cynader, M. S., 1988, Space-time separability of direction selectivity in cat striate cortex neurons, *Vis. Neurosci.* **28**:239–246.

Barlow, H. B., and Levick, W. R., 1965, The mechanism of directionally selective units in rabbit's retina, *J. Physiol.* (Lond.) **178**:477–504.

Bass, A. H., and Northcutt, R. G., 1981a, Retinal recipient nuclei in the painted turtle, *Chrysemys picta:* An autoradiographic and HRP study, *J. Comp. Neurol.* **199**:97–12.

Bass, A. H., and Northcutt, R. G., 1981b, Primary retinal targets in the Atlantic loggerhead sea turtle, *Caretta caretta, Cell Tiss. Res.* **218**:253–264.

Bauman, L. A., and Bonds, A. B., 1991, Inhibitory refinement of spatial frequency selectivity in single cells of the cat striate cortex, *Vis. Res.* **31**:933–944.

Baumfalk, U., and Albus, K., 1988, Phaclofen antagonizes baclofen-induced suppression of visually evoked responses in the cat's striate cortex, *Brain Res.* **463**:398–402.

Baylor, D. A., and Fettiplace, R., 1976, Transmission of signals from photoreceptors to ganglion cells in the eye of the turtle, *Cold Spring Harbor Symp. Quant. Biol.* **50**:529–536.

Baylor, D. A., and Fettiplace, R., 1977a, Transmission from photoreceptors to ganglion cells in turtle retina, *J. Physiol.* (Lond.) **271**:391–424.

Baylor, D. A., and Fettiplace, R., 1977b, Kinetics of synaptic transfer from receptors to ganglion cells in turtle retina, *J. Physiol.* (Lond.) **271**:425–448.

Belekhova, M. G., and Kosareva, A. A., 1971, Organization of the turtle thalamus: Visual, somatic and tectal zones, *Brain Behav. Evol.* **4**:337–375.

Belekhova, M. V., Kosareva, A. A., Veselkin, N. P., and Ermakova, T. V., 1979, Telencephalic afferent connections in the turtle *Emys orbicularis:* A peroxidase study, *J. Evol. Biochem. Physiol.* **15**:97–103.

Berman, N. J., Douglas, R. J., Martin, K. A. C., and Whitteridge, D., 1991, Mechanisms of inhibition in cat visual cortex, *J. Physiol.* (Lond.) **440**:697–722.

Blanton, M. G., and Kriegstein, A. R., 1991a, Morphological differentiation of distinct neuronal classes in embryonic turtle cerebral cortex, *J. Comp. Neurol.* **310**:550–570.

Blanton, M. G., and Kriegstein, A. R., 1991b, Appearance of putative amino acid neurotransmitters during differentiation of neurons in embryonic turtle cerebral cortex, *J. Comp. Neurol.* **310**:571–592.

Blanton, M. G., and Kriegstein, A. R., 1991c, Spontaneous action potential activity and synaptic currents in the embryonic turtle cerebral cortex, *J. Neurosci.* **11**:3907–3923.

Blanton, M. G., and Kriegstein, A. R., 1992, Properties of amino acid neurotransmitter receptors of embryonic cortical neurons when activated by exogenous and endogenous agonists, *J. Neurophysiol.* **67**:1185–1200.

Blanton, M. G., Shen, J. M., and Kriegstein, A. R., 1987, Evidence for the inhibitory neurotransmitter gamma-aminobutyric acid in aspiny and sparsely spiny nonpyramidal neurons of turtle dorsal cortex, *J. Comp. Neurol.* **259**:277–297.

Blanton, M. G., LoTurco, J., and Kriegstein, A. R., 1989, Whole cell recording from neurons in slices of reptilian and mammalian cerebral cortex, *J. Neurosci. Meth.* **30**:203–210.

Blomfeld, S., 1974, Arithmetical operations performed by nerve cells, *Brain Res.* **69**:115–124.

Bock, G. R., and Goode, J. A., 1994, *Higher-Order Processing in the Visual System,* Wiley, New York.

Boiko, V. P., 1980, Responses to visual stimuli in thalamic neurons of the turtle *Emys orbicularis, Neurosci. Behav. Physiol.* **10**:183–188.

Bonds, A. B., 1989, The role of inhibition in the specification of orientation selectivity of cells in the cat striate cortex, *Vis. Neurosci.* **2**:41–55.

Born, R. T., and Tootell, R. B. H., 1992, Segregation of global and local motion processing in primate middle temporal visual area, *Nature* **357**:497–499.

Borst, A., and Egelhaaf, M., 1989, Principles of visual motion detection, *Trends Neurosci.* **12**:297–306.

Bowling, D. B., 1980, Light responses of ganglion cells in the retina of the turtle, *J. Physiol.* (Lond.) **299**:173–196.

Boycott, B. B., and Wässle, H., 1974, The morphological types of ganglion cells of the domestic cat's retina, *J. Physiol.* (Lond.) **240**:397–419.

Braddick, O. J., 1974, A short-range process in apparent motion, *Vis. Res.* **14**:519–527.

Braddick, O. J., 1980, Low-level and high-level processes in apparent motion, *Phil. Trans. R. Soc. Lond. B* **290**:137–151.

Braddick, O., 1993, Segmentation versus integration in visual motion processing, *Trends Neurosci.* **7**:263–268.

Brenner, E., and Rauschecker, J. P., 1990, Centrifugal motion bias in the cat's lateral suprasylvian visual cortex is independent of early flow field exposure, *J. Physiol.* (Lond.) **423**:641–660.

Britten, K. H., Shadlen, M. N., Newsome, W. T., and Movshon, J. A., 1993, Responses of neurons in macaque MT to stochastic motion signals, *Vis. Neurosci.* **10:**1157–1169.

Brown, K. T., 1969, A linear area centralis extending across the turtle retina and stabilized to the horizon by non-visual cues, *Vis. Res.* **9:**1053–1062.

Buhl, E. H., Halasy, K., and Somogyi, P., 1994, Diverse sources of hippocampal unitary inhibitory postsynaptic potentials and the number of synaptic release sites, *Nature* **368:**823–828.

Buonomano, D. V., and Merzenich, M. M., 1995, Temporal information transformed into a spatial code by a neural network with realistic properties, *Science* **267:**1028–1030.

Burgi, P.-Y., and Grzywacz, N. M., 1994, Model for the pharmacological basis of spontaneous synchronous activity in developing retinas, *J. Neurosci.* **14:**7426–7439.

Burkhardt, D. A., 1995, The influence of center–surround antagonism on light adaptation in conces in the retina of the turtle, *Vis. Neurosci.* **12:**877–885.

Carandini, M., and Heeger, D. J., 1994, Summation and division by neurons in primate visual cortex, **264:**1333–1336.

Carr, C. E., 1993, Processing of temporal information in the brain, *Annu. Rev. Neurosci.* **16:**223–243.

Casagrande, V. A., and Kass, J. H., 1994, The afferent, intrinsic and efferent connections of primary visual cortex in primates, in: *Cerebral Cortex,* Vol. 10. *Primary Visual Cortex in Primates* (A. Peters and K. S. Rockland, eds.), Plenum Press, New York, pp. 201–260.

Celebrini, S., and Newsome, W. T., 1995, Microstimulation of extrastriate area MST influences performance on a direction discrimination task, *J. Neurophysiol.* **73:**437–448.

Colombe, J. B., and Ulinski, P. S., 1996, Temporal integration windows in the feedforward pathways of visual cortex, *Soc. Neurosci. Abstr.* **22:**284.

Connors, B. W., and Gutnick, M. J., 1990, Intrinsic firing patterns of diverse neocortical neurons, *Trends Neurosci.* **13:**99–104.

Connors, B. W., and Kriegstein, A. R., 1986, Cellular physiology of the turtle visual cortex: Distinctive properties of pyramidal and stellate neurons, *J. Neurosci.* **6:**164–177.

Connors, B. W., Gutnick, M. J., and Prince, D. A., 1982, Electrophysiological properties of neocortical neurons *in vitro, J. Neurophysiol.* **48:**1302–1320.

Cosans, C. E., and Ulinski, P. S., 1990, Spatial organization of axons in turtle visual cortex: Intralamellar and interlamellar projections, *J. Comp. Neurol.* **296:**548–558.

Creutzfeldt, O. D., Kuhnt, U., and Beneveto, L. A., 1974, An intracellular analysis of visual cortical neurons to moving stimuli: Responses in a cooperative neuronal network, *Exp. Brain Res.* **21:**251–274.

Davson, H., 1980, *Physiology of the Eye,* Academic Press, New York.

Davydova, T. V., and Goncharova, N. V., 1979, Comparative characterization of the basic forebrain cortical zones in *Emys orbicularis* (Linnaeus) and *Testudo horsefieldi* (Gray), *J. Hirnforsch.* **20:**245–262.

Davydova, T. V., Goncharova, N. V., and Boiko, V. P., 1975, Retinotectal system of the tortoise, *Testudo horsefieldi,* Gray (morpho-functional study in the norm and after enucleation), *J. Hirnforsch.* **17:**463–488.

DeAngelis, G. C., Ohzawa, I., and Freeman, R. D., 1993a, Spatiotemporal organization of simple-cell receptive fields in the cat's striate cortex. I. General characteristics and postnatal development, *J. Neurophysiol.* **69:**1091–1117.

DeAngelis, G. C., Ohzawa, I., and Freeman, R. D., 1993b, Spatiotemporal organization of simple-cell receptive fields in the cat's striate cortex. II. Linearity of temporal and spatial summation, *J. Neurophysiol.* **69:**1118–1135.

DeAngelis, G. C., Ohzawa, I., and Freeman, R. D., 1995, Receptive field dynamics in the central visual pathways, *Trends Neurosci.* **18:**451–458.

Dehay, C., Douglas, R. J., Martin, K. A. C., and Nelson, J. C., 1991, Excitation by geniculocortical synapses is not "vetoed" at the level of dendritic spines in cat visual cortex, *J. Physiol.* (Lond.) **440:**723–734.

Desan, P. H., 1984, The organization of the cerebral cortex of the pond turtle, *Pseudemys scripta elegans.* Ph.D. dissertation, Harvard University, Cambridge, MA.

Desan, P. H., 1985, Organization of cerebral cortex in turtle, in: *The Forebrain of Reptiles* (W. K. Schwerdtfeger and W. J. A. J. Smeets, eds.), Karger, Basel, pp. 1–11.

DeYoe, E. A., and Van Essen, D. C., 1988, Concurrent processing streams in monkey visual cortex, *Trends Neurosci.* **11:**219–226.

Douglas, R. J., Martin, K. A. C., and Whitteridge, D., 1991, An intracellular analysis of the visual responses of neurons in cat visual cortex, *J. Physiol.* (Lond.) **440:**659–696.

Dünser, K. R., Granda, A. M., Maxwell, J. H., and Fulbrook, J. E., 1981, Visual properties of cells in anterior dorsal ventricular ridge of turtle, *Neurosci. Lett.* **25**:281–285.

Ebner, F. F., and Colonnier, M., 1975, Synaptic patterns in the visual cortex of turtles: An electron microscopic study, *J. Comp. Neurol.* **160**:51–80.

Ebner, F. F., and Colonnier, M., 1978, A quantitative study of synaptic patterns in turtle visual cortex, *J. Comp. Neurol.* **179**:263–276.

Eccles, J. C., 1969, *The Inhibitory Pathways of the Central Nervous System*, Thomas, Springfield, IL.

Egelhaaf, M., and Borst, A., 1987, Dynamic response properties of movement detectors: Theoretical analysis and electrophysiological investigation in the visual system of the fly, *Biol. Cybernet.* **56**:69–97.

Egelhaaf, M., and Borst, A., 1993a, Motion computation and visual orientation in fields, *Comp. Biochem. Physiol.* **104A**:659–673.

Egelhaaf, M., and Borst, A., 1993b, Movement detection in arthropods, in: *Visual Motion and Its Role in the Stabilization of Gaze* (F. A. Miles and J. Wallman, eds.), Elsevier, Amsterdam, pp. 53–77.

Egelhaaf, M., and Borst, A., 1994, A look into the cockpit of the fly: Visual orientation, algorithms and identified neurons, *J. Neurosci.* **13**:4563–4574.

Emerson, R. C., and Citron, M. C., 1992, Linear and nonlinear mechanisms of motion selectivity in simple cells of the cats striate cortex, in: *Nonlinear Vision: Determination of Neural Receptive Fields, Function and Networks*, (R. Pinter and B. Nabet, eds.), CRC Press, Boca Raton, FL, pp. 39–74.

Emerson, R. C., and Gerstein, G. L., 1977, Simple striate neurons in the cat. II. Mechanisms underlying directional asymmetry and directional selectivity, *J. Neurophysiol.* **40**:136–155.

Emerson, R. C., Bergen, J. R., and Adelson, E. H., 1992a, Directionally selective complex cells and the computation of motion energy in cat visual cortex, *Vis. Res.* **32**:203–218.

Emerson, R. C., Korenberg, M. J., and Citron, M. C., 1992b, Identification of complex cell intensive nonlinearities in a cascade model of cat visual cortex, *Biol. Cybernet.* **66**:291–300.

Enroth-Cugell, C., 1993, The world of retinal ganglion cells, in: *Contrast Sensitivity* (R. Shapley and M.-K. Lam, eds.), MIT Press, Cambridge, MA, pp. 149–180.

Enroth-Cugell, C., and Robson, J. G., 1966, The contrast sensitivity of retinal ganglion cells of the cat, *J. Physiol.* (Lond.) **187**:517–552.

Ewert, J.-P., 1991, A prospectus for the fruitful interaction between neuroethology and neural engineering, in: *Visual Structures and Integrated Functions* (M. A. Arbib and J.-P. Ewert, eds.), Springer-Verlag, Berlin, pp. 3–30.

Ferrera, V. P., and Wilson, H. R., 1990, Perceived direction of moving two-dimensional patterns, *Vis. Res.* **30**:273–287.

Ferster, D., 1986, Orientation selectivity of synaptic potentials in neurons of cat primary visual cortex, *J. Neurosci.* **6**:1284–1301.

Ferster, D., 1987, Origin of orientation-selective EPSPs in simple cells of cat visual cortex, *J. Neurosci.* **7**:1780–1791.

Ferster, D. J., 1988, Spatially opponent excitation and inhibition in simple cells of the cat visual cortex, *J. Neurosci.* **8**:1172–1180.

Ferster, D., and Jagadeesh, B., 1992, EPSP–IPSP interactions in cat visual cortex studied with *in vivo* whole-cell patch recording, *J. Neurosci.* **12**:1262–1274.

Ferster, D., Wheat, H. S., and Jagadeesh, B., 1994, The linearity of synaptic mechanisms in simple cells of area 17 of cat visual cortex, in: *Structural and Functional Organization of the Neocortex* (B. Albowitz, K. Albus, U. Kuhnt, H.-Ch. Nothdruft, and P. Wahle, eds.), Springer-Verlag, Berlin, pp. 190–202.

Ferster, D., Chung, S., and Wheat, H., 1996, Orientation selectivity of thalamic input to simple cells of cat visual cortex, *Nature* **380**:249–252.

Fleet, D. J., and Langley, K., 1994, Computational analysis of non-Fourier motion, *Vis. Res.* **34**:3057–3079.

Fowler, M. H., and Ulinski, P. S., 1993, Synaptic organization of lateral pyramidal cells in visual cortex: Spatial distribution of $GABA_A$ receptors, *Soc. Neurosci. Abstr.* **19**:1577.

Fowler, M., Ulinski, P. S., Smith, L., and Larson-Prior, L. J., 1992, Evidence of a relationship between electrotonic properties and morphology of pyramidal cells in turtle visual cortex, *Soc. Neurosci. Abstr.* **18**:296.

Fowler, M., Nicolaus, J. M., and Ulinski, P. S., 1994, Integrative physiology of pyramidal cells in turtle visual cortex. III. Synapses of inhibitory neurons upon pyramidal cells, *Soc. Neurosci. Abstr.* **20**:626.

Franceschini, N., Riehle, A., and Le Nestour, A., 1989, Directionally selective motion detection by insect neurons, in: *Facets of Vision* (D. C. Stavenga and R. C. Hardie, eds.), Springer-Verlag, Berlin, pp. 361–390.

Fredericksen, R. E., Verstraten, F. A. J., and Van De Grind, W. A., 1993, Spatiotemporal characteristics of human motion perception, *Vis. Res.* **33:**1193–1205.

Fredericksen, R. E., Verstraten, F. A. J., and Van De Grind, W. A., 1994a, An analysis of the temporal integration mechanism in human motion perception, *Vis. Res.* **34:**3153–3170.

Fredericksen, R. E., Verstraten, F. A. J., and Van De Grind, W. A., 1994b, Spatial summation and its interaction with the temporal integration mechanism in human motion perception, *Vis. Res.* **34:**3171–3188.

Frost, B. J., 1982, Mechanisms for discriminating object motion from self-induced motion in the pigeon, in: *Analysis of Visual Behaviors* (D. J. Ingle, M. A. Goodale, and R. J. W. Mansfield, eds.), MIT Press, Cambridge, MA, pp. 177–198.

Frost, B. J., and Nakayama, K., 1983, Single visual neurons code opposing motion independent of direction, *Science* **220:**744–745.

Fulbrook, J. E., 1982, Motion sensitivity of optic nerve axons in turtle, *Pseudemys scripta elegans,* Ph.D. thesis, University of Delaware, Newark, DE.

Fuortes, M. G. F., and Hodgkin, A. L., 1964, Changes in the time scale and sensitivity in the ommatidia of *Limulus, J. Physiol.* (Lond.) **172:**239–263.

Gaidenko, G. V., 1978, Efferent connections of the dorsal cortex in tortoises, *J. Evol. Biochem. Physiol.* **13:**268–270.

Ganz, L., and Felder, R., 1984, Mechanism of directional selectivity in simple neurons of the cat's visual cortex analyzed with stationary flash sequences, *J. Neurophysiol.* **51:**294–324.

Geri, G. A., Kimsey, R. A., and Dvorak, C. A., 1982, Quantitative electron microscopic analysis of the optic nerve of the turtle, *Pseudemys, J. Comp. Neurol.* **207:**99–103.

Goodwin, A., Henry, G. H., and Bishop, P. O., 1975, Direction selectivity of simple striate cells: Properties and mechanism, *J. Neurophysiol.* **38:**1500–1523.

Gorea, A., 1995, Spatio-temporal characterization of a Fourier and non-Fourier motion system, *Vis. Res.* **35:**907–914.

Granda, A. M., and Dvorak, C. A., 1977, Vision in turtles, in: *Handbook of Sensory Physiology* (H. Autrum, R. Jung, R. Loewenstein, D. M. MacKay, and H. L. Teuber, eds.), Springer-Verlag, New York, Vol. VII/5, pp. 451–495.

Granda, A. M., and Fulbrook, J. E., 1989, Classification of turtle retinal ganglion cells, *J. Neurophysiol.* **62:**723–737.

Granda, A. M., and Sisson, D. F., 1992, Retinal function in turtles, in: *Biology of the Reptilia* Vol. 17, *Neurology C* (C. G. Gans and P. S. Ulinski, eds.), University of Chicago Press, Chicago, pp. 136–174.

Graziano, M. S. A., Andersen, R. A., and Snowden, R. J., 1994, Tuning of MST neurons to spiral motions, *J. Neurosci.* **14:**54–67.

Grzywacz, N. M., and Koch, C., 1987, Functional properties of models for direction selectivity in the retina, *Synapse* **1:**417–434.

Grzywacz, N. M., Amthor, F. R., and Mistler, L. A., 1990, Applicability of quadratic and threshold models to motion discrimination in the rabbit retina, *Biol. Cybernet.* **64:**41–49.

Guiloff, G. D., and Kolb, H., 1992, Ganglion cell types of the turtle retina that project to the optic tectum: Intracellular HRP fillings of retrogradely, rhodamine-marked cell bodies, *Vis. Neurosci.* **8:**295–313.

Gusel'nikov, V. I., and Pivovarov, A. S., 1978, Postsynaptic mechanism of habituation of turtle cortical neurons to moving stimuli, *Neurosci. Behav. Physiol.* **9:**1–7.

Gusel'nikov, V. I., Morenkov, E. D., and Pivovarov, A. S., 1972, Unit responses of the turtle forebrain to visual stimuli, *Neurosci. Behav. Physiol.* **5:**235–242.

Hall, J. A., Foster, R. E., Ebner, F. F., and Hall, W. C., 1977, Visual cortex in a reptile, the turtle (*Pseudemys scripta* and *Chrysemys picta*), *Brain Res.* **130:**197–216.

Hall, W. C., and Ebner, F. F., 1970a, Parallels in the visual afferent projections of the thalamus in the hedgehog (*Paraechinus hypomelas*) and the turtle (*Pseudemys scripta*), *Brain Behav. Evol.* **3:**135–154.

Hall, W. C., and Ebner, F. F., 1970b, Thalamotelencephalic projections in the turtle (*Pseudemys scripta*), *J. Comp. Neurol.* **140:**101–122.

Heeger, D. J., 1991, Nonlinear model of neural responses in cat visual cortex, in: *Computational Models of Visual Processing* (M. S. Landy and J. A. Movshon, eds.), MIT Press, Cambridge, MA, pp. 119–134.

Heeger, D. J., 1992, Normalization of cell responses in cat striate cortex, *Vis. Neurosci.* **9:**181–197.

Heggelund, P., and Hartveit, E., 1990, Neurotransmitter receptors mediating retinal input to cells in the cat lateral geniculate nucleus: I. Lagged cells, *J. Neurophysiol.* **63:**1347–1360.

Heller, S. B., and Ulinski, P. S., 1987, Morphology of geniculocortical axons in turtles of the genera *Pseudemys* and *Chrysemys*, *Anat. Embryol.* **175:**505–515.

Henry, G. H., 1985, Physiology of cat striate cortex, in: *Cerebral Cortex*, Vol. 3. *Visual Cortex* (A. Peters and E. G. Jones, eds.), Plenum Press, New York, pp. 119–156.

Hergueta, S., Lemire, M., Ward, R., Rio, J.-P., and Repérant, J., 1992, A reconsideration of the primary visual system of the turtle *Emys orbicularis*, *J. Hirnforsch.* **33:**515–544.

Hergueta, S., Lemire, M., Ward, R., Repérant, J., Rio, J.-P., and Weidner, C., 1994, Interspecific variation in the chelonian primary visual system, *J. Brain Res.* **36:**171–192.

Hildreth, E. C., and Koch, C., 1987, The analysis of visual motion: From computational theory to neuronal mechanisms, *Annu. Rev. Neurosci.* **10:**477–533.

Hodos, W., and Campbell, C. B. G., 1969, *Scalae naturae:* Why there is no theory in comparative psychology, *Psychol. Rev.* **76:**337–350.

Holmes, W. R., Segev, I., and Rall, W., 1992, Interpretation of time constant and electrotonic length estimates in multicylinder or branched neuronal structures, *J. Neurophysiol.* **68:**1401–1420.

Hounsgaard, J., and Nicholson, C., 1990, The isolated turtle brain and the physiology of neuronal circuits, in: *Preparations of Vertebrate Central Nervous System in Vitro* (H. Jahnsen, ed.), Wiley, New York, pp. 155–182.

Hubel, D. H., and Wiesel, T. N., 1962, Receptive fields of single neurons in cat's striate cortex, *J. Physiol.* (Lond.) **160:**106–154.

Hubel, D. H., and Wiesel, T. N., 1968, Receptive fields and functional architecture of monkey striate cortex, *J. Physiol.* (Lond.) **195:**215–243.

Humphrey, A. L., and Weller, R. E., 1988a, Functionally distinct groups of X-cells in the lateral geniculate nucleus of the cat, *J. Comp. Neurol.* **268:**429–447.

Humphrey, A. L., and Weller, R. E., 1988b, Structural correlates of functionally distinct X-cells in the lateral geniculate nucleus of the cat, *J. Comp. Neurol.* **268:**448–468.

Humphrey, A. L., and Saul, A. B., 1993, The temporal transformation of retinal signals in the lateral geniculate nucleus of the cat: Implications for cortical function, in: *Thalamic Networks for Relay and Modulation* (D. Minchiachi, M. Molinari, G. Macchi, and E. G. Jones, eds.), Pergamon Press, Oxford, pp. 81–89.

Innocenti, G. M., and Fiore, L., 1974, Post-synaptic inhibitory components of the responses to moving stimuli in area 17, *Brain Res.* **80:**122–126.

Jagadeesh, B., Wheat, H. S., and Ferster, D., 1993, Linearity of summation of synaptic potentials underlying direction selectivity in simple cells of the cat visual cortex, *Science* **262:**1901–1904.

Jensen, R. J., and DeVoe, R. D., 1983, Comparisons of directionally selective with other ganglion cells of the turtle retina: Intracellular recording and staining, *J. Comp. Neurol.* **217:**271–287.

Jin, Z. F., and Srinivasan, M. V., 1990, Neural gradient models for the measurement of image velocity, *Vis. Neurosci.* **5:**261–271.

Johansson, G., 1971, Studies on visual perception of locomotion, *Perception* **6:**365–376.

Johnston, D., and Wu, S. M., 1995, *Foundations of Cellular Neurophysiology*, MIT Press, Cambridge, MA.

Johnston, J. B., 1915, The cell masses in the forebrain of the turtle *Cistudo carolina*, *J. Comp. Neurol.* **25:**393–468.

Jones, E. G., 1985, *The Thalamus*, Plenum Press, New York.

Jones, J. P., and Palmer, L. A., 1987, The two-dimensional spatial structure of simple receptive fields in cat striate cortex, *J. Neurophysiol.* **58:**1187–1211.

Julesz, B., 1971, *Foundation of Cyclopean Perception*, University of Chicago Press, Chicago.

Kaas, J. H., and Krubitzer, L. A., 1992, Area 17 lesions deactivate area MT in owl monkeys, *Vis. Neurosci.* **9:**399–407.

Katz, B., and Miledi, R., 1968, The role of calcium in neuromuscular facilitation, *J. Physiol.* (Lond.) **195:**481–492.

Kenigfest, N. B., Repérant, J., Rio, J.-P., Belekhova, M. G., Tumanova, N. L., Ward, R., Vesselkin, N. P., Herbin, M., Chkeidze, D. D., and Ozirskaya, E. V., 1995, Fine structure of the dorsal lateral geniculate nucleus of the turtle, *Emys orbicularis:* A Golgi, combined HRP tracing and GABA immunocytochemical study, *J. Comp. Neurol.* **356:**595–614.

Killackey, H., Pellmar, T., and Ebner, F. F., 1972, Effects of general cortex ablation on habituation in the turtle, *Fed. Proc.* **31:**819.

Knapp, H., and Kang, D. S., 1968a, The retinal projection of the side-necked turtle (*Podocnemis unifilis*) with some notes on the possible origin of the pars dorsalis of the lateral geniculate body, *Brain Behav. Evol.* **1**:369–404.

Knapp, H., and Kang, D. S., 1968b, The visual pathways of the snapping turtle (*Chelydra serpentina*), *Brain Behav. Evol.* **1**:19–42.

Koch, C., and Poggio, T., 1983a, A theoretical analysis of electrical properties of spines, *Proc. R. Soc. Lond. B* **218**:455–477.

Koch, C., and Poggio, T., 1983b, Non-linear interaction in a dendritic tree: Localization, timing and a role in information processing, *Proc. Natl. Acad. Sci. USA* **80**:2799–2802.

Koch, C., Poggio, T., and Torre, V., 1986, Computations in the vertebrate retina: Gain enhancement, differentiation and motion discrimination, *Trends Neurosci.* **9**:204–211.

Koch, C., Douglas, R., and Wehmeier, U., 1990, Visibility of synaptically induced conductance changes: Theory and simulations of anatomically characterized pyramidal cells, *J. Neurosci.* **10**:1728–1744.

Kolb, H., 1982, The morphology of the bipolar cells, amacrine cells and ganglion cells in the retina of the turtle, *Pseudemys scripta elegans, Phil. Trans. R. Soc. Lond. B* **298**:355–393.

Kolb, H., Perlman, I., and Normann, R. A., 1988, Neural organization of the retina of the turtle *Mauremys caspica*: A light microscopic and Golgi study, *Vis. Neurosci.* **1**:47–72.

Kontsevich, L., 1995, The nature of inputs to cortical motion detectors *Vis. Res.* **35**:2785–2793.

Kosareva, A. A., 1967, Projection of optic fibers to visual centers in a turtle (*Emys orbicularis*), *J. Comp. Neurol.* **130**:263–276.

Kosslyn, S. M., and Andersen, R. A., 1995, *Frontiers in Cognitive Neuroscience,* MIT Press, Cambridge, MA.

Kriegstein, A. R., 1987, Synaptic responses of cortical pyramidal neurons to light stimulation in the isolated turtle visual system, *J. Neurosci.* **7**:2488–2492.

Kriegstein, A. R., and Connors, B. W., 1986, Cellular physiology of the turtle visual cortex: Synaptic properties and intrinsic circuitry, *J. Neurosci.* **6**:178–191.

Kuffler, S. W., 1953, Discharge patterns and functional organization of mammalian retina, *J. Neurophysiol.* **16**:37–68.

Kwon, Y. H., Nelson, S. B., Toth, L. J., and Sur, M., 1992, Effect of stimulus contrast and size on NMDA receptor activity in cat lateral geniculate nucleus, *J. Neurophysiol.* **68**:182–196.

Lamb, T. D., and Simon, E. J., 1976, The relation between intercellular coupling and electrical noise in turtle photoreceptors, *J. Physiol.* (Lond.) **263**:257–286.

Lamb, T. D., and Simon, E. J., 1977, Analysis of electrical noise in turtle cones, *J. Physiol.* (Lond.) **272**:435–468.

Landy, M. S., and Movshon, J. A., 1991, *Computational Models of Visual Processing,* MIT Press, Cambridge, MA.

Larson-Prior, L. J., and Slater, N. T., 1988, GABAergic inhibition and epileptiform discharges in the turtle hippocampus *in vitro, Brain Res.* **460**:369–375.

Larson-Prior, L. J., Ulinski, P. S., and Slater, N. T., 1991, Excitatory amino acid receptor-mediated transmission in geniculocortical and intracortical pathways within visual cortex, *J. Neurophysiol.* **66**:293–306.

Lee, D. N., 1980, The optic flow field: The foundation of vision, *Phil. Trans. R. Soc. Lond. B* **290**:169–179.

Lennie, P., 1993, Role of M and P pathways, in: *Contrast Sensitivity* (R. Shapley and M.-K. Lam, eds.), MIT Press, Cambridge, MA, pp. 201–215.

Lester, R. A., and Jahr, C. E., 1992, NMDA channel behavior depends on agonist affinity, *J. Neurosci.* **12**:635–643.

Lettvin, J. Y., Maturana, H. P., McCulloch, W. S., and Pitts, W. H., 1959, What the frog's eye tells the frog's brain, *Proc. Int. Radio Eng.* **47**:1940–1951.

Lipetz, L. E., and Hill, R. M., 1970, Discrimination characteristics of the turtle's retinal ganglion cells, *Experentia* **26**:373–374.

Llinás, R., 1993, Is dyslexia a dyschronia? *Ann. N. Y. Acad. Sci.* **682**:48–56.

Maex, R., and Orban, G. A., 1992, A model circuit for cortical temporal low-pass filtering, *Neural Computation* **4**:932–945.

Magelby, K. L., and Zengel, J. E., 1982, A quantitative description of stimulation-induced changes in transmitter release at the frog neuromuscular junction, *J. Gen. Physiol.* **80**:613–638.

Mallart, A., and Martin, A. R., 1967, An analysis of facilitation of transmitter release at the neuromuscular junction of the frog, *J. Physiol.* (Lond.) **193**:679–694.

Mancilla, J. G., and Ulinski, P. S., 1996a, Temporal structure of compound postsynaptic potentials in visual cortex, in: *Proceedings of the Fourth Computation and Neural Systems Conference* (J. M. Bower, ed.), Academic Press, New York, pp. 227–232.

Mancilla, J. G., Fowler, M., and Ulinski, P. S., 1998, Responses of regular spiking and fast spiking cells in turtle visual cortex to light flashes, *Vis. Neurosci.* **15**:979–993.

Marcar, V. L., and Cowey, A., 1992, The effect of removing superior cortical motion areas in macaque monkey on visual discrimination using random dot displays, *Eur. J. Neurosci.* **4**:1219–1227.

Marcar, V. L., Xiao, D. K., Raiguel, S. E., Maes, H., and Orban, G. A., 1995, Processing of kinetically defined boundaries in the cortical motion area MT of the macaque monkey, *J. Neurophysiol.* **74**:1258–1270.

Marchiafava, P. L., and Torre, V., 1977, Self-facilitation of ganglion cells in the retina of the turtle, *J. Physiol.* (Lond.) **268**:335–351.

Marchiafava, P. L., and Wagner, H. G., 1981, Interactions leading to colour opponency in ganglion cells of the turtle retina, *Proc. R. Soc. Lond. B* **211**:261–267.

Marchiafava, P. L., and Weiler, R., 1980, Intracellular analysis and structural correlates of the organization of inputs to ganglion cells in the retina of the turtle, *Proc. R. Soc. Lond. B* **208**:103–113.

Marchiafava, P. L., Weiler, R., and Strettoi, E., 1983, Intracellular recording with horseradish-peroxidase electrodes reveals distinct functional roles of retinal plexiform layers, in: *The Physiology of Excitable Cells* (A. D. Grinnell and W. J. Moody, Jr., eds.), Liss, New York, pp. 549–556.

Marmarelis, P. Z., and Marmarelis, V. Z., 1978, *Analysis of Physiological Systems: The White Noise Approach,* Plenum Press, New York.

Marr, D., 1982, *Vision. A Computational Investigation into the Human Representation and Processing of Visual Information,* Freeman, San Francisco.

Mastronarde, D. M., 1987a, Two classes of single-input X-cells in cat lateral geniculate nucleus. I. Receptive-field properties and classification of cells, *J. Neurophysiol.* **57**:357–380.

Mastronarde, D. M., 1987b, Two classes of single-input X-cells in cat lateral geniculate nucleus. II. Retinal inputs and the generation of receptive-field properties, *J. Neurophysiol.* **57**:381–413.

Mastronarde, D. N., Humphrey, A. L., and Saul, A. B., 1991, Lagged Y cells in the cat lateral geniculate nucleus, *Vis. Neurosci.* **7**:191–200.

Mather, G., and West, S., 1993, Evidence for second-order motion detectors, *Vis. Res.* **33**:1109–1112.

Maturana, H. R., Lettvin, J. Y., McCulloch, W. S., and Pitts, W. H., 1960, Anatomy and physiology of vision in the frog (*Rana pipiens*), *J. Gen. Physiol.* **43**:(*Suppl.* 2):129–171.

Mazurskaya, P. Z., 1972, Study of the projection of the retina to the forebrain of the tortoise *Emys orbicularis, J. Evol. Biochem. Physiol.* **8**:550–555.

Mazurskaya, P. Z., 1974, Organization of receptive fields in the forebrain of *Emys orbicularis, Neurosci. Behav. Physiol.* **7**:311–318.

McCormick, D. A., 1990, Possible ionic basis for lagged visual responses in cat LGNd relay neurons, *Soc. Neurosci. Abstr.* **16**:159.

McLean, J., and Palmer, L. A., 1989, Contribution of linear spatiotemporal receptive field structure to velocity selectivity of simple cells in area 17 of cat, *Vis. Res.* **29**:675–679.

McLean, J., and Palmer, L. A., 1994, Organization of simple cell responses in the three-dimensional (3-D) frequency domain, *Vis. Neurosci.* **11**:295–306.

McLean, J., Raab, S., and Palmer, L. A., 1994, Contribution of linear mechanisms to the specification of local motion by simple cells in areas 17 and 18 of the cat, *Vis. Neurosci.* **11**:271–294.

Merigan, W. H., and Maunsell, J. H. B., 1993, How parallel are the primate visual pathways? *Annu. Rev. Neurosci.* **16**:369–402.

Merzenich, M. M., Schreiner, C., Jenkins, W., and Wang, X., 1993, Neural mechanisms underlying temporal integration, segmentation and input sequence representation: Some implications for the origin of learning disabilities, *Ann. N. Y. Acad. Sci.* **682**:1–22.

Mikami, A., Newsome, W. T., and Wurtz, R. H., 1986a, Motion selectivity in macaque visual cortex. I. Mechanisms of direction and speed selectivity in extrastriate area MT, *J. Neurophysiol.* **55**:1308–1327.

Mikami, A., Newsome, W. T., and Wurtz, R. H., 1986b, Motion selectivity in macaque visual cortex. II. Spatiotemporal range of directional interactions in MT and V1, *J. Neurophysiol.* **55**:1328–1339.

Mizunumi, M., 1990, Synaptic rectification model equivalent to the correlation-type movement detector, *Biol. Cybernet.* **64**:1–6.

Mody, I. Y., DeKoninck, Y., Otis, T. S., and Soltesz, I., 1994, Bridging the cleft at GABA synapses in the brain, *Trends Neurosci.* **17**:517–525.

Motter, B. C., Steinmetz, M. A., Duffy, C. J., and Mountcastle, V. B., 1987, Functional properties of parietal visual neurons: Mechanisms of directionality along a single axis, *J. Neurosci.* **7**:154–176.

Movshon, J. A., Adelson, E. H., Gizzi, M. S., and Newsome, W. T., 1986, The analysis of moving visual patterns, in: *Pattern Recognition Mechanisms* (C. Chagas, R. Gattass, and C. Gross, eds.), Springer-Verlag, New York, pp. 117–151.

Movshon, J. A., Thompson, I. D., and Tolhurst, D. J., 1978a, Spatial summation in the receptive fields of simple cells in the cat's striate cortex, *J. Physiol.* (Lond.) **283:**53–77.

Movshon, J. A., Thompson, I. D., and Tolhurst, D. J., 1978b, Receptive field organization of complex cells in the cat's striate cortex, *J. Physiol.* (Lond.) **283:**79–99.

Mulligan, K. A., and Ulinski, P. S., 1990, Organization of geniculocortical projections in turtles: Isoazimuth lamellae in the visual cortex, *J. Comp. Neurol.* **296:**531–547.

Nakayama, K., 1985, Biological image motion processing: A review, *Vis. Res.* **25:**625–660.

Nelson, S., Toth, L., Bhavin, S., and Sur, M., 1994, Orientation selectivity of cortical neurons during intracellular blockade of inhibition, *Science* **265:**774–777.

Newsome, W. T., and Pare, E. B., 1988, A selective impairment of motion processing following lesions of the middle temporal visual area (MT), *J. Neurosci.* **8:**2201–2211.

Newsome, W. T., Mikami, A., and Wurtz, R. H., 1986, Motion selectivity in macaque visual cortex. III. Psychophysics and physiology of apparent motion, *J. Neurophysiol.* **55:**1340–1351.

Newsome, W. T., Britten, K. H., and Movshon, J. A., 1989, Neuronal correlates of a perceptual decision, *Nature* **341:**52–54.

Nicolaus, J. M., and Ulinski, P. S., 1991, Medial and lateral differences in populations of GABAergic neurons in layer 3 of turtle visual cortex, *Soc. Neurosci. Abstr.* **16:**114.

Nicolaus, J. M., and Ulinski, P. S., 1994, Inward rectifying conductances in inhibitory neurons in turtle visual cortex, *Neural Systems: Analysis and Modeling 3* (F. Eeckman, ed.), Kluwer, Boston, pp. 91–96.

Northmore, A. P. M., and Granda, A. M., 1991a, Refractive state, contrast sensitivity, and resolution in the freshwater turtle, *Pseudemys scripta elegans*, determined by tectal visual-evoked potentials, *Vis. Neurosci.* **7:**619–625.

Northmore, D. P. M., and Granda, A. M., 1991b, Ocular dimensions and schematic eyes of freshwater and sea turtles, *Vis. Neurosci.* **7:**627–635.

Orban, G. A., 1984, *Neuronal Operations in the Visual Cortex*, Springer-Verlag, Berlin.

Orban, G. A., 1986, Processing of moving images in the geniculocortical pathway, in: *Visual Neuroscience* (J. D. Pettigrew, K. J. Sanderson, and W. R. Levick, eds.), Cambridge University Press, Cambridge, pp. 121–141.

Orban, G. A., 1994, Motion processing in monkey striate cortex, in: *Cerebral Cortex*, Vol. 10, *Primary Visual Cortex in Primates* (A. Peters and K. Rockland, eds.), Plenum Press, New York, pp. 413–442.

Orrego, F., 1961, The reptilian forebrain. I. The olfactory pathways and the cortical areas in the turtle, *Arch. Ital. Biol.* **99:**425–445.

Oyster, C. W., 1968, The analysis of image motion by the rabbit retina, *J. Physiol.* (Lond.) **199:**613–635.

Pasternak, T., Schumer, R. A., Gizzi, M. S., and Movshon, J. A., 1985, Abolition of cortical directional selectivity affects visual behavior in cats, *Exp. Brain Res.* **61:**214–217.

Pasternak, T., Albano, J. E., and Harvitt, D. M., 1990, The role of directionally selective neurons in the perception of global motion, *J. Neurosci.* **10:**3079–3086.

Peiper, A. J., and Ulinski, P. S., 1996a, Activation of the fast sodium conductance in pyramidal cells by temporal patterns of geniculocortical synapses, in: *Proceedings of the Fourth Computation and Neural Systems Conference* (J. M. Bower, ed.), Academic Press, New York, pp. 119–124.

Peiper, A. J., and Ulinski, P. S., 1996b, Activation of the fast sodium conductance in pyramidal cells by temporal patterns of geniculocortical synapses, in: *Computational Neuroscience* (J. H. Bower, ed.), Academic Press, San Diego, CA, pp. 119–124.

Penn, A. A., Wong, R. O. L., and Shatz, C. J., 1994, Neuronal coupling in the developing mammalian retina, *J. Neurosci.* **14:**3805–3815.

Petersik, J. T., 1994, A comparison of varieties of "second-order" motion, *Vis. Res.* **35:**507–517.

Peterson, E. H., 1978, Size classes of ganglion cells which project to the optic tectum in the turtle, *Pseudemys scripta elegans*, **190:**509–510.

Peterson, E. H., 1992, Retinal structure, in: *Biology of the Reptilia*, Vol. 17, *Neurology C* (C. G. Gans and P. S. Ulinski, eds.), University of Chicago Press, Chicago, pp. 1–135.

Peterson, E. H., and Ulinski, P. S., 1979, Quantitative studies of retinal ganglion cells in a turtle, *Pseudemys scripta elegans*. I. Number and distribution of ganglion cells, *J. Comp. Neurol.* **186:**17–42.

Peterson, E. H., and Ulinski, P. S., 1982, Quantitative studies of retinal ganglion cells in a turtle, *Pseudemys scripta elegans*. II. Size spectrum of ganglion cells and its regional variation, *J. Comp. Neurol.* **208:**157–168.

Pettigrew, J. D., 1979, Binocular visual processing in the owl's telencephalon, *Proc. R. Soc. Lond. B* **204**:435–454.

Pfleger, B., and Bonds, A. B., 1995, Dynamic differentiation of $GABA_A$-sensitive influences on orientation selectivity of complex cells in the cat striate cortex, *Exp. Brain Res.* **104**:81–88.

Pinter, R. B., and Nabet, B., eds., 1992, *Nonlinear Vision: Determination of Neural Receptive Fields, Function and Networks,* CRC Press, Boca Raton, FL.

Poggio, T., and Reichardt, W., 1976, Visual control of orientation behaviour in the fly, *Q. Rev. Biophys.* **9**:377–438.

Powers, A. S., and Wojcik, L., 1992, Effects of dorsal cortex lesions on habituation to a looming stimulus in turtles, *Soc. Neurosci. Abstr.* **18**:1066.

Prechtl, J. C., 1994, Visual motion induces synchronous oscillations in turtle visual cortex, *Proc. Natl. Acad. Sci. USA* **91**:12467–12471.

Prechtl, J. C., 1995, Flutter-like response in visual cortex of the semi-isolated turtle brain, *Biol. Bull.* **189**:215–216.

Prechtl, J. C., and Bullock, T. H., 1995, Distinctive cortical potentials anticipate and follow spontaneous pupillary shifts in turtle, *Soc. Neurosci. Abstr.* **21**:127.

Qian, N., and Andersen, R. A., 1994, Transparent motion perception as detection of unbalanced motion signals. II. Physiology. *J. Neurosci.* **14**:7367–7380.

Qian, N., Andersen, R. A., and Adelson, E. H., 1994a, Transparent motion perception as detection of unbalanced motion signals. I. Psychophysics. *J. Neurosci.* **14**:7357–7366.

Qian, N., Andersen, R. A., and Adelson, E. H., 1994b, Transparent motion perception as detection of unbalanced motion signals. III. Modeling, *J. Neurosci.* **14**:7381–7392.

Raiguel, S., Van Hulle, M. M., Xiao, D.-K., Marcar, V. L., and Orban, G. A., 1995, Shape and spatial distribution of receptive fields and antagonistic motion surrounds in the middle temporal area (V5) of the macaque, *Eur. J. Neurosci.* **7**:2064–2082.

Rainey, W. T., and Ulinski, P. S., 1986, Morphology of neurons in the dorsal lateral geniculate nucleus in turtles of the genera *Pseudemys* and *Chrysemys, J. Comp. Neurol.* **253**:440–465.

Rall, W., 1977, Core conductor theory and cable properties of neurons, in: *Handbook of Physiology. The Nervous System. Cellular Biology of Neurons* (J. M. Brookhardt and V. B. Mountcastle, eds.), American Physiological Society, Bethesda, MD, pp. 39–96.

Rauschecker, J. P., van Grunau, M. W., and Poulin, C., 1987, Centrifugal organization of direction preferences in the cat's lateral suprasylvian visual cortex and its relation to flow field processing, *J. Neurosci.* **7**:943–958.

Regan, D., 1991a, *Spatial Vision,* CRC Press, Boca Raton, FL.

Regan, D., 1991b, A brief review of some of the stimuli and analysis methods used in spatiotemporal vision research, in: *Spatial Vision* (D. Regan, ed.), CRC Press, Boca Raton, FL, pp. 1–42.

Reichardt, W., 1961, Autocorrelation: a principle for the evaluation of sensory information by the central nervous system, in: *Sensory Communication* (W. A. Rosenblith, ed.), Wiley, New York, pp. 303–318.

Reichardt, W., and Poggio, T., 1976, A theory of pattern induced flight orientation of the fly *Musca domestica:* II. *Biol. Cybernet.* **18**:69–80.

Reid, C., Soodak, R., and Shapley, R. M., 1987, Linear mechanisms of directional selectivity in simple cells of cat striate cortex, *Proc. Natl. Acad. Sci. USA* **84**:8740–8744.

Reid, C. R., Soodak, R., and Shapley, R. M., 1991, Directional selectivity and spatiotemporal structure of receptive fields of simple cells in cat striate cortex, *J. Neurophysiol.* **66**:505–529.

Reiner, A., 1981, A projection of displaced ganglion cells and giant ganglion cells to the accessory optic nuclei in turtle, *Brain Res.* **204**:403–409.

Reiner, A., 1991, A comparison of neurotransmitter-specific and neuropeptide-specific neuronal cell types present in the dorsal cortex in turtles with those present in the isocortex in mammals: Implications for the evolution of isocortex. *Brain Behav. Evol.* **38**:53–91.

Repérant, J., Rio, J.-P., Ward, R., Heguerta, S., Miceli, D., and Lemire, M., 1992, Comparative analysis of the primary visual system of reptiles, in: *Biology of Reptilia,* Vol. 17, *Neurology C* (C. G. Gans and P. S. Ulinski, eds.), University of Chicago Press, Chicago, pp. 175–240.

Robson, J. G., 1966, Spatial and temporal contrast-sensitivity functions of the visual system, *J. Opt. Soc. Am.* **56**:1141–1142.

Rodieck, R. W., 1973, *The Vertebrate Retina. Principles of Structure and Function,* Freeman, San Francisco.

Rodieck, R. W., 1988, The primate retina, in: *Comparative Primate Biology,* Vol. 4. *Neurosciences,* Liss, New York, pp. 203–278.

Rodieck, R. W., and Stone, J., 1965, Analysis of receptive fields of cat retinal ganglion cells, *J. Neurophysiol.* **28:**819–832.

Rodieck, R. W., Brening, R. K., and Watanabe, M., 1993, The origin of parallel visual pathways, in: *Contrast Sensitivity* (R. Shapley and M.-K. Lam, eds.), MIT Press, Cambridge, MA, pp. 117–148.

Rosenberg, A. F., and Ariel, M., 1990, Visual-response properties of neurons in turtle basal optic nucleus *in vitro*, *J. Neurophysiol.* **63:**1033–1045.

Rosenberg, A. F., and Ariel, M., 1991, Electrophysiological evidence for a direct projection of direction-sensitive retinal ganglion cells to the turtle's accessory optic system, *J. Neurophysiol.* **65:**1022–1033.

Rosenquist, A. C., 1985, Connections of visual cortical areas in cats, in: *Cerebral Cortex*, Vol. 3. *Visual Cortex* (A. Peters and E. G. Jones, eds.), Plenum Press, New York, pp. 81–118.

Roy, J.-P., and Wurtz, R. H., 1990, The role of disparity-sensitive cortical neurons in signaling the direction of self-motion, *Nature* **348:**160–162.

Rudolph, K. H., Ferrera, V. P., and Pasternak, T., 1994, A reduction in the number of directionally selective neurons extends the spatial limit for global motion perception, *Vis. Res.* **34:**3241–3251.

Saito, H., Yukie, M., Tanaka, K., Hikosaka, K., Fukada, Y., and Iwai, E., 1986, Integration of direction signals of image motion in the superior temporal sulcus of the macaque monkey, *J. Neurosci.* **6:**145–157.

Salzman, C. D., Britten, K. H., and Newsome, W. T., 1990, Cortical microstimulation influences perceptual judgements of motion direction, *Nature* **346:**174–177.

Saul, A. B., and Humphrey, L., 1990, Spatial and temporal response properties of lagged and nonlagged cells in cat lateral geniculate nucleus, *J. Neurophysiol.* **64:**206–224.

Saul, A. B., and Humphrey, A. L., 1992a, Temporal frequency tuning of direction selectivity in cat visual cortex, *Vis. Neurosci.* **8:**365–372.

Saul, A. B., and Humphrey, A. L., 1992b, Evidence of input from lagged cells in the lateral geniculate nucleus to simple cells in cortical area 17 of the cat, *J. Neurophysiol.* **68:**1190–1207.

Schneeweis, D. M., and Green, D. G., 1995, Spectral properties of turtle cones, *Vis. Neurosci.* **12:**333–344.

Sejnowski, T. J., 1986, Open questions about computation in cerebral cortex, in: *Parallel Distributed Processing. Explorations in the Microstructure of Cognition*, Vol. 2. *Psychological and Biological Models* (J. L. McClelland, D. E. Rumelhart, and the PDP Research Groups, eds.), MIT Press, Cambridge, MA, pp. 372–389.

Sekuler, R., Anstis, S., Braddick, O. J., Brandt, T., Movshon, J. A., and Orban, G., 1990, The perception of motion, in: *Visual Perception. The Neurophysiological Foundations*, Academic Press, San Diego, CA, pp. 205–230.

Senseman, D. M., 1995a, Combined microelectrode and fast multisite optical recording of visually-evoked activity in the turtle cortex, *Invest. Ophthal. Vis. Sci.* **36:**S693.

Senseman, D. M., 1995b, High-speed imaging of neural activity in the turtle cortex evoked by moving stimuli, *Soc. Neurosci. Abstr.* **21:**1653.

Senseman, D. M., 1996, Correspondence between visually evoked voltage-sensitive dye signals and synaptic activity recorded in cortical pyramidal cells with intracellular microelectrodes, *Vis. Neurosci.* **13:**963–977.

Sereno, M. E., 1993, *Neural Computation of Pattern Motion*, MIT Press, Cambridge, MA.

Sernagor, E., and Gryzwacz, N. M., 1995, Emergence of complex receptive field properties of ganglion cells in the developing turtle retina, *J. Neurophysiol.* **73:**1355–1364.

Shapley, R., Reid, C., and Soodak, R., 1991, Spatiotemporal receptive fields and direction selectivity, in: *Computational Models of Visual Processing* (M. S. Landy and J. A. Movshon, eds.), MIT Press, Cambridge, MA, pp. 109–118.

Sherk, H., Kim, J., and Mulligan, K., 1995, Are the preferred directions of neurons in cat extrastriate cortex related to optic flow? *Vis. Neurosci.* **12:**887–894.

Sillito, A. M., 1975, The contribution of inhibitory mechanisms to the receptive field properties of neurons in the striate cortex of the cat, *J. Physiol.* (Lond.) **250:**305–329.

Sillito, A. M., 1977, Inhibitory processes underlying the directional specificity of simple, complex and hypercomplex cells in the cat's visual cortex, *J. Physiol.* **271:**699–720.

Sillito, A. M., 1992, GABA mediated inhibitory processes in the function of the geniculo-striate system, *Prog. Brain Res.* **90:**349–384.

Sjöstöm, A. M., and Ulinski, P. S., 1985, Morphology of retinogeniculate terminals in the turtle, *Pseudemys scripta elegans*, *J. Comp. Neurol.* **238:**107–120.

Smith, A. T., Snowden, R. J., and Milne, A. B., 1994, Is global motion really based on spatial integration of local motion signals? *Vis. Res.* **34**:2425–2430.

Smith, D. O., Franke, C., Rosenheimer, J. L., Zufall, F., and Hatt, H., 1991, Desensitization and resensitization rates of glutamate-activated channels may regulate motoneuron excitability, *J. Neurophysiol.* **66**:1166–1175.

Smith, L. M., Ebner, F. F., and Colonnier, M., 1980, The thalamocortical projection in *Pseudemys* turtles: A quantitative electron microscope study, *J. Comp. Neurol.* **190**:445–462.

Smith, R. D., Grzywacs, N. M., and Borg-Graham, L. J., 1996, Is the input to a GABA-ergic synapse the sole asymmetry in turtle's retinal directional selectivity? *Vis. Neurosci.* **13**:423–439.

Snowden, R. J., and Braddick, O. J., 1989, The combination of motion signals over time, *Vis. Res.* **29**:1621–1630.

Snowden, R. J., and Braddick, O. J., 1991, The temporal integration and resolution of velocity signals, *Vis. Res.* **31**:907–914.

Snowden, R. J., Treue, S., Erickson, R. G., and Andersen, R. A., 1991, The response of area MT and V1 neurons to transparent motion, *J. Neurosci.* **11**:2768–2785.

Somers, D. C., Nelson, S. B., and Sur, M., 1995a, An emergent model of visual cortical orientation selectivity, in: *The Neurobiology of Computation. The Proceedings of the Third Annual Computation and Neural Systems Conference* (J. M. Bower, ed.), Kluwer, Boston, pp. 311–316.

Somers, D. C., Nelson, S. B., and Sur, M., 1995b, An emergent model of orientation selectivity in cat visual cortical simple cells, *J. Neurosci.* **15**:5448–5465.

Spruston, N., Jaffe, D. B., and Johnston, D., 1994, Dendritic attenuation of synaptic potentials and currents: The role of passive membrane properties, *Trends Neurosci.* **17**:161–166.

Srinivasan, M. V., 1992, How flying bees compute range from optical flow: Behavioral experiments and neural models, in: *Nonlinear Vision: Determination of Neural Receptive Fields, Function and Networks* (R. B. Pinter and B. Nabet, eds.), CRC Press, Boca Raton, FL, pp. 353–377.

Steinmetz, M. A., Motter, B. C., Duffy, C. J., and Mountcastle, V. B., 1987, Functional properties of parietal visual neurons: Radial organization of directionalities within the visual field, *J. Neurosci.* **7**:177–191.

Stone, J., 1983, *Parallel Processing in the Visual System. The Classification of Retinal Ganglion Cells and Its Impact on the Neurobiology of Vision,* Plenum Press, New York.

Stoner, G. R., and Albright, T. D., 1992, Neural correlates of perceptual motion coherence, *Nature* **358**:412–414.

Stoner, G. R., Albright, T. D., and Ramachadran, V. S., 1990, Transparency and coherence in human motion perception, *Nature* **344**:153–154.

Sutter, A., Sperling, G., and Chubb, C., 1995, Measuring the spatial frequency selectivity of second-order texture mechanisms, *Vis. Res.* **7**:915–924.

Tanaka, K., Hikosaka, K., Saito, H., Yukie, M., Fukada, Y., and Iwai, E., 1986, Analysis of local and wide-field movements in the superior temporal visual areas of the macaque monkey, *J. Neurosci.* **6**:134–144.

Teller, D. Y., 1990, The domain of visual science, in: *Visual Perception. The Neurophysiological Foundations* (L. Spillman and J. S. Werner, eds.), Academic Press, San Diego, CA, pp. 11–22.

Troy, J. B., 1993, Modeling the receptive fields of mammalian retinal ganglion cells, in: *Contrast Sensitivity* (R. Shapley and D. M.-K. Lam, eds.), MIT Press, Cambridge, MA, pp. 85–102.

Trussell, L. O., and Fischbach, G. D., 1989, Glutamate receptor desensitization and its role in synaptic transmission, *Neuron* **3**:209–218.

Trussell, L. O., Thio, L. L., Zorumski, C. F., and Fischbach, G. D., 1988, Rapid desensitization of glutamate receptors in vertebrate central neurons, *Proc. Natl. Acad. Sci. USA* **85**:4562–4566.

Tsumoto, T., Eckart, W., and Creutzfeldt, O. D., 1979, Modification of orientation sensitivity of cat visual cortex neurons by removal of GABA-mediated inhibition, *Exp. Brain Res.* **34**:351–363.

Ulinski, P. S., 1980, Functional morphology of the vertebrate visual system: An essay on the evolution of complex systems, *Am. Zool.* **20**:229–246.

Ulinski, P. S., 1983, *Dorsal Ventricular Ridge. A Treatise on Forebrain Organization in Reptiles and Birds,* Wiley, New York.

Ulinski, P. S., 1986a, Ultrastructure of the dorsal lateral geniculate complex in turtles of the genera *Pseudemys* and *Chrysemys, Brain Behav. Evol.* **29**:117–142.

Ulinski, P. S., 1986b, Organization of the corticogeniculate projections in the turtle, *Pseudemys scripta, J. Comp. Neurol.* **254**:529–542.

Ulinski, P. S., 1987, Distribution of GABA accumulating neurons in the diencephalon of the turtle *Pseudemys scripta, Anat. Rec.* **218:**141A.

Ulinski, P. S., 1988, Functional architecture of turtle visual cortex, in: *The Forebrain of Reptiles* (W. K. Schwerdtfeger and W. J. A. J. Smeets, eds.), Karger, Basel, pp. 151–161.

Ulinski, P. S., 1990, Cerebral cortex in reptiles, in: *Cerebral Cortex,* Vol. 8A. *Comparative Structure and Evolution of Cerebral Cortex, Part I* (E. G. Jones and A. Peters, eds.), Plenum Press, New York, pp. 139–215.

Ulinski, P. S., 1997, The nervous system of vertebrates: A comparative perspective, in: *Handbook of Comparative Physiology* (W. H. Dantzler, ed.), Wiley, New York, pp. 17–53.

Ulinski, P. S., and Nautiyal, J., 1987, Organization of retinogeniculate projections in turtles of the genera *Pseudemys* and *Chrysemys, J. Comp. Neurol.* **276:**92–112.

Ulinski, P. S., Larson-Prior, L., and Slater, N. T., 1991, Cortical circuitry underlying visual motion analysis in turtles, in: *Visual Structures and Integrated Functions* (M. A. Arbib and J.-P. Ewert, eds.), Springer-Verlag, New York, pp. 307–324.

Ulinski, P. S., Larson-Prior, L. J., and Slater, N. T., 1993, Cellular and network properties in cortical neurons: Studies with an *in vitro* preparation of visual cortex, in: *Analysis and Modeling of Neural Systems* (F. Eeckman, ed.), Kluwer, Boston, pp. 211–218.

Ullman, S., 1984, On the measurement and use of visual motion: Computational considerations and some neurophysiological implications, in: *Dynamic Aspects of Neocortical Function* (G. M. Edelman, W. E. Gall, and W. M. Cowan, eds.), Wiley, New York, pp. 637–652.

van Santen, J. P. H., and Sperling, G., 1985, Elaborated Reichardt detectors, *J. Opt. Soc. Am. A* **2:**300–321.

Victor, J. D., 1992, Nonlinear systems analysis in vision: Overview of kernel methods, in: *Nonlinear Vision: Determination of Neural Receptive Fields, Function and Networks* (R. B. Pinter and B. Nabet, eds.), CRC Press, Boca Raton, FL, pp. 1–38.

Victor, J. D., and Conte, M. M., 1992, Evoked potential and psychophysical analysis of Fourier and non-Fourier motion mechanisms, *Vis. Neurosci.* **9:**105–123.

Wallach, H., and O'Connell, D. N., 1953, The kinetic depth effect, *J. Exp. Psychol.* **45:**205–217.

Wallman, J., and Fite, K. V., 1985, A survey of pretectal areas and the accessory optic system, *Brain Behav. Evol.* **26:**69–140.

Wandel, B. A., 1995, *Foundations of Vision,* Sinauer, Sunderland, MA.

Watamaniuk, S. N. J., Sekuler, R., and Williams, D. W., 1989, Direction perception in complex dynamic displays: The integration of direction information, *Vis. Res.* **29:**47–59.

Watamaniuk, S. N. J., McKee, S. P., and Grzywacs, N. M., 1995, Detecting a trajectory embedded in random-direction motion noise, *Vis. Res.* **35:**65–77.

Watson, A. B., and Ahumada, Jr., A. J., 1985, Model of human visual-motion sensing, *J. Opt. Soc. Am. A* **2:**322–341.

Welch, L., 1989, The perception of moving plaids reveals two motion-processing stages, *Nature* **337:**734–736.

Williams, D. W., and Sekuler, R., 1984, Coherent global motion percepts from stochastic local motions, *Vis. Res.* **24:**55–62.

Wilson, H. R., and Kim, J., 1992, Perceived motion in the vector sum direction. *Vis. Res.* **34:**1835–1842.

Wilson, H. R., and Kim, J., 1994, A model for motion coherence and transparency, *Vis. Neurosci.* **11:**1205–1220.

Woodbury, P. B., and Ulinski, P. S., 1986, Conduction velocity, size and distribution of optic nerve axons in the turtle, *Pseudemys scripta elegans, Anat. Embryol.* **174:**253–263.

Worgötter, F., and Holt, G., 1991, Spatiotemporal mechanisms in receptive fields of visual cortical simple cells: A model, *J. Neurophysiol.* **65:**494–510.

Wurtz, R. H., Yamasaki, D. S., Duffy, C. J., and Roy, J.-P., 1990, Functional specialization for visual motion processing in primate cerebral cortex, *Cold Spring Harbor Symp. Quant. Biol.* **54:**717–727.

Yuille, A. L., and Grzywacz, N. M., 1988, A computational theory for the perception of coherent visual motion, *Nature* **333:**71–74.

Zanker, J. M., 1990, On the directional sensitivity of motion detectors, *Biol. Cybernet.* **62:**177–183.

Zanker, J. H., 1995, Of models and men: Mechanisms of human motion perception, in: *Early Vision and Beyond* (T. V. Papathomas, ed.), MIT Press, Cambridge, MA, pp. 155–165.

Zeki, S. M., 1976, The functional organization of projections from striate to prestriate visual cortex in the rhesus monkey, *Cold Spring Harb. Symp. Quant. Biol.* **40:**591–600.

7

Linearity and Gain Control in V1 Simple Cells

MATTEO CARANDINI, DAVID J. HEEGER, and J. ANTHONY MOVSHON

1. Introduction

The primary visual cortex (V1) is arguably the most studied area in the mammalian cortex, and one of the very few for which we can say something sensible about the computations that it performs. V1 cells are selective for the position, shape, size, velocity, color, and eye of presentation of a visual stimulus. The mechanism of this selectivity as well as its rationale have recently begun to be understood, although some aspects still constitute an area of intense debate.

The receptive fields of V1 cells were first mapped by Hubel and Wiesel (1962) using flashing bars. They termed *simple cells* those cells for which they could find regions that responded either to the onset or to the offset of a bright bar, but not to both. Simple cells constitute around 50% of V1 neurons (De Valois *et al.*, 1982a; Schiller *et al.*, 1976). This chapter is devoted to simple cells, but it also includes ideas that can be useful in the understanding of the other major V1 cell type, the complex cells.

MATTEO CARANDINI and J. ANTHONY MOVSHON • Howard Hughes Medical Institute and Center for Neural Science, New York University, New York, New York 10003. *Present address for MC:* Institute of Neuroinformatics, ETH/UNIZ, Zurich 8057, Switzerland. DAVID J. HEEGER • Department of Psychology, Stanford University, Stanford, California 94305.

A number of researchers have proposed that simple cells behave linearly, and that their selectivity is determined by their linear weighting functions (Fig. 1A). This *linear model* of simple cells is attractive because if it were correct, it would be possible to predict the responses of a simple cell to any visual stimulus, based on a limited number of measurements. For example, any image can be approximated by a number of small pixels. Measuring the cell response by lighting each pixel one by one would allow the experimenter to predict the response to any visual stimulus.

We begin the chapter with the full definition of the linear model (Section 2). We explain its basic properties, and we summarize the vast number of studies that were devoted to testing it (Section 3). In these studies the model was found to be largely successful in explaining the selectivity of simple cells for stimulus shape, size, position, orientation, and direction of motion. We then propose a biophysical implementation of the linear model, and we discuss its plausibility (Section 4). In this implementation, simple cells receive both excitation and inhibition arranged in push–pull fashion, so that when one increases, the other decreases, and vice versa. The importance of this arrangement is that it makes it possible for the visual stimuli to result in perfect current injection in the cell, without any conductance increase. Conductance increases would result in nonlinear behavior.

While many aspects of simple cell responses are consistent with the linear model, there also are important violations of linearity (Section 5). For example, scaling the contrast of a stimulus would identically scale the responses of a linear cell. At high contrasts, however, the responses of simple cells show clear saturation (Maffei and Fiorentini, 1973). Moreover, simple cells are subject to cross-orientation inhibition: the responses to an optimally oriented stimulus can be diminished by superimposing an orthogonal stimulus, which would be ineffective in driving the cell when presented alone (Morrone *et al.,* 1982). These nonlinearities may be partially explained by contrast gain control mechanisms known to operate as early as in the retina (Shapley and Victor, 1978). There is evidence, however, suggesting that these nonlinearities have an important cortical component.

Our opinion is that the linear model is a powerful explanation of the behavior

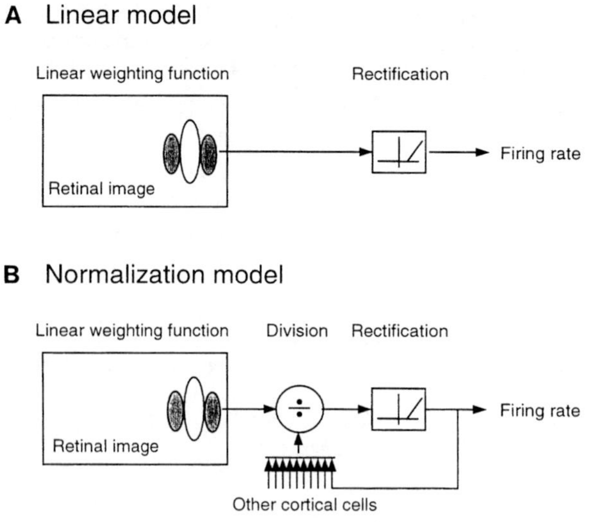

Figure 1. The two models of simple cell function discussed in this chapter. (A) The *linear model,* composed of a linear *weighting function* and a *rectification* stage. The cell performs a weighted average of the light intensities over local space and recent time. The rectification stage converts the output of this linear stage into firing rate. Note that rectification is a nonlinearity, so the "linear model" is not entirely linear. (B) The *normalization model,* which extends the linear model by adding a divisive stage. The output of each cell's linear stage is divided by the pooled output of a large number of other cells.

of simple cells. Part of the nonlinear behavior of simple cells can be attributed to the rectification (threshold) in the generation of action potentials. The many nonlinearities that are not accounted for by rectification can be explained by adding to the linear model a divisive inhibition stage (Fig. 1B). This stage controls the *gain* (responsiveness) of the neurons by means of intracortical feedback from a large group of other cortical cells. This extended linear model is called the *normalization* model. Response normalization was originally proposed by Robson (1988) to provide explanations for various failures of the linear model of simple cells responses. The model has been expanded and formalized by Heeger (1993, 1992a,b, 1991), Albrecht and Geisler (1991), and Carandini and Heeger (1994), who have shown that the normalization model is capable, in principle, of explaining a wide variety of empirical phenomena (see also Carandini *et al.*, 1997b; Nestares and Heeger, 1997; Tolhurst and Heeger, 1997a,b; DeAngelis *et al.*, 1992; Bonds, 1989). The overall motivation of the normalization model and its detailed synaptic mechanisms are surprisingly similar to Marr's (1970) general theory of cerebral neocortex, and to much of Grossberg's theoretical work on nonlinear neural networks (for review see Grossberg, 1988).

We propose (in Section 6) that normalization operates by shunting inhibition: cells inhibit each other by increasing each other's membrane conductance. This form of inhibition decreases the gain of the transformation of input currents into membrane potentials. The model explains response saturation and cross-orientation inhibition because increasing stimulus contrast or adding an orthogonal grating increases the activity of the network, resulting in increased conductance and decreased gain. Increasing the conductance also decreases the time constant of the membrane so the latency of the responses and the temporal filtering properties of the cells depend on the stimulus contrast. As a consequence, the model captures a number of temporal nonlinearities in the responses of V1 cells (Carandini *et al.*, 1997b; Albrecht, 1995; Hawken *et al.*, 1992; Reid *et al.*, 1992; Dean and Tolhurst, 1986; Holub and Morton-Gibson, 1981).

The normalization model is intentionally based on a simplified view of the cellular physiology. As a consequence, it makes strong quantitative predictions with few free parameters. We have recently tested some of these predictions by recording from simple cells in the primary visual cortex of paralyzed, anesthetized macaques, and presenting large sets of visual stimuli (Carandini *et al.*, 1997b; Carandini and Heeger, 1994). We derived closed-form equations for the model responses to such stimuli, and we found that these equations provide good fits to the neural responses. The results of these tests are presented in Section 7.

We conclude the chapter with a discussion of the biophysical plausibility of the normalization model (Section 8). Is shunting inhibition really the mechanism underlying gain control in the cortex? We tested the model using extracellular data, so we have no direct proof that the overall conductance grows with stimulus contrast. Large membrane conductance increases have been reported in some intracellular studies (Borg-Graham *et al.*, 1998; Carandini *et al.*, 1998), but not in others (Ahmed *et al* 1997; Ferster and Jagadeesh, 1992; Berman *et al.*, 1991). As a result, the true biophysical substrate of gain control is uncertain. The main advantage of shunting inhibition is that it constitutes the simplest possible way for the normalization pool to control both the gain and the dynamics of a cell's response. Until further data are available the model should be considered to lie between a phenomenological model and a true biophysical explanation.

2. The Linear Model of Simple Cells

2.1. Visual Stimuli in Space–Time

As a visual stimulus is projected on the retina it can be described by its intensity distribution $I(x, y, t)$, which varies in the two spatial dimensions, x, y, and in time t. This representation ignores the color of the stimulus and assumes monocular viewing, but is in all other respects complete. Consider, for example, a stimulus consisting of a dark vertical bar drifting from left to right on a white background. Figure 2A shows the bar at a particular instant in time. Figure 2B shows that as the bar drifts from left to right, it can be considered as a solid in the x–y–t space. Figure 2C shows a snapshot of the volume taken from above, a space–time (x–t) plot that ignores the y dimension.

Different velocities result in different orientations in space–time (Adelson and Bergen, 1985; van Santen and Sperling, 1985; Watson and Ahumada, 1985; Fahle and Poggio, 1981). For example, if the bar in Fig. 2A were going faster, its space–time (x–t) representation (Fig. 2C) would have been more tilted toward the horizontal. Had the bar been motionless, its x–t representation would have been vertical. Had the bar been going from right to left, the orientation of its x–t representation would have been opposite to the one in Fig. 2C.

2.2. Spatiotemporal Weighting Functions

A defining property of linearity is that of *superposition:* if L_1 is the response to stimulus I_1, and L_2 is the response to stimulus I_2, then the response of a linear system to the sum of the stimuli $I_1 + I_2$ is just the sum of the responses, $L_1 + L_2$. While the property of superposition may sound a little abstract, there is an equivalent statement that will make it concrete: simple cells are linear if and only if their responses are *a weighted sum of the light intensities* falling on their receptive fields.

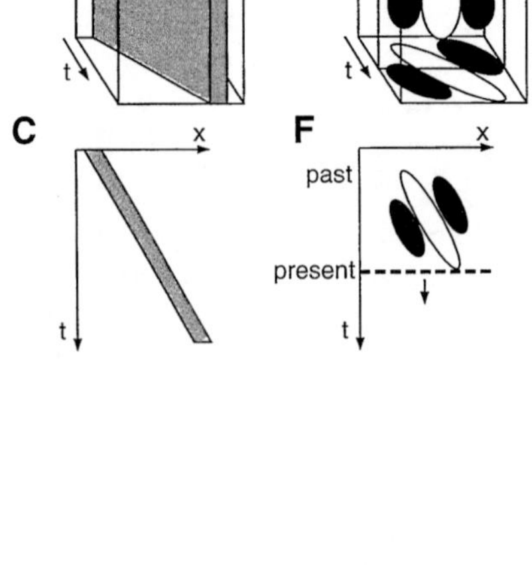

Figure 2. Space–time representations of a stimulus and of a receptive field. (A) A vertical bar translating to the right. (B) The space–time (x–y–t) volume of stimulus intensities corresponding to motion of the bar. (C) An x–t slice through the space–time volume. Orientation in the x–t slice represents motion of the bar in the x direction (D) x–y section of a spatiotemporal weighting function. Dark areas represent locations where the weighting function is negative, bright areas represent locations where it is positive. (E) The same x–y section together with a projection of the weighting function on the x–t plane. (F) The x–t projection, which ignores the y dimension. The receptive field travels in time t, from past to future. Panels A–C are based on an illustration by Adelson and Bergen (1985).

Figures 2D–2F show schematics of a spatiotemporal weighting function. Figure 2D shows a space–space (x–y) section of the weighting function. Figure 2F shows the space–time (x–t) projection of the weighting function, i.e., a snapshot of the weighting function taken from above. The relation between the two is shown in Fig. 2E. As we will see, real simple cell weighting functions do not look too different from this idealization (Fig. 6).

The response of a linear cell is simply obtained by weighting the stimulus intensity I at each location and time by the value of the cell's weighting function W at that location and at that time, and by summing the results:

$$L(t) = \int\int\int W(x, y, T) I(x, y, t - T) \; dx \; dy \; dT \tag{1}$$

The cell travels in time t from past to future (as we all do), while its retinal (x–y) location remains fixed. The weighting function is zero for any time that lies in the future, because the responses of the cell cannot depend on future events.

The spatiotemporal weighting function W of a linear cell determines its selectivity (e.g., for orientation or direction of motion). In particular, several researchers have pointed out that a linear cell is direction selective if and only if the subregions of its weighting function are tilted along an oblique axis in space–time (Adelson and Bergen, 1985; van Santen and Sperling, 1985; Watson and Ahumada, 1985; Fahle and Poggio, 1981).

Figure 3 illustrates how this selectivity arises, by showing the responses of a linear cell with a space–time oriented weighting function to a drifting grating. In Figs. 3A–3D the grating drifts from left to right, and the resulting space–time orientation is similar to the space–time orientation of the weighting function. This results in strong responses (Fig. 3A). In Figs. 3E–3H the grating drifts in the opposite direction, and the resulting space–time orientation is almost orthogonal to that of the weighting function. This results in small responses (Fig. 3E) because the weighting function averages out the variations in intensity present in the stimulus.

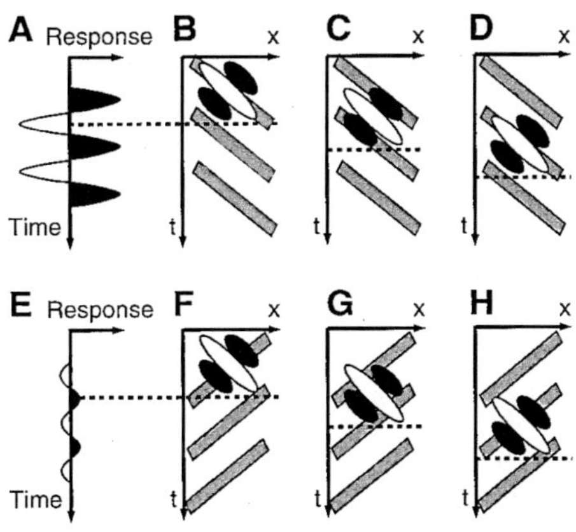

Figure 3. Direction selectivity in a linear cell. (A) Response to a grating drifting in the preferred direction. (B–D) Space–time representation of weighting function and stimulus at three instants in time. When the central excitatory region of the weighting function is aligned with a dark bar (B), the response is negative. When it is aligned with a bright bar, the response is positive (C), and so on. (E) Response to a grating drifting in the opposite direction. (F–H) Space–time representation of weighting function and stimulus at three instants in time. At any given time, each bar of the grating is covering both excitatory and inhibitory subregions of the weighting function. The resulting response is smaller than with the grating with the preferred direction. The filled areas in panels A and E show the portion of the responses that is visible after rectification.

If a linear cell's weighting function is not tilted along an oblique axis in space–time, then the cell will not have a preference for direction of motion. Figure 4, for example, shows the space–time projections of three different weighting functions. The weighting function in Fig. 4A prefers stationary objects, since it is vertical in the x–t plane. The weighting function in Fig. 4B prefers moving or flickering objects, but has no preference for the direction of motion. These two weighting functions cannot be direction selective because they are *space–time separable,* i.e., their weighting functions $W(x, y, t)$ can be expressed as the product of a function of space x, y and a function of time t. The weighting function in Fig. 4C is clearly tilted along an oblique axis in space–time, and is direction selective. Note that it is not space–time separable.

2.3. A Nonlinearity: Light Adaptation

In characterizing simple cells as spatiotemporal linear neurons, we have neglected an important nonlinearity: light adaptation in the retina (Shapley and Enroth-Cugell, 1984). We can, however, safely ignore light adaptation by restricting our choice of visual stimuli to luminance distributions $l(x, y, t)$ that modulate (transiently) about a fixed mean/background luminance m. Examples are drifting grating patterns and drifting or briefly flashed bars that are either brighter or darker than the mean. In these conditions the retina can be considered to be in a fixed state of adaptation, and its output is proportional to the "local contrast" $I(x, y, t) = [l(x, y, t) - m]/m$ of the stimulus (Shapley and Enroth-Cugell, 1984).

To avoid confusion we (improperly) use the term *intensity* to refer to the local contrast, and we reserve the term *contrast* for the maximum absolute value of the local contrast of a grating stimulus. The maximum contrast of a grating is 1, which is attained when the lowest intensity is zero and the highest intensity is twice the mean. Finally, we use the term *local energy* to denote the variance of the stimulus over local space and recent time, and within a band of spatiotemporal frequencies.

2.4. Another Nonlinearity: Rectification

Spatiotemporal linear weighting functions are intended to be models of the intracellular responses of simple cells. Most of the data discussed in this chapter,

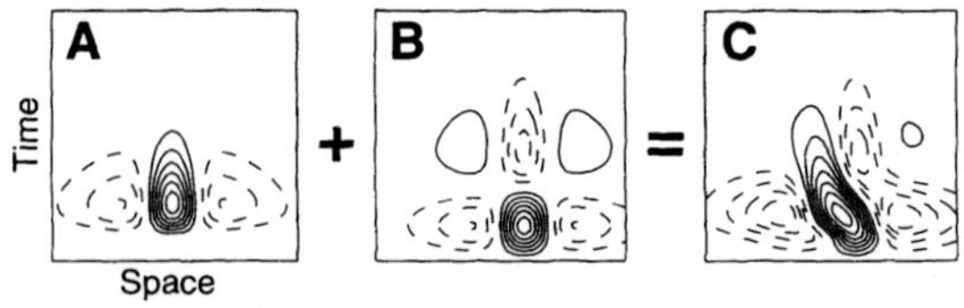

Figure 4. Contour plots of three spatiotemporal weighting functions, averaged across one spatial dimension. Continuous curves: positive contours. Dashed curves: negative contours. (A) A space–time separable weighting function, the product of a center–surround spatial weighting function and of a monophasic temporal weighting function. (B) Another space–time separable weighting function, spatially displaced with respect to that in panel A and with a biphasic temporal weighting function that is less delayed than the one in panel A. (C) Nonseparable weighting function obtained by summing those in panels A and B. The result is a space–time oriented weighting function resembling that of the direction-selective cell in Fig. 6.

however, consist of firing rates (recorded extracellularly). To model this, one is forced to consider also the transformation of membrane potentials into firing rates.

This transformation is bound to introduce a nonlinearity. The responses of a linear cell would assume both positive and negative values. Likewise, the membrane potential fluctuates above and below a cell's resting potential. Firing rates, on the other hand, are by definition positive. A linear cell with a high maintained firing rate could encode the positive and negative values by responding either more or less than the maintained rate. This is typical, for example, of retinal ganglion cells (Enroth-Cugell and Robson, 1966). Simple cells, however, have little maintained discharge. Since their negative responses cannot be encoded in their firing rate, simple cells cannot act truly linearly.

As we discuss in Section 2.4, the transformation of membrane potential into spike rate can be approximated by *rectification,* that is, by a function that is zero for membrane potentials below a threshold, and that grows linearly from there on. Figure 5 shows some examples of rectification. The three solid lines depict cases in which the firing threshold is 0, 5, and 10 mV away from the resting potential. Technically the first example of rectification, with a threshold at V_{rest}, is a case of "half-rectification." The other two are called "overrectification," since their threshold is above the resting potential. We use the term "rectification" to include all these cases.

Rectification is a static nonlinearity, that is, one that depends only on the instantaneous value of its input and not on its past history. Adding rectification after a spatiotemporal linear weighting function does not substantially alter the selectivity or other basic properties of the responses (Heeger, 1992a). In the following, whenever we refer to the *linear model* of simple cells, we tacitly assume it to be a *spatiotemporal linear weighting function followed by rectification,* as shown in Fig. 1A.

3. Some Linear Properties of Simple Cells

This section describes some experimental results that provide strong evidence in favor of the linear model of simple cells (see Heeger, 1993, 1992a, for a more thorough review). Most of the nonlinearities that are mentioned in this section are explained by the rectification stage that transforms intracellular responses into firing rates.

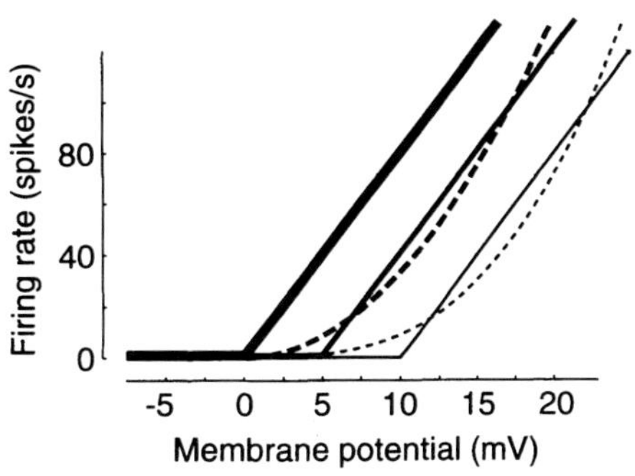

Figure 5. Three possible transformations of membrane potential into firing rate. For simplicity the resting potential is assigned the value $V_{rest} = 0$. The continuous lines represent *rectification* with threshold V_{thresh}. The thick, intermediate, and thin lines represent the cases in which the threshold is $V_{thresh} = 0, 5,$ and 10 mV. The dashed lines represent approximations to rectification. These approximations are useful to simplify the mathematics of the normalization model (Heeger, 1992a). They are power functions of the positive deviation from resting potential for different exponents n. The exponent n is 2 for the thicker dashed curve, 3 for the thinner dashed curve. After Carandini *et al.* (1997b).

3.1. Responses to Impulses

When Hubel and Wiesel (1962) first mapped the receptive fields of V1 cells, their stimuli were bright, flashing bars. Depending on whether a region responded positively to the onset or to the offset of a bright bar, they termed that region an ON or an OFF subregion. The linear model predicts the existence of these subregions (Heeger, 1992a; Emerson, 1988). This can be understood by considering the full space–time representation of a flashed bar, which is shown in Fig. 6A.

Since the bar does not change position in time, its space–time (x–t) projection is vertical. The top and bottom ends of the rectangle are the times at which the bar is turned on and off. The responses of a linear weighting function to such a stimulus are depicted in Figs. 6B and 6C. Figure 6B shows the case in which the bar is flashed in an OFF subregion. As the weighting function travels down in time, the first subregion of the weighting function that hits the stimulus is inhibitory; this gives a negative response. Later the stimulus overlaps both the excitatory and inhibitory subregions, so the response is about zero. Finally, when the stimulus overlaps only the excitatory subregion (right panel of Fig. 6B) the weighting function gives a positive response. In sum, the response is negative just after the bar is turned on and positive just after it is turned off. The opposite will happen when the bar is flashed in an ON subregion (Fig. 6C): the response is positive just after the bar is turned on and negative just after it is turned off. After the responses in the left panels of Figs. 6B and 6C are passed through a rectification stage that shows only their positive parts (shaded areas), they resemble the spike rate responses of a real simple cell.

Since Hubel and Wiesel's original work, the method for mapping a receptive field has been made more quantitative by having a computer show sequences of bars in many different positions and recording the correlation between firing rate and light intensity. Such a correlation depends on space and time: for each location x, y and *delay time* T, one can measure the correlation between the spike train $R(t)$

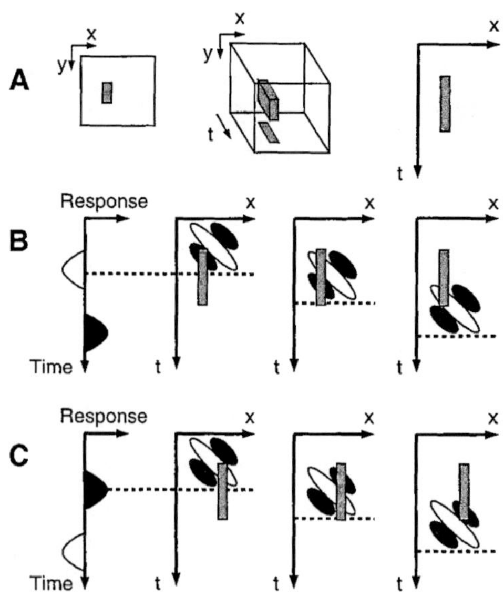

Figure 6. Responses of a linear cell to a flashed bar stimulus. (A) Spatiotemporal structure of a flashed bar stimulus. The x–t projection of a flashed bar is a vertical rectangle. (B) Response of a linear cell when the bar is flashed on an OFF subregion of the receptive field. The response is negative when the bar is turned on, positive when it is turned off. The left panel shows the responses as a function of time, before rectification. The other three panels represent receptive field and stimulus at three instants in time. (C) as in panel B, except that the bar is flashed on an ON subregion of its receptive field. The filled areas in the leftmost panel in parts B and C show the parts of the responses that would be visible after rectification.

and the sequence of stimulus intensities that occurred T seconds before at that x, y location, $I(x, y, t - T)$. The value of the correlation, which can be positive or negative, is taken as the strength of the weighting function at that position and time, $W(x, y, T)$. This method is called *reverse correlation,* and allows the measurement of full space–time (x–y–t) weighting functions (DeAngelis *et al.*, 1993a,b; Shapley *et al.*, 1991; McLean and Palmer, 1989, 1988; deBoer and Kuyper, 1968). Figure 7 shows the full space–time weighting function of a simple cell, measured with the reverse correlation technique. The four upper panels represent x–y snapshots of the weighting function measured at different times T in the past. A large number of snapshots like these are stacked to build a full space–time weighting function, whose x–t structure (averaged over the y axis) is shown at the bottom of the figure.

The reverse correlation method can be applied to any visual cell, linear or nonlinear, and it will always give a result, i.e., a space-time weighting function. For a linear cell, however, such a weighting function could then be used to predict the cell's responses to any visual stimulus [using Equation (1)]. This property of linear systems can be used to test whether simple cells are linear. For example, one can ask whether direction selectivity in simple cells is fully explained by an underlying linear stage.

The linear model predicts that simple cells are direction selective only if their weighting functions are oriented in space–time, and thus nonseparable (Fig. 4). This prediction was tested by McLean and Palmer (1989, 1988), Shapley *et al.*

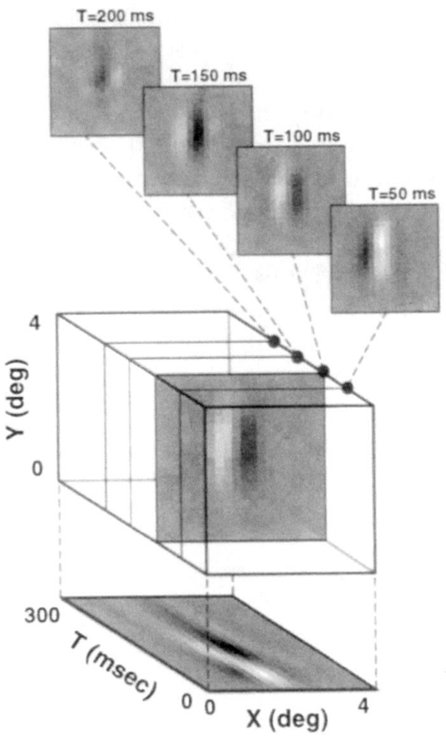

Figure 7. The full space–time receptive field of a simple cell, as obtained with the reverse correlation method. The four upper panels represent x–y snapshots of the receptive field measured at different times T in the past. Gray levels indicate the correlation between the appearance of a bar and the firing rate T milliseconds later. Zero correlation is indicated by mid-gray. Lighter grays indicate points of positive correlation with the appearance of a bright bar. Darker grays indicate points of positive correlation with the appearance of a dark bar. A large number of snapshots like these are stacked to build a full space–time receptive field, whose space–time projection is shown at the bottom. This previously unpublished figure is courtesy of Greg DeAngelis. Cell is part of sample published in DeAngelis *et al.* (1993a).

(1991), Emerson and Citron (1992), and DeAngelis *et al.* (1993b). Their findings are mostly consistent with the linear model. They found simple cells with weighting functions tilted along an oblique axis (inseparable) in space–time, like the one in Fig. 6. These cells were all direction selective, and the preferred direction of motion was always correctly predicted from the orientation of the weighting function. For example, the cell of Fig. 6 was highly direction selective and preferred stimuli moving from right to left in the x–y plane. A number of simple cells were found to have space–time separable weighting functions, like the ones depicted in Figs. 4A and 4B. Consistent with the linear model, most of the cells were not direction selective.*

3.2. Responses to Drifting Gratings

The stimulus of choice for linear systems analysis of the visual system is a stimulus whose luminance varies sinusoidally in space and time, the *sine grating.* There are many advantages to using sine gratings (reviewed in Enroth-Cugell and Robson, 1984), the most important being that linear systems are guaranteed to respond to sinusoidal modulation with a sinusoid. For example, had the modulation in luminance of the grating in Fig. 3 been sinusoidal, the responses in Figs. 3A and 3E would have been perfect sinusoids. The deviation of the responses from pure sinusoids can provide a quantitative measure of nonlinearity (Hochstein and Shapley, 1976). Sine gratings were first used to study the neurophysiology of the visual system by Enroth-Cugell and Robson (1966), who demonstrated the linearity of cat retinal X ganglion cells. One of their tests of linearity involved comparing the responses to gratings with the responses to luminance edges. The logic of this experiment is straightforward: since an edge is composed of the sum of a number of gratings, the responses of a linear cell to an edge would be predictable from its response to gratings.

A similar experimental paradigm was applied to the study of simple cells by Movshon *et al.* (1978b). They measured the sensitivity of the cells to drifting gratings of different spatial frequencies, and the sensitivity of different receptive field regions to flashing bars. Since a grating is composed of the sum of a number of bars, the response of a linear cell to a grating is predictable (via Fourier transform) from its response to bars. Likewise, since a bar can be thought of as the sum of a number of gratings, the response to a bar is predictable (via inverse Fourier transform) from the response to gratings. Movshon *et al.* found good agreement between the weighting function predicted by inverse Fourier transform of the grating sensitivity and the weighting function obtained from the flashing bar data (Fig. 8).This supports the linear model of simple cell responses.

Many other studies have compared grating responses to impulse responses (DeAngelis *et al.*, 1993b; Shapley *et al.*, 1991; Tadmor and Tolhurst, 1989; Jones and Palmer, 1987a,b; Jones *et al.*, 1987; Field and Tolhurst, 1986; Dean and Tolhurst,

*Some cells, however, are direction selective even though they have space–time separable weighting functions (McLean *et al.*, 1994; Emerson and Citron, 1992; McLean and Palmer, 1989). The behavior of these cells cannot be accounted for by models like those advocated in this chapter, in which direction selectivity is due to an underlying spatiotemporal linear stage.

1983; Kulikowski and Bishop, 1981a,b; Glezer *et al.*, 1980; Andrews and Pollen, 1979; Maffei *et al.*, 1979). In many cases, the inverse transform of the response to gratings gives a weighting function with additional sidebands beyond those measured directly. In addition, the measured response to gratings is often more narrowly tuned for spatial frequency than predicted from the Fourier transform of the response to impulses. This discrepancy between the grating responses and the impulse responses can be explained by overrectification, which conceals the impulse responses of the weaker receptive field regions, so it is consistent with the linear model (Tolhurst and Heeger, 1997b; Heeger, 1992a; Tadmor and Tolhurst, 1989).

Some of the above-mentioned studies, on the other hand, unveiled a serious failure of linearity: a discrepancy between the predicted and actual *sizes* of the responses. For example, when Movshon *et al.* (1978b) compared the observed and predicted weighting functions (Fig. 8), they did so only up to an arbitrary amplitude scaling factor. According to the linear model, this scaling factor should not be

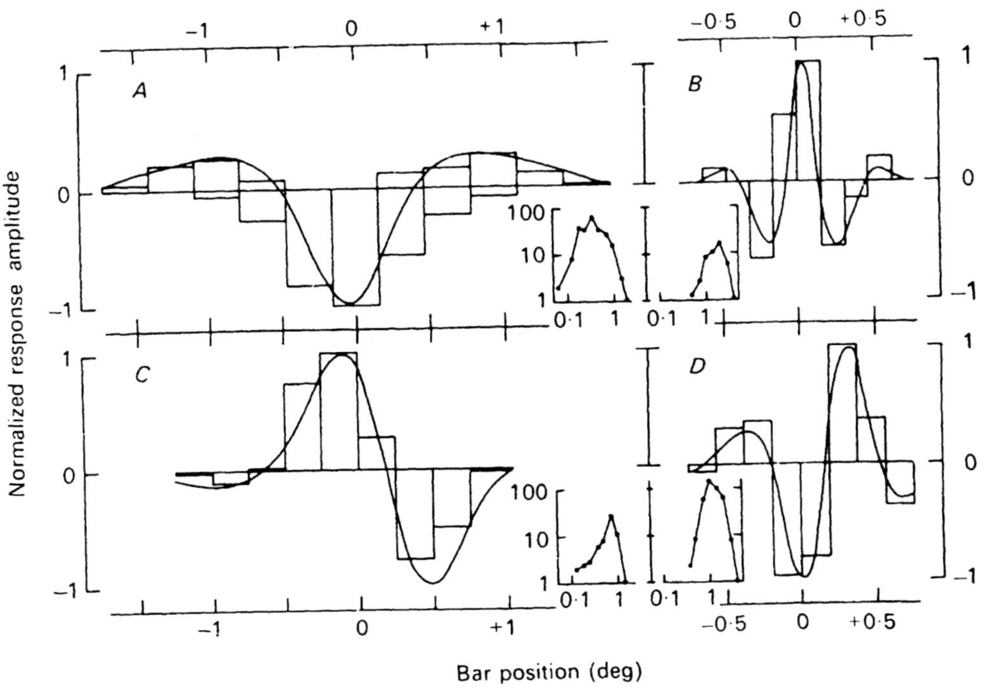

Figure 8. Linearity of spatial summation in four cat V1 simple cells. Spatial weighting functions as measured with flashing bars (histograms) and as predicted by inverse Fourier transformation of the spatial frequency tuning curves (*continuous curves*). For a linear cell the two would be identical. Plots show one spatial dimension (e.g., x), and collapse all information about the other two dimensions (y and t). Both the observed and predicted weighting functions were independently rescaled. Positive values in each weighting function represent incremental responses to the introduction of a bright bar; negative values represent incremental responses to the introduction of a dark bar. Insets: The spatial frequency tuning curves used to compute each predicted weighting function. The abscissa of these insets is spatial frequency (in cycles/degree) and the ordinate is contrast sensitivity, the inverse of the threshold contrast value for each spatial frequency. From Movshon *et al.* (1978b), with permission of the publisher.

necessary. We will see later that the normalization model predicts this failure of linearity: it predicts that the cell's gain is different when it is stimulated with flashed bars from when it is stimulated with drifting gratings.

3.3. Responses to Contrast-Modulated Gratings

A contrast-modulated grating is a standing sine grating whose intensity is modulated sinusoidally over time. Simple cell responses to drifting and contrast-modulated gratings are quite similar to rectified sinusoids (see, e.g., Fig. 11A). This is obviously consistent with the linear model: the spatiotemporal linear weighting function responds with a sinusoid and rectification hides everything that is below threshold.

A number of researchers (e.g., Reid *et al.*, 1991, 1987; Tolhurst and Dean, 1991; Kulikowski and Bishop, 1981b; Movshon *et al.*, 1978b; Maffei and Fiorentini, 1973) measured the responses of simple cells while varying the spatial phase of contrast-modulated gratings. Because these responses can be reasonably fit by a sinusoid, they can be described by just two numbers, the amplitude and phase of the sinusoid. A useful way to display both response amplitude and response phase at the same time is by a *polar plot* like the one in Fig. 9. Every point in the polar plot corresponds to a sinusoid whose amplitude is given by the distance from the origin and whose phase is given by the angle with the horizontal axis.

The linear model predicts that as the spatial phase of the contrast-modulated grating varies between 0° and 180°, the responses should describe a "wasp-waisted" ellipse in the polar plot (Movshon *et al.*, 1978b). In particular, for a linear cell a polar plot of the responses would be elliptical in shape. Overrectification distorts the ellipse, producing a wasp waist: if the neuron has to reach a certain level of excitation before any activity is seen, there will be a disproportionate decrease in small responses (Tolhurst and Heeger, 1997a; DeAngelis *et al.*, 1993b; Heeger, 1993; Albrecht and Geisler, 1991). The physiological results are in line with this prediction. Figure 9, for example, shows the responses of a simple cell to a contrast-modulated grating positioned at eight different phases over the receptive field, spanning the range from 0° to 180°. The raw data (filled symbols) describe a wasp-waisted ellipse since the amplitudes near the minor axes are smaller than they should be to fit an ellipse. When the distortion introduced by rectification is removed (in this case assuming a resting "firing rate" of −8 spikes/sec), the data fall on an ellipse (open symbols).

Because of superposition, the responses of a linear cell to contrast-modulated gratings would be easily predictable from the responses to drifting gratings and vice versa. Several researchers tested whether this was the case for simple cells. Ferster and collaborators performed intracellular *in vivo* recordings and found that the membrane potential responses were consistent with the output of a spatiotemporal linear weighting function (Jagadeesh *et al.*, 1997, 1993). Other researchers performed extracellular recordings (Albrecht and Geisler, 1991; Reid *et al.*, 1991, 1987; Tolhurst and Dean, 1991). These studies are generally consistent with the linear model in that a cell's preferred direction of motion for drifting gratings can be correctly predicted from its responses to contrast-modulated gratings. They uncovered, however, two nonlinearities in simple cell responses. First, the linear prediction from contrast-modulated grating responses underestimates the degree of di-

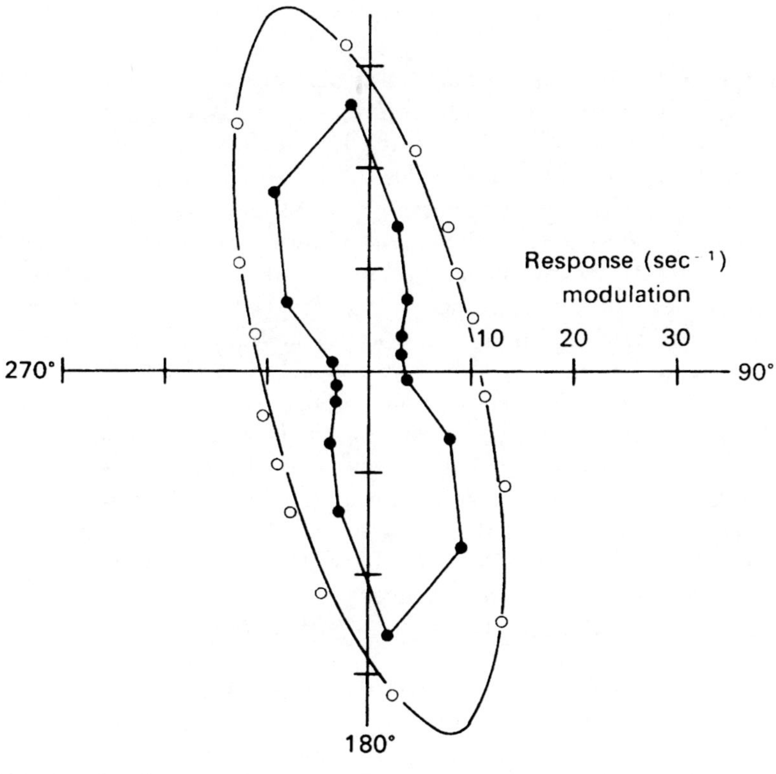

Figure 9. Linearity of spatiotemporal summation in a cat V1 simple cell. The polar plot shows the cell's responses to standing gratings whose contrasts were modulated sinusoidally in time at different spatial phases. The amplitude of the first harmonic sinusoid of each response is represented radially, while the angular coordinate indicates the temporal phase of each response. Units of sec^{-1} correspond to spikes/sec. The filled symbols represent the unaltered data from the experiment; the open symbols and the ellipse fitted to them represent the same data corrected for a resting "firing rate" of −8 spikes/sec. From Movshon *et al.* (1978b), with permission of the publisher.

rectional selectivity observed with drifting gratings. Second, the linear prediction overestimates the responses to gratings drifting in the nonpreferred direction. Albrecht and Geisler (1991) and Heeger (1993, 1991) showed that the first phenomenon can be explained by the rectification stage (which acts as an expansive nonlinearity), so it is consistent with the linear model of simple cells. The second phenomenon instead is not consistent with the linear model, but it can in most cases be explained by a gain control mechanism like that postulated by the normalization model. The normalization model predicts that the cells are less responsive in the presence of drifting gratings than in the presence of contrast-modulated gratings of equal contrast. This difference in gain can be shown to yield the observed discrepancy in the predicted and actual responses to gratings drifting in the nonpreferred direction (Tolhurst and Heeger, 1997a; Heeger, 1993).

3.4. Responses to Compound Stimuli

According to the linear model of simple cells, knowing a cell's responses to gratings would make it possible to predict the responses to any visual stimulus. This is because any visual stimulus can be expressed as the sum of many different gratings.

An important test of this prediction was performed by DeValois *et al.* (1979). They measured the orientation tuning of cat simple cells using individual gratings as well as checkerboard stimuli. The motivation for their use of checkerboards is very interesting. At the end of the 1970s the issue of whether simple cells were better modeled as linear weighting functions or as all-or-none edge detectors was the object of heated debate (e.g., Schumer and Movshon, 1984; Marr, 1982; Maffei and Fiorentini, 1973). DeValois *et al.* (1979) reasoned that the two models made different predictions of a cell's response to checkerboards. In checkerboards the strongest sine grating components are oriented along the diagonals, whereas the sharp edges are oriented along the rows and columns. According to the linear model, the cells will respond best when one of the diagonals is oriented in the cell's preferred orientation for gratings. According to the edge-detector model, on the other hand, a cell will respond best when either the rows or the columns are oriented in the cell's preferred orientation for gratings. The results of DeValois *et al.* (1979) were consistent with the linear model, and falsified the edge-detector model. The responses of the cells could be predicted by having knowledge of the location and orientation of the main sine gratings that compose the checkerboard. The precise location and orientation of the sharp edges was not relevant in predicting the cells' responses.

A similar approach was followed by other researchers (e.g., Gizzi *et al.*, 1990; Pollen *et al.*, 1988; DeValois *et al.*, 1982b; Pollen and Ronner, 1982; Maffei *et al.*, 1979), who tested linearity by comparing responses to sine gratings with responses to sums of sine gratings of different spatial frequencies or orientations. All of these results are qualitatively explained by the linear model (Heeger, 1992a).

The quantitative predictions of the linear model, however, are not always correct, and once again the discrepancy points to the existence of a gain control mechanism. For example, Gizzi *et al.* (1990) found that simple cell responses to plaids composed of two sine gratings with different orientations were on average only two thirds of the linear predictions based on the responses to the individual gratings. As we will see, the normalization model explains this behavior because it predicts that the gain of the cells is lower in the presence of plaids than in the presence of single gratings.

4. Biophysics of the Linear Model

We have been considering the linear model as little more than a mathematical abstraction. This Section describes how the model might be implemented physiologically.

4.1. Linearity of the Lateral Geniculate Nucleus

Unless some complicated linearization mechanism is invoked, simple cells can only be as linear as the inputs they get. Because the input to the visual cortex is constituted by the activity of lateral geniculate nucleus (LGN) cells, we must begin our task of modeling simple cell linearity by assuming that *the responses of LGN neurons are linear functions of the stimulus intensity distribution.*

This assumption is a better approximation in the monkey than in the cat. In the cat there are no LGN cells that are perfectly linear: the X cells are spatially and temporally linear, but they are subject to a (retinal) contrast gain control mechanism, which violates linearity (Victor, 1987; Enroth-Cugell *et al.*, 1983; Enroth-Cugell and Robson, 1966). The Y cells are extremely nonlinear (Victor, 1988; Troy, 1983; Hochstein and Shapley, 1976), but may not be contributing an important input to the primary visual cortex (Ferster, 1990a,b).

In the monkey, on the other hand, there is a geniculocortical channel, the P pathway, which is substantially linear. The other channel, the M pathway, is instead quite nonlinear, and its nonlinearity might be due to a gain control mechanism. The substantial linearity of the P pathway and the nonlinearity of the M pathway have been observed in the responses of retinal ganglion cells (Benardete and Kaplan, 1997; Lee *et al.*, 1994; Benardete *et al.*, 1992), and are reflected in the properties of LGN cells (Movshon *et al.*, 1994; Carandini et al., 1993; Derrington and Lennie, 1984; Sherman *et al.*, 1984). Even though P cells constitute around 90% of the monkey LGN (Dreher *et al.*, 1976), many simple cells also receive M inputs (Malpeli *et al.*, 1981). Indeed, while the two streams are segregated in layer 4C (Blasdel and Lund, 1983; Hendrickson *et al.*, 1978; Hubel and Wiesel, 1972), they are not segregated at all in the upper layers (Nealey and Maunsell, 1994; Yoshioka *et al.*, 1994; Lahica *et al.*, 1992). In particular, for those neurons that do receive M input, the first 7–10 msec of activation are due exclusively to the M signal (Maunsell and Gibson, 1992).

Besides the fact that monkey simple cells may receive some M input, there is another phenomenon that makes our assumption of linearity of the geniculate input an imperfect approximation: at high contrasts the responses of LGN cells show evidence of rectification. Resting firing rates in the LGN have been reported to be around 18 spikes/sec in monkey, and around 6 and 16 spikes/sec, respectively, in cat X and Y LGN cells (Kaplan *et al.*, 1987). An analysis of the 136 monkey LGN cells recorded by Sherman *et al.* (1984) reveals that their average resting firing rate was only around 7 spikes/sec. On average the modulation in spike rate due to a 50% contrast grating was twice as large as the resting firing rate for P cells (median: 2.1), and three times as large for M cells (median: 3.3). Rectification in the input will be ignored in the following, but should be considered in more detailed models.

4.2. Building Simple Cell Receptive Fields

In their 1962 paper, Hubel and Wiesel hypothesized that the receptive fields of simple cells are the result of an orderly arrangement of LGN inputs. Geniculate cells have center-surround receptive fields. When stimulated with a spot of light in their center they respond either to the onset (ON-center cells) or to the offset (OFF-center cells) of the stimulus. According to the scheme proposed by Hubel and Wiesel, an ON subregion of a simple cell receptive field would result from the sum of an aligned series of LGN ON-center inputs. Similarly, an OFF subregion would result from the sum of an aligned series of OFF-center inputs. This arrangement has been recently confirmed by Reid and Alonso (1995), who recorded simultaneously from simple cells and from LGN cells.

If some of the LGN inputs reach a simple cell before some others, the resulting receptive field can also display direction selectivity. In the cat LGN, for instance, there is evidence for the existence of two classes of cells, "lagged" and "nonlagged," whose responses have different latencies (Saul and Humphrey, 1992, 1990; Mastronarde, 1987). Figure 4 illustrates how the outputs of these two classes of cells could be summed to yield a direction-selective simple cell. The weighting functions in Figs. 4A and 4B were drawn to resemble those of idealized lagged and nonlagged LGN cells. These functions are space–time separable and have the same center—surround spatial structure, but slightly different spatial positions. They have different temporal structures, which result in different response latencies. Figure 4C depicts the weighting function obtained by summing the two LGN weighting functions. This weighting function is oriented in space–time, so it is direction selective. It is similar to that of a direction-selective V1 simple cell. In practice, V1 simple cell weighting functions would be the result of many, not just two, LGN inputs. In general, a nonseparable (direction-selective) weighting function can be obtained by simple addition of separable (not direction-selective) weighting functions (Adelson and Bergen, 1985; Watson and Ahumada, 1985; Fahle and Poggio, 1981), as long as the latter differ in their temporal structure.

4.3. Push–Pull Arrangement of Inputs

The linear combination of LGN inputs involves both sums and subtractions. Indeed, besides excitatory responses, simple cell receptive fields also exhibit inhibitory responses elicited by the onset of a light on an OFF region or the offset of a light on an ON region. Hubel and Wiesel (1962) pointed out that these inhibitory responses could originate either from the withdrawal of excitation or from actual inhibition.

There is now evidence that both mechanisms are at work. Ferster (1986) measured the selectivity of the inhibitory postsynaptic potentials (IPSPs) and found it to be identical to that of the excitatory postsynaptic potentials (EPSPs). Moreover, he found that in ON regions a light increase results in EPSPs and a light decrease results in IPSPs (Ferster, 1988). In other words, EPSPs and IPSPs are spatially overlapping. The inhibitory responses result from both withdrawal of excitation and actual inhibition, just as the excitatory responses result from both withdrawal of inhibition and actual excitation.

It is thus plausible that the inputs to a simple cell are arranged in *push–pull* fashion, i.e., they come from pairs of cells with opposite-signed receptive fields, one of which provides excitation and the other inhibition. For example, an ON subregion would be the result of excitatory ON-center inputs as well as of inhibitory OFF-center inputs.

This complementary arrangement of excitation and inhibition is also consistent with extracellular recording studies in cat V1 (Tolhurst and Dean, 1990, 1987; Heggelund, 1986; Glezer *et al.*, 1982; Heggelund, 1981; Palmer and Davis, 1981; Glezer *et al.*, 1980). A similar push–pull arrangement might also be used by retinal ganglion cells to integrate bipolar signals (Gaudiano, 1992).

For reasons of simplicity we assume that both the excitation and the inhibition are contributed by feedforward connections. In this we differ from a number of recent models that consider intracortical feedback crucial in sharpening the selectivity conferred by the inputs from the lateral geniculate nucleus (see also Carandini and Ringach, 1997; Ben-Yishai *et al.*, 1995; Douglas *et al.*, 1995; Somers *et al.*, 1995; Suarez *et al.*, 1995). While the feedforward view is supported by recent evidence (Ferster *et al.*, 1996; Reid and Alonso, 1995), the linear model should not necessarily be identified with a feedforward arrangement inputs. A linear receptive field could, in principle, be constructed with pure feedforward connections, pure feedback connections, or a combination of feedforward and feedback.

4.4. Linearity of Excitation and Inhibition

We now make the linking assumption that *simple cell synaptic conductances depend linearly on the responses of LGN neurons.* Strictly speaking, this assumption is simplistic for at least two reasons.

First, synaptic transmission can be well approximated by a linear transformation only in the absence of synaptic plasticity. In these conditions each presynaptic spike results in a stereotyped postsynaptic conductance increase (Koch and Poggio, 1987; Jack *et al.*, 1975), and synaptic transmission amounts to a linear transformation (whose impulse response is given by the shape of the stereotyped postsynaptic conductance increase). On the other hand, synaptic plasticity has been recently reported to be pervasive, and to have nonlinear effects (Abbott *et al.*, 1997; Markram and Tsodyks, 1996).

Second, while there is evidence for direct geniculocortical excitation (Reid and Alonso, 1995; Ferster and Lindstrom, 1983), the anatomical evidence for direct geniculocortical inhibition is contradictory (Einstein *et al.*, 1987; Garey and Powell, 1971), and the physiological evidence is overwhelmingly against it (Reid and Alonso, 1995; Ferster and Lindstrom, 1983; Tanaka, 1983; Toyama *et al.*, 1977a,b; Watanabe *et al.*, 1966). Most inhibitory inputs from the LGN to simple cells are disynaptic (Ferster and Lindstrom, 1983), so the linearity of inhibition would likely require an inhibitory cortical interneuron that performs a linear integration of LGN inputs and encodes them linearly into firing rate.

4.5. Simplified Model of a Cortical Cell

We adopt a very simplified model of a cortical cell (Fig. 10): a single-compartment circuit with only passive conductances. In particular we consider a leak conductance g_{leak} and two synaptic conductances, one excitatory (g_e) and one inhibitory (g_i). The membrane potential of a model cell then obeys

$$-C\frac{dV}{dt} = g_e(V - V_e) + g_i(V - V_i) + g_{\text{leak}}(V - V_{\text{leak}}) \tag{2}$$

where C is the membrane capacitance, and V_{leak}, V_e, and V_i are the equilibrium potentials of the leak, excitatory, and inhibitory channels.

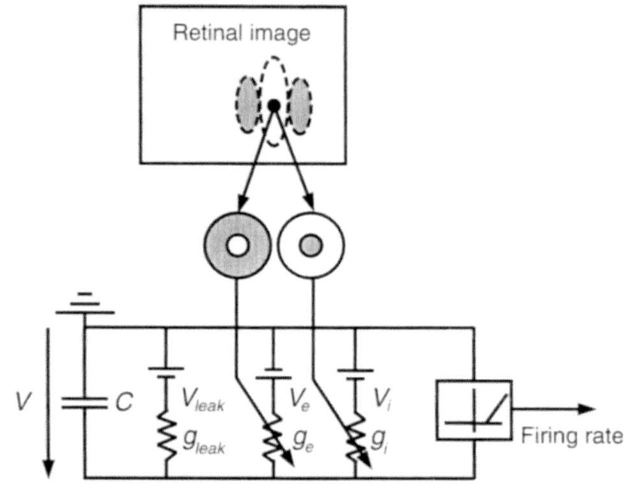

Figure 10. Simplified model of a cortical cell, and possible biophysical implementation of the linear model. The cell membrane is modeled as a single compartment with passive properties and two classes of synaptic inputs, excitatory and inhibitory. In the central excitatory subregion of the receptive field excitation is provided by ON-center cells and inhibition by OFF-center cells with superimposed receptive fields. The flanking inhibitory subregions are obtained by the opposite arrangement of excitation and inhibition (not shown). This *push–pull* arrangement of excitation and inhibition ensures the linearity of the membrane potential *V*. The membrane potential is encoded into firing rate by a rectifier. See text for explanation of symbols. After Carandini and Heeger (1994).

This view of the cellular physiology deliberately ignores many known aspects of neuronal biophysics such as voltage- and calcium-dependent channels, the possible nonlinear interactions between inputs caused by the dendritic structure, and the effects of electrotonic distance from the soma (Koch and Segev, 1989). Our model of a cortical cells is, however, in many respects a reasonable approximation. For example, Fig. 11A shows the responses of an intracellularly recorded cortical neuron to sinusoidal current injection at different temporal frequencies. The membrane potential responses are dominated by their sinusoidal (*first harmonic*) component. This means that if the generation of spikes is ignored, the membrane of cortical neurons can be reasonably modeled by passive conductances, which endow it with a linear behavior. In particular, when the first harmonic responses are plotted against the temporal frequency of the stimulus (filled circles in Fig. 11B), they are well fit by the predictions of a single-compartment model of the cell (dashed curve in Fig. 11B).

4.6. Linear Integration of the Synaptic Inputs

The push–pull arrangement of the LGN inputs to a simple cell can lead (through a balance of excitation and inhibition) to a perfectly linear integration of the synaptic conductances by the cell membrane (Carandini and Heeger, 1994).

For the sake of simplicity, consider the steady-state behavior of the membrane ($dV/dt = 0$). At steady state, Equation (2) can be rewritten as

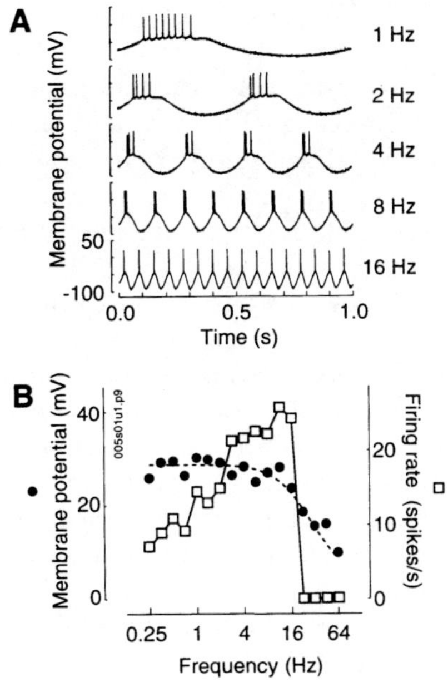

Figure 11. Encoding of input current in a visual cortical cell *in vitro*. Responses of an intracellularly recorded regular-spiking neuron in a slice of guinea pig cortex to sinusoidal current injection (0.8 nA). (A) Time course of the responses for five different frequencies of stimulation (1, 2, 4, 8, and 16 Hz). The traces are dominated by their first harmonic. This means that apart from the presence of the spikes, the membrane is acting linearly. (B) Temporal frequency tuning of spike rate and membrane potential. Filled symbols: Amplitude of the first harmonic of the membrane potential obtained by fitting a sinusoid to the raw membrane potential traces in panel A. Scale is on left axis. The dashed line is the prediction of a single compartment with only passive conductances. Open symbols: Amplitude of the first harmonic of the firing rate, obtained by fitting a sinusoid to the spike train. Scale is on right axis. Data from Carandini *et al.* (1996).

$$V = \frac{g_e V_e + g_i V_i + g_{\text{leak}} V_{\text{leak}}}{g_e + g_i + g_{\text{leak}}} \tag{3}$$

The push–pull arrangement (Section 4.3) guarantees that every increase in excitation will correspond to a decrease in inhibition, and vice versa. In particular, we assume that g_e and g_i are balanced so that the total conductance of the cell is constant:

$$g_e(t) + g_i(t) + g_{\text{leak}} = g_0 \tag{4}$$

Equation (3) can then be rewritten as $V = (g_e V_e + g_i V_i + g_{\text{leak}} V_{\text{leak}}) / g_0$, which is a linear function of g_e and g_i. In words, *the membrane potential V is a linear function of the synaptic conductances.*

If our model of the cell membrane as a single compartment is a good approximation, the exact balance of excitation and inhibition expressed in Equation (4) is an essential condition for the linear integration of the synaptic conductances. If, on the other hand, there is substantial electrotonic distance between synaptic sites on the membrane, there are other conditions in which linear integration of the synaptic inputs is possible. For example, Blomfield (1974), showed that if inhibition is located on the soma and excitation is electrotonically remote from it, there is a range of synaptic activations in which the membrane potential will be approximately a linear combination of the excitatory and inhibitory synaptic conductances. This approach would not require the strict balance of excitation and inhibition [Equation (4)], but it would require additional assumptions about the dendritic structure of the cell, the sites of the inputs, and the range of the synaptic conductances.

4.7. Spike Rate Encoding

If simple cells integrate their synaptic inputs linearly, if those inputs depend linearly on LGN activity, and if LGN activity is a linear function of the stimulus intensity distribution, then simple cells will integrate the stimulus intensity distribution linearly. This section discusses the final, nonlinear stage of the model, which is responsible for the encoding of input-driven membrane potential responses into spike trains.

Many characteristics of firing rate encoding are consistent with the view that the firing rate responses are a rectified copy of the membrane potential responses. An example of this can be seen in Fig. 11A. The spike responses closely mirror the membrane potential responses, and there is a threshold below which no spikes are generated. Once above threshold, the firing rate grows with the amplitude of the membrane potential modulation. There is in fact a large literature pointing to a linear or bilinear relation between injected current and firing rate, once the current is above a threshold level (Stafstrom *et al.*, 1984, and references therein).

There is, however, an additional experimental result that is not consistent with the view of firing rate encoding as rectification: the spike rate encoder has notable dynamic properties (Carandini *et al.*, 1996). For example, cortical neurons typically exhibit *spike frequency adaptation,* meaning that the firing rate response to steady depolarization decreases with time (Stafstrom *et al.*, 1984). Dynamic properties of spike encoding are also evident in Fig. 11. Figure 11B illustrates the temporal frequency tuning of the first harmonic of the spike train taken from records like the ones in Fig. 11A. It is clear that spike rate encoding is not at all independent of the temporal frequency of the stimulus, as would be the case for rectification. The middle temporal frequencies are transmitted much better than the low temporal frequencies, and the very high temporal frequencies are completely cut off.

Rectification cannot account for these behaviors because it is a *static* nonlinearity, i.e., it depends only on the instantaneous value of its argument. Strictly speaking, then, rectification is incorrect, because it would predict that the spike encoding properties would not depend on the past history of stimulation. Hence, we are forced to adopt a slightly more complicated model of spike rate encoding. In particular, we model the spike rate encoder as a band-pass filter followed by rectification.*

The behavior of this spike encoder model is actually quite simple when incorporated into the linear model of simple cells. In the full model, the spike rate encoder comes after a (synaptic)linear spatiotemporal weighting function. Since a chain of linear systems is itself a linear system, we can treat the band-pass (spike encoder) linear filter and the (synaptic) spatiotemporal linear weighting function as a single linear system. The weighting function of this final system is partly due to the synaptic inputs and partly to the band-pass properties of the firing rate encoder.

*An even better model for the spike rate encoder of visual cortical cells is given by the opposite arrangement, in which rectification is *followed* by a band-pass linear filter (Carandini *et al.*, 1996). This arrangement, however, would make the model very hard to deal with analytically. The two arrangements give identical first harmonic responses to sinusoidal stimulation at a fixed temporal frequency.

Having described the linear model of simple cells, and having discussed its numerous successes and its possible biophysical implementation, it is now the time to discuss its failures. We have already encountered a number of occasions in which the linear model fails to yield precise quantitative predictions. Indeed, the behavior of simple cells is in many ways nonlinear (for review see Heeger, 1993, 1992b). This section describes some of these nonlinearities, which will be discussed more quantitatively once we have introduced the normalization model.

5.1. Contrast Responses

Presented with a change in contrast, a linear neuron would scale its response by the same amount. The responses of simple cell, instead, are often not proportional to stimulus contrast. An example of this is illustrated in the central column of Fig. 12A, which shows the spike histograms of a simple cell in response to a drifting grating. The different rows correspond to different contrasts. As the contrast doubles from 50% to 100%, the response does not double. Instead it grows very little. This phenomenon is known as *response saturation* (Sclar *et al.*, 1990; Albrecht and Hamilton, 1982; Dean, 1981; Maffei and Fiorentini, 1973). Cells can even exhibit "supersaturation," in which increasing the contrast of the stimulus reduces the amplitude of the responses (Bonds, 1991; Li and Creutzfeldt, 1984).

Response saturation is not due to the high firing rates. This can also be seen in Fig. 12. The three columns in Fig. 12A show the responses of a cell to three gratings of different spatial frequency. Even though the left column and the right column stimuli elicit fewer spikes than the central column stimulus, there clearly is response saturation. This phenomenon thus depends on the contrast of the stimulus per se,

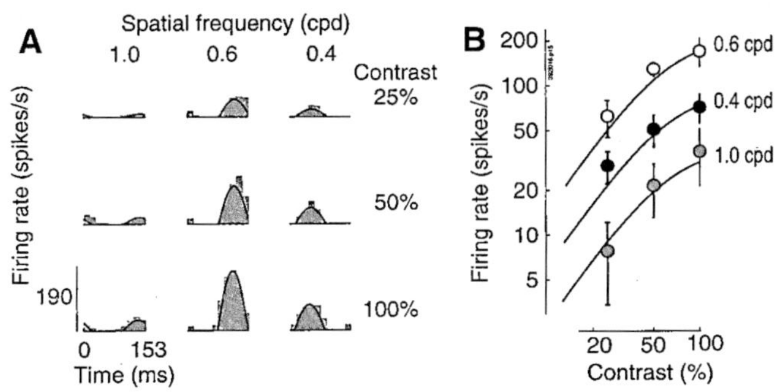

Figure 12. Responses of a monkey V1 simple cell to a drifting sine grating for different spatial frequencies and different contrasts. The curves are fits of the normalization model. (A) Spike histograms of one period of the responses, averaged over many presentations. The three columns show the responses to drifting gratings with a spatial frequency of 1, 0.6, and 0.4 cycles/degree (cpd). Each row corresponds to one of three different contrasts: 25%, 50%, and 100%. (B) Amplitude of the responses as a function of contrast. The ordinate represents the amplitude of the first harmonic of the responses in panel A. Error bars indicate the standard error of the mean (N = 3). After Carandini *et al.* (1997b).

not on the amplitude of the responses it elicits in the cell. This property of the contrast responses can be observed more precisely in Fig. 12B, which shows the amplitude of the responses shown in Fig. 12A, as a function of contrast. In spite of the amplitude saturation, the three contrast responses are vertical shifts of each other. Since the vertical axis is logarithmic, a vertical shift means that *the ratio of the responses to any two different spatial frequencies is constant, irrespective of the stimulus contrast.*

Another way to express this property is to say that the shape of the spatial frequency tuning curve is independent of the contrast at which it is measured. Changing the contrast of the stimuli just scales the tuning curve. This has been observed for both spatial frequency tuning and orientation tuning (Skottun *et al.*, 1987; Li and Creutzfeldt, 1984; Albrecht and Hamilton, 1982; Sclar and Freeman, 1982; Movshon *et al.*, 1978a). An example of the contrast invariance of the orientation tuning is shown in Fig. 13.

The contrast independence of the tuning curve shapes would be easy to explain if the contrast responses of simple cells were linear. The responses of a linear cell to two stimuli S_1 and S_2 with the same contrast c could be written as $cL(S_1)$ and $cL(S_2)$. Their ratio would be $L(S_1)/L(S_2)$, independent of the contrast c. We have seen, however, that the contrast responses of real simple cells are nonlinear, since they often saturate at high contrasts. The contrast independence of the tuning curves is thus by no means a trivial property.

5.2. Nonspecific Suppression

The response to a preferred stimulus can be suppressed by superimposing an additional stimulus that would not elicit any response when presented alone. This phenomenon is a violation of superposition, a defining property of linearity. We call it *nonspecific suppression,* as it has been found to be independent of direction of motion, largely independent of orientation, and broadly tuned for spatial and temporal frequency (Carandini *et al.*, 1997b; DeAngelis *et al.*, 1992; Geisler and

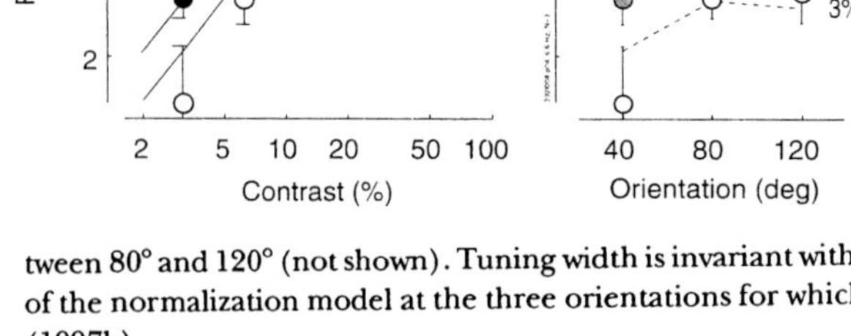

Figure 13. Responses of a monkey V1 simple cell to a drifting sine grating for different stimulus orientations and contrasts. The curves are fits of the normalization model. Error bars indicate the standard error of the mean ($N = 3$). (A) Contrast responses for two different stimulus orientations. Changing the orientation of a grating shifts the contrast responses up and down on a logarithmic scale (B) Effect of contrast on the orientation tuning. Data for 40° and 80° are the same as those in panel A. The responses peaked at orientation between 80° and 120° (not shown). Tuning width is invariant with contrast. Dashed lines join the predictions of the normalization model at the three orientations for which data were acquired. After Carandini *et al.* (1997b).

Albrecht, 1992; S. B. Nelson, 1991; Bonds, 1989; Gulyas *et al.*, 1987; De Valois *et al.*, 1985, 1982b; Kaji and Kawabata, 1985; Li and Creutzfeldt, 1984; Morrone *et al.*, 1982; Hammond and MacKay, 1981; Dean *et al.*, 1980; Bishop *et al.*, 1973). After some debate, there is now consensus that cross-orientation inhibition can be driven dichoptically (with one grating in each eye), although monoptic suppression (with both gratings in the same eye) is typically stronger than dichoptic suppression. (Walker *et al.*, 1996; Sengpiel *et al.*, 1995; Sengpiel and Blakemore, 1994; DeAngelis *et al.*, 1992; Freeman *et al.*, 1987; Ohzawa and Freeman, 1986; Ferster, 1981) Suppression is absent in monkey P LGN and cat LGN (Movshon *et al.*, 1994; Bonds, 1989), so its origins are most likely cortical. In particular, the temporal properties of suppression are consistent with the view that it originates from complex cells or from a large pool of simple cells (Bonds, 1989; Morrone *et al.*, 1982).

Figure 14 shows an example of cross-orientation inhibition. The stimulus was a plaid made of two gratings. One (the "test") drifted in the cell's preferred direction and evoked a large response when presented on its own. The other grating (the "mask") drifted at right angles with the test grating, and was ineffective in driving the cell. Its presence, however, clearly suppressed the responses. For example, when the mask contrast was 50% the cell responded only when the test grating had high contrast.

From Fig. 14B one can see that the presence of the mask shifts the contrast response to the right (Bonds, 1989). This corresponds to a scaling of contrast (Heeger, 1992b). We will see later that the contrast responses shift to the right only when the cell is *completely unresponsive to the mask.* If the cell gives even a minimal response to the mask, the effect is more complicated than just a rightward shift.

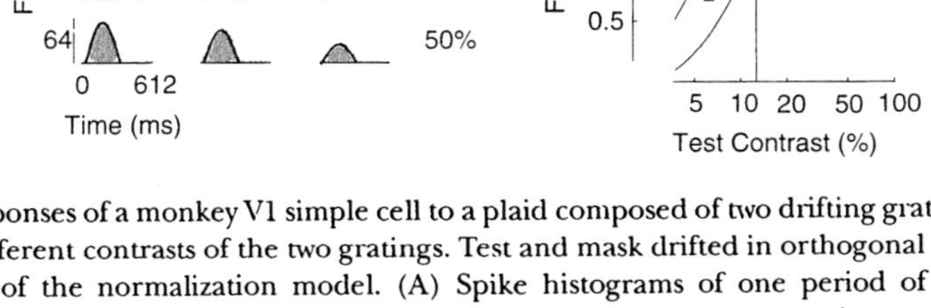

Figure 14. Responses of a monkey V1 simple cell to a plaid composed of two drifting gratings ("test" and "mask"), for different contrasts of the two gratings. Test and mask drifted in orthogonal directions. The curves are fits of the normalization model. (A) Spike histograms of one period of the responses, averaged over many presentations. Rows correspond to a fixed test contrast, columns to a fixed mask contrast. The mask does not elicit any overt response when presented alone (top row) and strongly inhibits the responses to the test (second and third columns). (B) Amplitude of the responses as a function of contrast. The ordinate represents the amplitude of the first harmonic of the responses in panel A. Error bars indicate the standard error of the mean ($N = 3$). The white, gray, and black circles refer to mask contrasts of 6%, 25%, and 50%, respectively. After Carandini *et al.* (1997b).

5.3. Temporal Nonlinearities

Simple cells display prominent temporal nonlinearities. Figure 15A provides a good example of this. As the stimulus contrast increases, the responses occur earlier in time. This is called *phase advance* (Carandini *et al.*, 1997b; Albrecht, 1995; Carandini and Heeger, 1994; Dean and Tolhurst, 1986). It is a nonlinearity because for a linear cell scaling the input would just scale the output, not change its timing. Phase advance is not entirely cortical in origin, but has a strong cortical component. Phase advance in cat LGN (measured with 2- to 5-Hz stimuli) is on average less than 20 msec (Carandini *et al.*, 1993), whereas phase advance in cat V1 (measured with 2-Hz stimuli) is on average around 47 msec (Dean and Tolhurst, 1986). In the monkey LGN, phase advance is present in M cells, but completely absent in P cells (Sherman *et al.*, 1984).

A second temporal nonlinearity of simple cells was uncovered by Reid *et al.* (1992). They measured the responses of cat simple cells to eight different stimuli and to the compound stimulus obtained by summing the eight stimuli. They found that the responses to the compound stimulus occur earlier in time than the linear prediction obtained from the responses to the individual stimuli. This decrease in "integration time" is quite prominent, in the range of 5–60 msec, and there is evidence that its origin is cortical (Reid *et al.*, 1992).

A third temporal nonlinearity of simple cell responses is given by the contrast dependence of their temporal frequency tuning (Holub and Morton-Gibson,

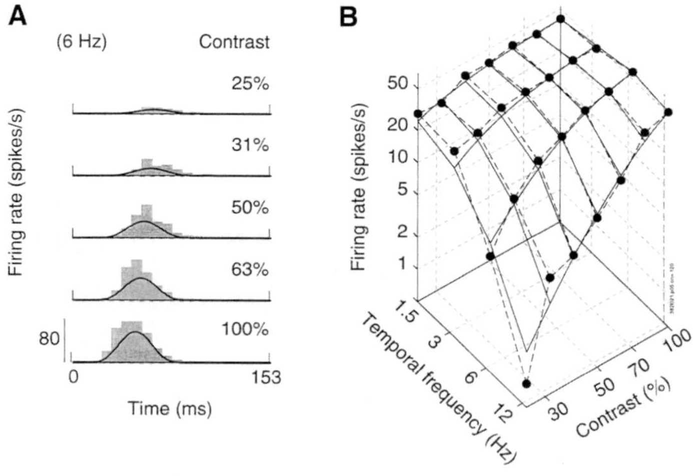

Figure 15. Responses of a monkey V1 simple cell to drifting sine gratings of different contrasts and temporal frequencies. The curves are fits of the normalization model. (A) Spike histograms of one period of the responses, averaged over many presentations. Each panel corresponds to a different contrast of the stimulus. Note the prominent phase advance with increasing contrast. (B) Amplitude of the responses as a function of contrast and temporal frequency. The ordinate represents the amplitude of the first harmonic responses such as those in panel A. Data points are joined by dashed lines, model predictions are solid curves. The temporal frequency tuning is strongly dependent on contrast: at low contrasts the high frequencies are more attenuated than at high contrasts. After Carandini *et al.* (1997b).

1981). According to the linear model, increasing the contrast should just scale the responses, with no effect on their temporal frequency tuning. Instead, as illustrated in Figure 15B, increasing stimulus contrast increases the responsivity of V1 cells to the high temporal frequencies (Hawken *et al.*, 1992). The origins of this nonlinear behavior are partially subcortical, since it was observed in the cat retina (Shapley and Victor, 1978) and in the monkey M LGN (Benardete *et al.*, 1992). There is evidence, however, that in the monkey this behavior is much stronger in V1 than in the LGN. Measurements by Hawken *et al.* (1992; M. J. Hawken, personal communication) indicate that on average the high-cutoff frequency of V1 cells changes from around 10 Hz at 8–16% contrast to around 30 Hz at 64% contrast. By contrast, the average change in high-cutoff frequency of LGN cells is negligible.

6. The Normalization Model of Simple Cells

We have now seen in many cases in which the linear model is an inadequate description of simple cells. Given its numerous successes, however, it would be unwise to dispense with the linear model altogether. The linear model can be extended by endowing it with a mechanism that controls the gain (sensitivity to input) and the time course of the cell responses (Carandini *et al.*, 1997b; Carandini and Heeger, 1994; Heeger, 1993, 1992b, 1991; Albrecht and Geisler, 1991; Robson, 1988). This mechanism ensures that the gain decreases when the contrast of a stimulus is increased, or when another stimulus is superimposed to it.

This is done in the *normalization* model (Heeger, 1992b) depicted in Fig. 1B. First comes a spatiotemporal linear weighting function. This is followed by a normalization stage, where each cell's linear response is divided by a quantity proportional to the pooled activity of a large number of other cells (the *normalization pool*). The normalization stage is followed by rectification.

Normalization is a nonlinear operation; one input (a cell's underlying linear response) is divided by another input (the activity of the normalization pool). The normalization pool is assumed to include cells tuned to all orientations, spatial frequencies, and temporal frequencies, so that its overall response is uniform across these parameters. The effect of normalization is that the response of each cell is rescaled with respect to local stimulus energy. The name "normalization" comes from the assumption that the gain of every cell in the pool is rescaled by the same amount ("normalized").

The normalization model explains all the nonlinearities described in the previous section, while retaining the main features of the linear model. According to the normalization model, a cell's selectivity is attributed to summation (the linear stage) and its nonlinear behavior is attributed to division (the normalization stage). For example, the model explains cross-orientation inhibition because a given cell is suppressed by many other cells including those with perpendicular orientation tunings. It explains response saturation because the divisive suppression increases with stimulus contrast.

The rest of this section is devoted to the description of a biophysical implementation of the normalization model. Once the details of the model are laid out, one can derive closed-form equations for the responses of the model to some types of visual stimulus. Section 7 will show how these closed-form equations perform in fitting the experimental data.

6.1. Shunting Inhibition

We hypothesize that normalization (division) acts by controlling the overall conductance of the cell membrane (Carandini *et al.*, 1997b; Carandini and Heeger, 1994). Figure 16 shows the biophysical structure of the model. The circuit depicted is identical to the linear model shown in Fig. 10, with the addition of a *shunting* synaptic conductance, g_{shunt}. The shunting conductance has the property that its equilibrium potential V_{shunt} is the same as the resting potential of the cell V_{rest} (Coombs *et al.*, 1955; Fatt and Katz, 1953). For ease of notation we pick this value as the origin of the membrane potential measurements:

$$V_{\text{shunt}} = V_{\text{rest}} = 0 \tag{5}$$

The membrane potenital in the circuit shown in Fig. 16 obeys the following equation:

$$-C\frac{dV}{dt} = g_e(V - V_e) + g_i(V - V_i) + g_{\text{shunt}}(V - V_{\text{shunt}}) + g_{\text{leak}}(V - V_{\text{leak}}) = gV - I_d \tag{6}$$

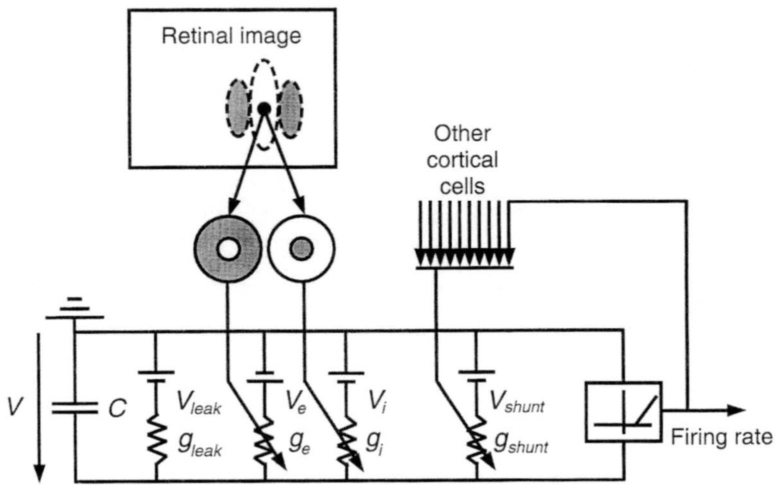

Figure 16. A possible biophysical implementation of the normalization model. The diagram is identical to the one for the linear model (Fig. 10), except for the presence of a shunting conductance g_{shunt}. The equilibrium potential for the shunt is $V_{\text{shunt}} = V_{\text{rest}}$. In the model the shunting conductance grows with the activity of a large number of cortical cells, the normalization pool, and is assumed to be the same for all the neurons in the pool. See Sections 6.1 and 6.2 for details. After Carandini and Heeger (1994).

where g is the *total conductance* and I_d is what we call the *driving current**:

$$g \equiv g_e + g_i + g_{\text{shunt}} + g_{\text{leak}}$$
$$I_d \equiv g_e V_e + g_i V_i + g_{\text{shunt}} V_{\text{shunt}} + g_{\text{leak}} V_{\text{leak}}$$

Since g_e and g_i are in push–pull arrangement [Equation (4) in Section 4.4], and the equilibrium potential of the shunt is equal to the resting potential of the cell [Equation (5)], the above equations can be rewritten as

$$g \equiv g_0 + g_{\text{shunt}}$$
$$I_d \equiv g_e V_e + g_i V_i + g_{\text{leak}} V_{\text{leak}}$$

As a result, *current and conductance are decoupled:* the driving current I_d is a linear function of the excitatory and inhibitory inputs g_e and g_i and does not depend on the shunt conductance g_{shunt}. The overall conductance g of the cell does not depend on the excitatory and inhibitory inputs g_e and g_i and is completely controlled by the shunt conductance g_{shunt}.

6.2. Role of the Membrane Conductance

We now make a crucial hypothesis: that *the shunt conductance g_{shunt} grows with the activity of the normalization pool.* The exact relation that we hypothesize between the two is depicted in Fig. 17A. This relation is entirely ad hoc: it was chosen so that it would allow us to solve the model mathematically, and it is not based on experimental data. Its principal consequence is that the conductance of a model cell will depend on the energy of the visual stimuli. Increasing the energy of a stimulus, e.g., by increasing its contrast, will increase the activity of the cells belonging to the normalization pool, and thus cause an increase in cell conductance.

The membrane conductance g in turn determines the cell's *gain,* which is the relation between input I_d and output V. At steady state the gain is $V/I_d = 1/g$, inversely proportional to the conductance. The conductance also has an effect on the time course of the response: the membrane capacitance takes time to charge and discharge, and this time is proportional to the membrane *time constant* $\tau = C/g$, which is also inversely proportional to the conductance. Figure 17B illustrates these concepts. It shows the responses of the membrane to a current step for three values of the conductance g. If the conductance is small, the response is slow and the cell has high gain (that is, the voltage response to a given current is high). If the conductance g is large (the membrane is leaky), the cell has small gain and it is fast in charging and discharging its membrane capacitance.

To summarize, increasing the stimulus energy increases the conductance of model cells. This has two effects: (1) it provides divisive inhibition, by reducing the cell's gain, and (2) it shortens the latency of the responses, by reducing the cell's time constant.

*The driving current depends on the cell's synaptic inputs, but it is independent of V, the cell's membrane potential. The driving current could only be measured by voltage clamping the cell. It is not the actual synaptic current, which depends on the membrane potential V.

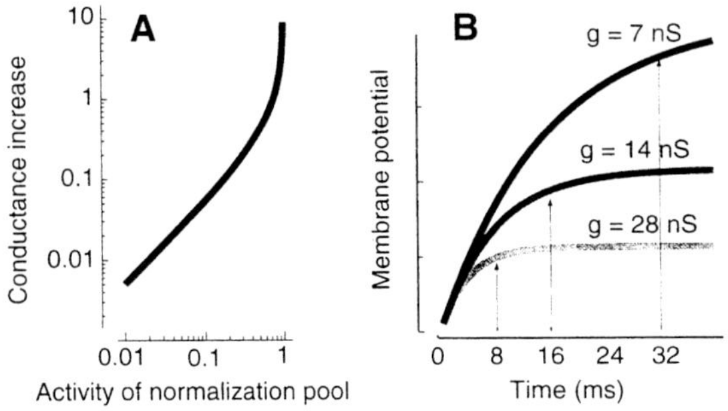

Figure 17. (A) Dependence of the shunt conductance on the activity of the normalization pool, as postulated by the normalization model. The ordinate represents the conductance increase g_{shunt}/g_0, the abscissa represents the overall firing rate of the normalization pool ΣR. The model postulates that the two are related by $g_{\text{shunt}}/g_0 = 1/\sqrt{1 - k \cdot \Sigma R} - 1$, where k is a constant. This function is entirely ad hoc and does not have any experimental support. It was chosen so that the model would be mathematically tractable and yield the desired equations. (B) Effect of the membrane conductance on the size and time course of the membrane potential responses. The three curves show the membrane potential responses of a model cell to a current step for three different values of the total conductance g. The current steps up at a time 0. The arrows point to twice the time constant of the cell. Increasing the conductance of the cell reduces both the gain and the latency. After Carandini *et al.* (1997b).

7. Testing the Normalization Model

The variables in the model depend on each other in a circular way: (1) each cell's firing rate depends on its membrane potential (Fig. 5); (2) each cell's membrane potential depends on its driving current and on its conductance [Equation (6)]; (3) each cell's conductance depends on the firing rates of the cells in the normalization pool (Fig. 17A). The model is a nonlinear neural network (Grossberg, 1988), and is in general quite complicated because both the driving current and the conductance vary over time. As a consequence, to predict its responses one is often forced to resort to computer simulations (Heeger, 1993, 1992b).

For drifting sine grating and plaid stimuli, however, we were able to solve the model and derive closed-form equations for its responses (Carandini *et al.*, 1997b; Carandini and Heeger, 1994). In fact the model was designed with these equations in mind. The advantage of closed-form equations is that they can be fitted to the data.

7.1. Fitting the Grating Responses

We derived two equations for the first harmonic responses to drifting gratings, one for response amplitude and one for response phase. These equations are detailed in (Carandini *et al.*, (1997b), which also contains a sketch of their derivation.

Here we concentrate on the expression for the response amplitude, which is

$$\text{amplitude}(R) \propto \left[\text{amplitude}(L) \frac{c}{\sqrt{\sigma(f)^2 + c^2}} \right]^n \tag{7}$$

where c is the contrast of the grating and f is its temporal frequency. The role of the quantities L, $\sigma(f)$, and n is easy to understand if one keeps in mind the structure of the model (Fig. 1B). $L(t)$ is the output of the cell's linear weighting function ([Equation (1)] when the grating has unit contrast. It is a sinusoid; here we are concerned only with its amplitude, which for a stimulus of contrast c is amplitude$(L)\,c$. The normalization stage divides that by a quantity that depends on the activity of a large number of neurons. In Carandini *et al.*, (1997b) we show that for drifting grating stimuli this quantity is $\sqrt{\sigma(f)^2 + c^2}$, where the function $\sigma(f)$ is related to the low-pass properties of the cell membrane. It grows with f, the temporal frequency of the stimulus. Finally, the exponent n is a constant and is related to the rectification stage that encodes the membrane potentials into firing rates.

Equation (7) is very similar to one that was empirically found to fit V1 contrast responses (Sclar *et al.*, 1990; Albrecht and Hamilton, 1982). In addition to stimulus contrast, the model predicts the dependence of the responses on all the other stimulus parameters: spatial frequency, temporal frequency, and orientation. To test these predictions we used the model to fit large data sets obtained by stimulating simple cells with drifting gratings of a variety of contrasts, spatial frequencies, temporal frequencies, and orientations (Carandini *et al.*, 1997b; Carandini and Heeger, 1994). Some of our results are shown in Figs. 12, 13, and 15. We found that the model provides good fits to the data, and accounts quantitatively for the linear and nonlinear behaviors described in this chapter.

Figure 12 illustrates the dependence of the responses of a simple cell on contrast and spatial frequency. The fits of the model capture the shape of the contrast responses and the fact that changing the stimulus spatial frequency simply scales these responses. This means that when the amplitude is plotted on a logarithmic scale, as in Fig. 12B, changes in spatial frequency shift the contrast responses vertically, without affecting their shape. Similar results are obtained when the gratings are varied in orientation instead of spatial frequency, as shown, for example, in Fig. 13A (Section 5.1).

The reason for this behavior can be understood by considering the expression for the response amplitude predicted by the model [Equation (7)]. The expression can be seen as the product of two factors, $[\text{amplitude}(L)]^n$ and $[c/\sqrt{\sigma(f)^2 + c^2}]^n$. The first factor depends on L, the response of the cell's linear receptive field to the grating at unit contrast. The second factor depends only on the contrast c and on the temporal frequency f of the grating. For a fixed temporal frequency the shape of the contrast responses is controlled by the second factor; hence, varying stimulus orientation or spatial frequency leads to a vertical shift of the contrast responses (Figs. 12B and 13A). Likewise, the shape of the orientation and spatial frequency tuning curves is controlled by the first factor; hence, varying stimulus contrast leads to a vertical shift of the orientation and spatial frequency tuning curves (Fig. 13B).

The normalization model also predicts the temporal nonlinearities of simple cell responses (Section 5.3). Figure 13A shows that the model captures the phase advance behavior observed with increasing stimulus contrast. Figure 13B shows that the model correctly predicts the dependence of the temporal frequency tuning on contrast: As the contrast increases, the cell is more responsive to the high temporal frequencies. The reason for the contrast dependence of the response phase and of the temporal frequency tuning can be understood in terms of the basic properties of the normalization model (Fig. 17). In model cells, increasing stimulus contrast increases the membrane conductance and thus decreases its time constant. A decreased time constant leads to a shorter response latency, thus explaining phase advance. Decreasing the time constant also enables the membrane potential to better follow the high temporal frequencies. This explains the fact that at high contrast the cell responds to higher temporal frequencies than at low contrast. This last argument can be made more quantitative by examining Equation (7). As a consequence of the low-pass properties of the membrane, the quantity $\sigma(f)$ grows with the temporal frequency f of the stimulus (Carandini *et al.*, 1997b). At low contrasts c, σ has a strong effect, considerably scaling down the responses. At high contrasts, when $c \gg \sigma(f)$, the effect is much weaker, so the high-frequency responses are relatively enhanced.

7.2. Fitting the Plaid Responses

Similar approximate equations can be derived for the amplitude and phase of the first harmonic response to plaid stimuli (Carandini *et al.*, 1997b). In particular, the expression for response amplitude is

$$\text{amplitude}(R) \propto \left[\frac{\text{amplitude}(c_1 L_1(t) + c_2 L_2(t))}{\sqrt{\sigma(f)^2 + c_1^2 + c_2^2}}\right]^n \tag{8}$$

where c_1 and c_2 are the contrasts of the two gratings, $L_1(t)$ and $L_2(t)$ are the sinusoidal responses of the linear weighting function to the individual gratings at unit contrast, and the remaining symbols have the same meaning as in the expression for the response to individual gratings [Equation (7)]. Since the spatiotemporal receptive field is linear, its response to the plaid is just a linear combination of its responses to the individual gratings, $c_1L_1(t) + c_2L_2(t)$. The normalization stage divides that by a quantity that depends on the activity of a large number of neurons. For plaids composed of two gratings this quantity is $\sqrt{\sigma(f)^2 + c_1^2 + c_2^2}$ (Carandini *et al.*, 1997b). Finally, the rectification stage is responsible for the exponent n (Fig. 5).

We tested the predictions of the model by recording the responses of simple cells to stimuli composed to two gratings whose contrasts c_1 and c_2 assumed a variety of different values (Carandini *et al.*, 1997b). The two gratings always had the same temporal frequency, but could differ in orientation and/or in spatial frequency.

Figure 14 shows an example of our results. In this case we chose the gratings to that one of them (the "test") would strongly drive the cell, while the other (the "mask") would not elicit any response when presented alone. The figure shows the responses of the cell for five different test contrasts and three different mask contrasts. The cell was actually tested with a wider variety of conditions, and the predictions of the model [Equation (8) for response amplitude, and an equation for response phase described in Carandini *et al.* (1997b)] were fit to all the responses at once. When we discussed nonspecific suppression (Section 5.2), we used this figure as an example of the rightward shift of the contrast responses that results from masking. It is clear from the quality of the fits that the normalization model captures this behavior. Indeed, the effect of a mask on the responses of a model cell can be seen directly in Equation (8). Let, for example, grating 2 be the "mask." If, as in Fig. 14, the mask alone does not elicit any response ($L_2 \approx 0$), then the mask contrast c_2 appears only in the denominator. The effect of an increase of c_2 in the denominator is to shift the contrast response to the right (Heeger, 1992b).

Things become more complicated when both gratings that compose the plaid are able to drive the cell (even minimally) when presented alone. In these conditions each component of the plaid acts both as a "test" and as a "mask." An example of this is shown in Fig. 18. In this experiment, the "mask" was not as effective as the "test" in driving the cell, but it did elicit some spikes when presented alone. Figure 18B shows that depending on the contrast of the stimulus, increasing the contrast of the mask can either enhance or suppress the responses. This behavior is predicted by the normalization model (continuous curves), as the contrasts of the two gratings c_1 and c_2 appear both in the numerator and in the denominator of Equation (8). Increasing one of the two can result either in an enhancement or in a reduction in the response, depending on the amplitude and phases of L_1 and L_2, the responses of the linear receptive field to the individual gratings, and on the size of the term $\sigma(f)$ in the denominator.

In summary, masking will cause the rightward shift in the contrast responses shown in Fig. 14 only if the mask elicits negligible responses when presented alone. The normalization model predicts that in the general case masking will deform the contrast responses in more complicated ways, like that shown in Fig. 18. Some researchers have indeed reported that suppression shifts the contrast responses downward on a logarithmic scale (Morrone *et al.*, 1982; Dean *et al.*, 1980), which is in apparent contradiction with the rightward shift reported by others (Section 5.2). The normalization model might explain this discrepancy: if the mask elicits even minimal (subthreshold) responses when presented alone, it may very well cause a downward shift in the contrast responses. The importance of "cross-talk" between different components of a stimulus has been previously recognized (Bonds, 1992;

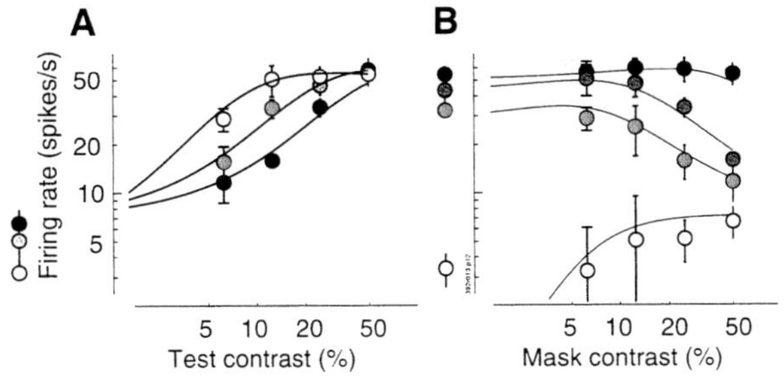

Figure 18. Responses of monkey V1 simple cell to a plaid composed of two drifting gratings ("test" and "mask"), for different contrasts of the two gratings. Test and mask drifted in orthogonal directions. The curves are fits of the normalization model. Error bars indicate the standard error of the mean ($N = 3$). (A) Dependence of the response amplitude on the test contrast, for three different values of the mask contrast (white: 6%, gray: 25%, black: 50%). Data points to the left of the ordinate correspond to test contrasts of zero. They indicate that the mask alone was able to elicit some responses. In these conditions the normalization model correctly predicts that increasing mask contrast does not simply shift the contrast responses to the right, as was the case in Fig. 14, but it also changes their shape. (B) Dependence of the response amplitude on the mask contrast, for four different values of the test contrast (white: 1%; light gray: 6%; dark gray: 12%; black: 50%). At low test contrasts the mask resulted in visual responses around 5 spikes/sec. At intermediate test contrasts the mask exerted a suppressive effect. At the highest test contrasts the mask did not much affect the responses. The prediction of the model for the data point in the lower left of panel B is off by 2 spikes/sec. After Carandini *et al.* (1997b).

Bauman and Bonds, 1991), and the normalization model may provide a quantitative framework to understand these interactions.

8. Biophysical Plausibility of the Normalization Model

The two models that we have described in this chapter call for the existence of two different kinds of inhibitory inputs to V1 simple cells. The linear model postulates hyperpolarizing inhibition with the same selective tuning properties as the excitation. This inhibition is needed to ensure the linearity of the simple cell responses (Section 4). The normalization model in addition postulates shunting inhibition originating from a large number of cortical units. This nonselective shunting inhibition is responsible for controlling the cell's gain (Section 6). While the existence of the first (hyperpolarizing) kind of inhibition in the visual cortex is widely accepted, the evidence in favor of the second (shunting) kind of inhibition is for now contradictory. This section discusses the effects of blocking inhibition in V1, the plausibility of nonselective inhibition, and the evidence in favor of membrane conductance increases in the visual cortex.

8.1. Effects of Blocking Inhibition

The importance of intracortical inhibition in the normal operation of the primary visual cortex was established by the work of Sillito and coworkers (Sillito, 1984, 1977, 1975; Sillito *et al.*, 1980). These authors blocked inhibition by iontophoresing a α-aminobutyric acid (GABA) antagonist on a region of cortex in the vicinity of the recording electrode. Blocking GABA strongly enhanced the responses of V1 cells and dramatically broadened their orientation tuning curves. Based on these results, it was concluded that the orientation tuning of a cell is substantially sharpened by cross-orientation inhibition. This conclusion, however, appears to be premature: blocking inhibition simultaneously in a large number of cortical cells greatly increases the responsiveness of all those cells, not just the one being recorded. This may bring that region of cortex close to epileptogenesis (Chagnac-Amitai and Connors, 1989). It is not entirely clear, therefore, how Sillito's results should be interpreted in the context of the tuning of single V1 cells.

A more focused blockade of inhibition was performed by Crook and Eysel (1992). These authors found that in cat area 18 a block of inhibition from cross-orientation sites broadens the orientation tuning, while a block of inhibition from iso-orientation sites increases response magnitude. This suggests that under normal conditions the cells receive inhibitory inputs from sites tuned to a wide variety of orientations. This nonspecific inhibition could be mediated by basket cells, which are GABAergic, and are considered to connect to cells tuned for all orientations (Kisvárday and Eysel, 1993; Kisvárday *et al.*, 1993).

S. Nelson *et al.*, (1994) have recently been able to perform an extremely focused blockade of inhibition. They recorded intracellularly from a neuron while blocking inhibition *to only that one cell.* Blocking inhibition under these circumstances does not lead to a substantial change in orientation tuning. The normalization model is generally consistent with this result. According to the normalization model, blocking inhibition in a simple cell would destroy both the underlying linearity and the gain control (normalization), but it would not have much of an effect on selectivity. This is because neither the shunting (normalization) inhibition nor the hyperpolarizing (linear) inhibition are critical for selectivity.

Removing the shunting inhibition corresponds to setting $g_{\text{shunt}} = 0$ in the model. This would turn a cell described by the normalization model into one described by the linear model. It would result in the complete loss of the gain control mechanism. But the cell's selectivity would be largely unaffected because the selectivity is set up by the underlying linear spatiotemporal weighting function.

Removing the hyperpolarizing inhibition corresponds to setting $g_i = 0$. This would interfere with the linearity of the cell: in our model the exact balance of excitation and inhibition [Equation (4)] is an essential condition for the linear integration of the synaptic inputs. On the other hand, removing the hyperpolarizing inhibition would have little influence on the cell's tuning curves. Note that the excitatory inputs alone would still provide differential responses for stimuli in the ON and OFF subregions of a simple cell's receptive field; flashing a light in an ON subregion would evoke increased excitation and flashing a light in an OFF subregion would cause a withdrawal of excitation. Instead of Equation (3), the membrane potential would obey $V = (g_e V_e + g_{\text{leak}} V_{\text{leak}}) / (g_e + g_{\text{leak}})$, which is a nonlinear (saturating) function of the synaptic excitation g_e. The cell's tuning would be a distorted version of that provided by the excitatory inputs. The distortion would only be noticeable if the excitatory inputs were large enough to approach saturation. But even then the distortion would be subtle; saturation would result in tuning curves with broader/flatter tops and steeper flanks.

8.2. Evidence for Shunting Inhibition

Shunting inhibition is a widely cited proposal for how neurons might perform division (Koch and Poggio, 1987; Coombs *et al.*, 1955; Fatt and Katz, 1953). Its defining property is that it does not introduce any current when the cell is at rest, thus affecting only the cell's overall conductance. Shunting inhibition is usually thought to operate through $GABA_A$ synaptic channels, permeable to chloride ions, because the equilibrium potential of chloride is close to the resting potential of a typical cell.

There is evidence that there are strong inhibitory circuits in the cortex, and that these circuits operate through GABA-mediated shunting inhibition. A seminal study by Krnjević and his colleagues (Dreifuss *et al.*, 1969) showed that electrical stimulation of the cortical surface produces very large (up to 300%) increases in membrane conductance, and that similar effects are obtained by iontophoretic application of GABA. In addition, Rose (1977) showed that iontophoresing GABA over V1 cells yields divisive effects on their visual responses. Cortical neurons are known to have $GABA_A$ receptors, which may be under a constant barrage of synaptic input originating from nearby cortical cells (Salin and Prince, 1996a, 1996b).

The normalization model predicts that a cell's conductance should increase with contrast. To evaluate how substantial these increases would be, we measured responses of a sample of simple cells for grating stimuli of a variety of temporal frequencies, contrasts, and orientations and/or spatial frequencies (Carandini *et al.*, 1997b). We fit the data with the normalization model and used the best-fit parameters to estimate the time constants predicted by the model for zero-contrast (uniform gray field) and full-contrast gratings. The estimated time constants for a zero-contrast stimulus generally varied between 10 and 100 msec. These time constants were estimated to drop by around a factor of four for a full-contrast grating. A fourfold drop in time constant corresponds to a fourfold increase in conductance. Even higher increases in conductance were predicted from other data sets in which plaids or flickering noise masks were employed (Carandini *et al.*, 1997b). These large conductance increases are unlikely to be realistic.

Our estimates of the conductance changes are based on extracellular data, and are surely inflated by our assumption that LGN cells are perfectly linear. As mentioned in Section 4.1, both cat LGN and monkey M LGN cells exhibit a gain control that is in some ways similar to what we have been ascribing entirely to simple cells. In addition, there is evidence for active, nonlinear processing in the dendrites (e.g., calcium spikes; Hirsch *et al.*, 1995). Including these factors in the model would most likely allow us to fit the simple cell responses while requiring smaller, possibly more realistic conductance increases.

Intracellular *in vivo* studies have yielded conflicting results for large conductance increases in V1 cells. Berman *et al.* (1991) reported visually driven conductance increases of less than 20%. These results were confirmed by Ferster and Jagadeesh (1992), who measured the conductance with synaptic current rather than with injected current. Ahmed *et al.* (1997) explicitly studied the dependence of conductance on stimulus contrast, and found conductance increases of up to 30%. Large conductance increases, as large as 300%, were reported only recently, by Borg-Graham *et al.* (1998), using a voltage-clamp approach, and by Carandini *et al.* (1998), using steady current injections in the current-clamp technique. The findings of this latter study, however, are only partially in agreement with the normalization model, as membrane conductance was found to depend on stimulus orientation as well as contrast, being maximal for optimal orientations. This is at odds with the normalization model, which predicts that conductance would not depend on stimulus contrast.

9. Conclusions

The ultimate goal of this research is to develop detailed, quantitative, predictive models of neural function in visual cortex. We will have succeeded when we can record from a neuron while presenting a basic set of visual stimuli (to measure model parameters) and then be able to predict the neuron's response to any visual stimulus.

We have presented two models of V1 simple cells. The first, the linear model, accounts for the cells' selectivity, but fails when the stimulus energy (or contrast) is varied. The second, the normalization model, overcomes this limitation by endowing the linear model with a feedback gain control mechanism. According to the normalization model, the gain of each cell is rescaled by a quantity that grows with the overall activity of a pool of cortical neurons. This effectively normalizes the outputs with respect to stimulus energy. We have recently tested the normalization model, using grating and plaid stimuli, with encouraging results (Carandini *et al.*, 1997b; Tolhurst and Heeger, 1997a,b).

Response normalization may be one of the primary roles of short-range suppression in cat and monkey V1. The V1 neurons have a limited dynamic range, a limit to how strong an output signal they can generate, and thus a limit to the range of contrasts over which they can respond differentially. The normalization operation preserves essential features of the neuron's responses (spatiotemporal frequency tuning, orientation, and direction selectivity in V1) while limiting the dynamic range of their outputs. For example, the response curves (for preferred and nonpreferred stimuli) in Figs. 12B and 13A are vertically shifted copies of one another. Because the data are plotted on a logarithmic response scale, this means that the ratio of responses is about the same at all stimulus contrasts, even in the face of response saturation. This invariance, which we attribute to normalization, is critical for encoding visual information (e.g., about motion, orientation, binocular disparity, etc.) independent of contrast.

The issue of limited dynamic range is, of course, not restricted to V1 neurons. Gain control has been measured and modeled in a variety of other neural systems including turtle photoreceptors (Baylor and Hodgkin, 1974), the vestibuloocular reflex (Lisberger and Sejnowski, 1992), and the velocity-selective neurons in the middle temporal (MT) area of the primate cortex (Simoncelli and Heeger, 1998). In particular, the normalization model of simple cell responses is analogous to models of retinal adaptation/normalization (Grossberg, 1988; Shapley and Enroth-Cugell, 1984; Tranchina *et al.*, 1984; Sperling and Sondhi, 1968), in which the stimulus intensity at a particular point is normalized with respect to the mean stimulus intensity. This makes the retinal response largely independent of the overall level of illumination, and allows the brain to proceed to process visual information without having to attend to the overall light level. Similarly, the normalization mechanism that we propose for V1 simple cells allows the brain to process visual information without having to attend further to contrast; the perceived orientation or motion direction of a stimulus is indeed largely invariant with respect to contrast.

A limitation of the normalization model is that it is local in space. It was not designed to account for the strong surround inhibition displayed by many cortical cells (Levitt and Lund, 1996, and references therein; DeAngelis *et al.*, 1994; Li and Li, 1994; Born and Tootell, 1991; De Valois *et al.*, 1985; Blakemore and Tobin, 1972). While surround suppression could in principle result from the same mechanism that provides masking, it is not clear that its nature is divisive. On the contrary, there is evidence that divisive gain control is highly spatially selective (DeAngelis *et al.*, 1992). In addition some V1 neurons exhibit center–surround phenomena that are significantly more complicated than divisive normalization: for some very specific stimulus configurations, introducing a stimulus in the surrounding field can facilitate a neuron's response (Polat *et al.*, 1998; Gilbert *et al.*, 1996; Kapadia *et al.*, 1995; Sillito *et al.*, 1995; Gilbert and Wiesel, 1990; van Essen *et al.*, 1989; J. I. Nelson and Frost, 1985; Maffei and Fiorentini, 1976). These issues are currently under investigation in our laboratory (Cavanaugh *et al.*, 1997).

Another limitation of the model is that it is local in time: it does not take into consideration the phenomenon of adaptation. To some extent, adaptation can be framed within the context of the normalization model: it can be treated as masking by assuming that gain control has a long memory (Heeger, 1992b). It is unlikely, however, that adaptation operates through the same mechanism that provides masking. First, adaptation was shown in the cat to result from a tonic hyperpolarization (Carandini and Ferster, 1997b), which is not observed during masking (Carandini and Ferster, 1997a). Second, there are some adaptation results that cannot be explained simply by changing a cell's gain. In particular, after long exposure to a high-contrast grating, the response to that grating is often reduced more than its response to other gratings, both in cats (Saul and Cynader, 1989a,b; Albrecht *et al.*, 1984; Movshon and Lennie, 1979) and monkeys (Carandini *et al.*, 1997a). It may very well be that there are additional neural circuits mediating these phenomena.

ACKNOWLEDGMENTS. We thank M. J. Hawken, C. S. Leonard, F. Mechler, L. P. O'Keefe, A. B. Poirson, D. Ringach, R. M. Shapley, and C. Tang for help and suggestions. Our work was supported by a CNR (Italy) grant to M.C., an NIMH grant (MH50228) and an Alfred P. Sloan Fellowship to D.J.H., and an NIH grant (EY2017) and a Howard Hughes Medical Institute Investigatorship to J.A.M.

10. References

Abbott, L. F., Varela, J. A., Sen, K., and Nelson, S. B., 1997, Synaptic depression and cortical gain control, *Science* **275**:220–224.

Adelson, E. H., and Bergen, J. R., 1985, Spatiotemporal energy models for the perception of motion, *J. Opt. Soc. Am. A* **2**:284–299.

Ahmed, B., Allison, J. D., Douglas, R. J., and Martin, K. A. C., 1997, An intracellular study of the contrast-dependence of neuronal activity in cat visual cortex, *Cerebral Cortex,* **7**:559–570.

Albrecht, D. G., 1995, Visual cortex neurons in monkey and cat: Effect of contrast on the spatial and temporal phase transfer functions, *Vis. Neurosci.* **12**:1191–1210.

Albrecht, D. G., and Geisler, W. S., 1991, Motion sensitivity and the contrast-response function of simple cells in the visual cortex, *Vis. Neurosci.* **7**:531–546.

Albrecht, D. G., and Hamilton, D. B., 1982, Striate cortex of monkey and cat: Contrast response function, *J. Neurophysiol.* **48**:217–237.

Albrecht, D. G., Farrar, S. B., and Hamilton, D. B., 1984, Spatial contrast adaptation characteristics of neurones recored in the Cat's visual cortex, *Physiol.* (Lond.) **347**:713–739.

Andrews, B. W., and Pollen, D. A., 1979, Relationship between spatial frequency selectivity and receptive field profile of simple cells, *J. Physiol.* (Lond.) **287**:163–176.

Bauman, L. A., and Bonds, A. B., 1991, Inhibitory refinement of spatial frequency selectivity in single cells of the cat striate cortex, *Vis. Res.* **31**:933–944.

Baylor, D. A., and Hodgkin, A. L., 1974, Changes in time scale and sensitivity in turtle photoreceptors, *J. Physiol.* (Lond.) **242**:729–758.

Benardete, E. A., and Kaplan, E., 1997, The receptive field of the primate P retinal ganglion cell. I: Linear dynamics, *Vis. Neurosci.* **14**:169–185.

Benardete, E. A., Kaplan, E., and Knight, B. W., 1992, Contrast gain in the primate retina: P cells are not X-like, some M cells are, *Vis. Neurosci.* **8**:483–486.

Ben-Yishai, R., Or, R. L. B., and Sompolinsky, H., 1995, Theory of orientation tuning in the visual cortex, *Proc. Natl. Acad. Sci. USA* **92**:3844–3848.

Berman, N. J., Douglas, R. J., Martin, K. A. C., and Whitteridge, D., 1991, Mechanisms of inhibition in cat visual cortex, *J. Physiol.* (Lond.) **440**:697–722.

Bishop, P. O., Coombs, J. S., and Henry, G. H., 1973, Receptive fields of simple cells in the cat striate cortex, *J. Physiol.* (Lond.) **231**:31–60.

Blakemore, C., and Tobin, E. A., 1972, Lateral inhibition between orientation detectors in the cat's visual cortex, *Exp. Brain Res.* **15**:439–440.

Blasdel, G. G., and Lund, J. S., 1983, Termination of afferent axons in macaque striate cortex, *J. Neurosci.* **3:**1389–1413.

Blomfield, S., 1974, Arithmetical operations performed by nerve cells, *Brain Res.* **69:**115–124.

Bonds, A. B., 1989, Role of inhibition in the specification of orientation selectivity of cells in the cat striate cortex, *Vis. Neurosci* **2:**41–55.

Bonds, A. B., 1991, Temporal dynamics of contrast gain in single cells of the cat striate cortex, *Vis. Neurosci* **6:**239–255.

Bonds, A. B., 1992, Spatial and temporal nonlinearities in receptive fields on the cat striate cortex, in: *Nonlinear Vision* (R. B. Pinter and B. Nabet, eds.), CRC Press, Boca Raton, FL, pp. 329–352.

Borg-Graham, L. J., Monier, C., and Frégnac, Y., 1998, Visual input evokes transient and strong shunting inhibition in visual cortical neurons, *Nature* **393:**369–373.

Born, R. T., and Tootell, R. B. H., 1991, Single-unit and 2-deoxyglucose studies of side inhibition in macaque striate cortex, *Proc. Natl. Acad. Sci. USA* **88:**7071–7075.

Carandini, M., and Ferster, D., 1997a, Intracellular correlates of adaptation and masking in cat simple cells, *Inv. Ophthalmol. Vis. Sci.* **38/4:**S16.

Carandini, M., and Ferster, D., 1997b, A tonic hyperpolarization underlying contrast adaptation in cat visual cortex, *Science* **276:**949–952.

Carandini, M., and Heeger, D. J., 1994, Summation and division by neurons in visual cortex, *Science* **264:**1333–1336.

Carandini, M., and Ringach, D. L., 1997, Predictions of a recurrent model of orientation selectivity, *Vis. Res.* **37:**3061–3071.

Carandini, M., Heeger, D. J., and Movshon, J. A., 1993, Amplitude and phase of contrast responses in LGN and V1, *Soc. Neurosci. Abstr.* **19:**628.

Carandini, M., Mechler, F., Leonard, C. S., and Movshon, J. A., 1996, Spike train encoding in regular-spiking cells of the visual cortex, *J. Neurophysiol.* **76:**3425–3441.

Carandini, M., Barlow, H. B., O'Keefe, L. P., Poirson, A. B., and Movshon, J. A., 1997a, Adaptation to contingencies in macaque primary visual cortex. *Proc. R. Soc. Lond. B* **352:**1149–1154.

Carandini, M., Heeger, D. J., and Movshon, J. A., 1997b, Linearity and normalization in simple cells of the macaque primary visual cortex, *J. Neurosci.* **17:**8621–8644.

Carandini, M., Anderson, J., and Ferster, D., 1998, Tuning of membrane conductance changes in simple cells of the cat striate cortex, *Soc. Neurosci. Abstr.* **24:**766.

Cavanaugh, J., Bair, W., and Movshon, J., 1997, Orientation-selective setting of contrast gain by the surrounds of macaque striate cortex neurons, *Soc. Neurosci. Abstr.* **23:**567.

Chagnac-Amitai, Y., and Connors, B. W., 1989, Horizontal spread of synchronized activity in neocortex and its control by GABA-mediated inhibition, *J. Neurophysiol.* **61:**747–758.

Coombs, J. S., Eccles, J. C., and Fatt, P., 1955, The inhibitory suppression of reflex discharges from motoneurones, *J. Physiol.* (Lond.) **130:**396–413.

Crook, J. M., and Eysel, U. T., 1992, GABA-induced inactivation of functionally characterized sites in cat visual cortex (area 18): Effects on orientation tuning, *J. Neurosci* **12:**1816–1825.

Dean, A. F., 1981, The relationship between response amplitude and contrast for cat striate cortical neurones, *J. Physiol.* (Lond.) **318:**413–427.

Dean, A. F., and Tolhurst, D. J., 1983, On the distinctness of simple and complex cells in the visual cortex of the cat, *J. Physiol.* (Lond.) **344:**305–325.

Dean, A. F., and Tolhurst, D. J., 1986, Factors influencing the temporal phase of response to bar and grating stimuli for simple cells in the cat striate cortex, *Exp. Brain Res.* **62:**143–151.

Dean, A. F., Hess, R. F., and Tolhurst, D. J., 1980, Divisive inhibition involved in direction selectivity, *J. Physiol.* (Lond.) **308:**84p–85p.

DeAngelis, G. C., Robson, J. G., Ohzawa, I., and Freeman, R. D., 1992, The organization of supression in receptive fields of neurons in cat visual cortex, *J. Neurophysiol.* **68:**144–163.

DeAngelis, G. C., Ohzawa, I., and Freeman, R. D., 1993a, Spatiotemporal organization of simple-cell receptive fields in the cat's striate cortex. I. General characteristics and postnatal development, *J. Neurophysiol.* **69:**1091–1117.

DeAngelis, G. C., Ohzawa, I., and Freeman, R. D., 1993b, Spatiotemporal organization of simple-cell receptive fields in the cat's striate cortex. II. Linearity of temporal and spatial summation, *J. Neurophysiol.* **69:**1118–1135.

DeAngelis, G. C., Freeman, R. D., and Ohzawa, I., 1994, Length and width tuning of neurons in the cat's primary visual cortex, *J. Neurophysiol.* **71:**347–374.

deBoer, E., and Kuyper, P., 1968, Triggered correlation. *IEEE Trans. Biomed. Eng.* **15**:169–179.

Derrington, A. M., and Lennie, P., 1984, Spatial and temporal contrast sensitivities of neurons in lateral geniculate nucleus of macaque, *J. Physiol.* (Lond.) **357**:219–240.

De Valois, K. K., De Valois, R. L., and Yund, E. W., 1979, Responses of striate cortex cells to grating and checkerboard patterns, *J. Physiol.* (Lond.), **291**:483–505.

De Valois, R. L., Albrecht, D. G., and Thorell, L. G., 1982a, Spatial frequency selectivity of cells in macaque visual cortex, *Vis. Res.* **22**:545–559.

De Valois, R. L., Yund, E. W., and Hepler, N., 1982b, The orientation and direction selectivity of cells in macaque visual cortex, *Vis. Res.* **22**:531–544.

De Valois, R. L., Thorell, L. G., and Albrecht, B. G., 1995, Periodicity of Striate-cortex-cell receptive fields, *J. Opt. Soc. Am.* **2**:1115–1123.

Douglas, R. J., Koch, C., Mahowald, M., Martin, K. A. C., and Suarez, H. H., 1995, Recurrent excitation in neocortical circuits, *Science* **269**:981–985.

Dreher, B., Fukuda, Y., and Rodieck, R. W., 1976, Identification, classification and anatomical segregation of cells with X-like and Y-like properties in the LGN of Old World monkeys, *J. Physiol.* (Lond.) **258**:433–452.

Dreifuss, J. J., Kelly, J. S., and Krnjević, K., 1969, Cortical inhibition and gamma-aminobutyric acid, *Exp. Brain Res.* **9**:137–154.

Einstein, G., Davis, T. L., and Sterling, P., 1987, Ultrastructure of synapses from the A-laminae of the lateral geniculate nucleus in layer IV of the cat striate cortex, *J. Comp. Neurol.* **260**:63–75.

Emerson, R. C., 1988, A linear model for symmetric receptive fields: Implications for classification test with flashed and moving images, *Spatial Vision* **3**:159–177.

Emerson, R. C., and Citron, M. C., 1992, Linear and nonlinear mechanisms of motion selectivity in simple cells of the cat's striate cortex. in: *Nonlinear Vision: Determination of Neural Receptive Fields, Function and Networks* (R. B. Pinter and B. Nabet, eds.), CRC Press, Boca Raton, FL, pp. 75–89.

Enroth-Cugell, C., and Robson, J. G., 1966, The contrast sensitivity of retinal ganglion cells of the cat, *J. Physiol.* (Lond.) **187**:517–552.

Enroth-Cugell, C., and Robson, J. G., 1984, Functional characteristics and diversity of cat retinal ganglion cells, *Inv. Ophthalmol. Vis. Sci.* **25**:250–267.

Enroth-Cugell, C., Robson, J. G., Schweitzer-Tong, D. E., and Watson, A. B., 1983, Spatiotemporal interactions in cat retinal ganglion cells showing linear spatial summation, *J. Physiol.* (Lond.) **341**:279–307.

Fahle, M., and Poggio, T., 1981, Visual hyperacuity: Spatiotemporal interpolation in human vision, *Proc. R. Soc. Lond. B* **213**:451–477.

Fatt, P., and Katz, B., 1953, The effect of inhibitory nerve impulses on a crustacean muscle fiber, *J. Physiol.* (Lond.) **121**:374–389.

Ferster, D., 1981, A comparison of binocular depth mechanisms in areas 17 and 18 of the cat visual cortex, *J. Physiol.* (Lond.) **311**:623–655.

Ferster, D., 1986, Orientation selectivity of synaptic potentials in neurons of cat primary visual cortex, *J. Neurosci.* **6**:1284–1301.

Ferster, D., 1988, Spatially opponent excitation and inhibition in simple cells of the cat visual cortex, *J. Neurosci* **8**:1172–1180.

Ferster, D., 1990a, Binocular convergence of synaptic potentials in cat visual cortex, *Vis. Neurosci.* **4**:625–629.

Ferster, D., 1990b, X- and Y-mediated current sources in area 17 and 18 of cat visual cortex, *Vis. Neurosci.* **4**:135–145.

Ferster, D., and Jagadeesh, B., 1992, EPSP–IPSP interactions in cat visual cortex studied with *in vivo* whole-cell patch recording, *J. Neurosci.* **12**:1262–1274.

Ferster, D., and Lindstrom, S., 1983, An intracellular analysis of geniculo-cortical connectivity in area 17 of the cat, *J. Physiol.* (Lond.) **342**:181–215.

Ferster, D., Chung, S., and Wheat, H. S., 1996, Orientation selectivity of thalamic input to simple cells of cat visual cortex, *Nature* **380**:249–252.

Field, D. J., and Tolhurst, D. J., 1986, The structure and symmetry of simple-cell receptive field profiles in the cat's visual cortex, *Proc. R. Soc. Lond. B* **228**:379–400.

Freeman, R. D., Ohzawa, I., and Robson, J. G., 1987, A comparison of monocular and binocular inhibitory processes in the visual cortex of cat, *J. Physiol.* (Lond.) **396**:69p.

Garey, L. J., and Powell, T. P. S., 1971, An experimental study of the termination of the lateral geniculocortical pathway in the cat and monkey, *Proc. R. Soc. Lond. B* **179:**41–63.

Gaudiano, P., 1992, A unified neural network model of spatiotemporal processing in X and Y retinal ganglion cells I: Analytical results, *Biol. Cybernet.* **67:**11–21.

Geisler, W. S., and Albrecht, D. G., 1992, Cortical neurons: Isolation of contrast gain control, *Vis. Res.* **8:**1409–1410.

Gilbert, C. D., and Wiesel, T. N., 1990, The influence of contextual stimuli on the orientation selectivity of cells in primary visual cortex of the cat, *Vis. Res.* **30:**1689–1701.

Gilbert, C. D., Das, A., Ito, M., Kapadia, M., and Westheimer, G., 1996, Spatial integration and cortical dynamics, *Proc. Natl. Acad. Sci. USA* **93:**615–622.

Gizzi, M. S., Katz, E., Schumer, R. A., and Movshon, J. A., 1990, Selectivity for orientation and direction of motion of single neurons in cat striate and extrastriate visual cortex, *J. Neurophysiol.* **63:**1529–1543.

Glezer, V. D., Tscherbach, T. A., Gauselman, V. E., and Bondarko, V. E., 1980, Linear and nonlinear properties of simple and complex receptive fields in area 17 of the cat visual cortex. *Biol. Cybernet.* **37:**195–208.

Glezer, V. D., Tscherbach, T. A., Gauselman, V. E., and Bondarko, V. E., 1982, Spatiotemporal organization of receptive fields of the cat striate cortex, *Biol. Cybernet.* **43:**35–49.

Grossberg, S., 1988, Nonlinear neural networks: Principles, mechanisms and architectures, *Neural Networks* **1:**17–61.

Gulyas, B., Orban, G. A., Duysens, J., and Maes, H., 1987, The suppressive influence of moving textured backgrounds on responses of cat striate neurons to moving bars, *J. Neurophysiol.* **57:**1767–1791.

Hammond, P., and MacKay, D. M., 1981, Modulatory influences of moving textured backgrounds on responsiveness of simple cells in feline striate cortex, *J. Physiol.* (Lond.) **319:**431–442.

Hawken, M. J., Shapley, R. M., and Grosof, D. H., 1992, Temporal frequency tuning of neurons in macaque V1: Effects of luminance contrast and chromaticity, *Inv. Ophthalmol. Vis. Sci. (Suppl.)* **33:**955.

Heeger, D. J., 1991, Nonlinear model of neural responses in cat visual cortex, in: *Computational Models of Visual Processing* (M. S. Landy and A. Movshon, eds.), MIT Press, Cambridge, MA, pp. 119–133.

Heeger, D. J., 1992a, Half-squaring in responses of cat simple cells, *Vis. Neurosci.* **9:**427–443.

Heeger, D. J., 1992b, Normalization of cell responses in cat striate cortex, *Vis. Neurosci.* **9:**181–197.

Heeger, D. J., 1993, Modeling simple cell direction selectivity with normalized, half-squared, linear operators, *J. Neurophysiol.* **70:**1885–1897.

Heggelund, P., 1981, Receptive-field organization of simple cells in cat striate cortex, *Exp. Brain Res.* **42:**89–98.

Heggelund, P., 1986, Quantitative studies of the discharge fields of single cells in cat striate cortex, *J. Physiol.* (Lond.) **373:**277–292.

Hendrickson, A. E., Wilson, J. R., and Ogren, M. P., 1978, The neuroanatomical organization of pathways between the dorsal lateral geniculate nucleus and visual cortex in Old World and New World primates, *J. Comp. Neurol.* **182:**123–136.

Hirsch, J. A., Alonso, J. M., and Reid, R. C., 1995, Visually evoked calcium action potentials in cat striate cortex, *Nature* **378:**612–616.

Hochstein, S., and Shapley, R. M., 1976, Quantitative analysis of retinal ganglion cell classifications, *J. Physiol.* (Lond.) **262:**237–264.

Holub, R. A., and Morton-Gibson, M., 1981, Response of visual cortical neurons of the cat to moving sinusoidal gratings: Response-contrast functions and spatiotemporal interactions, *J. Neurophysiol.* **46:**1244–1259.

Hubel, D. H., and Wiesel, T. N., 1962, Receptive fields, binocular interaction and functional architecture in the cat's visual cortex, *J. Physiol.* (Lond.) **160:**106–154.

Hubel, D. H., and Wiesel, T. N., 1972, Laminar and columnar distribution of geniculocortical fibers in macaque monkeys, *J. Comp. Neurol.* **146:**421–450.

Jack, J. J. B., Noble, D., and Tsien, R. W., 1975, *Electric Current Flow in Excitable Cells,* Oxford University Press, Oxford.

Jagadeesh, B., Wheat, H. S., and Ferster, D., 1993, Linearity of summation of synaptic potentials underlying direction selectivity in simple cells of the cat visual cortex, *Science* **262:**1901–1904.

Jagadeesh, B., Wheat, H. S., Kontsevich, L. L., Tyler, C. W., and Ferster, D., 1977, Direction selectivity of synaptic potentials in simple cells of the cat visual cortex, *J. Neurophysiol.* **78:**2772–2789.

Jones, J. P., and Palmer, L. A., 1987a, An evaluation of the two-dimensional Gabor filter model of simple receptive fields in cat striate cortex, *J. Neurophysiol.* **58:**1233–1258.

Jones, J. P., and Palmer, L. A., 1987b, The two-dimensional spatial structure of simple receptive fields in cat striate cortex, *J. Neurophysiol.* **58:**1187–1211.

Jones, J. P., Stepnoski, A., and Palmer, L. A., 1987, The two-dimensional spectral structure of simple receptive fields in cat striate cortex, *J. Neurphysiol.* **58:**1212–1232.

Kaji, S., and Kawabata, N., 1985, Neural interactions of two moving patterns in the direction and orientation domain in the complex cells of cat's visual cortex, *Vis. Res.* **25:**749–753.

Kapadia, M. K., Ito, M., Gilbert, C. D., and Westheimer, G., 1995, Improvement in visual sensitivity by changes in local context: Parallel studies in human observers and in V1 of alert monkeys, *Neuron* **15:**843–856.

Kaplan, E., Purpura, K., and Shapley, R., 1987, Contrast affects the transmission of visual information through the mammalian lateral geniculate nucleus, *J. Physiol.* (Lond.) **391:**267–288.

Kisvárday, Z., and Eysel, U. T., 1993, Functional and structural topography of horizontal inhibitory connections in cat visual cortex, *Eur. J. Neurosci.* **5:**1558–1572.

Kisvárday, Z. F., Beaulieu, C., and Eysel, U. T., 1993, Network of GABA-ergic large basket cells in cat visual cortex (area 18): Implication for lateral disinhibition, *J. Comp. Neurol.* **327:**398–415.

Koch, C., and Poggio, T., 1987, Biophysics of computation: Neurons, synapses and membranes, in: *Synaptic Function* (G. M. Edelman, W. E. Gall, and M. W. Cowan, eds.), Wiley, New York, pp. 637–698.

Koch, C., and Segev, I., eds., 1989, *Methods in Neuronal Modeling,* MIT Press, Cambridge, MA.

Kulikowski, J. J., and Bishop, P. O., 1981a, Fourier analysis and spatial representation in the visual cortex, *Experimentia* **37:**160–163.

Kulikowski, J. J., and Bishop, P. O., 1981b, Linear analysis of the responses of simple cells in the cat visual cortex, *Exp. Brain Res.* **44:**386–400.

Lahica, E. A., Beck, P. D., and Casagrande, V. A., 1992, Parallel pathways in macaque monkey striate cortex: Anatomically defined columns in layer III, *Proc. Natl. Acad. Sci. USA* **89:**3566–3570.

Lee, B. B., Pokorny, J., Smith, V. C., and Kremers, J., 1994, Responses to pulses and sinusoids in macaque ganglion cells. *Vis. Res.* **34:**3081–3096.

Levitt, J. B., and Lund, J. S., 1996, Contrast dependence of modulatory surround effects in macaque striate neurons, *Inv. Ophthalmol. Vis. Sci. (Suppl.)* **37:**S482.

Li, C., and Creutzfeldt, O., 1984, The representation of contrast and other stimulus parameters by single neurons in area 17 of the cat, *Pflugers Arch.* **401:**304–314.

Li, C., and Li, W., 1994, Extensive integration beyond the classical receptive field of cat's striate cortical neurons—Classification and tuning properties, *Vis. Res.* **34:**2337–2356.

Lisberger, S. G., and Sejnowski, T. J., 1992, Motor learning in a recurrent network model based on the vestibulo-ocular reflex, *Nature* **360:**159–161.

Maffei, L., and Fiorentini, A. 1973, The visual cortex as a spatial frequency analyzer, *Vis. Res.* **13:**1255–1267.

Maffei, L., and Fiorentini, A. 1976, The unresponsive regions of visual cortical receptive fields, *Vis. Res.* **13:**1255–1267.

Maffei, L., Morrone, C., Pirchio, M., and Sandini, G., 1979, Responses of visual cortical cells to periodic and nonperiodic stimuli, *J. Physiol.* (Lond.) **296:**27–47.

Malpeli, J. G., Schiller, P. H., and Colby, C. L., 1981, Response properties of single cells in monkey striate cortex during reversible inactivation of individual lateral geniculate laminae, *J. Neurophysiol.* **46:**1102–1119.

Markram, H., and Tsodyks, M., 1996, Redistribution of synaptic efficacy between neocortical pyramidal neurons, *Nature* **382:**807–10.

Marr, D., 1970, A theory for cerebral neocortex, *Proc. R. Soc. B* **176:**161–234.

Marr, D., 1982, *Vision,* Freeman, San Francisco.

Mastronarde, D., 1987, Two classes of single-input X-cells in cat lateral geniculate nucleus. I. Receptive-field properties and classification of cells, *J. Neurophysiol.* **57:**357–380.

Maunsell, J. H. R., and Gibson, J. R., 1992, Visual response latencies of striate cortex of the macaque monkey, *J. Neurophysiol.* **68:**1332–1344.

McLean, J., and Palmer, L. A., 1988, Contribution of linear mechanisms to direction selectivity of simple cells in area 17 and 18 of the cat, Presented at Annual Spring Meeting of the Association for Research in Vision and Ophthalmology, Sarasota, Florida, May 1988.

McLean, J., and Palmer, L. A., 1989, Contribution of linear spatiotemporal receptive field structure to velocity selectivity of simple cells in area 17 of the cat, *Vis. Res.* **29**:675–679.
McLean, J., Raab, S., and Palmer, L. A., 1994, Contribution of linear mechanisms to the specification of local motion by simple cells in areas 17 and 18 of the cat, *Vis. Neurosci.* **11**:295–306.
Morrone, M. C., Burr, D. C., and Maffei, L., 1982, Functional implications of cross-orientation inhibition of cortical visual cells. I. Neurophysiological evidence, *Proc. R. Soc. Lond. B* **216**:335–354.
Movshon, J. A., and Lennie, P., 1979, Pattern-selective adaptation in visual cortical neurones, *Nature* **278**:850–852.
Movshon, J. A., Thompson, I. D., and Tolhurst, D. J., 1978a, Spatial and temporal contrast sensitivity of neurones in areas 17 and 18 of the cat's visual cortex, *J. Physiol.* (Lond.) **283**:101–120.
Movshon, J. A., Thompson, I. D., and Tolhurst, D. J., 1978b, Spatial summation in the receptive fields of simple cells in the cat's striate cortex, *J. Physiol.* (Lond.) **283**:53–77.
Movshon, J. A., Hawken, M. J., Kiorpes, L., Skoczenski, A. M., Tang, C., and O'Keefe, L. P., 1994, Visual noise masking in macaque LGN neurons, *Inv. Ophthalmol. Vis. Sci. (Suppl.)*, **35**:1662.
Nealey, T. A., and Maunsell, J. H., 1994. Magnocellular and parvocellular contributions to the responses of neurons in macaque striate cortex, *J. Neurosci.* **14**:2069–2079.
Nelson, J. I., and Frost, B., 1985, Intracortical facilitation among co-oriented, co-axially aligned simple cells in cat striate cortex, *Exp. Brain Res.* **6**:54–61.
Nelson, S., Toth, L., Sheth, B., and Sur, M., 1994, Orientation selectivity of cortical neurons during intracellular blockade of inhibition, *Science* **265**:774–777.
Nelson, S. B., 1991, Temporal interactions in the cat visual system I. Orientation-selective suppression in visual cortex, *J. Neurosci.* **11**:344–356.
Nestares, O., and Heeger, D. J., 1997, Modeling the apparent frequency-specific suppression in simple cell responses, *Vis. Res.* **37**:1535–1543.
Ohzawa, I., and Freeman, R. D., 1986, The binocular organization of simple cells in the cat's visual cortex, *J. Neurophysiol.* **56**:221–242.
Palmer, L. A., and Davis, T. L., 1981, Receptive-field structure in cat striate cortex, *J. Neurophysiol.* **46**:260–276.
Polat, U., Mizobe, K., Pettet, M. W., Kasamatsu, T., and Norcia, A. M., 1998, Collinear stimuli regulate visual responses depending on cell's contrast threshold, *Nature* **391**:580–584.
Pollen, D., and Ronner, S., 1982, Spatial computation performed by simple and complex cells in the visual cortex of the cat, *Vis. Res.* **22**:101–118.
Pollen, D. A., Gaska, J. P., and Jacobson, L. D., 1988, Responses of simple and complex cells to compound sine-wave gratings, *Vis. Res.* **28**:25–39.
Reid, R. C., and Alonso, J. M., 1995, Specificity of monosynaptic connections from thalamus to visual cortex, *Nature* **378**:281–284.
Reid, R. C., Soodak, R. E., and Shapley, R. M., 1987, Linear mechanisms of direction selectivity in simple cells of cat striate cortex, *Proc. Natl. Acad. Sci. USA* **84**:8740–8744.
Reid, R. C., Soodak, R. E., and Shapley, R. M., 1991, Directional selectivity and spatiotemporal structure of receptive fields of simple cells in cat striate cortex, *J. Neurophysiol.* **66**:505–529.
Reid, R. C., Victor, J. D., and Shapley, R. M., 1992, Broadband temporal stimuli decrease the integration time of neuron in cat striate cortex, *Vis. Neurosci.* **9**:39–45.
Robson, J. G., 1988, Linear and nonlinear operations in the visual system, *Inv. Ophthalmol. Vis. Sci. (Suppl.)* **29**:117.
Rose, D., 1977, On the arithmetical operation performed by inhibitory synapses onto the neuronal soma, *Exp. Brain Res.* **28**:221–223.
Salin, P. A., and Prince, D. A., 1996a, Electrophysiological mapping of GABA-A receptor-mediated inhibition in adult rat somatosensory cortex, *J. Neurophysiol.* **75**: 1589–1599.
Salin, P. A., and Prince, D. A., 1996b, Spontaneous GABA-A receptor-mediated inhibitory currents in adult rat somatosensory cortex, *J. Neurophysiol.* **75**:1573–1588.
Saul, A. B., and Cynader, M. S., 1989a, Adaptation in single units in visual cortex: The tuning of aftereffects in the spatial domain, *Vis. Neurosci.* **2**:593–607.
Saul, A. B., and Cynader, M. S., 1989b, Adaptation in single units in visual cortex: The tuning of aftereffects in the spatial domain, *Vis. Neurosci.* **2**:609–620.
Saul, A. B., and Humphrey, A. L., 1990, Spatial and temporal response properties of lagged and nonlagged cells in cat lateral geniculate nucleus, *J. Neurophysiol.* **64**:206–224.
Saul, A. B., and Humphrey, A. L., 1992, Evidence of input from lagged cells in the lateral geniculate nucleus to simple cells in cortical area 17 of the cat, *J. Neurophysiol.* **68**:1190–1208.

Schiller, P. H., Finlay, B. L., and Volman, S. F., 1976, Quantitative studies of single-cell properties in monkey striate cortex. I. Spatiotemporal organization of receptive fields, *J. Neurophysiol.* **39:**1288–1319.

Schumer, R. A., and Movshon, J. A., 1984, Length summation in simple cells of cat striate cortex, *Vis. Res.* **24:**565–571.

Sclar, G., and Freeman, R. D., 1982, Orientation selectivity in the cat's striate cortex is invariant with stimulus contrast, *Exp. Brain Res.* **46:**457–461.

Sclar, G., Maunsell, J. H. R., and Lennie, P., 1990, Coding of image contrast in central visual pathways of the macaque monkey, *Vis. Res.* **30:**1–10.

Sengpiel, F., and Blakemore, C., 1994, Interocular control of neuronal responsiveness in cat visual cortex, *Nature* **368:**847–850.

Sengpiel, F., Blakemore, C., and Harrad, R., 1995, Interocular suppression in the primary visual cortex: A possible neural basis of binocular rivalry, *Vis. Res.* **35:**179–196.

Shapley, R. M., and Enroth-Cugell, C., 1984, Visual adaptation and retinal gain controls, *Prog. Retinal Res.* **3:**263–346.

Shapley, R. M., and Victor, J. D., 1978, The effect of contrast on the transfer properties of cat retinal ganglion cells, *J. Physiol.* **285:**275–298.

Shapley, R., Reid, R. C., and Soodak, R., 1991, Spatiotemporal receptive fields and direction selectivity, *Computational Models of Visual Processing* (M. S. Landy and J. A. Movshon, eds.), MIT Press, Cambridge, MA, pp. 109–118.

Sherman, S. M., Schumer, R. A., and Movshon, J. A., 1984, Functional cell classes in the macaque's LGN, *Soc. Neurosci. Abstr.* **10:**296.

Sillito, A. M., 1975, The contribution of inhibitory mechanisms to the receptive field properties of neurones in the cat's striate cortex, *J. Physiol.* (Lond.) **250:**304–330.

Sillito, A. M., 1977, Inhibitory processes underlying the directional specificity of simple, complex, and hypercomplex cells in the cat's visual cortex, *J. Physiol.* (Lond.) **271:**699–720.

Sillito, A. M., 1984, Functional considerations of the operation of GABAergic inhibitory processes in the visual cortex, in: *Cerebral Cortex,* Vol. 2. *Functional Properties of Cortical Cells,* (E. G. Jones and A. Peters, eds.), Plenum Press, New York, pp. 91–117.

Sillito, A. M., Kemp, J. A., Milson, J. A., and Berardi, N., 1980, A re-evaluation of the mechanisms underlying simple cell orientation selectivity, *Brain Res.* **194:**517–520.

Sillito, A. M., Grieve, K. L., Jones, H. E., Cudeiro, J., and Davis, J., 1995, Visual cortical mechanisms detecting focal orientation discontinuities, *Nature* **378:**492–496.

Simoncelli, E. P., and Heeger, D. J., 1998, A model of neuronal responses in visual area MT, *Vis. Res.* **38:**743–761.

Skottun, B. C., Bradley, A., Sclar, G., Ohzawa, I., and Freeman, R., 1987, The effects of contrast on visual orientation and spatial frequency discrimination: A comparison of single cells and behavior, *J. Neurophysiol.* **57,**773–786.

Somers, D. C., Nelson, S. B., and Sur, M., 1995, An emergent model of orientation selectivity in cat visual cortical simple cells, *J. Neurosci.* **15:**5448–5465.

Sperling, G., and Sondhi, M. M., 1968, Model for visual luminance discrimination and flicker detection, *J. Opt. Soc. Am. A* **58:**1133–1145.

Stafstrom, C. E., Schwindt, P. C., and Crill, W. E., 1984, Repetitive firing in layer V neurons from cat neocortex *in vitro, J. Neurophysiol.* **52:**264–277.

Suarez, H. H., Koch, C., and Douglas, R. J., 1995, Modeling direction selectivity of simple cells in striate visual cortex within the framework of the canonical microcircuit, *J. Neurosci.* **15:**6700–6719.

Tadmor, Y., and Tolhurst, D. J., 1989, The effect of threshold on the relationship between the receptive-field profile and the spatial-frequency tuning curve in simple cells of the cat's striate cortex, *Vis. Neurosci.* **3:**445–454.

Tanaka, K., 1983, Cross-correlation analysis of geniculostriate neuronal relationships in cats, *J. Neurophysiol.* **49:**1303–1318.

Tolhurst, D. J., and Dean, A. F., 1987, Spatial summation by simple cells in the striate cortex of the cat, *Exp. Brain Res.* **66:**607–620.

Tolhurst, D. J., and Dean, A. F., 1990, The effects of contrast on the linearity of spatial summation of simple cell in the cat's striate cortex, *Exp. Brain Res.* **79:**582–588.

Tolhurst, D. J., and Dean, A. F., 1991, Evaluation of a linear model of directional selectivity in simple cells of the cat's striate cortex, *Vis. Neurosci.* **6:**421–428.

Tolhurst, D. J., and Heeger, D. J., 1997a, Comparison of contrast-normalization and threshold models of the responses of simple cells in cat striate cortex, *Vis. Neurosci.* **14**:293–310.

Tolhurst, D. J., and Heeger, D. J., 1997b, Contrast normalization and a linear model for the directional selectivity of simple cells in cat striate cortex, *Vis. Neurosci.* **14**:19–26.

Toyama, K., Kimura, M., Shiida, T., and Takeda, T., 1977a, Convergence of retinal inputs onto visual cortical cells: II. A study of the cells disynaptically excited from the lateral geniculate body, *Brain Res.* **137**:221–231.

Toyama, K., Maikawa, K., and Tanaka, T., 1977b, Convergence of retinal inputs onto visual cortical cells: I. A study of the cells monosynaptically excited from the lateral geniculate body, *Brain Res.* **137**:207–220.

Tranchina, D., Gordon, J., and Shapley, R. M., 1984, Retinal light adaptation—Evidence for a feedback mechanism, *Nature* **310**:314–316.

Troy, J. B., 1983, Spatial contrast sensitivities of X and Y type neurones in the cat's dorsal lateral geniculate nucleus, *J. Physiol.* (Lond.) **344**:399–417.

van Essen, D., DeYoe, E. A., Olavarria, J. F., Knierim, J. J., Sagi, D., Fox, J. M., and Julesz, B., 1989, Neural responses to static and moving texture patterns in visual cortex of the macaque monkey, in: *Neural Mechanisms of Visual Perception* (D. M.-K. Lam and C. D. Gilbert, eds.), Portfolio, Woodlands, TX, pp. 137–156.

van Santen, J. P. H., and Sperling, G., 1985, Elaborated Reichardt detectors, *J. Opt. Soc. Am. A* **2**:300–321.

Victor, J., 1987, The dynamics of the cat retinal X cell centre, *J. Physiol.* (Lond.) **386**:219–246.

Victor, J., 1988, The dynamics of the cat retinal Y cell subunit, *J. Physiol.* (Lond.) **405**:289–320.

Walker, G. A., Ohzawa, I., and Freeman, R. D., 1996, interocular transfer of cross-orientation suppression in the cat's visual cortex, *Inv. Ophthalmol. Vis. Sci. (Suppl.)* **37**:S484.

Watanabe, S., Konishi, M., and Creutzfeldt, O. D., 1966, Postsynaptic potentials in the cat's visual cortex following electrical stimulation of afferent pathways, *Exp. Brain Res.* **1**:272–283.

Watson, A. B., and Ahumada, A. J., 1985, Model of human visual-motion sensing, *J. Opt. Soc. Am.* **2**:322–341.

Yoshioka, T., Levitt, J. B., and Lund, J., 1994, Independence and merger of thalamocortical channels within macaque monkey primary visual cortex: Anatomy of interlaminar projections, *Vis. Neurosci.* **11**:467–489.

8

Non-Fourier Cortical Processes in Texture, Form, and Motion Perception

HUGH R. WILSON

1. Introduction

1.1. Strategies in Neural Modeling

Modeling plays a unique role in visual neuroscience. On one hand, single-unit physiology catalogues neural responses to visual stimuli and perhaps correlates them with behavior; on the other hand, psychophysics measures the overall functional capabilities of the visual system. The goal of neural modeling is to produce theories that bridge the gap between activity of the neural elements and responses of the functioning system. Thus, an ideal neural model contains elements mimicking the response properties of neurons in relevant visual areas *and* makes predictions of system behavior that are psychophysically testable. Current neural theories of visual function take two very different approaches. The *connectionist* approach defines a set of inputs and a set of neurons representing desired outputs and then trains a

HUGH R. WILSON • Visual Sciences Center, University of Chicago, Chicago, Illinois 60637.

network to develop appropriate connections between these, typically with intervening layers of "hidden units." This approach has led to a number of biologically interesting results, such as the extraction of shape from shading (Lehky and Sejnowski, 1988). However, the connectionist approach frequently employs learning rules that are nonbiological and typically omits the feedback connections that are ubiquitous in the nervous system (Hertz *et al.*, 1991). The *anatomically motivated network* approach, on the other hand, aims at developing a network of model neurons with connections specified by available anatomy that will predict the system behavior. The component neurons may be described at various levels of generality, e.g., in terms of the dependence of spike rate on net postsynaptic potential, or in terms of individual ionic currents. The anatomically motivated approach can therefore be successfully applied to such complex problems as simulating human retinal function on a neuron by neuron basis (Wilson, 1997) or predicting responses of neurons in various visual cortical areas. In such tasks neuroanatomical data can usually be used to constrain network connectivity patterns. As noted by Hertz *et al.* (1991, p. 10), "[connectionist] neural networks are the second best way of doing just about anything. The *best* way is to find and use the right rules . . . for each particular problem." The networks described in this chapter represent an attempt to apply the latter approach to nonlinear aspects of cortical functioning.

1.2. Nonlinearities and Non-Fourier Processing

This chapter uses nonlinear filtering operations to elucidate various aspects of texture, form, and motion perception. In so doing it focuses on a class of nonlinear models having certain common processing features. Typically, a two-dimensional spatial image is first filtered using linear spatial filters with characteristics of orientation-selective cortical simple cells. These simulated neural responses are next either squared or rectified to ensure that all responses are nonnegative. In a visual system containing separate ON-center and OFF-center channels with thresholds (e.g., Schiller, 1992), rectification corresponds to summing ON and OFF responses from spatial arrays of units with otherwise identical receptive field characteristics. Due to this inherent nonlinearity, these models are able to extract stimulus elements that are not represented by any Fourier component or sum of components in the stimulus. In consequence, such processing is frequently termed "non-Fourier" processing (Chubb and Sperling, 1988, 1989), although it has also been referred to as "second-order" processing (Cavanagh and Mather, 1990).

The non-Fourier processing sequence of oriented filtering, rectification, and further pooling or filtering is exemplified in its simplest form by complex cell receptive fields in striate cortex. Following the discovery of complex cells by Hubel and Wiesel (1962, 1968), Movshon *et al.* (1978) showed that complex cell responses could be predicted by summation of rectified subunit responses, with the subunits being described by oriented linear filters. The subunit orientation and spatial frequency tuning imparts like tuning to the complex cell response, while the nonlinear rectification operation produces a response independent of phase and sign of contrast (i.e., light or dark). This represents the non-Fourier processing theme in its simplest form. As will be shown below, more sophisticated forms of non-Fourier processing produce units with characteristics of end-stopped or end-inhibited complex cells, which are common in visual cortex (Dreher, 1972; Henry, 1977). [These were originally termed "hypercomplex cells" by Hubel and Wiesel (1968).] Such units are important in texture boundary perception, and neurons responding to texture boundaries have been discovered in cortical area V2 (von der Heydt *et al.*, 1984; von der Heydt and Peterhans, 1989).

In sum, the filter–rectify–filter theme of non-Fourier processing describes complex cells in primary visual cortex as well as neurons at many higher cortical levels. It is to be emphasized that the nonlinearity in these non-Fourier schemes is an essential rather than an incidental element of the processing. As Malik (1994) has cogently observed, "Almost all systems will be nonlinear—very few will be linear. There is a fundamental computational reason for that. . . . A linear system followed by a second linear stage is equivalent to a single-stage linear system. Unless we have intervening non-linearities, nothing is gained by adding additional stages." Although the visual system certainly incorporates many more sophisticated nonlinearities than those in the non-Fourier scheme, the clear implication is that linear schemes of visual processing are doomed from the start.

The goal of this chapter is to demonstrate that non-Fourier, nonlinear processes can elucidate many higher levels of visual processing. Accordingly, I will first describe non-Fourier processing of texture boundaries and will compare this with simple cell responses to luminance edges. Next, extensions of the non-Fourier scheme will be shown to explain more complex form analysis by neurons in area V4 (Gallant *et al.*, 1993, 1996). A similar approach to motion perception demonstrates that a combination of Fourier and non-Fourier motion signals can predict the relevant psychophysics as well as much single-unit physiology in middle temporal cortex (MT). Finally, these non-Fourier models suggest an explanation for the multiple connections among cortical areas V1, V2, V4, and MT.

2. Analysis of Texture Boundaries by Non-Fourier Mechanisms

2.1. Simple Cells and Luminance Edges

Lines and edges defined by a change in luminance are easily extracted by cortical area V1 simple cells with the appropriate orientation selectivity. Psychophysics has led to the development of quantitative models for simple cell processing that have proved extremely accurate in predicting a wide range of discrimination abilities for luminance-defined patterns (Wilson, 1986, 1991a,b; Wilson and Richards, 1989). As shown in Fig. 1A, these models involve filtering by model receptive fields tuned for orientation and spatial frequency followed by a nonlinear response function incorporating threshold behavior at low stimulus contrasts and compression or saturation at high contrasts. Properties of these simple cell filters have been measured psychophysically using a pattern-masking paradigm (Wilson *et al.*, 1983; Phillips and Wilson, 1984), while the contrast nonlinearities have been measured in several laboratories (Nachmias and Sansbury, 1974; Legge and Foley, 1980; Wilson, 1980). One may describe the receptive field (*RF*) and contrast response nonlinearity (*R*) by the following equations:

$$RF(x,y) = A\left[\exp\left(-\frac{x^2}{\sigma_1^2}\right) - B\exp\left(-\frac{x^2}{\sigma_2^2}\right) + C\exp\left(-\frac{x^2}{\sigma_3^2}\right)\right]\exp\left(-\frac{y^2}{\sigma_y^2}\right) \tag{1}$$

$$R(S) = \begin{cases} \dfrac{S^2 + KS^{3-\varepsilon}}{K + S^2} & \text{for} \quad S \geq 0 \\ 0 & \text{for} \quad S < 0 \end{cases}$$

This receptive field is vertically oriented and consists of an elongated excitatory zone flanked on either side by inhibitory zones which are in turn flanked by weak secondary excitatory zones. S represents the result of convolving $RF(x,y)$ with the luminance pattern describing a stimulus. As the value of ϵ is generally around 0.5, the function R approximates a square-root compression for large values of S. Values of the various parameters have been tabulated elsewhere (Wilson, 1991a).

Predicting pattern discrimination based on these model simple cells requires convolving the stimulus with $RF(x,y)$ at several different sizes and orientations, passing the results through the response nonlinearity, and then comparing responses of neighboring units. In a task of discriminating line orientation, for example, the unit most strongly stimulated signals the approximate orientation of the stimulus, while comparison with responses of units tuned to adjacent orientations permits accurate interpolation of orientations to within about 0.4°, which agrees with human psychophysical performance (Westheimer *et al.*, 1976). A plausible network for performing neural interpolation is described elsewhere (Wilson *et al.*, 1992). Although current research indicates that simple cells interact via divisive gain controls (Heeger, 1991, 1992; Wilson and Humanski, 1993; Wilson, 1993), this does not significantly alter the basic picture of luminance pattern discrimination.

2.2. Non-Fourier Units, Texture Boundaries, and Curvature

Although simple cells respond optimally to oriented luminance edges, difficulties arise in patterns defined by edge discontinuities. For example, consider the pattern in Fig. 2, which is composed of parallel lines. One readily perceives an oblique contour running across the center of this figure where lines on the left are shifted relative to those on the right. However, there is no luminance border there, as the white regions are continuous across the pattern. Hence, these are often termed "illusory contours." They may equally well be regarded as very real texture boundaries, however, which could be produced by superimposing two pieces of obliquely cut, black and white-striped wallpaper with an offset between left and right. Such texture boundaries pose an insuperable problem for simple cell processing. As shown in Fig. 2, horizontally oriented simple cells will respond very well to the white bars on either side of the pattern, but they will not respond at all along the texture boundary! This is due to the balance of light and dark in the excitatory and inhibitory regions. In fact, it is easy to show mathematically that there are no components in the Fourier spectrum of these stimuli that are oriented along the texture boundaries. Thus, linear filters followed by the simple nonlinearity in Equation (1) simply cannot detect these texture boundaries.

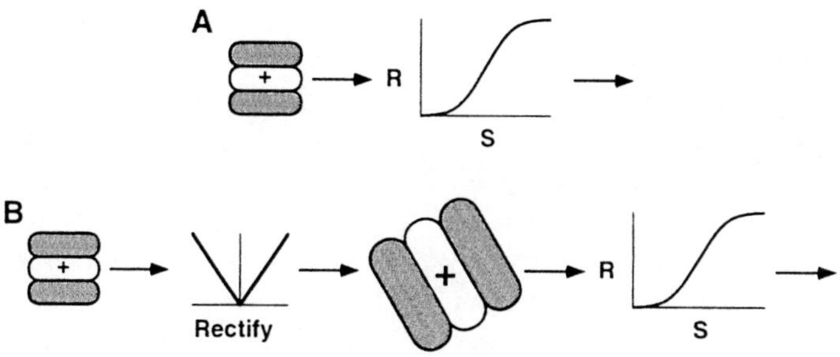

Figure 1. Schematic diagrams of Fourier (A) and non-Fourier (B) visual processing. Spatial weighting functions (filters or receptive fields) are all orientation selective and composed of parallel excitatory (+) and inhibitory (gray) zones. (A) Fourier pathways, which operate like V1 simple cells, employ orientation-selective filtering followed by a nonlinear stimulus–response function [Equation (1)] exhibiting both threshold and saturation behavior. (B) Non-Fourier pathways begin with similar oriented filtering, but the filter output is then rectified (easily accomplished by pooling of ON- and OFF-center neural responses), and the result is then processed by a second oriented filter that is about twice as large (half the peak spatial frequency). A nonlinear function generates the final response. Note that the first- and second-stage filters in non-Fourier pathways will involve many different combinations of relative orientations.

Consider now the sequence of neural processing diagrammed in Fig. 1B. Here oriented simple cell filtering is followed by full-wave rectification, and these neural responses are then pooled by a larger oriented receptive field with parallel excitatory and inhibitory zones. Applying this processing sequence to the pattern in Fig. 2, horizontally oriented simple cells are strongly activated except along the texture

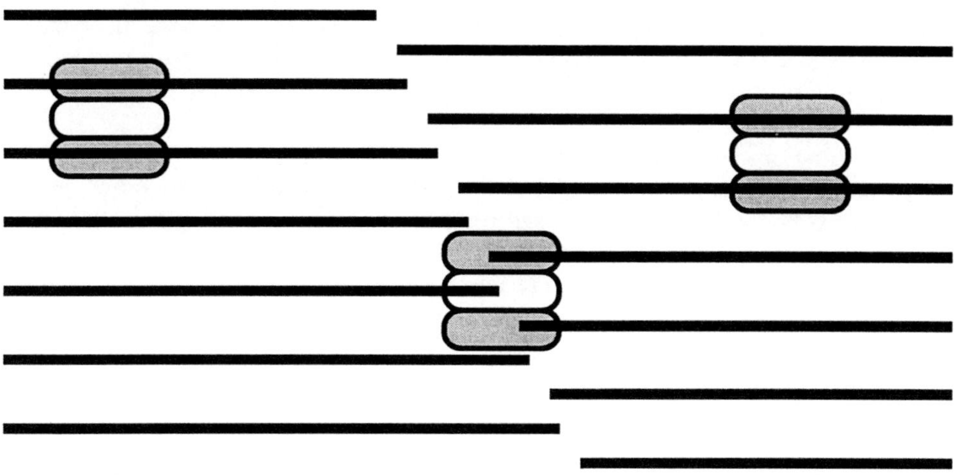

Figure 2. Example of an "illusionary contour" or texture boundary that cannot be detected by Fourier units. Three horizontal, on-center Fourier receptive fields are depicted with white excitatory and gray inhibitory zones. The left and right units will respond strongly to this pattern, as the black bars reduce the amount of light falling in the inhibitory zones. The unit along the texture boundary, however, will give no response, as the excitatory and inhibitory inputs are balanced. One can easily prove that this pattern contains *no* Fourier energy at the oblique orientation of the texture boundary. Such texture boundaries are readily extracted by the non-Fourier scheme in Fig. 1B.

boundary. Full-wave rectification is next accomplished by combining ON-center and OFF-center simple cell responses. Finally, a larger receptive field at the orientation of the texture boundary will detect the ridge along which simple cell activity drops to zero. The final receptive field is chosen to be larger (i.e., of lower spatial frequency) so that it will be insensitive to the very high spatial frequency responses produced by rectification of responses to the parallel lines themselves. A range of relative orientations is assumed to be paired as first- and second-stage receptive fields, although the exact number remains a topic of research.

In addition to encoding the location of texture boundaries and illusory contours, non-Fourier mechanisms play a key role in encoding boundary curvature. This was first demonstrated for end-stopped simple cells in the cortex (an instantiation of non-Fourier processing; see below) by Dobbins *et al.* (1987, 1989), and a theoretical basis was provided by Koenderink and Richards (1988). Curvature analysis by non-Fourier mechanisms was further generalized by Wilson and Richards (1992) using the scheme depicted in Fig. 1B. As shown in Fig. 3A, a curved contour will selectively activate oriented filters that are tangent to the curve. Subsequent rectification and larger scale filtering at right angles to this will produce a response that increases with curvature over a significant range (see Fig. 3B). The theoretical reason for this is that the center–surround organization of the second-stage filter approximates the second spatial derivative of the oriented responses extracted at the first stage, and curvature is directly proportional to this spatial second derivative (Koenderink and Richards, 1988). Thus, there is less surround inhibition generated in the second-stage filter as the contour curvature increases, so the non-Fourier response increases as shown in Fig. 3B. Finally, of course, the curvature becomes too great to activate the first-stage oriented filters effectively, so the response falls off at extreme curvatures, which also occurs for human curvature discrimination (Wilson, 1985; Wilson and Richards, 1989). Evidence will be presented below that curvature information extracted by non-Fourier mechanisms can be pooled to extract global aspects of stimulus shape. Further examples of non-Fourier image processing may be found elsewhere (Wilson and Richards, 1992; Wilson, 1993).

Just how precise is non-Fourier processing? This question has been answered by psychophysical comparison of discrimination ability at luminance and texture boundaries. For a range of tasks including orientation discrimination, spatial frequency discrimination, and curvature discrimination, thresholds for texture boundary stimuli are only about twice as large as those for luminance stimuli (Vogels and Orban, 1987; Wilson and Richards, 1992; Lin and Wilson, 1996). This small processing penalty is presumably due both to noise resulting from the extra neural processing and to some loss of accuracy inherent in the larger second-stage receptive fields of non-Fourier mechanisms. Interestingly, there is also evidence that non-Fourier pattern discrimination requires a longer stimulus duration than Fourier pattern discrimination (Lin and Wilson, 1996). This presumably reflects the extra time necessary for the additional rectification and pooling stages, a point to which we shall return in discussing non-Fourier motion processing.

The sequence of filtering followed by rectification and subsequent filtering at a lower spatial frequency represents the basic theme of non-Fourier processing, but it should be emphasized that this merely describes the simplest possible non-Fourier processor. The key aspect is the inherently nonlinear rectification stage, which

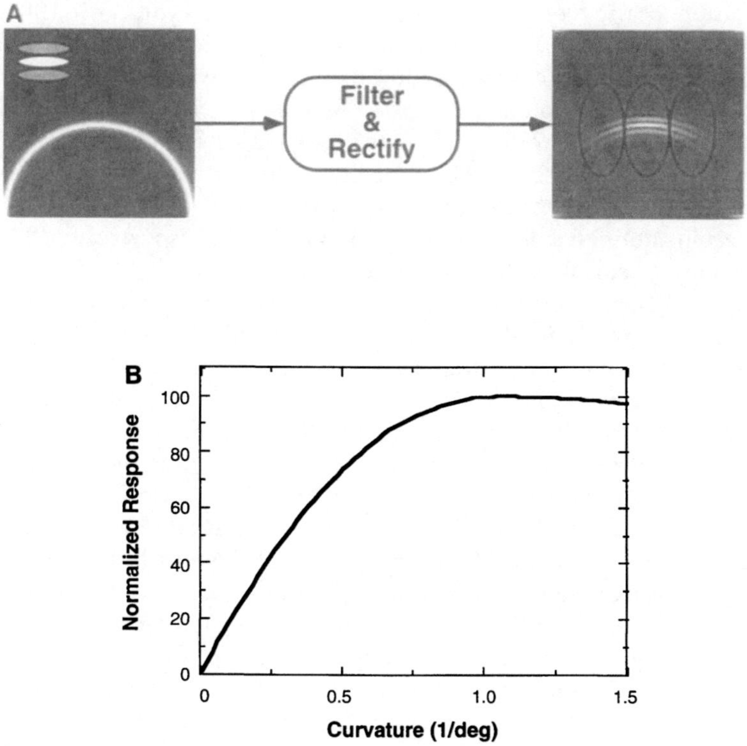

Figure 3. Non-Fourier processing of a curved contour. (A) Curved contour with inset of horizontal first-stage oriented filter. The result of first-stage filtering and rectification is depicted on the right, which shows that the neural response is largest at the point of horizontal tangency to the curve. The large vertical second-stage filter, shown in outline, will receive mainly excitatory input and little inhibition in response to the curved contour. (B) Computed model responses as a function of contour curvature (reciprocal of the radius of curvature). Zero curvature corresponds to a long, straight contour, which generates equal excitation and inhibition in the second-stage filter and therefore generates no response. As curvature increases, the neural response increases monotonically over a substantial range.

generates components not present in the Fourier description of the stimulus. Others have shown that some non-Fourier processes utilize half-wave rather than full-wave rectification (Malik and Perona, 1990; Sperling, *et al.,* 1994). In addition, several more complex models involve additional inhibitory or nonlinear stages before the final filtering (Malik and Perona, 1990; Bergen and Landy, 1991; Graham *et al.,* 1992). However, the simple non-Fourier scheme works extremely well with a wide range of texture boundaries generated by phase, orientation, or spatial frequency changes. As shown in Fig. 3, the same scheme is also effective in encoding boundary curvature.

2.3. Non-Fourier Mechanisms and Single-Cell Physiology

How do the properties of non-Fourier models relate to single-unit electrophysiology? The most direct comparison is with the work of von der Heydt and his colleagues, who reported that many neurons in cortical area V2 responded to illusory contours of the type depicted in Fig. 2 (von der Heydt, *et al.*, 1984; von der Heydt and Peterhans, 1989). Furthermore, they reported that no V1 neurons responded to these illusory contours, although recent evidence suggests that this conclusion may require some modification (Grosof *et al.*, 1993). In a striking behavioral study, Merigan *et al.* (1993) assessed the visual effects of V2 lesions in macaque monkeys. These lesions affected neither acuity nor orientation discrimination using simple lines. However, V2 lesions made it impossible for the monkeys to discriminate orientations defined solely by texture boundaries. This is strong evidence that V2 neurons play a crucial role in non-Fourier processing.

Model non-Fourier neurons respond to illusory contours in the same way as a subset of V2 neurons. It is also instructive to consider how model non-Fourier neurons respond to cosine gratings as a function of the height of the bars. As illustrated by the dashed curve in Fig. 4, the response of a vertically oriented simple cell to a vertical grating increases monotonically until the grating height exceeds the height of the receptive field, after which the response asymptotes. This is typical of the behavior of simple cells as reported by Hubel and Wiesel (1968). Consider now the response of a non-Fourier pathway neuron that employs the same vertical simple cell receptive field as its first-stage filter. Subsequent full-wave rectification followed by a horizontally oriented final pooling stage produces the response shown by the solid curve in Fig. 4. The response first increases until the height of the grating equals the width of the excitatory region of the final pooling process. Further increases in height impinge upon the inhibitory zones of this final receptive field, and the response then decreases back to zero. Thus, non-Fourier model neurons have response properties similar to complex, end-stopped cortical neurons (Dreher, 1972; Henry, 1977), which were termed "hypercomplex" by Hubel and Wiesel (1968). Indeed, there is evidence that the end-stopping originates with cells responding only to long bars, and these cells provide the inhibitory end-stopping for cells that respond to shorter bars (Bolz and Gilbert, 1986). This physiological evidence is fully compatible with the non-Fourier model in Fig. 1B if an interneuron were to perform the pooling that generates the inhibitory flanks of the final-stage receptive field. It has also been noted that end-stopped cells in the cortex will respond to boundary curvature (Dobbins *et al.*, 1987, 1989), and this characteristic is shared by non-Fourier model neurons as a result of their end-stopped characteristic (see Fig. 3).

3. Area V4 Neurons and Form Vision

3.1. Non-Cartesian Units in V4

Area V4 receives input from both V1 and V2, and it in turn projects to inferior temporal cortex (IT) (VanEssen, 1985; VanEssen *et al.*, 1992).* Although it was originally suggested that V4 was primarily devoted to color processing (Zeki, 1978),

*The direct projection from V1 is much smaller than that from V2 to V4, and the direct projection may be absent in the periphery. However, the focus here is on the fovea, and a direct V1-to-V4 projection in the fovea has been documented.

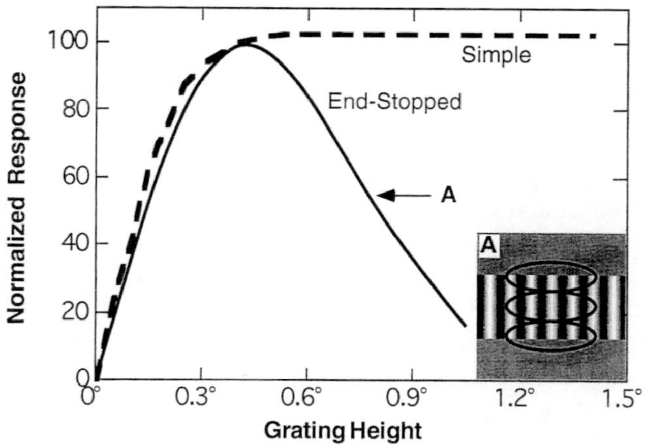

Figure 4. Normalized response of Fourier and non-Fourier units as a function of grating height. The non-Fourier unit responds to gratings in the same manner as end-stopped complex cortical neurons. All gratings were vertical, and they were centered on the receptive field. The dashed curve marked Simple plots the response of a Fourier unit like that in Fig. 1A. This response increases to an asymptotic level as the grating height increases to the length of the receptive field and thereafter remains constant. The solid curve marked End-Stopped illustrates responses of a unit like Fig. 1B. The second-stage filter was chosen to be orthogonal to the first stage and of twice the linear dimensions (half the peak spatial frequency). This non-Fourier response increases to a maximum as the grating height fills the excitatory region of the second-stage filter; thereafter it decreases to zero as the inhibitory zones of the second-stage filter are progressively invaded. The inset A depicts partial incursion of the grating into the inhibitory zones of the second-stage non-Fourier filter (point *A* on response curve).

recent evidence indicates that this is not the case. Following bilateral removal of V4 in macaques, Heywood *et al.* (1992) observed large deficits in form perception, but no loss of color discrimination capabilities. Merigan (1996) has also observed major deficits in form perception following V4 lesions.

As both the ablation data and its interconnections with IT indicate that V4 is part of the form vision pathway (Mishkin *et al.*, 1983; Desimone *et al.*, 1985), one might expect that V4 neurons would have spatial receptive fields that are more elaborate than those of V1 and V2 neurons. Gallant *et al.* (1993) explored this possibility using cosine gratings defined in polar or hyperbolic coordinates. To distinguish these from standard cosine gratings in Cartesian coordinates, these were termed non-Cartesian patterns. For example, concentric (Con) and radial (Rad)gratings are defined by the equations

$$\begin{aligned} \text{Con} &= \cos\,(2\pi\omega\,\sqrt{x^2 + y^2}) \\ \text{Rad} &= \cos\,[2\pi\omega\,\tan^{-1}(y/x)] \end{aligned} \tag{2}$$

where ω is the spatial frequency in cycles per degree (cpd) or cycles per 2π radians. A concentric function is plotted at the top left of Fig. 6. The striking result of their initial study was that 16% of V4 neurons responded much better to polar or hyperbolic cosines than to normal cosine gratings. Conversely, only 8% responded best to conventional gratings. In a subsequent, more extensive study of V4 neurons, these authors found that "The majority of cells fell into one of two groups that were selective for polar gratings; one selective for radial gratings, and one selective for concentric or spiral gratings. There was no group whose primary selectivity was for Cartesian gratings" (Gallant *et al.*, 1996).

If non-Cartesian units represent a new form of receptive field organization first appearing in V4, some form of pooling across orientations would be necessary. Striking anatomical support for such orientation convergence in V4 has recently been provided by Schoups *et al.* (1995). These investigators used a double deoxyglucose labeling approach to simultaneously map left oblique and right oblique orientations in cortical areas V1, V2, V3, and V4 of macaques. In V1, V2, and V3 the two labels marked separate, interdigitated orientation columns. In V4, however, the two labels were found to be superimposed, thus indicating anatomical convergence of orthogonal orientations! While superimposed deoxyglucose labeling from orthogonal orientations does not in itself guarantee convergence at the single-cell level, such superposition is to be expected if orientation pooling is an important organizational feature of V4 receptive fields.

3.2. A Model of Non-Fourier V4 Concentric Units

Evidence for the existence of non-Cartesian units in V4 raises the following question: what sort of neural circuitry might produce a concentric (or radial; see below) V4 unit? One such circuit is illustrated in Fig. 5. It consists of four parallel processing pathways that are subsequently summed and then passed through a nonlinearity such as $R(S)$ in Equation (1). Each individual pathway is nothing more than a non-Fourier pathway like that in Fig. 1B, except that the second-stage receptive field involves summation of two non-Fourier filters located symmetrically around the receptive field center. Following simple cell processing and rectification, each pathway thus pools responses using a weighting function of the form

$$W(x,y)=\left\{A\ \exp\left(-\frac{x^2}{\sigma_E^2}\right)-B\ \exp\left(-\frac{x^2}{\sigma_I^2}\right)\right\}\left\{\exp\left(-\frac{(y-\Delta)^2}{\sigma_y^2}\right)+\exp\left(-\frac{(y+\Delta)^2}{\sigma_y^2}\right)\right\} \quad (3)$$

This is just the sum of two oriented second-stage filters vertically separated by distance 2Δ. As depicted at the top of Fig. 5, this filter follows horizontally oriented first-stage processing. The formula for $W(x, y)$ in Equation (3) is rotated to other orientations for the remaining pathways in Fig. 5. Key to the operation of these filters is the observation that non-Fourier units will respond to contour curvature (see Fig. 3). In the first pathway in Fig. 5, therefore, responses of horizontally oriented simple cells above and below the center of $W(x, y)$ generate excitation as long as these horizontal units are tangent to curved rather than straight contours. Similarly, the second pathway performs the same sequence of operations, but with the simple cell receptive field and $W(x, y)$ both rotated by 45°. In this model only four parallel channels are depicted, although simulations have been implemented with up to 12 orientations incorporated.

It should be stressed that each of the parallel channels in Fig. 5 is simply a pair of non-Fourier units from Fig. 1B. Such units respond to concentric curved contours in virtue of their end-stopped characteristics (Dobbins *et al.*, 1987, 1989; Koenderink and Richards, 1988; Wilson and Richards, 1992). In addition, note that $W(x, y)$ must be significantly larger than the dimensions of the simple cell receptive fields if this scheme is to be successful. In this instance, the diameter of $W(x, y)$ was

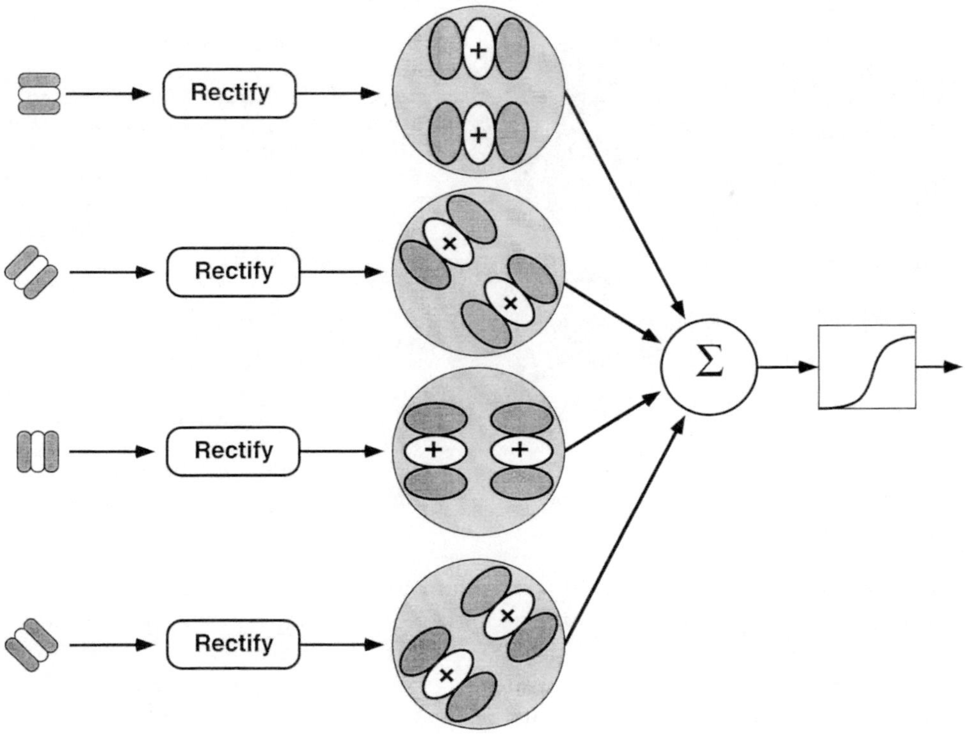

Figure 5. Schematic of a neural network for a V4 concentric unit. Each of the four parallel non-Fourier pathways consists of oriented filtering, rectification, and second-stage oriented filtering using the weighting function in Equation (3). These second-stage filters have a central excitatory lobe (white) and flanking inhibitory lobes (gray). Note that the second-stage filters are orthogonal to those in the first stage. Each non-Fourier pathway pools a different orientation (0°, ±45°, 90°), and all four are subsequently summed (Σ) and passed through a nonlinear response function with threshold and saturation behavior [see Equation (1)]. All second-stage filters are located on light gray circles that indicate the spatial extent of the resulting receptive field. Although only four orientations were used here, more could readily be incorporated.

chosen to be 5.0 times the length of the simple cell filter, which is in accord with the observation that V4 receptive fields have linear dimensions that are four to seven times larger than those of V1 neurons (Desimone and Schein, 1987).

The response of a concentric cell is produced by summing the four oriented pathway responses and passing this through an appropriate nonlinearity [see Equation (1)]. Responses of this model were simulated on a Macintosh computer using MatLab software. The first-stage simple cells had receptive fields described by $RF(x, y)$ in Equation (1), using parameters for the smallest of the receptive fields described in Wilson (1991a). Two examples of image processing by concentric units are shown in Fig. 6. In Fig. 6A a 128×128 array of concentric units was convolved with $\mathrm{Con}(x, y)$ [Equation (2)].* The diameter of this pattern was chosen to coin-

*The reference to convolution here is somewhat of a misnomer, as convolution is a linear process, while the neural architecture of the concentric units is highly nonlinear due to the rectification stage. Convolution in this context is used loosely to indicate that the first-stage and second-stage receptive fields were convolved with their appropriate inputs in order to generate an array of concentric unit responses.

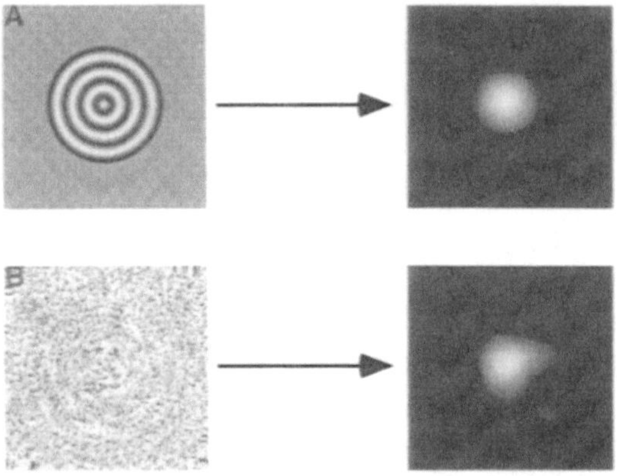

Figure 6. Responses of model V4 concentric units (Fig. 5) to two visual stimuli. In each case a 128 × 128 spatial array of concentric units was used to analyze the stimulus on the left, which yielded the V4 response on the right (active units depicted in gray to white, inactive and inhibited units in black). The stimulus in panel A is the concentric cosine [Con, Equation (2)] used to study V4 responses by Gallant *et al.* (1993). In agreement with physiology, the model V4 units respond well even when the target is displaced by several cycles. The model units also detect the circular structure evident in the Glass (1969) pattern in panel B. The V4 response here is not perfectly concentric, which results from noise in the stimulus. In addition, the response in panel B was weaker than that in panel A, reflecting the effective contrasts of the circles in the two stimuli.

cide with the diameter of the $W(x, y)$ function for the concentric units, but this is not necessary, as the concentric units respond to these patterns over a wide range of sizes. Responses of the array of model concentric units are plotted on the upper right of Fig. 6, with gray and white indicating the relative level of excitation of the unit centered at each point and black indicating units that are unresponsive or inhibited. The width of the activity pattern indicates that these units are insensitive to the positioning of $\text{Con}(x, y)$ within a range of about ±2.5 cycles in any direction. Furthermore, the model concentric units respond to $\text{Con}(x, y)$ patterns over a 1.5-octave spatial frequency range at half amplitude, and they respond to a wide range of pattern diameters. In all these respects the model neurons agree with properties reported for V4 concentric units by Gallant *et al.* (1993, 1996).

Further simulations indicate that these model concentric units do not respond to cosine gratings, radial [$\text{Rad}(x, y)$ in Equation (3)], or hyperbolic patterns. Subsequent work by Gallant *et al.* (1996), however, suggests that even those V4 neurons with the strongest preferences for concentric or radial patterns still responded weakly to conventional cosine gratings. This can be readily incorporated into the concentric unit model by simply weakening inhibition in the second-stage filters. For example, reducing the strength of the second-stage inhibitory zones by 50% produces a concentric unit that responds about 25% as strongly to a cosine grating as to a concentric grating.

3.3. Responses to Glass Patterns

To explore the responses of these model V4 neurons further, let us consider the random-dot patterns discovered by Glass (1969). An example is shown in Fig. 6B. To produce this image, a random-dot pattern with 16% black dots was generated. A copy of this was made, rotated about its center by 6.5°, and then added to the original pattern. As is evident from the figure, subjects readily perceive noisy concentric circles around the pattern center. Glass (1969) also demonstrated that these circles vanish if the angle of rotation is increased sufficiently, as subjects then perceive only random dots of double the initial density (due to the pattern addition). When the Glass pattern in Fig. 6B was used as a stimulus for the array of model V4 concentric units, the pattern of responses on the lower right of the figure resulted. In accord with the perception of human subjects, there was strong activity of concentric units centered on the pattern and inhibition virtually everywhere else. Due to the noisy structure of the Glass pattern, of course, the neural response is somewhat distorted relative to the precise response in Fig. 6A. To determine the relative strength of the model response, the neural computation was carried out for ten random Glass patterns and the average taken. If we arbitrarily set the maximum response to $\mathrm{Con}(x, y)$ in Fig. 6A to 100%, the maximum response to Glass patterns averaged 12% ±3% of this. This is in qualitative agreement with the observation that the circles in the Glass pattern are far less distinct than those in $\mathrm{Con}(x, y)$. In contrast, when the angle of rotation was large enough so that no circles could be perceived, the model response was zero on average!

Wilson *et al.* (1997) studied Glass pattern perception psychophysically and obtained direct evidence in support of the concentric unit model in Fig. 5. In particular, it was shown that human perception of concentric Glass patterns involves full-wave rectification followed by concentric orientation pooling. Simulations showed that the model in Fig. 5 (with slightly different second-stage filters) could quantitatively predict concentric Glass pattern thresholds for a variety of stimulus manipulations.

3.4. Non-Fourier Subunits and End-Stopping

The V4 concentric unit circuit depicted in Fig. 5 pools responses of non-Fourier subunits. As discussed above, these non-Fourier units respond to gratings in ways well documented for end-stopped complex cells, a characterization that fits many cells in both V2 (von der Heydt and Peterhans, 1989) and V4 (Desimone and Schein, 1987). In short, responses of the non-Fourier subunits comprising the concentric unit circuit in Figure 5 are well documented physiologically. What has been less well appreciated is the observation that end-stopped units will encode contour curvature (Dobbins *et al.*, 1987, 1989; Wilson and Richards, 1992). Koenderink and Richards (1988) have used observations from differential geometry to elucidate the theoretical basis for regarding end-stopped units as local curvature extractors. The concentric V4 unit model in Fig. 5 takes the next step in this analysis of stimulus geometry by integrating local curvature information in a globally structured fashion.

3.5. Non-Fourier Concentric Units and Face Processing

As V4 projects to area IT, and as there is overwhelming evidence that a subpopulation of IT neurons respond well to faces (Rolls, 1984; Perrett and Chitty, 1987; Gross, 1992), model V4 units were applied to face processing. Results of one simulation using a 256 × 256 image are depicted in Fig. 7, where the diameter of concentric unit receptive fields was one fourth the width of the image. The re-

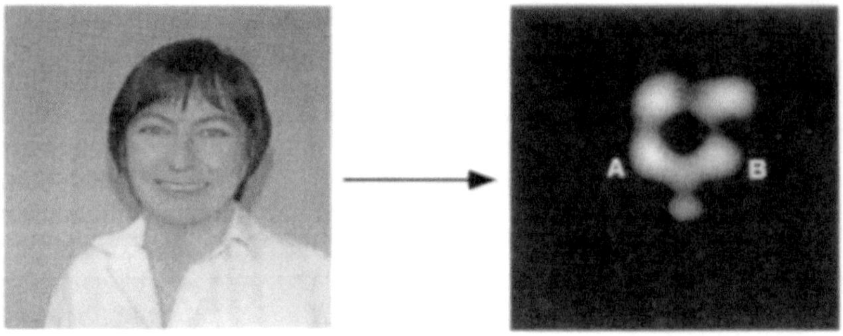

Stimulus V4 Response

Figure 7. Face processing by model V4 concentric units. The 256 × 256 pixel image on the left was processed by an array of model V4 units yielding the activity pattern on the right. The diameter of the V4 units was equal to one fourth of the image width. These units pick out quasicircular groupings of facial elements. For example, the two lower patches of high activity (labeled A and B) are centered on the cheeks and are responding to the curve through the eye and brow, side of the nose, shadow below the cheek, and finally the contour at the side of the head. Responses of larger model V4 units (not shown) encode the overall shape of the head.

sponse pattern on the right of Fig. 7 was characteristic of responses to several other faces as well. What these units do is to pick out the centers of regions that are approximately bounded by a smooth curve. For example, the activity peak labeled A is produced by the vertical curve of the nose, the horizontal curve of the left eye and eyebrow, the vertical curve of the left side of the face and hairline, and finally the lower horizontal curve of the jaw line and left part of the lips. Peak B responds to the same constellation of features around the other cheek. In contrast, concentric units centered between the eyes are inhibited, because the pattern of eyes, nose, and vertical hair on the forehead is roughly radial here. (Simulation of V4 units preferring radial gratings using a simple variant of the processing in Fig. 5 produced strong activation centered between the eyes.) V4 concentric units with larger receptive fields will, of course, respond to the overall shape of the head, which is roughly circular.

The point of this sojourn into the potential of V4 concentric units for face processing is definitely not to suggest that they are "face detectors." After all, the same units respond well to Con(x, y) and Glass patterns in addition to faces. The point is simply to indicate that human (and animal) faces typically have roughly circular regions delimited by characteristic facial features. Indeed, elementary drawing books suggest that the novice begin by constructing faces from circles that interlock in characteristic ways (Hamm, 1963), and the same is true for heads of animals (Hamm, 1969). In fact, Hamm (1963, p. 23) describes "hidden circles" in a female face, and their centers lie very close to the local maxima in the model V4 responses in Fig. 7. The suggestion is thus that V4 concentric units signal valuable structural information about faces, information that would be of value for face processing in IT.

Two striking observations support the proposition that V4 concentric units are involved in face processing. First, Yamane *et al.* (1988) have found that many cells in

IT "detect the combination of the distances between facial parts such as eyes, mouth, eyebrows, hair, and so on" (p. 209). This is exactly the response pattern of the model V4 units shown in Fig. 7 at A and B. Even more striking is the study by Rentschler *et al.* (1994), who compared the perceptual deficits of two human patients, one of whom was alexic and one of whom was prosopagnosic. The prosopagnosic had right temporal lobe damage and could not recognize faces, facial emotions, or the style of handwriting, although she could read perfectly well. In contrast, the alexic patient could not read, but had no difficulties with face perception. Rentschler *et al.* (1994) discovered that the prosopagnosic was virtually at chance in perceiving the global structure in concentric Glass (1969) patterns, while the alexic processed these patterns normally. This provides a direct association between the global, concentric pooling involved in Glass pattern perception (Wilson *et al.*, 1997) and face recognition.

3.6. Non-Fourier Form Units and V4 Physiology

A few further points deserve mention. First, it must be emphasized that the processing scheme in Fig. 5 represents a summation of several non-Fourier pathways. Simple modifications of the second-stage filters in this model produce units that respond best to hyperbolic cosines, and these units have been reported to occur in V4 (Gallant *et al.*, 1993, 1996). Indeed, Kobatake and Tanaka (1994) also reported that many V4 units respond best to either X-shaped or roughly circular stimuli. Thus, the generic non-Fourier scheme may be used to generate several types of units useful for V4 form analysis. Second, the final receptive fields in these pooled non-Fourier pathways must be significantly larger than those of the simple cells in the first stage to extract curves properly. In fact, one can estimate the size of receptive fields that would be optimal for this purpose given the orientations represented in V1, which are approximately 15° apart at the smallest (Hubel and Wiesel, 1977). Given 15° orientation differences, the maximum number of orientations around a circle processed by V1 cells would be $360°/15° = 24$. This would be approximated by a circle with a circumference equal to 24 times the length of V1 simple cell receptive fields, and its diameter would therefore be 7.6 times the simple cell length. At low spatial frequencies orientation bandwidths are about 30° (DeValois *et al.*, 1982; Phillips and Wilson, 1984), so there would be 12 orientations around a circle, which would therefore have a diameter 3.8 times greater than the length of V1 units. This simple calculation may account for the report that V4 units have receptive fields four to seven times larger than V1 units (Desimone and Schein, 1987).

In a similar vein, it is easy to show that the seemingly low percentage of concentric units in V4 is actually close to optimal. For a complete spatial array of perhaps 12 orientations at a single spatial frequency, V4 would require only one spatial array of concentric units (or perhaps two if both large and small units exist). One would therefore expect concentric units to represent between 8% and 17% of the V4 population, and this is what is found (Gallant *et al.*, 1993, 1996). The situation is analogous to that of M and P cells in the primate retina or X and Y cells in the cat retina, as the M/P and Y/X cell ratios are both around 10%. This does not reflect relative importance of the cell types, however, but relative receptive field size: M and Y cells have much larger receptive fields, and all cell types sample the central retinal image at their respective Nyquist frequencies (Wilson *et al.*, 1990).

Desimone and Schein (1987) have reported that V4 units have about the same spatial frequency bandwidths as V1 neurons. In the neural model this is a direct consequence of initial filtering by simple cell receptive fields, assuming them to reside in V1. Only frequencies to which the simple cells are sensitive can be processed further, and the spatial frequency tuning of the final receptive fields is too broad to change that significantly. Finally, most V4 units have been reported to have strong suppressive surrounds (Desimone *et al.*, 1985; Desimone and Schein, 1987). Such surrounds are not present in the scheme for concentric units developed here (although they are present in the form of end-stopping for the non-Fourier subunits; see Fig. 4). It is easy to include suppressive surrounds through recurrent lateral inhibition among spatially distributed model V4 concentric units (see discussion of spatial interactions in MT below).

3.7. Psychophysics of Non-Fourier Concentric Units

Several major results in the psychophysical literature can be given a unified explanation based on V4 concentric units. First, kittens can discriminate concentric patterns from radial patterns at a very early age, and they do this much better than discrimination of horizontal from vertical gratings (Wilkinson and Dodwell, 1980; Dodwell *et al.*, 1983). If concentric units were important for cat face processing, it would be extremely valuable for kittens to develop them early in life. Second, in a striking series of experiments, Olzak and Thomas (1991, 1992) discovered that orthogonal orientations are never processed independently. Rather, they showed that higher visual processes sum across all orientations linearly, but only at a single spatial frequency. Although their data do not speak to the spatial structure of this pooling process, if the pooling they documented is concentrically arranged (see below), this would be exactly the result predicted by V4 concentric units in human vision.

Field *et al.* (1993) studied the perception of contours defined by sequences of oriented Gabor functions. Subjects could detect these contours in a noise field of randomly oriented and positioned Gabors as long as the following conditions were met: (a) Gabors defining the contour had to be oriented along the contour and not perpendicular to it, (b) adjacent Gabors could not shift in orientation by more than 45° or performance suffered, and (c) the phase of the Gabors could be randomly varied with no detriment to performance. Reference to the model circuit will reveal that all of these conditions are optimal for responses by model V4 concentric units: points (a) and (b) reflect the relative orientations of first- and second-stage filters, and point (c) is a consequence of rectification.

Motivated by the success of the non-Fourier concentric unit model in explaining V4 responses to concentric patterns, we have conducted several human psychophysical experiments to characterize such units in human vision. As mentioned above, we discovered that the detection of circular Glass (1969) patterns in noise involves linear summation around the pattern circumference (Wilson *et al.*, 1997). This result is predicted by the model in Fig. 5. In addition, Wilkinson *et al.* (1998) have investigated pattern discrimination using stimuli that contain subtle distortions of circular shapes. For all subjects optimal discrimination falls in the hyperacuity range, with thresholds of 5.0 arc sec or less. Furthermore, subjects can identify certain characteristic shapes of distorted circles near threshold, even in 167 ms flashes, suggesting that certain quasi-circular shapes are particularly important in form vision.

A study using radial Glass patterns has provided evidence for the existence of a second class of V4 type unit in human vision that sums radially organized orientations (Wilson & Wilkinson, 1998). The data were fit quantitatively using a simple variant of the concentric model in Fig. 5, the only change being that the first and second stage filters were both radially oriented with respect to the respective field center. Experiments by Li and Westheimer (1997) also suggest that the visual system is very effective in processing X-shapes. Thus, psychophysics agrees with the V4 physiology of Gallant *et al.* (1993, 1996) in showing greater sensitivity to concentric and radial structures than to parallel ones. Models of the type depicted in Fig. 5 provide a unified account of both the physiology and psychophysics.

4. Two-Dimensional Motion

4.1. Parallel Fourier and Non-Fourier Motion Pathways

Let us now return to the basic non-Fourier scheme in Fig. 1B and explore its role in the analysis of two-dimensional motion. As details of the theoretical model for area MT discussed here have been recently published (Wilson *et al.,* 1992; Wilson and Kim, 1994a; Wilson, 1994a,b), the focus will be on the key qualitative features of that model and the similarities between model MT units and physiological data on MT neurons.

As shown in Fig. 8, the neural model for two-dimensional motion involves processing of moving stimuli by parallel Fourier and non-Fourier pathways followed by combination of these responses in a network that will be shown to have many of the properties of area MT. Psychophysical evidence that motion perception is a two-stage process has emerged from several studies (Adelson and Movshon, 1982; Welch, 1989; Derrington and Suero, 1991).

Initial stages of the Fourier and non-Fourier pathways are identical to those for static pattern analysis in Fig. 1, but each is followed by a motion extraction stage, which may be modeled equivalently by a Reichardt (1961) or an opponent motion energy mechanism (Adelson and Bergen, 1985). The existence of a non-Fourier motion pathway in human vision has been demonstrated in several laboratories (Badcock and Derrington, 1985, 1989; Chubb and Sperling, 1988, 1989; Turano and Pantle, 1989; Cavanagh and Mather, 1990), and Smith (1994) has provided an excellent contemporary synopsis of the literature. Recently, Ledgeway and Smith (1994) demonstrated psychophysically that Fourier and non-Fourier pathways do indeed operate in parallel in the human visual system.

It is natural for the visual system to contain both Fourier and non-Fourier motion pathways for the same reason that it contains parallel pathways for the analysis of static patterns: the Fourier motion pathway signals motion of luminance edges, while the non-Fourier pathway signals motion of texture boundaries and

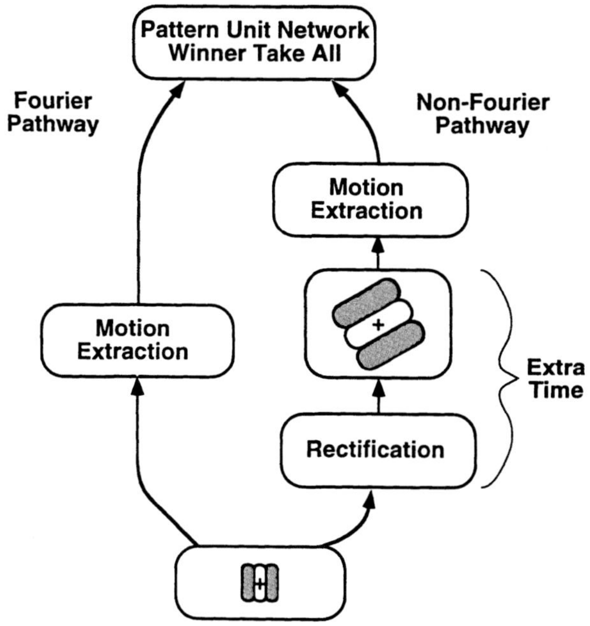

Figure 8. Schematic of a model of cortical motion processing. Following oriented filtering (bottom), motion information is extracted in parallel Fourier and non-Fourier pathways using Reichardt (1961) detectors. The non-Fourier motion pathway incorporates additional, time-consuming rectification and oriented filtering stages before motion extraction. The Fourier and non-Fourier pathways are subsequently combined as inputs to a pattern unit network (presumably in area MT) that calculates the vector sum direction using a modified "winner-take-all" architecture. As discussed in the text, the Fourier pathway may represent the V1 → MT pathway and the non-Fourier pathway the V1 → V2 → MT pathway.

illusory contours (see Fig. 2). Given that these two motion pathways process moving stimulus in parallel, it would be natural for them to be combined at a later stage of processing as illustrated in Fig. 8. Indeed, Albright (1992) showed that 87% of macaque area MT cells respond to both Fourier and non-Fourier motion, while Zhou and Baker (1993) obtained similar results in cat cortex. Derrington *et al.* (1992) showed directly that Fourier and non-Fourier motion signals are combined in human processing of two-dimensional motion. Thus, the combination of Fourier and non-Fourier motion signals extracted from the stimulus in parallel pathways is a well-established feature of human and primate motion processing.

4.2. Vector Summation of Fourier and Non-Fourier Motion Signals

There remains the issue of just how Fourier and non-Fourier motion signals are combined in the final stage of the motion model. Three key studies appeared almost simultaneously showing that the appropriate operation was likely to be vector summation. First, Yo and Wilson (1992) showed that two-dimensional plaids, which are composed of the sum of two cosine gratings moving in different directions (see Fig. 10), are perceived to move in the vector sum of grating directions at short durations. This is true even when the vector sum direction differs from the true direction of physical motion by up to 55°. Concurrently, Mingolla *et al.* (1992) and Rubin and Hochstein (1993) showed that patterns cleverly constructed to exclude non-Fourier motion were always seen to move in the vector sum of the component motion directions. A final critical test of the vector summation hypothesis showed that patterns composed solely of moving non-Fourier components are always perceived to move in the vector sum direction (Wilson and Kim, 1994b). So the final motion stage apparently involves a vector sum calculation.

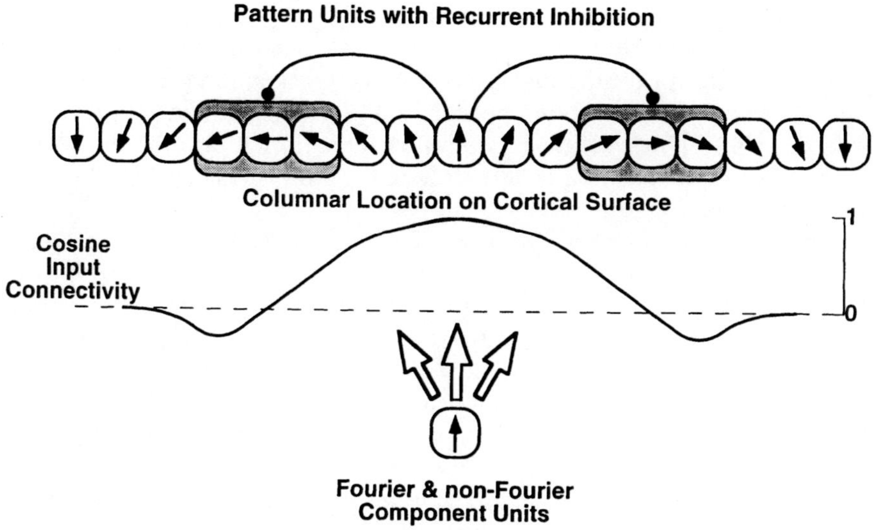

Figure 9. Synaptic connectivity function from Fourier and non-Fourier component units to model MT pattern units. The preferred direction of the depicted component unit is vertically upward, and this unit synapses with pattern units having the preferred directions indicated by the arrows with weights given by a cosine function of relative direction. As indicated, there are no inhibitory connections between units signaling directions differing by more than ±120°, which permits processing of transparent motion. Each pattern unit recurrently inhibits other pattern units (gray rectangles) except for its two nearest neighbors on either side and units beyond ±120°. The diagram is constructed in accord with physiological evidence for a regular, columnar organization of preferred direction in MT (Albright *et al.*, 1984). Given such a columnar arrangement, all connections represent forms of lateral excitation or lateral inhibition. The same connections are replicated for each pattern unit relative to its preferred direction.

A simple way in which a neural network can compute a vector sum is illustrated in Fig. 9. Let us assume that neural units signaling the velocities* of Fourier and non-Fourier components in the stimulus provide inputs to model MT pattern motion units, inputs that are weighted by the cosine of the component direction relative to the direction signaled by the pattern unit. That is, if a component unit is tuned to direction θ, its input to a pattern unit tuned to direction Ω will be weighted by $\cos(\Omega - \theta)$. This input connectivity function should be extremely easy to construct if direction-selective units in MT are arranged in regular columnar arrays as suggested by the data of Albright *et al.* (1984). As shown in the lower half of Fig. 9, this is nothing more than an excitatory center, inhibitory surround input distribution across the array of MT direction columns. Restriction of the inhibitory range of this cosine weighting will be discussed in a moment. Given this cosine weighting of Fourier and non-Fourier component inputs, we have proved that the MT unit receiving the maximum input will correctly signal the vector sum direction (Wilson *et al.*, 1992). In symbols:

$$\max_{\Omega}\left[\sum_{\theta=-120^\circ}^{120^\circ} S_\theta \cos(\Omega-\theta)\right] = \text{vector sum direction} \tag{4}$$

*Reichardt detectors respond to temporal frequency rather than velocity. We have shown, however, that simple gain controls operating on the outputs of Reichardt detectors suffice to provide a signal proportional to velocity over a considerable range (Wilson and Kim, 1994a).

where S_θ is the speed of the Fourier or non-Fourier component moving in direction θ, and Ω is the pattern-unit preferred direction. The maximum calculation is now readily implemented via competitive inhibitory feedback among the pattern units, which produces a "winner-take-all" network (Wilson and Cowan, 1973; Feldman and Ballard, 1982). The nonlinear differential equations that must be solved to implement this network can be found elsewhere (Wilson *et al.*, 1992; Wilson and Kim, 1994a; Wilson, 1994b).

The spread of inhibitory feedback in the winner-take-all network is diagramed above the pattern units in Fig. 9. As with the cosine weighting of inputs, this is nothing other than lateral inhibition across the pattern unit direction columns reported by Albright *et al.* (1984). In the traditional winner-take-all network each unit inhibits all other units except itself, with the result that there can only be one winner within the network. To tailor a winner-take-all network to characteristics of MT, two innovations must be made. First, pattern units do not inhibit either of their two nearest neighbors, which results in three neighboring units surviving the inhibition. This cluster of three neighboring responses contains a sparse code for direction of motion that is accurate to 0.5° even when the units themselves have preferred directions separated by 15°–30°, and this direction can easily be extracted by parabolic interpolation (Wilson *et al.*, 1992). The second embellishment of the winner-take-all concept is limitation of the recurrent inhibition among pattern units to a range of ±120°. This permits the network to signal both motion coherence and transparency accurately, as shown below (Wilson and Kim, 1994a).

4.3. MT Model Prediction of Coherent Motion

One illustrative example of motion network responses is illustrated in Fig. 10. The network was stimulated by the cosine plaid (Fig. 10A) moving vertically upward, which consisted of two superimposed cosine gratings moving toward 30° and 60° (mean direction of 45° shown by black arrow F). The Fourier motion pathway uses simple oriented receptive fields and will only detect the motion of the two cosine components, which produces the cosine-weighted Fourier input to the pattern network shown by the white area on the polar plot in Fig. 10B. The non-Fourier pathway, using the filter–rectify–filter sequence in Fig. 1B, detects the motion of the texture boundaries in the plaid (dashed line in Fig. 10A), which are moving in the direction of arrow NF. Cosine weighting then produces the hatched non-Fourier input shown in the polar plot in Fig. 10B. The embellished winner-take-all competition among the pattern units then produces the responses indicated by the gray region in Fig. 10B, and these signal a motion direction of 7.4° relative to vertical (arrow in Fig. 10B). This bias of perceived motion direction away from the true direction of physical motion has been shown experimentally to be an inherent characteristic of human visual motion perception for a wide variety of patterns (Ferrera and Wilson, 1990; Burke and Wenderoth, 1993), and model predictions of these biases are consistently accurate (Wilson and Kim, 1994a).

Examination of Fig. 10 reveals that the non-Fourier component inputs to the pattern unit network fall about 90° away from the Fourier inputs, so they pull the vector sum computation back toward the true direction of pattern motion, the small bias resulting because this non-Fourier compensation is not perfect. In fact,

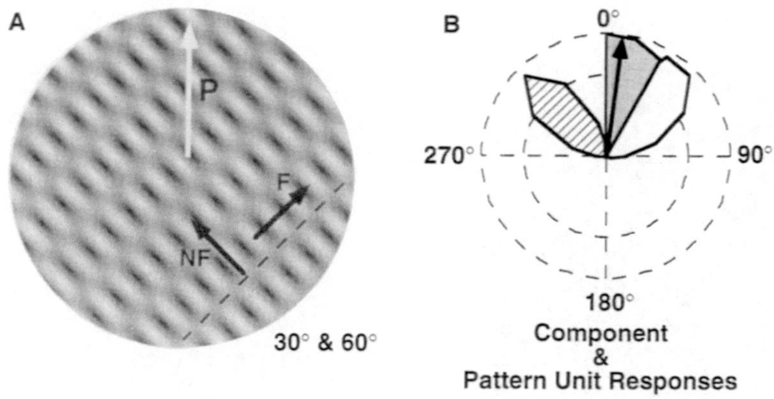

Figure 10. Depiction of a cosine plaid moving vertically upward (A) and a polar coordinate plot of neural responses in the MT model (B). Although the plaid in panel A is moving vertically upward (arrow P), the two component gratings are moving at 30° and 60° to the right, which generates the net Fourier component motion vector indicated by arrow F. Non-Fourier motion units extract the motion direction of the nodal lines or texture boundaries (dashed line), and these are moving upward and to the left as indicated by the NF arrow. Activity of Fourier component units is shown by the white lobe in panel B, and non-Fourier component activity is shown by the hatched lobe. Competitive interactions among the model MT pattern units generate the activity shown in gray, and parabolic interpolation (Wilson *et al.*, 1992) shows that this activity predicts a direction bias of 7.4° (arrow), which is in agreement with the psychophysics (Ferrera and Wilson, 1990).

much larger perceived direction biases are predicted and observed in peripheral vision, where non-Fourier motion signals are relatively attenuated (Yo and Wilson, 1992; Wilson *et al.*, 1992). As noted earlier, psychophysical evidence indicates that gratings like that in Fig. 10 are perceived to move in the vector sum of the Fourier component directions for short durations. For the pattern shown, this is a direction of 41° relative to the true vertical motion! The explanation for this effect is simple: the operations of rectification and subsequent filtering in the non-Fourier pathway require additional processing time, so the non-Fourier motion signals will have a delayed impact on the pattern unit network. Hence, processing will initially produce a vector sum of the Fourier component directions alone. A striking motion illusion discovered by the Japanese op artist Ouchi in 1977 will permit the reader to experience this consequence of slower non-Fourier motion processing directly. Figure 11 contains an adaptation of the Ouchi pattern plus instructions for generating the illusion.

4.4. MT Model Prediction of Motion Transparency

So far, we have only considered conditions under which the model signals the direction of rigid, coherent pattern motion. However, there are conditions under which the visual system fails to integrate local motions, and motion transparency results. For example, the multiaperture motion array in Fig. 12A consists of adjacent apertures, each containing a bar moving in one of two directions: obliquely upward to the left or right. Kim and Wilson (1993) showed that when the two directions were within about ±40° of vertical, the visual system fused them into a

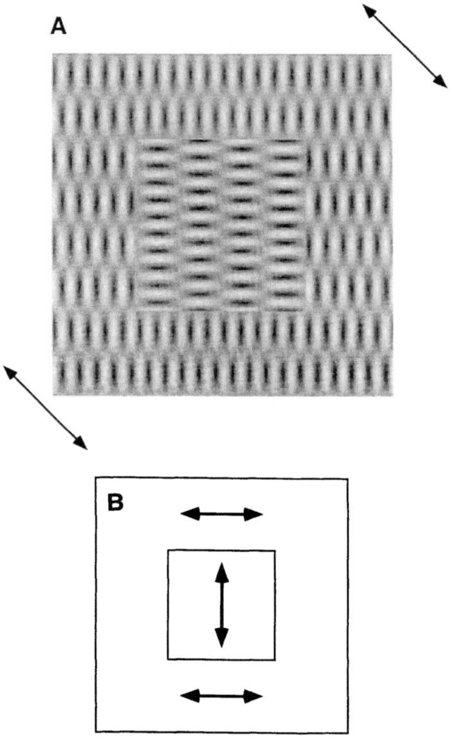

Figure 11. Modified version of the Ouchi (1977) motion illusion. If the pattern in panel A is viewed while oscillating the page back and forth diagonally as indicated by the double-headed arrows at the corners, the center of the pattern will be perceived to move up and down relative to side-to-side motion in the surround (perceived directions indicated schematically by arrows in panel B). For optimal perception of the illusion, the diagonal motion should have an amplitude of about 1 inch and a frequency of 1–3 Hz. The cortical motion model in Fig. 8 predicts this illusion as a result of slower processing in the non-Fourier motion pathway.

percept of rigid upward motion. For angles differing by ±60° or more, however, the patterns never cohered, and a percept of two transparent sheets sliding across each other was the result.* As shown in Fig. 12B, these widely separated input directions generate a bimodal input to the pattern unit network, and the network generates two winners indicating transparent motion in two directions (dark gray arrows). The use of multiple apertures in this case prevents the two sets of moving bars from overlapping and generating texture boundaries, so the only network inputs are Fourier component motions. It is the absence of recurrent inhibition beyond ±120° in the enhanced winner-take-all network that permits two winning directions for this stimulus, and this agrees with the data. Residual inhibition within the network causes the pattern units to signal directions of ±80° rather than the ±60° of the Fourier inputs, and this motion repulsion has been observed psychophysically (Marshak and Sekuler, 1979; Mather and Moulden, 1980; Kim and Wilson, 1996). MT model predictions of this motion repulsion as a function of bar contrast are quite accurate (Wilson and Kim, 1994a). Snowden *et al.* (1991) reported physiological evidence of inhibition in macaque MT neurons under conditions of motion transparency.

4.5. MT Model and Electrical Stimulation of MT

In an elegant experiment on motion perception in alert monkeys, Salzman and Newsome (1994) provided striking physiological support for the MT model

*Data at 45° varied from subject to subject, but were extremely consistent for each individual.

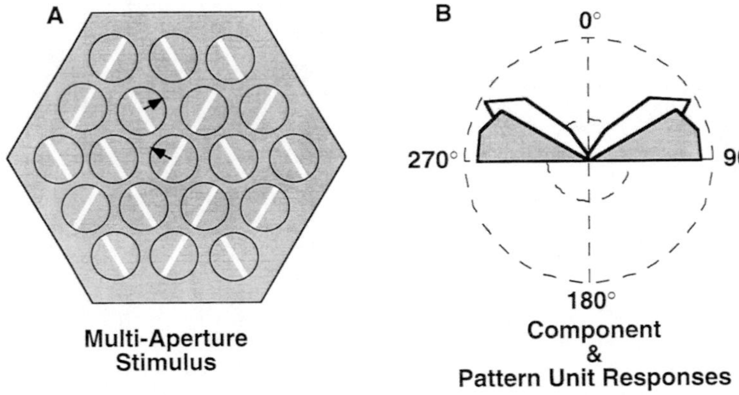

Figure 12. Motion transparency with multiaperture patterns. The pattern in panel A contains a hexagonal array of apertures through each of which a line is presented moving in one of two directions (arrows). When these two directions differ by more than 90° as illustrated, subjects perceive the lines as forming two transparent sheets moving almost in opposite directions horizontally (Kim and Wilson, 1996). A polar graph of model predictions is shown in panel B, where Fourier component responses are plotted in white. As the apertures prevent the lines from intersecting, no non-Fourier motion components occur in this stimulus. Given only the widely separated Fourier component inputs, the model predicts the two directions of pattern motion plotted in gray. Note that inhibition within the pattern unit network produced repulsion between these two response directions so that they are nearly at 90° and 270°. Such repulsion in motion transparency has been observed in several studies (see text).

developed here. They trained alert monkeys to signal the direction of motion of a random-dot pattern in order to receive a food reward. On some trials, they paired electrical stimulation of neurons in MT with the moving visual stimulus and thereby demonstrated that they could bias the monkey's response toward the direction signaled by the electrically stimulated cells. One major conclusion of this work was that the output of pattern units in MT reflected the operation of a winner-take-all network, which is precisely the computation implemented by the model. Indeed, their result was foreshadowed by the earlier psychophysical demonstration of hysteresis in motion perception (Williams *et al.*, 1986; Williams and Phillips, 1987), as hysteresis can only occur in a network with both cooperative and competitive interactions. In a winner-take-all network the cooperative aspect emerges from self-disinhibition by the winning units.

As a byproduct of their study, Salzman and Newsome (1994) also measured the direction tuning of both the visual inputs to MT direction columns and the direction tuning of the electrical stimulation as monitored by the behavioral response of the animal. As depicted in Fig. 13, the visual inputs spanned a range of greater than ±90°, while the MT output was much narrower. The cosine weighting of the Fourier and non-Fourier component units produces a very broad bandwidth for the visual inputs to the MT pattern network (heavy dashed line), in agreement with the multiunit data (open circles) of Salzman and Newsome (1994). The enhanced winner-take-all network among the model pattern units, however, produces a very narrow bandwidth centered on the direction of stimulus motion (heavy solid line), which agrees with the behavioral responses evoked by electrical stimulation (solid circles). In addition, Albright (1992) showed that the vast majority of MT cells respond to both Fourier and non-Fourier motion. Thus, key aspects of the cosine input weight-

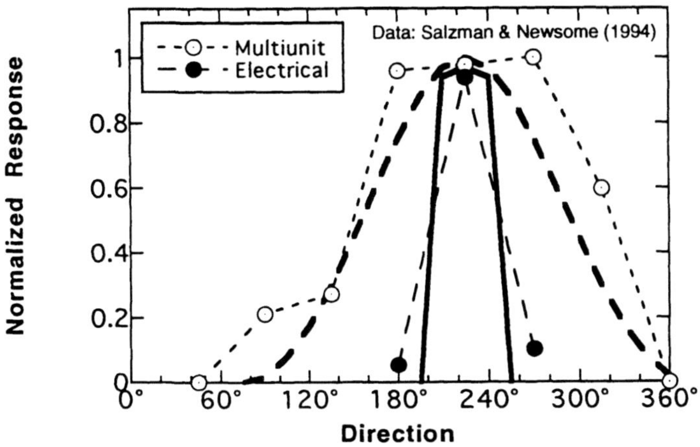

Figure 13. Comparison of cortical model with physiological responses in area MT of a behaving monkey. In both cases the stimulus was a moving sheet of random dots. Open circles show physiological data (Salzman and Newsome, 1994) on direction tuning of inputs to a single MT direction column. Note that they are nonzero over a range of more than 200°. As shown by the heavy dashed line, similarly broad tuning of component inputs occurs in the model due to the cosine weighting function (see Fig. 9). The solid circles show the narrow range over which electrical stimulation biased the monkey's behavioral response toward the preferred direction of this column. This narrow tuning is reflected in the model pattern-unit response tuning curve, which is plotted with a heavy solid line. Thus, the MT model correctly predicts that the broad cosine weighting of component direction inputs is transformed into a narrowly tuned pattern unit response by competition in the winner-take-all network.

ing and model MT network architecture have been experimentally corroborated since first publication of the model (Wilson *et al.*, 1992).

4.6. Spatial Interactions in MT

The motion model developed thus far has dealt with plaids, multiaperture patterns, and moving random dots: all of the stimuli are assumed to produce similar stimulation throughout the receptive field of all activated motion units. Clearly the next major step is to incorporate the spatial interactions that produce suppressive surrounds in many MT neurons (Allman *et al.*, 1985; Xiao *et al.*, 1995). Born and Tootell (1992) documented two categories of these neurones: for a given direction of motion in the receptive field center, one group is inhibited by the same direction of motion in its surround, while the other group is inhibited by the opposite direction of surround motion. We have suggested elsewhere how these units might easily be constructed from spatial arrays of the motion network described above (Wilson and Kim, 1994a; Kim and Wilson, 1997). In closing this section on motion, however, it is worth considering that motion of an object across a background that is moving in a very different direction will generate very different responses from the model MT motion network contingent upon the location of the receptive field. Motion networks with receptive fields covered entirely by the object or by the surround will accurately signal the directions of motion of object and surround, respectively (see Fig. 10). Motion-receptive fields lying along the object boundary, however, will be stimulated by two very different directions of motion and thus will signal transparency (Fig. 12). Thus, shape defined by motion discontinuities will be encoded into the locus of motion networks signaling transparency, and this information could be extracted by the MT receptive fields described by Born and Tootell (1992). Furthermore, Xiao *et al.* (1995) showed that the surrounds of many MT neurons consist of bilaterally opposed subfields that might usefully calculate motion-defined boundaries.

4.7. Motion and Marr's Algorithmic Approach

Two-dimensional motion processing forms an enlightening example in which Marr's (1982) algorithmic approach breaks down. A mathematically exact algorithm for computing the true direction of rigid translation, namely the "intersection of constraints" (IOC) construction (Adelson and Movshon, 1982), was known before the development of neural models. However, psychophysical research demonstrated that the IOC construction is *not* implemented in the human visual system (Ferrera and Wilson, 1990; Yo and Wilson, 1992; Mingolla *et al.*, 1992; Rubin and Hochstein, 1993; Wilson and Kim, 1994b; Victor and Conte, 1994). Instead, this psychophysics provided evidence for a vector sum computation and for the involvement of both Fourier and non-Fourier pathways. Psychophysics thus became a guide both to the algorithm employed in human motion perception as well as to the construction of a neural network that could predict the data. Thus, the existence of an exact mathematical algorithm to solve a visual problem is no guarantee that the visual system evolved the capacity to employ that algorithm.

5. Discussion

5.1. Importance of Nonlinearity

The focus of this chapter has been on non-Fourier processing of visual texture, form, and motion. The cardinal point to be emphasized is that non-Fourier processing is inherently nonlinear in consequence of the rectification stage sandwiched between the first- and second-stage filters (see Figs. 1B, 5, and 8). As a result, non-Fourier pathways extract spatial stimulus attributes that simply cannot be extracted by linear filtering such as that performed by cortical simple cells. This becomes obvious in the case of the texture boundaries in Fig. 2 and their moving counterparts in Fig. 10. In addition, it can also be proved that no linear filter can extract the perceptually salient circularity in Glass (1969) patterns (see Fig. 7B). Similarly, the fact that both real (Gallant *et al.*, 1993) and model V4 neurons respond well to concentric patterns even when the pattern is displaced from the receptive field center by up to 2.5 cycles indicates that linear filtering cannot explain the processing.

The evidence for non-Fourier processing in motion (Smith and Snowden, 1994) and texture perception (Malik and Perona, 1990; Bergen and Landy, 1991; Graham *et al.*, 1992; Wilson and Richards, 1992) is now overwhelming. The construction of model V4 concentric units above represents the first extension of non-

Fourier-type processing to the understanding of higher level form vision. While processing of Glass (1969) patterns might be regarded as a more complex form of texture analysis, the extraction of human facial characteristics in Fig. 7 by these units links them to form vision, as does the evidence that prosopagnosics are deficient in Glass pattern perception (Rentschler *et al.*, 1994). As pointed out above, neophyte artists are frequently taught to draw human faces by smoothly linking circles centered at virtually the same points as the model V4 unit responses in Fig. 7 (e.g., Hamm, 1963). While this correlation with art training might seem spurious, Gombrich (1972) has convincingly argued that the two millennium history of representational art may be regarded a long series of empirical experiments on depicting three-dimensional shapes in two dimensions. From this perspective art indeed provides empirical data of relevance to the vision scientist. Thus, it is likely that non-Fourier processing will begin to wedge open the door to an understanding of form vision.

5.2. A Role for Cortical Feedback

The ubiquity of non-Fourier processing in the visual system appears to raise one serious computational problem. As shown in Fig. 1B, the first- and second-stage receptive fields of a non-Fourier unit typically have different preferred orientations. In accord with this, Xiao *et al.* (1995) found all possible combinations of preferred motion direction and suppressive surround subfield axes among MT neurons. However, as the number of orientations represented in V1 is about 12, each separated by a 15° increment (Hubel and Wiesel, 1977), there are 144 possible combinations of first- and second-stage receptive field orientations possible for non-Fourier units. Such a monumental proliferation of non-Fourier units is simply not feasible, and it becomes even less plausible when one realizes that higher visual areas are typically ten times smaller than V1 or V2 (VanEssen, *et al.*, 1992). One elegant scheme by which this proliferation could be avoided involves the feedback circuits from areas such as V4 and MT back to V1 and V4 (VanEssen, 1985). Tsotsos (1990) devised a feedback scheme whereby higher visual units selectively activate winner-take-all networks operating on their own inputs. This would permit a non-Fourier unit to maintain connections with a wide range of first-stage orientations and then dynamically select the optimal one to process each stimulus via feedback. Though obviously speculative, this scheme appeals both by resolving the non-Fourier combinatoric problem and by providing a functional role for the ubiquitous feedback from higher to lower cortical areas.

5.3. Area V2, Cortical Connections, and Non-Fourier Pathways

As discussed above, it has been reported that non-Fourier units responsive to texture boundaries and illusory contours are to be found in V2, but not in V1 (von der Heydt *et al.*, 1984; von der Heydt and Peterhans, 1989). Grosof *et al.* (1993) contested this by claiming that some non-Fourier boundaries are processed in V1. However, the demonstration that monkeys with V2 lesions are blind to non-Fourier boundary orientation, but not to simple line orientation (Merigan *et al.*, 1993) seems definitive in showing that V2 represents a crucial stage in non-Fourier processing. As it has previously been suggested that the non-Fourier motion pathway in Fig. 8 involves V2 processing (Wilson *et al.*, 1992), let us now consider how V2 might simultaneously fit into the processing streams for both form and motion perception.

The models for texture, form, and motion perception developed above all begin with oriented simple cell processing, which is depicted in Fig. 14 as occurring in V1 (Hubel and Wiesel, 1968). Direct inputs from foveal V1 simple cells to V4 explain the psychophysics of pattern discrimination and hyperacuity with simple line and edge stimuli (Wilson, 1986; Wilson and Richards, 1989). This may be conceptualized as the Fourier form pathway. At the same time, Reichardt (1961) detectors operating on simple cell outputs in V1 comprise the Fourier motion pathway and provide direct inputs to MT. Non-Fourier processing of texture boundaries, illusory contours, and Glass (1969) patterns requires rectification and subsequent pooling in larger oriented receptive fields, and these operations are hypothesized to occur in V2 before the information is sent on to V4. Similarly, rectification and subsequent pooling in the non-Fourier motion pathway are also assumed to occur in V2, with the result being passed through a Reichardt (1961) detector before arriving at MT pattern units. Thus, the schematic in Fig. 14 associates simple pattern vision with the V1 → V4 pathway and non-Fourier texture and form vision with the V1 → V2 → V4 pathway. Similarly, Fourier motion processing is assumed to occur in the V1 → MT pathway and non-Fourier motion processing in the V1 → V2 → MT pathway.

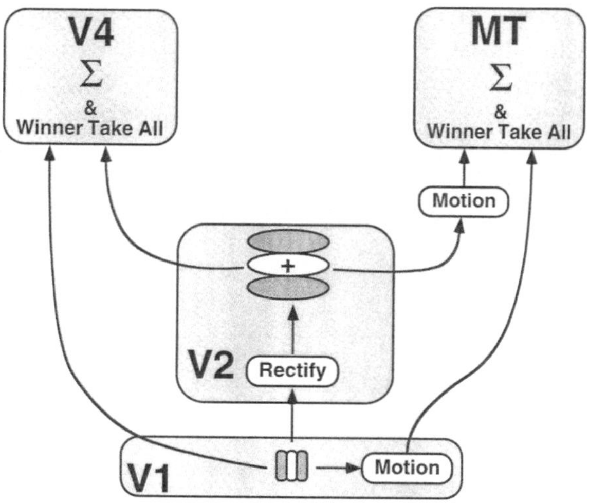

Figure 14. Summary diagram of Fourier and non-Fourier processing for both form and motion information. Outputs of oriented simple cells in V1 (bottom, only one orientation shown) are sent directly to foveal V4 (upper left) for higher form analysis, and they are also sent to MT (upper right) following motion extraction by Reichardt (1961) detectors. These are hypothesized to be the Fourier form and motion pathways, respectively. In addition, the V1 responses are sent to V2 (center), where rectification and second-stage oriented filtering is thought to occur (von der Heydt and Peterhans, 1989). Spatial profiles of second-stage V2 filters discussed above are illustrated with white excitatory and gray inhibitory zones. V2 unit responses are sent to MT following motion extraction and also to V4. The V1 → V2 → V4 and V1 → V2 → MT pathways are conjectured to subserve non-Fourier form and motion perception, respectively. The quantitative model for MT (Fig. 8) requires both summation (with cosine weighting) and recurrent inhibition within an embellished winner-take-all network. The model for V4 concentric units in Fig. 5 requires summation over component inputs, but implementation of a winner-take-all strategy within V4 has yet to be established.

Graziano *et al.* (1994) noted strong parallels between neural processing in the form and motion pathways. Morrone *et al.* (1995) went even further by showing that there is linear pooling of concentric motion direction information in human vision. This direction pooling, speculatively localized in medial superior temporal cortex, is based on motion experiments fully analogous to Glass pattern experiments showing concentric pooling of orientation information (Wilson *et al.*, 1997). Thus, the scheme in Fig. 14, indicating significant shared processing in motion and pattern vision, is consistent with physiology (von der Heydt and Peterhans, 1989), anatomy (VanEssen *et al.*, 1992), and psychophysics, and it offers an explanation for the fact that V1 projects both directly and via V2 to both V4 and MT.

At the highest levels of this model hierarchy, V4 and MT, it is proposed that there is appropriately weighted summation of inputs followed by an embellished winner-take-all network. In the case of motion processing, the combination of cosine-weighted summation and a winner-take-all network computes the vector sum of Fourier and non-Fourier component motions (see Fig. 9). Psychophysical and physiological evidence for this vector sum operation has been cited previously. To construct V4 concentric units, the summation over non-Fourier inputs shown in Fig. 5 is clearly necessary, so the parallel between form and motion processing clearly holds to this point. Although the network in Fig. 5 for V4 concentric units does not include a winner-take-all stage, an embellished winner-take-all network operating on concentric unit responses over appropriate two-dimensional spatial regions would pick out local maxima in the responses shown in Figs. 6 and 7. Indeed, there is evidence for such inhibition in IT, which is the major target of V4. First, the receptive fields of IT neurons in alert monkeys have been reported to be significantly smaller than in anesthetized preparations (Richmond *et al.*, 1983). Second, Sato (1989) and Miller *et al.*, (1993) reported direct evidence of inhibitory competition in recordings from IT neurons in alert macaques. Finally, Ferrera *et al.* (1994) reported strong memory- and attention-related modulation of V4 responses. Attentional effects in V4 may be plausibly interpreted as resulting from top-down biasing of winner-take-all networks in the manner envisioned by Tsotsos (1990). All these data are thus consistent with embellished winner-take-all networks operating among V4 inputs to IT and probably also among IT units themselves in alert animals.

The operations envisaged for V2 in Fig. 14 deserve two final comments. First, the Reichardt (1961) detector that extracts motion direction from the V2 output to MT might reside in either V2 or MT, an issue that will have to be resolved physiologically. Second, it will be noted that the same group of second-stage receptive fields can perform the requisite processing both for V4 concentric units and for MT texture boundary motion. This represents a clear example of neural processing economy.

5.4. Conclusion: Further Potential of Non-Fourier Processing

In conclusion, non-Fourier processing, which is inherently nonlinear, seems to be ubiquitous in the visual system. In addition to its role in texture, form, and motion perception, non-Fourier mechanisms are now known to be involved in stereopsis as well (Hess and Wilcox, 1994; Wilcox and Hess, 1996; Lin and Wilson,

1999). Non-Fourier processing is vastly more powerful and flexible than quasilinear processing by simple cells, and it is capable of predicting visual phenomena that simply cannot be explained by linear filtering (e.g., the perception of illusory contours in Fig. 2 or of Glass patterns in Fig. 6). Non-Fourier processing in form vision also provides a means of measuring the configuration of local "features" in complex images such as human faces. The networks described above have only begun to elucidate the potential of non-Fourier networks for extracting visually important information, and it now appears that further psychophysical and physiological exploration of non-Fourier processes will catapult us to a new level of sophistication in our understanding of visual function.

ACKNOWLEDGMENTS. This work was supported in part by NIH grant #EY02158, by a grant from the Brain Research Institute of Chicago, and by a grant to the University of Chicago from Research to Prevent Blindness. I am indebted to F. Wilkinson for many valuable discussions related to this work.

6. References

Adelson, E. H., and Bergen, J. R., 1985, Spatiotemporal energy models for the perception of motion, *J. Opt. Soc. Am. A* **2**:284–299.

Adelson, E. H., and Movshon, J. A., 1982, Phenomenal coherence of moving visual patterns, *Nature* **300**:523–525.

Albright, T. D., 1992, Form-cue invariant motion processing in primate visual cortex, *Science* **255**:1141–1143.

Albright, T. D., Desimone, R., and Gross, C. G., 1984, Columnar organization of directionally selective cells in visual area MT of the macaque, *J. Neurophys.* **51**:16–31.

Allman, J., Miezin, F., and McGuinness, E., 1985, Direction and velocity specific responses from beyond the classical receptive field in the middle temporal visual area (MT), *Perception* **14**:105–126.

Badcock, D. R., and Derrington, A. M., 1985, Detecting the displacement of periodic patterns, *Vision Res.* **25**:1253–1258.

Badcock, D. R., and Derrington, A. M., 1989, Detecting the displacements of spatial beats: No role for distortion products, *Vision Res.* **29**:731–739.

Bergen, J. R., and Landy, M. S., 1991, Computational modeling of visual texture segregation, in: *Computational Models of Visual Processing* (M. Landy and J. A. Movshon, eds.), MIT Press, Cambridge, MA, pp. 253–271.

Bolz, J., and Gilbert, C. D., 1986, Generation of end-inhibition in the visual cortex via interlaminar connections, *Nature* **320**:362–365.

Born, R. T., and Tootell, R. B., 1992, Segregation of global and local motion processing in primate middle temporal visual area, *Nature* **357**:497–499.

Burke, D., and Wenderoth, P., 1993, The effect of interactions between one-dimensional component gratings on two-dimensional motion perception, *Vision Res.* **33**:343–350.

Cavanagh, P., and Mather, G., 1990, Motion: The long and short of it, *Spatial Vision* **4**:103–129.

Chubb, C., and Sperling, G., 1988, Drift-balanced random stimuli: A general basis for studying non-Fourier motion perception, *J. Opt. Soc. Am. A* **5**:1986–2007.

Chubb, C., and Sperling, G., 1989, Two motion perception mechanisms revealed through distance-driven reversal of apparent motion, *Proc. Natl. Acad. Sci. USA* **86**:2985–2989.

Derrington, A., and Suero, M., 1991, Motion of complex patterns is computed from the perceived motions of their components, *Vision Res.* **31**:139–149.

Derrington, A. M., Badcock, D. R., and Holroyd, S. A., 1992, Analysis of the motion of 2-dimensional patterns: Evidence for a second-order process, *Vision Res.* **32**:699–707.

Desimone, R., and Schein, S. J., 1987, Visual properties of neurons in area V4 of the macaque: Sensitivity to stimulus form, *J. Neurophysiol.* **57**:835.

Desimone, R., Schein, S. J., Moran, J., and Ungerleider, L. G., 1985, Contour, color, and shape analysis beyond the striate cortex, *Vision Res.* **25:**441–452.

DeValois, R. L., Albrecht, D. G., and Thorell, L. G., 1982, Spatial frequency selectivity of cells in macaque visual cortex, *Vision Res.* **22:**545–559.

Dobbins, A., Zucker, S. W., and Cynader, M. S., 1987, Endstopped neurons in the visual cortex as a substrate for calculating curvature, *Nature* **329:**438–441.

Dobbins, A., Zucker, S. W., and Cynader, M. S., 1989, Endstopping and curvature, *Vision Res.* **29:**1371–1387.

Dodwell, P. C., Wilkinson, F. E., and von Grünau, M. W., 1983, Pattern recognition in kittens: Performance on Lie patterns, *Perception* **12:**393–410.

Dreher, B., 1972, Hypercomplex cells in the cat's striate cortex, *Invest. Ophthalmol.* **11:**355–356.

Feldman, J. A., and Ballard, D. H., 1982, Connectionist models and their properties, *Cognitive Sci.* **6:**205–254.

Ferrera, V. P., and Wilson, H. R., 1990, Perceived direction of moving two-dimensional patterns, *Vision Res.* **30:**273–287.

Ferrera, V. P., Rudolph, K. K., and Maunsell, J. H. R., 1994, Responses of neurons in the parietal and temporal visual pathways during a motion task, *J. Neurosci.* **14:**6171–6186.

Field, D. J., Hayes, A., and Hess, R. F., 1993, Contour integration by the human visual system: Evidence for a local "association field," *Vision Res.* **33:**173–193.

Gallant, J. L., Braun, J., and VanEssen, D. C., 1993, Selectivity for polar, hyperbolic, and Cartesian gratings in macaque visual cortex, *Science* **259:**100–103.

Gallant, J. L., Connor, C. E., Rakshit, S., Lewis, J. W., and Van Essen, D. C., 1996, Neural responses to polar, hyperbolic, and Cartesian gratings in area V4 of the macaque monkey, *J. Neurophys.* **76:**2718–2739.

Glass, L., 1969, Moiré effect from random dots, *Nature* **223:**578–580.

Gombrich, E. H., 1972, *Art and Illusion,* 4th ed., Phaidon, London.

Graham, N., Beck, J., and Sutter, A., 1992, Nonlinear processes in spatial frequency channel models of perceived texture segregation: Effects of sign and amount of contrast, *Vision Res.* **32:**719–743.

Graziano, M. S. A., Andersen, R. A., and Snowden, R. J., 1994, Tuning of MST neurons to spiral motions, *J. Neurosci.* **14:**54–67.

Grosof, D. H., Shapley, R. M., and Hawken, M. J., 1993, Macaque V1 neurons can signal 'illusory' contours, *Nature* **365:**550–552.

Gross, C. G., 1992, Representation of visual stimuli in inferior temporal cortex, *Phil. Trans. R. Soc. Lond.* **335:**3–10.

Hamm, J., 1963, *Drawing the Head and Figure,* Perigree Books, Putnam, New York.

Hamm, J., 1969, *How to Draw Animals,* Perigree Books, Putnam, New York.

Heeger, D. J., 1991, Nonlinear model of neural responses in cat visual cortex, in: *Computational Models of Visual Processing* (M. Landy and J. A. Movshon, eds.), MIT Press, Cambridge, MA, pp. 119–133.

Heeger, D. J., 1992, Normalization of cell responses in cat striate cortex, *Vis. Neurosci.* **9:**181–197.

Henry, G. H., 1977, Receptive field classes of cells in the striate cortex of the cat, *Brain Res.* **133:**1–28.

Hertz, J., Krogh, A., and Palmer, R. G., 1991, *Introduction to the Theory of Neural Computation,* Addison-Wesley, New York.

Hess, R. F., and Wilcox, L. M., 1994, Linear and nonlinear filtering in stereopsis, *Vision Res.* **34:**2431–2438.

Heywood, C. A., Gadotti, A., and Cowey, A., 1992, Cortical area V4 and its role in the perception of color, *J. Neurosci.* **12:**4056–4065.

Hubel, D. H., and Wiesel, T. N., 1962, Receptive fields, binocular interaction, and functional architecture in the cat's striate cortex, *J. Physiol.* **160:**106–154.

Hubel, D. H., and Wiesel, T. N., 1968, Receptive fields and functional architecture of monkey striate cortex, *J. Physiol.* **195:**215–243.

Hubel, D. H., and Wiesel, T. N., 1977, Functional architecture of macaque monkey visual cortex, *Proc. R. Soc. Lond. B* **198:**1–59.

Kim, J., and Wilson, H. R., 1993, Dependence of plaid motion coherence on component grating directions, *Vision Res.* **33:**2479–2489.

Kim, J., and Wilson, H. R., 1996, Direction repulsion between components in motion transparency, *Vision Res. 36:*1177–1187.

Kim, J., and Wilson, H. R., 1997, Motion integration over space: Interaction of center and surround motion, *Vision Res.* **37**:991–1005.

Kobatake, E., and Tanaka, K., 1994, Neuronal selectivities to complex object features in the ventral pathway of the macaque cerebral cortex, *J. Neurophys.* **71**:856–867.

Koenderink, J. J., and Richards, W. A., 1988, Two-dimensional curvature operators, *J. Opt. Soc. Am. A* **5**:1136–1141.

Ledgeway, T., and Smith, A. T., 1994, Evidence for separate motion-detecting mechanisms for first- and second-order motion in human vision, *Vision Res.* **34**:2727–2740.

Legge, G. E., and Foley, J. M., 1980, Contrast masking in human vision, *J. Opt. Soc. Am.* **70**:1458–1470.

Lehky, S., and Sejnowski, T. J., 1988, Network model of shape-from-shading: Neural function arises from both receptive and projective fields, *Nature* **333**:452–454.

Li, W., and Westheimer, G., 1997, Human discrimination of the implicit orientation of simple symmetrical patterns, *Vision Res.* **37**:565–572.

Lin, L.-M., and Wilson, H. R., 1996, Fourier and non-Fourier pattern discrimination compared, *Vision Res.* **36**:1907–1918.

Lin, L. M., and Wilson, H. R., 1999, Fourier and non-Fourier stereoacuities compared, submitted.

Malik, J., 1994, in: *Higher Order Processing in the Visual System* (J. Goode, ed.), p. 121, Wiley, New York.

Malik, J., and Perona, P., 1990, Preattentive texture discrimination with early vision mechanisms, *J. Opt. Soc. Am. A* **7**:923–932.

Marr, D., 1982, *Vision: A Computational Investigation into the Human Representation and Processing of Visual Information,* Freeman, San Francisco.

Marshak, W., and Sekuler, R., 1979, Mutual repulsion between moving visual targets, *Science* **205**:1399–1401.

Mather, G., and Moulden, B., 1980, A simultaneous shift in apparent direction: Further evidence for a "distribution shift" model of direction coding, *Q. J. Exp. Psychol.* **32**:325–333.

Merigan, W. H., 1996, Basic visual capabilities and shape discrimination after lesions of extrastriate area V4 in macaques, *Vis. Neurosci.* **13**:51–60.

Merigan, W. H., Nealey,, T. A., and Maunsell, J. H. R., 1993, Visual effects of lesions of cortical area V2 in macaques, *J. Neurosci.* **13**:3180–3191.

Miller, E. K., Gochin, P. M., and Gross, C. G., 1993, Suppression of visual responses of neurons in inferior temporal cortex of the awake macaque by addition of a second stimulus, *Brain Res.* **616**:25–29.

Mingolla, E., Todd, J. T., and Norman, J. F., 1992, The perception of globally coherent motion, *Vision Res.* **32**:1015–1031.

Mishkin, M., Ungerleider, L. G., and Macko, K. A., 1983, Object vision and spatial vision: Two cortical pathways, *Trends Neurosci.* **6**:414–417.

Morrone, M. C., Burr, D. C., and Vaina, L. M., 1995, Two stages of visual processing for radial and circular motion, *Nature* **376**:507–509.

Movshon, J. A., Thompson, I. D., and Tolhurst, D. J., 1978, Receptive field organization of complex cells in the cat's striate cortex, *J. Physiol.* **283**:79–99.

Nachmias, J., and Sansbury, R. V., 1974, Grating contrast: Discrimination may be better than detection, *Vision Res.* **14**:1039–1042.

Olzak, L. A., and Thomas, J. P., 1991, When orthogonal orientations are not processed independently, *Vision Res.* **31**:51–57.

Olzak, L. A., and Thomas, J. P., 1992, Configural effects constrain Fourier models of pattern discrimination, *Vision Res.* **32**:1885–1898.

Ouchi, H., 1977, *Japanese and Geometrical Art,* Dover, New York.

Perrett, D. I., and Chitty, A. J., 1987, Visual neurones responsive to faces, *Trends Neurosci.* **10**:358–364.

Phillips, G. C., and Wilson, H. R., 1984, Orientation bandwidths of spatial mechanisms measured by masking, *J. Opt. Soc. Am. A* **1**:226–232.

Reichardt, W., 1961, Autocorrelation, a principle for the evaluation of sensory information by the central nervous system, in: *Sensory Communication* (W. A. Rosenblith, ed.), Wiley, New York, pp. 303–317.

Rentschler, I., Treutwein, B., and Landis, T., 1994, Dissociation of local and global processing in visual agnosia, *Vision Res.* **34**:963–971.

Richmond, B. J., Wurtz, R. H., and Sato, T., 1983, Visual responses of inferior temporal neurons in awake rhesus monkey, *J. Neurophys.* **50**:1415–1432.

Rolls, E. T., 1984, Neurons in the cortex of the temporal lobe and in the amygdala of the monkey with responses selective for faces, *Hum. Neurobiol.* **3**:209–222.

Rubin, N., and Hochstein, S., 1993, Isolating the effect of one-dimensional motion signals on the perceived direction of moving two-dimensional objects, *Vision Res.* **33**:1385–1396.

Salzman, C. D., and Newsome, W. T., 1994, Neural mechanisms for forming a perceptual decision, *Science* **264**:231–237.

Sato, T., 1989, Interactions of visual stimuli in the receptive fields of inferior temporal neurons in awake macaques, *Exp. Brain Res.* **77**:23–30.

Schiller, P., 1992, The ON and OFF channels of the visual system, *TINS* **15**:86–92.

Schoups, A. A., Tootell, R. B., Vanduffel, W., and Orban, G. A., 1995, Use of the double-label deoxyglucose approach to map the orientation columnar system beyond area V1 and V2 in the macaque, and its plasticity, *Soc. Neurosci. Abstr.* **15**:3.

Smith, A. T., 1994, The detection of second-order motion, in: *Visual Detection of Motion* (A. T. Smith and R. J. Snowden, eds.), Academic Press, London, pp. 145–176.

Smith, A. T., and Snowden, R. J., 1994, eds., *Visual Detection of Motion*, Academic Press, London.

Snowden, R. J., Treue, S., Erickson, R. G., and Andersen, R. A., 1991, The response of area MT and V1 neurons to transparent motion, *J. Neurosci.* **11**:2768–2785.

Sperling, G., Chubb, C., Solomon, J. A., and Lu, Z.-L., 1994, Full-wave and half-wave processes in second order motion and texture, in: *Higher Order Processing in the Visual System* (J. Goode, ed.), Wiley, New York, pp. 287–303.

Tsotsos, J. K., 1990, Analyzing vision at the complexity level, *Behav. Brain Sci.* **13**:423–469.

Turano, K., and Pantle, A., 1989, On the mechanism that encodes the movement of contrast variations: Velocity discrimination, *Vision Res.* **29**:207–221.

VanEssen, D. C., 1985, Functional organization of primate visual cortex, in: *Cerebral Cortex*, Vol. 3. *Visual Cortex* (A. Peters and E. G. Jones, eds.), Plenum Press, New York, pp. 259–329.

VanEssen, D. C., Anderson, C. H., and Felleman, D. J., 1992, Information processing in the primate visual system: An integrated systems perspective, *Science* **255**:419–423.

Victor, J. D., and Conte, M. M., 1994, Investigation of a patient with severely impaired direction discrimination: Evidence against the intersection of constraints model, *Vision Res.* **34**:267–277.

Vogels, R., and Orban, G. A., 1987, Illusory contour orientation discrimination, *Vision Res.* **27**:453–467.

von der Heydt, R., and Peterhans, E., 1989, Mechanisms of contour perception in monkey visual cortex. I. Lines of pattern discontinuity, *J. Neurosci.* **9**:1731–1748.

von der Heydt, R., Peterhans, E., and Baumgartner, G., 1984, Illusory contours and cortical neuron responses, *Science* **224**:1260–1262.

Welch, L., 1989, The perception of moving plaids reveals two motion processing stages, *Nature* **337**:734–736.

Westheimer, G., Shimamura, K., and McKee, S. P., 1976, Interference with line orientation sensitivity, *J. Opt. Soc. Am.* **66**:332–338.

Wilcox, L. M., and Hess, R. F., 1996, Is the site of non-linear filtering in stereopsis before or after binocular combination? *Vision Res.* **36**:391–999.

Wilkinson, F., and Dodwell, P. C., 1980, Young kittens can learn complex visual pattern discriminations, *Nature* **284**:258–259.

Wilkinson, F., and Wilson, H. R., 1996, Discrimination of curvilinear patterns: Beyond simple cells, *Invest. Ophthalmol. Vis. Sci.* **37**:ARVO Abstracts 4370.

Wilkinson, F., Wilson, H. R., and Habak, C., 1998, Detection and recognition of radial frequency patterns, *Vision Res.* **38**:3555–3568.

Williams, D., and Phillips, G., 1987, Cooperative phenomena in the perception of motion direction, *J. Opt. Soc. Am. A* **4**:878–885.

Williams, D., Phillips, G., and Sekuler, R., 1986, Hysteresis in the perception of motion direction as evidence for neural cooperativity, *Nature* **324**:253–255.

Wilson, H. R., 1980, A transducer function for threshold and suprathreshold human vision, *Biol. Cybernet.* **38**:171–178.

Wilson, H. R., 1985, Discrimination of contour curvature: Data and theory, *J. Opt. Soc. Am. A* **2**:1191–1199.

Wilson, H. R., 1986, Responses of spatial mechanisms can explain hyperacuity, *Vision Res.* **26**:453–469.

Wilson, H. R., 1991a, Psychophysical models of spatial vision and hyperacuity, in: *Spatial Vision* (D. Regan, ed.), MacMillan, London, pp. 64–86.

Wilson, H. R., 1991b, Pattern discrimination, visual filters, and spatial sampling irregularity, in: *Computational Models of Visual Processing* (M. Landy and J. A. Movshon, eds.), MIT Press, Cambridge, MA, pp. 153–168.

Wilson, H. R., 1993, Nonlinear processes in visual pattern discrimination, *Proc. Natl. Acad. Sci. USA* **90**:9785–9790.

Wilson, H. R., 1994a, The role of second-order motion signals in coherence and transparency, in: *Higher Order Processing in the Visual System* (J. Goode, ed.), Wiley, New York, pp. 227–244.

Wilson, H. R., 1994b, A model for motion coherence and transparency, in: *Visual Detection of Motion* (A. T. Smith and R. J. Snowden, eds.), Academic Press, London, pp. 219–251.

Wilson, H. R., 1997, A neural model of foveal light adaptation and afterimage formation, *Vis. Neurosci.* **4**:403–423.

Wilson, H. R., and Cowan, J. D., 1973, A mathematical theory of the functional dynamics of cortical and thalamic nervous tissue, *Kybernetik* **13**:55–80.

Wilson, H. R., and Humanski, R., 1993, Spatial frequency adaptation and contrast gain control, *Vision Res.* **33**:1133–1149.

Wilson, H. R., and Kim, J., 1994a, A model for motion coherence and transparency, *Vis. Neurosci.* **11**:1205–1220.

Wilson, H. R., and Kim, J., 1994b, Perceived motion in the vector sum direction, *Vision Res.* **34**:1835–1842.

Wilson, H. R., and Richards, W. A., 1989, Mechanisms of contour curvature discrimination, *J. Opt. Soc. Am. A* **6**:106–115.

Wilson, H. R., and Richards, W. A., 1992, Curvature and separation discrimination at texture boundaries, *J. Opt. Soc. Am. A* **9**:1653–1662.

Wilson, H. R., and Wilkinson, F., 1998, Detection of global structure in Glass patterns: Implications for form vision, *Vision Res.*, **38**:2933–2947.

Wilson, H. R., McFarlane, D. K., and Phillips, G. C., 1983, Spatial frequency tuning of orientation selective units estimated by oblique masking, *Vision Res.* **23**:873–882.

Wilson, H. R., Levi, D., Maffei, L., Rovamo, J., and DeValois, R. L., 1990, The perception of form: Retina to striate cortex, in: *The Neurophysiological Foundations of Visual Perception* (L. Spillmann and J. S. Werner, eds.), Academic Press, New York, pp. 231–272.

Wilson, H. R., Ferrera, V. P., and Yo, C., 1992, Psychophysically motivated model for two-dimensional motion perception, *Vis. Neurosci.* **9**:79–97.

Wilson, H. R., Wilkinson, F., and Asaad, W., 1997, Concentric orientation summation in human form vision, *Vision Res.* **37**:2325–2330.

Xiao, D. K., Raiguel, S., Marcar, V., Koenderink, J., and Orban, G. A., 1995, Spatial heterogeneity of inhibitory surrounds in the middle temporal visual area, *Proc. Natl. Acad. Sci. USA* **92**:11303–11306.

Yamane, S., Kaji, S., and Kawano, K., 1988, What facial features activate face neurons in the inferotemporal cortex of the monkey? *Exp. Brain Res.* **73**:209–214.

Yo, C., and Wilson, H. R., 1992, Perceived direction of moving two-dimensional patterns depends on duration, contrast, and eccentricity, *Vision Res.* **32**:135–147.

Zeki, S. M., 1978, Functional specialization in the visual cortex of the rhesus monkey, *Nature* **274**:423–428.

Zhou, Y., and Baker, C. L., 1993, A processing stream in mammalian visual cortex neurons for non-Fourier responses, *Science* **261**:98–101.

9

Modeling Thalamocortical Oscillations

WILLIAM W. LYTTON and ELIZABETH THOMAS

1. Slow Thalamic Rhythms

The thalamic nuclei lie in the center of the brain and are thought to be functionally central to a variety of brain processes. In particular, the primary nuclei play two, possibly complementary, roles in brain function. The first role is that of a relay nucleus. In the relay mode a thalamic nucleus receives information from the sensory periphery and sends it on to the cortex. In this chapter, we will not be much concerned with this aspect of thalamic function. The second role involves the thalamus' involvement in slow oscillations. These are reflected in the cortex during various stages of sleep as well as in pathological processes associated with certain types of epilepsy. In this context the thalamus appears to be a sort of central pattern generator for the vertebrate brain. It is the slow oscillations associated with these rhythms that have been most intensively studied using modeling techniques.

Rhythmic oscillations involving the thalamus are seen in the human electroencephalogram (EEG) during sleep. During the early stage of sleep, shortly after the transition from the wake state, relatively high amplitude oscillation called sleep spindle are seen. The frequency of the oscillations is 12–14 Hz and the entire sleep

WILLIAM W. LYTTON and ELIZABETH THOMAS • Department of Neurology, University of Wisconsin, and William S. Middleton VA Hospital, Madison, Wisconsin 53706.

spindle has a duration of about 1 sec, during which time it grows in amplitude and subsequently diminishes. The word spindle refers to the shape of this envelope: the gradual increase in size followed by the decrease and cessation of the rhythm. Generically, a spindle has come to refer to any neural oscillation that arises spontaneously from a quiet or asynchronous background rhythm and then disappears. Spindles have been found in various *in vivo* and *in vitro* preparations and generally range in frequency from 7 to 14 Hz. The term spindle is therefore widely used and does not connote a common mechanism.

A separate, slower oscillation that takes place during deep sleep occurs within the delta range of 1–4 Hz and is sometimes referred to as the delta rhythm. This rhythm forms the major background of deep sleep and is ongoing rather than transient. In this case, similar frequencies found in a variety of animal and computer models may be taken as models of sleep delta. However, another phenomenon also is found within the same frequency range, namely the 3-Hz spike-and-wave oscillation of absence epilepsy. Absence epilepsy is a type of generalized epilepsy seen in children. "Generalized" means that the seizure appears to arise all over the brain at once instead of starting in a distinct location as is seen with focal epilepsy. Unlike the more common generalized epilepsies of adulthood, absence epilepsy does not involve a convulsion in the limbs. Instead the child simply loses touch with the surroundings for 2–10 sec before returning immediately and fully to consciousness. The episodes are so short and inapparent that they may be interpreted as absentmindedness, although on closer observation they will usually be associated with some twitching of the face or hands at 3 Hz.

Since both delta sleep and spike-and-wave epilepsy occur in the same frequency range, something other than a frequency criterion must be used to determine whether a given animal or computer model pertains to one or the other of them. Any oscillation that can be detected by EEG must necessarily involve a fair degree of synchrony between neurons in order to give rise to a signal summation large enough to be detected. The hallmark of absence epilepsy is the spike-and-wave, a very sharp EEG transient followed by a long, blunt one. The sharp spike must be caused by a very high degree of synchrony between neurons. An additional distinction between spike-and-wave oscillation and delta rhythm sleep is that the former is transient, while the latter persists over many minutes. Though spike-and-wave oscillations are, like spindles, transient, their frequency defines them separately from spindle phenomena.

2. Thalamus as Magnet for Modeling—Dynamics and Neural Systems

Because of the strong interconnections between thalamic relay nuclei, nucleus reticularis thalami, and cortex, it is inevitable that the various connections between different structures will alter the phase, frequency, and coherence of all oscillations. This being the case, it can be argued whether any given rhythm originates in one structure and is modified by others or whether the rhythm originates out of the interaction of multiple structures. Furthermore, it is often possible to produce a variety of rhythms of similar frequency that are actually due to different cellular interactions. A single term used to denote a particular oscillation frequency may in fact include a variety of rhythms that are associated with different intraneuronal ion channel activations and different interneuronal interactions. A central role of thalamic modeling has been to analyze how various dynamical interactions can produce the varieties of oscillations seen. This allows the physiologist to hone investigations to distinguish between them.

Though the complexity of the thalamus and its connections is a major factor underlying its resistance to our understanding, it has become clear that even seemingly simple dynamical systems can behave in complex ways. The word dynamics was originally used in classical physics to refer to the movements of mutually influential objects such as the planetary bodies in the solar system. A solar system can be described mathematically by a set of interlocking differential equations. By extension of the term, the word dynamics has come to be applied to any system that can be described by differential equations in this way. The underlying similarity of these systems is that they can all be described by "state variables" which identify the "motion" of the system's components. In a solar system these state variables would be the position and velocity of every plant. In a neural system the state variables include voltage as well as concentrations of calcium or other second messengers, depending on the degree of detail used in a particular model.

Parameters define the form of a dynamical system. Unlike state variables, parameters are fixed and do not change with the progression through time of the system. In the solar system the parameters are masses and separation distances of the heavenly bodies and the value of the acceleration due to gravity. A nervous system is much more heavily parametrized. Such things as the shapes of neurons and the concentration of various membrane-associated, voltage-dependent ion channels all must be given suitable parameters in order to explicitly define a network model. Another important parameters in neural systems is specific membrane resistance, a controversial parameter that will be discussed further below. In many cases, explicit parameter explorations must be performed because the correct value for a particular parameter has not or cannot be experimentally determined.

As long as the mathematical relationships in a system remain linear, it is relatively easy to predict and understand how a dynamical system will behave. This is because signals in linear systems add up neatly: if you know the response to one signal (say a sinusoid) and also know the response to another signal (say a decaying signal), then you know that the response to both signals combined (a decaying sinusoid) will simply be the superposition of the two individual responses. With nonlinear dynamical systems, however, all bets are off. The response to a decaying sinusoid may look nothing like the response to either one of the individual signals alone. One consequence of this is that mathematical analysis is of limited value in understanding these systems: they cannot be reduced to a single equation that express their behavior over time. For this reason, computer simulation is necessary to explore nonlinear dynamics. Another consequence is that strange things can happen. The formal study of bizarre behavior in nonlinear dynamical systems has been given the name "chaos theory."

One definition of a chaotic system is a dynamical system that shows sensitivity to initial conditions. Initial conditions are where you start. Models that are started with different conditions will in many cases show very different behavior. A model of a solar system that starts with all the planets extremely close to their sun will result in all the planets being swallowed up by the sun instead of rotating around it. While such gross alterations in initial conditions obviously give different outcome, the sensitivity to initial conditions seen with chaos was wholly unexpected, given classical notions of a neat and orderly universe. Given a chaotic system the tiniest nudge and it is likely to end up behaving in a completely different way than it would have without the nudge. The essential metaphorical "result" (metaphorical since impossible to test experimentally) is the effect of a butterfly on the weather: the orientation of a butterfly's wing in Brazil will determine a rainstorm next year in China. Because chaotic systems give such highly divergent responses, it can be difficult or impossible to distinguish chaos from random processes. Is the apparent noise of neural firing true noise or is it deterministic chaos?

Computer simulations are particularly convenient ways of studying dynamical systems. In a system that can be solved analytically, movement of each state variable is explicitly expressed in terms of time. Given the time, you can say exactly what position every piece of the system will be in. A nonlinear dynamical system cannot be solved in this way. It is instead expressed as a set of differential equations, each of which tells how a particular state variable's growth or diminution is effected by the parameters and by this and other state variables. Time is not explicitly mentioned in a differential equation. It only shows up as the denominator for the rate of change of each state variable. There is no way to calculate exactly the value of each state variable at a given time. Since we know how fast each state variable is changing, and since we also know its current value, we can simply add on the amount that the state variable would change in a brief time to get a new value for this much time in the future. But all the state variables are changing at the same time, and some state variables affect the rate of change of other state variables. Hence, the rate of change for one state variable is no sooner calculated than it is incorrect. This is the slippery slope that the digital computer manages to negotiate by virtue of taking very small steps forward and sometimes even by stepping back and trying again. if we recall the above discussion of chaotic systems, however, we realize that even these inevitable little errors can have profound effects on the evolution of the system.

The state of an entire dynamical system can be defined as a set of values for all of the state variables. This set can be thought of as a single point in a high-dimensional phase space, where each state variable is represented along a different dimension. Only one trajectory passes through a given point in phase space for a particular dynamical system: you cannot go off in two directions from one place. Therefore, if a system returns precisely to a particular point, it will follow the same path and come back there again. This is the nature of an orbit. An orbit is a type of attractor, a preferred state or sequence of states in which the system will tend to end up. The "strange attractors" of a chaotic system are represented by various shapes in phase space. Chaotic systems do not return to points through which they have previously passed, but instead pass through nearby points. A depiction of the formation of a strange attractor in phase space looks like a multidimensional pointillist portrait emerging over time. For this reason, modelers, like experimentalists, must be concerned with the general behavior of the system, rather than with precisely who spikes at what time.

An alternative to simulation is emulation. This involves putting together physical parts that are made to correspond to the system of interest. In the case of the solar system, orreries were built that illustrated planetary movements without explicitly representing gravitational forces. For the nervous system, electronic components such as translators, resistors, and capacitors are used to emulate the capacitance and resistance of neuronal membranes and cytoplasm as well as the active properties of ion channels. Done in this way, emulation has the advantage of directly embodying the differential equations and not then being prone to the simulation errors described above. Although neural emulation is currently a growing field of considerable interest, emulations are difficult to implement, and software simulation remains the modeling technique of choice for most researchers. It should be noted that, whether through emulation or simulation, such simulacra can only complement the direct experimental study of the nervous system.

3. Early Modeling Predicted a Role for Inhibitory Phasing

Thirty years ago, Andersen, Gillow, and Rudjord used an IBM 1620 computer to prepare a network model of the thalamus (Andersen and Andersson, 1968; Andersen *et al.*, 1966; Andersen and Rudjord, 1964). Their model was primitive in many ways and ran on a machine with the processing power of a modern-day calculator. Nonetheless, they were able to demonstrate the plausibility of their central hypothesis—the role of inhibitory phasing in producing synchronization of thalamocortical neurons. Inhibitory phasing was used to describe the ability of inhibitory postsynaptic potentials (IPSPs) to produce identical firing times in a set of postsynaptic neurons. In order to permit an IPSP to have this effect, they postulated that thalamocortical neurons would show "post-anodal exaltation," an increase in firing following hyperpolarization. In this very early neuronal network model, a probabilistic interpretation of spike firing was used rather than an explicit model of neuron dynamics. Nonetheless the model, along with their physiological studies, provided a central conceptual organization that continues in our present understanding of thalamic oscillations. Subsequent physiological studies and attendant modeling have filled in some of the blanks.

Figure 1 shows the original hand-drawn schematic of this model. The large circles are thalamocortical cells. The small circles are inhibitory cells, either thalamic interneurons or the reticularis neurons of the nucleus reticularis thalami. In their model, a single inhibitory cell projects to multiple excitatory thalamocortical cells. The firing of one thalamocortical cell is shown by the darkened cell in column 1 and corresponds to the trace of a burst shown below. This burst in turn produces excitation in associated inhibitory cells. The inhibitory cell with its relatively widespread divergence then serves to hyperpolarize a number of thalamocortical cells (column 2). The hyperpolarization is followed by a burst in multiple thalamocortical neurons (column 3). The ability of a large number of cells to burst simultaneously in response to divergent output from inhibitory cells permitted synchrony to develop and spread in the model. This tendency to burst after a hyperpolarizing input, dubbed postanodal exaltation by these researchers, is now called the "low-threshold spike." (The term low-threshold spike is also a bit confusing, however, since its prolonged duration makes it look decidedly nonspikey and since multiple sodium spikes generally ride on top of the single low-threshold spike.) The model also proposes a mechanism whereby a spindle could terminate through the desynchronizing influence of premature excitations.

The Andersen *et al.* model, although primitive by current standards, shows how modeling can play a valuable role in furthering our understanding of nervous organization. A critical aspect of models is that they enable the researcher to connect levels of organization whose relationship would otherwise not be immediately obvious. In this case, Andersen and colleagues observed a neural population phenomenon and utilized the model to make a predication about the expected behavior of single cells, connecting network phenomenon with single-cell behavior and ion channel properties. In experimental terms, this connected the *in vivo,* extracellular recording preparation with *in vitro,* intracellular recording techniques. Although the simplicity of their model is such that it can be readily expressed in a single figure, the computer model itself played a role in the conceptual development that led to this very concise expression of the hypothesis. It also provided a method of determining whether a mechanism that seemed plausible to the imagination would really work when scaled up to involve a large number of neuron-like elements. Just as experiments may only remain interesting for historical reasons after an overarching theory has explained the phenomena that they demonstrate, the conceptual model that arises from a computer simulation may make the computer model itself seem trivial and obsolete. The inhibitory phasing model

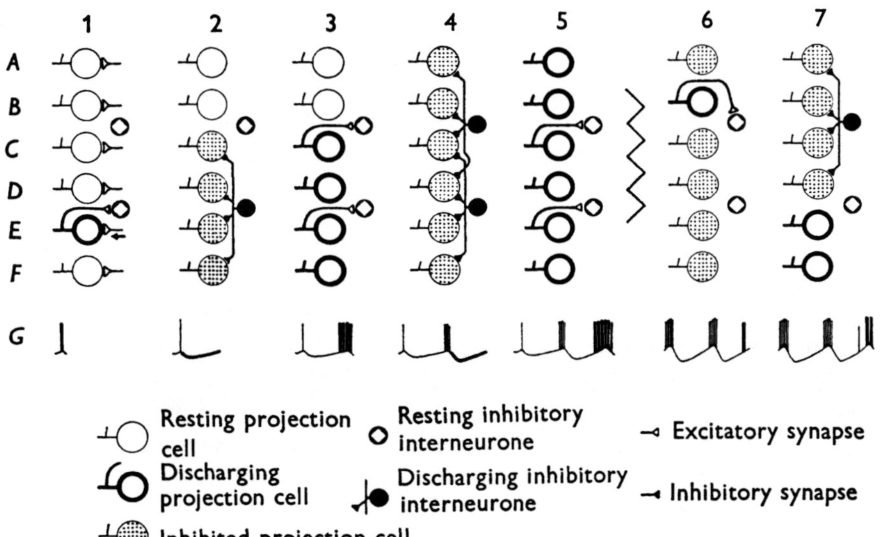

Figure 1. Original schematic showing the inhibitory phasing model of Andersen and colleagues. The large circles are thalamocortical neurons and the small ones are inhibitory neurons. These are here assumed to be interneurons, but could equally well be reticularis neurons. The discharge of a single thalamocortical neuron in column 1 (neuron E; also shown as trace in G), is followed by activation of the inhibitory neuron and hyperpolarization of several thalamocortical neurons (hatched in column 2). Shared input and projection allows the spread of the oscillation shown in columns 4 and 5. In the fullness of time (vertical sawtooth line), some thalamocortical neurons may discharge prematurely (cell B in column 6), leading to the breakdown of synchrony and termination of the spindle (column 7 and beyond). Reproduced with permission from Andersen and Sears (1964).

required that postanodal be an intrinsic property of the thalamocortical neuron. The existence of postanodal exaltation remained controversial for two decades until demonstrated *in vitro* by Jahnsen and Llinás (1984a, b). Further exploration of the low-threshold spike has remained a central focus of both physiology and modeling studies.

4. Modeling the T Channel

In the past decade, physiological investigation into the slow oscillations of the thalamus has clarified the source of the low-threshold spike and its role in thalamic oscillations. Modeling has worked hand in hand with electrophysiology to help us understand the genesis of the low-threshold spike and to start to appreciate its implications for sleep and epilepsy.

The molecular underpinning of the low-threshold spike is the T-type voltage-dependent calcium channel. Jahnsen and Llinás (1984a, b) showed that the thalamocortical cell could produce two very different firing patterns depending on the current-clamp protocol employed. Starting from the resting potential, a sustained depolarizing current resulted in rapid, repetitive firing. Starting instead at a hyperpolarized potential, a return either to rest or to a slightly depolarized potential produced the low-threshold spike: a prolonged depolarization with sodium spikes atop in a burst configuration. Blockade of sodium channels demonstrated that the low-threshold spike was not due to the influx of sodium, and replacements or blockade of calcium influx demonstrated that this was in fact a calcium-dependent mechanism. The requirement for a preceding hyperpolarization suggested the following scheme for activation of this channel. The channel would be inactivated at rest so that a depolarization from rest would not result in current passing through this channel. Hyperpolarization would serve to deinactivate the channel, thereby priming it for subsequent activation after the release from the hyperpolarized state.

This scheme was subsequently confirmed by voltage clamp by several groups (Coulter *et al.*, 1986b; Crunelli *et al.*, 1989; Hernandez-Cruz and Pape, 1989; Suzuki and Rogawski, 1989). As is generally done, they parametrized these results in the context of the Hodgkin–Huxley model, the mathematical model which underlies our understanding of transient neural potentials. This parametrization involves a pair of first-order differential equations that represent activation and inactivation, traditionally represented as m and h, respectively (m and h were originally used just for the Na^+ channel, n was used for the K^+ channel). The activation state-variable, m, proceeds toward its steady-state goal $m\infty$ at a rate dependent on time constant τm. Similarly, h is controlled by $h\infty$ and τh. Since $m\infty$, $h\infty$, τm, and τh are all functions of voltage and since the opening of a channel will change the voltage, this scheme creates a fairly complex dynamical system of feedback loops that can best be understood through modeling. (Note on nomenclature: deactivation is the reversal of activation, a decrease in the Hodgkin–Huxley m state-variable. Inactivation is a separate process given by the decrease of h. Deinactivation is the reversal of inactivation.)

Figure 2 shows the genesis of the low-threshold spike. In addition to the

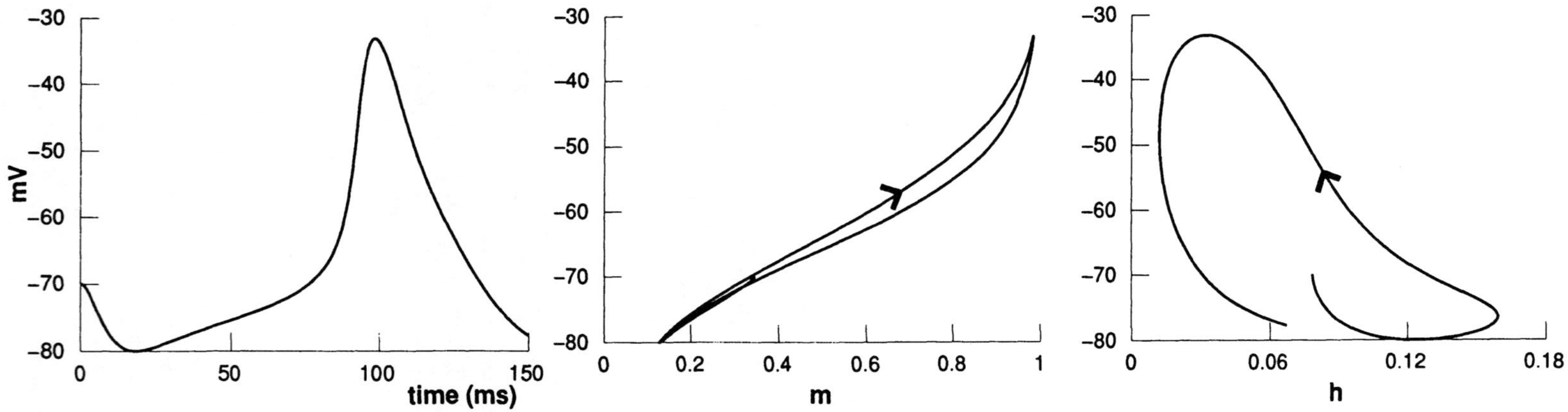

Figure 2. Low-threshold spike activation in response to an initial hyperpolarization. The left panel shows the low-threshold spike as it might be seen in an intracellular recording with sodium channel blockage. The phase plane graphs show the state variables m (middle panel) and h (right panel) on the abscissa and membrane potential on the ordinate. Arrows depict the direction of the trajectories. m initially declines but then increases to full activation (1.0), causing the inward calcium current that drives the potential. Note that the voltage lags behind at the top of the curve. Coming back down, the voltage drop precedes the reduction in m. The right-hand panel shows deinactivation as h goes from about 0.08 to 0.16 with the initial hyperpolarization. Inactivation then occurs as the voltage climbs up to the peak.

standard depiction of the change in cell potential with time, further insight can be gained by looking at specific state variables of the system in the phase-plane representation shown. Phase plane representations, used in dynamical modeling, represent an alternative to standard state variable versus time plots. The phase plane is a two-dimensional cut through the multidimensional phase space described above. In the phase plane, one state variable is graphed against another, yielding a trajectory that can provide insight into how the state variables are interacting. A classic phase plane representation is the view of the solar system from above the plane, where all the orbits are represented by sequential positions without an explicit representation of time. To look at the low-threshold spike, we follow either m or h on the x axis as they vary with voltage, depicted on the y axis. The state variables pursue their steady-state values, but fail to reach them because their own movement causes a current that changes the voltage, which in turn changes their own steady-state goal. In the case of a regenerative potential such as a sodium spike or low-threshold spike, an initial inward current produces a rising voltage which is associated with a steeply rising $m\infty$. Although m pursues this $m\infty$ in its upward course, the increase in m itself further depolarizes the membrane, making $m\infty$ rise still further. This gives m an ever-receding target to move toward. As seen in the middle panel, m follows a sigmoidal trajectory. This is similar to the parametrization of $m\infty$ (V), a function of voltage. Since the time constant, τm, is relatively short, m stays close to $m\infty$ (V) as it passes through different voltages. The role of h is apparent in the rightmost panel; its gradual increase with the initial hyperpolarization is the deinactivation that permits the subsequent low-threshold spike.

Our understanding of the electrodynamics of the single neuron has been repeatedly challenged by the finding of new ion channels and, more recently, by the finding that channel peptide subcomponents can be recombined in a myriad of ways. The simple picture of the Hodgkin–Huxley squid axon has been greatly modified as researchers have moved onto other neuron types. Fortunately, though many channels have been found, they have generally been amenable to approximation using the Hodgkin–Huxley parametrization or variations thereof. Since the Hodgkin–Huxley equations have become a neurophysiology standard, parameters for a new channel can be taken directly from physiology papers and inserted in a computer model. Although the Hodgkin–Huxley model is good enough to represent most ion channels, a more general channel model is the Markov model. In fact, the Hodgkin–Huxley model is an example of a Markov model with considerable state redundancy (Hille, 1984). Markov models represent ion channels as passing from one state to another according to kinetic variables that may have a voltage or substrate dependence. In some cases a complex Markov model is necessary as an alternative to a Hodgkin–Huxley model, in order to reproduce adequately the behavior of an ion channel. In the case of the T channel, for example, an alternative model with three states was used to represent activation more adequately (Fig. 3) (Wang *et al.*, 1991).

In the case of the thalamocortical neuron, at least two calcium channels are present in sizable amount so as to make a difference in the response of the cell. These are the T channel and the L channel, sometimes referred to as the low-threshold calcium current and the high-threshold calcium current, respectively. Experimentally, it is not always easy selectively to block ion channels pharmacologically so as to isolate a single ion-channel type of interest. In this case, both

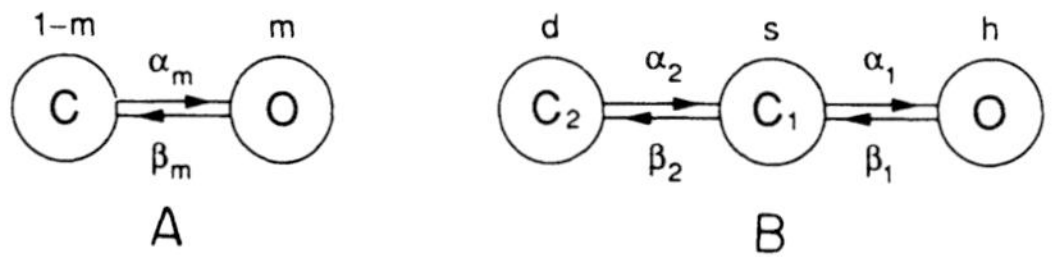

Figure 3. Two Markov models for the activation gate of the T calcium channel. Each of these alternative schemes only models a single gate. Four gates are used to model the entire channel; three identical activation gates and one inactivation gate. The four-gate scheme is the same as that originally used by Hodgkin and Huxley for the squid axon sodium channel. All four gates must be open for current to pass through the channel. This single *gate* can take on more than one *state* that can be either opened or closed. Note that in many ion-channel Markov models, the open and closed states would refer to the entire channel rather than to just a single gate. (A) The standard two-state model of the activation gate used by Hodgkin and Huxley and here adapted for the T channel. α and β are kinetic parameters that are functions of voltage. The values of α increases and the value of β decreases with more-depolarized voltages. (B) A more complex three-state model of the activation gate. Now there are two closed states as well as a single open state. Reproduced with permission from Wang *et al.* (1991).

T and L channels were demonstrated in voltage-clamp studies. Though the total current was made up of these two components, it was generally possible to disambiguate them due to the transient nature of the T current. Figure 4 summarizes the comparison between experimental and computer-modeled voltage clamp. Total current can be broken up into transient and persistent currents, which effectively distinguishes T from L components. The initial current is a result of ions passing through both types of channels. With time, the T channel inactivates, leaving a persistent current that is due to L alone. Since L shows minimal inactivation, this final current can be taken to be the L contribution to the current during the entire time. Subtracting this amount form the peak total current gives an estimate of the current that is passing through the T channel. This will be somewhat inaccurate, however, since it assumes that both current activate at approximately the same rate so that the peak represents a maximal contribution from the L current.

5. Modeling the Low-Threshold Spike

Having modeled the T channel at the voltage-clamp level, it is natural to use this model in order to better understand the role of the T channel in the genesis of the low-threshold spike. The simplest way to do this is simply to take the parameters from the voltage-clamp model and try to fit them into a current-clamp protocol similar to that used by Jahnsen and Llinás: a prolonged hyperpolarization followed by a return to rest. The question was how this channel would interact with other channels and with the capacitance and resistance of the cell to produce not only the low-threshold spike, but also repetitive firing.

When this was done, we found that it was surprisingly difficult to obtain a robust low-threshold spike. Figure 5 shows three different activation curves ($m\infty$) and indicates the low-threshold spikes obtained with two of them. A regenerative potential only occurs if the activation curve is steep enough for a voltage change to produce enough additional current entry to further augment the voltage, giving a positive feedback loop. In this case, none of the curves was sufficiently steep to produce a robust spike, although it was possible to get a rather anemic low-threshold spike using the steepest curve. The failure suggested that some critical aspect of the system was being overlooked or that some mechanism that appeared intuitively reasonable was perhaps not so reasonable after all. Ironically, one of the benefits of computer simulation is only realized when the simulations do not work. The failure compels one to look more closely at the data and the system to discover the missing links.

In this case, a variety of possibilities could account for the difficulties in generating the low-threshold spike using various voltage-clamp data. Technically, the voltage clamp may be somewhat inaccurate due to the difficulty in achieving suffi-

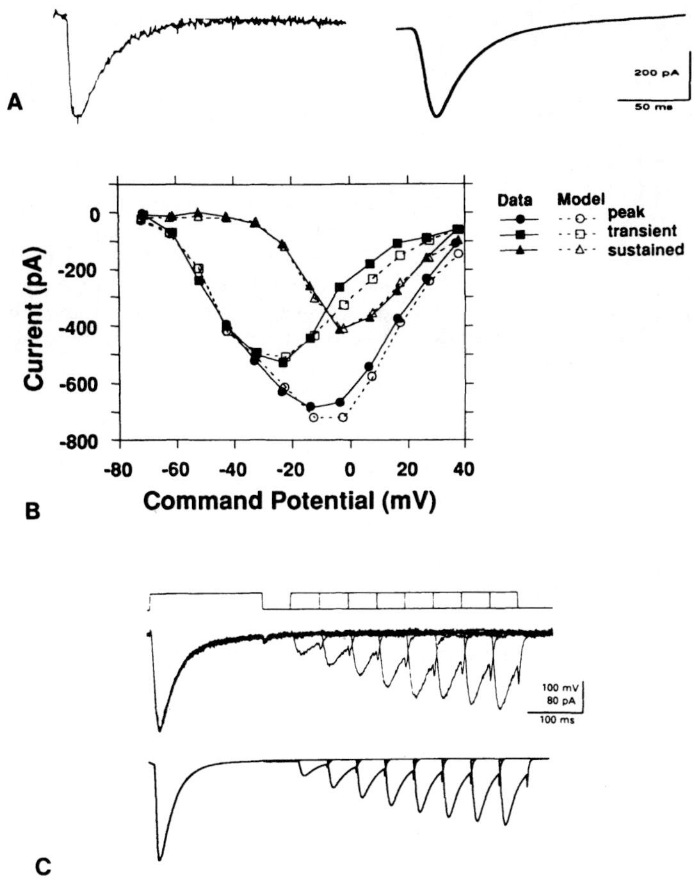

Figure 4. Comparison of experimental voltage clamp to computer model. (A) Neuron held at −100 mV and then stepped to −40 mV produces primarily the transient calcium current due to the T channel with little additional L-channel current. Experiment, left; model, right. (B) T-channel current can be estimated by subtracting the peak current that is due to both L and T components from the sustained current which is due to L alone. The existence of some slight L-channel inactivation as well as different rates of activation for the two channels will lead to some error in this estimate. Peak currents (ordinate) due to different command potentials (abscissa) are produced as in panel A. (C) The T channel recovers slowly from inactivation. Initial activation and inactivation similar to that in panel A is followed by prolonged hyperpolarization with subsequent attempts to reactivate the channel at different times in each of these superimposed traces. While an early reactivation (leftmost) gives only a small proportion of the original current, a late reactivation (rightmost) gives about 80%. This reflects the time course of deinactivation at the holding potential. Experiment, top: model, bottom. Reproduced with permission from Coulter *et al.* (1989a, b) and Lytton and Sejnowski (1992).

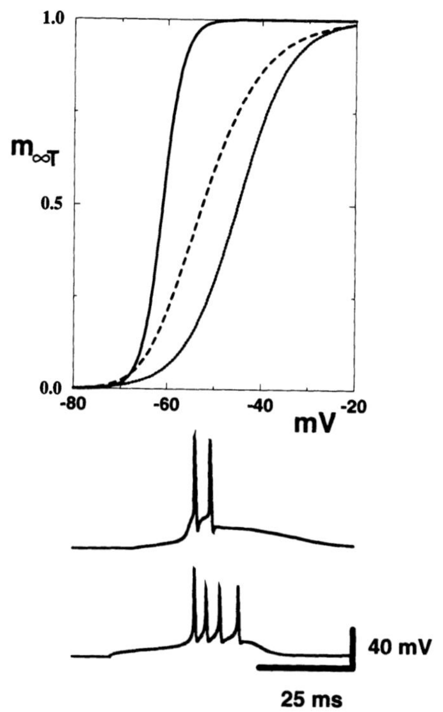

Figure 5. Different studies (left to right: Crunelli *et al.* 1989a; Coulter *et al.* 1989b; Hernandez-Cruz and Pape, 1989) gave very different voltage-clamp results for thalamocortical neuron T-channel activation. Although all three showed similar thresholds, there was a vast difference in steepness. A steeper activation curves implies a greater gain for positive feedback in creating a spike. Using the full morphology model, a low-threshold spikes was easiest to obtain using the steepest m_∞ (top trace), difficult, but possible to obtain using the next (lower trace), and impossible to obtain with the third (not shown).

ciently rapid switching between current injection and voltage measurement through a single electrode during the very rapid activation phase. As the voltage is stepped higher, the inward calcium current suddenly turns on. If the voltage clamp cannot sink this current as quickly as it enters, current escapes. The escape of current means, first, that some current is not being measured and, second, that this current will be altering the voltage, producing voltage escape. The consequence of voltage escape is that channels are seeing a higher voltage than has been selected for the clamp. Current escape tends to make the activation curve less steep, while voltage escape makes it more steep. Further artifact can be introduced if there is spatial as well as temporal failure of the clamp. In this case, some of the channels out in the dendrites may not be seeing the selected voltage because of their electrotonic distance. Additionally, the different techniques of whole-cell patch and electrode impalement may have different effects on accuracy.

These different studies were done in different animals and in different areas of thalamus. It may be that thalamocortical neurons in different species or areas have different T channels that are variably able to burst. Also, channel variation might be expected in response to hormonal or pharmacological influences. Conversely, it might be that neurons endowed with slightly different channels or different channel densities manage to compensate for such variability. Nature presumably finds robust solutions to its dynamical problems so that such minor variability still allows the neuron to function properly. Following this lead, it is important that the modeler find robust solutions that will not collapse with the slightest parameter tweak. In this case, the difficulty of obtaining the low-threshold spike with all parameters tested would suggest a systematic error, whether in the initial measurements or in the application of these measurements to the model.

A frequent critique of the realistic neural modeling enterprise is that, utilizing such a large number of parameters, it should be possible to recreate any desired neural firing pattern in many different ways. This concern arises from an analogy to curve fitting where large numbers of parameters permit ready duplication of a given curve in more than one way. With many possible solutions, any given solution has little value since it would be merely coincidence were it to correspond to physiological reality. We have found this concern to be largely unwarranted, at least a the cellular level. Firing patterns are characterized by their occurrence at certain a the cellular level. Firing patterns are characterized by their occurrence at certain frequencies or durations and within certain voltage ranges. Figure 6 gives a set of ion channel parameters in a display that illustrates time constants (ordinate) and the voltages of activation or inactivation (abscissa). By looking at such a diagram, one can pick out candidates for any given behavior. One can then pick out where there would be a chance for redundancy, one potassium channel substituting for another, for example. This type of redundancy may well be relevant to the functioning of the neuron, as it permits the robust response to minor alterations mentioned above.

Modeling difficulties, though frustrating, are in some ways the major contribution that modeling can make. If our understanding is perfect, then we can readily build a model, and it will work. This model may be emotionally, and perhaps even intellectually satisfying, but it has not told us anything that we did not already know. When a model fails, on the other hand, it is at least telling us that we do not know. The efforts to rectify its failure will generate new ideas as to what may be missing. The transformation of hypothesis into explicit computer model entails accounting

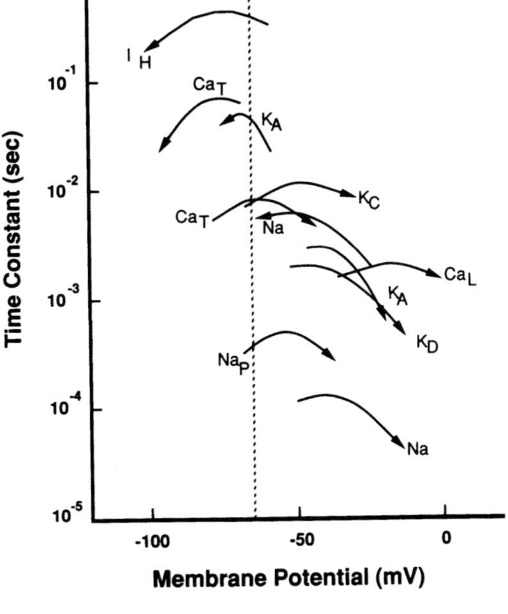

Figure 6. Conductance time constants for a typical thalamocortical neuron model. Parametrization for each time constant is shown over a voltage range corresponding to activation (rightward arrow) or inactivation (leftward arrow) for that conductance. Observations of neuron behavior in response to current injection (current clamp) will reveal activity that occurs in a certain voltage range and over a characteristic amount of time. Reference to the corresponding location in this type of figure will reveal which channels could be involved. RMP, Resting membrane potential; Na, fast sodium channel; Na_P, persistent sodium channel; Ca_T, T-type calcium current; Ca_L, L-type calcium current; K_D, delayed rectifier; K_A, A current; K_H, H current; K_C, C current at 500 nM intracellular Ca^{2+}. Reproduced with permission from Lytton and Sejnowski (1992).

for all system parameters. In itself, compiling this list may reveal basic, critical aspects of the system that are not known.

In this case, the difficulty in obtaining the low-threshold spike was most likely due to uncertainties as to location and density of the T channels. The simulations shown in Fig. 5 were done using a full compartment model with all of the channels localized to the soma. It may be more realistic to assume that there are T calcium channels in the dendrites as well.

5.1. Compartment Models versus the Point Neuron

In addition to physiological parameters, anatomical parameters also play a role in electrophysiological manifestation such as the low-threshold spike. Most models of thalamocortical neurons have been done using a simple "point neuron." A point neuron is a neuron represented as a single resistor–capacitor parallel circuit which represents the cell membrane. The name refers to the fact that the model only simulates a single point in the neuron, which would generally be considered to be the soma. To the extent that a neuron is electrophysiologically compact, all points on the neuron will have the same potential as all other points and the point neuron will be an adequate model. However, the neurons of the thalamus do not appear to be electrotonically compact.

To model electrotonically complex cells, a compartment model is used. This approach, pioneered by Wilfred Rall (see Segev *et al.*, 1995) links the membrane model utilized for a point neuron into chains. Since dendrites branch, these chains can become very complex (Fig. 7). Anatomically accurate fall compartment models can be produced with the aid of a translation program for reading in the output of a neuron tracing system. In the compartment model, every dendrite is represented by a series of resistor–capacitor (*RC*) circuits connected by a resistance that represents the axial (cytoplasmic) resistance within the cell. By stringing these together and branching them appropriately, one can get a model of the full cell as is shown here. Each of the *RC* membrane models is regarded as a compartment which represents one point along the membrane of the dendrite or soma of the neuron. Usually axons are not modeled since they are believed to generally communicate action potentials without alteration along their length.

This classical view of dendrites as summators of incoming signals depended on two assumptions. The first assumption was that we knew the specific membrane resistance of neural membranes. The second assumption was that most dendritic trees are passive, that they do not have the voltage-sensitive sodium, calcium, and potassium channels (active channels) that characterize neuron soma membranes and axon membranes. The development of whole-cell patch recording has allowed much better intracellular access than the earlier technique of cell impalement and has led to the idea that membrane resistance might be far higher than previously believed (Storm, 1990; Gola and Niel, 1993). Higher membrane resistance would lead to less leakage of current out of the neuron and would tend to make the neuron more electrotonically compact, hence more like a point neuron. This would suggest that physically distant dendritic domains would tend to influence each other and might not be capable of independent voltage processing.

A still more revolutionary complexity has been added by the discovery of active elements in the dendrites of a variety of neuron types (Spruston *et al.*, 1995; Magee and Johnston, 1995). Properties of dendrites are difficult to assess experimentally because neuron access is generally through the soma, whose larger size offers a far better target. However, new techniques in voltage imaging as well as refinement of electrode access has led to the ability to monitor these elusive structures. It is now accepted that a variety of cell types may use active properties to promulgate signals along their dendrites. Initial results suggest that this communication may be primarily antidromic (from soma to dendrite). However, involvement of dendritic channels intradendritic processing may also be important.

The three major neuron types of thalamus have all been investigated with compartment models. The thalamic interneuron was looked at several years ago by Bloomfield and Sherman (1989). Thalamic interneurons are of particular interest because input to the thalamus comes directly to their dendrites. Incoming afferents synapse onto unusual trisynaptic structures that are made up of synapses from the afferent to both interneuron and thalamocortical dendrites as well as a dendrodentritic synapse from interneuron to thalamocortical neuron. Such complex dendritic interactions suggest the hypothesis of local processing occurring out in the

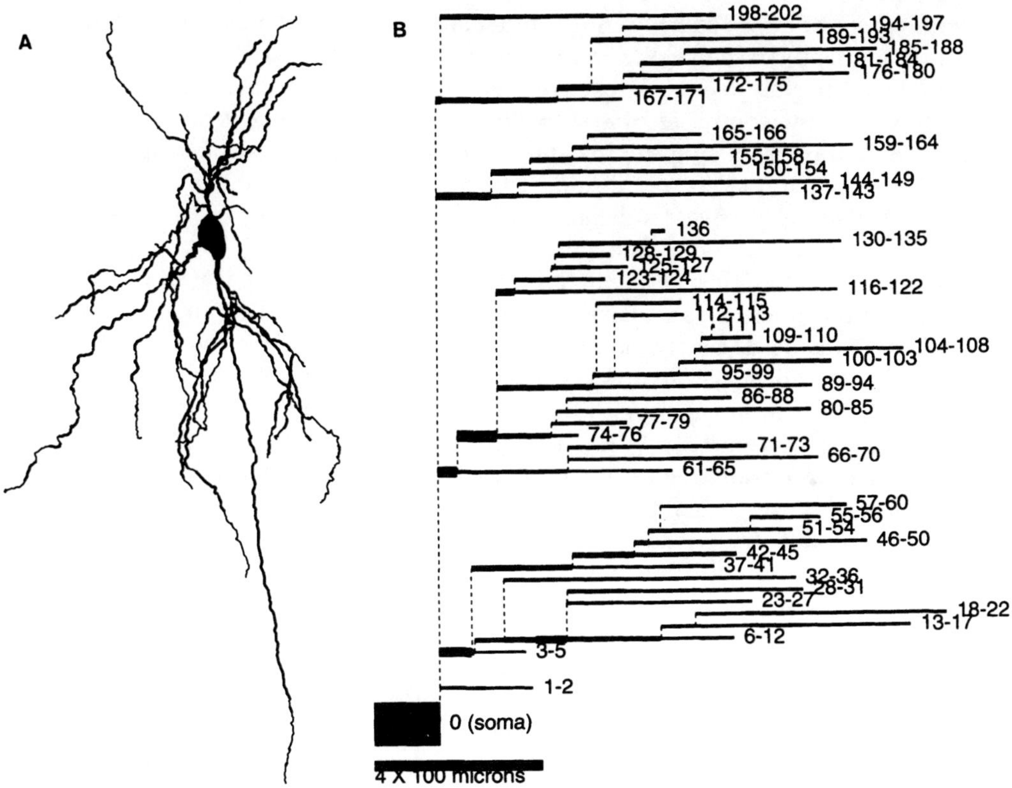

Figure 7. A compartment model of a neuron has multiple resistor–capacitor circuits strung together to match the dendritic tree of a neuron. In this case the thalamocortical neuron (A) (digitized using a Eutectics system by Drs. Chie-Fang Hsiao and Mark Dubin, University of Colorado) corresponds to the compartment model (B). The numbers indicate the range of compartments represented by each apposed line.

dendrites, making the interneuron a multiprocessor that could do several calculations at the same time in different parts of its dendritic tree. This Bloomfield and Sherman study supported this hypothesis, concluding that the interneurons were likely to be fairly large electronically. Hence, voltages in one part of the neuron would not have much effect on other parts, allowing local processing to occur. However, these conclusions were dependent on the aforementioned assumptions of low membrane resistance and lack of dendritic active channels. They will need to be reevaluated as new data become available for this neuron type.

Most modeling of thalamocortical neurons has been done using simple point-neuron models. By contrast, the model used in Fig. 5 used the full compartment model of Fig. 7. Because so little is known about the densities, kinetics, and distribution of voltage-sensitive channels in the dendrites, we performed this modeling with ion channels in the soma alone. These somatic channels had to drive the low impedance represented by a passive dendritic tree. (The word impedance, taken from electrical engineering, is simply a generalization of the concept of resistance). The dendrites produce a low impedance because they are so large that a lot of current has to flow into them in order to get a voltage change. It was considerably harder to generate a low-threshold spike in the full compartment model than in the point neuron. The difficulty can be traced to the failure of positive feedback in the presence of the low dendritic input impedance. Current flowing through the T channels disappears down the dendrites instead of altering the voltage locally. There is less positive feedback from voltage to cause regenerative T-channel opening. In the absence of this feedback, the low-threshold spike fizzles out. A large part of the difficulty producing the low-threshold spike with the various experimental voltage-clamp parameters was undoubtedly due to this low impedance. Most likely, the assumption of passive dendrites is inadequate and will need to be corrected as more information becomes available.

Destexhe *et al.* (1996) evaluated the role of active conductances in the dendritic tree of the nucleus reticularis neurons. They postulated that different results obtained in dissociated cells and in slice might be due to the removal of these active dendrites in the dissociation process. They were able to show that differences in the density of dendritic ion channels would produce specific alterations in the response to voltage and current clamp that matched experimental findings. The apparently slower channel kinetics seen in slice with voltage clamp appeared in the model to be an artifact of the distant location of the channels in the intact neuron. This sort of analysis is valuable because it points up recording artifacts that are not made obvious by notching or other obvious inconsistencies in the current trace.

This study also explained a difference in the response to current clamp *in vivo* and in slice (Fig. 8). In slice, traces generally show an all-or-none burst response with a characteristic "accelerando–decelerando" pattern (increasing action potential frequency followed by decreasing frequency). By contrast, *in vivo* responses were graded with increasing number of action potentials depending on the depth of the preceding hyperpolization. Simulations with dendritic channels present showed the all-or-none pattern characteristic of the slice. The accelerando–decelerando pattern was found to be due to the gradual onset of the dendritic low-threshold spike followed by its gradual termination as current leaked out into the soma. The graded response of *in vivo* recording could be reproduced by assuming that the dendrites are exposed to ongoing depolarization through synaptic input.

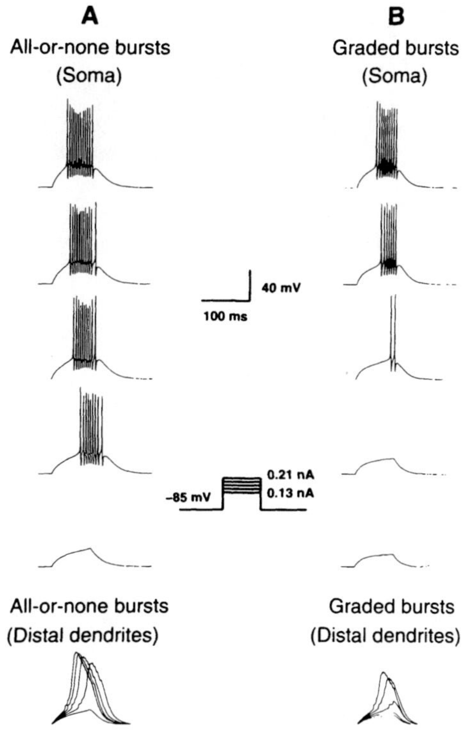

Figure 8. Models of reticularis neurons with (B) and without (A) dendritic current injection explain different responses found *in vivo* and *in vitro.* Current injections from a hyperpolarized state were used to elicit bursts. (A) A high density of T-type calcium channel in the dendrites leads to all-or-none bursts with the accelerando–decelerando pattern. (B) Bursts are graded when the same model is subjected to an ongoing dendritic current as might be seen with constant synaptic barrage *in vivo.* This current tends to partially inactivate the dendritic T channel, causing it to play less of a role relative to the soma in producing the low-threshold spike. Reproduced with permission from Destexhe *et al.* (1996).

This input served to partially inactive the T channel, leading to a less robust, hence more gradual, burst onset.

5.2. Beyond the Low-Threshold Spike

In addition to the single low-threshold spike burst, the thalamocortical neuron can show repetitive bursting, potentially making it a pacemaker neuron. Isolated thalamocortical neurons have been shown to produce a variety of rhythms (Fig. 9) (Leresche *et al.,* 1991; Toth and Crunelli, 1992). The demonstration of the variety of rhythms in isolated neurons validated the Jahnsen and Llinás claim that their slice observations were due to intrinsic and not network properties. The single neuron, and the single-neuron model, could produce fast, repetitive firing reminiscent of the rapid thalamocortical neuron firing of the relay mode (top trace). They could also produce rhythmic low-threshold spikes above or below the threshold for sodium channel activation (fourth and third traces, respectively). In addition, neuronal quiescence was observed due to either excessive hyperpolarization (bottom trace) or to an intermediate potential between sodium channel activation and T-channel deinactivation.

The form of the slow oscillation, a low-threshold spike followed by a gradual depolarizing interburst interval, was hypothesized by McCormick and Pape (1990) to be due to the interplay between the T channel and the H channel (Fig. 10).The H channel is a mixed-cation channel that passes both sodium and potassium and therefore has an intermediate reversal potential of about -40 mV. This means that

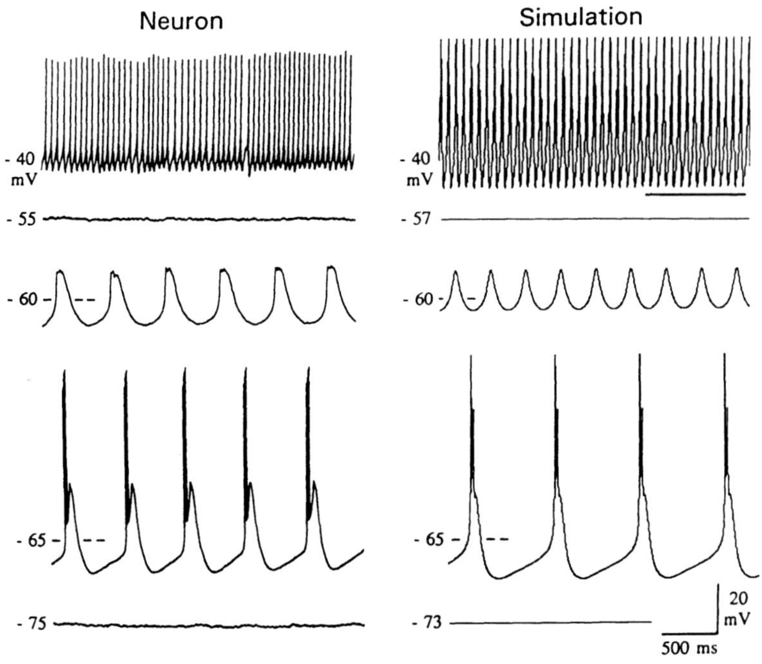

Figure 9. Comparison between experiment and simulation for an isolated cat lateral geniculate nucleus thalamocortical neuron at different potentials determined by the amount of injected current. Tonic firing is seen at top and slow, repetitive low-threshold spikes are second from bottom. Excessive hyperpolarization (bottom) precludes activity, as does an intermediate level of depolarization that is too high to permit T-channel deinactivation and too low to cause sodium channel activation. Reproduced with permission from Toth and Crunelli (1992).

when it is activated it will tend to depolarize the thalamocortical neuron. In contrast to most channels, which tend to activate with depolarization, I_H turns on with hyperpolarization. The hyperpolarization that follows the low-threshold spike activates I_H producing an inward current that progressively depolarizes the cell. In this way, a sustained cycle can be set up. The hyperpolarization of the interburst interval is sufficient to deinactivate the T channel, arming it for its next burst. The I_H-mediated depolarization gradually increases the potential until the T-channel threshold is reached, producing the low-threshold spike. During the low-threshold spike the T channel inactivates and the H channel deactivates.

This basic interplay between the two channel types has been subsequently confirmed in many models. Additional detail has been added as more channels have been included, producing models of increasing complexity. However, these additions have not obscured the pattern described above and many modelers primarily utilize these two channels, electing not to include the potassium and sodium channels that produce fast spiking atop the low-threshold spike. Huguenard and McCormick (1992) have utilized their own voltage-clamp data to produce perhaps the most detailed model of the thalamocortical neuron (McCormick and Huguenard, 1992). They identified a variety of potassium channels as well as the major calcium channels T and L, the sodium channel, and the H channel. They were able to investigate the role of each current at various phases of the oscillation.

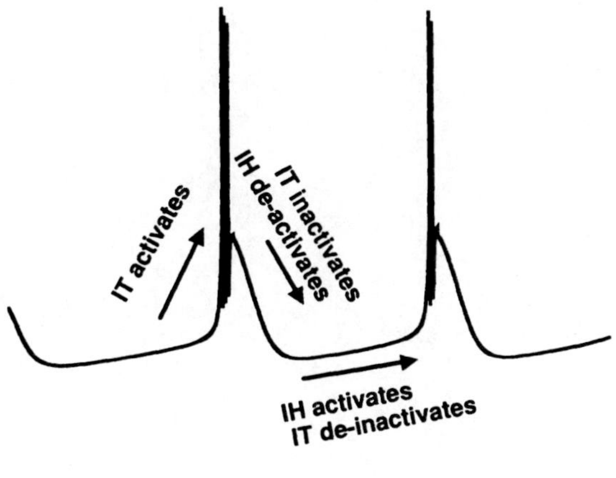

Figure 10. Two channel types interact to produce repeated bursts in a thalamocortical neuron. The burst itself is produced by inward current through the T calcium channel and is predicated on a prior hyperpolarization as has been described. In addition to deinactivating I_T, the hyperpolarization activates I_H, leading to a gradual depolarization up to the threshold for a subsequent I_T-mediated burst. Note that this model also includes many other channels, such as the Na^+ channels producing the fast spikes atop the bursts, but that these channels do not contribute substantially to the form of the low-threshold spike itself. Adapted from McCormick and Pape (1990).

The manner in which the critical currents I_H, I_T, and I_{leak} influenced the oscillations was studied by altering the maximum amplitude of each of these currents. As I_H density was increased, the oscillation first underwent an increase in frequency due to more rapid interburst depolarization mediated by the greater inward current. With further increase, however, there was a critical value beyond which I_H acted to dampen further oscillations since its depolarizing current prevented the interburst hyperpolarization required for I_T deinactivation. Altering the voltage dependence of I_H by shifting its activation curve had a similar effect (Fig. 11). Deeper and slower oscillation was possible with the curve in a hyperpolarized position. Shifting the curve to more depolarized position produces a shallower and less sustained oscillation. This observation is consistent with a model of the spindle envelope to be discussed below. Altering the maximum amplitude of I_T also affected the oscillation. A certain minimum I_T was required in order to obtain sustained oscillations. As I_T strength was doubled from this minimum, the frequency and amplitude of the oscillations increased.

The leak current was also found to play a significant role in the oscillations. The phrase "leak current" is used to refer to any current mediated by a non-voltage-sensitive channel. This is typically thought to be a hyperpolarizing current. In the Huguenard and McCormick model, a minimal hyperpolarizing leak was required to permit adequate interburst T-channel deinactivation. Further increases in the leak produced higher amplitude and lower frequency of the oscillation. Still further increase eliminated the oscillation since I_H was no longer powerful enough to overcome the hyperpolarization. The other currents in the model (the I_L calcium channel and the I_A, I_C, and I_{K2} potassium channels) were found to play smaller roles in the generation of rhythmic activity. Alterations made to these currents were found to change the shape and repolarization of the low-threshold spike. The high-threshold calcium current, I_L, came on at a similar time to I_T, but made a smaller contribution to the Ca^{2+} current. Reduction of the potassium current, I_{K2}, lead to a broadening of the low-threshold spike. It was found that the effects of decreasing I_{K2} were partially compensated by an increase in current through the calcium-sensitive potassium channel, I_C. This is a possible example of electrophysiological

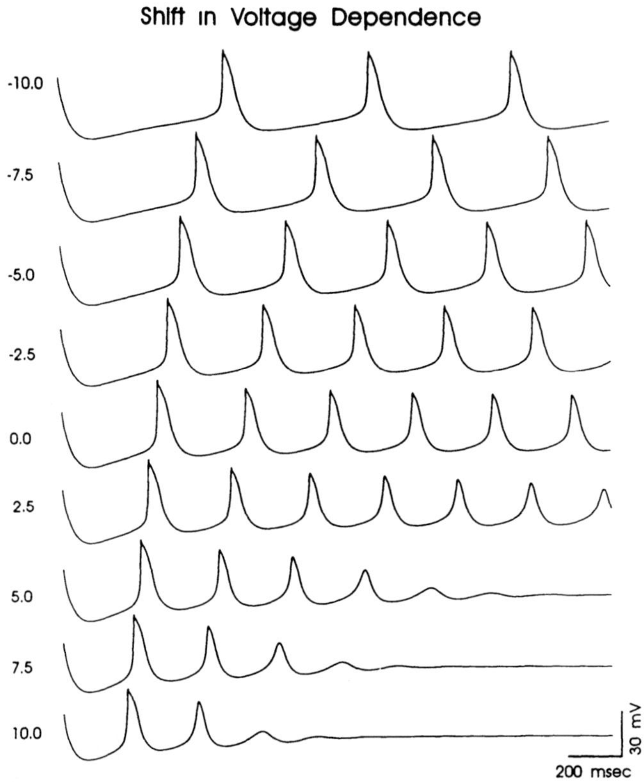

Figure 11. Change in thalamocortical neuron model slow oscillation with shift in voltage dependence of I_H. Note that this simple model does not include sodium spikes. Reproduced with permission from McCormick and Huguenard (1992).

homeostasis maintaining the activity of the neuron in the face of perturbations (Lemasson *et al.*, 1993).

5.3. Waxing and Waning as an Intrinsic Property

Many thalamocortical models have sought to reproduce frequency and coherence characteristics of spindlelike oscillations, but have ignored the characteristic wax and wane of activity. Of course, the actual "spindle" of the spindle is the waxing-and-waning envelope surrounding the underlying oscillation. Spindles generally last about 1 sec in human sleep. Thalamocortical neurons in slice can show a waxing-and-waning oscillation which is resistant to tetrodotoxin and therefore intrinsic to the thalamocortical neuron itself (Leresche *et al.*, 1991).

Based on previous proposals (McCormick, 1992; Soltesz *et al.*, 1991), Destexhe *et al.* (1993) were able to produce waxing and waning in their model thalamocortical neuron using a hypothesized effect of calcium on I_H (Fig. 12). Since there are no experimental studies showing an influence of Ca^{2+} on I_H channels for thalamocortical neurons, models were based on cardiac sinoatrial node cells (Hagiwara and Irisawa, 1989). In the cardiac cells, a very similar ion channel called I_F shows a

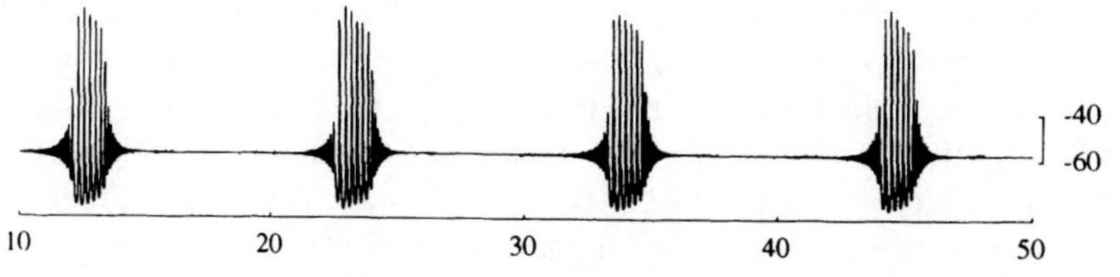

Figure 12. Postulating that Ca^{2+} may control I_H voltage dependence in thalamocortical neurons as it does in the heart suggests a mechanism for the waxing and waning of the spindle envelope. As shown in Fig. 11, I_H will only sustain oscillations when its activation curve is situated at a relatively hyperpolarized potential. Making the activation curve position dependent on Ca^{2+} allows it to shift back and forth, shaping the oscillation. Reproduced with permission from Destexhe *et al.* (1993).

positive shift in activation function with Ca^{2+} binding. The new I_H model was also more detailed than previous ones due to the inclusion of two activation gates with different time constants.

With low density of H channels an oscillation with a frequency of about 3.5 Hz was seen. With an increase in channel density waxing and waning was found to occur. The decline in amplitude of the spindle (the wane) occurred because the entry of Ca^{2+} during the spindle caused I_H to turn on at less-hyperpolarized potentials, thereby permitting progressively less hyperpolarization during the interburst interval (Fig. 11). This lack of hyperpolarization prevented full deinactivation of the T channel, reducing the low-threshold spike amplitudes progressively until they no longer occurred. During the silent phase that followed, calcium was removed and I_H shifted back, permitting the baseline to become more hyperpolarized so that I_T was again deinactivated. Still further hyperpolarization allowed I_H to activate despite its shift, permitting a new spindle to occur.

Further explorations indicated that spindles could also be obtained using a slow depolarization-activated potassium channel in lieu of a calcium-dependent I_H. In contrast to the calcium-dependent mechanism, a progressive depolarization in the oscillatory period is caused by increasing activation of I_H and inactivation of I_{K2}. The depolarization again prevents I_T deinactivation. The intraspindle frequency in this model was 10–14 Hz, much faster than that seen in the Ca^{2+}-sensitive I_H model and more comparable to the frequencies seen in human sleep and in *in vitro* slice preparations (Leresche *et al.*, 1991).

Waxing and waning in these models takes place due to intrinsic properties of the neurons. Alternatively, network properties could produce spindling as a population response. Increasing coherent activity would give rise to higher amplitudes in the electroencephalogram followed by a reduction in amplitude with reduced coherence as individual neurons drop out.

5.4. Frequency and Phase Control

An important and difficult question that frequently arises in the study of the nervous system is at what level a particular phenomenon can best be described. In general, we can anticipate that no level is primary and that both tertiary protein structures and network circuitry, to name two examples, are critical in generating brain physiology. Modeling should be able to help us determine to what extent something observed can be said to be due to this particular ion channel or this particular circuit diagram. In modeling thalamic slow rhythms, the distinction between intrinsic versus network behavior that arises with regard to spindles is a recurring theme.

A simple way to look at the interplay of extrinsic and intrinsic influences is by explorations of the response of the neuron's native properties to various externally imposed driving protocols. Most simply, it is possible to assess the range of frequencies that can be obtained with and without external driving. Since network effects will be realized at the single-neuron level through the instantaneous driving of the neuron by synaptic influences, this type of analysis can provide valuable insight into how the network influence single-neuron behavior. However, it must also be appreciated that the complexity of network interactions resides in the existence of feedback. In the network the driving of the single neuron alters the driving that subsequently occurs so that highly complex rhythms can emerge that stray very far from the simple single-frequency driving protocols that are usually assessed.

Frequency of bursting in thalamocortical neurons can be readily altered by changing the cell potential either by injecting current or by altering the leak as described by McCormick and Huguenard (1992). Figure 13 shows the progressive slowing with increasing hyperpolarizing current injection. Frequency control can also be produced by simulating alterations in potassium conductance that would occur with the changes in aminergic transmission from brainstem sites that might occur during sleep (McCormick, 1992). This differs from current clamp since an increase in membrane conductance is associated with the current flow and since the amount of current flow depends on the changing membrane potential. In either case, the lowest frequency possible occurs at the point when the current injected overwhelms the ability of I_H to repolarize the neuron to the point where a low-threshold calcium spike can occur.

Different parameters allowed slightly different ranges of frequencies to be obtained, but all were generally in the delta range of 1–4 Hz, consistent with *in vitro* results and supporting the hypothesis that these slowest rhythms may be primarily intrinsic. Driving these rhythms with simulated cortical input allowed entrainment of oscillation frequency to higher, but not lower values. A simulated cortical shock, resulting in a triphasic postsynaptic potential with initial excitatory postsynaptic potential (EPSP) followed by $GABA_A$ and $GABA_B$ IPSPs, was modeled after those recorded *in vivo* (Pare *et al.,* 1991). The initial EPSP component produced a low-threshold spike, allowing repetitive stimulation to approximately double the frequency (Fig. 14), as demonstrated experimentally (Steriade *et al.,* 1991). Faster driving did not allow a sufficiently long interstimulus interval to permit I_T to recover. Therefore, the time constant of deinactivation for I_T predicts the fastest driving frequency possible. With insufficient time for I_T recovery, it was not possible to start a new low-threshold spike. However, the initial EPSP could confound interpretation, since a large enough EPSP produced a burst, with sodium spikes riding on top of synaptic depolarization. These burst EPSPs could be driven to very high frequencies since they did not involve the T channel at all. At frequencies above about 15 Hz, this mechanism also failed since the feedforward IPSP from the

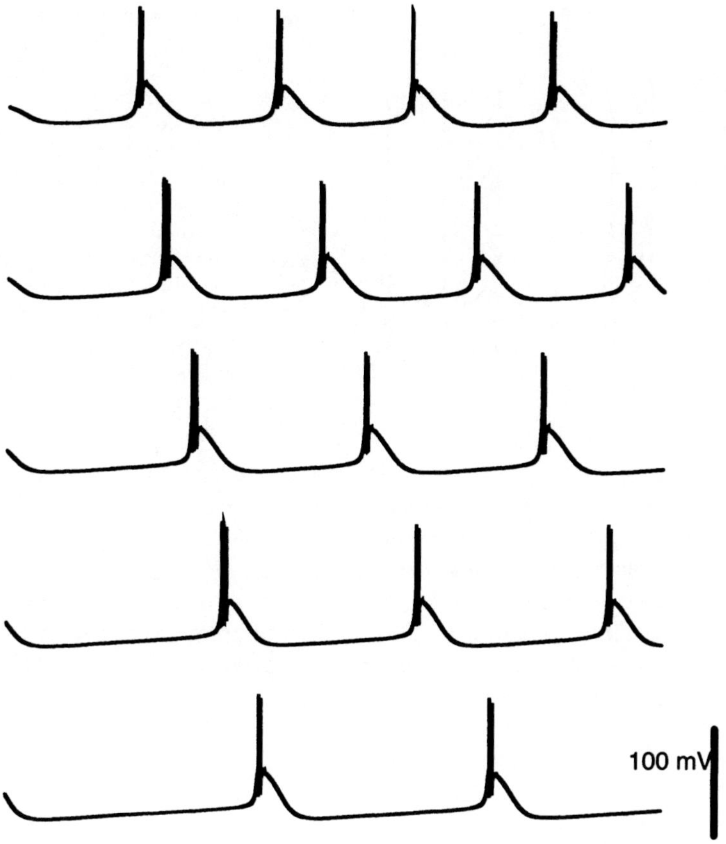

Figure 13. Increasing hyperpolarization by negative current injection progressively reduces the rate of bursting of thalamocortical neurons. The range of frequencies seen is generally between 2 and 6 Hz depending on the precise parameters used.

reticularis neurons interfered with the subsequent EPSP, eliminating the burst. There were also intermediate cases where the distinction between low-threshold spike and EPSP burst was not obvious, since the EPSP provided part of the depolarization, but the T channel augmented it. In such cases the distinction between intrinsic and extrinsic causes is blurred.

6. The Basic Two-Neuron Network

The nucleus reticularis is a thin layer that encases principal thalamic nuclei like a pillowcase around a pillow. These inhibitory neurons see all the signal traffic passing from thalamocortical neurons to cortex and back, and contribute their projections into the principal thalamic nuclei. They are not the only inhibitory neuron type in the thalamus. The thalamic interneurons are also inhibitory and

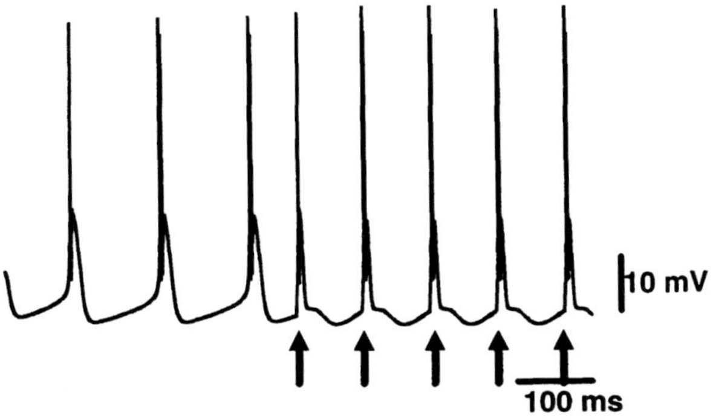

Figure 14. Entrainment by repetitive cortical activation can readily increase the rate of bursting of the thalamocortical neuron. This activation results in a multiphase postsynaptic potential which includes the direct corticothalamic EPSP as well as two or more feedforward IPSPs mediated by nucleus reticularis and interneurons. The EPSP can itself produce a burst, while the following IPSPs may produce further T-channel deinactivation to prime the following burst.

appear to be strategically placed at the interface between the afferents and thalamocortical neurons. While the interneurons may be expected to play an interesting role in relay-mode processing of incoming information, reticularis neurons are likely to be important in the slow oscillations that we have been discussing. In particular, they seem well situated to participate in the inhibitory phasing scheme put forth by Andersen and Rudjord (1964) and Anderson and Sears (1964) before the inhibitory nature of reticularis neurons was known.

As extensively explored by Traub *et al.* (1987) in another system, population oscillations need not denote substantial involvement of individual neurons in individual cycles. Instead groups of neurons may coalesce in synchronous firing and then disappear. On the next cycle of the oscillation, another group of neurons comes together. A group that fires together at any one time might never fire together again. If the individual neurons have different relative refractory periods, it is obvious that this would occur through some neurons being ready to join in the oscillation before others are. However, this pattern of shifting alliances may even be seen in the setting of very similar refractory periods. In the thalamus, it has been observed that the reticularis neurons usually fire with a much greater frequency than do the thalamocortical neurons (Steriade *et al.*, 1990; von Krosigk *et al.*, 1993). The thalamocortical neurons therefore probably engage only infrequently in the population oscillation in varying combinations. Surprisingly, recent *in vivo* intracellular recordings have suggested that thalamocortical neuron activity might be even more restricted, with many thalamocortical neurons completely silent despite coordinated activity all around them (Steriade and Contreras, 1995). Such neuronal silence would never be detected with extracellular recording since there is no way to detect the presence of a silent neuron extracellularly. Thus widespread quiescence belies the large population activity recorded during slow thalamic oscillations.

Since the thalamocortical neuron can burst in response to hyperpolarization and the reticularis neuron can burst in response to the depolarization, they are nicely matched to engage in the reciprocal oscillatory activity postulated in the classical inhibitory phasing model (Fig. 15). The thalamocortical burst produces an EPSP in the reticularis cell whose rapid depolarization results in a burst. This in turn produces a prolonged combination $GABA_A$ and $GABA_B$ IPSP in the thalamocortical neuron. The termination of this IPSP allows the thalamocortical neuron to

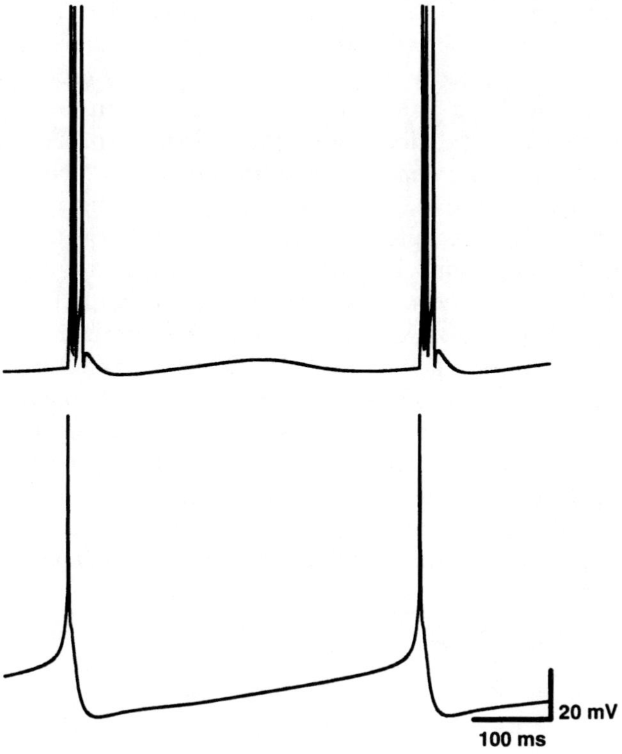

Figure 15. The thalamocortical neuron and reticularis neuron are reciprocally connected neurons that are well suited to engage each other in a mutual oscillation. Since the thalamocortical neuron is excitatory and the reticularis neuron inhibitory, the thalamocortical neuron can drive the reticularis neuron directly. The reticularis neuron can in turn drive the thalamocortical neuron indirectly by causing a hyperpolarization which primes the low-threshold spike.

produce a low-threshold spike, finishing the cycle. This can be considered a symmetrical-mode oscillation, since both neuron types are firing. An increase in the inhibitory strength from reticularis to thalamocortical neuron will instead produce an asymmetrical mode where ongoing reticularis activity tonically hyperpolarizes the thalamocortical neuron so that it is unable to burst (Fig. 16). The reticularis neuron in this setting shows continued intrinsic bursting.

The switch between symmetrical and asymmetrical modes of firing may provide insight into the limited involvement of thalamocortical neurons in slow oscillations. Perhaps many thalamocortical neurons are tonically hyperpolarized in the asymmetrical mode most of the time, and only become briefly involved in the symmetrical mode on occasion. Switching between modes in the model occurred either spontaneously or in response to simulated cortical stimulation. Extrapolating to a larger network, one could imagine that incidents of strong coincident corticothalamic feedback could cause particular thalamocortical neurons to either join or leave the ongoing oscillation.

7. The Search for Origins: Whence Spindling?

A common heuristic in many areas of scientific endeavor is the search for the ultimate source of observed phenomena. In most cases, this is readily seen as a simplifying tactic rather than a statement about reality. The limitations of such causal hypotheses are particularly apparent when it comes to understanding complex systems such as thalamic slow rhythms. Complex systems often show "emergent properties" that cannot be traced to the actions of one molecule or one neuron or even be conceptualized through one easily expressed dictum. Instead, the property emerges from complex interactions that have many causes.

Despite the likely difficulty in ever pinning a single cause or site or source on something as complex as a brain rhythm, the search for a site of spindle origin has yielded important insights and raised intriguing possibilities. A debate has taken place between advocates of the nucleus reticularis thalami as the spindling center and those who have suggested that the interaction between thalamocortical neurons and reticularis neurons is critical for spindling to occur. Early *in vivo* studies showed that the isolated nucleus reticularis thalami could produce oscillations in the spindle frequency range with some homology to cat sleep spindles (Steriade *et al.*, 1987). These experiments, performed in cats, involved extremely difficult surgical efforts to isolate a single brain structure without killing the animal. An additional caveat, previously mentioned, is that the existence of physiological homology, like that of anatomical homology, does not always connote common origin.

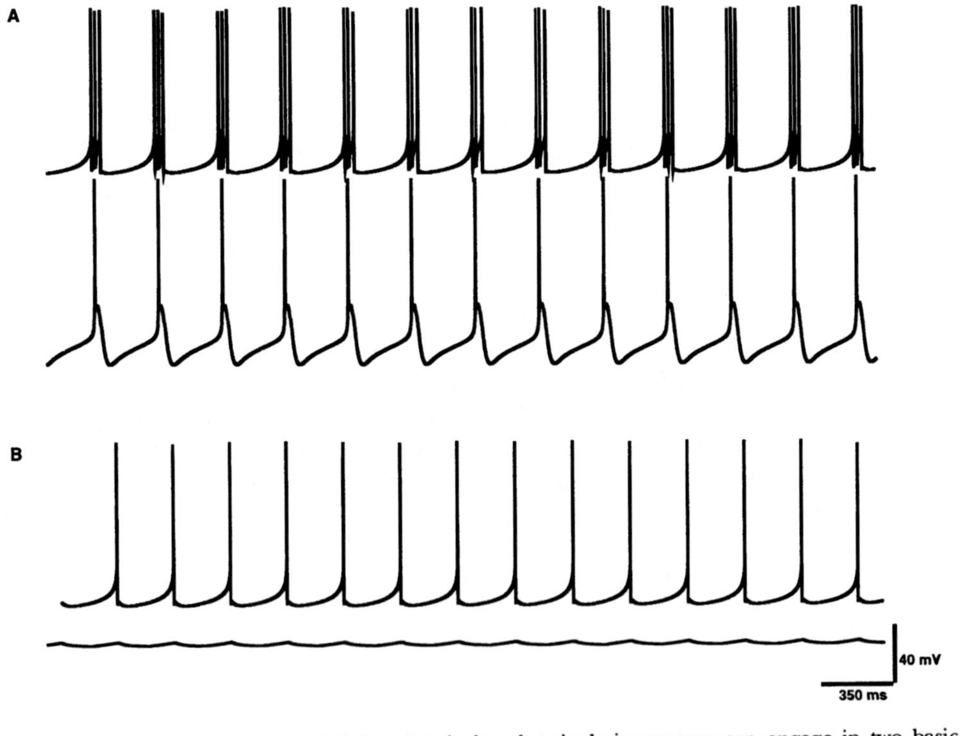

Figure 16. Reciprocally connected thalamocortical and reticularis neurons can engage in two basic oscillation types depending on the strength of synaptic connectivity between them. (A) Thalamocortical neuron and reticularis neuron oscillate together. (B) With stronger inhibition the reticularis neuron dominates with an intrinsic oscillation that produces a tonic hyperpolarization in the thalamocortical neuron.

Nonetheless, something spindlelike was going on in the isolated nucleus reticularis thalami. This seemed surprising since the nucleus reticularis thalami is made up exclusively of inhibitory neurons. It was no surprise that these inhibitory neurons could excite one another since the reticularis neurons have a T channel only slightly different from that found in thalamocortical neurons. It was surprising that they could be synchronous, since mutually inhibitory neurons will fall most naturally into an asynchronous mode of firing in antiphase (Perkel and Mulloney, 1974). This is easily understood by considering the elementary two-neuron circuit. Activity in neuron 1 actively inhibits neuron 2, causing neuron 2 to be silent. If each neuron's activity is self-limiting due to a process of adaptation, then the conclusion of firing will terminate the inhibition onto neuron 2. This will allow neuron 2 to fire, whether through an intrinsic process such as the low-threshold spike, or through extrinsic activation via excitatory inputs. Now neuron 2's activity will cause neuron 1 to be silent. In this way, the firing of each neuron is associated with the absence of firing in the other: an antiphase oscillation.

Wang and Rinzel (1993) showed that it is possible to get in-phase oscillations using mutually inhibitory neurons. Using a model reticularis neuron, they found that IPSPs of long duration would permit synchrony. Again considering the two-neuron network, a simultaneous burst in both cells would produce simultaneous inhibitory activation of both of these neurons. If these IPSPs are completely identical, then the next burst will once again occur at the same time in both neurons, allowing for synchrony. However, with many parameters, this will be an unstable equilibrium, equivalent to standing a pencil on its point. The network will tend to desynchronize in response to even slight perturbation. With long IPSPs the oscillation was more stable since the IPSP duration itself set and dominated the oscillation frequency. Antiphase oscillations were not favored since the long, deep IPSPs substantially deinactivated the T channel, allowing it to escape early, rather than waiting for the end of the burst in the other neuron.

Another possibility for the genesis of synchrony in isolated nucleus reticularis thalami arises from the relationship of many neurons in a large population (Destexhe *et al.*, 1993). With many pairs of neurons bursting in antiphase, two such antiphase partners would be in phase due to their individual antiphase relationships with a shared partner. Such networks have been shown to alternate between periods of high coherence and low coherence following the chaotic dynamics of the network as it evolves through time (Golomb *et al.*, 1994). Similarly, various clusters made up of different neurons can form and break up. The alternation between greater and lesser coherence in the network would be seen externally as a spindle.

Oddly, spindles spontaneously emerge in thalamic slices made from ferrets, but not in slices made from other animals. A set of *in vitro* studies of ferret thalamic slice yielded results that differed from the earlier *in vivo* study (von Krosigk *et al.*, 1993; Bal *et al.*, 1995a, b). When a knife cut was made in the slice, severing the nucleus reticularis thalami remnant from the relay nucleus remnant, the spindles were eliminated. Of course, the slice preparation is subject to its own set of experimental artifacts, exemplified by the word "remnant" in the previous sentence. Even if the entire nucleus reticularis thalami in the intact animal can generate spindles, a small piece of the nucleus might not be able to do so.

Simulations have suggested that it is indeed far easier to produce synchronous activity in a network made up of both reticularis neurons and thalamocortical neurons (Golomb *et al.,* 1994). Adding small amounts of thalamocortical neuron feedback to a nucleus reticularis thalami model eliminated clustering of the reticularis neurons and produced full synchronization that was resistant to parameter changes or to externally imposed noise. Adding still more thalamocortical neuron feedback would allow an otherwise quiescent nucleus reticularis thalami to produce synchronized oscillations.

A model of neuromodulation of a reticularis network provided a hypothesis for the difference in oscillatory activity of the reticular thalamic nucleus *in vivo* and *in vitro* (Destexhe *et al.,* 1994a). Serotonin and norepinephrine are known to block I_{leak} in reticularis neurons and hence depolarize them. The nucleus reticularis thalami network model showed synchronized rhythmic activity with these neuromodulators present, but not otherwise. The depolarization was sufficient to allow spontaneous rebound bursts in the reticularis neurons and permit oscillations to emerge. Since the neuromodulators are presumably present in higher concentrations *in vivo,* this could produce the oscillations seen under these conditions.

8. Synchrony and Spread of Network Activity

Recent studies have looked at larger networks involving both thalamocortical neurons and reticularis neurons, in order to establish factors that might lead to changes in synchrony and spread of synchronous activity. In addition to the many parameters that we have discussed in looking at individual neurons, these large-scale networks must also determine parameters for synaptic strength and time course as well as for divergence and convergence between and among cellular components. One simplifying anatomical observation is that there appear to be no connections between the thalamocortical neurons themselves.

Wang *et al.* (1995) have systematically explored the effects of convergence on such a network using a relatively simplified simple model of the individual neuron and assuming that any neuron could connect to any other. Using equal numbers of reticularis neurons and thalamocortical neurons, they found that coherent synchronization could occur even with sparse connections, such that each neuron received input from only 1% of the presynaptic population. These coherent population oscillations did not depend on the synchronous activation of all of the neurons. Instead, individual thalamocortical neurons would fire only intermittently as I_{H} gradually recovered from deactivation. Since thalamocortical neurons would fire only relatively rarely, a certain critical degree of convergence from thalamocortical neurons onto reticularis neurons was required for population activity. Network oscillation frequency could be changed by altering the density of I_{H} or by hyperpolarizing the individual thalamocortical neurons by changing I_{leak}. The interconnected reticularis neurons were not active by themselves in this network, so that oscillations were dependent on the excitatory feedback. Blockade of $\mathrm{GABA_A}$ receptors would cause significant slowing of the oscillation. This is consistent with experiments using $\mathrm{GABA_A}$ blockers that have been done in ferret slice. In contrast to the situation with sparse connectivity, all-to-all connections did not produce network-wide synchrony, but instead showed the breakup of the population into synchronously firing subgroups, similar to what has been seen in the reticularis-only networks (Destexhe *et al.*, 1993; Golomb *et al.*, 1994, Wang and Rinzel, 1993).

Utilizing a similar network with one dimension of spatial extent, the same group has shown propagation of population activity from one end of the network to the other (Golomb *et al.*, 1996). They discovered an unusual type of lurching wave propagation (Fig. 17). This occurred because thalamocortical neurons required a certain duration of hyperpolarization to deinactivate the T channel before they began oscillating. This temporal requirement is translated into a spatial phenomenon since a group of thalamocortical neurons will hold up the wavefront while they wait to receive sufficient hyperpolarization to engage in the oscillation. In order to obtain the propagation velocity of 1 mm/sec observed in ferret slice, the distance of lateral connectivity was constrained. In this way, the model used physiological data to make an anatomical prediction concerning total lateral projection from thalamocortical neurons to reticularis neurons and back again.

9. Summary and Conclusions

The thalamocortical circuit has attracted modeling attention primarily because of its involvement in the slow oscillations of sleep and epilepsy. Such oscillations promise to be relatively simple and thus provide a good entree into more general issues of thalamocortical interaction, including the coding of memory and cognition.

Neural systems, like other physiological processes, are dynamical systems that evolve through time in complex ways. Since such complex systems are not amenable to simple, single-equation, mathematical solution, computer simulations must be employed to investigate them. There has been increasing appreciation in recent years that even very simple dynamical systems can show astoundingly diverse behav-

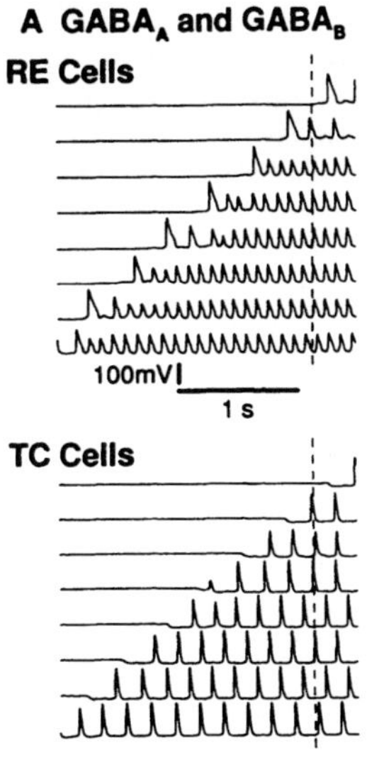

Figure 17. Spatial and temporal aspects of an oscillation spreading across a one-dimensional thalamic–reticular network. Individual voltage traces are shown for neurons of each type that become progressively involved as the oscillation spreads from one side (trace at bottom) to the other (trace at top). The wave spreads with a lurching quality that can be partially appreciated by noting the two to three cycles required before the next thalamocortical neuron in line joins the oscillation. Also note the highly variable phase relationships among the neurons engaged in the oscillation (relative to dashed line). Reproduced with permission from Golomb *et al.* (1996).

iors depending on seemingly inconsequential alterations in initial conditions. It is not yet clear how these results can be applied to the brain.

In this review, we started with a classical model of three decades ago and showed how its critical prediction, the low-threshold spike, could be explored from the molecular to the network level. As computer power becomes greater and as interest in complex systems grows, we will be able to learn more about how the entire thalamocortical network influences slow oscillations. The tools for evaluating such enormous systems remain primitive, but we can expect them to improve as further research compels their development.

ACKNOWLEDGMENTS. This work was supported by the National Institute for Neurological Disease and Stroke and by the Veterans Administration.

10. References

Andersen, P., and Andersson, S. A., 1968, *Physiological Basis of the Alpha Rhythm,* Appleton-Century-Crofts, New York.

Andersen, P., and Rudjord, T., 1964, Simulation of a neuronal network operating rhythmically through recurrent inhibition, *Nature* **204:**289–290.

Andersen, P., and Sears, T., 1964, The role of inhibition in the phasing of spontaneous thalamo-cortical discharge, *J. Physiol.* **173:**459–480.

Andersen, P., Gillow, M., and Rudjord, T., 1966, Rhythmic activity in a simulated neuronal network, *J. Physiol.* **185:**418–428.

Bal, T., von Krosigk, M., and Mccormick, D. A., 1995a, Synaptic and membrane mechanisms underlying synchronized oscillation in the ferret LGNd *in vitro, J. Physiol.* (Lond.) **483:**641–663.

Bal, T., Von Krosigk, M., and Mccormick, D. A., 1995b, Role of the ferret perigeniculate nucleus in the generation of synchronized oscillation *in vitro, J. Physiol.* (Lond.) **483:**665–685.

Bloomfield, S., and Sherman, S., 1989, Dendritic current flow in relay cells and interneurons of the cat's lateral geniculate nucleus, *Proc. Natl. Acad. Sci. USA* **86:**3911–3914.

Coulter, D. A., Huguenard, J. R., and Prince, D. A., 1989a, Characterization of ethosuximide reduction of low-threshold calcium current in thalamic neurons, *Ann. Neurol.* **25:**582–593.

Coulter, D. A., Huguenard, J. R., and Prince, D. A., 1989b, Calcium currents in rat thalamocortical relay neurones: Kinetic properties of the transient low-threshold current, *J. Physiol.* (Lond.) **414:**587–604.

Crunelli, V., Lightowler, S., and Polland, C. E., 1989, A T-type Ca^{2+} current underlies low-threshold Ca^{2+} potentials in cells of the cat and rat lateral geniculate nucleus, *J. Physiol.* (Lond.) **413:**543–561.

Destexhe, A., McCormick, D. A., and Sejnowski, T. J., 1993, A model for 8–10 Hz spindling in interconnected thalamic relay and reticularis neurons, *Biophys. J.* **65:**2473–2477.

Destexhe, A., Contreras, D., Sejnowski, T. J., and Steriade, M., 1994a, Modeling the control of reticular thalamic oscillations by neuromodulators, *Neuroreport* **5:**2217–2220.

Destexhe, A., Contreras, D., Sejnowski, T. J., and Steriade, M., 1994b, A model of spindle rhythmicity in the isolated thalamic reticular nucleus, *J. Neurophysiol.* **72:**803–818.

Destexhe, A., Contreras, D., Steriade, M., Sejnowski, T. J., and Huguenard, J. R., 1996, *In vivo, in vitro* and computational analysis of dendritic calcium currents in thalamic reticular neurons, *J. Neurosci.* **16:**169–185.

Gola, M., and Niel, J. P., 1993, Electrical and integrative properties of rabbit sympathetic neurones re-evaluated by patch clamping non-dissociated cells, *J. Physiol.* **460:**327–349.

Golomb, D., Wang, X. J., and Rinzel, J., 1994, Synchronization properties of spindle oscillations in a thalamic reticular nucleus model, *J. Neurophysiol.* **72:**1109–1126.

Golomb, D., Wang, X. J., and Rinzel, J., 1996, Propagation of spindle waves in in a thalamic slice model, *J. Neurophysiol.* **75:**750–769.

Hagiwara, N., and Irisawa, H., 1989, Modulation by intracellular Ca^{2+} of the hyperpolarization-activated inward current in rabbit single sino-atrial node cells, *J. Physiol.* **409:**121–141.

Hernandez-Cruz, A., and Pape, H. C., 1989, Identification of two calcium currents in acutely dissociated neurons from the rat lateral geniculate nucleus, *J. Neurophysiol.* **61:**1270–1283.

Hille, B., 1984, *Ionic Channels of Excitable Membranes,* 2nd ed., Sinauer, Sunderland, MA.

Huguenard, J. R., and McCormick, D. A., 1992, Simulation of the currents involved in rhythmic oscillations in thalamic relay neurons, *J. Neurophysiol.* **68:**1373–1383.

Jahnsen, H., and Llinás, R., 1984a, Electrophysiological properties of guinea pig thalamic neurones: An *in vitro* study, *J. Physiol.* (Lond.) **349:**205–226.

Jahnsen, H., and Llinás, R., 1984b, Ionic basis for the electroresponsiveness and oscillatory properties of guinea-pig thalamic neurons *in vitro, J. Physiol.* (Lond.) **349:**227–247.

Lemasson, G., Marder, E., and Abbott, L. F., 1993, Activity-dependent regulation of conductances in model neurons, *Science* **259:**1915–1917.

Leresche, N., Lightowler, S., Soltesz, I., Jassik-Gerschenfeld, D., and Crunelli, V., 1991, Low-frequency oscillatory activities intrinsic to rat and cat thalomocortical cells, *J. Physiol.* **441:**155–174.

Lytton, W. W., and Sejnowski, T. J., 1992, Computer model of ethosuximide's effect on a thalamic neuron, *Ann. Neurol.* **32:**131–139.

Magee, J. C., and Johnston, D., 1995, Synaptic activation of voltage-gated channels in the dendrites of hippocampal pyramidal neurons, *Science* **268:**301–304.

McCormick, D. A., 1992, Neurotransmitter actions in the thalamus and cerebral cortex and their role in neuromodulation of thalamocortical activity, *Prog. Neurobiol.* **39:**337–388.

McCormick, D. A., and Huguenard, J. R., 1992, A model of the electrophysiological properties of thalamocortical relay neurons, *J. Neurophysiol.* **68:**1384–1400.

McCormick, D. A., and Pape, H. C., 1990, Properties of a hyperpolarization-activated cation current and its role in rhythmic oscillations in thalamic relay neurones, *J. Physiol.* (Lond.) **431:**291–318.

Pare, D., Dossi, R. C., and Steriade, M., 1991, Three types of inhibitory postsynaptic potentials generated by interneurons in the anterior thalamic complex of cat, *J. Neurophysiol.* **66:**1190–1204.

Perkel, D. H., and Mulloney, B., 1974, Motor pattern production in reciprocally inhibitory neurons exhibiting postinhibitory rebound, *Science* **185:**181–183.

Segev, I., Rinzel, J., and Shepherd, G. M., eds., 1995, *The Theoretical Foundation of Dendritic Function: Selected Papers of Wilfrid Rall with Commentaries,* MIT Press, Cambridge, MA.

Soltesz, I., Lightowler, S., Leresche, N., Jassik-Gerschenfeld, D., Pollard, C. E., and Crunelli, V., 1991, Two inward currents and the transformation of low-frequency oscillations of rat and cat thalamocortical cells, *J. Physiol.* **44:**175–197.

Spruston, N., Schiller, Y., Stuart, G., and Sakmann, B., 1995, Activity-dependent action potential invasion and calcium influx into hippocampal CA1 dendrite, *Science* **8:**297–300.

Steriade, M., and Contreras, D., 1995, Relations between cortical and thalamic cellular events during transition from sleep patterns to paroxysmal activity, *J. Neurosci.* **15:**623–642.

Steriade, M., Domich, L., Oakson, G., and Deschenes, M., 1987, The deafferented reticular thalamic nucleus generates spindle rhythmicity, *J. Neurophysiol.* **57:**260–273.

Steriade, M., Jones, E. G., and Llinas, R. R., 1990, *Thalamic Oscillations and Signaling,* Wiley, New York.

Steriade, M., Curró Dossi, R., and Nunez, A., 1991, Network modulation of a slow intrinsic oscillation of cat thalamocortical neurons implicated in sleep delta waves: Cortically induced synchronization and brainstem cholinergic suppression, *J. Neurosci.* **11:**3200–3217.

Storm, J. F., 1990, Why is the input conductance of hippocampal neurones impaled with microelectrodes so much higher than when giga-seal patch pipettes are used? *Soc. Neurosci. Abstr.* **16:**506.

Suzuki, S., and Rogawski, M. A., 1989, T-type calcium channels mediate the transition between tonic and phasic firing in thalamic neurons, *Proc. Natl. Acad. Sci. USA* **86:**7228–7232.

Toth, T., and Crunelli, V., 1992, Computer simulation of the pacemaker oscillations of thalamocortical cells, *Neuroreport* **3:**65–68.

Traub, R. D., Miles, R., and Wong, R. K. S., 1987, Models of synchronized hippocampal bursts in the presence of inhibition. II. Ongoing spontaneous population events, *J. Neurophysiol.* **58:**752–764.

von Krosigk, M., Bal, T., and McCormick, D., 1993, Cellular mechanisms of a synchronized oscillation in the thalamus, *Science* **261:**361–364.

Wang, X. J., and Rinzel, J., 1993, Spindle rhythmicity in the reticularis–thalami nucleus—Synchronization among mutually inhibitory neurons, *Neuroscience* **53:**899–904.

Wang, X. J., Rinzel, J., and Rogawski, M. A., 1991, A model of the T-type calcium current and the low-threshold spike in thalamic neurons, *J. Neurophysiol.* **66:**839–850.

Wang, X. J., Golomb, D., and Rinzel, J., 1995, Emergent spindle oscillations and intermittent burst firing in a thalamic model: Specific neuronal mechanisms, *Proc. Natl. Acad. Sci. USA* **92:**5577—5581.

10

Realistic Network Models of Synchronized Oscillations in Visual Cortex

PAUL BUSH

1. Introduction

Stimulus-dependent synchronous oscillatory firing of populations of cortical neurons at frequencies around 40 Hz has been observed within and between many different cortical areas in primates and cats both awake and anaesthetized (for review see Singer, 1993). It has been proposed that this synchronous activity is used to group separated parts of single objects (Engel *et al.*, 1991a; Sporns *et al.*, 1991) and even signal visual awareness (Crick and Koch, 1990). Although there is still much debate as to the role and significance of synchrony, it is generally agreed that many cortical neurons can fire synchronously under some conditions. Therefore, it is worthwhile to discover exactly how cortical tissue generates and sustains synchronized oscillatory firing of its component neurons, both within a single column and between columns that may be located in different hemispheres (Engel *et al.*, 1991b).

PAUL BUSH • Computational Neurobiology Laboratory, Salk Institute, La Jolla, California 92037; *present address:* Department of Physiology, University of California at San Francisco, San Francisco, California 94143-0444.

Synchronized oscillations have been demonstrated in a variety of models based on coupled intrinsic oscillators (Winfree, 1967; Schuster and Wagner, 1990; Sompolinsky *et al.*, 1990; Williams, 1992). These models use simplified representations of generic neurons which permit thorough analysis, but cannot be used for detailed, realistic simulation of actual cortical networks. Realistic simulations are currently being used to investigate the biophysical mechanisms underlying cortical microcuitry dynamics and their dependences on various physiological parameters. The results of these simulations can be compared directly with recordings from cortical neurons.

One of the first realistic simulations of synchronized oscillatory firing in visual cortex was a model from G. Edelman's group (Sporns *et al.*, 1989). This model consisted of a layer of excitatory and inhibitory neurons organized into groups. Each group received orientation and direction selective inputs from an 'LGN' and connected to other groups with similar response properties. The 'orientation/direction' layer was connected in a divergent reciprocal fashion to a 'movement' layer consisting of groups similar to those in the first layer. The connections to the movement layer were organized so as to allow each group to respond to stimulus motion in only one direction. When the model was stimulated with coherent bar stimuli, each responding group displayed oscillatory discharge, synchronized with zero average phase delay. The coherency of the oscillations was ascribed to the reciprocal, excitatory connections between each group. Single units did not discharge regularly, as in the experimental data, leading to the observation that oscillatory behavior is not easy to detect without the use of statistical techniques. This model also showed that groups of neurons need not be directly coupled to be synchronized—multiple and even indirect paths will suffice. In addition, it was found that transmission (axonal) delays greater than one quarter of the oscillation period produced phase shifts between the connected groups.

As to the biophysical mechanism of synchronized oscillatory firing within one group/column, this model could only show that "oscillations are generated by activity of relatively sparse local excitatory and inhibitory connections and occur independently of intergroup coupling." Each neuron in the model was a simple integrate-and-fire leaky integrator with Gaussian noise and a refractory period added. The refractory period of the excitatory cells contributed to the oscillatory activity of the network. It was left to later models to illuminate the details of the biophysical mechanisms underlying synchronized oscillatory firing.

A more biophysically realistic simulation based on a previous model of oscillatory activity in piriform cortex used compartmental representations of the pyramidal neurons (Wilson and Bower, 1991), although they had somewhat arbitrary dimensions and passive properties (e.g., time constant τ_m of 2 m sec). Action potentials were triggered by threshold crossing and followed by an absolute refractory period; thus these model neurons did not display the burst firing seen in experimental data (Gray *et al.*, 1992). The network (1125 cells) was a generic representation of cortical tissue, consisting of long-range excitatory and short-range inhibitory connections with 100 topographically organized input fibers. When stimulated with a continuous bar, responding cells showed synchronized oscillatory activity that became desynchronized if a section of the bar was removed in a position between the receptive fields of the synchronized groups, as in experimental recordings (Gray *et al.*, 1989). The synchrony was found to be dependent on the long-range horizontal connections between pyramidal cells. The importance of long-range connections from pyramidal cells to inhibitory interneurons (see below) was not noted. Interestingly, zero-phase difference synchrony required relatively weak long-range connections, indicating that coherence was maintained through activation of spatially intermediate cells. This study also found that adding connections between the inhibitory cells improved frequency locking and increased the strength of the oscillation (see below).

The mechanism generating synchronized oscillatory firing in a local group of cells was not addressed in detail, except to state that the frequency of the oscillation was inversely proportional to the amplitude of the inhibitory feedback to the pyramidal cells. This result contrasts with the results of the more physiologically accurate model presented below.

A pilot study for this more detailed model (Bush and Douglas, 1991) directly addressed the question of synchronization within a small group of cells. This model was composed of ten pyramidal and one inhibitory compartmental neurons, incorporating voltage- and ligand-dependent Hodgkin–Huxley-type conductances that produced burst firing in the pyramidal cells in response to depolarizing input. In the network (every cell connected to every other one), constant (or noisy) depolarizing input to the pyramidal cells produced synchronized oscillatory burst firing that was very similar to that seen in intracellular *in vitro* recordings. It is proposed that the following mechanism is the basis of this activity. Pyramidal cell burst discharges rapidly excite other pyramidal cells, producing a large compound excitatory postsynaptic potential (EPSP) in all cells in the network, including the inhibitory basket cells. The basket cells are driven to fire simultaneously; their feedback inhibition onto the pyramidal cells then terminates the population burst and, together with intrinsic potassium conductances, produces a postburst hyperpolarization. Since the input resistance of the pyramidal cells is greatly reduced during this hyperpolarization (Bush and Sejnowski, 1994), there is less chance of the cell spiking during this time (out of phase with the oscillation). Cells with intrinsic bursting frequencies that vary over an octave (15–30 Hz) can be made to synchronize together at one frequency (e.g., 45 Hz) by this mechanism. Cells with lower intrinsic frequencies tend to 'miss' cycles of the oscillation rather than fire out of phase and they tend to fire single spikes instead of bursts.

Although this model demonstrated a plausible mechanism for generating synchronized oscillatory firing in a column of cortical neurons, it was subject to many simplifications: It consisted of only 11 fully connected cells and each synapse had to be spread out in time as well as increased in amplitude to compensate. The network oscillation showed frequency variability within a single trial, but this was an artifact of having only one inhibitory neuron in the model: A single extra spike from this neuron would significantly delay the onset of the next population burst.

The model presented below is a much more faithful simulation of a piece of cortical tissue, incorporating a more realistic model neuron, many more cells, sparse connectivity, and connections between as well as within columns. This more accurate model not only produced results that were better fits to the experimental data, but also allowed the systematic determination of how and in what way each physiological variable affected synchronization.

2. Model Structure

2.1. Model Neurons

The primary neuron used in network simulations was a layer 5 pyramidal neuron modeled by nine compartments. This model neuron was reduced from a 400-compartment model of a reconstructed cat visual cortex pyramidal cell (Koch *et al.*, 1990) using a method that preserves essential electrotonic parameters (Bush and Sejnowski, 1993). Since approximately 20% of neocortical cells are inhibitory (Douglas and Martin, 1990), 20% of the cells were seven-compartment inhibitory (basket) neurons. Each model neuron had Hodgkin–Huxley-type active conductances at the soma only. These conductances produced burst firing and high-frequency, nonadapting spiking in pyramidal and basket cells, respectively, in response to steady depolarizing input. The pyramidal cells had differing rates of internal calcium elimination, resulting in different intrinsic bursting frequencies. Noise in the form of current injection at the soma was used in each cell to provide a varying resting membrane potential and a low background firing rate.

2.2. Synaptic Connectivity

Synaptic conductances were modeled using an alpha-function conductance change (1 msec to peak) activated on the postsynaptic dendrite with some delay after the presynaptic spike (Bernander *et al.*, 1991). Synapses were placed at appropriate positions on the dendrites and somas of the model cells. Pyramidal cells received external driving input from uncorrelated Poisson spike trains. Within a model column composed of 80 pyramidal and 20 basket cells, every cell had an equal chance of contacting every other cell (but no self-connections were allowed) with a probability of 10%.

In order to view rapidly and easily the average activity of the whole network an analog of the local field potential (LFP) was calculated for each simulation, called the local averaged potential (LAP). This was a running average of all the membrane potentials of all the pyramidal cell somas. The presence of oscillations in this LFP analog was an indicator of synchronized activity in the pyramidal cell population.

3. Model Results

3.1. Synchronization within a Single Column

Figure 1 shows the output of the model network when the neurons were driven with stationary Poisson spike trains. Oscillations were visible in the LAP indicating synchronized firing of the pyramidal cells. The period of oscillation was 22.2 msec, giving a frequency of 45 Hz. The basket cells showed evidence for bursting at the same frequency as the pyramidal cells even though they have no intrinsic oscillatory properties; they are directly driven by the pyramidal cells, so the fact that they fired in bursts is further evidence of synchronized population discharge. The mechanism of synchronization was detailed in the introduction.

As observed in experimental recordings *in vivo* (Gray *et al.*, 1992), the synchronization sometimes spontaneously ceased (in this case at 100 and 350 msec) and then reappeared. The basket cell bursts in the model became less clumped at these times. Since the driving input to the network was stationary throughout, these changes were not due to changes in the nature of the external input. The synchronization was a statistical phenomenon, as observed in the experimental recordings; it was often difficult to see a regular oscillation in the spike train of a single cell, especially if it did not fire many spikes, and of course it was impossible to see synchronization. When examining all seven single-cell traces, one observes moments when the spikes all 'lined up,' but even in these cases there was significant jitter in individual spike times. Comparing two traces spike by spike, such as the top two pyramidal cell traces in Fig. 1, one often sees few instances of simultaneous firing, although on average both cells were locked to the underlying oscillation of the population (shown by the LAP). In other cases (such as the third and fourth

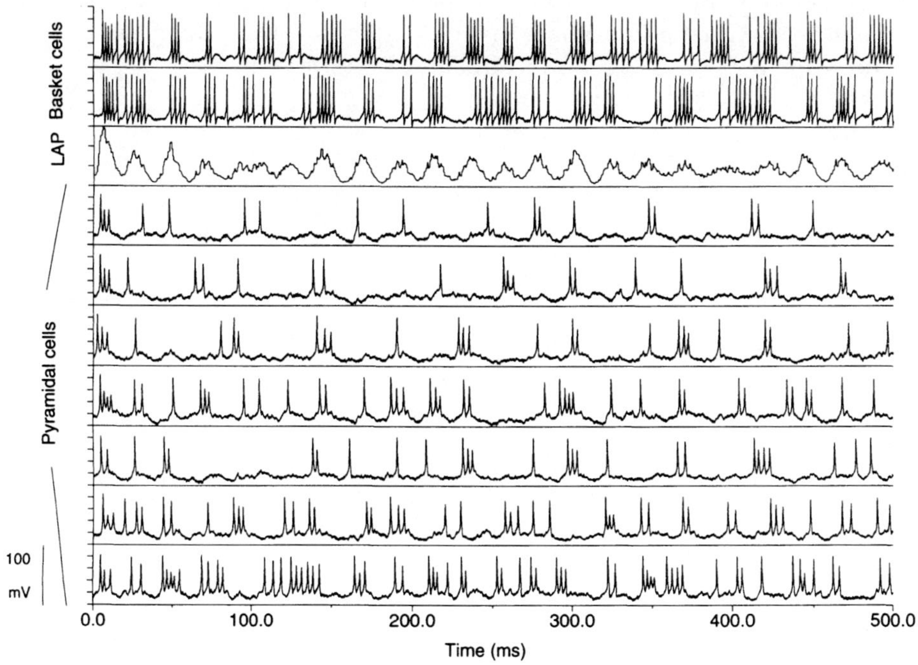

Figure 1. Intracolumnar synchronization in a model network of neurons connected with a probability of 10%. The top two traces show the somatic membrane potential of 2 basket cells from a population of 20 making up one column. The next trace is an analog of the local field potential, called the local averaged potential (LAP). The bottom seven traces show the somatic membrane potential of 7 sample pyramidal cells from the total population of 80. The oscillations in the LAP indicate that the neurons in the column fired synchronously. This is confirmed by the firing of the basket cells, which have no intrinsic bursting dynamics and only fire bursts in response to synchronized input from the pyramidal cells. There is considerable 'jitter' in an individual pyramidal cell's output; the network oscillation is a statistical property of the population and not at all 'clock like.' The synchrony spontaneously disappears and reappears at 100 and 350 msec. This rapid shift has been seen in real experimental data.

traces) the synchronization of spike firing was clearer. It is important to have an averaged measure of the activity of the whole population (such as the LAP shown here) to determine if the component cells are collectively oscillating in synchrony.

Experimental recordings show a frequency variability of synchronized activity within and between trials ranging from 40 to 60 Hz (Gray *et al.*, 1992). In these simulations different frequencies of oscillation could only be produced by changing the strength of the external driving input. The external input was kept constant within a trial and a change in oscillation frequency was never observed. Therefore, I suggest that the frequency variability observed in experimental recordings is due to a concomitant variability in the strength of th external input. External here means external to the synchronized group, coming from the thalamus or other areas of cortex. For a pyramidal cell in one of the simulations, approximately half of its excitatory input was external stationary Poisson spike trains and half was excitatory feedback from other pyramidal cells in the network, the sole source of synchronized inputs.

The membrane potential trace of one of the pyramidal cells is shown at a higher temporal resolution in Fig. 2. The large compound EPSP present on each cycle of the population oscillation is clearly visible. Spikes only arise from the top of these compound EPSPs, but not every compound EPSP causes a spike. Intracellular recordings from visual cortical neurons responding to physiological stimuli often show these features (e.g., Fig. 3C in Bush and Douglas, 1991).

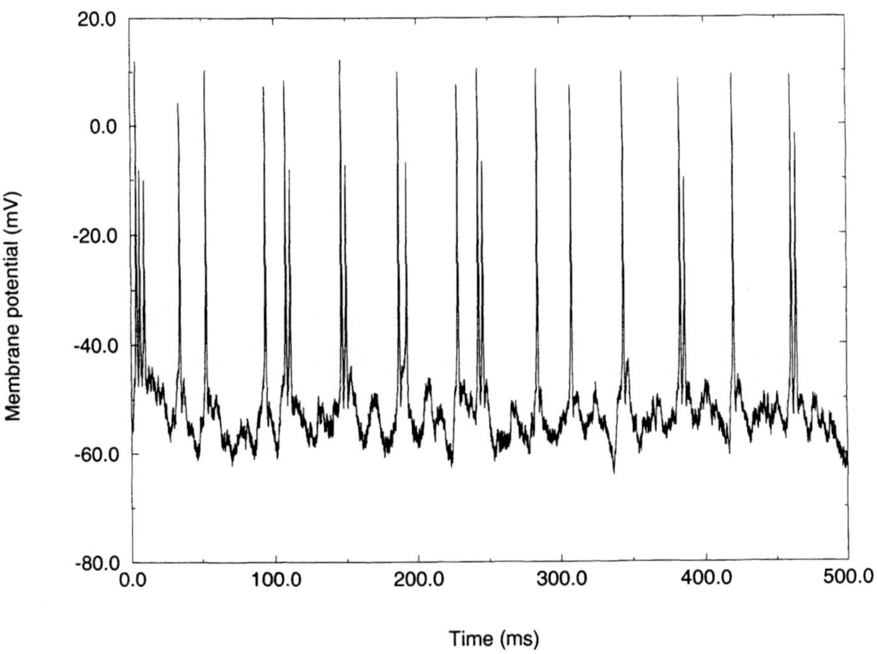

Figure 2. Higher resolution plot of the membrane potential of a pyramidal cell. There is a series of rhythmic compound EPSPs, some of which have spikes arising from them. Spikes do not occur at other times (in the 'troughs') because the membrane resistance is substantially reduced by a combination of inhibitory feedback and intrinsic potassium currents.

Sensitivity to Parameter Variation

The distributions of synaptic conductances and time delays in all of the simulations had large standard deviations, but the synchronized firing of the network was not affected by this variability, and was in general a robust phenomenon: All simulations were repeated with a noise level four times higher than that shown here and with a resting membrane potential of −65 mV instead of −55 mV, but synchronized oscillation was still present. With some sets of parameters (such as higher connectivity, stronger inhibitory synapses, and lower resting membrane potentials) even stronger synchronization occurred, but the values of parameters used for the simulations presented here are probably closer to those in the real cortex.

Increasing the mean synaptic delay within a column did not disrupt synchronization, even when the mean delay was as large as 7.2 rather than 1.2 msec. In this case, the oscillation became more regular, with every cell firing on every cycle and the cells with the highest intrinsic bursting frequency always leading the others. The frequency of oscillation of the network was also decreased due to the long duration of each burst. The only physiologically valid way found to alter the frequency of oscillation of the network was to vary the strength of the external driving input. By changing this parameter it was possible to vary the oscillation frequency of the network in the range 35–60 Hz.

Network synchronization was generally resistant to variations in synaptic strengths. However, inhibitory synapses had to be stronger than excitatory ones. A ratio of approximately 4:1, the value initially chosen on the basis of physiological results, was optimal for realistic synchronization. In networks without inhibition, simulating the effect of bicuculine, excitatory feedback operates unconstrained, producing large paroxysmal burst discharges, with the membrane potentials of the highest intrinsic frequency cells latching up past spike threshold (not shown). The only inhibitory/hyperpolarizing force in these networks was the intrinsic potassium conductance of the pyramidal cells, which was not strong enough to control the excitatory feedback.

The network was also somewhat sensitive to the strength of reciprocal inhibition between the basket cells. Although it is known from anatomical studies that local inhibitory interneurons make synaptic connections on each other (Douglas and Martin, 1990), very little is known about the function of these connections beyond the vague concept of 'disinhibition.' Figure 3B shows the same network LAP as in Fig. 1 and one of the basket cell traces. Figure 3A is the LAP from the same network with inhibitory contacts between basket cells removed. Synchrony was slightly weaker because the basket cell bursts were no longer terminated by inhibitory feedback and as a result became less discrete, sometimes continuing into the next cycle. This produced a lower frequency of oscillation (38 Hz compared to 45 Hz in Fig. 3B). This higher level of inhibitory activity also resulted in less activity in the pyramidal cell population. When synapses between baskets cells were made very strong (Fig. 3C), synchronization was severely disrupted. In this case basket cells were inhibited before they could provide effective inhibition to the pyramidal cell population, so the pyramidal cells began to fire continuously instead of in synchronous bursts. Thus, mutual inhibition between inhibitory interneurons within a single column is important for producing synchronized population oscillations, but the strength of this inhibition should not be as great as that between inhibitory cells and their pyramidal targets.

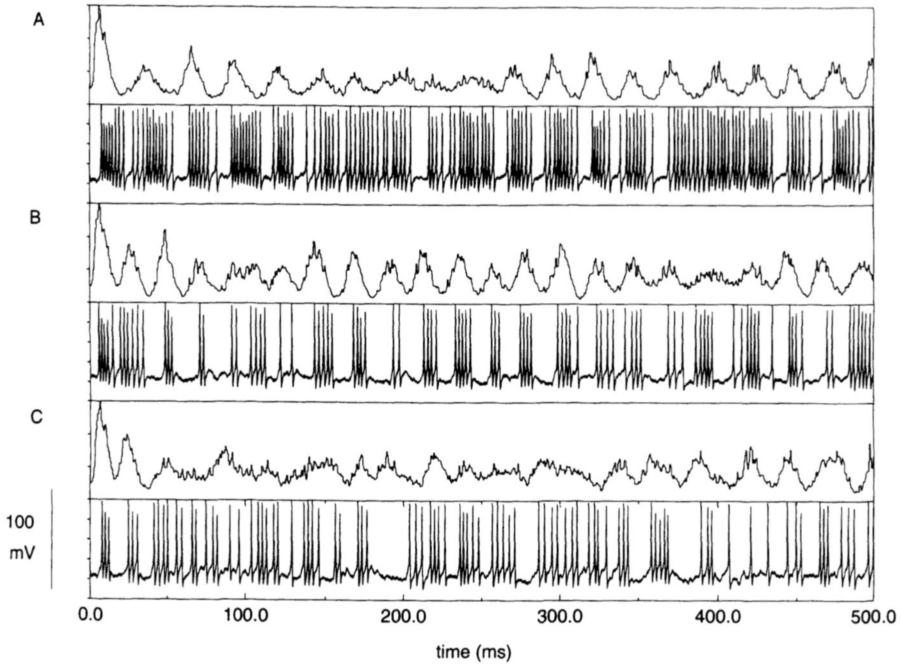

Figure 3. Effect of increasing the strength of reciprocal inhibition between the basket cells in a 100-neuron network. Inhibition increases from top to bottom. Each trace shows the population LAP and a sample basket cell trace. (A) No inhibition between basket cells. Synchronization can still occur, but is not optimal because the discharges of the basket cells on each cycle perseverate, sometimes running into the next cycle. (B) Basket cells connected by 1-nS synapses. These traces are taken from Fig. 1. This is the optimal amount of mutual inhibition between basket cells in the model column. (C) Basket cells connected by 2-nS synapses. Synchronization is degraded because the basket cells inhibit each other too strongly before they can fire a coherent burst and effectively terminate the burst of firing in the pyramidal cell population.

3.2. Synchronization between Columns

Since synchronization has been observed between different cortical columns and even different cortical areas (Eckhorn *et al.*, 1988; Gray *et al.*, 1989; Engel *et al.*, 1991b; Kreiter and Singer, 1992; Koenig *et al.*, 1995) simulations were performed to examine how synchronization arises between two columns, both internally synchronized. Two columns of 100 neurons each were simulated, each column connected as in Fig. 1. Connections between the columns were then added to see if synchronization between them could be established. Long-range connections in cortex are mediated by pyramidal cell axons, and these axons make most of their synapses on other pyramidal cells, although at a significantly lower density than within their own column (Hirsch and Gilbert, 1991; Martin, 1988; Kisvarday and Eysel, 1992). Therefore, pyramidal cells between the two columns were connected with a density of 2.5% (each pyramidal cell received a synapse from two pyramidal cells in the other column). In this case and all other simulations using pyramidal–pyramidal connections only, good synchronization between the two columns could not be obtained. The internal synchrony of each column appeared to be degraded and the only clear tendency was for the columns to sometimes oscillate 180 deg out of phase with each other. With some parameter choices this was a very strong effect, but in no case did clear synchronization with zero phase lag develop between the columns.

Long-range pyramidal cell axons make some of their synapses on dendritic shafts, many presumably belonging to inhibitory neurons (Douglas and Martin, 1990; White, 1989). Thus intercolumnar connections were added from pyramidal cells to basket cells to see if this would synchronize the two columns. Figure 4 shows the results of a simulation with two columns connected together at a density of 4% (each pyramidal cell received three synapses and each basket cell received four synapses from pyramidal cells in the other column). When the intercolumnar connections were turned on at 100 msec, the two columns immediately began to synchronize and maintained near-zero average phase difference (1.2 msec) for the duration of the simulation. When one column spontaneously desynchronized (e.g., at 350 msec) the other did, too, then both rapidly resynchronized. Intercolumnar synchronization could not be obtained by simply increasing the number of pyramidal–pyramidal connections. Thus the pyramidal–basket intercolumnar connection, while numerically small, was vital for intercolumnar synchronization. The greater effectiveness of inhibitory input in producing synchronization (Lytton and

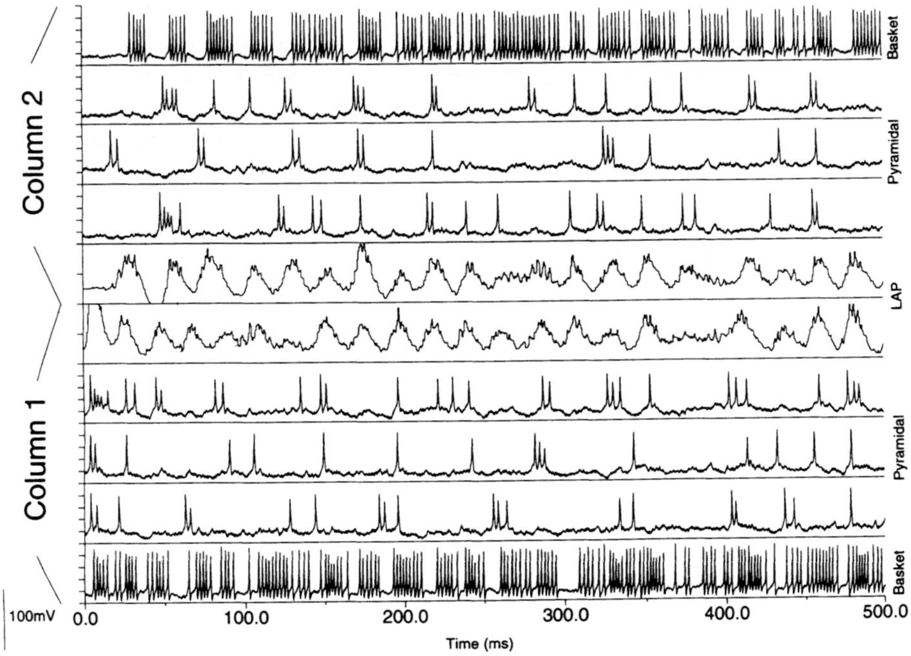

Figure 4. Zero-phase-lag synchrony between two columns of 100 neurons connected with pyramidal-to-pyramidal and -basket cell synapses at a probability of 4%. For each column a basket cell trace, three pyramidal cell traces, and the LAP of all the pyramidal cells in that column are shown. Synchrony is rapidly established when the intercolumnar connections turn on at 100 msec and remains for the duration of the simulation. When the columns temporarily desynchronize and then resynchronize (e.g., at 375 msec), they do so together.

Sejnowski, 1991) may explain the need for pyramidal–basket intercolumnar connections.

Zero-phase-lag synchrony between the two populations was established within one or two cycles of the intercolumnar connections being activated. Often the oscillation in one column would continue unperturbed while a cycle in the other would be suppressed or a new one prematurely initiated to get the two columns in phase. Reciprocal connections were not strictly necessary: Two columns could be synchronized with unidirectional connections from one to the other, although the synchrony obtained with this connectivity was not as strong as with reciprocal connections between the columns.

Increasing synaptic time delays between neurons within a column did not disrupt synchronization. However, increasing the synaptic delay of the intercolumnar connections adversely affected the synchronization of the two columns. As the delay was increased to 3–4 msec a phase shift of a few milliseconds developed between the two populations. When the intercolumnar delay was greater than 5 msec, synchronization was severely disrupted. Thus, the model predicts that two cortical areas cannot be synchronized by direct connections if the delay in those connections is significantly greater than 5 msec (~1/4 period), in agreement with previous models based on coupled oscillators (Schuster and Wagner, 1989; Sompolinsky *et al.*, 1990). Interestingly, in the cat this does not appear to exclude any areas of cortex directly connected to each other, even if they are in opposite hemispheres. Time delays between cells in the same column, separated by tens of micrometers, are 1–2 msec due to conduction along thin, unmyelinated axons (Mason *et al.*, 1991; Thomson *et al.*, 1988). Time delays between cells in opposite hemispheres, separated by several centimeters, have been recorded at or below 5 msec due to conduction along thick, myelinated axons (McCourt *et al.*, 1990). It would appear that the axons are organized to put cortical cells functionally next to each other. Thus, cells coupled by synchronization throughout the extent of the visual system can work on the same task simultaneously.

However, the same is not true for the cortex of the rabbit, where callosal conduction times are tens of milliseconds (Swadlow, 1991), too long for synchronization to work by direct connections. Thus, for this animal either there is some other mechanism of synchronization, or it does without synchronized 40-Hz activity of its two hemispheres.

Function of Long-Range Horizontal Axons

The results of the model suggest that intercolumnar connections mediated by long-range horizontal axons are modulatory in function rather than directly excitatory, primarily because of inhibition directly evoked by long-range axonal stimulation. Due to the relatively low numbers of intercolumnar axons and the lower threshold of the target basket cells, the excitatory intercolumnar fibers do not necessarily have a strong excitatory effect on their target columns. This is demonstrated in Fig. 5, which shows the output of two connected column, the bottom one driven normally by external input,, the top receiving no external input until 350 msec. Although the top column received input from the bottom one that was sufficient to synchronize the two populations when they were both being driven (after 350 msec), this input was not strong enough to cause significant firing before

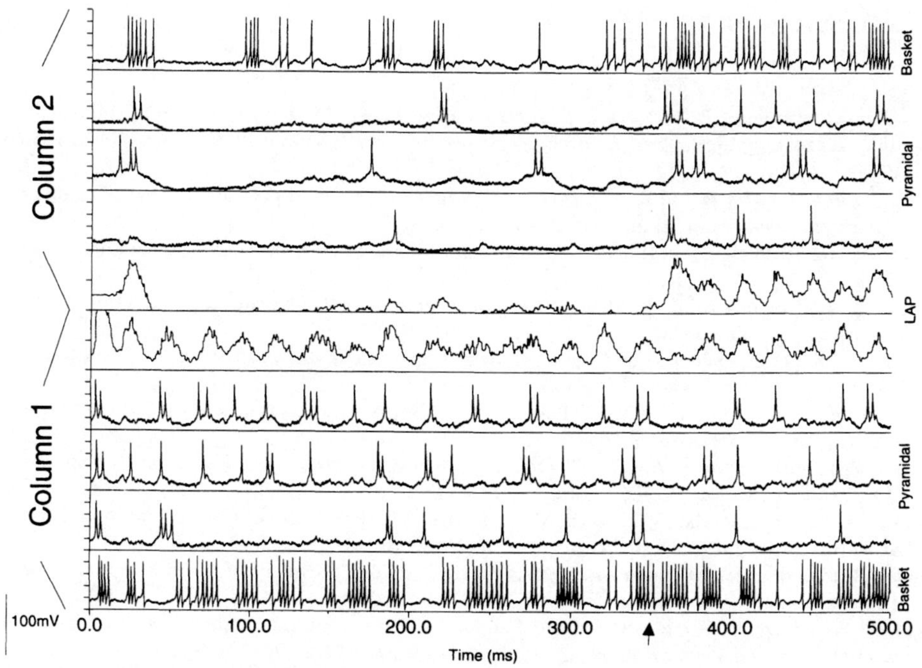

Figure 5. Intercolumnar connections have little effect on an undriven column. The two columns shown here are connected as in Fig. 4, with the intercolumnar connections turned on at 0 msec. The upper column receives no external driving input until 350 msec (arrow). After this time the two columns are synchronized by the intercolumnar connections, but before this they have very little effect on the undriven column, producing a few subthreshold depolarizations, but few spikes. The basket cells have a lower threshold and fire a few spikes, which helps to suppress the pyramidal cells.

350 msec. The low firing rate in the top column was not significantly higher than the spontaneous rate due to noise.

The strong effect of surround stimulation on a neuron that is also receiving direct stimulation of its receptive field (Gilbert and Wiesel, 1990) is also consistent with a modulatory role for the long-range horizontal connections. It should be noted that recent work has suggested that these connections can be strengthened under some conditions (perhaps by sprouting of extra axon collaterals) to the point where they can indeed drive cells in the absence of other input (Pettet and Gilbert, 1992; Darian-Smith and Gilbert, 1994).

4. Conclusion

Synchronization in model cortical networks is a robust phenomenon resistant to variation in parameters that are known to show a wide degree of variation in cortex. As long as time delays are short enough, which in the cat seems to hold even across hemispheres, cells in any number of directly connected cortical regions could fire in synchrony with each other regardless of where they are located physically.

5. References

Bernander, O., Douglas, R. J., Martin, K. A. C., and Koch, C., 1991, Synaptic background activity influences spatiotemporal integration in single pyramidal cells, *Proc. Natl. Acad. Sci. USA* **88:**11569–11573.

Bush, P. C., and Douglas, R. J., 1991, Synchronization of bursting action potential discharge in a model network of neocortical neurons, *Neural Computation* **3:**19–30.

Bush, P. C., and Sejnowski, T. J., 1993, Reduced compartmental models of neocortical pyramidal cells, *J. Neurosci. Meth.* **46:**159–166.

Bush, P. C., and Sejnowski, T. J., 1994, Effects of inhibition and dendritic saturation in simulated neocortical pyramidal cells, *J. Neurophysiol.* **71:**2183–2193.

Crick, F., and Koch, C., 1990, Towards a neurobiological theory of consciousness, *Semin. Neurosci.* **2:**263–275.

Darian-Smith, C., and Gilbert, C. D., 1994, Axonal sprouting accompanies functional reorganization in adult cat striate cortex, *Nature* **368:**737–740.

Douglas, R. J., and Martin, K. A. C., 1990, Neocortex, *Synaptic Organization of the Brain* (G. Shepherd, ed.), Oxford University Press, Oxford, pp. 220–248.

Eckhorn, R., Bauer, R., Jordan, W., Brosch, M., and Kruse, W., 1988, Coherent oscillations: A mechanism for feature linking in the visual cortex, *Biol. Cybernet.* **60:**121–130.

Engel, A. K., Koenig, P., and Singer, W., 1991a, Direct physiological evidence for scene segmentation by temporal coding, *Proc. Natl. Acad. Sci. USA* **88:**9136–9140.

Engel, A. K., Konig, P., Kreiter, A. K., and Singer, W., 1991b, Interhemispheric synchronization of oscillatory neuronal responses in cat visual cortex, *Science* **252:**11177–11179.

Gilbert, C. D., and Wiesel, T. N., 1990, The influence of contextual stimuli on the orientation selectivity of cells in primary visual cortex of the cat, *Vis. Res.* **30:**1689–1701.

Gray, C. M., Konig, P., Engel, A. K., and Singer, W., 1989, Oscillatory responses in cat visual cortex exhibit inter-columnar synchronization which reflects global stimulus properties, *Nature* **338:**334–337.

Gray, C. M., Engel, A. K., Koenig, P., and Singer, W., 1992, Synchronization of oscillatory neuronal responses in cat striate cortex: Temporal properties, *Vis. Neurosci.* **8:**337–347.

Hirsch, J. A., and Gilbert, C. D., 1991, Synaptic physiology of horizontal connections in the cat's visual cortex, *J. Neurosci.* **11:**1800–1809.

Kisvarday, Z. F., and Eysel, U. T., 1992, Cellular organization of reciprocal patchy networks in layer III of cat visual cortex (area 17), *Neuroscience* **46:**275–286.

Koch, C., Douglas, R., and Wehmeier, U., 1990, Visibility of synaptically induced conductance changes: Theory and simulations of anatomically characterized cortical pyramidal cells, *J. Neurosci.* **10:**1728–1744.

Koenig, P., Engel, A. K., and Singer, W., 1995, Relation between oscillatory activity and long-range synchronization in cat visual cortex, *Proc. Natl. Acad. Sci. USA* **92:**290–294.

Kreiter, A. K., and Singer, W., 1992, Oscillatory neuronal responses in the visual cortex of the awake macaque monkey, *Eur. J. Neurosci.* **4:**369–375.

Lytton, W. W., and Sejnowski, T. J., 1991, Simulations of cortical pyramidal neurons synchronized by inhibitory interneurons, *J. Neurophysiol.* **66:**1059–1079.

Martin, K. A. C., 1988, From single cells to simple circuits in the cerebral cortex, *Q. J. Exp. Physiol.* **73:**637–702.

Mason, A., Nicoll, A., and Stratford, K., 1991, Synaptic transmission between individual pyramidal neurons of the rat visual cortex *in vitro*, *J. Neurosci.* **11:**72–84.

McCourt, M. E., Thalluri, G., and Henry, G. H., 1990, Properties of area 17/18 border neurons contributing to the visual transcallosal pathway in the cat, *Vis. Neurosci.* **5:**83–98.

Pettet, M. W., and Gilbert, C. D., 1992, Dynamic changes in receptive-field size in cat primary visual cortex, *Proc. Natl. Acad. Sci. USA* **89:**8366–8370.

Schuster, H. G., and Wagner, P., 1989, Mutual entrainment of two limit cycle oscillators with time delayed coupling, *Prog. Theor. Phys.* **81:**939–945.

Schuster, H. G., and Wagner, P., 1990, A model for neuronal oscillations in the visual cortex. 1. Mean-field theory and derivation of the phase equations, *Biol. Cybernet.* **64:**77–82.

Singer, W., 1993, Synchronization of cortical activity and its putative role in information processing and learning, *Annu. Rev. Physiol.* **55:**349–374.

Sompolinsky, H., Golomb, D., and Kleinfeld, D., 1990, Global processing of visual stimuli in a neural network of coupled oscillators, *Proc. Natl. Acad. Sci. USA* **87:**7200–7204.

Sporns, O., Gally, J. A., Reeke, G. N., and Edelman, G. M., 1989, Reentrant signalling among simulated neuronal groups leads to coherency in their oscillatory activity, *Proc. Natl. Acad. Sci. USA* **86:**7265–7269.

Sporns, O., Tononi, G., and Edelman, G. M., 1991, Modeling perceptual grouping and figure–ground separation by means of active reentrant connections, *Proc. Natl. Acad. Sci. USA* **88:**129–133.

Swadlow, H. A., 1991, Efferent neurons and suspected interneurons in second somatosensory cortex of the awake rabbit: Receptive fields and axonal properties, *J. Neurophysiol.* **66:**1392–1408.

Thomson, A. M., Girdlestone, D., and West, D. C., 1988, Voltage-dependent currents prolong single-axon postsynaptic potentials in layer III pyramidal neurons in rat neocortical slices, *J. Neurophysiol.* **60:**1896–1907.

White, E. L., 1989, *Cortical Circuits,* Birkhauser, Boston.

Williams, T. L., 1992, Phase coupling by synaptic spread in chains of coupled neuronal oscillators, *Science* **258:**662–665.

Wilson, W. A., and Bower, J. A., 1991, A computer simulation of oscillatory behaviour in primary visual cortex, *Neural Computation* **3:**498–509.

Winfree, A. T., 1967, Biological rhythms and the behavior of populations of coupled oscillators, *J. Theor. Biol.* **16:**15–42.

11

Modeling the Piriform Cortex

MICHAEL E. HASSELMO and
CHRISTIANE LINSTER

1. Introduction

The primary olfactory cortex, or piriform cortex, has been the focus of considerable modeling research. These models range from early work focused on electroencephalographic (EEG) data to more recent work constructing networks from compartmental biophysical simulations of individual neurons. Ultimately, effective models of the cortex must link different levels of analysis; they must account for the function of the region in a behavioral context, without violating the constraints of physiological and anatomical data. Section 2 provides an overview of the experimental data available for constraining models of the piriform cortex. Models of the piriform cortex have addressed both physiological data and functional hypotheses, and will here be evaluated from both perspectives. In Section 3, this review focuses on the use of models to simulate specific physiological data—particularly EEG and field potential data demonstrating oscillatory dynamics. In Section 4, this review focuses on the use of models to evaluate specific functional hypotheses. The chapter concludes in Section 5 with a discussion of important issues to be addressed in further modeling research.

MICHAEL E. HASSELMO • Department of Psychology, Boston University, Boston, Massachusetts 02215. CHRISTIANE LINSTER • Department of Psychology and Program in Neuroscience, Harvard University, Cambridge, Massachusetts 02138.

2. Summary of Data Being Modeled

Detailed anatomical and physiological data can be used to guide the development of computational models of the piriform cortex. Different types of models incorporate different sets of data, but it is useful to maintain awareness of the full range of available data in evaluating the validity of specific models.

2.1. Anatomy of Piriform Cortex

The anatomy of the piriform cortex has been reviewed extensively elsewhere (Haberly, 1985, 1990) and will only be briefly summarized here. Figure 1 shows the location of the piriform cortex in a schematic representation of the ventral surface of the rat brain. The location of the piriform cortex is only two synapses removed from the sensory periphery. Receptor cells in the olfactory epithelium project into the olfactory bulb, where they synapse on the dendrites of mitral and tufted cells. These mitral and tufted cells project along the lateral olfactory tract (LOT) to the piriform cortex. Most models of the piriform cortex contain two primary elements of the anatomical circuitry of the piriform cortex, as summarized in Fig. 2.This includes the excitatory afferent input from the lateral olfactory tract, terminating in a broadly distributed manner on the distal dendrites of pyramidal cells in the superficial layer of piriform cortex, layer Ia. Excitatory association and intrinsic

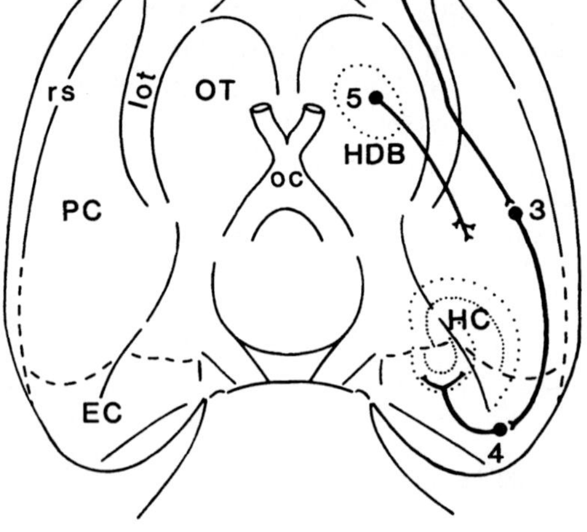

Figure 1. Schematic representation of the ventral surface of the rat brain, showing steps in the processing of olfactory information. (1) Receptor cells in olfactory epithelium; (2) olfactory bulb mitral cells; (3) olfactory cortex pyramidal cell; (4) entorhinal cortex pyramidal cell; (5) cholinergic neuron in horizontal limb of the diagonal band of Broca. OB, Olfactory bulb; AON, anterior olfactory nucleus; OT, olfactory tubercle; lot, lateral olfactory tract; rs, rhinal sulcus; oc, optic chiasm; PC piriform cortex; EC, entorhinal cortex; HDB, horizontal limb of the diagonal band of Broca; HC, hippocampus.

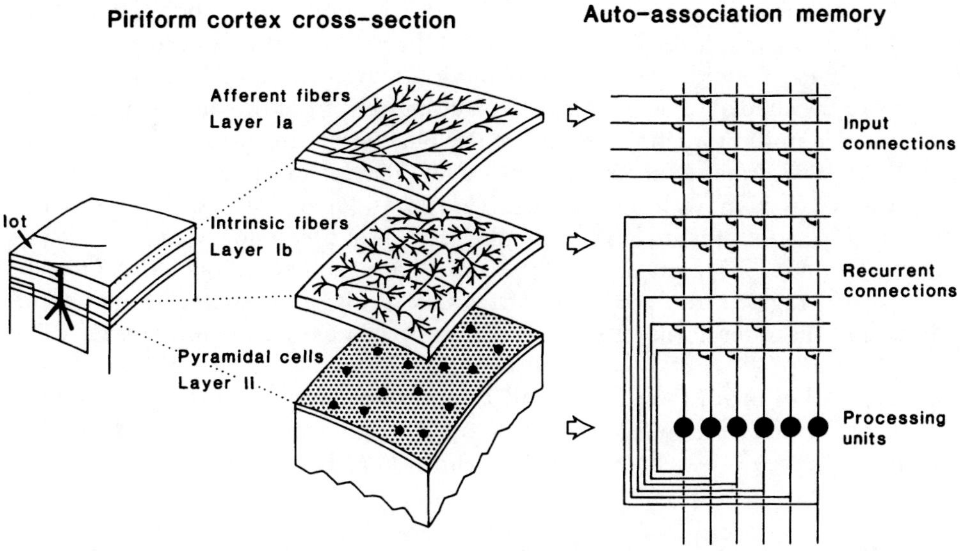

Figure 2. Anatomical structure of the piriform cortex compared to the structure of an associative memory model of this region. In layer Ia, afferent input from the LOT forms synapses on the distal dendrites of the piriform cortex pyramidal cells in a diffuse, nontopographic manner (Haberly and Price, 1977). This input is represented with distributed excitatory afferent connections in computational models. In layer Ib, intrinsic fibers arising from piriform cortex pyramidal cells in layers II and III synapse on proximal portions of other pyramidal cell dendrites in a diffuse, nontopographic manner. These connections are represented with distributed excitatory feedback connections between neurons in computational models.

fibers arise from other regions of the piriform cortex and adjacent cortical areas and terminate in a broadly distributed manner on the proximal dendrites of pyramidal cells in layer Ib and on basal dendrites in layer III. Models differ in their representation of excitatory association and intrinsic connections, particularly with regard to the relative predominance of feedforward versus feedback connectivity.

A number of different types of interneurons have been described in the piriform cortex (Haberly, 1985, 1990), which are presumed to be GABAergic, based on gamma-amino decarboxylase (GAD) immunoreactivity. The effects of inhibition have been incorporated into existing models, with a division of these effects according to two primary parameters. First, inhibition has been described as being either feedforward or feedback: Feedforward inhibition results from direct activation of inhibitory interneurons by the afferent input from the lateral olfactory tract; feedback inhibition results from activation of inhibitory interneurons by excitatory input from piriform cortex pyramidal cells. In addition, inhibitory influences have been divided according to the involvement of $GABA_A$ and $GABA_B$ receptor subtypes: activation of $GABA_A$ receptors increases chloride conductances with a rapid time course (Scholfield, 1978; Satou *et al.*, 1982; Tseng and Haberly, 1988), whereas activation of $GABA_B$ receptors causes increased potassium conductances with a slower time course (Satou *et al.*, 1982; Tseng and Haberly, 1988). In addition, evidence shows that activation of $GABA_B$ receptors causes suppression of synaptic transmission in the piriform cortex (Tang and Hasselmo, 1994).

Other than these primary excitatory and inhibitory connections, few other afferent, efferent, and intrinsic connections of the piriform cortex have been incorporated in computational models. Some models have included the excitatory connections of the medial olfactory tract, projecting back from the piriform cortex to synapse on the inhibitory granule cells of the olfactory bulb (Bressler, 1987a; Freeman, 1986, 1987; Ambros-Ingerson *et al.*, 1990). The piriform cortex has extensive reciprocal connections with the entorhinal cortex and neocortical structures as well as subcortical structures including the amygdala and mediodorsal nucleus of the thalamus (Price, 1985), but these connections have not been modeled. Similar to most cortical regions, the piriform cortex also receives modulatory innervation, including noradrenergic axons from the locus coeruleus (Fallon and Moore, 1978; Macrides *et al.*, 1981), cholinergic innervation from the horizontal limb of the diagonal band of Broca (Wenk *et al.*, 1977), and serotonergic innervation from the median raphe. This modulatory innervation has only recently been addressed in modeling research (Hasselmo, 1993; Hasselmo *et al.*, 1992, 1994).

2.2. EEG and Field Potential Data *in Vivo*

Much of the modeling research in piriform cortex has focused on EEG and field potential data obtained from the piriform cortex (Bressler and Freeman, 1980; Bressler, 1984, 1987a; Freeman, 1968b, 1978; Haberly, 1973a; Ketchum and Haberly, 1993b, c), primarily in anesthetized preparations. A major focus of many of the models discussed here concerns the oscillatory dynamics observed in EEG recordings. In the following sections, different aspect of these oscillatory dynamics will be summarized.

2.2.1. Gamma and Theta Frequency Oscillations

EEG recordings from the piriform cortex show the characteristics of fast, high-amplitude "bursts" of oscillations in the gamma range (30–90 Hz) superimposed on slower background oscillations in the theta frequency range (3–10 Hz) (Bressler and Freeman, 1980; Bressler, 1984; Freeman, 1978). The primary frequency in this band is lower in cats (around 38 Hz) as compared to rabbits and rats (around 52 Hz) (Bressler and Freeman, 1980). Some data suggest a bimodal distribution of the higher frequency oscillations in the rat, with a larger peak at 50 Hz and a smaller peak at about 80 Hz (Eeckman and Freeman, 1990). These higher frequency oscillations resemble the bursts of oscillations which appear in the olfactory bulb on each cycle of inhalation during sniffing (Bressler, 1984, 1987b). The higher frequency oscillations are superimposed on lower frequency oscillations in the theta range from 3 to 10 Hz (Bressler and Freeman, 1980). These slower theta oscillations appear to be synchronized with the sniffing cycle during performance of specific behavioral tasks (Macrides *et al.*, 1982). This synchronization with sniffing also appears in the olfactory bulb (Freeman, 1978).

The high-frequency gamma oscillations are also apparent in evoked potential data. In response to a weak electrical shock to the olfactory bulb or LOT, the piriform cortex evoked potential shows multiple oscillations at about 50 Hz (Freeman, 1968b; Ketchum and Haberly, 1993c). These oscillations die out gradually over a number of cycles. In contrast, after a strong shock to the bulb or LOT, the evoked potential shows a briefer characteristic waveform, with a divided peak proposed to reflect afferent fiber and associations fiber activation, respectively, followed by a rebound trough and a long, slow wave (Haberly, 1973a; Haberly and Shepherd, 1973; Ketchum and Haberly, 1993b, 1993c; Stripling *et al.*, 1988; Patneau and Stripling, 1992). It is not clear if the slow oscillatory dynamics influence the evoked potential, though the time course of these oscillations may be related to the time course of the decay of fast oscillations after a weak electrical shock.

Models have extensively explored the physiological basis of these characteristic frequencies in the EEG and evoked potentials. Some models suggest that gamma and theta oscillations may depend upon fast $GABA_A$ and slow $GABA_B$ inhibitory influences, respectively (Liljenstrom, 1991; Liljenstrom and Hasselmo, 1995; Wilson and Bower, 1989, 1992). Other models have emphasized the driving force of oscillatory input from the olfactory bulb in eliciting piriform cortex oscillations (Bressler, 1987b; Freeman, 1987; Ketchum and Haberly, 1993c). Experimental evidence suggests that though the main driving force may be bulbar input, intrinsic cortical dynamics makes some contribution to the oscillations (Bressler, 1987b), since recordings from surgically isolated cortex show fast oscillations in the evoked potentials (Freeman, 1968a) and EEG (Becker and Freeman, 1968).

2.2.2. Spatial Distribution of Field Potentials

Two important features of field potential data from the piriform cortex are the spatial distribution of activity across the width and length of the cortex, and the laminar distribution of activity across the depth of the cortex. Spatial data have been obtained using multiple recording electrodes spaced across the cortex (Bressler, 1984; Freeman, 1978). This recording demonstrates particular spatial distributions of activity during individual bursts of high-frequency oscillations. Maximal amplitude of bursts occurred on both sides of the LOT. These bursts appear to show some specificity of pattern in response to particular odors which are relatively stable unless the animal undergoes conditioning associated with specific odors. However, use of spatial distributions from EEG data for the identification of which individual odor was presented to the animal required extensive statistical manipulation of the data sets (Freeman and Skarda, 1985). This suggests that the representations determined by low-resolution recording of field potentials are not highly accurate, but if higher resolution recording of single-unit activity from a broad range of the cortex were available it might indicate the encoding of odors as a pattern of neuronal activity. Note that these are dynamic spatial distributions, reflected by the amplitude of oscillations in different cortical subregions, which can vary even within a burst.

2.2.3. Laminar Distribution of Field Potentials

In addition to explorations of the spatial distribution of activity, the laminar distribution of activity has been explored using current source density analysis of data gathered at multiple depths within the piriform cortex after electrical stimulation of the LOT (Haberly and Shepherd, 1973; Ketchum and Haberly, 1993a–c; Rodriguez and Haberly, 1989). These recordings demonstrate characteristic temporal relationships between synaptic activity in the different layers of the piriform cortex, with activity appearing successively in layer Ia, distal layer Ib, proximal layer Ib, and layer II. This suggests very accurate temporal relationships between the various synaptic influences on piriform cortex pyramidal cells.

2.3. Field Potential Recording *in Vitro*

Field potential data can also be gathered from brain slice preparations for investigation of synaptic transmission. In brain slice preparations cut tangential to the surface of the cortex, stimulation of the LOT results in a field potential with some similarity to the evoked potential observed *in vitro* (Collins and Howlett, 1988; Williams and Constanti, 1988), but the synaptic basis of its subcomponents is not clear. In contrast, slices cut perpendicular to the cortical surface allow separate stimulation of afferent and intrinsic fiber layers and more detailed analysis of these synaptic potentials (Hasselmo and Bower, 1991, 1992; Hori *et al.*, 1988; Jung *et al.*, 1990; Kanter and Haberly, 1990; Tang and Hasselmo, 1994). In particular, certain modulatory substances cause selective suppression of synaptic transmission in slice preparations. Acetylcholine selectively suppresses intrinsic fiber synaptic potentials (elicited in layer Ib), while having little effect on afferent fiber potentials in layer Ia (Hasselmo and Bower, 1992). Similar selectivity has been observed in the hippocampal formation (Hasselmo and Schnell, 1994). The selective suppression of intrinsic fiber synaptic potentials also appears with the $GABA_B$ agonist baclofen (Tang and Hasselmo, 1994) and noradrenergic agonists (Hasselmo *et al.*, 1997; Vanier and Bower, 1992). In contrast, the metabotropic glutamate agonist L-AP4 causes stronger suppression of afferent fiber than intrinsic fiber synaptic potentials (Hasselmo and Bower, 1991; Hori *et al.*, 1988).

2.4. Intracellular Recording

The construction of compartmental biophysical simulations can be guided by intracellular recording, which has been performed both *in vivo* (Haberly and Bower, 1984) and more extensively *in vitro* (Barkai and Hasselmo, 1994; Constanti and Bagetta, 1991; Constanti and Galvan, 1983; Constanti and Sim, 1987; Hasselmo and Barkai, 1995; Tseng and Haberly, 1988, 1989). Intracellular recording provides data showing the membrane potential of individual piriform cortex pyramidal cells during intracellular current injection or synaptic stimulation. Replication of these traces allows the formation of reasonably accurate representations of individual cellular response characteristics, which give an additional constraint on network properties when these single-cell simulations are assembled into network simulations (Barkai and Hasselmo, 1994; Hasselmo *et al.*, 1994).

2.5. Extracellular Unit Activity

Electrophysiological recording in piriform cortex has tended to focus on evoked potentials, with single- or multiunit spiking activity primarily discussed in association with evoked potentials (Eeckman and Freeman, 1990; Freeman, 1968b;

Haberly, 1973b). Recordings of multiple-unit activity can be obtained during recording of field potential data. The collection of field potential data utilizes a band-pass filter set at lower frequency ranges (e.g., 3–300 Hz) in comparison to the higher frequency bands used for multiunit recording (e.g., 300–3000 Hz). This allows analysis of the relationship of the probability of neuronal firing relative to field potentials at a particular moment. This has been used to analyze the validity of models of oscillatory dynamics (Eeckman and Freeman, 1990). Very few studies have examined single-unit responses during performance of behavioral tasks. Early work in the monkey showed that neurons responded to a moderate range of odors in the piriform cortex (Tanabe *et al.*, 1975). A study in the rat shows that neuronal responses during the presentation of odors are very brief and often habituate rapidly (McCollum *et al.*, 1991). However, more recent work shows that in addition to the response of neurons to specific odors, piriform cortex neurons also respond to a variety of other components of a task including trial initiation, discriminative response, and reward consumption (Schoenbaum and Eichenbaum, 1994).

2.6. Behavioral Data

One difficulty for extending modeling to behavioral data concerns the shortage of experimental data focusing on the role of the piriform cortex in behavior. Possibly due to its large size and ventral location, few studies have attempted to determine the effect of lesioning the piriform cortex on olfactory perception. Incomplete lesions of the piriform cortex have been demonstrated to impair the learning of complex olfactory discrimination problems (Staubli *et al.*, 1987), but such lesions do not prevent the learning of simple discriminations or detection problems. Thus, the role of the piriform cortex in olfactory perception appears to be necessary only for the formation of abstract relationships between individual odor percepts. This is further supported by the fact that cutting the lateral olfactory tract, and thereby depriving the piriform cortex of most of its afferent innervation from the olfactory bulb, does not prevent the learning of a go, no-go task using odor cues (Slotnick and Berman, 1980).

3. Modeling of Physiological Data

In this section we will review how models have been used to simulate electrophysiological data obtained from the piriform cortex. These data include both evoked potentials recorded in response to electrical stimulation of the olfactory bulb or lateral olfactory tract as well as EEG recordings of ongoing activity in the cortex during a range of behaviors. Both types of field potential recording demonstrate specific oscillatory characteristics as summarized in Section 2. Many models have focused on the oscillatory dynamics observed in field potential recordings.

3.1. Continuous-Firing-Rate Representations

Many models that focus on describing the oscillatory dynamics of piriform cortex field potential recordings use an abstraction of cortical activity in which the firing rate or probability of firing of a population of neurons is represented as a continuous variable. Models which use this abstraction will be referred to here under the category of continuous-firing-rate representations. These models have the advantage of greater mathematical tractability in contrast to models which realistically represent the spiking activity of neurons, and are therefore quite common in the neural network literature. Since the field potential data already have the characteristic of a continuous variable, these types of models prove useful for simulating such data. However, as we will discuss below, their use of a continuous firing rate raises the additional problem of relating their properties to the activity of individual neurons.

3.1.1. Lumped Models

The simplest representations of oscillatory dynamics do not attempt to address the spatial distribution of activity, but use lumped representations of whole populations of neurons to model field potentials recorded at a single location or averaged over the full region. In this manner, the activity of an entire region can be reduced to a single differential equation or system of coupled differential equations. Most of the modeling of olfactory cortex by Freeman's research group has utilized lumped representations of this region (Bressler, 1987a; Freeman, 1962, 1979b, 1986, 1987). However, the cortex is often proposed to have similar properties to those of the olfactory bulb, for which distributed models have been presented (Freeman *et al.*, 1988; Yao and Freeman, 1990).

3.1.1a. Modeling of Evoked Potentials. In early work on the piriform cortex, the fast gamma frequency oscillation in evoked potentials was modeled using a lumped representation of the cortex as a linear second-order differential equation, with the properties of a damped harmonic oscillator (Freeman, 1962). This allowed accurate simulation and characterization of different evoked potential traces according to phase of onset, frequency, amplitude, and a damping parameter Q. Later models utilized a control theory representation, with transfer functions for cortical pyramidal cells and for inhibitory feedback, and simulations were performed on an analog computer (Freeman, 1964). This led to later models which utilized complex lumped representations of two populations of neurons—a population of excitatory neurons representing piriform cortex pyramidal neurons, and a population of inhibitory neurons representing feedback inhibition activating GAB-A_A receptors (Freeman, 1979b, 1987). The intrinsic properties of each set of neurons are represented by second-order differential equations for neuronal activity. Coupled differential equations representing the activity of these populations have the capability of generating damped oscillations with an amplitude and time course dependent on the magnitude of the coupling coefficients. For example, different oscillatory dynamics associated with increased arousal or with habituation in experimental work can be modeled by increasing or decreasing the strength of excitatory interactions in the equations (Freeman, 1979b). These models use a nonlinear input–output function for each population of neurons (Freeman, 1979c). This input–output function was based on experiments in the bulb in which the same electrode was used to record multiunit spiking activity and field potentials, allowing the spiking probability to be plotted against the field potential amplitude. This provides a function which usually has a sigmoid shape. Naturally, this reflects the input–output function of an averaged heterogeneous population of excitatory and inhibitory neurons, rather than the actual input–output function of any individual neuron in that region. In addition, this reflects the steady-state output of a region, neglecting properties such as the adaptation of neurons dependent upon previous activity. In general, the structure of these models is very closely linked to the field potential data being described, which makes it difficult to relate parameters of the models to specific properties of individual neurons.

3.1.1b. Modeling of EEG Data. Lumped model representations of the piriform cortex were combined with lumped or distributed representations of the olfactory bulb and anterior olfactory nucleus in models of EEG data (Freeman, 1986, 1987). In these extended models, the impulse responses of each subregion were tuned to approximate experimental data, and then combined with longer range connections representing the feedforward and feedback connections of the lateral and medial olfactory tracts. Presentation of sustained input from a representation of olfactory receptor cells provides the basis for the sustained oscillatory activity resembling EEG recordings. Transmission delays on the feedback connections from the anterior olfactory nucleus to the periglomerular cells resulted in the generation of chaotic dynamics in the extended system model, which approximated the properties of EEG recordings in the olfactory bulb. The transition between the fast bursting oscillations of inhalation and the lower amplitude EEG during exhalation result from differences in the response of the network when input is present and when it is absent. The network also demonstrated activity resembling the EEG recorded during seizures, provided the strength of LOT input to the piriform cortex was reduced and the strength of feedback from the anterior olfactory nucleus to the bulb was increased (Freeman, 1986). This model could not account for the generation of seizure activity in the piriform cortex after isolation from the bulb.

3.1.2. Spatially Distributed Models

Modeling the spatial pattern of EEG activity in olfactory structures requires the use of distributed models with multiple sets of equations. A number of models have focused on the distribution of EEG activity in the olfactory bulb (Baird, 1986; Freeman, 1979a, 1988; Li and Hopfield, 1989; Wang *et al.*, 1991; Yao and Freeman, 1990), using multiple interacting populations of excitatory and inhibitory neurons. In models from the Freeman group, the heterogeneous distribution of oscillations depends upon the heterogeneity of the excitatory coupling between sets of coupled differential equations. There is not extensive evidence for excitatory connections between mitral cells in the olfactory bulb, which presents a major problem for these models. However, similar function has been demonstrated with inhibitory coupling between mitral cells (Li and Hopfield, 1989), and the dynamics of this network has been explored with a range of both inhibitory and excitatory couplings (Erdi *et al.*, 1993).

The continuous-firing-rate representation of neurons has also been utilized in distributed models of piriform cortex oscillatory activity (Liljenstrom, 1991; Liljenstrom and Hasselmo, 1993, 1995). These models of the piriform cortex incorporate the distributed excitatory afferent input and excitatory intrinsic and association connections. In contrast to the bulb, extensive experimental data demonstrate strong excitatory connections between cortical pyramidal cells. This model allowed simulation of the evoked potentials in response to strong and weak stimulation of the LOT, as well as the gamma and theta frequency bands in EEG recording (Liljenstrom, 1991). Despite its use of a continuous-firing-rate representation, this model utilizes a number of features from the more detailed compartmental biophysical simulation developed by Wilson and Bower (1988, 1989, 1992). In particular, these models contain a reasonably realistic representation of the anatomical connectivity of excitatory and inhibitory neurons in the cortex, with diffuse statistical patterns of connectivity which incorporate overlap between the connections of nearby neurons, and random variability in connection strength. This contrasts with many distributed models of the bulb, which combine excitatory and inhibitory neurons into lumped oscillators before linking them together into a full network with excitatory and inhibitory connections between the oscillators (Baird, 1986; Freeman *et al.*, 1988; Yao and Freeman, 1990).

Initial versions of the Liljenstrom model used traditional sigmoid input–output functions with fixed asymptotes which do not change dependent upon previous activity (Liljenstrom, 1991), but more recent versions of this model have used an input–output function which does change dependent upon previous output (Liljenstrom and Hasselmo, 1995). Most piriform cortex pyramidal cells show strong adaptation, causing a decrease in firing frequency during sustained input (Barkai and Hasselmo, 1994; Tseng and Haberly, 1989). The input–output function was modified so that the gain would decrease in proportion to previous output, giving model units the property of adaptation. Models using this modification have addressed a number of additional dynamical issues. In collaborative research, this model was utilized to explore the influence of cholinergic modulation on the dynamics of the piriform cortex (Liljenstrom and Hasselmo, 1993, 1995). This model was modified to incorporate the effects of acetylcholine within the piriform cortex, including suppression of synaptic transmission at excitatory intrinsic, but not afferent fiber synapses (Hasselmo and Bower, 1992) and the suppression of adaptation (Barkai and Hasselmo, 1994). Suppression of adaptation in the model meant that the slope and final asymptote remained larger, causing neurons to fire more strongly in response to a given synaptic input. These changes allowed accurate simulation (Liljenstrom and Hasselmo, 1993, 1995) of experimental data showing the effects of the cholinergic agonist carbachol on evoked and spontaneous field potentials recorded from the piriform cortex of cats (Biedenbach, 1966), as shown in Fig. 3. This model suggests how the cholinergic enhancement of excitability may influence the dynamics of cortical networks, but a more detailed account of this influence on network dynamics may require the additional realism provided by detailed compartmental biophysical models.

While many models focus on the intrinsic source of fast oscillations, these oscillations might also be the result of the driving force of oscillatory input from the olfactory bulb. Experimental data suggest that while this driving force may not be the only source of oscillations, it appears to be quite important. Comparison of EEG recordings from the bulb and cortex indicates that the correlation of burst onset, phase, and frequency in these recordings suggest much of the piriform cortex EEG is driven by input from the bulb (Bressler, 1987a). In current–source density analy-

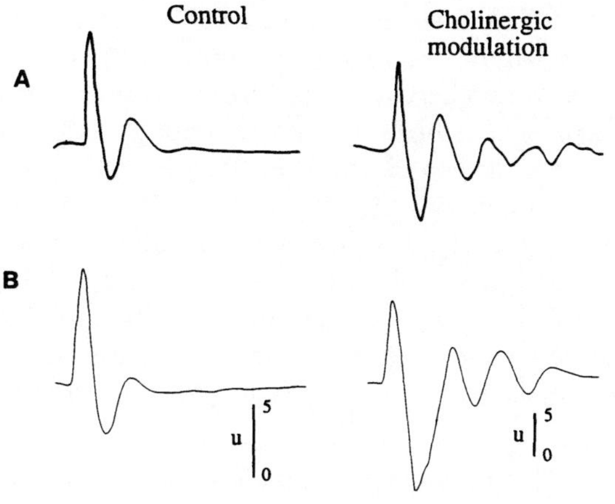

Figure 3. Experimental data and computational modeling showing the evoked potential elicited by stimulation of the lateral olfactory tract *in vivo* in control conditions and after application of the cholinergic agonist carbachol. Each trace is 125 msec in length. (A) Experimental data (Biedenbach, 1966) showing evoked potential in response to stimulation of the LOT in control condition and 45 min after application of solid carbachol into the superficial pyramidal cell layer (cholinergic modulation). (B) Simulation (Liljenstrom and Hasselmo, 1995) showing the activity evoked in the model by a strong afferent input pulse. Decreasing the adaptation of modeled neurons as a representation of the effects of cholinergic modulation causes the model to show slower damping of oscillations similar to that observed physiologically.

sis of the multiple damped oscillations resulting from weak electrical shock of the LOT, it is clear that each oscillation includes initial activity in the afferent fiber layer (Ketchum and Haberly, 1993c), suggesting that they arise due to input from the bulb rather than from purely intrinsic feedback. However, when the cortex is isolated from the bulb, it continues to demonstrate fast oscillatory dynamics in the evoked potential (Freeman, 1968a) and EEG (Becker and Freeman, 1968). Bressler utilized transfer functions for the cortex based on the Freeman model to derive the predicted shift in phase between cortex and bulb given the frequency of oscillation in each region (Bressler, 1987b). This model of interaction included the effects of transmission delay along the LOT as well as differences in the frequency between the two regions (both increased delay and greater oscillatory frequency in the bulb result in increased lag of the cortical oscillations in the model). This model could account for the phase lag at a number of the paired bulb–cortex recording sites which showed high correlation of bursting. Discrepancies between the model and data were attributed to the distributed input from the bulb to each cortical site, the intrinsic generation of oscillatory dynamics in the cortex, feedback from the cortex to the bulb, or the possible source of correlated bursts in a structure external to the system.

3.2. Compartmental Biophysical Simulations

The continuous-firing-rate models presented above can replicate many of the basic features of oscillatory dynamics in the piriform cortex. However, it remains somewhat difficult to link the parameters of these models to the specific physiological parameters of cortical neurons, due to the representation of large populations of neurons as units with continuous firing rate. Thus, many modelers have found it useful to develop more detailed biophysical simulations of cortical neurons to utilize in generating simulations.

Compartmental biophysical simulations contrast with the continuous-firing-rate representations in containing more detailed representations of individual neurons. The term compartmental refers to the modeling strategy developed by Rall and others (for review see Rall, 1989; Bower and Beeman, 1995) in which activity distributed across the different structural components of a neuron could be modeled with a combination of compartments representing subunits of a neuron. This allows analysis of how the structure of cortical neurons, and the laminar input to a neuron, could influence the distribution of field potentials in this region. In addition, these types of models normally contain considerably more detailed representations of membrane currents, using the Hodgkin–Huxley framework for representing voltage-dependent currents and dual-exponential representations of the time course of synaptic currents. These types of models have allowed more detailed analysis of specific physiological data, as described below.

3.2.1. Distributed Oscillations in a Network Biophysical Simulation

Working with James Bower at Caltech, Matthew Wilson developed a large scale and extensively detailed biophysical simulation of the piriform cortex (Wilson and Bower, 1988, 1989, 1992; Bower, 1995). This modeling project formed the basis for development of the GENESIS simulation package, which has been used extensively by other researchers (Bower and Beeman, 1995). In contrast to models using the continuous-firing-rate representation, this biophysical network simulation directly incorporates a wide range of physiological parameters. The simulation included 1500 pyramidal cells with five compartments: one compartment for the portion of the dendritic tree in layer Ia, one each for the portions of the dendritic tree in the distal and proximal layer Ib, one for the soma, and one for the basal dendrites. The soma of these neurons contained the traditional Hodgkin–Huxley currents underlying action potential generation, and a spike detection threshold was used to trigger synaptic currents at the postsynaptic targets. These synaptic currents consisted of excitatory sodium currents in other pyramidal cells with a dual-exponential time course. In addition, the model incorporated single-compartment representations of interneurons mediating feedforward and feedback inhibition, with 1500 neurons of each type. Feedback interneurons activated chloride currents in the soma, whereas feedforward interneurons activated potassium currents in the apical dendrites (these neurons also received some feedback connections from pyramidal cells).

This detailed biophysical network could be used to simulate the evoked potential in response to strong and to weak stimulation, as well as the oscillatory characteristics of the EEG (Wilson and Bower, 1992). Similar to previous models, the fast oscillatory dynamics of evoked potentials and the gamma frequency oscillations in the EEG resulted primarily from the interaction of excitatory neurons with the fast chloride currents mediated by feedback inhibition. Thus, if the time constant of feedback inhibition was increased, the fast oscillations were decreased considerably. In addition, the amplitude of the high-frequency oscillations would increase with stronger excitatory intrinsic connectivity, and would disappear if excitatory intrinsic connectivity was sufficiently weakened. This is of interest with regard to the cholinergic suppression of excitatory synaptic transmission (Hassselmo and Bower, 1992), since it suggests that the increased postsynaptic excitability must be strong enough to offset the effect of suppression on fast oscillations, since experiments demonstrate increased gamma oscillations in the presence of acetylcholine (Biedenbach, 1966). The model also replicated slow theta frequency oscillations as partly due to the activation of slow potassium currents, but elimination of this influence did not complete remove theta rhythm oscillations, leading the authors to suggest that population effects such as spike synchronization might play a role in these oscillations. These authors did not incorporate the potassium currents underlying adaptation in this model.

This model provided an extremely detailed view of the spatial distribution of activity during network oscillations. This might prove useful for testing the relationship between single-unit recording and field potential recording in this structure. However, the large-scale simulation used to explore oscillatory dynamics was not utilized to explore functional hypotheses. An earlier and smaller version of the model was used to test associative memory function (Wilson and Bower, 1988), and will be discussed in the following section. Later, this model was also adapted for analysis of the role of cholinergic modulation in associative memory function (Barkai *et al.,* 1994; Hasselmo *et al.,* 1994), as discussed below.

3.2.2. Modeling the Laminar Distribution of Field Potentials

The use of multiple compartments in the Wilson and Bower model allowed that model to incorporate the temporal relationships between activity induced in different layers. Thus, the activity induced by simulated stimulation of the LOT-induced sequential activation in different layers as activity arrived in the cortex at the afferent fiber synapses in layer Ia progressed along excitatory feedforward connections into distal layer Ib in more caudal regions, and then finally activated proximal layer Ib with activity spreading rostrally from neurons active in more caudal regions. This matches the temporal distribution of activity found in current source density analysis, with the significant difference that after weak stimulation the Wilson and Bower model contains only a single activation of afferent fiber synaptic currents—as connections between bulb and cortex were not explicitly modeled. For this reason, subsequent oscillations activated currents only in layer Ib. However, experimental data show activity in layer Ia at the start of each damped oscillation (Ketchum and Haberly, 1993c) suggesting a coordination in the oscillatory properties of these two networks via their reciprocal connections.

Detailed modeling of current source density data has been performed (Ketchum and Haberly, 1993a; Wilson, 1989). The Ketchum and Haberly model contains a single neuron with multiple compartments to provide a detailed description of the local currents evoked by synaptic activation. The use of a single neuron allowed much more detailed representation of the laminar distribution of excitatory and inhibitory synaptic inputs and the interaction of passive currents, but naturally does not address the topographic distribution of activity. The correct temporal representation of synaptic events was explored extensively, based on estimates of the dispersion of action potential arrival times at different distances from the location of stimulation. Similar techniques were employed to estimate the statistical distribution of synaptic input generated by feedback from other pyramidal cells activated by afferent input. This extensive detail allowed a very accurate simulation of the current source density data and explorations of questions about the basis for some components of this data. In particular, the model suggested that the excitatory currents in layer Ib did not arise from activation of voltage-dependent sodium conductances in this region of dendrite, as had been suggested. In addition, it showed that inhibitory chloride currents could not shift the location of source current induced by afferent input, and that these currents could only exert a strong shunting action on currents at substantially depolarized potentials. Questions at this level of detail require correspondingly greater detail in the simulations utilized to answer them. A detailed model of a single piriform pyramidal cell has also been used to determine the effect of fluctuations in synaptic timing around the values determined by current source density analysis (Protopapas and Bower, 1994). Results showed that neuron output was relatively unaffected by small temporal fluctuations near physiological values.

3.2.3. Modulatory Influences on Single Pyramidal Cells

In the previous section, a single-cell simulation was utilized to explore field potential phenomena involving network synaptic interactions. Single-cell models can also be used to simulate intracellular recordings of membrane potentials from single piriform cortex pyramidal cells. this provides additional physiological constraints on the intrinsic properties of these modeled neurons. The representations of pyramidal cells used in other compartmental models (Ketchum and Haberly, 1993b; Wilson and Bower, 1992) were not explicitly based on simulations of intracellular data.

Recent work has utilized the GENESIS simulation package (Bower and Beeman, 1995) to develop single-cell simulations of intracellular recordings obtained from piriform cortex pyramidal cells in brain slice preparations of the piriform cortex (Hasselmo and Barkai, 1992; Hasselmo *et al.*, 1994; Barkai and Hasselmo, 1994; Protopapas and Bower, 1994). The results of this single-cell modeling work were then used to develop a modified version of the Wilson and Bower network model. As shown in Fig. 4, these simulations incorporated the fast sodium and delayed rectifier potassium current and the various synaptic currents used in the previous model. However, these simulations also added potassium currents which were necessary to simulate the adaptation properties of single pyramidal neurons. These simulations included two voltage-dependent potassium currents (the $I_{K(A)}$ current and the $I_{K(M)}$ current), a high-threshold, voltage-dependent calcium current (I_{Ca}), and the calcium-dependent potassium current underlying the slow afterhyperpolarization ($I_{K(AHP)}$). Simulation of the calcium-dependent potassium current required a representation of intracellular calcium concentrations. The change in membrane potential of individual pyramidal cell compartments was computed with the following equation:

$$\frac{dV_m}{dt} = \frac{1}{C_m}\left[\frac{V_m - E_{\text{leak}}}{R_m} + (g_{\text{Na}} + g_{\text{Na(GLUT)}})(V_m - E_{\text{Na}}) + g_{\text{Cl(GABA}_\text{A})}(V_m - E_{\text{Cl}}) \right.$$
$$\times (g_{\text{K(DR)}} + g_{\text{K(A)}} + g_{\text{K(M)}} + g_{\text{K(AHP)}} + g_{\text{K(GABA}_\text{A})})(V_m - E_{\text{K}})$$
$$\left. + \sum_x \frac{V_{m(x)} - V_m}{R_a} + I_{in}\right]$$

This allowed a reasonably accurate representation of the adaptation characteristics of a piriform cortex pyramidal cell. In addition, this allowed simulation of the

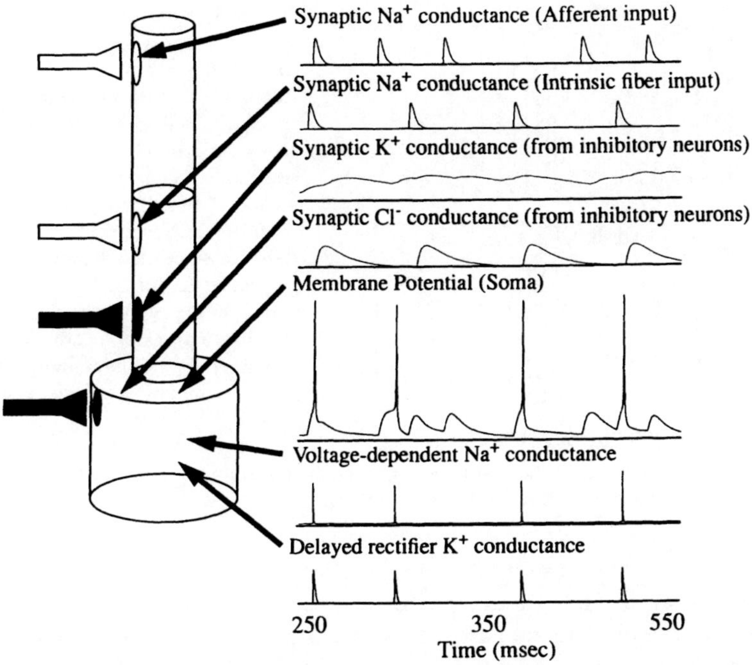

Figure 4. Biophysical simulation of a single piriform cortex pyramidal cell. The neuron contains a distal dendritic compartment, a proximal dendritic compartment, and a somatic compartment. On the right, a 200-msec component of a small network simulation is presented with traces from a single pyramidal cell showing (1) sodium conductance due to afferent synaptic input in the distal compartment, (2) sodium conductance due to intrinsic synaptic input in the proximal compartment, (3) potassium and chloride conductances due to synaptic input from inhibitory interneurons in the proximal dendritic and somatic compartment respectively, (4) the membrane potential at the soma (showing four spikes), (5) the voltage-dependent sodium conductance underlying action potential generation, and (6) the delayed-rectifier potassium conductance.

decrease in adaptation associated with perfusion of 20 μM carbachol through the slice chamber. Effects of acetylcholine were represented by decreasing $g_{K(AHP)}$ and $g_{K(M)}$ as well as by increasing the membrane resistance R_m. This resulted in simulated suppression of adaptation resembling the suppression of adaptation observed experimentally (Barkai and Hasselmo, 1994; Hasselmo *et al.*, 1994). The comparison of biophysical simulation and experimental data is shown in Fig. 5. These simulations of the response characteristics of piriform cortex pyramidal cells were utilized in network biophysical simulations exploring the role of cholinergic modulation in cortical associative memory function, as discussed in the following section.

4. Modeling of Functional Hypotheses

Section 3 focused on the use of models to simulate physiological data obtained from the piriform cortex. Many of the models discussed in that section have also been utilized to address possible functional implications of these physiological data.

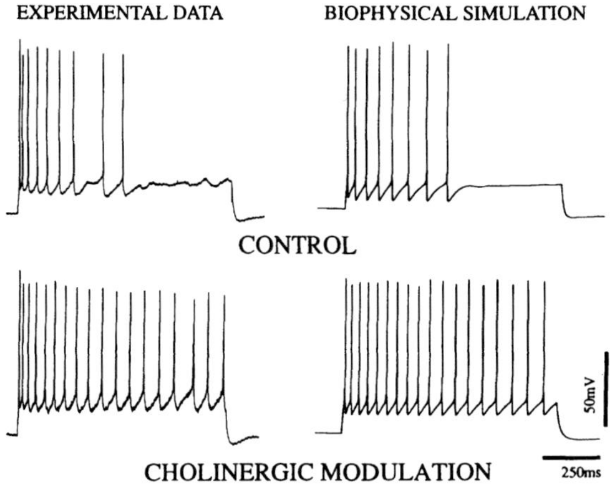

Figure 5. Experimental data and biophysical simulation of the suppression of neuronal adaptation by cholinergic modulation. On the left, results are shown from intracellular recording of a piriform cortex pyramidal cell during a 1-sec current injection at 0.7 nA (Barkai and Hasselmo, 1994). Perfusion with the cholinergic analogue carbachol (20 μM) causes a decrease in the adaptation of neuronal response to intracellular current injection. On the right, in the biophysical simulation, comparable adaptation can be obtained by implementation of voltage-dependent and calcium-dependent potassium conductances. The suppression of adaptation is simulated as a decrease in the maximum conductance of these currents. This allows the simulation to fire a larger number of action potentials in response to simulated current injection.

In addition, other models of the piriform cortex have been developed with the primary focus of exploring specific functional hypotheses. In some cases, the focus on functional hypotheses results in less effective implementation of physiological constraints.

Neural network models which have physiological features relevant to the functioning of cortex can be divided into two broad classes: associative memory models and self-organizing networks (Hasselmo, 1995). Models of the piriform cortex have focused on both of these properties, but in most cases these models have remained distinct from each other, despite the fact that final understanding of the piriform cortex will probably involve a unification of these two properties. This section will first discuss models of the piriform cortex as an associative memory, and subsequently discuss models of the piriform cortex as a self-organizing network.

4.1. Associative Memory Function

It has been noted that the anatomical and physiological characteristics of the piriform cortex strongly resemble a class of neural network models termed associative memories (Bower, 1991, 1995; Haberly, 1985; Haberly and Bower, 1989; Hasselmo *et al.*, 1992; Hasselmo and Bower, 1993). In particular, the broadly distributed excitatory afferent input arriving from the olfactory bulb (Haberly and Price, 1977) and the broadly distributed excitatory recurrent connections within the piriform cortex (Haberly and Price, 1978; Luskin and Price, 1983) resemble the distributed input and recurrent feedback present in autoassociative memory models. This suggests that the piriform cortex could be analyzed in relation to theoretical work on associative memory function (Amit, 1988; Anderson, 1972, 1983; Hopfield, 1984; Kohonen, 1972, 1984). An important aspect of modeling research concerns the linking of physiologically motivated models to more abstract models for which mathematical analysis has been more extensive. The primary feature of an associative memory model is the storage of associations between the components of a specific input pattern, allowing incomplete or noisy versions of this pattern to recall the full originally learned pattern. Associative memory function has been defined as resulting from the modification of synapses that are not the predominant influence on postsynaptic activity during modification (Hasselmo, 1995).

Associative memory function in all the models discussed below requires modifiability of the excitatory intrinsic synapses between pyramidal cells in the piriform cortex. The modifiability of these synapses is supported by the demonstration of long-term potentiation of synaptic potentials elicited in brain slice preparations of the piriform cortex (Jung *et al.*, 1990; Kanter and Haberly, 1990; Patil *et al.*, 1998). Interestingly, long-term potentiation is more easily obtained at the synapses of excitatory intrinsic fibers in layer Ib of piriform cortex than at the synapses of afferent fibers in layer Ia. Synaptic modifications has also been examined in whole-animal preparations. In these preparations, repetitive stimulation of the afferent fibers in the LOT commonly does not elicit modification of monosynaptic potentials (Racine *et al.*, 1983; Stripling *et al.*, 1988; Patneau and Stripling, 1992). However, this stimulation does alter the later components of the field potential, which may be due to feedback inhibition (Stripling *et al.*, 1988). In addition, modulatory influences activated by reinforcement contingencies have been proposed to result in modification of afferent input (Roman *et al.*, 1987).

4.1.1. Continuous-Firing-Rate Models

The simplest models of associative memory function in the piriform cortex utilize continuous-firing-rate representations. As described above, this involves representing the mean firing rate of a neuron or group of neurons as a continuous variable. This representation has been used in the vast majority of associative memory models. However, these models have a range of different dynamical characteristics, and will here be subdivided according to the dynamics of recall within the network. In very simple models, the dynamics of recall involves just a single step of matrix multiplication. In more complex models, the dynamics of recall involves a longer period of settling into a stable pattern of activity (a fixed-point attractor). Many recent models incorporate recall dynamics in which the network settles into an active oscillation with specific spatial distribution (a limit-cycle attractor). Finally, some models demonstrate chaotic dynamics in certain regions of parameter space.

4.1.1a. Single-Step Spread of Activity. The simplest class of associative memory models consist of nothing more sophisticated than matrix multiplication, and are commonly referred to as linear associative memories (Anderson, 1972; Kohonen, 1972). These have been classified as autoassociative or heteroassociative, depending upon whether they involve storage of associations between different components of the same pattern or between components of two different patterns. The feedback connectivity of the piriform cortex is more suggestive of the autoassociative version of these models, though heteroassociative function could take place between different subregions, for instance, the anterior and posterior portions of the cortex.

In these abstract models (Anderson, 1972; Kohonen, 1972), during a learning stage synaptic connectivity is modified such that the matrix of intrinsic synaptic connections resembles the outer product of the patterns being stored. Subsequently, during a recall stage a degraded version of one stored pattern is presented to the network, multiplied by the synaptic connectivity matrix, and the resulting vector is considered the output of the model. Note that this dynamic only allows for a single cycle of spreading activity within the network. Thus, each cycle of the model only addresses a brief period of activation within the cortex. However, it has been suggested that this single-step matrix multiplication could correspond to the spread of excitatory activity within the cortex during each of the fast gamma oscillations, after which feedback inhibition sets network activity back to a starting baseline (Hasselmo *et al.*, 1992).

A matrix multiplication model of this type was used to simulate the piriform cortex (Hasselmo *et al.*, 1992), with two primary features different from linear associative memories. The first feature was a nonlinear input–output function which tapered off to an asymptote, and the second was the addition of feedback and feedforward inhibition. These features enhanced the capability of a single-step associative memory to remove the noise from afferent input during recall (Hasselmo, 1993; Hasselmo *et al.*, 1992; Hasselmo and Bower, 1993). In addition, these models explored the requirement for different dynamics during learning and recall. Effective learning of new stored patterns required that recall of previously stored patterns be suppressed, by decreasing the strength of the excitatory feedback matrix. Effective recall required that these excitatory feedback synapses be increased to their full strength again. All associative memory models have this requirement, which is usually met by clamping activity to the input pattern during learning. Without the suppression of recall during learning, interference due to recalled patterns causes all stored patterns to be associated with each other, preventing discrimination between the stored patterns. The requirement of associative memory models for clamping of activity to the afferent input pattern closely matches the dynamical effects of cholinergic modulation. Acetylcholine causes selective suppression of synaptic transmission at excitatory feedback synapses in layer Ib, while leaving synaptic transmission at afferent fiber synapses in layer Ia almost unaffected (Hasselmo and Bower, 1992). In models, implementing this cholinergic suppression of intrinsic synaptic transmission during learning allows the network to store new patterns without interference from the recall of previously stored patterns. Thus, a simple associative memory model of the piriform cortex allowed the development of a functional hypothesis for the role of acetylcholine in learning of new information.

4.1.1b. Fixed-Point Attractor Dynamics. Many associative memory models have focused on attractor dynamics, with an emphasis on fixed-point attractors. In these models, recall involves setting initial conditions within the network, after which the activity state settles into a stable equilibrium which consists of a stored pattern. These models have an advantage over the matrix multiplication models in that the final equilibrium states do not depend upon the initial conditions. Thus, the network can recall a pattern with no interference from elements of the input used to cue that pattern.

The piriform cortex has been modeled as a Hopfield type network, with sigmoid input–output functions limiting the maximum output of individual neurons (Hasselmo *et al.*, 1991). This model differed from a classical Hopfield network in that it used separate populations of excitatory and inhibitory neurons, and utilized afferent input as the recall cue, rather than setting initial conditions directly. The model focused on the use of cholinergic modulation to set appropriate dynamics for learning. In contrast to the standard Hopfield network, in which the connectivity matrix is determined by an external calculation independent of the network, in this model patterns of afferent input were presented to the network during learning. Effective storage of these patterns required suppression of synaptic transmission at intrinsic fibers, similar to the effect of cholinergic modulation.

A significant problem with this and other attractor neural networks consists in their use of sigmoid input–output functions to constrain activity at maximal output. Piriform cortex pyramidal cells can fire at a rate of hundreds of spikes per second in response to current injection, but normally do not fire at much above 10 spikes/sec *in vivo* (Tanabe *et al.*, 1975; McCollum *et al.*, 1991; Schoenbaum and Eichenbaum, 1994). Thus, the constraint on maximal activity should not be intrinsic—i.e., should not involve use of an input–output function. Instead, network dynamics such as feedback inhibition should provide this constraint on spiking activity. Most models with fixed-point attractors do not incorporate separate populations of inhibitory neurons (Amit and Treves, 1989; Hopfield, 1984). However, in a recent model of recurrent excitation and feedback inhibition, stable fixed-point attractors were obtained using only threshold linear input–output functions, with no artificial constraint on maximal output (Hasselmo *et al.*, 1995). This model used simple subtractive inhibition, but a similar effect has been obtained with models of shunting inhibition (Abbott, 1991, 1992). Models of this type may more effectively represent cortical attractor states.

4.1.1c. Limit-Cycle Attractor Dynamics. Despite the considerable focus on fixed-point attractor dynamics in cortical models described above, physiological evidence suggests that these attractors are only approached very briefly. Thus, similar to the matrix multiplication models, the fixed-point-attractor models described above may only be useful for modeling a short period of cortical activity, such as a single cycle of gamma oscillations. Extensive field potential data and limited single-unit data suggest that cortical neurons undergo rapid increases and decreases in activity, particularly during the gamma frequency bursts (Bressler, 1984; Eeckman and Freeman, 1990; Freeman, 1978). Thus, physiologically realistic models must incorporate these rapid changes in activity. It is still possible to consider these dynamical states within the framework of associative memory models.

In the models of spatially distributed oscillations within the olfactory bulb, it was noted that Hebbian synaptic modification of connections between neurons provided the capacity for completion of individual stored patterns of oscillation (Baird, 1986; Freeman, 1979a; Freeman *et al.*, 1988; Yao and Freeman, 1990). Thus, in these models it was proposed that degraded versions of stored patterns could evoke the full, learned version of a pattern, described as a Hebbian cell assembly. This notion of a cell assembly has similarly guided work on fixed-point attractors. Thus, the primary difference between models of spatially distributed oscillation patterns and fixed-point attractors lies in the dynamical state of individual neurons. Neurons do not attain a stable state of activity, but membership in a particular assembly is determined by the relative amplitude of oscillations in different groups of neurons. Variants of these types of models can store sequences of patterns, with changes in pattern across successive individual cycles (Wang *et al.*,

1991). This focus on associative memory function in the bulb suffers from the fact that excitatory interactions between mitral cells have not been conclusively demonstrated in the bulb, where most intrinsic connections involve feedback inhibition. However, similar function has been shown in a network with inhibitory interactions (Li and Hopfield, 1989). In addition, the level of abstraction of these models makes them just as suitable for describing associative memory function in the piriform cortex.

More recent work has explicitly modeled the piriform cortex as an oscillatory network with the capacity for storage of spatially distributed patterns of oscillatory activity using physiologically realistic distributed inhibitory feedback (Liljenstrom, 1994, 1995; Liljenstrom and Hasselmo, 1995). In these models, Hebbian synaptic modification of connections between subsets of neurons allows these neurons to oscillate in synchrony, while other neurons which are not components of the stored pattern do not oscillate. Figure 6 shows an example of the oscillatory patterns stored in a network of this type. In particular, Fig. 6 demonstrates how the property of adaptation in cortical pyramidal cells might allow the network to make transitions between limit-cycle attractors. With this model it was demonstrated that learning and recall in an oscillatory mode are faster and more robust than these functions in a corresponding model with fixed-point attractor dynamics (Liljenstrom and Wu, 1995; Liljenstrom, 1995). Again, the storage of these oscillatory patterns requires different dynamics during learning, in which the recall of previously stored patterns is suppressed (Liljenstrom and Hasselmo, 1995). Cholinergic suppression of synaptic transmission suppresses the recall of old patterns during storage of new patterns in this model, while the cholinergic suppression of adaptation allows the network to remain in an individual attractor for a longer period during learning than during recall.

Most of the models using interactions of oscillatory units have focused on patterns of amplitude distribution which extend across full sets of fast oscillations. That is, in most models of bulb and cortex, the spatial pattern being stored is defined across the full burst of gamma oscillations. However, it has been suggested that stored representations might involve a different pattern of activity on each gamma oscillation, with excitatory connections mediating transitions between successive patterns (Ketchum and Haberly, 1991; Wang *et al.*, 1991). The capability of storing such temporal sequences has been demonstrated in an abstract mathematical representation of coupled oscillators (Wang *et al.*, 1991). In addition, models of current source density data have been analyzed to explore whether associations across different gamma oscillations would be possible, and what physiological properties storage of such associations across cycles would require (Ketchum and Haberly, 1991). These models demonstrated that the faster progression of afferent activity across the cortex would allow associations with the slower moving association-fiber activity, and that Hebbian modification would result in selective strengthening of faster association fibers in more-caudal regions of the cortex. Experimental data support the prediction that association fibers should be faster in more-caudal regions (Ketchum and Haberly, 1991). The evidence for single-unit responses to a number of different stages in an odor discrimination task (Schoenbaum and Eichenbaum, 1994) supports the notion that the piriform cortex may store transitions between different odor-related sensory information, rather than storing static attractors representing individual odors.

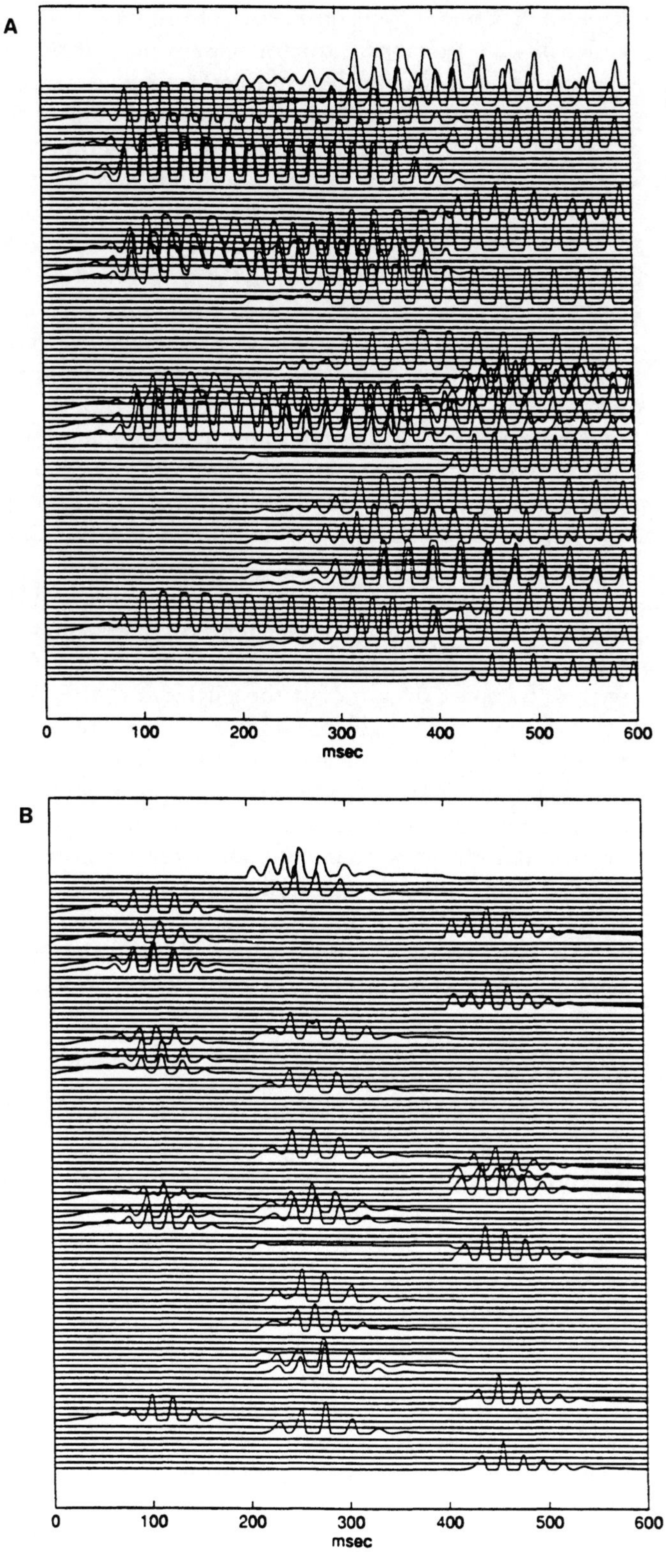

Figure 6. Patterns of oscillatory activity stored in a network model of the piriform cortex. The figure shows the activation of a set of neurons during sequential presentation of three different stored input patterns. In response to each input, the network shows a distribution of oscillatory activity reflecting the stored representation of that input pattern. (A) Without adaptation of the output of modeled neurons, a previously active pattern can interfere with subsequent patterns. (B) With adaptation of a neuronal output, the activity associated with one stored pattern gradually disappears before a subsequent pattern is activated.

4.1.1d. Chaos. The complexity of models of the olfactory system provides dynamics in which chaotic attractors can appear, though this chaotic behavior often depends upon higher order interactions between regions, rather than within regions (Freeman, 1987). More detailed modeling of the role of chaotic attractors versus limit-cycle attractors in network function will be required to determine whether chaotic dynamics gives functional advantages which can be demonstrated quantitatively. However, chaotic dynamics has been proposed to play a number of specific functional roles (Freeman, 1987; Skarda and Freeman, 1987). First, it has been proposed that activity during the interburst interval (between high-amplitude gamma bursts) shows chaotic properties which are well suited to the rapid transition to any of a number of stored attractor states. Thus, chaotic dynamics is proposed to provide a resting state which can rapidly enter various stored recall states due to a phase transition. Second, it has been proposed that some of the attractor states entered during the gamma bursts associated with inhalation are themselves chaotic. These chaotic states are observed experimentally during a subset of the inhalations to learned odors, and in response to almost all novel odor presentations. This chaotic state is seen as an escape from all established limit-cycle attractors, allowing unstructured activity during the development of a new limit-cycle attractor (Freeman, 1987; Skarda and Freeman, 1987). An additional chaotic state consists of a breakdown of the system into a low-dimensional attractor, consisting of seizure activity dominated by feedback inhibition (Freeman, 1986). It is possible that the complex dynamics of the piriform cortex is under neuromodulatory control. By varying a single parameter corresponding to neuronal excitability in a model of the olfactory cortex (Wu and Liljenstrom, 1994), the dynamics can switch from point-attractor to limit-cycle to chaotic-attractor behavior. Transient chaos during an initial input phase has been demonstrated to give rapid convergence to any stored limit-cycle memory state. Neuronal noise can also result in faster memory recall or transition between different memory states (Liljenstrom and Wu, 1995). All of these models of chaotic dynamics use the continuous-firing-rate representation. It remains to be shown whether networks of spiking neurons demonstrate dynamical properties which are accurately represented by these simplified models.

4.1.2. Biophysical Models of Associative Memory Function

The use of continuous-firing-rate models allows exploration of a wide range of dynamical properties. However, these models all suffer from the simplifications which make it difficult to relate their dynamical properties to the intrinsic properties of real cortical neurons. Thus, it has proven useful to explore possible associative memory function in realistic biophysical network simulations of the piriform cortex.

The Wilson and Bower model of piriform cortex was utilized initially to explore possible associative memory function in this structure (Wilson and Bower, 1988). This early model used 100 neurons, in contrast to the 1500-neuron version used for analysis of oscillations. In that model, it was shown that different patterns of afferent input from the olfactory bulb could be stored in the model with synaptic modification of the excitatory intrinsic synapses connecting pyramidal neurons. In response to degraded versions of these stored patterns, the network showed activity more closely resembling the originally stored version of the pattern, as evaluated by a dot product between the response to the degraded pattern and the response to the full learned version of the pattern.

This original work has been extended more recently, with a model analyzing storage of a larger number of patterns in a network with some additional modifications (Hasselmo *et al.,* 1994; Barkai *et al.,* 1994; Barkai and Hasselmo, 1994; Hasselmo and Barkai, 1995). This model utilized the *in vitro* data discussed in the previous section to constrain the development of the biophysical simulations of piriform cortex pyramidal cells. As discussed above, these modeled neurons contained additional potassium currents mediating adaptation of spiking response. This allowed effective modeling of the cholinergic suppression of neuronal adaptation, as shown in Fig. 5. In addition, the model incorporated other cholinergic effects including the suppression of excitatory intrinsic synaptic transmission, as shown in Fig. 7, and the depolarization of pyramidal cell membrane potential.

In a network of 240 pyramidal cells, five patterns were stored, which activated 20 neurons each. Figure 8 shows simulated neuronal activity during recall in response to complete or degraded versions of one of the five afferent input patterns. This particular pattern was chosen for clarity—actual patterns of physiological activity would not be expected to show such regular characteristics. Different-sized black squares represent the different numbers of spikes fired during recall by individual pyramidal cells within the situation. Membrane potential traces of individual pyramidal cell simulations (numbers 31, 111, and 190) are shown to illustrate the time course of activity during recall.

In Fig. 8A, the full pattern causes spiking activity in 20 of the 240 neurons during the 600-msec simulation period. To test the storage of this pattern, a degraded version containing only 12 active input lines was presented during recall. Before Hebbian modification of the excitatory intrinsic synapses, the network shows poor recall performance in response to the degraded pattern (Fig. 8B). None of the neurons normally activated by the missing elements of the full pattern shows spiking activity (including neuron 190). However, if Hebbian synaptic mod-

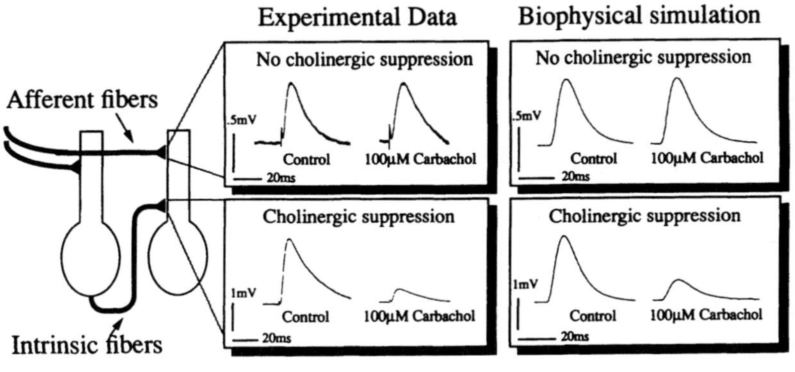

Figure 7. Experimental data and biophysical simulation of the selective cholinergic suppression of intrinsic fiber synaptic transmission. On the left, perfusion with the cholinergic analogue carbachol suppresses synaptic potentials elicited by stimulation of intrinsic, but not afferent fibers in brain slice preparations of rat piriform cortex (Hasselmo and Bower, 1992). On the right, in the biophysical simulation this effect is represented as a decrease in the number of synaptic sodium channels activated by intrinsic fiber synaptic transmission.

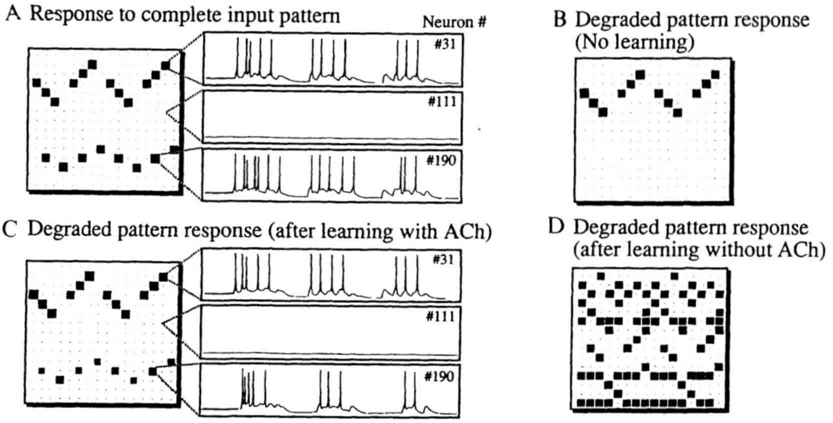

Figure 8. Associative memory function with a single learned pattern in a 240-neuron network biophysical simulation. For each section, the size of black squares represents the number of spikes fired by each of the 240 neurons during a 600-msec period of recall. In the center, membrane potential traces are shown for three simulated pyramidal cells (numbers 31, 111, and 190) over a period of 600 msec (A) Response of the network to the complete version of the pattern. Twenty neurons show spiking activity due to activation by the afferent input pattern. On the right, two neurons receiving afferent input generate action potentials, while one neuron not receiving input shows only slight inhibitory influences. (B) This example shows what happens when no learning has taken place (when no modification of excitatory intrinsic synapses has occurred). A degraded pattern is presented to the network, containing only 12 of the usual 20 active afferent input lines. Because learning has not occurred, activity does not spread to the other 8 neurons normally activated by the input pattern. (Normalized dot product with response to the complete pattern in panel A is 0.775.) On the right, one of the neurons which previously received afferent input now does not fire (neuron 190). (C) This example shows completion after learning with cholinergic suppression of synaptic transmission (synaptic modification of intrinsic synapses is applied during presentation of the full pattern, but intrinsic synaptic transmission is suppressed). The same degraded pattern is presented as in panel B, with 12 of 20 active afferent lines. However, note that the 8 neurons not receiving afferent input now fire in response to the spread of activity across strengthened excitatory intrinsic synapses. On the right, the spiking activity of the neuron not receiving afferent input has been restored (neuron 190). (D) Response of the network after learning without cholinergic suppression of synaptic transmission (intrinsic synapses were modified during presentation of the full pattern, but intrinsic synaptic transmission was not suppressed). The same degraded pattern is presented as in panels B and C, but now activity occurs in neurons activated by every pattern stored in the network. Interference during learning has caused runaway synaptic modification—this makes discrimination between patterns during recall impossible.

ification of the excitatory synapses between pyramidal cells is applied during presentation of the full pattern, the network responds to the degraded version of the pattern with completion of the missing elements. In Fig. 8C this completion results in activity more closely resembling the response to the full learned pattern. The spread of activity across modified intrinsic fiber synapses causes spiking activity in neurons which are not receiving afferent input. Spiking activity in neuron 190 is entirely elicited by transmission at excitatory intrinsic synapses which were modified during learning.

The effective learning shown in Fig. 8C requires cholinergic suppression of synaptic transmission during learning. Without this cholinergic suppression of synaptic transmission, the spread of activity across previously modified synapses recalls components of previously stored patterns during learning of a new pattern. This interference during learning results in strengthening of additional synapses between elements of the stored patterns. As learning progresses, this interference becomes worse, until presentation of the degraded version of one pattern recalls elements of all the patterns stored in the network, as shown in Fig. 8D. The progressive strengthening of additional undesired synaptic connections in the network is referred to as runaway synaptic modification. Figure 9 shows the progression of runaway synaptic modification in the network during learning without cholinergic suppression of synaptic transmission. In addition, it shows the strengthening of only the desired synaptic connections in the network if cholinergic suppression of intrinsic synaptic transmission is applied during learning. Runaway synaptic modification has been proposed as a possible synaptic mechanism underlying some of the pathological features of Alzheimer's disease (Hasselmo, 1994).

These principles are similar to those explored in earlier simulations (Hasselmo, 1993, 1994; Hasselmo *et al.*, 1991, 1992). However, the use of a detailed biophysical simulation allowed analysis of two additional issues. One concerned the role of neuronal adaptation and the modulation of this adaptation in learning (Hasselmo *et al.*, 1994; Barkai *et al.*, 1994). Adaptation was shown to result in greater stability of the network in certain cases, allowing effective sequential recall of different patterns. Cholinergic suppression of this adaptation during learning allowed greater activity to drive more strongly synaptic modification, resulting in a more complete strengthening of synaptic connections, as shown in Fig. 10. Thus, modulation of adaptation sets appropriate activation dynamics in the network during learning and recall. In addition, the network allowed exploration of the dynamics of recall with spiking neurons. It proved very difficult to balance the dynamics of the network to allow the appearance of either fixed-point or limit-cycle attractor dynamics. Instead, each cycle of fast $GABA_A$-mediated inhibition allowed activity to

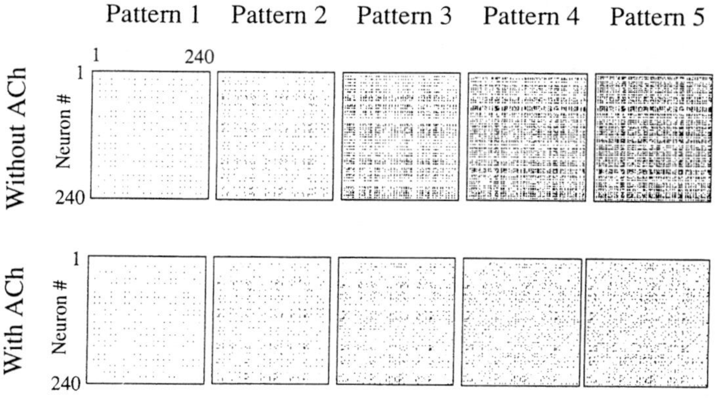

Figure 9. Progression of runaway synaptic modification in the 240-neuron network. The synaptic connectivity matrix of the 240-neuron network is shown after learning of each of five patterns. The size of the black squares represents the strength of synaptic connections. Bottom: With cholinergic suppression of synaptic transmission during learning, the synaptic connectivity matrix shows stable growth, with strengthening of synapses only between neurons directly activated by afferent input. Top: As learning progresses without cholinergic suppression of synaptic transmission, the synaptic connectivity matrix shows runaway synaptic modification. Note that growth is reasonably stable for the first two patterns, but during learning of the third and subsequent patterns, the spread of activity across previously modified synapses causes a rapid increase in the number of strengthened synapses.

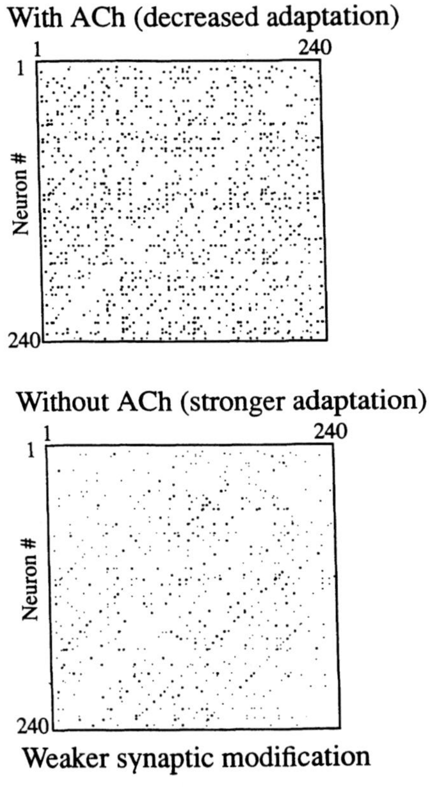

Figure 10. Final connectivity matrix after the same period of learning with and without cholinergic suppression of neuronal adaptation. Top: After learning with cholinergic suppression of neuronal adaptation, synapses reach the full strength necessary for effective recall of stored patterns. Bottom: After learning without cholinergic suppression of neuronal adaptation, many of the intrinsic excitatory synapses have not reached the full strength necessary for effective recall of stored patterns.

spread along modified synapses for a limited period, after which activity was suppressed until the arrival of additional spikes on the afferent input fibers. These models need to be explored to determine whether spiking neurons can demonstrate the complex attractor dynamics described in models using continuous-firing-rate representations.

4.2. Short-Term Memory Function

In order to explore the interplay between preprocessing in the olfactory bulb and associative memory function in the piriform cortex, Linster and Hassselmo (1997) have developed a combined model of both structures. This model is based on previous models of the olfactory bulb (Linster and Gervais, 1996; Linster and Hasselmo, 1996) as well as on previously described models of olfactory cortex (Wilson and Bower, 1992; Liljenstroem, 1991; Barkai *et al.*, 1994). Models of both structures are spatially distributed models, and reproduce single-cell data as well as field potential recordings. In this model, connections from the olfactory bulb to the piriform cortex are made by way of mitral cell axons projecting onto the piriform cortex with a connection probability and strength that decrease from rostral to caudal piriform cortex (Haberly, 1985; Wilson and Bower, 1992). As a consequence, rostral pyramidal cells and interneurons can be directly driven by mitral cell input, whereas more-caudal pyramidal cells are not driven by mitral cell input alone. Using a Hebbian learning rule (Barkai *et al.*, 1994), the authors show that the representation of olfactory stimuli spreads from rostral toward more-caudal regions with increasing stimulus familiarity. Thus, associative learning during successive sniff cycles leads to a spread of the olfactory information from rostral to caudal piriform cortex pyramidal cells (Fig. 11A). After learning during several sniff cycles, the synaptic connections in the rostral region are mostly saturated, and any additional learning will result in strengthening of synaptic connections in the caudal region. In a delayed match-to-sample task, one of several highly familiar stimuli (O1) is presented for a short duration (one cycle) (Fig. 11B). After a delay, a second stimulus, also highly familiar, is presented, which is either the same stimulus (O1 match) or a different stimulus (O2 nonmatch). During the first presentation, intrinsic synaptic connections between caudal pyramidal cells are strengthened. This short presentation of one of the stimuli results in short-term synaptic changes, strongly activating caudal pyramidal cells. A comparison between the pyramidal cell responses in the caudal region to the sample (match) or to one of the other stimuli (nonmatch) yields over 85% percent correct choices after learning of the sample stimulus when number of spikes is counted. This model, based on short-term synaptic potentiation, is an interesting alternative to other models of short-term memory (Zipser, 1991; Linster and Masson, 1996) proposing sustained neural activity during the delay.

4.3. Self-Organizing Models

Another broad class of models are here referred to as self-organizing networks. They are also referred to as competitive or winner-take-all networks. In these models, modifications of connections between neurons serve to form new representations of input activity. For example, these types of models have been used extensively to model the formation of feature detectors and topographic maps in primary neocortical regions such as the striate cortex (Miller *et al.*, 1989; von der Malsburg, 1973). They differ from associative memory models, which focus on the learning of associations between components of representations, not on the formation of new representations. Thus, the representations stored in associative memory models are either fixed by the initial structure of a network or formed by separate self-organizing connections providing afferent input to the associative memory. The function of different synapses can be defined according to the influence of modifiable synapses on postsynaptic activity during learning (Hasselmo, 1995). Synapses mediating associative memory are not the predominant influence on postsynaptic activity during learning (preventing them from influencing the representation formed). Synapses undergoing self-organization *are* the predominant influence on postsynaptic activity during learning (allowing them to influence the formation of a new representation). Obviously, self-organization is very important to the function of cortical networks, and must be incorporated into any model which focuses on the formation and modification of internal representations of sensory stimuli.

4.3.1. Hierarchical Classification

Considering the importance of self-organization for cortical function, surprisingly few models of the piriform cortex have addressed this issue. The models of

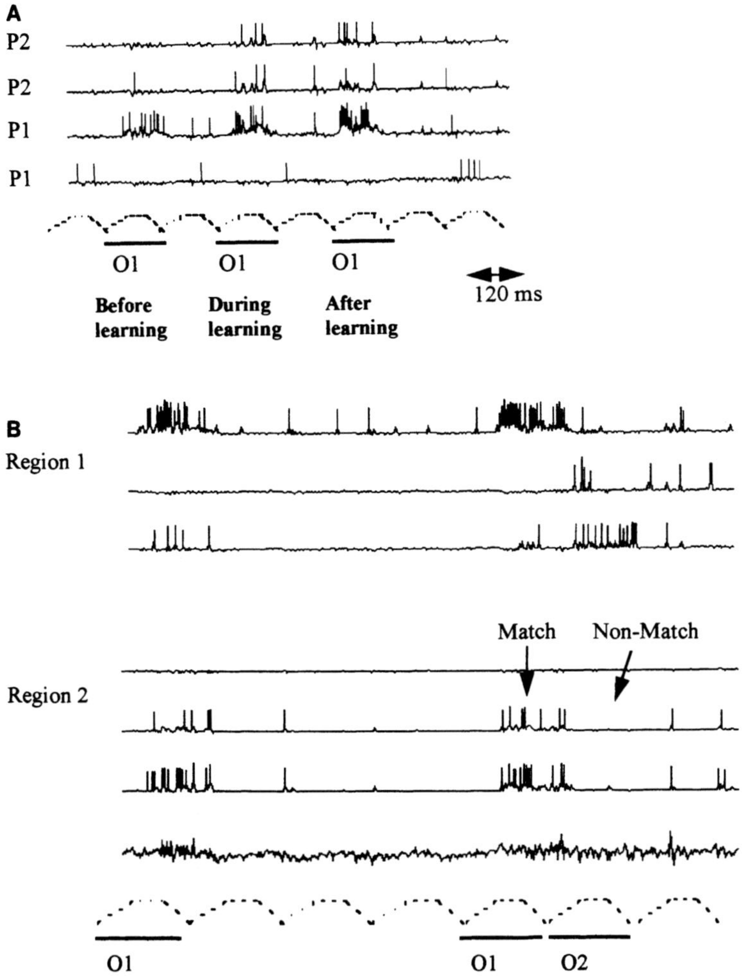

Figure 11. Short-term memory function in a combined model of olfactory bulb and cortex. (A) The neural representation of an olfactory stimulus spreads from rostral (P1) to caudal (P2) pyramidal cells during learning. Before learning, only P1 cells respond with increased spiking frequencies to input (O1). After learning, intrinsic connections between pyramidal cells are strengthened and P2 cells start responding. Membrane potential fluctuations and action potentials of pyramidal cells in the model are shown in response to olfactory stimuli to olfactory receptor cells. Dashed lines show time course of low noise mimicking respiration, to which olfactory stimuli (solid lines) are added. (B) After exposure to a highly familiar pattern (O1) during a single sniff cycle, a subsequent presentation of O1 results in increased spiking activity in rostral (region 1) and in caudal (region 2) pyramidal cells (match) while the presentation of a different, also highly familiar stimulus (O2) does not lead to increased spiking in caudal pyramidal cells (nonmatch).

olfactory cortex described above focus on modification of excitatory intrinsic synapses between pyramidal cells. In these models, the pattern induced by afferent input is usually set by the initial connectivity from the bulb to the cortex. Afferent synapses are either not modifiable or are less modifiable than intrinsic synapses. However, it is the modification of these afferent synapses which determines the nature of the representation being stored. Any modification of these synapses will constitute self-organization.

A separate body of modeling research has focused on how the modification of excitatory afferent and feedforward synapses within the piriform cortex allows determination of neuronal representations of olfactory stimuli (Ambros-Ingerson *et al.*, 1990; Granger *et al.*, 1989). These models focus on the response of layer II pyramidal neurons to a randomly distributed and sparse afferent input from the lateral olfactory tract. Strong lateral inhibition within patches of pyramidal neurons allows winner-take-all activation dynamics in which a small number of neurons respond to each afferent input pattern, and thereafter become selectively responsive to that afferent input pattern (Ambros-Ingerson *et al.*, 1990; Coultrip *et al.*, 1992; Granger *et al.*, 1989). Modification of the strength of afferent synapses changes the representation such that for the initial response to an odor, similarity between odors leads them to be lumped in the same category. Incorporation of a special refractory period for neurons which fire causes this initial categorization to become dormant, allowing subsequent activity to reflect a discrimination between different subcomponents within a category. Thus, the network ultimately learns to respond to odors with a temporal sequence of activity states which start with broad categories lumping odors together on the basis of similarity, and progresses to narrow categories discriminating odors on the basis of small differences (Ambros-Ingersen *et al.*, 1990; Granger *et al.*, 1989). The representation of individual odors is presumed to be very sparse, which is consistent with single-unit data showing sparse responsiveness of piriform cortex neurons to individual odors during a behavioral task (McCollum *et al.*, 1991).

While this model shows interesting functional characteristics, it obtains these by making very specific physiological assumptions. In particular, the refractory period of 1 sec is very long even if intended to represent the effect of a strong calcium-dependent potassium current. As an alternative to this refractory period, it has been proposed that modification of feedback connections to granule cells in the bulb might allow selective suppression of shared components of the input from the bulb (Ambros-Ingerson *et al.*, 1990). Another strong assumption concerns the use of separate "learning" and "performance" modes, with synaptic modification only during learning, and very different temporal patterns of afferent input during the two periods. These modes might correspond to the influence of cholinergic or noradrenergic modulation, but the properties of these modes have not been discussed with regard to data from the piriform cortex showing that these substances cause suppression of intrinsic synaptic transmission (Hasselmo and Bower, 1992; Vanier and Bower, 1992) and suppression of neuronal adaptation (Barkai and Hasselmo, 1994). To prevent certain categories from dominating the response of the network, these networks require a novel "no-history" assumption of the learning rule, in which learning does not affect the influence of afferent synapses during subsequent "learning" modes, but only during subsequent "performance." While these assumptions give the network very interesting functional properties, they require further experimental work to demonstrate that they do not violate physiological constraints.

Simple aspects of a competitive neural network have been tested experimentally to determine its validity for the learning of odors (Granger *et al.*, 1991b). Testing of the response to odor mixtures with a strong consistent component A and weaker varying components (B, C, D, . . .) demonstrated that rats did respond on the basis of a large category which lumped these odors together. However, they demonstrated the ability to make a final discrimination between AE and AH, demonstrating that they maintained the ability to discriminate these odors. While this behavioral test may support the general notions of clustering and differentiation of odors, these results could be supported by a wide range of different types of models. The model makes a more specific prediction that pretraining on component A would allow better discrimination of mixtures AB and AC than without pretraining, which is supported by data which have not yet been published (Granger *et al.,* 1991a).

4.3.2. Dynamical Self-Organization

Self-organization can refer to formation of representations which are not just stable patterns of activity. Thus, dynamical attractor states have been described as developing from self-organization (Skarda and Freeman, 1987). The definition utilized here, that modifiable synapses influence postsynaptic activity during learning, could include the formation of attractors as self-organization, even if they only depend upon modification of intrinsic connections, provided that the intrinsic connections determine the postsynaptic activity sufficiently to influence the learning. In fact, learning of dynamical states could be described as self-organization of subsequent responses on the basis of previous responses.

5. Summary and Future Directions

A variety of models of the piriform cortex have simulated physiological data and addressed functional hypotheses about this region. Different levels of analysis have proven useful. Mathematical abstractions such as the continuous-firing-rate representation of neuronal output provide simpler representations which aid the conceptualization of network function. However, these simplifications may not accurately portray the dynamics of real cortical networks. Biophysically detailed simulations are necessary to demonstrate that functions described in simpler models can be implemented with real neurons. In addition, these more detailed models allow the parameters of models to be more directly related to physiological data.

Modeling of evoked potentials supports the notion that the cortex can generate fast gamma oscillations on the basis of feedback inhibition, but the input from the olfactory bulb may be the predominant driving force of these fast oscillations. The basis for the slower oscillations observed in EEG recordings is not clear yet, with a number of hypotheses from modeling work. These slow oscillations have been suggested to arise from slower potassium currents activated by inhibition, from spike synchronization effects, from neuronal adaptation, and from the driving influence of cycles of afferent input due to inhalation.

The final representation of odors in the piriform cortex remains unclear, and has taken a number of forms in different models of the piriform cortex. A major difference between models results from whether they are developed with an emphasis on EEG data or have an implicit focus on single-neuron responses. Models based on EEG data (e.g., Freeman *et al.*, 1988) tend to focus on the continuous distribution of activity across space, with distinctions between patterns on the basis of rather subtle differences of distribution, or differences in temporal evolution of activity. The smoothing effect of EEG techniques may actually obscure important distinctions between patterns on a smaller spatial scale. In contrast, other models focus on representations at more of a single-neuron level (e.g., Hasselmo *et al.*, 1994; Wilson and Bower, 1988), with an emphasis on sparse distributions of spiking activity. These representations are not necessarily incompatible, but further effort must be focused on reconciling these different types of models.

The predominant function explored in models of the piriform cortex (and olfactory bulb) is associative memory function, in which modification of excitatory recurrent connections allows the network to store new patterns of activity and respond to noisy versions of these patterns with the originally stored pattern. This basic property can be used to describe models with a wide range of dynamical characteristics regardless of whether the final stored pattern state is a fixed-point attractor, limit-cycle attractor, or chaotic attractor. Recent formulations have suggested that activity does not necessarily approach a single stable pattern within each gamma frequency burst, but may involve a transition between each of a series of stored patterns on every gamma frequency oscillation. This may prove the most physiologically valid representation.

Associative memory models assume representation dependent upon an initial connectivity of the afferent inputs, which may actually undergo self-organization during development or learning. A final description of the function of this region will probably require a unification of self-organization on the afferent fibers and associative memory function on intrinsic fibers. This combination of self-organization and associative memory function may prove to be a basic operating principle of cortical function (Hasselmo, 1995). Further modeling must also focus on the interaction of the piriform cortex with the earlier stages of the olfactory system, as proposed by Linster *et al.* (1995) and Linster and Hasselmo (1996), and with neocortical and subcortical structures, to integrate the representations in the olfactory cortex with the guidance of behavior. Modeling provides a powerful technique for linking behavioral function to physiological substrates, but this work has only just begun.

6. References

Abbott, L. F., 1991, Realistic synaptic inputs for model neural networks, *Network* **2**:245–258.

Abbott, L. F., 1992, Firing-rate models for neural populations, in: *Neural Networks: From Biology to High-Energy Physics* (O. Benhar, C. Bosio, P. D. Giudice, and E. Tabet, eds.), ETS, Pisa, Italy.

Ambros-Ingerson, J., Granger, R., and Lynch, G., 1990, Simulation of paleocortex performs hierarchical clustering, *Science* **247**:1344–1348.

Amit, D. J., 1988, *Modeling Brain Function: The World of Attractor Neural Networks,* Cambridge University Press, Cambridge.

Amit, D. J., and Treves, A., 1989, Associative memory neural networks with low temporal spiking rates, *Proc. Natl. Acad. Sci. USA* **86**:7671–7673.

Anderson, J. A., 1972, A simple neural network generating an interactive memory, *Math. Biosci.* **14**:197–220.

Anderson, J. A., 1983, Cognitive and psychological computation with neural models, *IEEE Trans. Systems Man Cybernet.* **SMC-13:**799–815.

Baird, B., 1986, Nonlinear dynamics of pattern formation and pattern recognition in the rabbit olfactory bulb, *Physica D* **22:**150–175.

Barkai, E., and Hasselmo, M. E., 1994 Modulation of the input/output function of rat piriform cortex pyramidal cells, *J. Neurophysiol.* **72:**644–658.

Barkai, E., Bergman, R. E., Horwitz, G., and Hasselmo, M. E., 1994, Modulation of associative memory function in a biophysical simulation of rat piriform cortex, *J. Neurophys.* **72:**659–677.

Becker, C. J., and Freeman, W. J., 1968, Prepyriform electrical activity after loss of peripheral or central input, or both, *Physiol. Behav.* **3:**597–599.

Biedenbach, M. A., 1966, Effects of anaesthetics and cholinergic drugs on prepyriform electrical activity in cats, *Exp. Neurol.* **16:**464–479.

Bower, J. M., 1991, Piriform cortex and olfactory object recognition, in: *Olfaction: A Model System for Computational Neuroscience* (J. L. Davis and H. Eichenbaum, eds.), MIT Press, Cambridge, MA, pp. 265–286.

Bower, J. M., 1995, Reverse engineering the nervous system: An *in vivo, in vitro* and computational approach to understanding the mammalian olfactory system, in: *An Introduction to Neural and Electronic Networks* S. Zornetzer, J. Davis, and C. Lau, eds.), Academic Press, New York, pp. 3–24.

Bower, J. M., and Beeman, D., 1995, *The Book of GENESIS: Exploring Realistic Neural Models with the GEneral NEural SImulation System,* Springer-Verlag, New York.

Bressler, S. L., 1984, Spatial organization of EEGs from olfactory bulb and cortex, *Electroencephalogr. Clin. Neurophysiol.* **57:**270–276.

Bressler, S. L., 1987a, Relation of olfactory bulb and cortex. I. Spatial variation of bulbocortical interdependence, *Brain Res.* **409:**285–293.

Bressler, S. L., 1987b, Relation of olfactory bulb and cortex. II. Model for driving of cortex by bulb, *Brain Res.* **409:**294–301.

Bressler, S., and Freeman, W., 1980, Frequency analysis of olfactory system EEG in cat, rabbit and rat, *Electroencephalogr. Clin. Neurophysiol.* **50:**19–24.

Collins, G. G. S., and Howlett, S. J., 1988, The pharmacology of excitatory transmission in the rat olfactory cortex slice, *Neuropharmacology* **27:**697–705.

Constanti, A., and Bagetta, G., 1991, Mascarinic receptor activation induces a prolonged post-stimulus after depolarization with a conductance decrease in guinea-pig of factory cortex neurons in intro, *Neurosci. Lett.* **131:**27–32.

Constanti, A., and Galvan, M., 1983, M-current in voltage clamped olfactory cortex neurones, *Neurosci. Lett.* **39:**65–70.

Constanti, A., and Sim, J. A., 1987, Calcium-dependent potassium conductance in guinea-pig olfactory cortex neurones *in vitro, J. Physiol.* **387:**173–194.

Coultrip, R., Granger, R., and Lynch, G., 1992, A cortical model of winner-take-all competition via lateral inhibition, *Neural Networks* **5:**47–54.

Eeckman, F. H., and Freeman, W. J., 1990, Correlations between unit firing and EEG in the rat olfactory system, *Brain Res.* **528:**238–244.

Erdi, P., Grobler, T., Barna, G., and Kaski, K., 1993, Dynamics of the olfactory bulb: Bifurcations, learning and memory, *Biol. Cybernet.* **69:**57–66.

Fallon, J. H., and Moore, R. Y., 1978, Catecholamine innervation of the basal forebrain (3): Olfactory bulb, anterior olfactory nucleus, olfactory tubercle, and pyriform cortex, *J. Comp. Neurol.* **180:**533–544.

Freeman, W. J., 1962, Linear approximation of prepyriform evoked potential in cats, *Exp. Neurol.* **5:**477–499.

Freeman, W. J., 1964, A linear distributed feedback model of prepyriform cortex, *Exp. Neurol.* **10:**525–547.

Freeman, W. J., 1968a, Effects of surgical isolation and tetanization on prepyriform cortex in cats, *J. Neurophysiol.* **31:**349–357.

Freeman, W. J., 1968b, Relations between unit activity and evoked potentials in prepyriform cortex of cats, *J. Neurophysiol.* **31:**337–348.

Freeman, W. J., 1978, Spatial properties of an EEG event in the olfactory bulb and cortex, *Electgroencephalogr. Clin. Neurophysiol.* **44:**586–605.

Freeman, W. J., 1979a, EEG analysis gives model of neuronal template-matching mechanism for sensory search with olfactory bulb, *Biol. Cybernet.* **35**:221–234.

Freeman, W. J., 1979b, Nonlinear dynamics of paleocortex manifested in the olfactory EEG, *Biol. Cybernet.* **35**:21–37.

Freeman, W. J., 1979c, Nonlinear gain mediating cortical stimulus–response relations, *Biiol. Cybernet.* **33**:237–247.

Freeman, W. J., 1986, Petit mal seizure spikes in olfactory bulb and cortex caused by runaway inhibition after exhaustion of excitation, *Brain Res. Rev.* **11**:259–284.

Freeman, W. J., 1987, Simulation of chaotic EEG patterns with a dynamic model of the olfactory system, *Biol. Cybernet.* **56**:139–150.

Freeman, W. J., and Skarda, C. A., 1985, Spatial EEG patterns, non-linear dynamics and perception: The neo-Sherringtonian view, *Brain Res. Rev.* **10**:147–175.

Freeman, W. J., Yao, Y., and Burke, B., 1988, Central pattern generating and recognizing in olfactory bulb: A correlation learning rule, *Neural Networks* **1**:277–288.

Granger, R., Ambros-Ingerson, J., and Lynch, G., 1989, Derivation of encoding characteristics of layer II cerebral cortex, *J. Cognitive Neurosci.* **1**:61–87.

Granger, R., Staubli, U., Ambros-Ingerson, J., and Lynch, G., 1991a, Specific behavioral predictions from simulations of the olfactory system, in: *Olfaction: A Model System for Computational Neuroscience* (J. L. Davis and E. Eichenbaum, eds.), MIT Press, Cambridge, MA, pp. 251–264.

Granger, R., Staubli, U., Powers, H., Ambros-Ingerson, J., Otto, T., and Lynch, G., 1991b, Behavioral tests of a prediction from a cortical network simulation, *Psychol. Sci.* **2**:116–118.

Haberly, L. B., 1973a, Summed potentials evoked in opossum prepyriform cortex, *J. Neurophysiol.* **36**:775–788.

Haberly, L. B., 1973b, Unitary analysis of the opossum prepyriform cortex, *J. Neurophysiol.* **36**:762–774.

Haberly, L. B., 1985, Neuronal circuitry in olfactory cortex: Anatomy and functional implications, *Chem. Senses* **10**:219–238.

Haberly, L. B., 1990, Olfactory cortex, in: *Synaptic Organization of the Brain* (G. M. Shepherd, ed.), Oxford University Press, Oxford, pp. 317–345.

Haberly, L. B., and Bower, J. M., 1984, Analysis of association fiber system in piriform cortex with intracellular recording and staining techniques, *J. Neurophysiol.* **51**:90–112.

Haberly, L. B., and Bower, J. M., 1989, Olfactory cortex: Model circuit for study of associative memory? *Trends Neurosci.* **12**:258–264.

Haberly, L. B., and Price, J. L., 1977, The axonal projection patterns of the mitral and tufted cells of the olfactory bulb in the rat, *Brain Res.* **129**:152–157.

Haberly, L. B., and Price, J. L., 1978, Association and commissural fiber systems of the olfactory cortex of the rat. I. Systems originating in the piriform cortex and adjacent areas, *J. Comp. Neurol.* **178**:711–740.

Haberly, L. B., and Shepherd, G. M., 1973, Current density analysis of summed evoked potentials in opossum prepyriform cortex, *J. Neurophysiol.* **36**:789–802.

Hasselmo, M. E., 1993, Acetylcholine and learning in a cortical associative memory, *Neural Computation* **5**:32–44.

Hasselmo, M. E., 1994, Runaway synaptic modification in models of cortex: Implications for Alzheimer's disease, *Neural Networks* **7**:13–40.

Hasselmo, M. E., 1995, Neuromodulation and cortical function, *Behav. Brain Res.* **67**:1–27.

Hasselmo, M. E., and Barkai, E., 1992, Cholinergic modulation of the input/output function of rat piriform cortex pyramidal cells, *Soc. Neurosci. Abstr.* **18**:521.

Hasselmo, M. E., and Barkai, E., 1995, Cholinergic modulation of activity-dependent synaptic plasticity in the piriform cortex: Brain slice physiology and computational modeling, *J. Neurosci.* **15**:6529–6604.

Hasselmo, M. E., and Bower, J. M., 1991, Selective suppression of afferent but not intrinsic fiber synaptic transmission by 2-amino-4-phophonobutyric acid (AP4) in piriform cortex, *Brain Res.* **548**:248–255.

Hasselmo, M. E., and Bower, J. M., 1992, Cholinergic suppression specific to intrinsic not afferent fiber synapses in rat piriform (olfactory) cortex, *J. Neurophysiol.* **67**:1222–1229.

Hasselmo, M. E., and Bower, J. M., 1993, Acetylcholine and memory, *Trends Neurosci.* **16**:218–222.

Hasselmo, M. E., and Schnell, E., 1994, Laminar selectivity of the cholinergic suppression of synaptic transmission in rat hippocampal region CA1: Computational modeling and brain slice physiology, *J. Neurosci.* **14**:3898–39914.

Hasselmo, M. E., Anderson, B. P., and Bower, J. M., 1991, Cholinergic modulation may enhance cortical associative memory function, in: *Advances in Neural Information Processing Systems* (R. P. Lippman, J. Moody, and D. S. Touretsky, eds.), Morgan Kaufmann, San Mateo, CA, pp. 46–52.

Hasselmo, M. E., Anderson, B. P., and Bower, J. M., 1992, Cholinergic modulation of cortical associative memory function, *J. Neurophysiol.* **67**:1230–1246.

Hasselmo, M. E., Barkai, E., Horwitz, G., and Bergman, R. E., 1994, Modulation of neuronal adaptation and cortical associative memory function, in: *Computation in Neurons and Neural Systems* (F. Eeckman and J. M. Bower, eds.), Kluwer, Norwell, MA, pp. 287–292.

Hasselmo, M. E., Schnell, E., and Barkai, E., 1995, Dynamics of learning and recall at excitatory recurrent synapses and cholinergic modulation in hippocampal region CA 3, *J. Neurosci.* **15**:5249–5252.

Hasselmo, M. E., Linster, C., Patil, M., Ma, D., and Cekic, M., 1997, Noradrenergic suppression of synaptic transmission may influence corical signal-to-noise ratio, *J. Neurophysiol.* **77**:3326–3339.

Hopfield, J. J., 1984, Neurons with graded responses have collective computational properties like those of two-state neurons, *Proc. Natl. Acad. Sci. USA* **81**:3088–3092.

Hori, N., Akaike, N., and Carpenter, D. O., 1988, Piriform cortex brain slices: Techniques for isolation of synaptic inputs, *J. Neurosci. Meth.* **25**:197–208.

Jung, M. W., Larson, J., and Lynch, G., 1990, Long-term potentiation of monosynaptic EPSPs in rat piriform cortex *in vitro, Synapse* **6**:279–283.

Kanter, E. D., and Haberly, L. B., 1990, NMDA-dependent induction of long-term potentiation in afferent and association fiber systems of piriform cortex *in vitro, Brain Res.* **525**:175–179.

Ketchum, K. L., and Haberly, L. B., 1991, Fast oscillations and dispersive propagation in olfactory cortex and other cortical areas: A functional hypothesis, in: *Olfaction: A Model System for Computational Neuroscience* (J. L. Davis and H. Eichenbaum, eds.), MIT Press, Cambridge, MA pp. 679–100.

Ketchum, K. L., and Haberly, L. B., 1993a, Membrane currents evoked by afferent fiber stimulation in rat piriform cortex. II. Analysis with a system model, *J. Neurophysiol.* **69**:261–281.

Ketchum, K. L., and Haberly, L. B., 1993b, Membrane currents evoked by afferent fiber stimulation in rat piriform cortex. I. Current source-density analysis, *J. Neurophysiol.* **69**:248–260.

Ketchum, K. L., and Haberly, L. B., 1993c, Synaptic events that generate fast oscillations in piriform cortex, *J. Neurosci.* **13**:3980–3985.

Kohonen, T., 1972, Correlation matrix memories, *IEEE Trans. Computers* **C-21**:353–359.

Kohonen, T., 1984, *Self-Organization and Associative Memory*, Springer-Verlag, Berlin.

Li, Z., and Hopfield, J. J., 1989, Modeling the olfactory bulb and its neural oscillatory processings, *Biol. Cybernet.* **61**:379–392.

Liljenstrom, H., 1991, Modeling the dynamics of olfactory cortex using simplified network units and realistic architecture, *Int. J. Neural Syst.* **2**:1–15.

Liljenstrom, H., 1994, Oscillations and associative memory: Brain and model, *Biol. Trans. R. Dan. Acad. Sci. Lett.*

Liljenstrom, H., 1995, Autonomous learning with complex dynamics, Int. J. Intell. Syst. **10**:119–153.

Liljenstrom, H., and Hasselmo, M. E., 1993, Acetylcholine and cortical oscillatory dynamics, in *Computation and Neural Systems* (F. Eeckman and J. M. Bower, eds.), Kluwer, Boston, pp. 523–530.

Liljenstrom, H., and Hasselmo, M. E., 1995, Cholinergic modulation of cortical oscillatory dynamics, *J. Neurophysiol.* **74**:288–297.

Liljenstrom, H., and Wu, X., 1995, Noise-enhanced performance in a cortical associative memory model, *Int. J. Neural Syst.* **6**:19–29.

Linster, C., and Gervais, R., 1996, Investigation of the role of interneurons and their modulation by centrifugal fibers in a neural model of the olfactory bulb, *J. Computational Neurosci.* **3**:225–246.

Linster C., and Hasselmo, M., 1996, Modulation of inhibition in a model of olfactory bulb reduces overlap in the neural representation of olfactory stimuli, *Behv. Brain. Res.* **84**:117–127.

Linster, C., and Hasselmo, M. E., 1997, Olfactory delayed-match-to-sample in a combined model of olfactory bulb and cortex, in: *Computational Neuroscience* (J. M. Bower, ed.), Plenum Press, New York.

Linster, C., and Masson, C., 1996, A neural model of olfactory sensory memory in the honeybee's antennal lobe, *Neural Computation* **8**:94–114.

Linster, C., Hasselmo, M. E., and Gervais, R., 1995, Interactions between olfactory bulb and piriform cortex in a neural model of olfactory processing, *Soc. Neurosci. Abst.* **21**:1747.

Luskin, M. B., and Price, J. L., 1983, The topographic organization of associational fibers of the olfactory system in the rat, including centrifugal fibers to the olfactory bulb, *J. Comp. Neurol.* **216**:264–291.

Macrides, F., Davis, B. J., Youngs, W. M., Nadi, N. S., and Margolis, F. L., 1981, Cholinergic and catecholaminergic afferents to the olfactory bulb in the hamster: A neuroanatomical, biochemical and histochemical investigation, *J. Comp. Neurol.* **203**:495–514.

Macrides, F., Eichenbaum, H. B., and Forbes, W. B., 1982, Temporal relationships between sniffing and limbic theta rhythm during odor discrimination reversal learning, *J. Neurosci.* **2**:1705–1717.

McCollumn, J., Larson, J., Otto, T., Schottler, F., Granger, R., and Lynch, G., 1991, Short-latency single unit processing in olfactory cortex, *J. Cognitive Neurosci.* **3**:293–299.

Miller, K. D., Keller, J. B., and Stryker, M. P., 1989, Ocular dominance column development—Analysis and simulation, *Science* **245**:605–615.

Patil, M. M., Linster, C., Lubenov, E., and Hasselmo, M. E., 1998, The cholinergic agonist carbachol enables associative long-term potentiation in piriform cortex slices, *J. Neurophysiol.*, in press.

Patneau, D. K., and Stripling, J. S., 1992, Functional correlates of selective long-term potentiation in the olfactory bulb and olfactory cortex, *Brain Res.* **585**:219–228.

Price, J. L., 1985, Beyond the primary olfactory cortex: Olfactory-related areas in the neocortex, thalamus and hypothalamus, *Chem. Senses* **10**:239–258.

Protopapas, A., and Bower, J. M., 1994, Sensitivity in the response of piriform pyramidal cells to fluctuations in synaptic timing, in: *Computation in Neurons and Neural Systems* (F. H. Eeckman, and J. M. Bower, eds.), Kluwer, Boston, pp. 185–190.

Racine, R. J., Milgram, N. W., and Hafner, S., 1983, Long-term potentiation phenomena in the rat limbic forebrain, *Brain Res.* **260**:217–231.

Rall, W., 1989, Cable theory for dendritic neurons, in: *Methods in Neuronal Modeling: From Synapses to Networks* (C. Koch and I. Segev, eds.), MIT Press, Cambridge, MA, pp. 9–62.

Rodriguez, R., and Haberly, L. B., 1989, Analysis of synaptic events in the opossum piriform cortex with improved current source-density techniques, *J. Neurophysiol.* **61**:702–718.

Roman, F., Staubli, U., and Lynch, G., 1987, Evidence for synaptic potentiation in a cortical network during learning, *Brain Res.* **418**:221–226.

Satou, M., Mori, K., Tazawa, Y., and Takagi, S. F., 1982, Two types of postsynaptic inhibition in pyriform cortex of the rabbit: Fast and slow inhibitory postsynaptic potentials, *J. Neurophysiol.* **48**:1142–1156.

Schoenbaum, G. M., and Eichenbaum, H., 1994, Information coding in prefrontal cortex of the behaving rodent: Single unit and ensemble data in orbital prefrontal and piriform cortex, *Soc. Neurosci. Abstr.* **20**:808 (335.4).

Scholfield, C. N., 1978, A depolarizing inhibitory potential in neurons of the olfactory cortex in vitro, *J. Physiol. Cond.* **279**:547–557.

Skarda, C. A., and Freeman, W. J., 1987, How brains make chaos in order to make sense of the world, *Behav. Brain Sci.* **10**:161–195.

Slotnick, B. M., and Berman, E. J., 1980, Transection of the lateral olfactory tract does not produce anosmia, *Brain Res. Bull.* **5**:141–145.

Staubli, U., Schottler, F., and Nejat-Bina, D., 1987, Role of dorsomedial thalamic nucleus and piriform cortex in processing olfactory information, *Behav. Brain Res.* **25**:117–129.

Stripling, J. S., Patneau, D. K., and Gramlich, C. A., 1988, Selective long-term potentiation in the pyriform cortex, *Brain Res.* **441**:281–291.

Tanabe, T., Iino, M., and Takagi, S. F., 1975, Discrimination of odors in olfactory bulb, pyriform-amygdaloid areas, and orbitofrontal cortex of the monkey, *J. Neurophysiol.* **38**:1248–1296.

Tang, A. C., and Hasselmo, M. E., 1994, Selective suppression of intrinsic but not afferent fiber synaptic transmission by baclofen in the piriform (olfactory) cortex, *Brain Res.* **659**:75–81.

Tseng, G.-F., and Haberly, L. B., 1988, Characterization of synaptically mediated fast and slow inhibitory processes in piriform cortex in an *in vitro* slice preparation, *J. Neurophysiol.* **59**:1352–1376.

Tseng, G.-F., and Haberly, L. B., 1989, Deep neurons in piriform cortex. II. Membrane properties that underlie unusual synaptic responses, *J. Neurophysiol.* **62**:386–400.

Vanier, M. C., and Bower, J. M., 1992, Noradrenergic suppression of synaptic transmission in rat piriform (olfactory) cortex, *Soc. Neursci. Abstr.* **18**:1353.

von der Malsburg, C., 1973, Self-organization of orientation sensitive cells in the striate cortex, *Kybernetik* **14**:85–100.

Wang, D., Buhmann, J., and Von der Malsburg, C., 1991, Pattern segmentation in associative memory, in: *Olfaction* (J. L. Davis and H. Eichenbaum, eds.), MIT Press, Cambridge, MA, pp. 213–224.

Wenk, H., Meyer, U., and Bigl, V., 1977, Centrifugal cholinergic connections in the olfactory system of rats, *Neuroscience* **2**:797–800.

Williams, S. H., and Constanti, A., 1988, Quantitative effects of some muscarinic agonists on evoked surface-negative field potentials recorded from the guinea-pig olfactory cortex slice, *Br. J. Pharmacol.* **93**:846–854.

Wilson, M. A., 1989, An analysis of olfactory cortical behavior and function using computer simulation techniques, Ph.D. thesis, California Institute of Technology.

Wilson, M. A., and Bower, J. M., 1988, A computer simulation of olfactory cortex with functional implications for storage and retrieval of olfactory information, in: *Neural Information Processing Systems:* (D. Z. Anderson, ed.), AIP Press, New York, 114–126.

Wilson, M. A., and Bower, J. M., 1989, The simulation of large-scale neuronal networks, in: *Methods in Neuronal Modeling: From Synapses to Networks* (C. Koch and I. Segev, eds.), MIT Press, Cambridge, MA, pp. 291–334.

Wilson, M. A., and Bower, J. M., 1992, Cortical oscillations and temporal interactions in a computer simulation of piriform cortex, *J. Neurophysiol.* **67**:981–995.

Wu, X., and Liljenstrom, H., 1994, Regulating the nonlinear dynamics of olfactory cortex, *Network Computation Neural Syst.* **5**:47–60.

Yao, Y., and Freeman, W. J., 1990, Model of biological pattern recognition with spatially chaotic dynamics, *Neural Networks* **3**:153–170.

Zipser, D., 1991, Recurrent network model of the neural mechanism of short-term active memory, *Neural Computation* **3**:179–193.

Index

SPRINGER NATURE

GPSR Compliance

The European Union's (EU) General Product Safety Regulation (GPSR) is a set of rules that requires consumer products to be safe and our obligations to ensure this.

If you have any concerns about our products, you can contact us on ProductSafety@springernature.com

In case Publisher is established outside the EU, the EU authorized representative is:

Springer Nature Customer Service Center GmbH
Europaplatz 3
69115 Heidelberg, Germany

MIX
Papier | Fördert
gute Waldnutzung
FSC® C083411

Kolibri 360 GmbH
Ferdinand-Jühlke-Straße 7
99095 Erfurt, Deutschland
produktsicherheit@kolibri360.de